Localization Algorithms and Strategies for Wireless Sensor Networks

Guoqiang Mao
University of Sydney, Australia

Barış Fidan
National ICT Australia, Australia & Australian National University, Australia

INFORMATION SCIENCE REFERENCE
Hershey · New York

Director of Editorial Content: Kristin Klinger
Senior Managing Editor: Jamie Snavely
Managing Editor: Jeff Ash
Assistant Managing Editor: Carole Coulson
Typesetter: Jeff Ash
Cover Design: Lisa Tosheff
Printed at: Yurchak Printing Inc.

Published in the United States of America by
Information Science Reference (an imprint of IGI Global)
701 E. Chocolate Avenue
Hershey PA 17033
Tel: 717-533-8845
Fax: 717-533-8661
E-mail: cust@igi-global.com
Web site: http://www.igi-global.com/reference

and in the United Kingdom by
Information Science Reference (an imprint of IGI Global)
3 Henrietta Street
Covent Garden
London WC2E 8LU
Tel: 44 20 7240 0856
Fax: 44 20 7379 0609
Web site: http://www.eurospanbookstore.com

Library of Congress Cataloging-in-Publication Data

Localization algorithms and strategies for wireless sensor networks / Guoqiang Mao and Baris Fidan, editors.
p. cm.

Includes bibliographical references and index.
Summary: "This book encompasses the significant and fast growing area of wireless localization technique"--Provided by publisher.

ISBN 978-1-60566-396-8 (hardcover) -- ISBN 978-1-60566-397-5 (ebook) 1. Wireless sensor networks. 2. Proximity detectors. 3. Location problems (Programming) I. Mao, Guoqiang, 1974- II. Fidan, Baris.

TK7872.D48L63 2009
621.382'1--dc22

2008052196

British Cataloguing in Publication Data
A Cataloguing in Publication record for this book is available from the British Library.

List of Reviewers

Brian Anderson, *Australian National University and National ICT Australia, Australia*
Adrian Bishop, *KTH Royal Institute of Technology, Sweden*
Chun Tung Chou, *University of New South Wales, Australia*
Soura Dasgupta, *University of Iowa, USA*
Kutluyıl Doğançay, *University of South Australia, Australia*
Jia Fang, *Yale University, USA*
Tolga Girici, *TOBB University of Economics and Technology, Turkey*
Fredrik Gustafsson, *Linköping University, Sweden*
Hatem Hmam, *Defence Science and Technology Organisation, Australia*
Julien Hendrickx, *Université catholique de Louvain, Belgium*
Tibor Jordán, *Eötvös University, Hungary*
Anushiya Kannan, *University of Sydney, Australia*
Emre Köksal, *Ohio State University, USA*
Ullrich Köthe, *University of Hamburg, Germany*
Anthony Kuh, *University of Hawaii at Manoa, USA*
Lavy Libman, *National ICT Australia, Australia*
Sarfraz Nawaz, *University of New South Wales, Australia*
Michael L. McGuire, *University of Victoria, Canada*
Garry Newsam, *Defence Science and Technology Organisation, Australia*
M. Özgür Oktel, *Bilkent University, Turkey*
Neal Patwari, *University of Utah, USA*
Parastoo Sadeghi, *Australian National University, Australia*
Yi Shang, *University of Missouri-Columbia, USA*
Qinfeng Shi, *Australian National University and National ICT Australia, Australia*
Bülent Tavlı, *TOBB University of Economics and Technology, Turkey*

Table of Contents

Detailed Table of Contents

Chapter IX focuses on localization in indoor wireless local area network (WLAN) environments and presents a RSS-based localization system for indoor WLAN environments. The localization problem is formulated as a multi-hypothesis testing problem and an algorithm is developed using this algorithm to identify in which region the sensor resides. A solid theoretical discussion of the problem is provided, backed by experimental validations.

Chapter X first presents an analytical framework for ascertaining the attainable accuracy of RSS-based localization techniques. It then summarizes the issues that may affect the design and deployment of RSS-based localization systems, including deployment ease, management simplicity, adaptability and cost of ownership and maintenance. With this insight, the authors present the "LEASE" architecture for localization that allows easy adaptability of localization models.

Chapter XI surveys and compares several RSS-based localization techniques from two broad categories: point-based and area-based. It is demonstrated that there are fundamental limitations for indoor localization performance that cannot be transcended without using qualitatively more complex models of the indoor environment, e.g., modelling every wall, desk or shelf, or without adding extra hardware in the sensor node other than those required for communication, e.g., very high frequency clocks to measure the time of arrival.

Chapter XII presents a machine learning approach to localization. The applicability of two learning methods, the classification method and the regression model, to RSS-based localization is discussed.

Chapter XIII presents another paradigm for robust localization based on the use of identifying codes, a concept borrowed from the information theory literature with links to covering and superimposed codes. The approach is reported to be robust and suitable for implementation in harsh environments.

Chapter XIV introduces a methodological approach to the evaluation of localization algorithms. The authors argue that algorithms should be simulated, emulated (on test beds or with empirical data sets) and subsequently implemented in hardware, in a realistic WSN deployment environment, as a complete test of their performance.

Chapter XV looks at evaluation of localization algorithms from a different perspective and takes an analytical approach to performance evaluation. In particular, the authors advocate the use of the Weinstein-Weiss and extended Ziv-Zakai lower bounds for evaluating localization error, which overcome the problem in the widely used Cramer-Rao bound that the Cramer-Rao bound relies on some idealizing assumptions not necessarily satisfied in real systems.

Chapter XVI discusses algorithms and solutions for signal processing and filtering for localization and tracking applications. The authors explain some practical issues for engineers interested in implementing tracking solutions and their experiences gained from implementation and deployment of several such systems.

Chapter XVII presents an experimental study on the integration of Wi-Fi based wireless mesh networks and Bluetooth technologies for detecting and tracking travelling cars and measuring their speeds for road traffic monitoring in intelligent transportation systems.

Chapter XVIII discusses an interesting aspect of the geographic routing problem. The authors propose the use of virtual coordinates, instead of physical coordinates, of sensors for improved geographic routing performance. This chapter motivates us to think beyond the horizon of localization and invent smarter ways to label sensors and measurement data from sensors to facilitate applications that do not rely on the knowledge of physical locations of sensors.

Preface

Distributed sensor networks have been discussed for more than 30 years, but the vision of wireless sensor networks has been brought into reality only by the recent advances in wireless communications and electronics, which have enabled the development of low-cost, low-power and multi-functional sensors that are small in size and communicate over short distances. Today, cheap, smart sensors, networked through wireless links and deployed in large numbers, provide unprecedented opportunities for monitoring and controlling homes, cities, and the environment. In addition, networked sensors have a broad spectrum of applications in the defence area, generating new capabilities for reconnaissance and surveillance as well as other tactical applications.

Localization (location estimation) capability is essential in most wireless sensor network applications. In environmental monitoring applications such as animal habitat monitoring, bush fire surveillance, water quality monitoring and precision agriculture, the measurement data arc meaningless without an accurate knowledge of the location from where the data are obtained. Moreover, the availability of location information may enable a myriad of applications such as inventory management, intrusion detection, road traffic monitoring, health monitoring, reconnaissance and surveillance.

Wireless sensor network localization techniques are used to estimate the locations of the sensors with unknown positions in a network using the available *a priori* knowledge of positions of, typically, a few specific sensors in the network and inter-sensor measurements such as distance, time difference of arrival, angle of arrival and connectivity. Sensor network localization techniques are not just trivial extensions of the traditional localization techniques like GPS or radar-based geolocation techniques. They involve further challenges in several aspects: (1) a variety of measurements may be used in sensor network localization; (2) the environments in which sensor networks are deployed are often complicated, involving urban environments, indoor environments and non-line-of-sight conditions; (3) wireless sensors are often small and low-cost sensors with limited computational capabilities; (4) sensor network localization techniques are often required to be implemented using available measurements and with minimal hardware investment; (5) sensor network localization techniques are often required to be suitable for deployment in large scale multi-hop networks; and (6) the choice of sensor network localization techniques to be used often involves consideration of the trade-off among cost, size and localization accuracy to suit the requirements of a variety of applications. It is these challenges that make localization in wireless sensor networks unique and intriguing.

This book is intended to cover the major techniques that have been widely used for wireless sensor network localization and capture the most recent developments in the area. It is based on a number of stand-alone chapters that together cover the subject matter in a fully comprehensive manner. However, despite its focus on localization in wireless sensor networks, many localization techniques introduced in the book can be applied in a variety of wireless networks beyond sensor networks.

The targeted audience for the book includes professionals who are designers and/or planners for wireless localization systems, researchers (academics and graduate students), and those who would like to learn about the field. Although the book is not exactly a textbook, the format and flow of information have been organized such that it can be used as a textbook for graduate courses and research-oriented courses that deal with wireless sensor networks and wireless localization techniques.

ORGANIZATION

This book consists of 18 chapters. It begins with an introductory chapter that covers the basic principles of techniques involved in the design and implementation of wireless sensor network localization systems. A focus of the chapter is on explaining how the other chapters are related to each other and how topics covered in each chapter fit into the architecture of this book and the big picture of wireless sensor network localization. The other chapters are organized into three parts: measurement techniques, localization theory, and algorithms, experimental study and applications.

Measurement techniques are of fundamental importance in sensor network localization. It is the type of measurements employed and the corresponding precision that fundamentally determine the estimation accuracy of a localization system and the localization algorithm being implemented by this system. Measurements also determine the type of algorithm that can be used by a particular localization system. The part on *Measurement Techniques* includes Chapters II-V, which discuss various aspects of measurement techniques used in sensor network localization. Chapter II introduces a common framework for analysing the information content of various measurements, which can be used to derive localization bounds for integration of any combination of measurements in the network. Chapter III discusses challenges in time-of-arrival measurement techniques and methods to overcome these challenges. A focus of the chapter is on the identification of non-line-of-sight conditions in time-of-arrival measurements and the corresponding mitigation techniques. Chapter IV gives a detailed discussion on the impact of various factors, that is, noise, clock synchronization, signal bandwidth and multipath, on the accuracy of signal propagation time measurements. Chapter V features a thorough discussion on a number of practical issues involved in the use of received signal strength (RSS) measurements. In particular, it focuses on the device calibration problem and its impact on localization.

Chapters VI-XV give an in-depth discussion of the fundamental theory underpinning sensor network localization and various localization approaches. Chapter VI gives a detailed overview of various tools in graph theory and combinatorial rigidity, many of which are just recently developed, to characterize uniquely localizable networks. A network is said to be uniquely localizable if there is a unique set of locations consistent with the given data, that is, location information of a few specific sensors and inter-sensor measurements. Chapter VII presents a class of computationally efficient sequential algorithms based on graph theory for estimating sensor locations using inaccurate distance measurements. Chapter VIII presents several centralized and distributed localization algorithms based on multidimensional scaling techniques for implementation in regular and irregular networks. Chapters IX-XI feature a thorough discussion on theoretical and practical issues involved in the design and implementation of RSS-based localization algorithms. Chapter IX focuses on localization in indoor wireless local area network (WLAN) environments and presents a RSS-based localization system for indoor WLAN environments. The localization problem is formulated as a multi-hypothesis testing problem and an algorithm is developed using this algorithm to identify in which region the sensor resides. A solid theoretical discussion of the problem is provided, backed by experimental validations. Chapter X first presents an analytical framework for ascertaining the attainable accuracy of RSS-based localization techniques. It then summarizes the issues that may affect the design and deployment of RSS-based localization systems, including deployment ease, management simplicity, adaptability and cost of ownership and maintenance. With this insight, the authors present the "LEASE" architecture for localization that allows easy adaptability of localization models. Chapter XI surveys and compares several RSS-based localization techniques from two broad categories: point-based and area-based. It is demonstrated that there are fundamental limitations for indoor localization performance that cannot be transcended without using qualitatively more complex models of the indoor environment, for example, modelling every wall, desk or shelf, or without adding extra hardware in the sensor node other than those required for communication, e.g., very high frequency clocks to measure the time of arrival. Chapter XII presents a machine learning approach to localization. The applicability of two learning methods, the classification method and the regression model, to RSS-based localization is discussed. Chapter XIII presents another paradigm for robust localization based on the use of identifying codes, a concept borrowed from the information theory literature with links to covering and superimposed codes. The approach is reported to be robust and suitable for implementation in harsh environments. Chapters XIV and XV consider the evaluation of localization algorithms. Chapter XIV introduces a methodological approach to the evaluation of localization algorithms. The authors argue that algorithms should be simulated, emulated (on test beds or with empirical data sets) and subsequently implemented in hardware, in a realistic WSN deployment environment, as a complete test of their performance. Chapter XV looks at evaluation of localization algorithms from a different perspective and takes an analytical approach to performance evaluation. In particular, the authors advocate the use of the Weinstein-Weiss and extended Ziv-Zakai lower bounds for evaluating localization error, which overcome the problem in the widely used Cramer-Rao bound that the Cramer-Rao bound relies on some idealizing assumptions not necessarily satisfied in real systems.

Chapters XVI, XVII, and XVIII discuss the applications of localization techniques in tracking and sensor network routing. Chapter XVI discusses algorithms and solutions for signal processing and filtering for localization and tracking applications. The authors explain some practical issues for engineers interested in implementing tracking solutions and their experiences gained from implementation and deployment of several such systems. Chapter XVII presents an experimental study on the integration of Wi-Fi based wireless mesh networks and Bluetooth technologies for detecting and tracking travelling cars and measuring their speeds for road traffic monitoring in intelligent transportation systems. Chapter XVIII discusses an interesting aspect of the geographic routing problem. The authors propose the use of virtual coordinates, instead of physical coordinates, of sensors for improved geographic routing performance. This chapter motivates us to think beyond the horizon of localization and invent smarter ways to label sensors and measurement data from sensors to facilitate applications that do not rely on the knowledge of physical locations of sensors.

Guoqiang Mao
University of Sydney, Australia

Barış Fidan
National ICT Australia, Australia & Australian National University, Australia

Acknowledgment

This book would not have been possible without the expertise and commitment of our contributing authors. The editors are grateful to all the authors for their contributions to the quality of this book.

The editors also greatly appreciate the reviewers of all the chapters for their constructive and comprehensive reviews. The list of reviewers is provided separately in the book. We are immensely indebted to them.

We want to thank the publishing team at IGI Global, whose contributions throughout the whole process from inception of the initial idea to final publication have been invaluable, in particular to Rebecca Beistline, Julia Mosemann and Christine Bufton, who continuously provided valuable support via e-mail.

Our special thanks go to Brian D.O. Anderson, whose collaborative studies with us in the last four years have helped provide the foundation and motivation for us to edit this book. He is a person of great character, and he has been a selfless mentor, a brilliant research partner and a precious friend during these stimulating collaborative studies. We have enjoyed collaboration with him enormously.

Guoqiang Mao
University of Sydney, Australia

Barış Fidan
National ICT Australia, Australia & Australian National University, Australia

Chapter I
Introduction to Wireless Sensor Network Localization

Guoqiang Mao
University of Sydney, Australia

Barış Fidan
National ICT Australia, Australia & Australian National University, Australia

ABSTRACT

Localization is an important aspect in the field of wireless sensor networks that has attracted significant research interest recently. The interest in wireless sensor network localization is expected to grow further with the advances in the wireless communication techniques and the sensing techniques, and the consequent proliferation of wireless sensor network applications. This chapter provides an overview of various aspects involved in the design and implementation of wireless sensor network localization systems. These can be broadly classified into three categories: the measurement techniques in sensor network localization, sensor network localization theory and algorithms, and experimental study and applications of sensor network localization techniques. This chapter also gives a brief introduction to the other chapters in the book with a focus on explaining how these chapters are related to each other and how topics covered in each chapter fit into the architecture of this book and the big picture of wireless sensor network localization.

INTRODUCTION

Distributed sensor networks have been discussed for more than 30 years, but the vision of wireless sensor networks (WSNs) has been brought into reality only by the recent advances in wireless communications and electronics, which have enabled the development of low-cost, low-power and multi-functional sensors that are small in size and communicate over short distances. Today, cheap, smart sensors, networked through wireless links and deployed in large numbers, provide unprecedented opportunities for monitoring and controlling homes, cities, and the environment. In addition, networked sensors have a broad spectrum of applications in the defence area, generating new capabilities for reconnaissance and surveillance as well as other tactical applications (Chong & Kumar, 2003).

Localization (location estimation) capability is essential in most WSN applications. In environmental monitoring applications such as animal habitat monitoring, bush fire surveillance, water quality monitoring and precision agriculture, the measurement data are meaningless without an accurate knowledge of the location from where the data are obtained. Moreover, the availability of location information may enable a myriad of applications such as inventory management, intrusion detection, road traffic monitoring, health monitoring, reconnaissance and surveillance.

WSN localization techniques are used to estimate the locations of the sensors with initially unknown positions in a network using the available *a priori* knowledge of positions of a few specific sensors in the network and inter-sensor measurements such as distance, time difference of arrival, angle of arrival and connectivity. Sensors with the *a priori* known location information are called *anchors* and their locations can be obtained by using a global positioning system (GPS), or by installing anchors at points with known coordinates, etc. In applications requiring a global coordinate system, these anchors will determine the location of the sensor network in the global coordinate system. In applications where a local coordinate system suffices (e.g., in smart homes, hospitals or for inventory management where knowledge like in which room a sensor is located is sufficient), these anchors define the local coordinate system to which all other sensors are referred. Because of constraints on the cost and size of sensors, energy consumption, implementation environment (e.g., GPS is not accessible in some environments) and the deployment of sensors (e.g., sensors may be randomly scattered in the region), most sensors do not know their own locations. These sensors with unknown location information are called *non-anchor* nodes and their coordinates need to be estimated using a sensor network localization algorithm. In some other applications, e.g., for geographic routing in WSN, where there are no anchor nodes and also knowledge of the physical location of a sensor is unnecessary, people are more interested in knowing the position of a sensor *relative* to other sensors. In that case, sensor localization algorithms can be used to estimate the relative positions of sensors using inter-sensor measurements. The obtained estimated locations are usually a reflected, rotated and translated version of their global coordinates.

In this chapter, we provide an overview of various aspects of WSN localization with a focus on the techniques covered in the other chapters of this book. These chapters can be broadly classified into three categories: the *measurement techniques* in sensor network localization, sensor network *localization theory and algorithms*, and *experimental study and applications of sensor network localization* techniques.

The rest of the chapter is organized as follows. In Section MEASUREMENT TECHNIQUES, measurement techniques in WSN localization and the basic principle of localization using these measurements are discussed. These measurements include *angle-of-arrival (AOA) measurements, distance related measurements* and *received signal strength (RSS) profiling techniques.* Distance related measurements are further classified into *one-way propagation time* and *roundtrip propagation time* measurements, the *lighthouse approach* to distance measurements, *RSS-based distance measurements, time-difference-of-arrival (TDOA)* measurements and *connectivity* measurements. In Section LOCALIZATION THEORY AND ALGORITHMS, fundamental theory underpinning WSN localization algorithms and some fundamental problems in WSN localization are discussed with a focus on the use of graph theory in WSN localization. Later in this section, a set of major localization algorithms are discussed. Section EXPERI-

MENTAL STUDIES AND APPLICATIONS OF WSN LOCALIZATION discusses implementation of WSN localization techniques and their use in a number of areas, e.g., intelligent transportation and WSN routing. The aim of each of these three later sections is to provide an overall review of its topic and to give brief introduction of the relevant chapters of the book.

MEASUREMENT TECHNIQUES

WSN localization relies on measurements. There are many factors that affect the choice of the algorithm to be used for a specific application and the accuracy of the estimated locations, to name but a few, the network architecture, the average node degree (i.e., the average number of neighbours per sensor), the geometric shape of the network area and the distribution of sensors in that area, sensor time synchronization and the signalling bandwidth among the sensors. However, it is the type of measurements employed and the corresponding precision that fundamentally determine the estimation accuracy of a localization system and the localization algorithm being implemented by this system. Measurements also determine the type of algorithm that can be used by a particular localization system.

In a typical WSN localization system, the available measurements can often be related to the coordinates of sensors using the following generic formula:

$$\boldsymbol{Y} = \boldsymbol{h}(\boldsymbol{X}) + \boldsymbol{e}$$

where $\boldsymbol{Y}$ is the vector of all measurements, $\boldsymbol{X}$ contains the true coordinate vectors of sensors whose locations are to be estimated and $\boldsymbol{e}$ is the vector of measurement errors. If the distribution of measurement errors f_e is known, the estimated locations of sensors can be obtained using the maximum likelihood approach by minimizing an optimization criterion:

$$\hat{X} = \arg\min\left(\log f_e\left(Y - \boldsymbol{h}\left(\hat{X}\right)\right)\right)$$

A particular cost function related to this optimization criterion is the *Fisher Information Matrix*

$$\boldsymbol{J}(\boldsymbol{X}) = E\left(\nabla_X^T \log f_e\left(\boldsymbol{Y} - \boldsymbol{h}(\boldsymbol{X})\right)\nabla_X \log f_e\left(\boldsymbol{Y} - \boldsymbol{h}(\boldsymbol{X})\right)\right)$$

where $\nabla_X \log f_e\left(Y - \boldsymbol{h}(\boldsymbol{X})\right)$ is the partial derivative of $\log f_e\left(Y - \boldsymbol{h}(\boldsymbol{X})\right)$ with respect to X evaluated at $\boldsymbol{X}$.

A common technique that has been widely used to evaluate the location accuracy that can be expected from measurements is the Cramer-Rao bound. The Cramer-Rao lower bound is given by

$$Cov\left(\hat{X}\right) = E\left(X - \hat{X}\right)\left(X - \hat{X}\right)^T \geq \boldsymbol{J}^{-1}(X)$$

The Cramer-Rao bound is valid for any unbiased estimator of sensor locations and gives the best performance that can be achieved by an unbiased location estimator. Therefore it is a valuable tool for analysing the information content of various measurements. ***Chapter II - Measurements Used in Wireless Sensor Networks Localization*** features a thorough discussion on this topic. It establishes a common framework for analysing the information content of various measurements, which can be used to derive localization bounds for integration of any combination of measurements in the network.

Measurement techniques in WSN localization can be broadly classified into three categories: *AOA measurements*, *distance related measurements* and *RSS profiling techniques*. Next, we introduce these three categories in more detail.

Angle-of-Arrival Measurements

The *AOA* measurements are also known as the *bearing* measurements or the *direction of arrival* measurements. The AOA measurements can usually be obtained from two categories of techniques: those making use of the receiver antenna's amplitude response and those making use of the receiver antenna's phase response. In addition to the directivity of the antenna (Cheng, 1989), the accuracy of AOA measurements are affected by other environmental factors like shadowing and multipath, and the later effect may make the transmitter look like located at a different direction of the receiver.

The first category of AOA measurements is widely known as *beamforming* and it is based on the anisotropy in the reception pattern (Cheng, 1989) of an antenna. The size of the measurement unit can be comparatively small with regards to the wavelength of the signals. Figure 1 shows the beam pattern of a typical anisotropic antenna. When the beam of the receiver antenna is rotated electronically or mechanically, the direction corresponding to the maximum signal strength is taken as the direction of the transmitter. The accuracy of the measurements is determined by the sensitivity of the receiver and the beam width. Using a rotating beam has the potential problem that the receiver cannot differentiate the signal strength variation caused by the varying amplitude of the transmitted signal and the signal strength variation caused by the anisotropy in the reception pattern. This problem can be dealt with by using a second non-rotating and omnidirectional antenna at the receiver. The impact of varying signal strength can be largely removed by normalizing the signal strength received by the rotating anisotropic antenna with respect to the signal strength received by the non-rotating omnidirectional antenna. Alternatively, one may also use multiple stationary antennas with known, anisotropic antenna patterns to overcome the difficulty caused by the varying signal strength problem. Comparing the signal strength received from each antenna at the same time, together with the knowledge of their antenna patterns, leads to an estimate of the transmitter direction, even when the signal strength changes (Koks, 2005).

Figure 1. The horizontal antenna pattern of a typical anisotropic antenna in polar coordinates

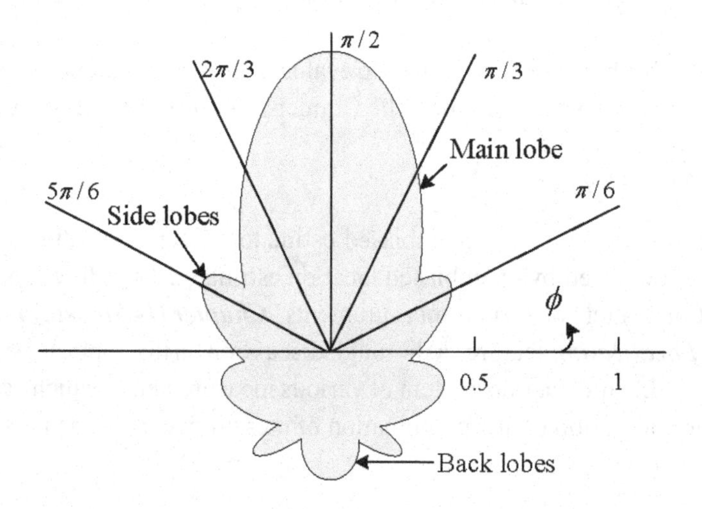

The other category of AOA measurement techniques is widely known as *phase interferometry* and it derives the AOA measurements from the measurements of the phase differences in the arrival of a wave front (Rappaport, Reed, & Woerner, 1996). A large receiver antenna (relative to the wavelength of the transmitter signal) or an antenna array is typically required when using this technique. Figure 2 shows an antenna array of N elements. The adjacent antennas are separated by a fixed distance d. For a transmitter far away from the antenna array, its distance to the kth antenna can be approximated by

$$R_k \approx R_0 - kd\cos\theta \tag{1}$$

where R_0 is the distance between the transmitter and the 0th antenna and θ is the direction of the transmitter viewed from the antenna array. The transmitter signal received by the adjacent antennas will have a phase difference of $2\pi\frac{d\cos\theta}{\lambda}$ with λ being the wavelength of the transmitter signal. Therefore the AOA of the transmitter with respect to the antenna array can be derived from the measurements of the phase differences. The accuracy of the AOA measurements obtained using this approach is usually not affected by high signal-to-noise-ratio (SNR) but this approach may fail in the presence of strong co-channel interference and/or multipath signals (Rappaport, Reed, & Woerner, 1996).

The accuracy of AOA measurements is limited by the directivity of the antenna and the measurements are further complicated by the presence of *shadowing* and *multipath* in the measurement environment. A major challenge in AOA measurements is therefore the accurate estimation of AOA in the presence of multipath and shadowing. AOA measurements rely on a direct *line-of-sight (LOS)* path between the transmitter and the receiver. A multipath component from the transmitter signal may appear as a signal coming from an entirely different direction and consequently causes a very large error in the AOA measurement.

Multipath problems in AOA measurements have been usually addressed using *maximum likelihood (ML)* algorithms (Rappaport, Reed, & Woerner, 1996). Depending on the assumptions being made about the statistical characteristics of the transmitter signals, i.e., whether the structure of the transmitter signal is known or unknown to the receiver, these ML algorithms can be further classified into deterministic (Agee, 1991; Halder, Viberg, & Kailath, 1993; Jian, Halder, Stoica, & Viberg, 1995) and stochastic (Biedka, Reed, & Woerner, 1996; Bliss & Forsythe, 2000; Ziskind & Wax, 1988) ML algorithms.

Yet another class of AOA estimation techniques, which relies on the presence of a multi-antenna array that is composed of, say, N antennas at the receiver, is based on the so-called *subspace-based algorithms* (Paulraj, Roy, & Kailath, 1986; Roy & Kailath, 1989; Schmidt, 1986; Tayem & Kwon, 2004). The most well known methods in this category are *MUSIC (multiple signal classification)* and *ESPRIT (estimation of signal parameters by rotational invariance techniques)* (Paulraj et al., 1986; Roy & Kailath, 1989). The measured transmitter signal received at the N antennas of the receiver antenna array is considered as a vector in N dimensional space. A correlation matrix is formed utilizing the N signals received at the antennas of the receiver antenna array. By using an eigen-decomposition of the correlation matrix, the vector space is separated into signal and noise subspaces. Then the MUSIC algorithm searches for nulls in the magnitude squared of the projection of the direction vector onto the noise subspace. The nulls are a function of angle-of-arrival, from which AOA can be estimated. Other techniques that have been developed based on the MUSIC algorithms include *Root-MUSIC* (Barabell, 1983), a polynomial rooting version of MUSIC which improves the resolution capabilities of MUSIC, *WMUSIC* (Kaveh & Bassias, 1990), a weighted norm version of MUSIC which also gives an extension in the resolution capabilities to the original MUSIC. ESPRIT (Paulraj et al., 1986; Roy & Kailath, 1989) is based on the estimation of signal parameters via rotational invariance techniques. It uses two displaced subarrays of matched sensor doublets to exploit an underlying rotational invariance among signal subspaces for such an array. A comprehensive experimental evaluation of MUSIC, Root-MUSIC, WMUSIC, Min-Norm (Kumaresan & Tufts, 1983) and ESPRIT algorithms can be found in (Klukas & Fattouche, 1998). A significant number of AOA measurement techniques have been developed which are based on MUSIC and ESPRIT, to cite but two, see e.g., (Klukas & Fattouche, 1998; Paulraj et al., 1986). Readers may refer to (Schell & Gardner, 1993) for a detailed discussion on AOA measurement techniques.

Chapter III - Overview of RF Localization Sensing Techniques and TOA-Based Positioning for WSNs provides further discussion on AOA measurements using antenna arrays, and gives the Cramer-Rao lower bound on AOA estimation error. The lower bound is determined by the SNR of the received signal from the transmitter, the carrier frequency of the transmitter and the number of antenna elements of the antenna array.

In $\Re^2$, AOA measurements from a minimum of two receivers can be used to estimate the location of the transmitter. However in the presence of measurement errors, more than two AOA measurements will be needed for accurate location estimate. In the presence of measurement errors, AOA measurements from more than two receivers will not intersect at the same point. This is illustrated in Figure 3.

Denote by $\boldsymbol{X}_t = [x_t, y_t]^T$ the true coordinate vector of the transmitter whose location is to be estimated from AOA measurements $\alpha = [\alpha_1, \ldots, \alpha_N]^T$, where N is the total number of receivers. Let $\boldsymbol{X}_i = [x_i, y_i]^T$ be the known coordinate vector of the ith receiver associated with the ith AOA measurement α_i. Denote by $\theta(\boldsymbol{X}_t) = [\theta_1(\boldsymbol{X}_t), \ldots, \theta_N(\boldsymbol{X}_t)]$ the AOA vector of the transmitter located at x_t from the receiver locations, i.e., $\theta_i(\boldsymbol{X}_t)$ $(i \in \{1, \ldots, N\})$ is related to x_t and x_i by

$$\tan\theta_i(\boldsymbol{X}_t) = \frac{y_t - y_i}{x_t - x_i} \quad (2)$$

Figure 2. An illustration of AOA measurements using an antenna array of N antennas

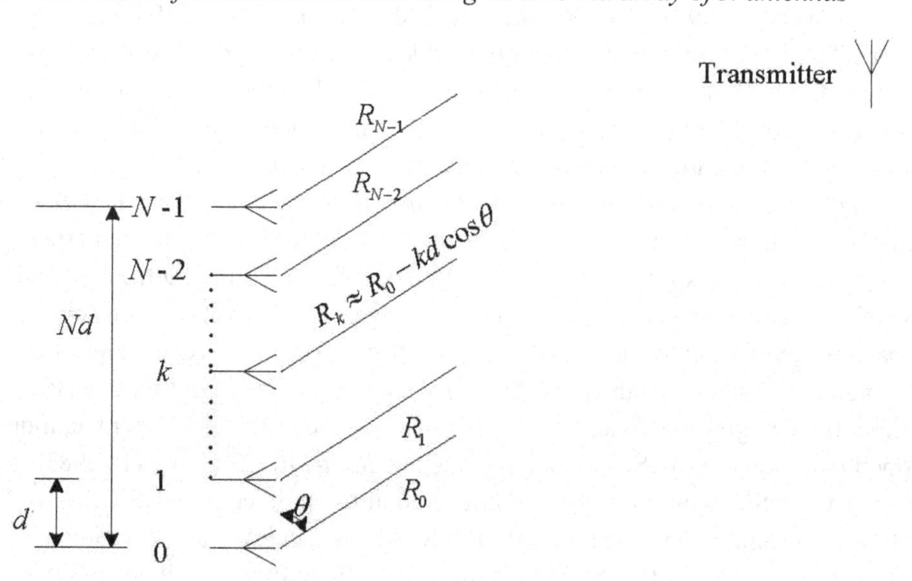

Figure 3. In the presence of measurement errors, AOA measurements from three receivers will not intersect at the same point

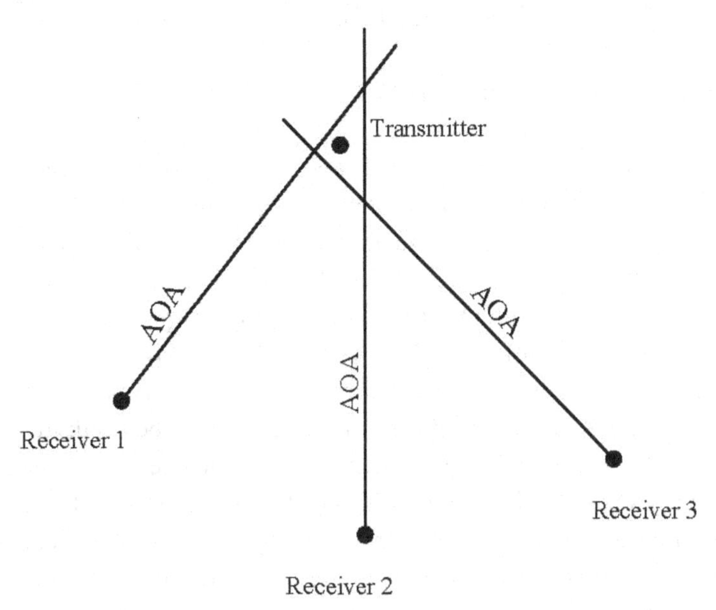

In the presence of measurement errors, the measured AOA vector $\boldsymbol{\alpha}$ consists of the true bearing vector corrupted by noise $\boldsymbol{e} = [e_1, \ldots, e_N]^T$, which is usually assumed to be additive zero mean Gaussian noise with covariance matrix $\boldsymbol{S} = diag\{\sigma_1, \ldots, \sigma_N\}$, i.e.,

$$\alpha = \theta(\boldsymbol{X}_t) + \boldsymbol{e} \tag{3}$$

The transmitter location can then be estimated using an ML estimator as follows:

$$\hat{\boldsymbol{X}}_t = \arg\min\left[\theta\left(\hat{\boldsymbol{X}}_t\right) - \alpha\right]^T \boldsymbol{S}^{-1}\left[\theta\left(\hat{\boldsymbol{X}}_t\right) - \alpha\right] \tag{4}$$

When the receivers are identical and much closer to each other than to the transmitter, the variances of AOA measurement errors can be considered as equal, i.e., $\sigma_1^2 = \cdots = \sigma_N^2 = \sigma^2$. The nonlinear optimization problem in Equation (4) can be solved by a Newton-Gauss iteration (Gavish & Weiss, 1992; Torrieri, 1984), which requires an initial estimate of the transmitter location close to its true location. If additional information, such as the measurement errors being small or rough estimates of the distances between the transmitter and the receivers, is available *a priori*, techniques like the Stanfield approach (Stanfield, 1947) can be used to simplify the optimization problem in Equation (4) and an analytical solution to $\hat{\boldsymbol{X}}_t$ can be obtained directly. We refer the readers to (Gavish & Weiss, 1992; Torrieri, 1984) for more detailed discussions on this topic.

Distance Related Measurements

Measurements that can be classified into the category of distance related measurements include *propagation time based measurements*, i.e., *one-way propagation time measurements*, *roundtrip propagation time measurements* and *TDOA* measurements; *RSS based measurements*; and *connectivity measurements*. Another interesting approach to distance measurements, which does not fall into any of the above categories, is the *lighthouse approach* (Romer, 2003).

One-Way Propagation Time Measurements

The principle of *one-way propagation time measurements* is straightforward: measuring the difference between the sending time of a signal at the transmitter and the receiving time of the signal at the receiver. Given this time difference measurement and the propagation speed of the signal in the media, the distance between the transmitter and the receiver can be obtained. Time delay measurement is a relatively mature field. The most widely used method for obtaining time delay measurement is the *generalized cross-correlation method* (Carter, 1981, 1993; Knapp & Carter, 1976).

A major challenge in the implementation of one-way propagation time measurements is that it requires the local time at the transmitter and the local time at the receiver to be accurately synchronized. Any difference between the two local times will become the bias in the one-way propagation measurement. At the speed of light, a very small synchronization error of 1ns will translate into a distance measurement error of 0.3m. The accurate synchronization requirement may add to the cost of sensors, by demanding a highly accurate clock, or increase the complexity of the sensor network, by demanding a sophisticated synchronization algorithm. This disadvantage makes one-way propagation time measurements a less attractive option in WSNs.

In addition to using an accurate clock for each sensor or using a sophisticated synchronization algorithm, an interesting approach has been proposed in the literature which overcomes the synchronization problem (Priyantha, Chakraborty, & Balakrishnan, 2000) based on the observation that the speed of sound in the air is much smaller than the speed of light or radio-frequency (RF) signal in the air. A combination of RF and ultrasound hardware is used in the technique. On each transmission, a transmitter sends an RF signal and an ultrasonic pulse at the same time. The RF signal will arrive at the receiver earlier than the ultrasonic pulse. When the receiver receives the RF signal, it turns on its ultrasonic receiver and listens for the ultrasonic pulse. The time difference between the receipt of the RF signal and the receipt of the ultrasonic signal is used as an estimate of the one-way acoustic propagation time. This method gives fairly accurate distance estimate at the cost of additional hardware and complexity of the system because ultrasonic reception suffers from severe multipath effects caused by reflections from walls and other objects. This method is referred to as *time-difference-of-arrival (TDOA)* measurement, i.e., measurement of the difference between the arrival times of RF signal and ultrasonic signal, in some papers as well as some chapters in this book. However it should be noted that it is different from the TDOA measurements discussed later in this chapter and in most papers on geolocation.

Roundtrip Propagation Time Measurements

Roundtrip propagation time measurements measure the difference between the time when a signal is sent by a sensor and the time when the signal returned by a second sensor comes back to the original sensor. Since the same local clock is used to compute the roundtrip propagation time, there is no synchronization problem. The major error source in roundtrip propagation time measurements is the delay required for handling the signal in the second sensor. This internal delay is either known via *a priori* calibration, or measured and sent to the first sensor to be subtracted. A technique that can be used to overcome the above internal delay problem involves the cooperation of the two sensors in the measurements. First sensor A sends a signal to sensor B at sensor A's local time t_{A1}, the signal arrives at sensor B at sensor B's local time t_{B1}. After some delay, sensor B sends a signal to sensor A at sensor B's local time t_{B2}, together with the time difference $t_{B2} - t_{B1}$. The signal arrives at sensor A at sensor A's local time t_{A2}. Then sensor A is able to compute the round-trip-time using $(t_{A2} - t_{A1}) - (t_{B2} - t_{B1})$. Because the computation only needs the difference between two local time measurements at sensor A and the difference between two local time measurements at sensor B, no synchronization problem exists. The internal delay in the second sensor B is also removed in the round-trip time measurements. A detailed discussion on circuitry design for roundtrip propagation time measurements can be found in (McCrady, Doyle, Forstrom, Dempsey, & Martorana, 2000).

In addition to the synchronization error, the accuracy of both one-way and roundtrip propagation time measurements is affected by noise, signal bandwidth, non-line-of-sight (NLOS) and multipath. Recently, ultra-wide band (UWB) signals have started to be used for accurate propagation time measurements (Gezici et al., 2005; Lee & Scholtz, 2002). A UWB signal is a signal whose bandwidth to centre frequency ratio is larger than 0.2 or a signal with a total bandwidth of more than 500 *MHz*. In principle, UWB can achieve higher accuracy because its bandwidth is very large and therefore its pulse has a very short duration. This feature makes fine time resolution of UWB signals and easy separation of multipath signals possible.

Chapter III - Overview of RF Localization Sensing Techniques and TOA-Based Positioning for WSNs first discusses time of arrival (TOA) measurement techniques and challenges in the measurements. The chapter then focuses on the identification of NLOS conditions in TOA measurements and techniques that can be used to mitigate the performance impact of NLOS conditions.

Chapter IV - RF Ranging Methods and Performance Limits for Sensor Localization gives a detailed discussion on the impacts of various factors, including noise, clock synchronization, signal bandwidth and multipath, on the accuracy of propagation time measurements. The chapter also features a discussion on the characteristics of some deployed systems.

In $\Re^2$, measured distances from a non-anchor node to three non-collinear anchors determine three circles whose centres are at the three anchors and radii are the associated measured distances respectively. When there is no measurement error, the three circles intersect at a single point which is the location of the non-anchor node. In the presence of measurement errors, the three circles do not intersect at a single point. A large number of approaches have been developed to estimate the location of the non-anchor node in such noisy cases. Assuming the measurement errors are additive zero mean Gaussian noises, for a non-anchor node at unknown location X_t with noise-contaminated distance measurements $\tilde{\boldsymbol{d}} = \left[\tilde{d}_1, \ldots, \tilde{d}_N\right]^T$ to N anchors at known locations $\boldsymbol{X}_1, \ldots, \boldsymbol{X}_N$, an ML formulation of the location estimation problem is given by

$$\hat{\boldsymbol{X}}_t = \arg\min\left[\boldsymbol{d}\left(\hat{\boldsymbol{X}}_t\right) - \tilde{\boldsymbol{d}}\right]^T \boldsymbol{S}^{-1}\left[\boldsymbol{d}\left(\hat{\boldsymbol{X}}_t\right) - \tilde{\boldsymbol{d}}\right] \tag{5}$$

where $\boldsymbol{d}\left(\hat{\boldsymbol{X}}_t\right) = \left[\| \hat{\boldsymbol{X}}_t - \boldsymbol{X}_1 \|, \ldots, \| \hat{\boldsymbol{X}}_t - \boldsymbol{X}_N \|\right]^T$ and $\boldsymbol{S}$ is the covariance matrix of the distance measurement errors. This minimization problem can be solved using ML techniques similar to those discussed in the previous section.

In real applications the situation is much more complicated. Some challenges that can be encountered in distance-based localization include: the distance measurement error may be neither additive nor Gaussian noises; the measured distances may be biased; a non-anchor node may have to derive its location from the estimated locations (containing errors) of its neighbouring non-anchor nodes instead of anchors; if a non-anchor node is a neighbour of a set of nodes which are almost collinear, the non-anchor node may not be able to uniquely determine its location estimate; the network topology may be irregular, not to mention the challenge of designing a computationally efficient localization algorithm for large scale networks. It is these challenges that make distance-based localization problem both challenging and intriguing. The other chapters of this book explore various aspects of distance-based localization problems and lead readers to establish a solid understanding in both distance-based localization and localization using other types of measurements.

Time-Difference-of-Arrival Measurements

Time-difference-of-arrival (TDOA) measurements measure the difference between the arrival times of a transmitter signal at two receivers respectively. In $\Re^2$, denote the coordinates of the two receivers by X_i and X_j, and the coordinates of the transmitter by X_t. The measured TDOA Δt_{ij} is related to the locations of the two receivers by

$$\Delta t_{ij} = t_i - t_j = \frac{1}{c}\left(\| X_t - X_i \| - \| X_t - X_j \|\right) \tag{6}$$

where t_i and t_j are the arrival times of the transmitter signal at receivers i and j respectively and c is the propagation speed of the transmitter signal. Assuming the receiver locations are known and the two receivers are perfectly synchronized, Equation (6) defines one branch of a hyperbola on which the transmitter must lie. The foci of the hyperbola are at the locations of the receivers i and j. In a system of N receivers, there are N–1 linearly independent TDOA measurements, hence N–1 linearly independent equations like (6). In $\Re^2$, TDOA measurements from a *minimum* of three receivers are required to uniquely determine the location of the transmitter. This is illustrated in Figure 4.

The accuracy of TDOA measurements is affected by the synchronization error between receivers and multipath. The accuracy and temporal resolution capabilities of TDOA measurements will improve when the separation between receivers increases because this increases differences between times of arrival. Readers are referred to (C. K. Chen & Gardner, 1992; Rappaport, Reed, & Woerner, 1996; Schell & Gardner, 1993) for more detailed discussion.

In the presence of measurement errors and assuming that the errors are in the form of additive zero mean Gaussian noise, in a system of N receivers, the TDOA equations can be written compactly in matrix form as

$$\Delta\tilde{t} = \Delta X + e \tag{7}$$

where $\Delta\tilde{t} = \left[\Delta\tilde{t}_{21}, \Delta\tilde{t}_{31}, \ldots, \Delta\tilde{t}_{N1}\right]^T$, $\Delta X = \frac{1}{c}\left[\|X_t - X_1\| - \|X_t - X_2\|, \ldots, \|X_t - X_1\| - \|X_t - X_N\|\right]^T$ and $e = [e_{21}, \ldots, e_{N1}]$ with e_{j1} being the measurement error of $\Delta\tilde{t}_{j1}$. Defining $f(\hat{X}) = \frac{1}{c}\left[\|\hat{X} - X_1\| - \|\hat{X} - X_2\|, \ldots, \|\hat{X} - X_1\| - \|\hat{X} - X_N\|\right]^T$, an ML formulation of the location estimation problem using TDOA measurements is:

$$\hat{X}_t = \arg\min\left[\Delta\tilde{t} - f(\hat{X})\right]^T S^{-1}\left[\Delta\tilde{t} - f(\hat{X})\right] \tag{8}$$

Figure 4. Two intersecting branches of two hyperbolas obtained by TDOA measurements from three receivers uniquely determine the location of the transmitter

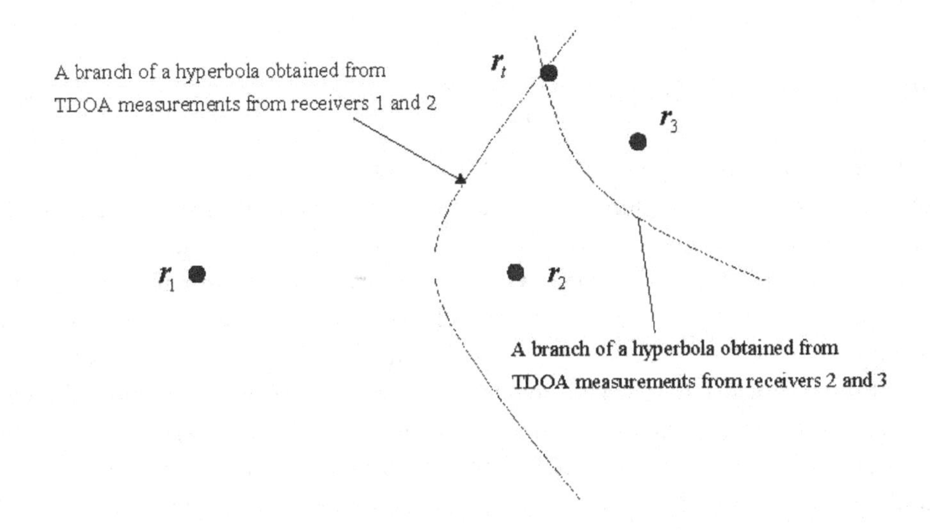

where S is the covariance matrix of TDOA measurement errors. Equation (8) however is in a very complicated form. In order to obtain a reasonably simple estimator, $f(X)$ can be linearized around a reference point X_0 using Taylor series:

$$f(X) \approx f(X_0) + f'(X_0)(X - X_0) \tag{9}$$

where $f'(X_0)$ is the partial derivative of $f(X)$ with respect to X evaluated at X_0. A recursive solution to the maximum likelihood estimator can then be obtained (Torrieri, 1984):

$$\hat{X}_{t,k+1} = \hat{X}_{t,k} + \left(f'\left(\hat{X}_{t,k}\right)^T S^{-1} f'\left(\hat{X}_{t,k}\right)\right)^{-1} f'\left(\hat{X}_{t,k}\right)^T S^{-1}\left(\Delta\tilde{t} - f\left(\hat{X}_{t,k}\right)\right) \tag{10}$$

This method obviously relies on a good initial guess of the transmitter location. Furthermore, the method can result in significant location estimation errors in some situations due to geometric delusion of precision (GDOP) effects. GDOP describes situation in which a relatively small measurement error can cause a large location estimation error because the transmitter is located on a portion of the hyperbola far away from the receivers (Bancroft, 1985; Rappaport, Reed, & Woerner, 1996). There are many other approaches presented in the literature on TDOA based location estimation and we refer readers to (Abel, 1990; Chan & Ho, 1994; Crippen & Havel, 1988; Dogancay, 2005; B. T. Fang, 1990; Smith & Abel, 1987)

Received Signal Strength Measurements

Received signal strength (RSS) measurements estimate the distances between neighbouring sensors from the received signal strength measurements between the two sensors (Bergamo & Mazzini, 2002; Elnahrawy, Li, & Martin, 2004; Madigan et al., 2005; Niculescu & Nath, 2003; Patwari et al., 2005). Most wireless devices have the capability of measuring the received signal strength.

The wireless signal strength received by a sensor from another sensor is a monotonically decreasing function of their distance. This relationship between the received signal strength and distance is popularly modelled by the following log-normal model:

$$P_r(d)[dBm] = P_0(d_0)[dBm] - 10n_p \log_{10}\left(\frac{d}{d_0}\right) + X_\sigma \quad (11)$$

where $P_0(d_0)[dBm]$ is a reference power in dB milliwatts at a reference distance d_0 from the transmitter, n_p is the path loss exponent that measures the rate at which the received signal strength decreases with distance, and X_σ is a zero mean Gaussian distributed random variable with standard deviation σ and it accounts for the random effect caused by shadowing. Both n_p and σ are environment dependent. The path loss exponent n_p is typically assumed to be a constant however some measurement studies suggest the parameter is more accurately modelled by a Gaussian random variable or different path loss exponent should be used for a receiver in the far-field region of the transmitter or in the near-field region of the transmitter. Given the model and model parameters, which are obtained via *a priori* measurements, the inter-sensor distances can be estimated from the RSS measurements. Localization algorithms can then be applied to these distance measurements to obtain estimated locations of sensors.

Chapter V - Calibration and Measurement of Signal Strength for Sensor Localization features a thorough discussion on a number of practical issues involved in the use of RSS measurements for distance estimation. The chapter focuses on device effects and modelling problems which are important for the implementation of RSS-based distance estimation but are not well covered in the literature. These include transceiver device manufacturing variations, battery effects on transmit power, nonlinearities in the circuit, and path loss model parameter estimation. Measurement methodologies are presented to characterize these effects for wireless sensors and suggestions are made to limit impact of these effects.

Note that in addition to the log-normal model many other models have also been proposed in the literature which can better describe the wireless signal propagation characteristic for signals within a specific frequency spectrum in a specific environment, for example Longley-Rice model, Durkin's model, Okumuran model, Hata model and wideband PCS microcell model for outdoor environments, and Ericsson multiple breakpoint model, attenuation factor model and the combined use of site specific propagation models and graphical information system databases for radio signal prediction in indoor environments (Rappaport, 2001).

Yet another interesting technique to estimate the distance between an optical receiver and an optical transmitter is the *lighthouse approach* reported in (Romer, 2003). The lighthouse approach estimates the distance between an optical receiver and a transmitter of a parallel rotating optical beam by measuring the time duration that the receiver dwells in the beam. A parallel optical beam is a beam whose beam width is constant with respect to the distance from the rotational axis of the beam. It is the characteristic of the parallel beam that the time the optical receiver dwells in the beam is inversely proportional to the distance between the optical receiver and the rotational axis of the beam enables the distance measurements. A major advantage of the lighthouse approach is the optical receiver can be of a very small size and low cost, thus making the idea of "smart dust" possible. However the transmitter may be large and expensive. The approach also requires a direct LOS between the optical receiver and the transmitter.

Connectivity Measurements

Connectivity measurements are possibly the simplest measurements. In connectivity measurements, a sensor measures which sensors are in its transmission range. Such measurements can be interpreted as binary distance measurements, i.e., either another particular sensor is within the transmission range of a given sensor or it is outside the transmission range of that sensor.

A sensor being in the transmission range of another sensor defines a proximity constraint between these two sensors, which can be exploited for localization. In its simplest form, a non-anchor sensor being a neighbour of three anchors means the non-anchor sensor is very close to the three anchors and many algorithms then use the centroid of the three anchors as the estimated location of the non-anchor sensor. In the later section, we shall give a more detailed discussion of connectivity-based localization algorithms in large scale networks.

RSS Profiling Measurements

Above, we have mentioned some techniques to estimate the distances between sensors from RSS measurements. Localization algorithms can then be applied to these distance measurements to obtain estimated locations of sensors. The implementation of such localization techniques however faces two major challenges: first the wireless environments, especially indoor wireless environments, are very complicated. It is often difficult to determine the best model for RSS-based distance estimation. Second, the determination of model parameters is also a difficult task. Such difficulties can be overcome using another category of localization techniques, namely the *RSS profiling-based localization techniques* (Bahl & Padmanabhan, 2000; Krishnan, Krishnakumar, Ju, Mallows, & Gamt, 2004; Prasithsangaree, Krishnamurthy, & Chrysanthis, 2002; Ray, Lai, & Paschalidis, 2005; Roos, Myllymaki, & Tirri, 2002), which estimate sensor location from RSS measurements directly.

The RSS profiling-based localization techniques works by first constructing a form of map of the signal strength behaviour of anchor nodes in the coverage area. The map is obtained either offline by *a priori* measurements or online using sniffing devices (Krishnan et al., 2004) deployed at known locations. The RSS profiling-based localization techniques have been mainly used for location estimation in wireless local area networks (WLANs), but they would appear to be attractive also for WSNs.

In RSS profiling-based localization systems, in addition to anchor nodes (e.g., access points in WLANs) and non-anchor nodes, a large number of sample points, e.g., sniffing devices or *apriori* chosen locations at which the RSS measurements from anchors are to be obtained before the localization of non-anchor nodes starts, are distributed throughout the coverage area of the sensor network. At each sample point, a vector of signal strengths is obtained, with the kth entry corresponding to the signal strength received from the kth anchor at the sample point. Of course, many entries of the signal strength vector may be zero or very small, corresponding to anchor nodes at larger distances (relative to the transmission range) from the sample point. The collection of all these vectors provides (by extrapolation in the vicinity of the sample points) a RSS map of the whole region. The collection constitutes the RSS map, and it is unique with respect to the anchor locations and the environment. The model is stored in a central location. By referring to the RSS map, a non-anchor node can estimate its location using the RSS measurements from anchors by either choosing the location of the sample point, whose signal strength vector is the closest match of that of the non-anchor node, to be its location, or derive its estimated location from the locations of a set of sample points whose signal strength vectors better match that of the non-anchor node than other sample points.

In this section, a number of measurement techniques and the basic principles of location estimation using these measurements are discussed. Which measurement technique to use for location estimation will depend on the requirements of the specific application on localization accuracy, cost and complexity of localization algorithms. Typically, localization algorithms based on AOA and propagation time measurements are able to achieve better accuracy than localization algorithms based on RSS measurements. However, that improved accuracy is achieved at the expense of higher equipment cost. Also the high nonlinearity and complexity in the observation model, i.e., the equation relating the coordinates of sensors to measurements, of AOA and TDOA measurements make them a less attractive option than distance measurements for location estimation in large scale multi-hop wireless sensor networks.

SENSOR NETWORK LOCALIZATION THEORY AND ALGORITHMS

In this section, we give a brief introduction to some fundamental theories in sensor network localization and major sensor network localization algorithms as well as introducing the relevant chapters of the book.

Graph Theory and its Applications in Sensor Network Localization

The task of WSN localization algorithms is to estimate the locations of sensors with initially unknown location information, i.e., the non-anchors, by using *a priori* knowledge of the locations of a few sensors, i.e., anchors, and inter-sensor measurements such as distance, AOA, TDOA and connectivity. A fundamental question in sensor network localization is whether a solution to the localization problem is unique. The network, with the given set of anchors, non-anchors and inter-sensor measurements, is said to be *uniquely localizable* if there is a unique set of locations consistent with the given data. Graph theory has been found to be particularly useful for solving the above problem of unique localization. Graph theory also forms the basis of many localization algorithms, especially for the category of distance-based localization problem, noting that it has been used to study the localization problem using other types of measurements, e.g., TDOA and AOA measurements, as well.

The task of distance-based localization problem is to estimate the locations of non-anchors using the known locations of anchors and inter-sensor distance measurements. A graphical model for distance-based localization problem can be built by representing each sensor in the network uniquely with a vertex and vice versa. An edge exists between two vertices if the distance between the corresponding sensors is known. Note that there is always an edge between two vertices representing two anchors as the distance between two anchors can be obtained from their known locations. The obtained graph $G(V,E)$ with V being the set of vertices and E being the set of edges is called the *underlying graph* of the sensor network. Details of graph theoretical representations of WSNs and their use in localization can be found in ***Chapter 6- Graph Theoretic Techniques in the Analysis of Uniquely Localizable Sensor Networks.***

In rigid graph theory, a mapping $p : V \rightarrow \Re^d$ $(d \in \{2,3\})$, assigning a location in $\Re^d$ to each vertex of graph $G = (V, E)$, is called a d–dimensional *representation* of G. With this definition the localization problem can be seen as finding the *correct representation* of the underlying graph of the WSN that is consistent with the given data. Given a graph $G = (V, E)$ and a representation p of it, the pair (G, p) is called a *framework*. A particular graph property associated with unique localizability of sensor networks is *global rigidity*: A framework (G, p) is called *globally rigid* if every framework (G, p_1) satisfying $\|p_1(i) - p_1(j)\| = \|p(i) - p(j)\|$ for any vertex pair $i, j \in V$, which are connected by an edge in E, also satisfies the same equality for any other vertex pairs that are not connected by an edge. A relaxed form of global rigidity is *rigidity*: A framework (G, p) is *rigid* if there exists a sufficiently small positive constant e_p such that every framework (G, p_1) satisfying $\|p_1(i) - p(i)\| < \varepsilon_p$ for all $i \in V$ and $\|p_1(i) - p_1(j)\| = \|p(i) - p(j)\|$ for any vertex pair $i, j \in V$, which are connected by an edge in E, satisfies $\|p_1(i) - p_1(j)\| = \|p(i) - p(j)\|$ for any other vertex pairs that are not connected by a single edge as well. If the framework (G, p) formed by the underlying graph G of a WSN and its correct representation p is not rigid, there are an infinite number of solutions to the localization problem that are consistent with the given data.

If the framework (G, p) formed by the underlying graph G of a WSN and its correct representation p is globally rigid, the sensor network with at least three non-collinear anchors in $\mathfrak{R}^2$ or four non-coplanar anchors in $\mathfrak{R}^3$ is uniquely localizable. If a framework (G, p) is rigid but not globally rigid, there exist two types of discontinuous deformations that can prevent finding a unique representation of G consistent with the information of anchor node positions and distance measurements: *flip ambiguities* and *discontinuous flex ambiguities*. In *flip ambiguities* in $\mathfrak{R}^d$ $(d \in \{2,3\})$, a vertex (sensor) v has a set of neighbours which span a $(d-1)$-dimensional subspace, e.g., v has only d neighbours, in $\mathfrak{R}^2$ v has a set of neighbours located on a line, or in $\mathfrak{R}^3$ v has a set of neighbours located on a plane, which leads to the possibility of the neighbours forming a mirror through which v can be reflected. In *discontinuous flex ambiguities* in $\mathfrak{R}^d$ $(d \in \{2,3\})$, the removal of an edge or a set of edges allows the remaining part of the graph to be flexed to a different realization (which cannot be obtained from the original realization by translation, rotation or reflection) such that the removed edge can be reinserted with the same length. Figure 5 shows an example of flip ambiguity and discontinuous flex ambiguity in $\mathfrak{R}^2$. Note that in Figure 5.(a) and 5.(b), both the figure on the left side and the figure on the right side satisfy the same set of distance constraints but the locations of vertices are different, which means the associated sensor network is not uniquely localizable.

Using graph theory, we can identify necessary conditions as well as sufficient conditions that need to be satisfied by the underlying graph of a sensor network in order for the network to be uniquely localizable. ***Chapter VI*** gives a detailed overview of this topic, providing various results in graph theory to characterize uniquely localizable networks in two dimensions. Conditions required for the sensor network to be uniquely localizable are discussed and techniques to test the unique localizability are introduced. While the focus of the chapter is 2-dimensional distance-based localization, the authors also consider sensor networks with mixed distance and AOA measurements as well as unique localizability of 3-dimensional networks.

Note that the unique localizability conditions mentioned above are independent of the specific localization algorithm being used. Furthermore, the above discussion has been carried out without considering measurement errors. The problem becomes more complicated when the effects of measurement errors are considered. For example, it has become a common knowledge that in $\mathfrak{R}^2$ in the presence of measurement errors, a non-anchor node connected to a set of two or more anchors which are exactly or almost collinear, the non-anchor node is *likely* to have *flip ambiguity* problem. However we are yet to establish an accurate knowledge in the area, i.e., given the measurement error distribution and anchor locations, how to compute the probability that the non-anchor's location estimation be contaminated by

Figure 5. An illustration of the flip and discontinuous flex ambiguity in 2D: (a) Flip ambiguity: The neighbours of vertex v_4, v_1, v_2 and v_3 are on the same line. Vertex v_4 can be reflected across the line on which vertices v_1, v_2 and v_3 locate to a new position without violating the distance constraints. (b) Discontinuous flex ambiguity: Removing the edge between v_3 and v_4, the vertices v_1, v_2, v_3 and v_4 can be moved continuously to other positions while maintaining the length of the edges between them. When these vertices move to positions such that the edge between v_3 and v_4 can be reinserted with the same length, we obtain a new graph. Both the graph on the left side and the graph on the right side satisfy the same set of distance constraints.

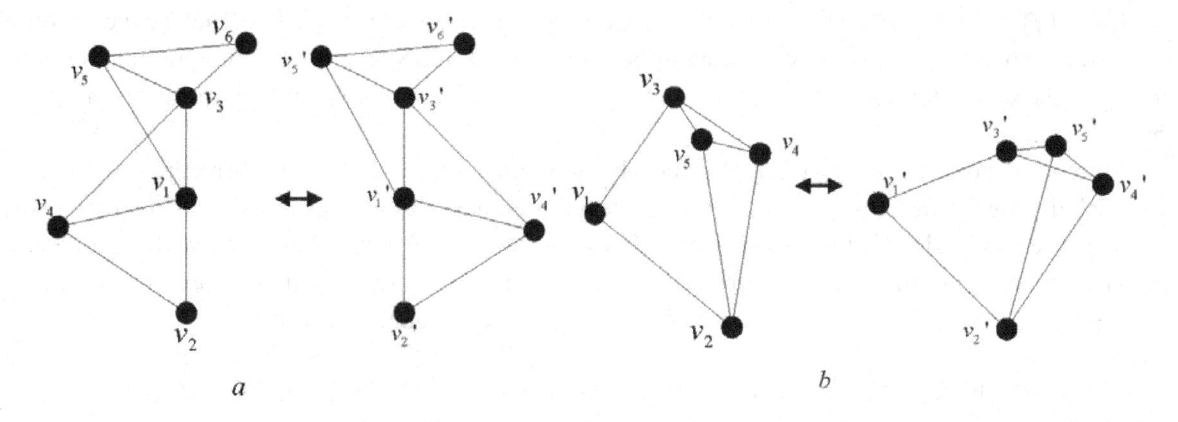

flip ambiguity error? The problem is further complicated in a large scale network where the non-anchor node may have to rely on the inaccurate location estimates of its non-anchor neighbours to estimate its own location. Therefore the analysis on unique localizability can be used to label those sensors with large errors in their location estimates so that those errors do not propagate to the rest of the network.

It is worth noting that flip ambiguity and discontinuous flex ambiguity problems do not necessarily occur in every sensor network. The probability of occurrence of ambiguities is generally smaller in dense networks where the average number of neighbours per node is high. However when such ambiguities occur, they generally cause a large error in the location estimate of a non-anchor node. This error may further propagate to other non-anchor nodes when they use the estimated location of the non-anchor node to determine their own locations. Therefore the performance impact of flip ambiguity and discontinuous ambiguity on sensor network localization may be significant. This has been validated by a number of analytical and simulation studies including some of our own work.

Graph theory has also been used to characterize large scale networks in which the design of an efficient localization algorithm is possible. The computational complexity of localization algorithms is an important consideration in the localization of large scale networks and the computational complexity of distance-based localization algorithms in large scale networks has been investigated in the literature (Aspnes et al., 2006; Eren et al., 2004; Saxe, 1979). In general, the computational complexity of localization algorithms is exponential in the number of sensor nodes (Saxe, 1979). Nevertheless, there is a category of networks where the design of efficient localization algorithms is possible. Specifically, if the underlying graph of the network is a *bilateration*, *trilateration* or *quadrilateration* graph, it is possible to design localization algorithms whose computational complexity is polynomial (and on occasions linear) in the number of sensor nodes (Aspnes et al., 2006; Cao, Anderson, & Morse, 2005; Eren et al., 2004).

A graph $G = (V,E)$ is called a *bilateration graph* if there exists an ordering of vertices $v_1, v_2, \ldots, v_{|V|}$, termed *bilaterative ordering*, such that (i) the edges (v_1, v_2), (v_1, v_3), (v_2, v_3) are all in E, (ii) each vertex v_i for $i = 4,5,\ldots,|V|-1$ is connected to (at least) two of the vertices in $v_1, v_2, \ldots, v_{i-1}$, and (iii) the vertex $v_{|V|}$ is connected to (at least) three of the vertices $v_1, v_2, \ldots, v_{|V|-1}$. The symbol $|V|$ denotes the cardinality of set V. If the underlying graph of a network is a bilateration graph, an efficient sequential localization algorithm can be designed for the network (J. Fang, Cao, Morse, & Anderson, 2006). The concepts of *trilateration graphs* and *quadrilateration graphs* are defined analogously. Note that trilateration and quadrilateration graphs are necessarily bilateration graphs as well. We refer readers to the above reference and ***Chapter VII - Sequential Localization with Inaccurate Measurements*** for more detailed discussions on this topic. ***Chapter VII*** further presents an efficient sequential algorithm for estimating sensor locations using inaccurate distance measurements. The algorithm is based on the above graph theory concepts; the authors have further developed existing work by demonstrating that it is possible to design a computationally efficient sequential localization algorithm for networks whose underlying graphs are not necessarily bilateration graphs.

Sensor Network Localization Algorithms

Centralized vs. Distributed Localization

Based on the approach of processing the individual inter-sensor measurement data, localization algorithms can be broadly classified into two categories: *centralized algorithms* and *distributed algorithms.* In centralized algorithms, all the individual inter-sensor measurements are sent to a single central processor where the estimated locations of non-anchor nodes are computed; while in distributed algorithms each node (or a group of nodes in close proximity to each other) estimate its (their) own location(s) using inter-sensor measurements and the location information collected from its (their) neighbours. Major approaches for designing centralized algorithms include *multidimensional scaling (MDS), linear programming* and *stochastic optimization* approaches. Some well-known distributed localization algorithms include the "*DV-hop*" and "*DV-distance*" algorithms (Niculescu & Nath, 2001), a number of other algorithms based on the above two algorithms (Chris Savarese & Rabaey, 2002; C. Savarese, Rabaey, & Beutel, 2001), and the *nonparametric belief propagation* algorithms (Ihler, Fisher, Moses, & Willsky, 2005) and its variants (Fox, Hightower, Lin, Schulz, & Borriello, 2003). The "*sweep*" category of *sequential algorithms* reported in ***Chapter VII*** also represents a promising direction in the development of distributed algorithms, which may offer an optimum balance between localization accuracy and computational efficiency in large scale sensor networks.

Centralized and distributed distance-based localization algorithms can be compared from several perspectives, including location estimation accuracy, implementation and computational complexities, and energy consumption.

Distributed localization algorithms are generally considered to be more computationally efficient and easier to implement in large scale networks. However in certain networks where centralized information architecture already exists, such as road traffic monitoring and control, environmental monitoring, health monitoring, and precision agriculture monitoring networks, the measurement data of all the nodes in the network need to be collected and sent to a central processor unit. In such a network the individual sensors may be of limited computational capability; it is convenient to piggyback localization related measurements to other measurement data and send them together to the central processing unit. Therefore a centralized localization algorithm appears to be a natural choice for such networks with existing centralized information architecture.

In terms of location estimation accuracy, centralized algorithms are likely to provide more accurate location estimates than distributed algorithms. One of the reasons is the availability of global information in centralized algorithms. However centralized algorithms suffer from the scalability problem and generally are not feasible to be implemented for large scale sensor networks. Other disadvantages of centralized algorithms, as compared to distributed algorithms, are their requirement of higher computational complexity and lower reliability due to accumulated information inaccuracies/losses involved in multihop transmission from individual sensors to the centralized processor over a WSN.

On the other hand, distributed algorithms are more difficult to design because of the potentially complicated relationship between local behaviour and global behaviour. That is, algorithms that are locally optimal may not perform well globally. Optimal distribution of the computation of a centralized algorithm in a distributed implementation in general remains an open research problem. Error propagation is another potential problem in distributed algorithms. Moreover, distributed algorithms generally require multiple iterations to arrive at a stable solution. This may cause the localization process to take longer time than the acceptable in some cases.

From the perspective of energy consumption, the individual amounts of energy required for each type of operation in centralized and distributed localization algorithms in the specific hardware and the transmission range setting needs to be considered. Depending on the setting, the energy required for transmitting a single bit could be used to execute 1,000 to 2,000 instructions (Chen, Yao, & Hudson, 2002). Centralized algorithms in large networks require each sensor's measurements to be sent over multiple hops to a central processor, while distributed algorithms require only local information exchange between neighbouring nodes. Nevertheless, in distributed algorithms, many such local exchanges may be required, depending on the number of iterations needed to arrive at a stable solution. A comparison of the communication energy efficiencies of centralized and distributed algorithms is provided in (Rabbat & Nowak, 2004), where it is concluded that in general, if in a given sensor network and distributed algorithm, the average number of hops to the central processor exceeds the necessary number of iterations, then the distributed algorithm will be more energy-efficient than a typical centralized algorithm.

Finally it is worth noting that the separation between distributed localization algorithms and centralized localization algorithms can sometimes be blurred. Any algorithm for distributed localization can always be applied to centralized problems. Distributed versions of centralized algorithms can also be designed for certain applications. A typical way of designing distributed versions of centralized algorithms involves dividing the entire network into several overlapping regions; implementing centralized localization algorithms in each region; then stitching these local maps for each region together by using common nodes between overlapping regions to form a global map (Capkun, Hamdi, & Hubaux, 2001; Ji & Zha, 2004; Oh-Heum & Ha-Joo, 2008). Such techniques may offer an optimum tradeoff between the advantages and disadvantages of centralized and distributed algorithms discussed above. A particular example of such techniques is multidimensional scaling-based localization, which is discussed further in the next subsection.

In the rest of this section, we give a brief introduction to each major localization technique.

Multidimensional Scaling Algorithms

The *Multidimensional Scaling (MDS)* technique can find its basis in graph theory and was originally used in psychometrics and psychophysics. It is often used as part of exploratory data analysis or information visualization technique that displays the structure of distance-like data as a geometric picture. The typical goal of MDS is to create a configuration of points in one, two, or three dimensions, whose inter-point distances are "close" to the known (and possibly inaccurate) inter-point distances. Depending on the criteria used to define "close", many variants of the basic MDS exist. MDS has been applied in many fields, such as machine learning and computational chemistry. When used for localization, MDS utilizes connectivity or distance information between sensors for location estimation.

Typical procedure of MDS algorithms involves first computing the shortest paths (i.e., the least number of hops) between all pairs of nodes. If distances between all pairs of sensors along the shortest path connecting two nodes are known, the distance between the two nodes along the shortest path can be computed. This information is used to construct a distance matrix for MDS, where the entry (i, j) represents the distance along the shortest path between nodes i and j. If only connectivity information is available, the entry (i, j) then represents the least number of hops between nodes i and j. Then MDS is applied to the distance matrix and an approximate value of the relative coordinates of each node is obtained. Finally, the relative coordinates are transformed to the absolute coordinates by aligning the estimated relative coordinates of anchors with their absolute coordinates. The location estimates obtained using earlier steps can be refined using a least-squares (LS) minimization.

The basic form of MDS is a centralized localization technique and may only be used in a regular network where the distance between two nodes along the shortest path is close to their Euclidean distance. However several variants of the basic MDS algorithm are proposed which allow the implementation of MDS technique in distributed environment and in irregular networks.

Chapter VIII - MDS-Based Localization provides a more detailed discussion on MDS localization techniques and presents several network localization methods based on these techniques. The chapter first introduces the basics of MDS techniques, and then four algorithms based on MDS: *MDS-MAP(C), MDS-MAP(P), MDS-Hybrid* and *RangeQ-MDS*. MDS-MAP(C) is a centralized algorithm. MDS-MAP(P) is a variant of MDS-MAP(C) for implementation in distributed environment. It has better performance than MDS-MAP(C) in irregular networks. MDS-Hybrid considers relative location estimation in an environment without anchors. RangeQ-MDS uses a quantized RSS-based distance estimation technique to achieve more accurate localization than algorithms using binary measurements of connectivity only (i.e., two nodes are either connected or not connected).

Linear Programming Based Localization Techniques

Many distance-based or connectivity-based localization problems can be formulated as a convex optimization problem and solved using linear and semidefinite programming (SDP) techniques (Doherty, Pister, & El Ghaoui, 2001). Semidefinite programs are a generalization of the linear programs and have the following form

$$\begin{aligned} &\text{Minimize} && \boldsymbol{c}^T \boldsymbol{X} \\ &\text{Subject to} && \boldsymbol{F}(\boldsymbol{X}) = \boldsymbol{F}_0 + \boldsymbol{X}_1 \boldsymbol{F}_1 + \cdots + \boldsymbol{X}_N \boldsymbol{F}_N \\ &&& \boldsymbol{AX} < \boldsymbol{B} \\ &&& \boldsymbol{F}_k = \boldsymbol{F}_k^T \end{aligned} \tag{12}$$

where $\boldsymbol{X} = [\boldsymbol{X}_1, \boldsymbol{X}_2, \ldots, \boldsymbol{X}_N]^T$ and $\boldsymbol{X}_k = [x_k, y_k]^T$ represents the coordinate vector of node k. The quantities $\boldsymbol{A}$, $\boldsymbol{B}$, $\boldsymbol{c}$ and $\boldsymbol{F}_k$ are all known. The inequality in (12) is known as a *linear matrix inequality (LMI)*.

If only connectivity information is available, a connection between nodes i and j can be represented by a "radial constraint" on the node locations: $\left\|\boldsymbol{X}_i - \boldsymbol{X}_j\right\| \leq R$ with R being the transmission range of wireless sensors. This constraint is a convex constraint and can be transformed into an LMI to be used in (12). A solution to the coordinates of the non-anchor nodes satisfying the "radial constraints" can be obtained by leaving the objective function $\boldsymbol{c}^T\boldsymbol{X}$ blank and solving the problem. Obviously there may be many possible coordinates of the non-anchor nodes satisfying the constraints, i.e., the solution may not be unique. If we set the entry of $\boldsymbol{c}$ corresponding to x_k (or y_k) to be 1 (or -1) and all other elements of $\boldsymbol{c}$ to be zero, the problem becomes a constrained maximization (or minimization) problem, which gives respectively the maximum (or minimum) value of x_k (or y_k) satisfying the constraints in (12). A rectangular box bounding the location estimates of the non-anchor node k can be obtained from these lower and upper bound on x_k and y_k. The detailed connectivity-based localization algorithm is reported in (Doherty et al., 2001).

The above SDP formulation of the connectivity-based localization problem can be readily extended to incorporate distance measurements (Doherty et al., 2001). In (Biswas & Ye, 2004) the distance-based localization problem is used in a quadratic form and solved using SDP. In (Liang, Wang, & Ye, 2004) gradient search is used to fine tune the initial estimated locations obtained using SDP and improves the accuracy of localization.

Note that different linear programming techniques have been used in various chapters of this book.

Stochastic Optimization Based Localization Techniques

The stochastic optimization approach provides an alternative formulation and solution of the distance-based localization problem using combinatorial optimization notions and tools. One of the most widely used tools in this approach is the *simulated annealing (SA)* technique (Kannan, Mao, & Vucetic, 2005).

SA is a technique for combinatorial optimization problems. The SA algorithm exploits an analogy between the way in which a metal cools and freezes into a minimum energy crystalline structure (the annealing process) and the search for a minimum in a more general system. It is a generalization of the Monte Carlo method. It transforms a poor, unordered solution into a highly optimized, desirable solution. This principle of SA technique with an analogous set of "controlled cooling" operations was used in the combinatorial optimization problems, such as minimizing functions of multiple variables, to obtain a highly optimized, desirable solution (Kirkpatrick, Gelatt, & Vecchi, 1983). We refer the readers to (Kannan et al., 2005; Kannan, Mao, & Vucetic, 2006) for a more detailed description of the design of a SA algorithm for distance-based localization problems.

A properly designed SA has the advantage that it is robust against being trapped into a false local minimum. However SA is also well-known to be very computationally demanding.

The DV-Hop and DV-Distance Localization Algorithms

The *DV(distance vector)-hop* algorithm (Niculescu & Nath, 2001) utilizes the connectivity measurements to estimate locations of non-anchor nodes. The algorithm starts with all anchors broadcasting their locations to other nodes in the network. The messages are propagated hop-by-hop and there is a hop-count in the message. Each node maintains an anchor information table and counts the least number of hops that it is away from an anchor. When an anchor receives a message from another anchor, it estimates the average distance of one hop using the locations of both anchors and the hop-count, and sends it back to the network as a correction factor. When receiving the correction factor, a non-anchor node is able to estimate its distance to anchors and performs trilateration to estimate its location if its distances to at least three anchors are available.

The *DV-distance* algorithm is similar to the DV-hop algorithm except that it includes measured distances into the localization process. The main idea in the DV-distance algorithm is the propagation of measured distance among neighbouring nodes instead of hop count.

Since the proposal of the DV-hop and DV-distance algorithms, many other algorithms based on essentially the same principle were proposed which aims to improve the performance of the basic DV-hop and DV-distance algorithms under various conditions, e.g., in irregular networks or when there are additional information such as node distribution available. We refer interested readers to (Chris Savarese & Rabaey, 2002; Shang, Ruml, Zhang, & Fromherz, 2004) for more detailed discussion.

Statistical Location Estimation Techniques

In the early part of this chapter, we have mentioned in a number of places the use of the ML estimator for localization under various types of measurements. Denote the coordinator vectors of non-anchor nodes by $\boldsymbol{X}$ and the vector of all inter-sensor measurements by $\boldsymbol{Z}$. Denote by $f(\boldsymbol{Z})$ the distribution of $\boldsymbol{Z}$ so that $f\left(\boldsymbol{Z} \mid \boldsymbol{X}\right)$ is the conditional probability of $\boldsymbol{Z}$ when the non-anchor nodes are at $\boldsymbol{X}$. The ML estimator is given by

$$\hat{\boldsymbol{X}} = \arg\max_{\hat{\boldsymbol{X}}} f\left(\boldsymbol{Z} \mid \hat{\boldsymbol{X}}\right) \tag{13}$$

When the inter-sensor measurements can be modelled by the sum of their respective true values and additive Gaussian noises with zero mean and the same variance, the ML estimator is equivalent to an LS estimator. When the variances of additive Gaussian noises are different, the ML estimator is equivalent to a weighted LS estimator. All three estimators, i.e., the ML estimator, the LS estimator and the weighted LS estimator, have been widely used in both centralized and distributed localization algorithms.

Occasionally we may have prior knowledge on the possible locations of non-anchor nodes. In that case, the *maximum a posteriori (MAP) estimator* can be used, which utilizes the prior knowledge on non-anchor nodes' locations to obtain a more accurate estimate. Denote the *a priori* known distribution of the non-anchor nodes by $g(X)$. The MAP estimator is given in the following:

$$\hat{\boldsymbol{X}} = \arg\max_{\hat{\boldsymbol{X}}} f\left(\boldsymbol{Z} \mid \hat{\boldsymbol{X}}\right) g\left(\hat{\boldsymbol{X}}\right) \tag{14}$$

Note that the MAP estimator of $\boldsymbol{X}$ coincides with the ML estimator when the non-anchor nodes have equal probability to be distributed anywhere in the sensor network area, i.e., $g(\boldsymbol{X})$ is a constant function.

The above estimators have often been used to obtain a point estimate of the non-anchors' locations. In some applications, we are interested in knowing in which region a non-anchor node is located. Such knowledge is often useful in asset management for example. Both the ML estimator and the MAP estimator can be altered to generate such location information. Assume that the entire network area is divided into M regions and each region is labelled by $L_k, 1 \le k \le M$. Denote by $g(L_k)$ the *a priori* known probability that a non-anchor node is located in L_k. Denote by $f(\boldsymbol{Z} \mid L_k)$ the conditional probability of $\boldsymbol{Z}$ when the non-anchors node is in L_k. The region in which the non-anchor node is located given the measurements $\boldsymbol{Z}$ can be estimated using the MAP estimator as:

$$L_k = \arg \max_{L_i, 1 \le i \le M} f(\boldsymbol{Z} \mid L_i) g(L_i) \tag{15}$$

An ML estimate of the region in which the non-anchor is located can be obtained analogously. ***Chapter IX - Statistical Location Detection*** provides more detailed discussions on the topic and presents a localization algorithm in indoor WLAN environment based on the same principle as that in Equation (15).

A recent statistical approach in distributed sensor network localization is the use of *Bayesian filter-based localization* techniques (Kwok, Fox, & Meila, 2004). Different from other localization techniques whose outputs are deterministic estimates of non-anchors' locations, Bayesian filters probabilistically estimate sensors' locations from noisy measurements. The outputs of Bayesian filters are probability distributions of the estimated locations conditioned on all available sensor data. Such probability distribution is known as *belief* representing uncertainty in estimated locations. Bayesian filter-based localization techniques are often implemented as iterative algorithms which iteratively update and improve such beliefs as localization process proceeds and more accurate knowledge about the neighbouring sensors become available. This process is known as *belief propagation.* In (Ihler et al., 2005), based on the Bayesian filters, the sensor network localization problem is formulated as an inference problem on a graphical model and a variant of *belief propagation (BP)* techniques, the so-called *nonparametric belief propagation (NBP)* algorithm, is applied to obtain an approximate solution to the sensor locations. The NBP idea is implemented as an iterative local message exchange algorithm, in each step of which each sensor node quantifies its "belief" about its location estimate, sends this belief information to its neighbours, receives relevant messages from them, and then iteratively updates its belief using Bayes' formula. The iteration process is terminated only when some convergence criterion is met about the beliefs and location estimates of the sensors in the network. Because of the difficulty both in obtaining an analytical expression of the belief function and in updating the belief function analytically, particle filters (Kwok et al., 2004) are often used to represent beliefs numerically by sets of samples, or particles. The main advantages of the NBP algorithm and the use of particle filters are its easy implementation in a distributed fashion and sufficiency of a small number of iterations to converge. Furthermore it is capable of providing information about location estimation uncertainties and accommodating non-Gaussian measurement errors. These advantages make the approach particularly attractive in non-linear systems with non-Gaussian measurement errors.

RSS-Based Localization Techniques

Chapters IX-XI of this book give a thorough discussion on various aspects involved in the design and implementation of RSS-based localization systems. The number of chapters in this book, the number of research papers in the area and the number of deployed systems on RSS-based localization techniques properly reflects the huge interest in the research community and industry on the techniques. As mentioned previously in this chapter, RSS-based localization techniques can only provide a coarse-grained estimate of sensor locations. However almost every wireless device has the capability of performing

RSS measurements and RSS-based localization techniques meet the exact demand from industry on localization solutions with minimal hardware investment. It is this feature of RSS-based localization techniques that drives the tremendous interest in their research and developments.

As mentioned above, ***Chapter IX*** presents an RSS-based localization system for indoor WLAN environments. The entire network area is divided into several regions and the algorithm identifies the region in which the non-anchor node resides. The localization problem is formulated as a multi-hypothesis testing problem and the authors provide an asymptotic performance guarantee of the system. The authors further investigate the optimal placement of anchor nodes in the system. The optimal placement problem is formulated as a mixed integer *linear programming* problem and a fast algorithm is presented for solving the problem. Finally the proposed techniques are validated using testbed implementations involving MICAz motes manufactured by Crossbow.

Chapter X - Theory and Practice of Signal Strength-Based Localization in Indoor Environments starts with a brief overview of *indoor localization techniques* and then focuses on RSS-based techniques for indoor wireless deployments using 802.11 technology. The authors present an analytical framework that aims to ascertain the attainable accuracy of RSS-based localization techniques. It provides answers to questions like "Is there any theoretical limit to the localization accuracy using techniques based on signal strength?". The approach is based on the analysis of *α-regions* in location space: If the probability that the observed signal strength at the receiver is due to a transmitter located inside a certain region is α, then this certain region is called an *α-region.* The definition of α-region leads to an analytical approach for characterizing uncertainties in RSS-based localization. Several properties of the uncertainties are established, including that uncertainty is proportional to the variance in signal strength. This observation has resulted in several algorithms which aim at improving localization performance by reducing the variance. The authors also summarize issues that may affect the design and deployment of RSS-based localization systems, including deployment ease, management simplicity, adaptability and cost of ownership and maintenance. With this insight, the authors present the "LEASE" architecture for localization that allows easy adaptability of localization models. The chapter concludes with a discussion of some open issues in the area.

Chapter XI - On a Class of Localization Algorithms Using Received Signal Strength surveys and compares several RSS-based localization techniques from two broad categories: *point-based* and *area-based.* In point-based localization, the goal is to return a single point estimate of the non-anchor node's location while in area-based localization the goal is to return the possible locations of the non-anchor node as an area or a volume. The authors find that individual RSS-based localization techniques have similar limited performance in localization error (i.e., the distance between the estimated location and the true location) and reveal the empirical law that using 802.11 technology, with dense sampling and a good algorithm, one can expect a median localization error of about 3 m; with relatively sparse sampling, every 6 m, one can still get a median localization error of 4.5 m. Therefore it can be concluded that there are fundamental limitations in indoor localization performance that cannot be transcended without using qualitatively more complex models of the indoor environment, e.g., models considering every wall, desk or shelf, or by adding extra hardware in the sensor node above that required for communication, e.g., very high frequency clocks to measure the TOA. The authors also briefly describe a sample core localization system called *GRAIL (General purpose Real-time Adaptable Localization)*, which can be integrated seamlessly into any application that utilizes radio positioning via simple Application Program Interfaces (APIs). The system has been used to simultaneously localize multiple devices running 802.11 (WiFi), 802.15.4 (ZigBee) and special customized RollCallTM radios.

Localization Techniques Based on Machine Learning and Information Theory

In the earlier part of this section, we have mentioned some widely used WSN localization approaches and introduced the relevant chapters of this book. There exist other less conventional approaches in the literature as well, which complement the above widely used approaches, especially by providing alternative localization solutions suitable for various specific application domains and settings. ***Chapters XII and XIII of this book*** present two such approaches.

Chapter XII - Machine Learning Based Localization presents a machine learning approach to localization. *Machine learning* is an information science field, studying algorithms that improve automatically through experience. It is concerned with the design and development of algorithms and techniques that allow computers or computing systems to "learn" rules and patterns out of massive data sets automatically, using certain computational and statistical tools of regression, detection, classification, pattern recognition, and data cleaning as well as convex optimization techniques. Two key concepts used in machine learning are *kernels,* which can be considered as systems that describe similarities between objects, and *support vector machines*, supervised learning methods used for regression and classification. Machine learning has been used in a number of areas including syntactic pattern recognition, search engines, medical diagnosis, bioinformatics, object recognition in computer vision, game playing and robot locomotion.

Chapter XII discusses the application of machine learning methods to WSN localization based on formulation of the localization problem (i) as a classification problem and (ii) as a regression problem. Both problem definitions are RSS-based, and RSS measurements from anchors at various sample points distributed inside the sensor network area are used as training data for the support vector machines. In the classification problem based approach, the sensor network area is partitioned into (overlapping or non-overlapping) geographical regions, and a set of classes are defined to represent membership to these regions. Using RSS measurements received from anchors at the non-anchor node and rules established from the training data, the classes attached to the non-anchor node location estimate, which represent the regions where the non-anchor node is estimated to lie, are found. If the found classes are more than one then the localization algorithm returns the centroid of the intersection of the regions corresponding to these classes as the location estimate of the non-anchor node. If only a single class is found, then the location estimate is determined as the centroid of the corresponding region. The regression problem based approach exploits the correlation between the RSS measurements from anchors at the non-anchor node and the RSS measurements from anchors at sampling points. The non-anchor node is estimated to be at the centroid of the sampling points whose RSS measurements have the highest correlation with those of the non-anchor node.

Chapter XIII - Robust Localization Using Identifying Codes presents a different paradigm for robust WSN localization based on *identifying codes*, a concept borrowed from the *information theory* literature with links to covering and superimposed codes. The approach involves choosing a set of discrete sampling points and transmitters in a given region such that each discrete sampling point is covered by a distinct set of transmitters. The location of a non-anchor node is estimated to be at the location of the discrete sampling point, which is covered by the same set of transmitters as the non-anchor node. The major challenges involved in using this approach are choosing the set of transmitters and finding good and robust identifying codes. The chapter presents the basics of robust identifying codes, use of these codes in WSN localization, design and analysis of an identifying code based algorithm, and implementation of the proposed algorithm on a test bed at Boston University involving a 33mx76m indoor region (fourth floor of the Photonics building) and four transmitters (anchors). The identifying codes-based approach has the simplifying advantage that a non-anchor node only needs to know the set of transmitters it can detect in order to infer its location. This feature makes the approach robust to spurious connections or sensor failures and suitable for implementation in harsh environments, at the expense of reduced localization accuracy.

Evaluation of Localization Algorithms

It is often the case that a number of solutions exist for solving the same localization problem. A question naturally arises is how to evaluate and compare the performance of various localization solutions.

Evaluating the performance of localization algorithms is important for both researchers and practitioners, either when validating a new algorithm against the previous state of the art, or when choosing existing algorithms which best fit the requirements of a given WSN application. However, there is currently no agreement in the research and engineering community on the criteria and performance metrics that should be used for the evaluation and comparison of localization algorithms. Neither there exists a standard methodology which takes an algorithm through modelling, simulation and emulation stages, and into real deployment. Part of the problem lies in the large number of factors that may affect the performance of a localization algorithm, including but not limited to: the type of measurements being used and measurement errors, the distributions of anchor and non-anchor nodes, the density of network nodes which is usually measured by the average node degree, the geometric shape of the network area, whether or not there is any prior knowledge of the network, the wireless environment in which the localization technique is being deployed, the presence of NLOS conditions. Quite often a localization algorithm performing well in one scenario, e.g., in regular networks, does not deliver a good performance in another scenario, e.g., in irregular networks. A localization algorithm delivering an excellent performance in simulation environment may also not perform satisfactorily in real deployment. All these phenomena highlight the importance of building a scientific methodology for the evaluation of localization algorithms.

Chapter XIV - Evaluation of Localization Algorithms addresses the above challenges by introducing a methodological approach to the evaluation of localization algorithms. The chapter contains a discussion of evaluation criteria and performance metrics, which is followed by statistical/empirical simulation models and parameters that affect the performance of the algorithms and hence their assessment. Two contrasting localization studies are presented and compared with reference to the evaluation criteria discussed throughout the chapter. The chapter concludes with a localization algorithm development cycle overview: from simulation to real deployment. The authors argue that algorithms should be simulated, emulated (on test beds or with empirical data sets) and subsequently implemented in hardware, in a realistic WSN deployment environment, as a complete test of their performance. It is hypothesised that establishing a common development and evaluation cycle for localization algorithms among researchers will lead to more realistic results and viable comparisons.

Chapter XV - Accuracy Bounds for Wireless Localization Methods looks at evaluation methods for localization systems from a different perspective and takes an analytical approach to performance evaluation. The authors argue that evaluation methods for localization systems serve two purposes. First, they allow a network designer to determine the achievable performance of a localization system from a given network configuration and available measurements prior to the deployment of the system. Second, these tools can be used to evaluate the performance of an existing localization system to see if the potential location accuracy is being achieved or if further improvements are possible.

The authors present several methods for calculating performance bounds for node localization in WSNs. The authors point out that the widely used *Cramer-Rao bound* relies on several assumptions: (i) The environment is an LOS radio propagation environment; (ii) The location estimator is unbiased; (iii) No prior information on node's location is available. Obviously, not all these assumptions are valid in real applications. Indeed, most distance-based, AOA-based and TDOA-based location estimators are biased which makes the second assumption invalid. The authors advocate the use of the *Weinstein-Weiss* and *extended Ziv-Zakai lower bounds* to address the above problems. These bounds remain valid under NLOS conditions and can also use all available information for bound calculations. It is demonstrated that these bounds are tight to actual estimator performance and may be used to determine the available accuracy of location estimation from survey data collected in the network area.

EXPERIMENTAL STUDIES AND APPLICATIONS OF WSN LOCALIZATION

The earlier sections of this chapter and correspondingly ***Chapters II-XV*** have largely focused on measurement techniques, theoretical backgrounds, and algorithm design for WSN localization. Nevertheless, there exist various other issues to consider in order to guarantee that an actual real-time WSN localization system works properly and performs well. The amount and the type of these issues in general differ for different application domains and tasks. ***Chapters XVI-XVIII*** of this book present three different WSN localization application studies exemplifying such further issues.

Chapter XVI - Experiences in Data Processing and Bayesian Filtering Applied to Localization discusses algorithms and solutions for signal processing and filtering for localization and *location tracking* applications. Here, the term *location tracking* is used for estimation of the trajectory of an object based on sequential measurements. As opposed to localization in static networks in which sensor locations do not change with time, location tracking techniques are developed to meet the demand (in a large number of application domains) for knowledge of the time-varying location of a moving object, which can be a vehicle, a robot, a mobile sensor unit, a human operator, etc.

Chapter XVI explains some practical issues for engineers interested in implementing location tracking solutions and their experiences gained from implementation and deployment of several such systems. In particular, the chapter introduces the data processing solutions found appropriate for commonly used sensor types, and discusses the use of Bayesian filtering for solving position tracking problem. The use of particle filters is recommended as a flexible solution appropriate for tracking in non-linear systems with non-Gaussian measurement errors. Finally the authors also give a detailed discussion on the design of some of the indoor and outdoor position tracking systems they have implemented, highlighting major design decisions and experiences gained from test deployments. Note that, the basics of Bayesian filters and particle filters and their use in location estimation in static networks have been introduced in the subsection *Stochastic Optimization Based Localization Techniques* above, and Chapter ***XVI*** features a more detailed introduction to Bayesian and particle filters as well as *Kalman filters*, focusing more on their application in location tracking.

Chapter XVII - A Wireless Mesh Network Platform for Vehicle Positioning and Location Tracking presents an experimental study on the integration of Wi-Fi based wireless mesh networks and Bluetooth technologies for detecting and tracking travelling cars and measuring their speeds. The authors propose a wireless platform for these purposes and deploy a small-scale network of four access points to validate the proposal. The platform employs RSS measurements and is shown to be able to track cars travelling at speeds of 0 to 70 km/h. The platform is found to be cost-effective and is envisaged to be a significant contribution to intelligent transportation systems for road traffic monitoring.

The availability of physical locations enables a myriad of applications, as exemplified extensively throughout this book. A particular application domain that benefits from the availability of location information is sensor network routing. Specifically the prospects brought by recent developments in WSN localization have sparked interest on a category of routing algorithms, known as geographical routing (D. Chen & Varshney, 2007). Geographic routing utilizes the location information of sensors to make routing decisions. It does not require the establishment or maintenance of routes from sources to destinations. Sensor nodes do not need to store routing tables. These features make geographic routing an attractive option for routing in large scale sensor networks.

Chapter XVIII - Beyond Localization: Communicating Using Virtual Coordinator discusses an interesting aspect of the *geographic routing* problem and question: for the purpose of improved *geographic routing*, whether it would be more efficient to label sensors by information other than their physical locations. Specifically the chapter advocates labelling sensors by their *virtual coordinates*, which are not related to their physical coordinates, and let the geographic routing algorithm use these virtual coordinates for routing. The concept of virtual coordinates is based on the notion of *greedy embedding.* A greedy embedding of a geometric graph (G,p) is the geometric graph (G,p') that has the same underlying graph G, i.e., the same edges interconnecting the same set of vertices, but having the vertices placed at different coordinates (p') such that greedy routing always functions when sending a message between arbitrarily chosen nodes. *Greedy routing* refers to a simple geographic routing scheme in which a node always forwards a packet to the neighbour that has the shortest distance to the destination. The use of virtual coordinates greatly facilitates geographic routing and removes void areas which have been a major hurdle in the implementation of geographic routing algorithms. The authors then present an algorithm that assigns virtual coordinates to sensors and the algorithm has been validated by both simulations and experiment.

Chapter XVIII reveals some insight that may be of interest for some applications currently using the physical location information of sensors. Physical locations of sensors can, to a large extent, be considered as a means to label sensors. It is possibly the most intuitive and useful way of labelling the sensors so that people know where the sensors are located and where the measured information by sensors comes from. Location information cannot be replaced by other information in many applications. However, in some applications which do not necessarily need to know the physical location of sensors but rely on some sort of sensor labels for identification of sensors or supporting the correct functioning of the application, there may be more efficient ways to label sensors that facilitate the application. It is in this sense that ***Chapter XVIII*** motivates us to think beyond the horizon of localization.

REFERENCES

Abel, J. S. (1990). A divide and conquer approach to least-squares estimation. *IEEE Transactions on Aerospace and Electronic Systems, 26*(2), 423-427.

Agee, B. G. (1991). Copy/DF approaches for signal specific emitter location. *the Twenty-Fifth Asilomar Conference on Signals, Systems and Computers* (pp. 994-999).

Aspnes, J., Eren, T., Goldenberg, D. K., Morse, A. S., Whiteley, W., Yang, Y. R., et al. (2006). A theory of network localization. *IEEE Transactions on Mobile Computing, 5*(12), 1663-1678.

Bahl, P., & Padmanabhan, V. N. (2000). RADAR: An in-building RF-based user location and tracking system. *IEEE INFOCOM* (pp. 775-784).

Bancroft, S. (1985). Algebraic solution of the GPS equations. *IEEE Transactions on Aerospace and Electronic Systems AES-21*(1), 56-59.

Barabell, A. (1983). Improving the resolution performance of eigenstructure-based direction-finding algorithms. *IEEE International Conference on Acoustics, Speech, and Signal Processing* (pp. 336-339).

Bergamo, P., & Mazzini, G. (2002). Localization in sensor networks with fading and mobility. *The 13th IEEE International Symposium on Personal, Indoor and Mobile Radio Communications* (pp. 750-754).

Biedka, T. E., Reed, J. H., & Woerner, B. D. (1996). Direction finding methods for CDMA systems. *Thirteenth Asilomar Conference on Signals, Systems and Computers* (pp. 637-641).

Biswas, P., & Ye, Y. (2004). Semidefinite programming for ad hoc wireless sensor network localization. *Third International Symposium on Information Processing in Sensor Networks* (pp. 46-54).

Bliss, D. W., & Forsythe, K. W. (2000). Angle of arrival estimation in the presence of multiple access interference for CDMA cellular phone systems. *Proceedings of the 2000 IEEE Sensor Array and Multichannel Signal Processing Workshop* (pp. 408-412).

Cao, M., Anderson, B. D. O., & Morse, A. S. (2005). Localization with imprecise distance information in sensor networks. *Proc. Joint IEEE Conf on Decision and Control and European Control Conf.* (pp. 2829-2834).

Capkun, S., Hamdi, M., & Hubaux, J. (2001). GPS-free positioning in mobile ad-hoc networks. *34th Hawaii International Conference on System Sciences* (pp. 3481-3490).

Carter, G. (1981). Time delay estimation for passive sonar signal processing. *IEEE Transactions on Acoustics, Speech, and Signal Processing, 29*(3), 463-470.

Carter, G. (1993). *Coherence and time delay estimation*. Piscataway, NJ: IEEE Press.

Chan, Y. T., & Ho, K. C. (1994). A simple and efficient estimator for hyperbolic location. *IEEE Transactions on Signal Processing, 42*(8), 1905-1915.

Chen, C. K., & Gardner, W. A. (1992). Signal-selective time-difference of arrival estimation for passive location of man-made signal sources in highly corruptive environments. Ii. Algorithms and performance. *IEEE Transactions on Signal Processing, 40*(5), 1185-1197.

Chen, D., & Varshney, P. K. (2007). A survey of void handling techniques for geographic routing in wireless networks. *IEEE Communications Surveys & Tutorials, 9*(1), 50-67.

Chen, J. C., Yao, K., & Hudson, R. E. (2002). Source localization and beamforming. *IEEE Signal Processing Magazine, 19*(2), 30-39.

Cheng, D. K. (1989). *Field and wave electromagnetics* (2nd ed.): Addison-Wesley Publishing Company, Inc.

Chong, C.-Y., & Kumar, S. P. (2003). Sensor networks: Evolution, opportunities, and challenges. *Proceedings of the IEEE, 91*(8), 1247-1256.

Crippen, G. M., & Havel, T. F. (1988). *Distance geometry and molecular conformation.* New York: John Wiley and Sons Inc.

Dogancay, K. (2005). Emitter localization using clustering-based bearing association. *IEEE Transactions on Aerospace and Electronic Systems, 41*(2), 525-536.

Doherty, L., Pister, K. S. J., & El Ghaoui, L. (2001). Convex position estimation in wireless sensor networks. *IEEE INFOCOM* (pp. 1655-1663).

Elnahrawy, E., Li, X., & Martin, R. P. (2004). The limits of localization using signal strength: A comparative study. *First Annual IEEE Conference on Sensor and Ad-hoc Communications and Networks* (pp. 406-414).

Eren, T., Goldenberg, D., Whiteley, W., Yang, R. Y., Morse, A. S., Anderson, B. D. O., et al. (2004). Rigidity and randomness in network localization. *IEEE INFOCOM* (pp. 2673-2684).

Fang, B. T. (1990). Simple solutions for hyperbolic and related position fixes. *IEEE Transactions on Aerospace and Electronic Systems, 26*(5), 748-753.

Fang, J., Cao, M., Morse, A. S., & Anderson, B. D. O. (2006). Sequential localization of networks. *The 17th International Symposium on Mathematical Theory of Networks and Systems- MTNS 2006.*

Fox, V., Hightower, J., Lin, L., Schulz, D., & Borriello, G. (2003). Bayesian filtering for location estimation. *IEEE Pervasive Computing, 2*(3), 24-33.

Gavish, M., & Weiss, A. J. (1992). Performance analysis of bearing-only target location algorithms. *IEEE Transactions on Aerospace and Electronic Systems, 28*(3), 817-828.

Gezici, S., Tian, Z., Giannakis, G. B., Kobayashi, H., Molisch, A. F., Poor, H. V., et al. (2005). Localization via ultra-wideband radios: A look at positioning aspects for future sensor networks. *IEEE Signal Processing Magazine, 22*(4), 70-84.

Halder, B., Viberg, M., & Kailath, T. (1993). An efficient non-iterative method for estimating the angles of arrival of known signals. *The Twenty-Seventh Asilomar Conference on Signals, Systems and Computers* (pp. 1396-1400).

Ihler, A. T., Fisher, J. W., III, Moses, R. L., & Willsky, A. S. (2005). Nonparametric belief propagation for self-localization of sensor networks. *IEEE Journal on Selected Areas in Communications, 23*(4), 809-819.

Ji, X., & Zha, H. (2004). Sensor positioning in wireless ad-hoc sensor networks using multidimensional scaling. *IEEE INFOCOM* (pp. 2652-2661).

Jian, L., Halder, B., Stoica, P., & Viberg, M. (1995). Computationally efficient angle estimation for signals with known waveforms. *IEEE Transactions on Signal Processing, 43*(9), 2154-2163.

Kannan, A. A., Mao, G., & Vucetic, B. (2005). Simulated annealing based localization in wireless sensor network. *The 30th IEEE Conference on Local Computer Networks* (pp. 513-514).

Kannan, A. A., Mao, G., & Vucetic, B. (2006). Simulated annealing based wireless sensor network localization with flip ambiguity mitigation. *63rd IEEE Vehicular Technology Conference* (pp. 1022-1026).

Kaveh, M., & Bassias, A. (1990). Threshold extension based on a new paradigm for music-type estimation. *International Conference on Acoustics, Speech, and Signal Processing* (pp. 2535-2538).

Kirkpatrick, S., Gelatt, C. D., & Vecchi, M. P. (1983). Optimization by simulated annealing. *Science, 220*(4598), 671–680.

Klukas, R., & Fattouche, M. (1998). Line-of-sight angle of arrival estimation in the outdoor multipath environment. *IEEE Transactions on Vehicular Technology, 47*(1), 342-351.

Knapp, C., & Carter, G. (1976). The generalized correlation method for estimation oftime delay. *IEEE Transansaction on Acoustics, Speech, Signal Processing, 24*(4), 320–327.

Koks, D. (2005). *Numerical calculations for passive geolocation scenarios* (No. DSTO-RR-0000). Edinburgh, SA, Australiao. Document Number)

Krishnan, P., Krishnakumar, A. S., Ju, W.-H., Mallows, C., & Gamt, S. N. (2004). A system for lease: Location estimation assisted by stationary emitters for indoor RF wireless networks. *IEEE INFOCOM* (pp. 1001-1011).

Kumaresan, R., & Tufts, D. W. (1983). Estimating the angles of arrival of multiple plane waves. *IEEE Transactions on Aerospace and Electronic Systems, AES-19*, 134-139.

Kwok, C., Fox, D., & Meila, M. (2004). Real-time particle filters. *Proceedings of the IEEE, 92*(3), 469-484.

Lee, J.-Y., & Scholtz, R. A. (2002). Ranging in a dense multipath environment using an UWB radio link. *IEEE Journal on Selected Areas in Communications, 20*(9), 1677-1683.

Liang, T.-C., Wang, T.-C., & Ye, Y. (2004). *A gradient search method to round the semidefinite programming relaxation for ad hoc wireless sensor network localization*: Standford Universityo. Technical Report.

Madigan, D., Einahrawy, E., Martin, R. P., Ju, W.-H., Krishnan, P., & Krishnakumar, A. S. (2005). Bayesian indoor positioning systems. *IEEE INFOCOM 2005* (pp. 1217-1227).

Niculescu, D., & Nath, B. (2001). Ad hoc positioning system (APS). *IEEE GLOBECOM* (pp. 2926-2931).

Niculescu, D., & Nath, B. (2003). Localized positioning in ad hoc networks. *IEEE International Workshop on Sensor Network Protocols and Applications* (pp. 42-50).

Oh-Heum, K., & Ha-Joo, S. (2008). Localization through map stitching in wireless sensor networks. *IEEE Transactions on Parallel and Distributed Systems, 19*(1), 93-105.

Patwari, N., Ash, J. N., Kyperountas, S., Hero, A. O., III, Moses, R. L., & Correal, N. S. (2005). Locating the nodes: Cooperative localization in wireless sensor networks. *IEEE Signal Processing Magazine, 22*(4), 54-69.

Paulraj, A., Roy, R., & Kailath, T. (1986). A subspace rotation approach to signal parameter estimation. *Proceedings of the IEEE, 74*(7), 1044-1046.

Prasithsangaree, P., Krishnamurthy, P., & Chrysanthis, P. (2002). On indoor position location with wireless lans. *The 13th IEEE International Symposium on Personal, Indoor and Mobile Radio Communications* (pp. 720-724).

Priyantha, N. B., Chakraborty, A., & Balakrishnan, H. (2000, August). The cricket location-support system. *Proc. of the Sixth Annual ACM International Conference on Mobile Computing and Networking* (pp. 32-43).

Rabbat, M., & Nowak, R. (2004). Distributed optimization in sensor networks. *Third International Symposium on Information Processing in Sensor Networks* (pp. 20-27).

Rappaport, T. S. (2001). *Wireless communications: Principles and practice* (2nd ed.): Prentice Hall PTR.

Rappaport, T. S., Reed, J. H., & Woerner, B. D. (1996). Position location using wireless communications on highways of the future. *IEEE Communications Magazine, 34*(10), 33-41.

Ray, S., Lai, W., & Paschalidis, I. C. (2005). Deployment optimization of sensornet-based stochastic location-detection systems. *IEEE INFOCOM 2005* (pp. 2279-2289).

Romer, K. (2003). The lighthouse location system for smart dust. *Proceedings of MobiSys 2003 (ACM/USENIX Conference on Mobile Systems, Applications, and Services)* (pp. 15-30).

Roos, T., Myllymaki, P., & Tirri, H. (2002). A statistical modeling approach to location estimation. *IEEE Transactions on Mobile Computing, 1*(1), 59-69.

Roy, R., & Kailath, T. (1989). ESPRIT-estimation of signal parameters via rotational invariance techniques. *IEEE Transactions on Acoustics, Speech, and Signal Processing, 37*(7), 984-995.

Savarese, C., & Rabaey, J. (2002). Robust positioning algorithms for distributed ad-hoc wireless sensor networks. *Proceedings of the General Track: 2002 USENIX Annual Technical Conference* (pp. 317-327).

Savarese, C., Rabaey, J. M., & Beutel, J. (2001). Locationing in distributed ad-hoc wireless sensor networks. *IEEE International Conference on Acoustics, Speech, and Signal Processing* (pp. 2037 - 2040).

Saxe, J. (1979). Embeddability of weighted graphs in k-space is strongly NP-hard. *17th Allerton Conference in Communications, Control and Computing* (pp. 480-489).

Schell, S. V., & Gardner, W. A. (1993). High-resolution direction finding. *Handbook of Statistics, 10*, 755-817.

Schmidt, R. (1986). Multiple emitter location and signal parameter estimation. *IEEE Transactions on Antennas and Propagation, 34*(3), 276-280.

Shang, Y., Ruml, W., Zhang, Y., & Fromherz, M. (2004). Localization from connectivity in sensor networks. *IEEE Transactions on Parallel and Distributed Systems, 15*(11), 961-974.

Smith, J., & Abel, J. (1987). The spherical interpolation method of source localization. *IEEE Journal of Oceanic Engineering, 12*(1), 246-252.

Stanfield, R. G. (1947). Statistical theory of DF finding. *Journal of IEE, 94*(5), 762 - 770.

Tayem, N., & Kwon, H. M. (2004). Conjugate esprit (C-SPRIT). *IEEE Transactions on Antennas and Propagation, 52*(10), 2618-2624.

Torrieri, D. J. (1984). Statistical theory of passive location systems. *IEEE Transactions on Aerospace and Electronic Systems, AES-20*(2), 183-198.

Ziskind, I., & Wax, M. (1988). Maximum likelihood localization of multiple sources by alternating projection. *IEEE Transactions on Acoustics, Speech, and Signal Processing, 36*(10), 1553-1560.

Chapter II
Measurements Used in Wireless Sensor Networks Localization

Fredrik Gustafsson
Linköping University, Sweden

Fredrik Gunnarsson
Linköping University, Sweden

ABSTRACT

Wireless sensor networks (WSN) localization relies on measurements. Availability of, and the information content in, these measurements depend on the network architecture, connectivity, node time synchronization and the signaling bandwidth between the sensor nodes. This chapter addresses wireless sensor networks measurements in a general framework based on a set of nodes, where each node either emits or receives signals. The emitted signal can for example be a radio, acoustic, seismic, infrared or sonic wave that is propagated in a certain media to the receiver. This general observation model does not make any difference between localization of sensor network nodes or unknown objects, or whether the nodes or objects are stationary or mobile. The information available for localization in wireless cellular networks (WCN) is in literature classified as direction of arrival (DOA), time of arrival (TOA), time difference of arrival (TDOA) and received signal strength (RSS). This chapter generalizes these concepts to the more general wireless sensor networks.

INTRODUCTION

Distributed sensor networks have been widely discussed for more than 30 years, but it is with more recent advances in hardware and processing capabilities that the vision of the wireless sensor networks can become a reality. In general, the ambition is to perform fusion of the information provided by the distributed sensors (Luo and Kay, 1989, Jayasimha et al., 1991, Wesson et al., 1981, Yemini, 1978, Akyildiz et al., 2002). One important wireless sensor networks research area concerns localization of objects moving within the coverage, or monitoring, area of the network. Applications include surveillance of both military and civilian areas and passages. Sensor networks are often deployed without knowledge of the exact sensor locations. Furthermore, sensor nodes might be mobile to some extent. Therefore, localization of individual nodes is also important. The scope of this chapter is measurements useful for all kinds of localization.

Wireless sensor network localization is in many ways similar to positioning in wireless cellular networks. This is particularly true when it comes to network-assisted positioning, where the network elements in the cellular network observe measurements related to the position of the mobile terminal. Some related surveys include (Caffery and Stuber, 1998, Zhao, 2002, Drane et al., 1998, Sun et al., 2005, Sayed et al., 2005, Pahlavan et al., 2002, Gustafsson and Gunnarsson, 2005, Deblauwe, 2008). In recent years, however, cellular network positioning has become more and more mobile-centric, where the mobiles perform the measurements, and the network either provides necessary information, or completes parts of the positioning calculations. Despite this fact, the generic measurements are still similar, which is evident after a direct comparison between this chapter and (Gustafsson and Gunnarsson, 2005). One can also consider wireless cellular networks as a special case of wireless sensor networks when it comes to localization in networks. Therefore, wireless cellular networks are used to some extent to exemplify wireless sensor network concepts.

The outline of the chapter is as follows. We start with explaining the main principles of WSN measurements in terms of a simple but generic transmitter-receiver model. This leads to a separation into waveform, time and power observations. The availability and limitations of these are discussed in terms of network architecture. Then more specific signal models are defined such as direction of arrival (DOA), time of arrival (TOA) and received signal strength (RSS), and properties in practice and typical performance values are discussed. Error modeling is discussed, and Gaussian mixture models (GMM) are considered as a generally applicable model. The potential estimation accuracy in terms of position root mean square error is discussed and related to fundamental and computable lower bounds, which can be used to evaluate network configuration. Aspects like routing protocols, information reporting protocols, signaling protocols, radio interface design, power consumption, etc. are all challenging parts of wireless sensor networks, but are not within the scope of this chapter.

GENERAL OBSERVATION MODELS

We start with a high level characterization of a wireless sensor network (WSN), and proceed with more detailed sensor models. An as generic vocabulary as possible will be used, and the particular application dependent notation will be introduced later on.

The WSN consists in general of nodes $i = 1, \ldots, M$ positioned at p^i_t at time t. A general observation model of a signal $s^j(t)$ emitted from node j and received at node i with an amplification $a^{ij}(t)$ and a delay τ_{ij} observed in noise $w^{ij}(t)$ is:

$$r^{ij}(t) = a^{ij}(t)s^j(t - \tau_{ij}) + w^{ij}(t) \tag{0.1}$$

A node can in this context be characterized with the following properties:

- The node either emits (target or emitter) a signal $s^j(t)$ or receives (sensor or receiver) a signal $r^j(t)$ at some frequency in some medium, or both. Examples include radio waves propagated with the speed of light, acoustic waves propagated with the speed of sound, seismic waves propagated in the earth or sonic waves propagated in water. The delay τ_{ij} depends on the propagation time among other things.
- A node is either a sensor node, which can act as both a target and a sensor, or an unknown object, which can only act as a target.
- The node can be known to be stationary or moving implying a constant or time-varying p^i_t.
- The node can have known or unknown position p^i_t at each time instant.
- One emitting and one receiving node can be anything between perfectly synchronized and completely unsynchronized. The same applies to two emitting nodes or two receiving nodes. The time index t refers to the time reference of the receiving node i, and synchronization errors between node i and j are included in τ_{ij}.
- The communication between two nodes (two emitters or two receivers or one of each) can be wired or wireless, and the bandwidth can be low or high.
- The emitted wave may propagate directly to the receiver (line of sight) or be subject to multipath effects. These effects have impact on both the amplification $a^{ij}(t)$ and the delay τ_{ij}.

These properties mean that the problems of determining the location of the sensor node itself, or of a different sensor node or an unknown object are equivalent and can be described with the same framework. Figure 1 shows an example of a WSN, where each sensor node consists of a microphone, a geophone and a magnetometer. The sensor nodes are stationary with known position. They have a fairly low bandwidth in their wireless communication links, and they are synchronized to an accuracy of a few milliseconds. The target node, on the other hand, is a moving vehicle with unknown position

Figure 1. The sensor nodes '+' in a WSN overlayed on an aerial photo and sampled positions 'o' of the unknown object (target node)

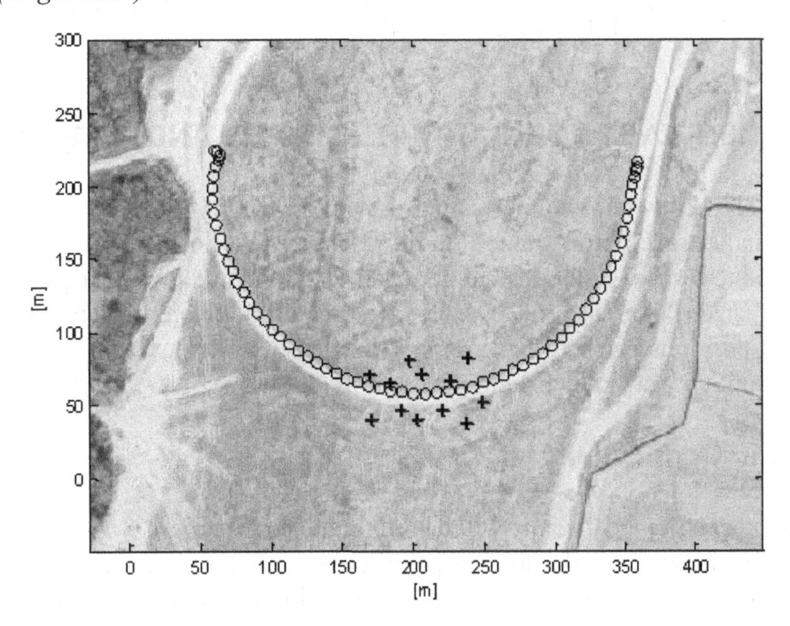

and on its way it is emitting acoustic and seismic waves. Depending on the type of vehicle, the range of these waves differs between some ten meters to some hundred meters.

The difference of a wireless sensor network and a sensor network in general is the assumption that at least two nodes exchange information using a wireless link and thus are subject to bandwidth limitations and non-trivial synchronization. Depending on sensor capabilities and the nodes' synchronization and communication capabilities, three different types of observations can be distinguished:

- **Waveform observations.** A highly capable sensor node is able to operate on the signal waveform $s^j(t)$, and this observation can be shared with other nodes if bandwidth allows. This is only meaningful when these nodes are synchronized at an accuracy comparable to the inverse waveform frequency. Sensors very close to each other (in the order of half a wavelength) can form a sensor array and correlate the phase of the emitted signal to get a direction of arrival estimates. For sensors further separated, there will be an integer problem in the ambiguity of the number of periods that may be resolved by merging other information.
- **Timing observations.** If a known or easily distinguished signature is embedded in the signal, the sensor can correlate the signal with the signature to accurately estimate time of arrival t-τ_{ij}. The timing estimation accuracy depends on the signature as well as the sensor capability. This is meaningful only if the sensor is synchronized either with the emitting node or another receiving node. In the former case, an absolute distance can be computed and in the latter case a relative distance follows. Geometrically, these two cases constrain the position to a circle or a hyberbolic function.
- **Power observations.** Another possibility is that the sensor estimates the received signal power (received signal strength - RSS) $||r^{ij}(t)||^2$. In essence, this means integrating the received signal power within a certain frequency band during an integration interval to estimate the received signal energy during the time interval. If the emitted power is known, RSS provides coarse range information. Otherwise, two or more sensors can compare their RSS observations to eliminate the unknown emitted power.

These three types are discussed and exemplified further below. However, first implications from the sensor network architecture and classification are addressed.

WIRELESS SENSOR NETWORK ARCHITECTURES AND CLASSIFICATIONS

The characteristics and behavior of a wireless sensor network is very much dependent on the architecture, organization and connectivity. A rather loosely connected ad-hoc network typically has different properties compared to a strictly hierarchical network. Handling and processing of time-related measurements are facilitated if the sensor nodes have a common time reference. The localization problem of unknown object positions is more challenging if the sensor node positions are unknown and time varying compared to if they are known and fixed.

Architecture and Connectivity

A network with a rather strict architecture will be referred to as a *hierarchical* network as illustrated by Figure 2a. Conversely, networks with a loose architecture as in Figure 2b are denoted *anarchical* (Wesson et al., 1981). The target node for the aggregated sensor information is referred to as the *sink*.

Figure 2. Two wireless sensor network architecture examples. (a) Hierarchical network with clusters of sensors associated to cluster heads. (b) Anarchical network with sensors loosely organized in a sensor field.

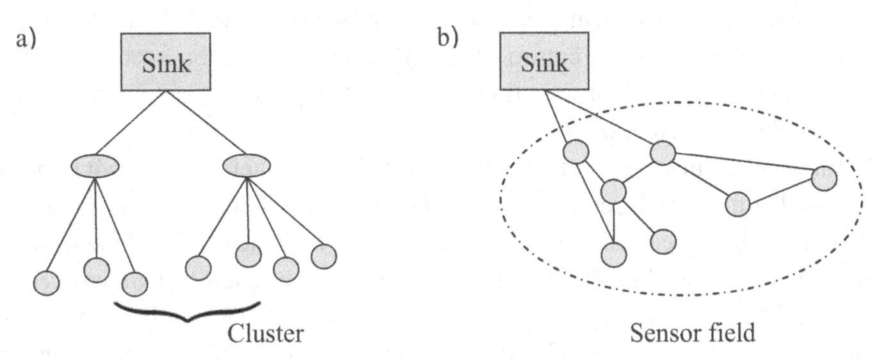

In a hierarchical network sensor nodes may be associated to more potent nodes known as *cluster heads* (C.-C. Shen et al., 2001) or *processing elements* (Jayasimha et al., 1991). Depending on the size of the sensor network, such clustering may be organized over multiple levels in a hierarchical fashion. The many low complexity, low power sensors with little processing power are placed in the bottom, while sensors with increasing complexity and processing power are found higher in the hierarchy. Information from child nodes may be aggregated, fused and refined by parent nodes before being passed upwards in the hierarchy. Furthermore, the cluster heads on the highest level distribute the information to the sink, where the final analysis and visualization take place. Connections are primarily established between child nodes and associated parent node, even though connections within the cluster is also possible. The strict hierarchy enables a prompt and reliable information transport. Wireless cellular networks are typically strictly hierarchical networks.

In an anarchical network, the sensors are loosely organized, and the sensor capability is typically more equal. The set of the sensors are popularly denoted the *sensor field* (Akyildiz et al., 2002), or *smart dust* (Kahn et al., 1999). The information may be aggregated and fused in some or all of the sensor nodes, before passing it on in the sensor field. Also the sink may be a node in the sensor field, but typically with somewhat better processing capabilities. Essentially all connections can be seen as volatile, and a connection between two nodes may be a multi-hop connection with several node to node links in sequence. Loosely organized networks with volatile communications links are often referred to as *mesh* networks (Akyildiz and Wang, 2005). Mechanisms such as self-configuration and self-organization (a new node finds its place in the network, and the network-wide connectivity establishes automatically), and self-healing (the connectivity is re-established after some disruptive event) are important components of the mesh network.

Sharing waveform observation between nodes is most probably only plausible in hierarchical networks. Distribution of timing and power observations should on the other hand be possible in any type of network, provided that the connection bandwidth between the nodes is sufficient.

Time Synchronization

In a distributed environment such as wireless sensor networks there may be no central synchronized time reference that regulates the activities of each node. Instead, each node manages its own time reference.

Most of the sensor measurements, both with respect to localization and relative distances, but also with respect to a general sensor observation, need to be associated with a common time reference in order to be properly fused together. If the time synchronization in the wireless sensor network is acceptable, then node-pairs can exchange information and measurements and readily use the information for localization. Waveform observations require the highest synchronization accuracy, timing observation intermediate accuracy and power observation the crudest accuracy. Synchronization inaccuracies can be included in the delay distribution in (0.1).

Network time synchronization protocols used over the fixed network are not applicable, due to the volatile and possibly time varying connectivity and the clock drift of the low complexity sensor devices. One tractable approach is to consider one node as the node with the time reference, and use adjustment mechanisms to gradually synchronize the sensor network. Such schemes are surveyed in (Jayasimha et al., 1991, Elson and Römer, 2003, Sivrikaya and Yener, 2004) and the results indicate that they provide accurate synchronization in the order of 10 μs. Recent advances provide algorithms that operate without a specific node as time reference (Solis et al., 2006, Schenato and Gamba, 2007) something that is very important for robustness reasons.

Sensor Node Location Information

Some nodes in the wireless sensor network may have known positions, either constant or time-varying. For example, the location information can be provided locally by some external location system such as GPS. Nodes with known position are referred to as *anchor* nodes, and nodes with unknown position consequently *non-anchor* nodes (Luo et al., 2006).

The wireless sensor network may consist of only non-anchor nodes. Then some sort of cooperative localization scheme (Patwari et al., 2005) is a necessity. The communication to support self-localization of sensor nodes can be either *interactive* or *non-interactive* (Luo et al., 2006). The interactive support means that all the nodes actively participate in the localization by signaling and sharing information (bidirectional connectivity), while in case of non-interactive support, most of the nodes passively observe other objects and beacons (unidirectional connectivity). These mechanisms may be supported by signals (e.g. pilot signals) broadcasted continuously or at least regularly from some nodes, which then act as *beacons* (Sun et al., 2005). Another way of exciting the self-localization mechanisms in non-interactive sensor networks is by observing the same mobile and unknown object both in time and in space and share the information with other nodes. This could be seen as a calibration phase.

In hierarchical networks, the existence of anchor nodes is quite plausible. Furthermore, an anchor node that acts as a beacon is denoted a *landmark* node (similar to a lighthouse) (Niculescu, 2004). Cluster head nodes become central and important, and this motivates that such nodes are relatively potent and deployed with care. This could include a determination of the exact node position. When some sensor locations are known as is likely in hierarchical networks, methods based on relative position measurements are available (N. Patwari et al., 2003). Other self-localization schemes include (Niculescu, 2004, Langendoen and Reijers, 2003, Savarese et al., 2001, Coates, 2004, Galstyan et al., 2004).

In some applications, the signal propagation speed is significantly higher than the velocity of an unknown object or a moving sensor node. This is the case for radio frequency signals. However, when the measurement signal is seismic or audible then the delay between signal emission to signal observation cannot be neglected.

SPECIFIC OBSERVATION MODELS

The notation assumes a two-dimensional position, $p^i_t = (X^i_t, Y^i_t)^T$, but can be extended to higher dimensions. Depending on whether one, two or three nodes have to collaborate to form an observation, the following notation will be used:

$$\begin{aligned} y_t^i &= h_{type}(p_t^i) + e_t^i \\ y_t^{ij} &= h_{type}(p_t^i, p_t^j) + e_t^{ij} \\ y_t^{ijk} &= h_{type}(p_t^i, p_t^j, p_t^k) + e_t^{ijk} \end{aligned} \quad (0.2)$$

Waveform Observations

Observing the full frequency band of a signal requires a quite capable sensor. For example, a message sent over radio is typically encoded to a base band signal, which then is modulated onto a carrier frequency of much higher frequency than the base band signal. The typical processing of the measurements includes correlation of waveform observations from different, spatially close, sensors. This is popularly referred to as sensor arrays. It could be instructive to consider observations from the different sensors as one sensor observation from a sensor array. As already mentioned, one application of the sensor array measurement at node i is *direction-of-arrival* estimation relative another node j.

Essentially all signal frequencies and media are possible. The nonlinear measurement function is given by

$$y_t^{ij} = h_{DOA}(p_t^i, p_t^j) + e_t^{ij} \quad (0.3)$$

Furthermore, the direction-of-arrival angle can be calculated as

$$h_{DOA} = \text{angle}(p_t^i - p_t^j) \quad (0.4)$$

Timing Observations

For signals with a known or detectable signature or fingerprint, timing information can be obtained by correlating the signal with the signature. The signature can be a pilot or a training sequence transmitted from a radio base station or node. Compared to a waveform observation, the signature is typically part of the base band signal. It can also be a detectable impulse in the signal, for example from a gun shot in an acoustic signal or from an explosion in a seismic signal. A similar signature can be the arrival of a message sent from a different location.

The accuracy depends mainly on the signature properties, but also on the processing capabilities of the sensor. One typical application of timing observations is *time-of-arrival* estimation

$$y_t^{ij} = h_{TOA}(p_t^i, p_t^j) + e_t^{ij} \quad (0.5)$$

If both nodes i and j have the same time reference, the time of signal transmission t^j and reception t^i combined can be related to the relative distance if the signal propagation speed of the medium v is known:

$$\left\| p_t^i - p_t^j \right\| = v(t^i - t^j) \tag{0.6}$$

In a message-oriented implementation, it could be easier to instead measure the round-trip time from time of transmission t^i_{tran} and time of response reception t^i_{resp}. If the processing time t^j_{proc} at the other node j is included in the response message, the following holds

$$\left\| p_t^i - p_t^j \right\| = \frac{v}{2}(t^i_{\text{resp}} - t^i_{\text{tran}} - t^j_{\text{proc}}) \tag{0.7}$$

The uncertainty in (0.6) is dominated by the difference in time references at the two locations, while the uncertainty in (0.7) depends on the time of arrival estimation of the response at node i and the accuracy of the processing time estimation at the other node j.

Hence, the nonlinear time-of-arrival measurement modeling can be summarized as

$$h_{TOA} = \left\| p_t^i - p_t^j \right\|. \tag{0.8}$$

As discussed for time-of-arrival measurements, a different or uncertain time reference can be troublesome. If the ambiguity in time-references t^j_{ref} and t^k_{ref} between two nodes j and k is known or can be resolved at some node or location, then *time-difference-of-arrival* measurements can become useful

$$y_t^{ijk} = h_{TDOA}(p_t^i, p^j, p^k) + e_t^{ijk} \tag{0.9}$$

A signal emitted (received) from node i and received (emitted in parallel) at nodes j and k at times t^{ij} and t^{ik}, respectively, leads to the following observation

$$\left\| p_t^i - p_t^j \right\| - \left\| p_t^i - p_t^k \right\| = \left(t^{ij} - t^{ik} + t^j_{ref} - t^k_{ref}\right)v \tag{0.10}$$

The uncertainty e_t^{ijk} depends both on the time-of-arrival estimation accuracies e_t^{ij} and e_t^{ik} as well as the time synchronization accuracy between the locations.

The nonlinear time-difference-of-arrival measurement is thus characterized by

$$h_{TDOA} = \left\| p_t^i - p_t^j \right\| - \left\| p_t^i - p_t^k \right\|. \tag{0.11}$$

If two of these positions are known, the position of the third is constrained to a hyperbolic function, whose asymptotes define a direction of arrival. This indicates a close relation of relative range information and angle information. Note the generality of this setup. Any of the three nodes can be the one to be located, and there is a duality between emitting and receiving. The distinguishing feature here is that exactly two nodes are synchronized, or the difference in time references is known somewhere in the sensor network.

Similar equations based upon relative distances also appear when considering measurements based on interferometrics (M. Maroti et. al, 2005). In such a case, two nodes A and B transmit pure sine waves at two close frequencies $f_A > f_B$, and two other nodes C and D measure and filter the total received signal power and estimate the absolute phase offset $v = \phi_A - \phi_B$. It has been shown that the relative phase offset

v_C - v_D is proportional to a combination of relative distances, provided that the frequencies are close enough and the relative node distances are short enough, compared to the travel time of the signals:

$$y_t^{ABCD} = h_{IF}\left(p_t^A, p_t^B, p_t^C, p_t^D\right) + e_t^{ABCD} \tag{0.12}$$

$$h_{IF}\left(p_t^A, p_t^B, p_t^C, p_t^D\right) = \left\|p_t^A - p_t^D\right\| - \left\|p_t^B - p_t^D\right\| + \left\|p_t^B - p_t^C\right\| - \left\|p_t^A - p_t^C\right\| \tag{0.13}$$

Power Observations

Power observations are typically obtained by integrating the signal power over a certain time interval. The signal may be low-pass or band-pass filtered before calculating the signal power to capture the power in the frequency band where the signal is expected to reside. This means that the power signal sample interval is orders of magnitude longer than the signature sequences.

The observed power is related to the relative distance between two nodes

$$y_t^{ij} = h_{RSS}(\|p_t^i - p_t^j\|) + e_t^{ij} \tag{0.14}$$

Radio frequency signals, acoustic signals, seismic signals, radar signals etc have in common that the emitted signal strength $\bar{P}_0^i$ (bar denotes here and in the sequel power in linear scale) at node i approximately decays exponentially with distance in linear scale. For example, this is observed for radio signals in e.g. (Okumura et al., 1968, Hata, 1980). Hence, the received signal strength $\bar{P}^{ij}$ at node j can be modeled as

$$\bar{P}^{ij} = \bar{P}_0^i \left\|p_t^i - p_t^j\right\|^{-n_{ij}}. \tag{0.15}$$

The corresponding relation in logarithmic scale is

$$P^{ij} = P_0^i - n_{ij}\log(\|p_t^i - p_t^j\|) \tag{0.16}$$

where 'log' denotes the natural logarithm. It is a matter of taste whether a model in linear or in logarithmic scale is selected. One aspect is that the uncertainty e^{ij}_t with good approximation can be modeled as Gaussian when the received signal strength measurement equation is in logarithmic scale (Okumura et al., 1968, Hata, 1980). Equations (0.15) and (0.16) yield two alternative non-linear functions

$$\begin{aligned} h_{RSS,log} &= P_0^i - n_{ij}\log(\|p_t^i - p_t^j\|) \\ h_{RSS,lin} &= \bar{P}_0^i \left\|p_t^i - p_t^j\right\|^{-n_{ij}}. \end{aligned} \tag{0.17}$$

Figure 3 illustrates acoustic and seismic sensor signals, whose propagation can be described with (0.17) at good accuracy. Furthermore, Figure 4 shows WiMax radio signal RSS measurements as a function of logarithmic distance. Again, the observation model in (0.17) fits data well.

Scanning radar was mentioned as a means of estimating directions above. It is also an example of power measurements where the received radar signal power is determined. In this case, the transmitted power is known, the distance is round-trip, but the uncertainty comes from the propagation properties and the radar cross section of the object.

In wireless cellular networks, received signal strength (or quality) measurements are integral parts of the mobility management, which aims at ensuring that the mobiles are connected to the most favorable base station over time. The pilot signal strengths are estimated, and it is possible in some standards to inform the mobiles about the pilot transmission power.

Observations Related to Digital Maps

The power observations model in the previous section only models the exponential decay with distance of the power. The main reason for deviations from this model is due to diffraction and reflection phenomena. If a reliable signal propagation tool is available, or if some effort is spent scanning the observation area, this information could be aggregated into a database, which can be seen as a digital map with a measurement model

$$y_t^i = h_{MAP}(p_t^i) + e_t^i \tag{0.18}$$

The observed or predicted values associated with a specific position p^i and different nodes j can be seen as a fingerprint for this position. The map information could be received signal strength measurements at all possible locations in the area as discussed for indoor localization in (Chen and Kobayashi, 2002, Kaemarungsi and Krishnamurthy, 2004). In this case, the nonlinear measurement function depends on at least two node positions $h_{MAP}^j(p_t^i, p_t^j)$. As another example, an object could travel along the trajectory in Figure 2 during a measurement campaign, while all sensor nodes measure the received signals as is emitted from the object. This information could be stored in a database and subsequent sensor observations can be correlated to the map information.

Figure 3. Received sensor energy in log scale Pij versus log range $\log(\| p_t^i - p_t^j \|)$, *together with a fitted linear relation as modeled in (0.14). The estimated path loss exponent nij is 2.3 and 2.6 respectively, and the model error standard deviation 0.56 and 0.60 respectively. If a dB scale is used, then these values scale by a factor of 10/log(10), which yields model error standard deviations of 2.4 and 2.6 dB.*

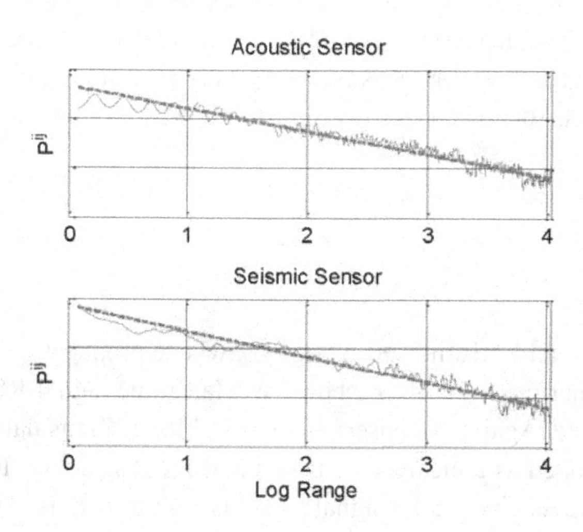

Figure 4. Radio signal received signal strength in log scale versus log range, together with a fitted linear relation as modeled in (0.14). The estimated path loss exponent n^{ij} is 2.3, and the estimated error distribution is Gaussian with standard deviation 0.70 (corresponding to 3 dB). Data from a Wimax radio network deployment in Brussels kindly provided by Mussa Bshara.

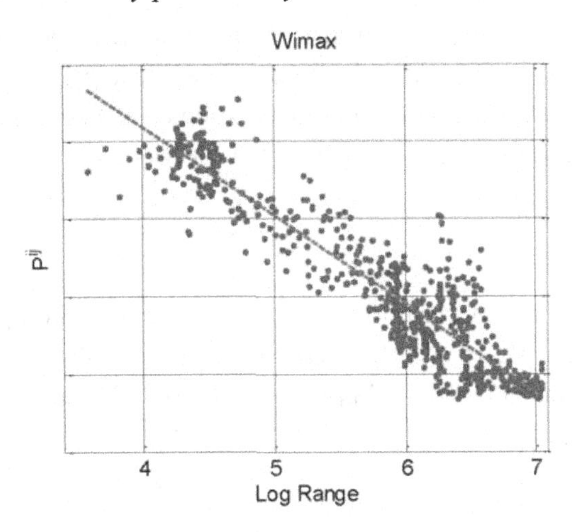

It could also be a digital map of walls in an area, and the distance to the walls in different directions are measured at node *i*. Similarly, it could be a digital road map (Gustafsson et al., 2002), where the measurement is the distance to the road which should be zero for road-bound mobility. Another example is a map of sea depths (bathymetry map), and the corresponding measurement is a depth measurement (Karlsson and Gustafsson, 2003). This means that the nonlinear measurement function depends only on one node position $h_{MAP}(p_t^i)$.

Position and Attitude Observations

Sensor observations for self-localization include local sensors measurements directly related to the position of the same node. A direct position estimate can simply be expressed as

$$y_t^i = p_t^i + e_t^i \tag{0.19}$$

and may be available from an external system such as the Global Positioning System. Typical GPS accuracy without differential support is in the order of 5-10 m. However, this could be significantly improved with differential support of some kind. The non linear measurement equation for position estimates is thus

$$h_{POS} = p_t^i \tag{0.20}$$

Note that GPS positioning can be incorporated either as a position estimate with estimated error statistics, or as separate timing estimates relative to the GPS satellites.

It is also possible that inertial measurement units (IMU) are available at certain nodes, providing attitude and acceleration information. Such measurements can be modeled as

$$y_t^i = h_{IMU}(\dot{p}_t^i, \ddot{p}_t^i) + e_t^i \tag{0.21}$$

Measurement Error Modeling

In all specific measurement models, the measurement noise is additive. A first and convenient approximation is that the measurement is unbiased and the noise is white and Gaussian with a standard deviation σ_e. Appropriate accuracy levels depend on both the type of measurement, as well as the network architecture and classification as discussed in the previous sections.

Waveform Observations

Geometrically, the spatial resolution of the intersection of two perfectly complementing AOA measurements is limited to $2d_A\sin(\alpha/2)$, where α is the angular resolution and d_A is the distance between the antennas. For $\alpha = 30°$, this means 36% of d_A. Approximately 5° - 10° can be considered appropriate.

Timing Observations

One key contributor to the measurement noise is the time reference inaccuracy between two nodes. Very exact time reference can be obtained by using a GPS receiver. Otherwise, one has to rely on different distributed time synchronization mechanisms. Furthermore, the timing determination accuracy depends on the resolution of the reference signal. In GSM, one training sequence bit corresponds to 554 m, and the timing accuracy is a fraction in the same order. In WCDMA the accuracy may be down to 0.5 chip of the scrambling code corresponding to an accuracy of 20 m. For radio frequency signals, the time reference accuracy needs to be less than 1 μs, but for audible signals can be far cruder since it is the accuracy relative the travel time of the signal that matters.

Power Observations

The received signal strength model describes signal propagation in a desert environment without interfering objects. The latter may result in diffraction and reflections, depending on the relation between the signal wavelength and the size of the objects. This shadow fading gives an additive component with 4-12 dB standard deviations. As has been discussed above, the accuracy of the power measurements can be improved by the use of well-predicted or measured power level maps. In such a case, the error standard deviation can be less than 3 dB.

Error Distributions and Correlations in More Detail

A generally applicable assumption is that the measurement error is Gaussian with a probability density function

$$p_E(e_t) = N(0, \sigma) \tag{0.22}$$

This may be valid, provided that the measurement is based on a signal that has traveled along the line-of-sight (LoS) between the nodes. In a non-LoS situation, the measurement will get a positive bias μ and probably another (larger) variance σ_{NLOS}

$$p_E(e_t) = N(\mu, \sigma_{NLOS}) \tag{0.23}$$

The problem here is that we cannot easily detect NLOS. One solution is to use robust algorithms. Another approach is to model the error distribution with a mixture, for instance the two-mode Gaussian mixture model (GMM)

$$p_E(e_t) = \alpha N(0, \sigma_{LOS}) + (1-\alpha) N(\mu_{NLOS}, \sigma_{NLOS}) \tag{0.24}$$

where (α, μ_{NLOS}, σ_{LOS}, σ_{NLOS}) are free parameters in the distribution. Here, e_t falls in the LOS distribution with probability α and the NLOS distribution with probability $1-\alpha$. Algorithms based on this distribution will automatically be more robust than algorithms that do not model NLOS.

Furthermore, the errors may be correlated over time and space. For example, the diffraction and reflection phenomena that cause most of the line-of-sight errors for power measurements are strongly correlated to the terrain. One modeling approach (M. Gudmundson 1991) considers distance dependent measurement errors and is based on the introduction of a decorrelation distance d_{dc} after which the error correlation has dropped to e^{-1}. As an approximation, the correlation is only based on the position of one node i, typically the one of an unknown object

$$corr\left(e(p^i), e(p^i + \Delta)\right) = e^{-\|\Delta\|/d_{dc}} \tag{0.25}$$

Sensor Observations in Summary

Table 1 summarizes the discussed sensor observations. Note that the quality of the sensor information depends not only on the noise variance, but also on the size and variation in $h(p)$. The sensor information included in the measurements is further discussed in the next section.

FUNDAMENTAL PERFORMANCE BOUNDS

Consider now the set of all available measurements as a vector y, and all positions to be estimated as a vector p. The total signal model can then be written

$$y = h(p) + e \tag{0.26}$$

This is the basic measurement relation for localization services.

Firstly, a nonlinear least squares (NLS) optimization routine can be applied to get a position estimate $\hat{p}_{NLS}$. Secondly, if the noise covariance R=cov(e) is known, the weighted nonlinear least squares (WNLS) estimate $\hat{p}_{WNLS}$ can be computed. Finally, if the noise distribution p_E is known, the maximum likelihood (ML) approach applies, which gives $\hat{p}_{ML}$. In general, the position estimate can be obtained by minimizing an optimization criterion

Table 1. Mathematical notation of available sensor observations in wireless sensor networks described by the non-linear location dependency h_{type}, such that $y = h_{type} + e$

Type of Measurement	Nonlinear Measurements	Accuracy
Direction of arrival	$h_{DOA} = \text{angle}(p_t^i - p_t^j)$	5° - 10°
Time of arrival	$h_{TOA} = \lVert p_t^i - p_t^j \rVert$	5 – 100 m
Time difference of arrival	$h_{TDOA} = \lVert p_t^i - p_t^j \rVert - \lVert p_t^i - p_t^k \rVert$	10 – 60 m
Interferometrics	$h_{IF} = \lVert p_t^A - p_t^D \rVert - \lVert p_t^B - p_t^D \rVert + \lVert p_t^B - p_t^C \rVert - \lVert p_t^A - p_t^C \rVert$	0.1 – 1 m
Received signal strength	$h_{RSS,log} = P_0^i - n_{ij} \log(\lVert p_t^i - p_t^j \rVert)$ $h_{RSS,lin} = \bar{P}_0^i \lVert p_t^i - p_t^j \rVert^{-n_{ij}}$	4 – 12 dB
Digital map information	$h_{MAP}^j(p_t^i, p_t^j)$	(RSS MAP 3dB)
Position estimates Inertial sensors	$h_{POS} = p_t^i$ $h_{INS}(\dot{p}_t^i, \ddot{p}_t^i)$	5 – 20 m (GPS)

$$\hat{p} = \arg\min_{p} V(p) \qquad (0.27)$$

where the different optimization routines are obtained by the following criteria

$$\begin{aligned} V_{NLS}(p) &= (y - h(p))^T (y - h(p)) \\ V_{WNLS}(p) &= (y - h(p))^T R^{-1} (y - h(p)) \\ V_{ML}(p) &= \log p_E (y - h(p)) \end{aligned} \qquad (0.28)$$

The estimates come with an estimation uncertainty, whose second order moment is represented by a covariance matrix P. Increasing the amount of considered knowledge in the estimation step, improves the estimation accuracy and reduces the estimation error covariance. In general

$$P_{NLS} > P_{WNLS} > P_{ML} \qquad (0.29)$$

For the user, the position root mean square error (RMSE) in meters is a useful measure. It includes estimation errors both due to the covariance and bias errors. RMSE can be lower bounded by the covariance matrix using the inequality

$$\text{RMSE} = \sqrt{\text{E}\left[(X^o - \hat{X})^2 + (Y^o - \hat{Y})^2\right]} \geq \sqrt{\text{tr Cov}(\hat{p})} \qquad (0.30)$$

where p^o denotes the true position. It should be noted that the covariance matrix may not be feasible to compute, because it is either not straightforward to express, or it may need an excessive number of computations.

A more challenging question is how small the covariance, and thus the RMSE, can be for a given sensor network constellation and measurement characteristics. Such a bound is provided by the Cramer-Rao Lower Bound (CRLB) for unbiased estimators. This bound is expressed in terms of the Fisher

Information Matrix (FIM) (Kay, 1993). CRLB has been analyzed thoroughly in the literature, primarily for DOA, TOA and TDOA (Botteron et al., 2004b, Botteron et al., 2004a, Botteron et al., 2002, Botteron et al., 2001, Patwari et al., 2003, Koorapaty et al., 1998), but also for RSS (Weiss, 2003, Koorapaty, 2004) and with specific attention to the impact from non-line-of-sight (Qi and Kobayashi, 2002b, Qi and Kobayashi, 2002a).

The 2 × 2 Fisher Information Matrix $J(p)$ is defined as

$$\begin{aligned} J(p) &= \mathrm{E}\left(\nabla_p^T \log p_E(y-h(p))\nabla_p \log p_E(y-h(p))\right) \\ \nabla_p \log p_E(y-h(p)) &= \begin{pmatrix} \dfrac{\mathrm{d}\log p_E(y-h(p))}{\mathrm{d}X} & \dfrac{\mathrm{d}\log p_E(y-h(p))}{\mathrm{d}Y} \end{pmatrix} \end{aligned} \tag{0.31}$$

where $p=(X,Y)^T$ the two-dimensional position vector and $p_E(y\text{-}h(p))$ the likelihood given the error distribution p_E of the measurement uncertainty.

In case of Gaussian measurement error distributions $p_E(e)= N(0,R(p))$, the FIM equals

$$\begin{aligned} J(p) &= H^T(p)R(p)^{-1}H(p) \\ H(p) &= \nabla_p h(p) \end{aligned} \tag{0.32}$$

Table 2 summarizes the involved expressions for $h(p)$ and its gradient for range and direction measurements with focus on location i.

In the general case with non-Gaussian error distributions, numerical methods are needed to evaluate the CRLB. The larger the gradient $H(p)$, or the smaller the measurement error, the more information is provided from the measurement, and the smaller potential estimation error.

Information is additive, so if two measurements are independent, the corresponding information matrices can be added. This is easily seen for instance from (0.32) for $H^T = (H_1^T, H_2^T)$ and R being block

Table 2. Analytical expressions related to some measurements related to the location i *and corresponding gradients, where* $\tilde{X}^{ij} = X^i - X^j$, $\tilde{Y}^{ij} = Y^i - Y^j$ $D^{ij} = \sqrt{(\tilde{X}^{ij})^2 + (\tilde{Y}^{ij})^2}$ $\varphi^{ij} = \mathrm{angle}\left(p_t^i - p_t^j\right)$.

Method	$h(p)$	$H(p)=\dfrac{\partial h(p)}{\partial p}$
DOA	φ^{ij}	$\left(\dfrac{-\tilde{Y}^{ij}}{(D^{ij})^2}, \dfrac{\tilde{X}^{ij}}{(D^{ij})^2}\right)$
TOA	D^{ij}	$\left(\dfrac{\tilde{X}^{ij}}{D^{ij}}, \dfrac{\tilde{Y}^{ij}}{D^{ij}}\right)$
TDOA	$D^{ij} - D^{ik}$	$\left(\dfrac{\tilde{X}^{ij}}{D^{ij}} - \dfrac{\tilde{X}^{ik}}{D^{ik}}, \dfrac{\tilde{Y}^{ij}}{D^{ij}} - \dfrac{\tilde{Y}^{ik}}{D^{ik}}\right)$
RSS	$P_0^i - n_{ij}\log(\lVert p^i - p^j \rVert)$ $\bar{P}_0^i \lVert p_t^i - p_t^j \rVert^{-n_{ij}}$	$\left(\dfrac{-n_{ij}\tilde{X}^{ij}}{(D^{ij})^2}, \dfrac{-n_{ij}\tilde{Y}^{ij}}{(D^{ij})^2}\right)$ $\left(\dfrac{-\bar{P}_0^i n_{ij}\tilde{X}^{ij}}{(D^{ij})^{n_{ij}+2}}, \dfrac{-\bar{P}_0^i n_{ij}\tilde{Y}^{ij}}{(D^{ij})^{n_{ij}+2}}\right)$

diagonal, in which case we can write $J = J_1 + J_2$. Plausible approximate scalar information measures are the trace of the FIM and the smallest eigenvalue of FIM

$$J_{tr}(p) \triangleq \text{tr } J(p), \quad J_{\min}(p) \triangleq \min \text{eig } J(p) \tag{0.33}$$

The former information measure is additive as FIM itself, while the latter is an under-estimation of the information useful when reasoning about whether the available information is sufficient or not. Note that in the Gaussian case with a diagonal measurement error covariance matrix, the trace of FIM is the squared gradient magnitude.

The Cramer-Rao Lower Bound is given by

$$\text{Cov}(\hat{p}) = \text{E}(p^o - \hat{p})(p^o - \hat{p})^T \geq J^{-1}(p^o) \tag{0.34}$$

The CRLB holds for any unbiased estimate of $\hat{p}_t$. The lower bound may not be an attainable bound. It is known that asymptotically in the information, the ML estimate is $\hat{p} \sim \mathbb{N}(p^o, J^{-1}(p^o))$ (Lehmann, 1991) and thus reaches this bound, but this may not hold for finite amount of inaccurate data.

The right hand side of (0.34) gives an idea of how suitable a given sensor node configuration is for localization e.g. of an unknown object. It can also be used for *deployment design*, e.g. where to place the sensor nodes in order to enable pre-determined localization accuracy, given the sensitivity of the sensors. However, it should always be kept in mind though that this lower bound is quite conservative and relies on many assumptions. In practice, the root mean square error (RMSE) is perhaps of more importance. This can be interpreted as the achieved position error in meters. The CRLB implies the following bound:

$$\text{RMSE} = \sqrt{\text{E}\left[(X^o - \hat{X})^2 + (Y^o - \hat{Y})^2\right]} \geq \sqrt{\text{tr Cov}(\hat{p})} \geq \sqrt{\text{tr } J^{-1}(p^o)} \tag{0.35}$$

The first inequality becomes an equality for unbiased estimates. If RMSE requirements are specified, it is possible to make the sensor network denser until (0.35) indicates that the amount of information provided from the sensor nodes is enough.

A Geometric Example

Consider the scenario in Figure 5, where four sensor nodes are placed in the positions (-1,0), (1,0), (0,-1) and (0,1), respectively. Each node measures the arrival time of a transmitted signal from an unknown position (either a sensor node or an unknown object), using accurate and synchronized clocks. If the transmitter is also synchronized, the signal propagation time can be computed, which leads to a TOA measurement. Propagation time corresponds to a distance, which leads to the distance circles around each receiver in Figure 5 (upper left). If the transmitter is unsynchronized, each pair of receivers can compute a time-difference of arrival TDOA. This leads to a hyperbolic function where the transmitter can be located (Fang, 1990, Spirito and Mattioli, 1998).

The RMSE lower bounds from (0.35) for the TOA and TDOA measurements in this example are plotted in Figure 5 (lower plots). They indicate estimation accuracy limits spatially depending on the actual position of a sensor node or an unknown object. The level curves are scaled by σ_e, so a range error with standard deviation of 100 meter will in the most favorable position lead to a position estimation error of 100 meter. A bit counterintuitive, TDOA and TOA give the same performance close to the

Figure 5. First row: Example scenario with four sensor node placed in a square, and there is one transmitter at (1.2,1). With TOA measurements, each receiver measurement constrains the transmitter position to a circle, while with TDOA measurements, each pair of receiver measurements constrains the transmitter position to a hyperbola. Second row: RMSE lower bound implied by the CRLB for TOA and TDOA, respectively, measurements. The unit is scaled to the measurement standard deviation.

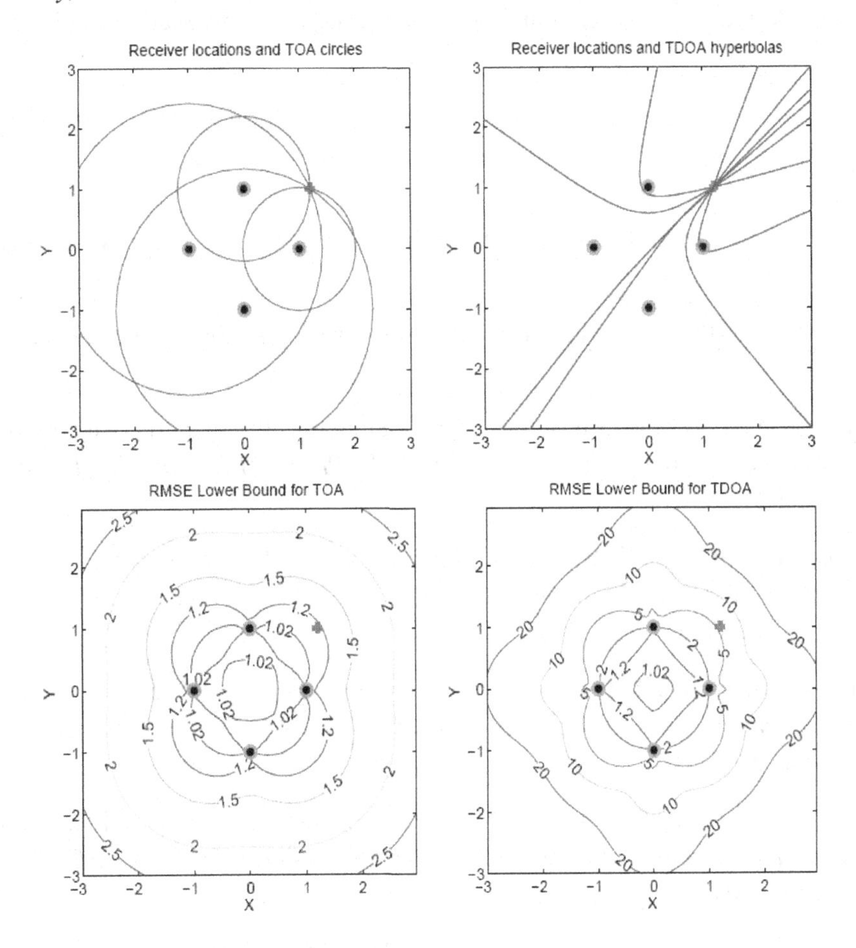

origin. To explain this, note that if all signals arrive simultaneously to all receivers, the transmission time does not add any information.

SUMMARY

This chapter addresses generic models of sensor observations in a wireless sensor network (WSN) by dividing these into detailed waveform observations, less detailed timing observations and much less detailed power observations. This view is well established in wireless cellular networks (WCN), though a more general framework was presented that besides radio waves also covers WSN with acoustic, seismic, sonic, infrared waves in different media as air, water and ground. A careful investigation associates these observation types to the more traditional measurement related to range, relative range and angle between nodes. A WSN is characterized by its architecture, connectivity, interactivity, the degree of synchronization and the available bandwidth in the wireless links, and these are also the main factors for which sensor observations that are plausible and can be shared between nodes and the kind of information that can be computed from the received signals.

A framework for analyzing the information content of various observations was also presented, that can be used to derive localization bounds for integrating any combination of observations in the network.

REFERENCES

Akyildiz, I., Su, W., Sankarasubramaniam, Y., & Cayirci, E. (2002). A survey on sensor networks. *IEEE Communications Magazine*, 40(8).

Akyildiz, I., & Wang, X. (2005). A survey on wireless mesh networks. *IEEE Communicatios Magazine*, 43(9).

Botteron, C., Fattouche, M., & Host-Madsen, A. (2002). Statistical theory of the effects of radio location system design parameters on the position performance. In *Proc. IEEE Vehicular Technology Conference*, Vancouver, Canada.

Botteron, C., Host-Madsen, A., & Fattouche, M. (2001). Cramer-Rao bound for location estimation of a mobile in asynchronous DS-CDMA systems. In *Proc. IEEE Conference on Acoustics,Speech and Signal Processing*, Salt Lake City, UT, USA.

Botteron, C., Host-Madsen, A., & Fattouche, M. (2004a). Cramer-Rao bounds for the estimation of multipath parameters and mobiles' positions in asynchronous DS-CDMA systems. *IEEE Transactions on Signal Processing, 52*(4).

Botteron, C., Host-Madsen, A., & Fattouche, M. (2004b). Effects of system and environment parameters on the performance of network-based mobile station position estimators. *IEEE Transactions on Vehicular Technology, 53*(1).

C.-C. Shen, C., Srisathapornphat, C., & Jaikaeo, C. (2001). Sensor information networking architecture and applications sensor information networking architecture and applications. *IEEE Personal Communications, 8*(4).

Caffery, J. J., & Stuber, G. L. (1998). Overview of radiolocation in CDMA cellular systems. *IEEE Communications Magazine, 36*(4).

Chen, Y., & Kobayashi, H. (2002). Signal strength based indoor geolocation. In *Proc. IEEE International Conference on Communications*, New York, NY, USA.

Coates, M. (2004). Distributed particle filters for sensor networks. In *Proc. IEEE Information Processing in Sensor Networks*, Berkeley, CA, USA.

Deblauwe, N. *GSM-based Positioning: Techniques and Application*. PhD Dissertation, Vrije University, Brussels, Belgium, 2008.

Drane, C., Macnaughtan, M., & Scott, C. (1998). Positioning GSM telephones. *IEEE Communications Magazine*, *36*(4).

Elson, J., & Römer, K. (2003). Wireless sensor networks: A new regime for time synchronization. *ACM Computer Communication Review*, *33*(1).

Fang, B. (1990). Simple solutions for a hyperbolic and related position fixes. *IEEE Transactions on Aerospace and Electronic Systems*, *26*(5).

Galstyan, A., Krishnamachari, B., Lerman, K., & Pattem, S. (2004). Distributed particle filters for sensor networks. In Proc. *IEEE Information Processing in Sensor Networks*, Berkeley, CA, USA.

Gudmundson, M. (1991). Correlation model for shadow fading in mobile radio systems. *IEE Electronics Letters*, 27(23).

Gustafsson, F., & Gunnarsson, F. (2005). Possibilities and fundamental limitations of positioning using wireless communications networks. *IEEE Signal Processing Magazine*, *22*(7).

Gustafsson, F., Gunnarsson, F., Bergman, N., Forssell, U., Jansson, J., Karlsson, R., & Nordlund, P.-J. (2002). Particle filters for positioning, navigation and tracking. *IEEE Transactions on Signal Processing*, *50*(2), 425–437.

Hata, M. (1980). Empirical formula for propagation loss in land mobile radio services. *IEEE Transactions on Vehicular Technology*, *29*(3).

Jayasimha, D., Iyengar, S., & Kashyap, R.(1991). Information integration and synchronization in distributed sensor networks. *IEEE Transactions Systems, Man, and Cybernetics*, *21*(5).

Kaemarungsi, K., & Krishnamurthy, P. (2004). Modeling of indoor positioning systems based on location fingerprinting. In *Proc. IEEE INFOCOM*, Hong Kong, P.R. China.

Kahn, J. M., Katz, R. H., & Pister, K. S. J. (1999). Mobile networking for smart dust. In *Proc. ACM/IEEE MobiCom*, Seattle, WA, USA.

Karlsson, R., & Gustafsson, F. (2003). Particle filter and Cramer-Rao lower bound for underwater navigation. In *IEEE Conference on Acoustics, Speech and Signal Processing (ICASSP)*, Hongkong, China.

Kay, S. (1993). *Fundamentals of signal processing – estimation theory*. Prentice Hall.

Koorapaty, H. (2004). Barankin bound for position estimation using received signal strength measurements. In *Proc. IEEE Vehicular Technology Conference*, Milan, Italy.

Koorapaty, H., Grubeck, H., & Cedervall, M. (1998). Effect of biased measurement errors on accuracy of position location methods. In *Proc. IEEE Global Telecommunications Conference*, Sydney, Australia.

Langendoen, K., & Reijers, N. (2003). *Distributed Localization in Wireless Sensor Networks: A Quantitative Comparison*. Elsevier Science.

Lehmann, E. (1991). *Theory of point estimation*. Statistical/Probability series. Wadsworth & Brooks/Cole.

Luo, J., Shukla, H., and Hubaux, J.-P. (2006). Noninteractive location surveying for sensor networks with mobility-differentiated toa. In *Proc. IEEE INFOCOM*, Barcelona, Spain.

Luo, R., & Kay, M. G. (1989). Multisensor integration and fusion in intelligent systems. *IEEE Transactions Systems, Man, and Cybernetics, 19*(5).

Maroti, M, Kusy, B., Balogh, G., Volgyesi, P., Nadas, A., Molnar, K., Dora, S., & Ledeczi, A. (2005, November). Radio Interferometric Geolocation. *In Proc. ACM 3rd Conference on Embedded Networked Sensor Systems (SenSys'05).*

Niculescu, D. (2004). Positioning in ad hoc sensor networks. *IEEE Network, 50*(4).

Okumura,Y., Ohmori, E., Kawano, T., & Fukuda, K. (1968). Field strength and its variability in VHF and UHF land-mobile radio service. *Review of the Electrical Communication Laboratory, 16*(9-10).

Pahlavan, K., Xinrong, L., & Mäkelä, J.-P. (2002). Indoor geolocation science and technology. *IEEE Communications Magazine*, 40(2).

Patwari, N., Hero III, A., Perkins, M., Correal, N., & O′ Dea, R. (2003). Relative location estimation in wireless sensor networks. *IEEE Trans. Signal Processing, 51*(8).

Patwari, N., Hero III, A., Perkins, M., Correal, N., & O'Dea, R. (2003). Relative location estimation in wireless sensor networks. *IEEE Transactions on Signal Processing, 51*(8).

Patwari, N., Hero III, A. O., Ash, J., Moses, R. L., Kyperountas, S., & Correal, N. S. (2005). Locating the nodes – cooperative localization in wireless sensor network. *IEEE Signal Processing Magazine, 22*(7).

Qi, Y., & Kobayashi, H. (2002a). Cramer-Rao lower bound for geolocation in non-line-of-sight environment. In *Proc. IEEE Conference on Acoustics, Speech and Signal Processing*, Orlando, FL, USA.

Qi, Y., & Kobayashi, H. (2002b). On geolocation accuracy with prior information in non-line-of-sight environment. In *Proc. IEEE Vehicular Technology Conference*, Vancouver, Canada.

Savarese, C., Rabaey, J., & Beutel, J. (2001). Locationing in distributed ad-hoc wireless sensor networks. In *Proc. IEEE Conference on Acoustics, Speech, and Signal Processing*, Salt Lake City, UT, USA.

Sayed, A. H., Tarighat, A., & Khajehnouri, N. (2005). Network-based wireless location. *IEEE Signal Processing Magazine, 22*(7).

Schenato, L., & Gamba, G. (2007). A distributed consensus protocol for clock synchronization in wireless sensor network. In *Proc. IEEE Conference on Decision and Control*, New Orleans, LA, USA.

Sivrikaya, F., & Yener, B. (2004). Time synchronization in sensor networks: A survey. *IEEE Network, 18*(4).

Solis, R., Borkar, V., and Kumar, P. (2006). A new distributed time synchronization protocol for multihop wireless networks. In *Proc. IEEE Conference on Decision and Control*, San Diego, CA, USA.

Spirito, M., & Mattioli, A. (1998). On the hyperbolic positioning of GSM mobile stations. In *Proc. International Symposium on Signals, Systems and Electronics.*

Sun, G., Chen, J., Guo, W., & Ray Liu, K. J. (2005). Signal processing techniques in network-aided positioning: A survey *IEEE Signal Processing Magazine*, *22*(7).

Weiss, A. J. (2003). On the accuracy of a cellular location system based on received signal strength measurements. *IEEE Transactions on Vehicular Technology*, *52*(6),1508–1518.

Wesson, R., Hayes-Roth, F., Burge, J. W., Stasz, C., & Sunshine, C. A. (1981). Network structures for distributed situation assessment. *IEEE Transactions Systems, Man, and Cybernetics*, *11*(1).

Yemini, Y. (1978). Distributed sensors networks (dsn): An attempt to define the issues. In *Proc. Distributed Sensor Networks Workshop*, Carnegie Mellon, Pittsburgh, PA, USA.

Zhao, Y. (2002). Standardization of mobile phone positioning for 3G systems. *IEEE Communications Magazine*, *40*(7).

Chapter III
Localization Algorithms and Strategies for Wireless Sensor Networks:
Monitoring and Surveillance Techniques for Target Tracking

Ferit Ozan Akgul
CWINS, Worcester Polytechnic Institute, USA

Mohammad Heidari
CWINS, Worcester Polytechnic Institute, USA

Nayef Alsindi
CWINS, Worcester Polytechnic Institute, USA

Kaveh Pahlavan
CWINS, Worcester Polytechnic Institute, USA

ABSTRACT

This chapter discusses localization in WSNs specifically focusing on the physical limitations imposed by the wireless channel. Location awareness and different methods for localization are discussed. Particular attention is given to indoor TOA based ranging and positioning systems. Various aspects of WSN localization are addressed and performance results for cooperative schemes are presented.

INTRODUCTION

Wireless sensor networks are ideal candidates for data gathering and remote sensing purposes in various environments, where the communications between the sensors mostly take place in a distributed manner (Akyildiz et al., 2002). Data obtained by individual sensors are relayed to a central station for further processing and logging. Usually, information obtained through a sensor needs to be associated with the location of each sensor which necessitates the localization of sensors within certain accuracy (Patwari et al., 2005). Since primary purpose of sensor networks is to gather information on environmental changes such as temperature, pressure or humidity, it is almost always required to determine the coordinates of a specific sensor so that the appropriate steps can be taken in a more effective way in the case of emergencies. As a matter of fact, about 13.3% of the recent scientific WSN publications focus on target tracking and localization aspects as mentioned by Sohraby et al. (2007).

Owing to their small form factor and low-power consumption, WSNs have found numerous applications in both civil and military use. In the civil domain, applications can be further divided into various categories like environmental, health related, commercial and public safety applications. Environmental applications may include fire/flood detection/prevention (Pathan et al., 2006), crop quality detection, field surveying; health related applications may be listed as patient/doctor/instrument tracking inside hospitals, elderly care and remote monitoring of biological data (Cypher et al., 2006); commercial applications might be inventory control, product tracking in warehouses (Rohrig & Spieker, 2008) and remote product quality assessment. For military applications, the use of WSNs is also important in rough terrain conditions where a centralized communication system may be too costly to build (Merrill et al., 2003). Soldier and mine tracking, as well as intelligence gathering can be some applications in this domain.

As it can be seen from a sample of applications in each field, location and tracking capability is an important aspect of WSNs that need to be considered and developed further. A number of researchers studied positioning using sensor nodes and investigated the performance of algorithms and presented theoretical bounds in WSNs (Bulusu et al., 2000; Niculescu et al., 2001; Savarese et al., 2001; Savvides et al., 2001; Doherty et al., 2001; Chang & Sahai, 2004; Kanaan et al., (2006a, 2006c)). The positioning capability can be implemented using different sensing technologies like ultrasonic waves as in the Active Bat system proposed by Ward et al. (1997) or using RF or both as in the Cricket system (Priyantha et al., 2000). By using sound waves, researchers have been able to obtain cm. accuracy; however, these systems are only suitable for very small areas like a single office environment and are not intended for outdoor or indoor/outdoor hybrid node localization. In the latter case, RF solutions are generally preferred since quick deployment is possible and hardware and various ranging/localization algorithms that can be directly applied are widely available. Nevertheless, due to the nature of RF propagation, ranging/localization accuracy is not on par with solutions using sound waves. Later in this chapter we will cover the basics of RF channel and how it affects the performance of localization in various environments.

Figure 1 shows a typical setup for a WSN with location capability. Here, the sensor nodes are able to communicate with each other as well as pre-deployed anchor nodes whose coordinates are known in advance. At this point, it might be appropriate to present the two methods of WSN localization. Sensor node localization can take place in a centralized or a distributed manner.

In centralized localization, individual sensors relay their ranging estimates from the anchors to a central processing station where the localization algorithm is implemented. In the centralized approach, the processing station has the knowledge of each location requesting node and hence network topology can easily be associated with the geographical positions of the sensor nodes. Drawbacks of this method include traffic congestion and computational complexities, especially for larger sensor networks.

In distributed positioning, the sensors get ranging estimates from the anchors and try to localize themselves. After they have performed localization they might act as pseudo-anchors so that other nodes

Figure 1. Typical WSN setup

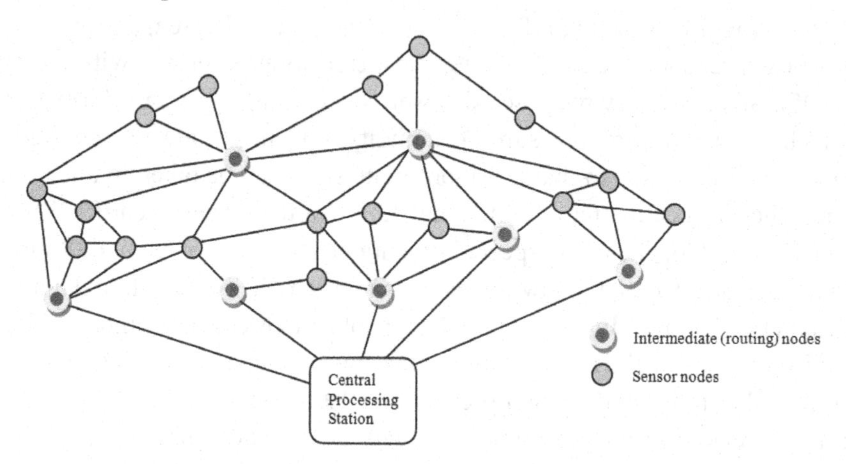

might benefit from these nodes for their own position fixing. Most of the time, pseudo-anchors will have estimation quality based on how many anchors they captured during localization and link quality for each individual anchor connection (Kanaan, 2006b). The estimation quality might be used in order to inform the end user of the overall quality factor of a particular node position. Distributed localization is most suited for large networks where central processing might be a limiting factor for the application being considered and when efficient energy utilization is required.

The performance of localization for WSNs depends mostly on the availability and quality of individual links between sensors and the anchors. The geographical distribution of the nodes with respect to each other also plays an important role in positioning accuracy. As the *connectivity* of a WSN increases, more reliable estimates can be obtained. The connectivity may be considered as a measure of how robust the WSN against node failures is. The number of alternative routes for relaying information from a specific node to the destination is directly proportional to connectivity. A high node density, as described by nodes per unit area, would primarily lead to high connectivity since nodes will be able to detect RF signals from their neighbors. On the other hand, if the nodes are sparsely connected, connectivity will be less hence nodes might not be able to obtain accurate position fixes due to lack of ranging information. Hence a high node density will provide better localization performance. A detailed analysis will be presented relating the node density to localization performance as well as performance bounds for the cooperative localization.

The overall organization of the chapter is as follows: The second section gives an overview of location awareness and a typical localization system. The third section presents the distance/position estimation metrics along with their achievable performance bounds and also the mapping techniques. Also part of this section is a brief overview of dynamic tracking and monitoring. The fourth section focuses on the TOA based ranging and localization and the challenges associated with it. The fifth section introduces and discusses the methods that can be used in the absence of DP conditions, particularly non-direct-path exploitation and cooperative localization. The sixth section presents the two studies showing the effect of various parameters on WSN localization. The seventh presents the practical implementation issues related to the implementation of WSN localization and finally the eighth section provides a conclusion for the chapter.

LOCATION AWARENESS

The question "where" might seem simple at first but the answer might not. Throughout the centuries, mankind has always tried to find the right answer to this question in his quest for exploring new lands and navigating the unknown seas. The first sailors relied on particular water currents, landmarks and positions of the celestial bodies to navigate through the waters. With the discovery of compass about 700 years ago, mariners were able to identify their directions. However, the need to get precise position and navigation, primarily for military reasons, led the nations and researchers to develop systems closer to achieving this goal. After the first developments in radio navigation starting in the first half of the 20th century (Schroer, 2003), the first successful implementation of such a system came in the form of a global positioning system or GPS, developed by the US military. In its 40 years of development and maturity, GPS has become a reliable location finding and tracking system for use not only by military but also by civilian world. Today, after various advancements in the field such as differential-GPS (DGPS) and Wide Area Augmentation System (WAAS) typical commercial grade GPS receivers can achieve accuracies of 1-5m with DGPS and 3 meters with WAAS. Study by Blomenhofer et. al (1994) reports DGPS accuracies in the centimeter range. Higher end geodetic and surveying GPS units using carrier phase, dual frequency methods and sophisticated algorithms can achieve centimeter and even sub-centimeter accuracy through GPS ambiguity resolution techniques (Kim & Langley, 2000; Poling & Zatezalo, 2002). Following GPS, other countries also started their own satellite positioning projects (EU's Galileo and Russia's GLONASS (Zaidi & Suddle, 2006)). Owing to its accurate positioning capability, most industries rely on GPS and the position information obtained via GPS (such as anchor node position information) serves as reference for other small scale localization systems.

Although GPS is a proven and reliable technology, it falls short of expectations for some terrestrial applications where the GPS signals cannot be detected due to obstructions. Satellite signals are attenuated heavily through the atmosphere and further obstruction by trees, heavy fog or manmade structures such as building tops prevent this system to be useful especially for densely populated urban settlements and inside buildings. In order to overcome this issue, researchers turned their attention to land-based positioning and tracking systems for situations that cannot make use of satellite signals.

After the proliferation of cellular based radio communications systems, FCC mandated mobile phones be located within a certain accuracy (FCC, 1999). According to this report, mobile operators should be able to locate phones with 50m accuracy 67% of the time and 150m accuracy 95% of the time for handset based positioning, and 100m accuracy 67% of the time and 300m 95% of the time for network based positioning.

The fundamentals of locating and tracking RF emitting devices differ greatly from those of data communications. In communications, information is transferred from one entity to another and the information carrier might be RF, sound or light. A single link, as long as it is reliable, will be enough to transfer data between the entities. However, locating a device whose location is completely unknown requires a completely different approach than transmitting data.

Localization might be realized in two ways:

- Geometric methods (Trilateration, triangulation, hyperbolic methods)
- Fingerprinting methods (Signal mapping)

Geometric methods include techniques that can locate or track devices based on signal properties that are estimated. TOA/TDOA, RSS and AOA are examples of geometric measurement techniques.

Fingerprinting methods require a two-phase approach. In the first phase (also called the off-line phase), a database is formed based on signal parameters and this database is utilized in the second phase to estimate the location. Nearest-neighbor mapping and artificial neural networks are examples of such methods (Dasarathy, 1991; Heidari et al., 2007b; Nerguizian et al., 2006; Kanaan & Pahlavan, 2004a).

Overview of a Localization System

A typical localization system consists of mobile terminals (nodes in case of a WSN) that need to be located/tracked, beacon or anchors serving as reference points, a central processing station that implements the positioning algorithm and keeps track of all the terminals as well coordinates data communications and a higher layer system that shows the results of positioning or tracking like an LCD panel. Figure 2 shows the components of such a localization system. The system might use different ranging metrics for obtaining the position information. The most common of these metrics are RSS, TOA/TDOA and AOA. RSS and TOA might be considered as ranging metrics since ranging information can be obtained from these signal parameters. The nodes will need at least three ranging estimates from different anchors to be able to obtain a position fix. In the case of AOA, two different AOA estimates from two anchors will suffice to obtain a location fix. More details will be given for each of these techniques in the coming sections.

DISTANCE/POSITION ESTIMATION METRICS

As mentioned in the previous section, a localization system needs to obtain range estimates from fixed anchors or reference points in order to estimate the location of a node. Ranges estimates can be obtained using different metrics. RSS and TOA/TDOA are examples of such techniques.

RSS: After the RF signal is transmitted by a transmitter, its energy experiences loss that is proportional to the distance signal travels. A common model based on single-path radio propagation is given by

Figure 2. High level architecture of a typical positioning system

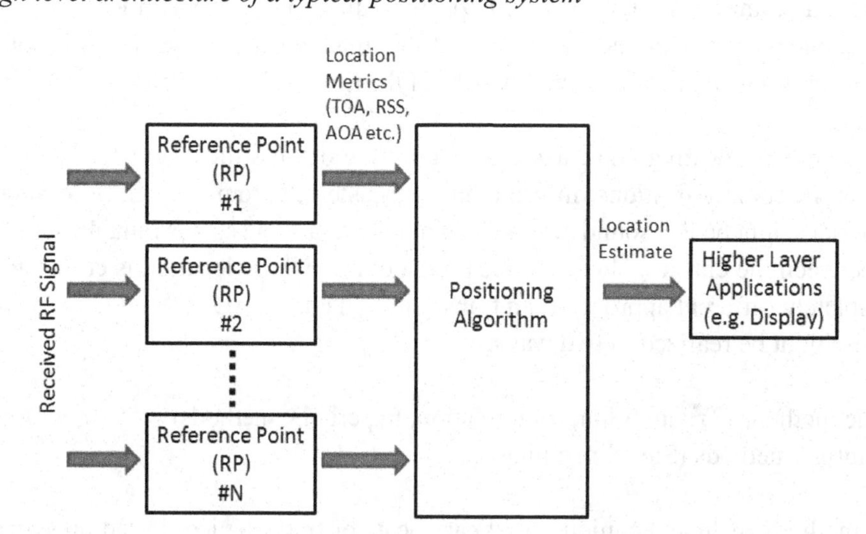

$$P_r(dB) = P_t(dB) - 10\alpha \log_{10} d \qquad (1)$$

where P_r(dB) and P_t(dB) denote the received and transmitted signal powers in dB. α is the distance-power gradient and is dependent on the propagation environment. For free space, $\alpha = 2$. A wide range of values are possible for α, i.e. for a brick construction office environment α is reported to be 3.9 or for a laboratory environment with metal-faced partitioning it is found to be 6.5 (Pahlavan & Levesque, 2005).

Other empirical models have also been developed based on extensive measurements in various environments. Motley (1988) proposed a path-loss model for multi-floor buildings. Technical working group of TIA/ANSI JTC recommended an indoor path loss model (JTC, 1994) for PCS applications. Apart from the indoor model, the same group proposed micro and macro-cellular models for outdoor applications. Other popular models for outdoor environments are the Okumura (1968), Hata (1980) and COST231 (1991) models.

Either by using the simple radio-propagation or the more complicated empirical models, distance information can be obtained from the received signal power given the transmitted signal power. Although this method can be easily applied since almost all RF wireless devices can report received signal strength, its accuracy is not always acceptable due to the stochastic variation of the channel. The path loss models discussed in this section are deterministic models that do not consider the fading and shadowing effects. At any time instant, the signal level experiences slow and fast fading caused by local scatterers and the movement of the receiver node. Slow fading is also called Shadow fading and it is generally modeled as a zero-mean normal variable, *X(dB),* in the logarithmic scale. Hence shadows are generally log-normally distributed. The probability density function (pdf) for the log-normal distribution is given as:

$$f(x) = \frac{1}{x\sigma\sqrt{2\pi}} e^{-\frac{(\ln(x)-\mu)^2}{2\sigma^2}} \qquad (2)$$

where μ and σ are the mean and standard deviation respectively.

Hence the received power can be given as

$$P_r(dB) = P_t(dB) - 10\alpha \log_{10} d + X \qquad (3)$$

Due to this fluctuating behavior of received signal power, accurate ranging measurements are not always possible hence leading to lower accuracy position estimation. In fact, the accuracy of such estimation is lower bounded by its Cramer-Rao lower bound (CRLB). CRLB basically specifies the lower bound on the variance of estimation. For the simplistic RSS model this bound has been given by Qi and Kobayashi (2003) as:

$$\sigma_{RSS} \geq \frac{(\ln 10)}{10} \frac{\sigma_{sh}}{\alpha} d \qquad (4)$$

Here, σ_{RSS} is the standard deviation of RSS estimation, σ_{sh} is the variance for shadow fading, d is the actual distance between the transmitter and the receiver and α is the power-distance gradient.

AOA: AOA information from two different anchors might be used to determine the position of a node by using triangulation as shown in Figure 3. AOA estimation is also referred to as direction-of-arrival (DOA) estimation, direction finding or bearing estimation in many contexts and has been researched

Figure 3. Triangulation of a node by two anchors

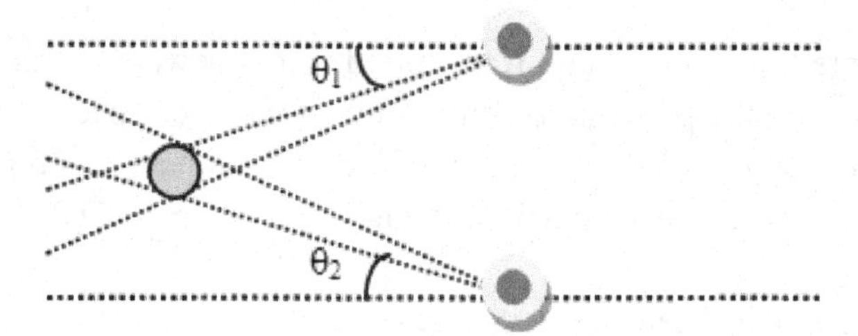

extensively in the literature (Cedervall & Moses, 1997; Krim & Viberg, 1996; Kumaresan & Tufts, 1983; Li et al., 1995; Stoica & Sharman, 1990; Tewfik & Hong, 1992). A common method for AOA estimation is by using special structures called uniform linear arrays (ULA) (Schelkunoff, 1943). The n elements of an ULA with spacing d can be used to estimate the direction of arrival of a RF signal based on the following relation:

$$\theta = \arcsin\left(\frac{c\Delta t}{d}\right) \quad (5)$$

where θ is the angle at which the signal is impinging upon the ULA, c is the speed of light, Δt is the time difference between the arrivals of the signal at consecutive array elements and d is the spacing between consecutive elements.

To achieve finer results by using a certain antenna array configuration one can employ super-resolution techniques. Although various methods are available in literature, most common ones are MUSIC (Schmidt, 1986) and ESPRIT (Roy & Kailath, 1989) and their variations (Rao & Hari, 1989; Ottersten et al., 1991). Kuchar et al. (2002) report angular estimation variance of 1° with 0dB SNR and down to 0.01° with SNRs of about 40 dB.

The performance bounds for AOA estimation can also be studied to derive the CRLB. The bound for AOA is formulated to be (Tingley, 2000)

$$\sigma_{AOA} \geq \frac{c\sqrt{2}\sqrt{BN_0}}{N\Delta 2\pi f_c \sin(\theta) A A_T} \quad (6)$$

where σ_{AOA} is the standard deviation for AOA estimation, c is the speed of light, B is the bandwidth of the signal, N_0 is the noise spectral density, N is the number of elements of the ULA, Δ is the spacing between the elements, A is the channel coefficient. A_T and f_c are respectively the amplitude and carrier frequency of the source signal $x(t)$ denoted as

$$x(t) = A_T \cos 2\pi f_c t \quad (7)$$

The *SNR* of the signal can be expressed as

$$SNR = \frac{A^2 A_T^{\ 2}}{BN_0} \quad (8)$$

so we can rewrite (6) as

$$\sigma_{AOA} \geq \frac{c\sqrt{2}}{N\Delta 2\pi f_c \sin(\theta)\sqrt{SNR}} \tag{9}$$

From (9), it can be seen that the CRLB is inversely proportional to the number of elements N, f_c and SNR. Thus having a high SNR and high frequency signal like an Ultra-Wide-Band (UWB) signal as well as a high number of array elements give higher resolution AOA estimation. Oppermann et al., 2004 give an overview of UWB signals and their applications.

TOA: Another distance estimation method is the TOA method in which the range is estimated based on the time the signal spends traveling from the transmitter to the receiver. Since the speed of RF propagation is very well known in both free space and air, it gives a direct estimation of the distance between the transmitter and the receiver once the travel time is estimated. When TOA systems are considered, the only important parameter that needs to be estimated correctly in a multipath propagation environment is the TOA of the LOS path or the direct-path (DP). Other multipath components are not important for ranging and localization purposes except for the cases when the DP is not available. This condition will be investigated in detail in the section *Challenges for the TOA based Systems.* The basic equation needed to obtain the distance is given as

$$d = \tau_{DP} c \tag{10}$$

where d is the distance estimate, τ_{DP} is the TOA of the DP and c is the speed of light. Accurate TOA estimation needs perfect synchronization between the clocks of the transmitter and the receiver. Clock synchronization might be achieved by regular data exchange between the transmitter and the receiver or an additional anchor for correcting the clock bias. Although 3 anchors are necessary to obtain position, a 4th anchor will be needed for time correction. This method is readily applied for the GPS in which a 4th satellite is used to compensate for the receiver clock bias. The TOA location estimation is depicted

Figure 4. Trilateration of a node by three anchors (RSS and TOA)

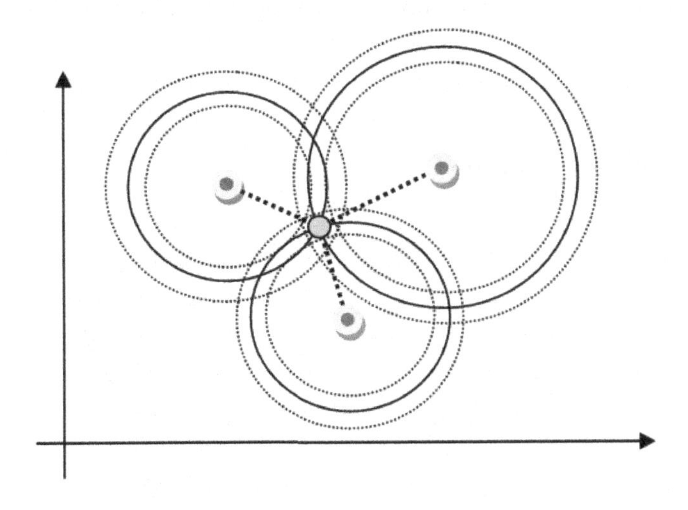

in Figure 4 where a perfect synchronization is assumed between the transmitters and the receiver. Same procedure also applies to the RSS method in which individual distance estimates are also used for position fixing. The dotted circles denote the uncertainty in range estimation hence leading to an area for the possible location of the receiver between the three estimation circles, rather than a single point.

Several methods are available to estimate the TOA. The traditional methods of estimation are the inverse Fourier transform (IFT) and maximum-likelihood (ML) estimations. The latter one is also called the cross-correlation method.

In the IFT method, observed frequency domain channel response is transformed into time-domain to obtain the time-response of the channel (Figure 5). The delay value of the DP is then used to calculate the distance.

The ML method assumes the following signal model for the estimation of TOA (Li, 2003).

$$r(t) = s(t-\tau) + w(t) \tag{11}$$

Here r(t) is the received signal, s(t) is the transmitted signal, τ is the delay and w(t) is noise modeled as AWGN. The signal at the receiver is basically a delayed version of the signal plus noise. ML dictates that maximum possible cross-correlation of the transmitted and the received signal occurs at the actual delay of the signal shown as (Li, 2003; Knapp & Carter, 1976)

$$\frac{d}{d\tau}\left(\int_{T_0} r(t)s(t-\tau)dt\right)\Bigg|_{\tau=\hat{D}_{ML}} = 0 \tag{12}$$

To obtain the delay estimate, τ is varied over a range of delay values and the value of τ that gives the maximum of the cross-correlation (or equivalently makes the derivative of the cross-correlation equal to 0) becomes the distance estimate. A block diagram is also given in Figure 6 to show the implementation of this method.

In the case of single path TOA estimation as applied to (11), CRLB is computed to be

$$\sigma_{\hat{D}}^2 \geq \frac{1}{8\pi^2}\frac{1}{SNR}\frac{1}{BT_0}\frac{1}{f_0^2 + B^2/12} \tag{13}$$

where $\sigma_{\hat{D}}^2$ is the variance of TOA estimate, $B = f_2 - f_1$, $f_0 = (f_2 + f_1)/2$ and T_0 is the observation time. From (13), it is easy to see that the bound is inversely proportional to the SNR, the signal bandwidth and observation time.

Figure 5. IFT operation for obtaining CIR

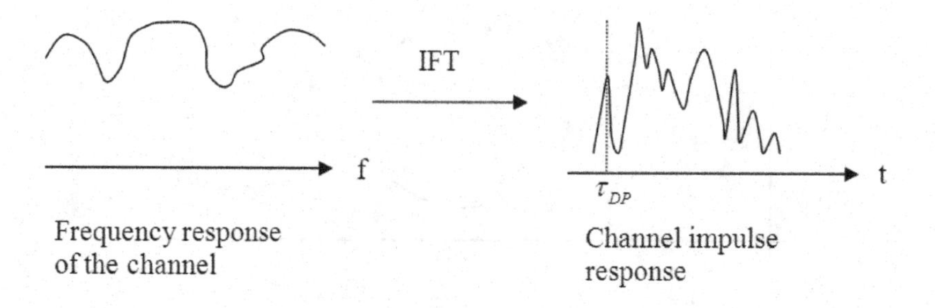

Figure 6. ML TOA estimation (reproduced from (Li, 2003))

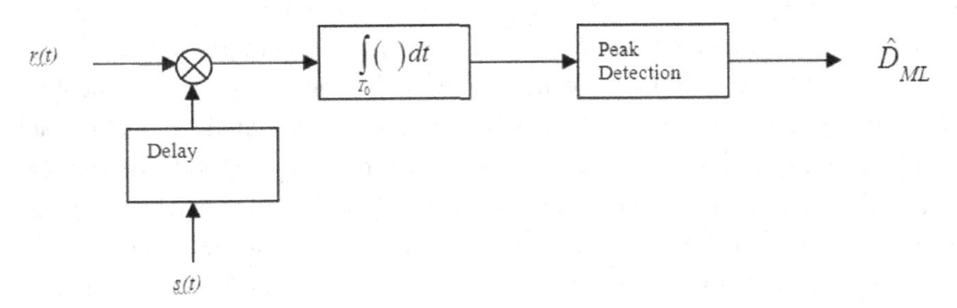

Figure 7. Hyperbolic positioning of a node by three anchors (TDOA)

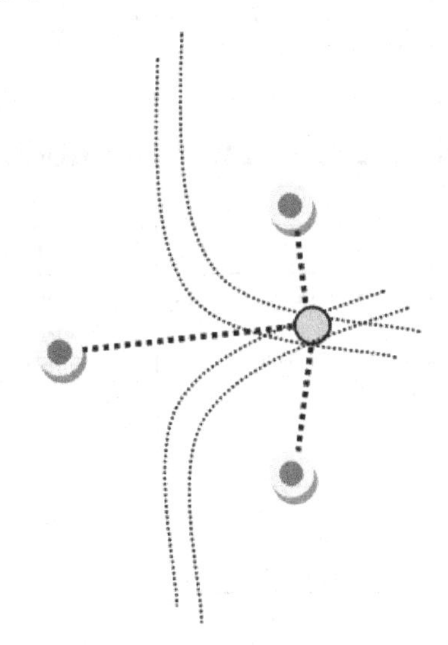

Similar to the AOA case, super-resolution methods might be used to extract indistinguishable peaks from the channel response in time-domain. This method has been shown to be effective in both wide-band and narrow band TOA estimation methods (Li & Pahlavan, 2004; Dumont et al., 1994) making these methods superior to traditional ML and cross-correlation methods.

TDOA: TDOA, also known as hyperbolic positioning, is a method whereby the receiver calculates the differences in the TOAs from different RPs. By using this method the clock biases between the transmitters and receivers are automatically removed, since only the differences between the TOAs from two transmitters are only considered. The estimation using TDOA is shown in Figure 7.

Mapping Techniques

Fingerprinting or mapping techniques are also widely studied for their robustness in terms of performance and some advantages in comparison to geometric methods. The fingerprinting methods employ a two-step approach. The first step involves the construction of the signal map for a desired environment (also called the offline phase), the second step is the actual positioning step (online phase). These techniques inherently capture all environment related propagation effects like multipath, shadowing, scattering etc and hence might be used for applications where geometric methods fall short of expectations. However due to extensive measurement and mapping involved in this approach, it might not be preferred for large areas where it may not be feasible. Additionally the structural changes in the environment might necessitate remapping for the affected regions of the database (Bahl & Padmanabhan, 2000; Nerguizian et al., 2006; Steiner et al., 2008). The two most commonly used mapping methods are the RSS mapping and CIR mapping. In RSS mapping, a receiver terminal is taken to almost every feasible part of an area that is intended to be mapped and signal power from multiple anchors are recorded into the database. Once mapping is done, actual positioning is obtained by comparing the RSS in online phase to one of the mapped points. The algorithms employed for this purpose are mostly k-NN algorithms (Dasarathy, 1991; Hatami & Pahlavan, 2005). Another application is the mapping of CIR for desired locations. The unique characteristics of the CIR, such as rms delay spread, average power… etc might then be used for comparison (Heidari et al., 2007b; Nerguizian et al., 2006). The same paper also discusses the use of neural networks for position estimation.

Overall Comparison for Different Localization Methods

Geometric Methods	**Advantages**	**Disadvantages**
RSS	- Simple to implement (most wireless devices report power) - Not sensitive to timing and RF bandwidth	- Not accurate - Requires models specific to application case and environment
AOA	- Only requires 2 anchors for localization	- DP blockage and multipath affects accuracy - Requires use of antenna arrays/smart antennas - Accuracy is dependent on RF bandwidth
TOA/TDOA	- Accurate ranging/localization can be obtained - Can be scaled to a multitude of applications	- Accuracy is dependent on RF bandwidth - DP blockage might cause large errors
Mapping Methods	- Captures all channel related parameters hence resilient to DP blockage	- Requires extensive database construction/training

Tracking and Dynamic Monitoring

Most of the time, the nodes to be located are mobile and hence it becomes essential to determine the location of these mobile nodes periodically. This real-time periodic location update is also called "tracking". Tracking keeps a history of the location information and hence is considered as a dynamic methodology for the localization problem. As opposed to blind positioning, which is based either on geometric or fingerprinting method and which basically requires locating the node without any prior position information, tracking makes use of location history to estimate the future positions. This might be obtained by various methods. The most popular of these approaches is to employ a Kalman Filter (Kalman, 1960), which is a recursive filter that estimates the state of a system in the presence of noisy measurements. However, for most systems that do not exhibit linear behavior, Kalman filtering is not an effective solution. For these non-linear systems other types of filtering such as Extended Kalman Filtering (EKF) (Haykin, 2002) and Unscented Kalman Filtering (UKF) (Julier & Uhlmann, 1997) are preferred. EKF is particularly useful for nonlinear but differentiable systems. It is a first order approximation for the nonlinear filtering problem. UKF, on the other hand is applicable to highly nonlinear systems and produces more accurate results than EKF. Another advantage of UKF over EKF is that

UKF does not require the computation of Jacobians that are needed for EKF. Hence it is more practical from an implementation point of view.

Another method is to use dead reckoning (Beauregard, 2006; Randell et al., 2003) which estimates the future position based on current speed, bearing and elapsed time. Inertial navigation systems are based on this principle. Even though these systems might obtain estimates for incomplete measurements, error propagation is a major concern for prolonged duration of information absence. Hence these tracking methods should be complemented with other true positioning approaches for a more complete positioning system design.

TOA-BASED RANGING AND LOCALIZATION

Due to recent advances in UWB signaling and hardware and its potential for accurate ranging, TOA based ranging systems utilizing UWB signals have gained particular popularity (Lee & Scholtz, 2002; Gezici et al., 2005; Qi et al.; 2004). On the other hand, the knowledge gained by the implementation and challenges of the now widely used GPS system has been instrumental in the advancement of these TOA systems. The following parts of this section will particularly focus on the TOA-based systems.

Peak Detection Strategies for TOA Based Systems

In this part, we present the two most commonly used methods for obtaining ranging measurements, which have also been outlined by Denis et al. (2003). In the following, τ_{sel} denotes the estimated TOA.

Detection of the First Peak

This method relies on the detection of the first available peak in the CIR. As long as the first path power is above the detection threshold of the system, the method gives the best possible results for ranging. However, accurate detection depends on high SNR (mostly DDP conditions) which is not always possible. The path decision can be expressed as

$$\tau_{sel} = \{\tau_i \mid i = \arg\min_p \tau_p\} \tag{14}$$

where τ_p is the TOA for the pth path.

Detection of the Strongest Peak

In this method, the path with the strongest power is detected and its TOA is considered as the ranging estimate between the transmitter and the receiver. Detecting the strongest peak is easily implementable when compared with the first method; however ranging accuracy is not always acceptable since the strongest path may not always be the DP. Most practical receivers implementing this method are S-rake receivers (Oppermann et al., 2004). Path decision in this case is

$$\tau_{sel} = \{\tau_i \mid i = \arg\max_p P_p\} \tag{15}$$

where P_p is the power of the pth path.

Challenges for the TOA-Based Systems

One major challenge facing the high precision TOA systems is the obstruction of the DP in the channel profile. Since the DP is the true indicator of the range between the transmitter and the receiver, its obstruction by various means such as metallic or thick concrete walls will lead to substantial ranging errors. This particular obstruction of the DP leads to a specific channel impairment that has been named as the undetected direct path (UDP) condition (Pahlavan et al., 1998; Wylie & Holtzman, 1996). To understand the effects of DP obstruction, it is convenient to consider the commonly used mathematical expression for channel impulse response at this point. This model takes into account the multipath components (MPCs) that arrive at the receiver via different propagation mechanisms such as reflection, transmission or scattering and is given as:

$$h(\tau,t)=\sum_{i=1}^{L}\beta_i\delta(t-\tau_i)e^{j\phi_i} \qquad (16)$$

where h denotes the CIR, L is the number of MPCs, β_i is the gain(amplitude), τ_i is the TOA and φ_i is the angle (phase) of the ith arriving path respectively. The DP might be characterized as the path having gain β_1, TOA τ_1 and phase φ_1. In this case the range between the receiver and the transmitter will be

$$d=\tau_1 c \qquad (17)$$

When the DP is blocked or cannot be detected, the indirect paths will be detected giving rise to substantial ranging errors. Figure 8 and Figure 9 below, show real world example channel profiles for both DDP and UDP obtained using a 1GHz bandwidth UWB signal. In the DDP case there is only 50cm of ranging error that can be attributed to the limited bandwidth (multipath error) of the signal, whereas the UDP case (by inserting a metallic shield in between the transmitter and receiver) introduces more than 2m of ranging error for the same setup.

Figure 8. Sample DDP channel profile

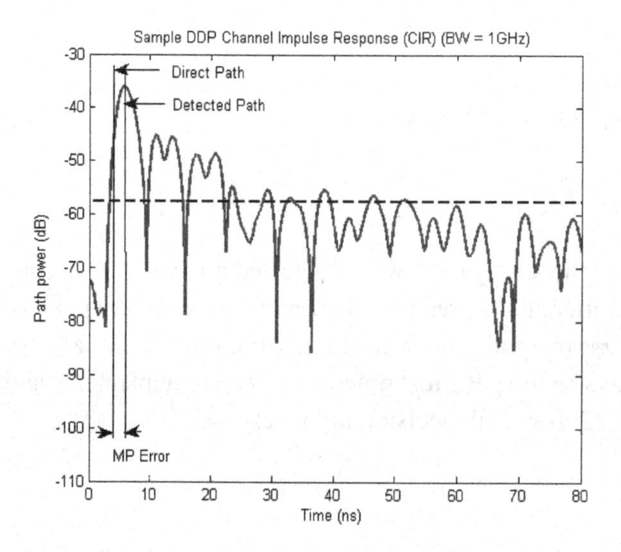

Figure 9. Sample UDP channel profile

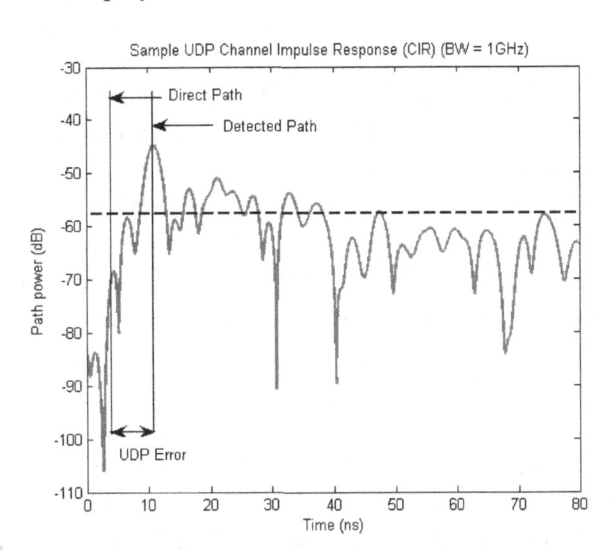

Next, we will present a new study on the detection of UDP conditions. The method presented here tries to estimate the occurrence of UDP condition hence necessary actions might be taken to mitigate the errors associated with UDP regions by employing proper adjustments to ranging data.

UDP Identification

Another open problem in the field of localization and positioning is the problem of identification of the channel profiles which exhibit unexpected large ranging errors. In traditional localization the distance measurements from different RPs were incorporated into the localization algorithm without mitigating the effects of ranging error. Therefore, the accuracy of the localization system would degrade drastically when problem of UDP occurred. This part investigates different techniques for identification of these channel profiles (UDP identification) with large ranging errors. We have used propagation parameters of the wireless channel and observed channel profile at the receiver side to decide whether a channel profile is in UDP conditions (Heidari et al., (2007a, 2007b)). To illustrate the effectiveness of such identification of channel profiles with UDP conditions, we set up an experiment with very limited number of RPs and compared the performance of the traditional algorithm with the proposed algorithm based on UDP identification. In the proposed algorithm we suggest to adjust the value of the distance measurement once a channel profile is identified to be in UDP conditions. The adjustment is performed by subtracting a correction value from the distance measurement (Heidari et al., (2007a, 2007b)). The amount of the correction value is determined based on predefined distributions of the ranging error in the typical environments similar to the building under study.

As discussed in the previous sections, the most important and distinguishing parameter in TOA-based localization system is the presence of the DP component. Detecting the DP component results in accurate ranging estimate of the true distance between the antenna pair. On the other hand, the erroneous estimation of the distance of the antenna pair results in large ranging error observed by the localization system and drastically degrades the performance of such systems. Therefore, we face two hypotheses:

$$\begin{cases} H_0 : DDP \mid d_{FDP} \approx d_{DP}, \varepsilon \approx 0 \\ H_1 : UDP \mid d_{FDP} \gg d_{DP}, \varepsilon \gg 0 \end{cases} \tag{18}$$

where H_0 denotes the DDP hypothesis, which indicates that the channel profile can effectively be used for localization, and H_1 denotes the UDP hypothesis, which indicates that the channel profile is not appropriate for being used for localization purposes.

There are two types of metrics being extracted from channel profile which can be utilized in identification of UDP conditions. The first class of metrics, is the time delay characteristics of the channel profile, while the second class deals with power characteristics of the channel profile. We can also utilize a hybrid metric, consists of time and power, in order to classify the receiver location.

Time Metrics

Delay information encrypted in the channel profile is our first time metric to investigate. Amongst all of the delay metrics the mean excess delay of the channel profile is the easiest to find and perhaps the most effective metric, relatively, to efficiently identify the UDP conditions. We used a Ray-Tracing (RT) database for modeling the different distributions of mean excess delay in this section. Mean excess delay is defined as the

$$\tau_m = \frac{\sum_{i=1}^{L_p} \hat{\tau}_i |\alpha_i|^2}{\sum_{i=1}^{L_p} |\alpha_i|^2} \tag{19}$$

where $\hat{\tau}_i$ and α_i represent the TOA and complex amplitude of the ith detected path, respectively, and L_p represents the number of detected peaks. Conceptually, it can be observed that profiles with higher mean excess delay are more likely to be UDP conditions.

Power Metrics

RSS is a simple metric that can be measured easily and it is measured and reported by most wireless devices. For example, the MAC layer of IEEE 802.11 WLAN standard provides RSS information from all active access points (APs) in a quasi-periodic beacon signal that can be used as a metric for localization (Kanaan et al., 2006b)

$$-P_{tot} = r = 10\log_{10}\left(\sum_{i=1}^{L_p} |\alpha_i|^2\right) \tag{20}$$

For identification, we used $r = -P_{tot}$ which is referred to as power loss. It can be observed that profiles with higher power loss are more likely to be UDP conditions.

Hybrid Time/Power Metric

Although, each time or power metric can be used individually to identify the class of receiver locations, but one can form a hybrid metric to achieve better results in identification of the UDP conditions. Here, we propose to use a hybrid metric consisting of TOA of DP component and its respective power as the metric to identify the UDP conditions. Mathematically

$$\zeta_{hyb} = -P_{FDP}\tau_{FDP} \tag{21}$$

where ζ_{hyb} represents the metric being extracted. It can be shown that the desired metric can be best modeled with Weibull distribution.

Binary Hypothesis Testing

Knowledge of the statistics of τ_m, r, and ζ_{hyb} enables us to identify the UDP conditions. In order to do so binary likelihood ratio tests can be performed to select the most probable hypothesis (Heidari et al., 2007a). For this purpose, we picked a random profile and extracted its respective metrics. The likelihood function of observed RMS delay spread,τ_{mi}, for DDP condition can then be described as

$$L(H_0 \mid \tau_{m_i}) = \mathrm{P_r}(\tau_{m_i} \mid H_0) \tag{22}$$

Similarly, the likelihood function of observed mean excess delay, τ_{mi}, for UDP condition can be described as

$$L(H_1 \mid \tau_{m_i}) = \mathrm{P_r}(\tau_{m_i} \mid H_1) \tag{23}$$

The likelihood ratio function of τ_m can then be determined as

$$\Lambda(\tau_{m_i}) = \frac{\sup\{L(H_0 \mid \tau_{m_i})\}}{\sup\{L(H_1 \mid \tau_{m_i})\}} \tag{24}$$

The defined likelihood ratio functions are the simplified Bayesian alternative to the traditional hypothesis testing. The outcome of the likelihood ratio functions can be compared to a certain threshold, i.e. unity for binary hypothesis testing, to make a decision

$$\Lambda(\tau_{m_i}) \underset{H_1}{\overset{H_0}{\gtrless}} \eta_\tau \tag{25}$$

Similar procedure takes place to obtain the other likelihood ratio functions. Each of the above likelihood ratio tests can individually be applied for UDP identification of an observed channel profile. The outcome of the likelihood ratio test being greater than unity indicates that the receiver location is more likely to be a DDP condition and can appropriately be used in localization algorithm while the outcome less than unity indicates that the profile is, indeed, more likely to belong to UDP class of receive location; hence, the estimated d_{FDP} has to be remedied before being used in the localization algorithm.

To use the likelihood functions more effectively, we can combine the functions and form a joint likelihood function. Using the distributions obtained from RT database for UDP identification we can exploit τ_m,r and ζ_{hyb} distributions and their respective parameters obtained from RT channel profiles to form the joint density function. Assumption of the independence of the likelihood functions leads to a suboptimal combined likelihood function defined as

$$\Lambda(\tau_m, r, \zeta_{hyb}) = \Lambda(\tau_m)\Lambda(r)\Lambda(\zeta_{hyb}) \quad (26)$$

For the simulation of the accuracy of the UDP identification we set up a small experiment that consisted all the RT channel profiles existing in our database. We used channel profiles, including the ones used for obtaining the parameters, for UDP identification and recorded the percentage of accuracy of each method. The results of the accuracy of the likelihood hypothesis tests, individually and as a joint distribution, are summarized in Table 1.

It can be observed that the accuracy of using individual metrics for identification of UDP conditions is about 70% while combining the metrics for UDP identification can achieve 90% of accuracy.

LOCALIZATION AND TRACKING IN THE ABSENCE OF DIRECT PATH

The previous part introduced the concept of UDP condition and its effect on the TOA estimation problem. Furthermore, a study describing the identification of UDP has also been given in order to predict the occurrence of such conditions in order to mitigate the errors associated with it. In this part we will focus on methods that try to alleviate the UDP problem by the exploitation of indirect paths in regions of UDP and the concept of cooperative localization for a WSN.

Exploiting Non-Direct Paths

The existence and detectability of DP is essential for accurate ranging estimates in TOA based systems as stated earlier. However, since the availability of DP cannot be guaranteed for typical indoor localization systems, other alternatives should be considered to improve ranging accuracy. One of them is to exploit the multipath components in the channel.

Multipath components might be used to remedy UDP conditions if they exhibit steady behavior in the region of interest. Figure 10 illustrates the basic principle underlying the relationship between the TOA of the direct path and a path reflected from a wall, for a simple two path scenario. As the mobile receiver moves along the x-axis, the change in the distance in that direction is related to the length of the DP by $dx\cos\alpha = dl_{DP}$. As the geometry of the shows, for the reflected path we also have $dx\cos\beta = dl_{P_n}$. Therefore, we can calculate the change in the length of the direct path from the change in the reflected path, using

$$dl_{DP} = dl_{P_n}\frac{\cos\alpha}{\cos\beta} \quad \text{or} \quad d(TOA_{DP}) = d(TOA_{P_n})\frac{\cos\alpha}{\cos\beta} \quad (27)$$

Table 1. Accuracy of the likelihood hypothesis test

Likelihood Ratio	*Correct Decision*
τ_m	70.85 %
r	67.06 %
ζ_{hyb}	69.73 %
$\Lambda(\tau_m, r, \zeta_{hyb})$	89.29 %

In other words, knowing the angle, β, between the arriving path and direction of movement and the angle, α, between the direction of movement and the DP, we can estimate the changes in the TOA of the DP from changes in the TOA of the reflected path. A study based on this principle is introduced by Akgul & Pahlavan (2007).

In order to use a path other than the DP for tracking the location, we should be able to identify that path among all other paths, and the number of reflections for that path should remain the same in the region of interest. In the simple two path model shown in Figure 10, the second path consistently reflects from one wall as we move along the region and hence we can identify that path easily because it is the only path other than the DP. Since both conditions hold for the second path, the behavior of the TOA of that path, shown in Figure 10-b, is smooth and we can use it for tracing the DP. In realistic indoor scenarios, in the absence of direct path, we have numerous other paths to use and the simplest paths to track are the first detected path (FDP) and the strongest path (SP).

With regard to channel behavior, we need to look into the principles underlying this behavior to learn how to remedy the situation. The basic problem is path-indexing changes, and the rate of path indexing exchange is a function of number of paths in the impulse response. The number of paths can be reduced by restricting the AOA of the received signal using a sectored antenna. Using sectored antennas to restrict the AOA provides two benefits: (I) it reduces the number of multipath components and hence reduces the path index crossing rate, facilitating improved tracking of specific paths in the channel profile; (II) it allows a means for estimating the angle of the arriving path needed in (27). Figure 11 shows the behavior of the SP at a receiver using sectored antenna with a variety of aperture angles. Details of this setup have been outlined by Pahlavan et al., 2006.

Figure 11 shows the ideal behavior of different paths without bandwidth considerations as a receiver moves along the left segment of the rectangular route. The blue line shows the actual distance and the blue line with star marker shows the behavior of the FDP, which in this case is also the strongest path. The receiver starts in a DDP condition, then moves to a UDP region, and then returns to another DDP area. In the DDP regions the DP, FDP and the SP are the same and the range estimate is accurate and consistent (steady). In the UDP region, the FDP, which is also the SP, remains steady for short periods but due to the path index changes of the FDP it cannot maintain its steadiness and it experiences about ten transitions of the path index or reflection scenario for the FDP. This high rate of transitions is due

Figure 10. (a) Basic two path reflection environment; (b) Relation between the TOA of multipath and DP

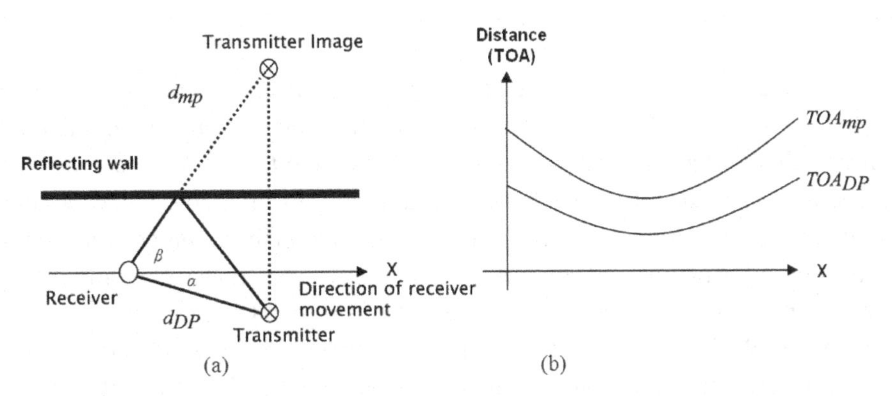

Figure 11. UDP region

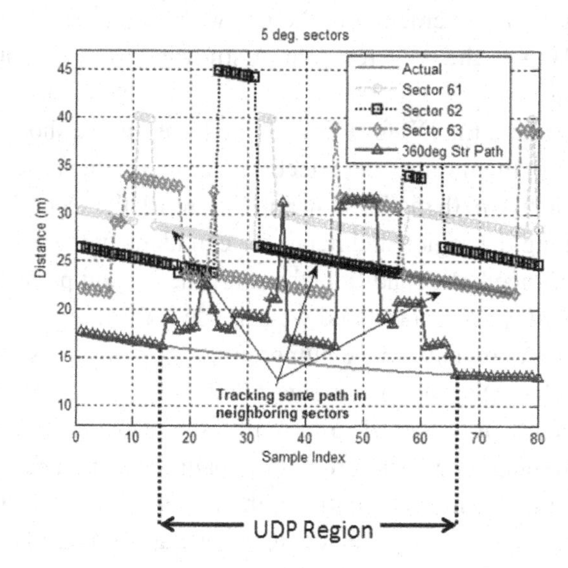

to the large number of multipath components and we can reduce these components by using sectored antennas to limit the AOA of the paths.

Figure 11 also shows the behavior of the SP in three neighboring 5 degree sectors along a UDP region. These are three of the 72 ideal 5 degree sectors assumed in this example. The SP at the start of the UDP region is in sector 61, and then it moves to sectors 62 and 63. As the SP moves among these sectors it has a steady behavior with no change in path index, which we can use for the detection of the TOA of the DP.

The discussion above shows the potential for the implementation of ranging using non-direct paths with an ideal sectored antenna with 5 degrees aperture angle for each sector and a simple algorithm which traces the strongest path as it moves from one sector to the next neighboring sector. Development of more practical algorithms to implement this concept with finite bandwidth and realistic antennas will require significant additional research.

Cooperative Localization

Background

The previous sections have provided an understanding of the different traditional approaches to the localization problem. It is evident that the localization accuracy depends on the ranging metric, deployment environment (which affects the ranging error statistics), and the relative geometry of the sensor node to the anchors. The major difference between traditional localization and WSN localization is cooperative localization. Cooperative localization refers to the collaboration between sensor nodes to estimate their location information. In traditional wireless networks, nodes can only range to specific anchors. As a result, nodes that are beyond the coverage of sufficient anchors fail to obtain a location estimate. In a cooperate WSN, however, nodes do not need to have a single-hop connection to anchors in order to localize. Cooperative localization makes propagating range information throughout the network possible. Due to random dcployment in a WSN, some parts of the network may still be isolated or may not have the required number of connections to nearby nodes (hence ill-connected), which further introduces limitations in position estimation. Obviously, increasing the sensor node density can reduce the probability of isolated sub-networks, but this approach has its own limitations. With increased range information cooperative localization has the following advantages: The first is that the *coverage* of the anchor nodes to the sensor nodes increases substantially compared the traditional counterpart. Second, the increased range information between the nodes allows for improvements in localization accuracy.

In two-dimensional localization we need at least three links or connections to reference terminals with known locations. These links may have different qualities of estimate for the distance between the reference and target terminals, depending on the availability of direct path in the channel. In cooperative precise localization in multipath-rich environments, we simply avoid ranging estimates reported from the links with UDP conditions. In other words the redundant information provided by the additional reference points is used to reduce the localization error. This situation is common in ad-hoc and sensor networks where we have a fixed infrastructure of known reference points for positioning and a number of mobile users in the area. When we want to locate a mobile terminal, in addition to the distances from the respective fixed reference points, we can also use the relative distances from other mobile users. We refer to this approach as cooperative localization since the localization is conducted through a cooperative method. A similar approach is also used for general localization in sensor and ad-hoc network when we have a limited number of dispersed references and a number of ad-hoc sensor terminals with less than an adequate number of connections to reference points (Savares et al., 2001; Savvides et al., 2005). For general localization, we only need the whereabouts of the terminals and the literature in that field does not address the large error caused by UDP conditions. The concept introduced in this part uses the redundancy of the links embedded in the sensor and ad-hoc network environments to achieve precise indoor localization.

Figure 12 shows a positioning scenario with three reference transmitters in a selected office building and a loop-route scenario. The transmitter TX-1 located in a large laboratory on the left side of the building has UDP conditions caused by the RF-isolation chamber in 40% of locations around the loop, transmitter TX-2 located in the small office on upper parts of the building layout, covers the entire loop without any UDP location, and transmitter TX-3 located in the lower corridor has around 50% UDP conditions around the loop caused by the elevator. The route estimation-1 in Figure 12 shows the results of location estimation using the traditional least square algorithm (Pahlavan et al., 2006) with the three known reference transmitters along the loop. Whenever the direct path is present, for example in the lower and right hand routes, the DME is small. As we have one or two UDP conditions for our three links to the references, for example in the upper route, the DME is substantially large. This observation suggests that whenever direct path is available for all links we can achieve precise localization, but as soon as one of the links loses the direct path we have large localization errors. In other words, if we avoid UDP conditions we can achieve precise positioning. Therefore, if we have more than the minimum number of references, assuming we can detect the UDP conditions, we can avoid links with UDP and achieve precise localization. A method of UDP detection and identification will also be presented at the end of this chapter.

To demonstrate the effectiveness of this approach, we consider an example where we have two other RPs, transmitter TX-4 and transmitter TX-5, which are located in good positions, where each has three direct path connections to the main reference transmitters. As shown in Figure 12, when we use the three main reference transmitters to estimate the location of transmitter TX-4 and transmitter TX-5 we

Figure 12. Cooperative localization

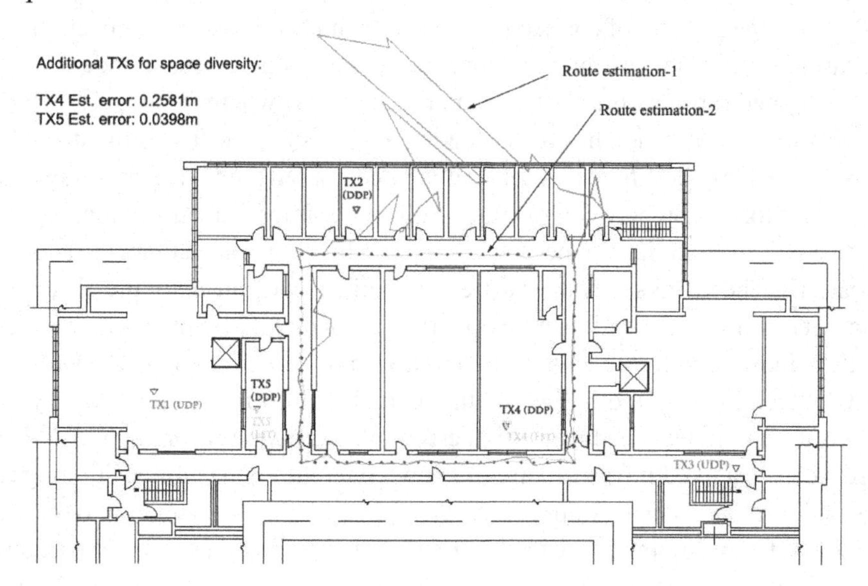

have very good estimated locations for them. In an ad-hoc sensor environment we can assume that our target receiver moving along the loop route can also measure its distances from transmitter TX-4 and transmitter TX-5. In this particular example, as shown in the figure, these ranging estimates are also very accurate because they are based on the availability of the direct path. The route estimation-2 in Figure 12 shows the estimate of location for the mobile terminal as it moves along the loop when it uses the estimated locations of transmitter TX-4 and transmitter TX-5 and the actual location of transmitter TX-2 to locate itself with the traditional least squares algorithm. As shown in Figure 12, our estimates are now substantially more accurate. The drastic improvement in the accuracy of localization is a result of avoiding UDP conditions and taking advantage of the redundancy of the ad-hoc sensor networks to achieve precise cooperative localization.

In the above example, we showed the potential advantage of using the redundancy in sensor and ad-hoc networks to achieve precise cooperative positioning. In practice, we need to develop algorithms for implementation of this concept. These algorithms need the intelligence to discover the quality of ranging estimates and possibly occurrence of the UDP conditions to use them for positioning. The algorithms for general cooperative localization first suggested in (Savares et al., 2001) and later on discussed in the follow up literature (Gezici et al., 2005; Savvides et al., 2005) are not applicable to our approach. We need new algorithms to address specific methods to handle the behavior of the DME errors in the absence of direct path, which is reported by Alavi and Pahlavan (2006). We have to find techniques for relating a quality of estimate to each ranging and positioning estimate in order to develop precise cooperative localization algorithms for sensor and ad-hoc networks. These algorithms should take advantage of redundancy to avoid unreliable reference sources and achieve robust precise localization. More research in this area is needed for the design of algorithms that take account of different radio propagation conditions.

Challenges Facing WSN Localization Algorithms

The major challenges facing WSN localization can be categorized into network and channel parameters. When considering network parameters, localization is usually constrained by the size (i.e., the number of nodes and *anchors*), the topology, and the connectivity of the network. Anchors or *beacons* are sensor nodes that are aware of their locations (usually through GPS or pre-programmed during setup) and they are necessary for WSN applications that require localization with respect to an absolute global frame of reference, e.g., GPS. Network connectivity is determined by node density, which is usually defined as the number of nodes in a meter square (nodes/m^2). A network with a high node density exhibits improved localization performance compared to a sparse network. A study related to localization performance vs. node density will be presented after this section. Furthermore, in sparse WSNs there is a high probability of ill-connected or isolated nodes and in such cases localization accuracy can be degraded substantially. Therefore, it is always favorable to increase the node density (higher connectivity information means a lower probability of ill-connected networks) to improve the accuracy of localization. However, with increased sensor nodes, the error propagates and accumulates from one hop to the next, which can be a serious problem in WSN localization algorithms. Error propagation in WSN localization is the accumulation of errors in estimated sensor node positions in each iterative step. When a node *transforms* into an anchor, the error in the range estimates used in the localization process impacts its position estimate. When other nodes in the network use this newly transformed anchor, the position error will *propagate* to the new node. Therefore, in several iterative steps, error propagation can substantially degrade the localization performance. Finally, the topology and geometric relation between nodes will further add limitations to the localization performance.

The second and most limiting factor affecting WSN localization is the wireless RF channel. Effective cooperative localization hinges on the RF ranging technology and its behavior in the deployed environment. For example, deploying hundreds of nodes in outdoor environments faces different challenges as opposed to trying to locate sensors inside a building. WSNs in indoor areas, particularly, face severe multipath fading and harsh radio propagation environments, which causes large ranging estimation errors that impact localization performance directly. To develop practical and accurate cooperative localization algorithms, the behavior of the wireless channel must be investigated and incorporated. Specifically, the localization algorithms used to determine the position must be able to assess the quality of the ranging estimates and integrate that information into the localization process.

Performance

The performance of WSN localization algorithms can be determined through very well established CRLB analysis that has recently attracted attention from different scholars and researchers. The definitions of the bound and thus the analytical derivation involved are similar, but they usually differ in their assumptions about the characteristics of the corrupting noise. The details of these different approaches for computing the CRLB are beyond the scope of the chapter and the interested reader can find more details in (Larsson, 2004; Savvides et al., 2005; Chang & Sahai, 2006). Due to its simplicity and applicability to sensor networks, we now provide an overview of CRLB analysis provided by Savvides et al. (2005) for unbiased estimate of the sensor positions. Although this is not the case in certain environments, such as indoors, it provides, nonetheless, a very important analytical foundation for analyzing the localization performance in WSNs.

For known anchor locations $\boldsymbol{\varphi} = [x_{N+1},\ldots,x_{N+M},y_{N+1},\ldots,y_{N+M}]^T$, we wish to estimate the unknown locations of sensor nodes, $\boldsymbol{\theta} = [x_1,\ldots,x_N,y_1,\ldots,y_N]^T$. The CRLB provides a lower bound on the error covariance matrix for an unbiased estimate of è (Savvides et al., 2005). For a given estimate of the sensor locations $\hat{\boldsymbol{\theta}}$ and Gaussian range measurement noise *z*, the Fisher Information Matrix (FIM) can be represented by (Savvides et al. 2005)

$$J(\boldsymbol{\theta}) = E\left\{\left[\nabla_{\boldsymbol{\theta}} \ln f_z(Z;\boldsymbol{\theta})\right]\left[\nabla_{\boldsymbol{\theta}} \ln f_z(Z;\boldsymbol{\theta})\right]^T\right\} \tag{28}$$

where $f_z(Z;\boldsymbol{\theta})$ is the joint Gaussian PDF given by

$$f_z(Z;\boldsymbol{\theta}) = \frac{1}{(2\pi)^K |\Sigma|^{\frac{1}{2}}} \exp\left\{-\frac{1}{2}\left[z - \mu(\boldsymbol{\theta})\right]^T \Sigma^{-1}\left[z - \mu(\boldsymbol{\theta})\right]\right\} \tag{29}$$

where $\mu(\boldsymbol{\theta})$ is the vector of the actual distances between the sensor nodes corresponding to available K measurements. FIM for the specific PDF in can be written as

$$J(\boldsymbol{\theta}) = \left[G(\boldsymbol{\theta})\right]^T \Sigma^{-1}\left[G(\boldsymbol{\theta})\right] \tag{30}$$

where $G(\boldsymbol{\theta})$ contains the partial derivatives of $\mu(\boldsymbol{\theta})$. The CRLB is then given by

$$CRLB = \left[J(\boldsymbol{\theta})\right]^{-1} \tag{31}$$

More detailed implementation of the CRLB expression can be found in Savvides et al. (2005).

WSN algorithms should then compare the localization performance to the widely available CRLB analysis in literature. One important note here is that both the bound and the algorithm performance rely mainly on the statistics of the ranging error. Although sensor density and geometry have an impact on the performance, the statistics of the ranging error specifically provides the main challenge for accurate localization. If the ranging error assumptions taken into the algorithm and the CRLB analysis do not reflect the actual behavior of the propagation channel, both the algorithm performance and the bound will be non-realistic. For the indoor ranging scenarios where range estimates are not just corrupted by zero mean Gaussian noise (a positive bias can corrupt the TOA measurements) we will need to analyze CRLB for biased estimates. In the presence of such biases, the Generalized-CRLB (G-CRLB) can be obtained instead and it has been derived for traditional wireless networks by Van Trees (1968) and for indoor WSNs by Qi et al. (2006). A detailed treatment of this problem is available by Alsindi & Pahlavan (2008).

A STUDY OF RP DENSITY AND MSE PROFILING ON WSN LOCALIZATION

In this section, we will be investigating the effects of various parameters that are related to localization in sensor networking.

In the first part, we will be showing how RP density affects the localization performance by using two different localization algorithms.

Effects of RP Density on TOA-Based UWB Indoor Positioning Systems

The performance of UWB indoor positioning systems based on TOA techniques is generally affected by the density of RPs, as well as UDP conditions. For a fixed number of RPs, the performance of some indoor positioning algorithms tends to degrade as the size of the area is increased, i.e. the RP density is decreased. In this part, we evaluate the effects of RP density on the performance of different positioning algorithms in the presence of empirical distance measurement error (DME) models derived from UWB measurements in typical indoor environments. We then present functional relationships between RP density and positioning mean-square error (MSE) for these algorithms. These relationships can be used for more effective indoor positioning system design and deployment. Finally, we investigate the effects of bandwidth with respect to improving the performance of these algorithms.

In addition to the inherent stochastic variations of the channel (which can induce distance measurement error, or DME), the indoor environment itself also does not necessarily stay static. Indoor areas can be remodeled, made larger, or portions of it can be rebuilt with different building materials. This will change the RP density and by extension, the estimation accuracy that we can obtain from the network used for positioning. The RP density, denoted by ρ, can be viewed as a measure of the number of RPs per unit area, and is defined as:

$$\rho = N / A \tag{32}$$

where N is the number of RPs covering a given indoor area, and A is the size of the area, generally given in m^2. It is noted in a prior work by Kanaan and Pahlavan (2004b) that given a *fixed* number of RPs, the performance of certain positioning algorithms tends to degrade as the size of the area to be covered is increased (i.e the RP density is decreased). This observation makes intuitive sense since the DP will be attenuated more as the distance between the RP and the sensor is increased. This will give rise to more DME which, in turn, will lead to degraded location estimation performance. Although the effects of RP density on location estimation accuracy has been known, the exact nature of the functional relationship between these two quantities has not, to the best of our knowledge, been formulated to date. This raises a valid question: why is it important to characterize this relationship? The answer fundamentally lies in the fact that different indoor positioning applications have different requirements for estimation accuracy. For example, in a commercial application (such as inventory tracking in a warehouse), low accuracy might be acceptable. However, in a public-safety or military application (such as keeping track of the locations of firefighters or soldiers within a building), much higher accuracy would be needed. This implies that the RP densities required for these two application domains would be different. Knowledge of the functional relationship between RP density and estimation accuracy enables a system designer to figure out how many RPs are required to meet a given accuracy target, thereby results in a cost-effective network deployment.

The manner in which RP density affects positioning accuracy depends principally on two factors: the particular algorithm used for the location estimation, and the DME model. The basic contribution of this paper is to explore these kinds of relationships for different positioning algorithms, both to get an insight into their performance, and also to provide a useful tool for designers of indoor positioning networks. In addition to addressing the above-mentioned issues, this part also extends the study reported by Kanaan and Pahlavan (2004b) in two important ways. First, the performance evaluations we undertake are based on DME models obtained from empirical UWB measurements in a typical indoor area, rather than models derived from ray-tracing simulations. Second, since the DME also depends on bandwidth (Alavi & Pahlavan, 2003), we also explore the impact of bandwidth on the performance of a given algorithm.

The relationship between RP density and positioning accuracy has been studied, mainly for ad-hoc sensor networks. Savarese et al. (2001) have studied positioning in distributed ad-hoc sensor networks through cooperative ranging. The paper by Chintalapudi et al. (2004) studies the effects of density of RPs on ad-hoc positioning algorithms employing both distance and bearing measurements. While these works have identified the relationship between positioning accuracy and RP density, they have not explicitly presented that relationship mathematically. Also, the DME models used in these studies are generally very simple. Here, we seek to explore the functional dependency of the positioning accuracy (as expressed by the *mean square error* or *MSE*) on RP density in the presence of DME models based on empirical measurements within actual indoor environments.

UDP conditions generally occur in cases where there are multiple walls and/or metallic objects between the transmitter and the receiver. As a result, the DP can experience severe fading (Pahlavan et al., 2002). It has also been observed that UDP conditions tend to occur along coverage boundaries or areas with coverage deficiencies (Alavi et al., 2005). As mentioned above, UDP occurs only on occasion and when it does, it is the dominant source of error for distance measurements. However, the DDP error is always present. Based on extensive UWB measurements, a DME model is introduced by Alavi et al. (2005). The model is given by:

$$\hat{d}_i = \begin{cases} d_i + \xi_{DDP,w} \log(1+d_i) & \text{DDP case} \\ d_i + \xi_{DDP,w} \log(1+d_i) + \xi_{UDP,w} & \text{UDP case} \end{cases} \tag{33}$$

where $\hat{d}_i$ is the observed distance measurement from the *i*-th RP, and $\xi_{DDP,w}$ and $\xi_{UDP,w}$ are random variables that characterize the DDP and UDP-based DME respectively. The distributions of $\xi_{DDP,w}$ and $\xi_{UDP,w}$ have been observed to be Gaussian and dependent on bandwidth, i.e, $\xi_{DDP,w} = N(m_{DDP,w}; \sigma^2_{DDP,w})$ and $\xi_{UDP,w} = N(m_{UDP,w}; \sigma^2_{UDP,w})$ where the means (denoted by $m_{DDP,w}$ for $\xi_{DDP,w}$ and $m_{UDP,w}$ for $\xi_{UDP,w}$) and standard deviations (denoted by $\sigma^2_{DDP,w}$ for $\xi_{DDP,w}$ and $\sigma^2_{UDP,w}$ for $\xi_{UDP,w}$) are a function of the system bandwidth, *w*, used to make the TOA-based distance measurements (hence the subscript *w*). The parameters for the distributions, as a function of the bandwidth *w*, are listed in Table 2.

Additionally, it has been determined through measurements that the probability of UDP occurrence $P_{UDP,w}$ increases as the bandwidth is increased (Alavi et al., (2005). Observations indicate that $P_{UDP,w}$ values also depend on the actual distance between the transmitter and the receiver. Specifically,

$$P_{UDP,w} = \begin{cases} P_{UDP\_close,w} & d \le 10\text{m} \\ P_{UDP\_far,w} & \text{otherwise} \end{cases} \tag{34}$$

The error modeling introduced here is detailed by Alavi et al. (2005). This model can be compared to the work by Alavi & Pahlavan (2003) and has a fundamentally different approach. Alavi & Pahlavan's model (2003) was developed based on Ray-Tracing results and categorized the conditions of the channel as Line-of-sight / Obstructed line-of-sight (LOS/OLOS), assuming that in the OLOS case, we always have the UDP case. However, the introduced DME model is based on UWB measurement data and classification of the channel as DDP and UDP, as this approach reflects the behavior of the indoor channel in a more realistic manner.

Table 2. Parameters for DDP and UDP-based DME

w(MHz)	500	1000	2000	3000
$m_{DDP,w}$ (m)	0.21	0.09	0.02	0.004
$\sigma_{DDP,w}$ (cm)	26.9	13.6	5.2	4.5
$m_{UDP,w}$ (m)	1.62	0.96	0.76	0.88
$\sigma_{DDP,w}$ (cm)	80.87	60.45	71.53	152.21
$P_{UDP\_far,w}$	0.33	0.62	0.74	0.77
$P_{UDP\_close,w}$	0.06	0.06	0.07	0.12

Algorithms

We investigate the effects of RP density on the performance of two algorithms using simulations: the Closest-Neighbor with TOA Grid (CN-TOAG) (Kanaan & Pahlavan, 2004a), and the Davidon LS algorithm (Davidon, 1968).

CN-TOAG Algorithm

In essence, the CN-TOAG algorithm estimates the location of the sensor S, by minimizing the following objective function:

$$f(x,y)=\sqrt{\sum_{k=1}^{N}\left(d_k-\sqrt{(x-X_k)^2+(y-Y_k)^2}\right)^2} \qquad (35)$$

where $(X_k; Y_k)$ is the location of the k-th RP, dk is the observed distance measurement from the k-th RP and $(x; y)$ is the unknown location of the sensor to be estimated. The estimated location is the one that minimizes (35). In order to find the minimum, one needs to solve the following partial differential equation:

$$\left(\frac{\delta f(x,y)}{\delta x},\frac{\delta f(x,y)}{\delta y}\right)=0 \qquad (36)$$

Due to the complexity of $f(x; y)$ in (35), it is not feasible to solve (36) analytically. Therefore, the CN-TOAG algorithm tries to solve it numerically using the concept of a TOA grid (Kanaan & Pahlavan, 2004a). The size of the grid, as given by the spacing between grid points, h, is a major determinant of performance for this algorithm. Specifically, values of h below a certain value can result in better performance than the LS algorithm (Kanaan & Pahlavan, 2004a).

Davidon Least-Squares Algorithm

The particular instance of the LS algorithm that has been used for our evaluations is the one by Davidon (1968), which attempts to minimize the objective function:

$$f(\mathbf{x}) = f(x,y) = \sum_{k=1}^{N}\left(d_k - \sqrt{(x-X_k)^2 + (y-Y_k)^2}\right) \tag{37}$$

in an iterative manner using the following relation:

$$\mathbf{x}_{k+1} = \mathbf{x}_k - \mathbf{H}_k g(\mathbf{x}_k) \tag{38}$$

where $\mathbf{H}_k$ represents an approximation to the inverse of the Hessian of f(x), G(x), which is defined as:

$$\mathbf{G}(\mathbf{x}) = \begin{pmatrix} \frac{\partial^2 f}{\partial x^2} & \frac{\partial^2 f}{\partial x \partial y} \\ \frac{\partial^2 f}{\partial y \partial x} & \frac{\partial^2 f}{\partial y^2} \end{pmatrix} \tag{39}$$

and g(x) is the gradient of f(x), defined as:

$$g(\mathbf{x}) = \left(\frac{\delta f(\mathbf{x})}{\delta x}, \frac{\delta f(\mathbf{x})}{\delta y}\right) \tag{40}$$

The following relation defines when the computations will be terminated:

$$\eta_k = (g(\mathbf{x}_{k+1}))^T \mathbf{H}_k (g(\mathbf{x}_{k+1})) \tag{41}$$

so that the iterations will stop when $\eta_k \leq \varepsilon$, where ε is a small tolerance value.

Simulation Platform

Figure 13 shows the general system scenario, where a regular grid arrangement of RPs is assumed to be available. The use of the regular grid arrangement for the RPs is common in indoor wireless networks, as this approach often provides adequate coverage (Unbehaun, 2002). We also note that this system scenario is a fundamental building block for certain indoor ad-hoc sensor networks, and could be a considered a realistic deployment scenario for such scenarios (Stoleru & Stankovic, 2004). The important parameter that determines performance is not the absolute number of RPs, but the ratio of the number of RPs to the area, as given by (32).

Here, we consider varying sizes of L for each algorithm. By varying the room size while keeping the number of RPs fixed at each of the four corners, we evaluate the performance of positioning algorithms as a function of RP density in the scenario and also show the effect of system bandwidth on overall performance. Synchronization mismatch between the transmitter and receiver is assumed to be small. For each algorithm, a total of 5000 uniformly distributed random sensor locations are simulated for different bandwidth values and for varying room dimensions. In line with the FCC's formal definition of UWB signal bandwidth as being equal to or more than 500 MHz (FCC, 2002), we will present our results for bandwidths of 500, 1000, 2000, and 3000 MHz. Once a sensor is randomly dropped in the

Figure 13. General system scenario

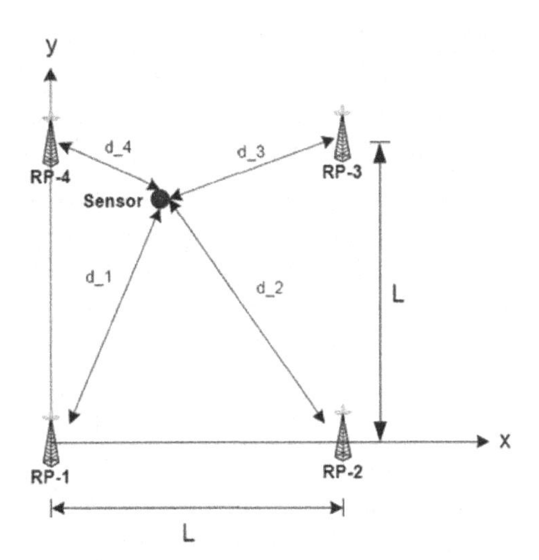

area, the actual distance measurements, d_i, from each RP at the corners are individually corrupted with DME, as given in (33). The corrupted distance measurements are then fed into the positioning algorithm to get the position estimate. As noted by Kanaan & Pahlavan (2004a), the performance of the CN-TOAG algorithm is dependent on the size of the TOA grid, as determined by the bin size, h, which for the purposes of this study, was fixed at 0.3125 m.

Results

The results are shown in Figure 14, Figure 15, and Figure 16 . From the Figure 14 and Figure 15 we can immediately see that as the node density is increased, the MSE decreases. This is an expected result, since a finer installation of the RPs will reduce the probability of the occurrence of UDP conditions and hence will result in better positioning accuracy. Another important observation is that as the bandwidth of the system is increased, the estimation accuracy is also increased with the exception of 3 GHz. Increasing the system bandwidth provides a better time resolution, thereby ensuring accurate estimation of the TOA of the DP. However, increasing the bandwidth beyond a certain point (2000 MHz in this case) also gives rise to increased probability of UDP conditions due to the faster attenuation of higher bandwidth signals. In Figure 16, we compare the performance of LS and CN-TOAG as a function of RP density using a system bandwidth of 2000 MHz. This bandwidth was arbitrarily selected, since it appears to be the bandwidth where both algorithms perform best. The results clearly indicate that CN-TOAG has better performance, particularly for higher values of ρ. Since CN-TOAG is based on the concept of a TOA grid, increasing ρ (i.e. decreasing the area) for a fixed bin size h brings the grid points closer together. This, in turn, places a much tighter bound on the positioning error for CN-TOAG. By examining the results of the simulation, we can derive a mathematical relation between RP density and MSE by applying a third order polynomial fit to the results. Our choice of the third order polynomial was simply influenced by the fact such a fit showed better agreement with simulation results than, say, a second-order fit. We have chosen to derive these relations on the basis of Monte-Carlo simulations,

Figure 14. Performance of LS algorithm

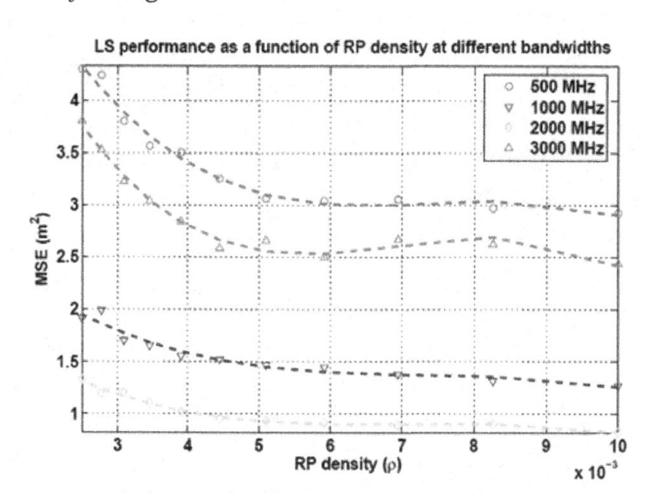

Figure 15. Performance of CN-TOAG algorithm

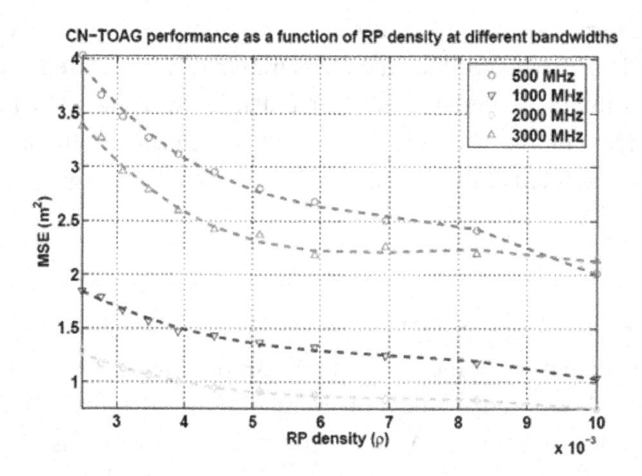

rather than analytically, in order to be able to compare and contrast the performance of the LS and CN-TOAG algorithms. It is certainly possible to derive these relations analytically for the LS algorithm, but not necessarily for CN-TOAG due to the complexity of the objective function of (35). These relations can be a valuable tool in determining the RP density for a required positioning accuracy. The 3*rd* order polynomial is given as:

$$MSE = a_3 \rho^3 + a_2 \rho^2 + a_1 \rho + a_0 \qquad (42)$$

where a_i ($i \in \{1,2,3,4\}$) denote the polynomial coefficients. Table 3 and Table 4 show the coefficient values for the two algorithms. These values are dependent on the DME model used; however, we note that the DME model parameters are still representative of typical indoor environments. A simple numerical

Figure 16. Performance of CN-TOAG and LS for 2000 MHz

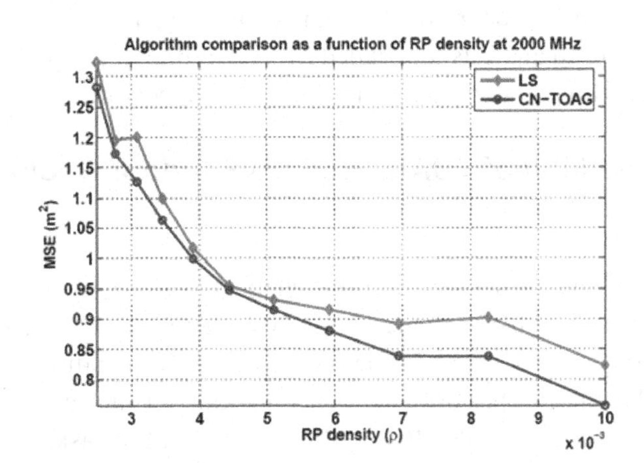

Table 3. Coefficients of the 3rd degree polynomial fit for LS algorithm

w(MHz)	a_3	a_2	a_1	a_0
500	-1.20e+07	2.69e+05	-1.98e+03	7.7776
1000	-4.53e+06	1.00e+05	-749.52	3.2647
2000	-4.36e+06	93645	-662.95	2.4484
3000	-1.72e+07	3.60e+05	-2427.4	7.8446

Table 4. Coefficients of the 3rd degree polynomial fit for CN-TOAG algorithm

w(MHz)	a_3	a_2	a_1	a_0
500	-1.15e+07	2.42e+05	-1771	7.0203
1000	-4.61e+06	97736	-725.22	3.1171
2000	-3.51e+06	76142	-557.93	2.2317
3000	-1.04e+07	2.34e+05	-1728.7	6.4152

example illustrates how these relations could be used. Suppose we have a 900 m^2 indoor area where we would like to implement a positioning system using CN-TOAG at a bandwidth of 1 GHz, and we would like the MSE to be no more than 1.5 m^2. Referring to Figure 15, we see that the corresponding value of ρ should be no less than 0.004. Using our knowledge of the size of the area, the value of ρ, and (32), we see that we need to have a minimum of 4 RPs in order to ensure satisfactory performance.

Conclusions

In this part, we have investigated the performance of two indoor positioning algorithms and compared their performance as a function of RP density and system bandwidth. We also presented mathematical relations between RP density and achievable MSE and showed how they can be used to ensure the required performance with a given indoor positioning network scenario.

The second part focuses on an analysis tool called MSE profiling developed to benchmark the performance of an indoor positioning system in the presence of UDP conditions.

Performance Benchmarking of TOA-Based UWB Indoor Geolocation Systems Using MSE Profiling

The presence of UDP conditions presents significant challenges for TOA-based indoor geolocation, since it introduces major errors into distance measurements. Therefore, it is critical to characterize the performance of indoor geolocation systems in the presence of UDP conditions. Until now, however, there has been no standard method of performance benchmarking for such cases. Towards that end, we present an analysis tool, known as the MSE Profile, that can aid in this task. We use the MSE Profile to analyze the performance of two TOA-based geolocation algorithms and show how the MSE Profile can be used to gain insight into their performance, in the presence of DME models derived from UWB measurements.

The accuracy of the location estimate can be viewed as a measure of the Quality of Service (QoS) provided by the geolocation system. Different location-based applications will have different requirements for accuracy. In a military or public-safety application (such as keeping track of the locations of fire-fighters or soldiers inside a building), high accuracy is desired. In contrast, lower accuracy might be acceptable for a commercial application (such as inventory control in a warehouse). In such cases, it is essential to be able to answer questions like: "what is the probability of being able to obtain an MSE of 1 m^2 from an algorithm x in any building configuration?" or "what algorithm should be used to obtain an MSE of 0.1 cm^2 in any building configuration?". Answers to such questions will heavily influence the design, operation and performance of indoor geolocation systems. Here, we propose the use of the *MSE Profile* to answer these kinds of questions and illustrate its use with examples.

Although indoor geolocation is a relatively new area, there is a large body of literature on performance analysis of geolocation systems in general. A number of researchers have studied geolocation systems intended primarily for outdoor deployments. A number of different performance metrics have been defined and used. Foy (1976) used the covariance matrix of the position estimation error as a performance metric for the evaluation of Taylor-Series algorithm for geolocation. Torrieri (1984) formally defined and used the circular error probability (CEP), which is a measure of the uncertainty in the location estimate $\hat{x}$, relative to its mean, $E\{\hat{x}\}$. The calculation of the CEP is, in general, quite complicated. This issue can be alleviated by making suitable approximations. However, from a QoS perspective, the most we can say after calculating the CEP is that the estimate is likely to be within $\hat{x}+CEP$ with probability 1/2. The CEP, therefore, will only be of limited use in answering the types of questions given in the previous section. The work of Deng and Fan (2000) and others working in the *E-911* field bears the closest resemblance to our work in the sense that it considers the CDF of the MSE in order to assess the performance of outdoor cellular positioning systems in relation to E-911 requirements outlined by the FCC. However, this cannot be directly applied to our work, as we specifically consider the effect of varying UDP conditions on UWB indoor geolocation system performance. Therefore, to the best of our knowledge, our approach to performance analysis of such systems is unique.

Given the variability of the indoor propagation conditions, it is possible that the distance measurements performed by some of the RPs will be subject to DDP errors, while some will be subject to UDP-based errors. The DDP/UDP errors can be observed in various combinations. For example, the distance measurements performed by RP-1 in Figure 13 may be subject to UDP-based DME, while the measurements performed by the other RPs may be subject to DDP-based DME; we can denote this combination as *UDDD*. Other combinations can be considered in a similar manner. Since the occurrence of UDP conditions is random, the performance metric used for the location estimate (such as the MSE) will also vary stochastically and depends on the particular combination observed. For the four-RP case shown in Figure 13, it is clear that we will have the following distinct combinations: *UUUU*, *UUUD*, *UUDD*, *UDDD*, and *DDDD*. Each of these combinations can be used to characterize a different type of building environment. The occurrence of each of these combinations will give rise to a certain MSE value in the location estimate. This MSE value will also depend on the specific algorithm used. There may be more than one way to obtain each DDP/UDP combination. If UDP conditions occur with probability P_{udp}, then the overall probability of occurrence of the *i*-th combination, P_i can be generally expressed as:

$$P_i = \binom{N}{N_{udp,i}} P_{udp}^{N_{udp,i}} (1 - P_{udp})^{N-N_{udp,i}} \tag{43}$$

where N is the total number of RPs (in this case four), and $N_{udp,i}$ is the number of RPs where UDP-based DME is observed. Combining the probabilities, P_i, with the associated MSE values for each combination we can obtain a discrete CDF of the MSE. We call this discrete CDF the *MSE Profile*. In the next section, we will illustrate the use of the MSE Profile with examples.

The algorithms used in this part have already been introduced in the first part. Hence, the results will be presented after introducing the simulation platform.

Simulation Platform

We consider the system scenario in Figure 13 with $L = 20$ m for each algorithm. A total of 1000 uniformly distributed random sensor locations are simulated for different bandwidth values. Similar to the previous study, we will present our results for bandwidths of 500, 1000, 2000, and 3000 MHz. For each bandwidth value we also simulate different combinations of UDP and DDP-based DMEs for each RP, specifically *UUUU*, *UUUD*, *UUDD*, *UDDD*, *DDDD*. Once a sensor is randomly placed in the simulation area, each RP calculates TOA-based distances to it. The calculated distances are then corrupted with UDP and DDP-based DMEs in accordance with (33). The positioning algorithm is then applied to estimate the sensor location. Based on 1000 random trials, the MSE is calculated for each bandwidth value and the corresponding combinations of UDP and DDP-based DMEs. The probability of each combination is also calculated. For example, take the combination *UUUU* for a bandwidth of 3000 MHz, where two of the RPs are assumed to be far from the sensor, and the other two are assumed to be close. Using the values for $P_{udp,close}$, and $P_{udp,far}$, we can obtain the probability of the combination as 0.0085. The means (denoted by $m_{DDP,w}$ for $\xi_{DDP,w}$ and $m_{UDP,w}$ for $\xi_{UDP,w}$) and standard deviations (denoted by $\sigma^2_{DDP,w}$ for $\xi_{DDP,w}$ and $\sigma^2_{UDP,w}$ for $\xi_{UDP,w}$) are a function of the system bandwidth used to make the TOA-based distance measurements. The parameters for the distributions, as a function of the bandwidth w, are listed in Table 2. As noted by Kanaan & Pahlavan (2004a), the performance of the CN-TOAG algorithm is dependent on the size of the TOA grid, as determined by the the bin size, h, which for the purposes of this study, was varied between 1.25 m down to 0.3125 m for a total of three different values.

Results

The results are shown in Figure 17, Figure 18, Figure 19 and Figure 20. Figure 17 and Figure 18 show the MSE Profiles for the LS and CN-TOAG algorithms respectively. From these plots, we observe that as the bandwidth increases from 500 MHz to 2000 MHz, the range of MSE Profile values gets smaller. This correlates with the findings of Alavi and Pahlavan (2006), where it has been observed that the overall DME goes down over this specific range of bandwidths. Above 2000 MHz, however, the MSE Profile becomes wider as a result of increased probability of UDP conditions (Alavi & Pahlavan, 2006), which increases the overall DME. This, in turn, translates into an increase in the position estimation error for both algorithms. Using the MSE Profile, we can gain insight into the MSE behavior of a given

Figure 17. MSE profile for the LS algorithm

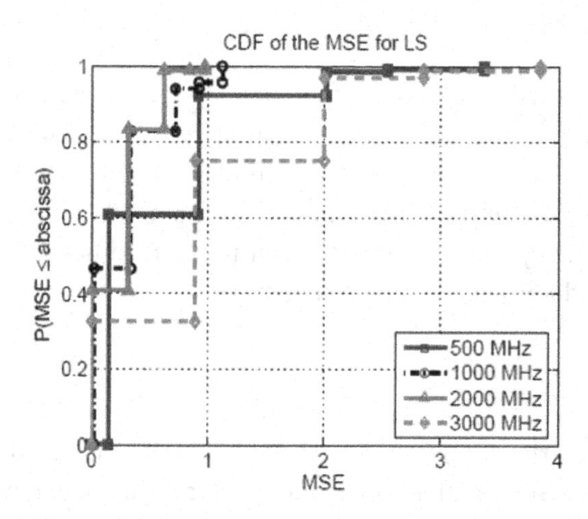

Figure 18. MSE profile for the CN-TOAG algorithm

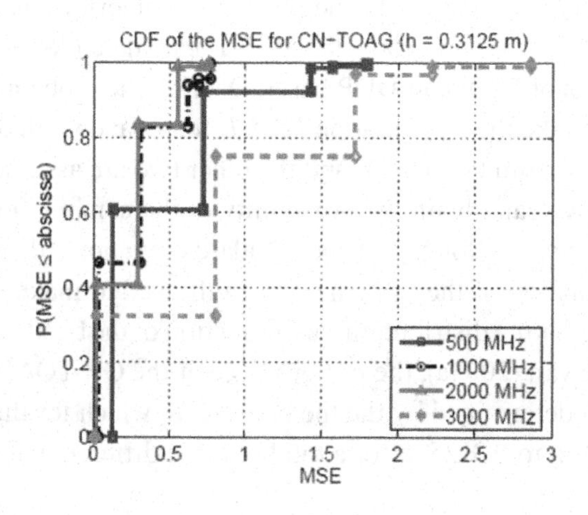

Figure 19. Average MSE comparison: LS vs CN-TOAG

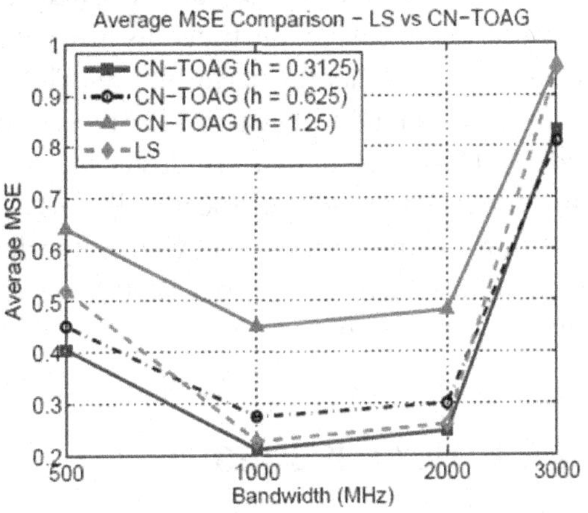

Figure 20. Variance comparison: LS vs CN-TOAG

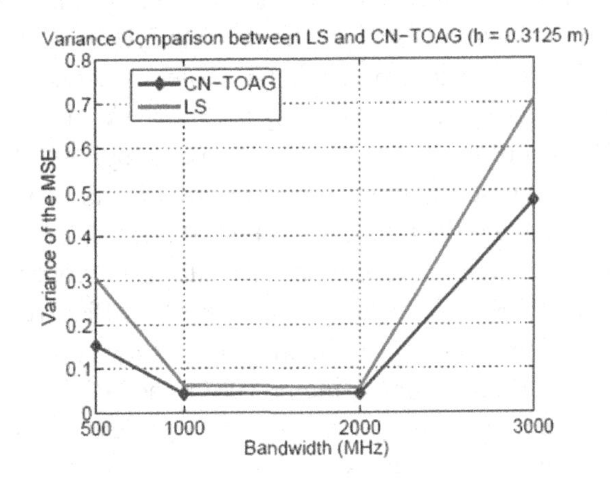

algorithm under varying amounts of UDP (i.e. different building configurations) by calculating the mean and the variance of the MSE for a given bandwidth value. The results of these calculations are shown as a function of bandwidth in Figure 19 and Figure 20. These results clearly indicate that CN-TOAG can outperform LS as long as $h \leq 0.3125$ m. In addition, there appears to be an optimal bandwidth for both algorithms where the average MSE is minimum. Our results indicate that this bandwidth value is 1000 MHz.

Conclusions

In the second part, we proposed the use of the MSE Profile to gauge the performance of any indoor geolocation algorithm under a variety of building conditions. The MSE Profile has been defined as the CDF of the MSE given the varying severity of UDP conditions across different building environments. We also showed that the MSE Profile can be used for performance benchmarking of different TOA-based indoor geolocation algorithms. We have illustrated its use in analyzing the performance of two algorithms: CN-TOAG and LS. We found that the performance exhibited by both algorithms is in line with previously reported observations on DME behavior. For the scenario and system bandwidths considered, we demonstrated that CN-TOAG can outperform LS as long as the number of points in the grid (as determined by the parameter h) is large enough. Specifically, we noted that h needs to be about 0.3125 m for the case of a 20m x 20m area in order for CN-TOAG to outperform LS. We also showed that bandwidth of operation of both algorithms needs to be about 1000 MHz in order to guarantee optimal performance across different building configurations.

IMPLEMENTATION AND PRACTICAL ISSUES

Even though no particular emphasis is given to a certain type of technology for the implementation of previous approaches in order to keep the methodology robust and applicable to a broad range of systems, certain aspects of practicality and candidate technologies need to be considered. The success of TOA/TDOA based methods depend primarily on the availability and the quality of RF detection hardware. Although most common systems such as the IEEE 802.11 WLANs and 802.15.4 are not primarily designed for ranging and localization applications, studies exist that show the feasibility of using TOA/TDOA positioning using these systems. The study by Yamasaki et al. (2005) report a positioning accuracy of 2.4m in the 67th percentile with a 802.11b system employing TDOA. The important aspect of AP synchronization is also discussed in this work. Similarly, Duan et al. (2007) report a range estimation error of less than 30cm using the 802.15.4a WPAN standard. The choice of 802.15.4a is also suitable for applications requiring low-power and low cost. Ma et al. (2005) discuss the power aspects of a 802.15.4 based WSN. The UWB PHY layer of 802.15.4a also offers very precise ranging. Hence, 802.15.4a would be a good candidate system for the implementation of TOA based localization using WSNs. However, considering the limited computing power on the 802.15.4 nodes, direct implementation of various algorithms may not be feasible. For this reason, a central computing station (a regular laptop or a PC) with the required technical specs such as memory and CPU speed might be needed to complement the 802.15.4 WSN in order to evaluate the algorithms discussed previously.

CONCLUSION

In this chapter, we presented various aspects of RF and TOA-based localization as applied to WSNs. The challenges of the RF channel have been presented along with their impact on localization algorithms. Particular attention has been given to wireless channel impairments and indoor TOA based systems. The general characteristics of a WSN localization system are discussed and a CRLB expression is also given for the performance of cooperative localization. Methods of remedying underlying wireless channel impairments have been discussed in the context of multipath exploitation along with cooperative localization and UDP detection.

REFERENCES

Akgul, F. O., & Pahlavan, K. (2007). AOA Assisted NLOS Error Mitigation for TOA-based Indoor Positioning Systems. *IEEE MILCOM.* (pp. 1-5). Orlando, FL.

Akyildiz, I., Su, W., Sankarasubramaniam,Y., & Cayirci, E. (2002). A survey on sensor networks, *IEEE Commun. Mag.*, *40*(8), 102-114.

Alavi, B., & Pahlavan, K. (2003). Bandwidth effect of distance error modeling for indoor geolocation. In *IEEE Personal Indoor Mobile Radio Communications Conference (PIMRC)*, *3*, 2198-2202.

Alavi, B., Pahlavan, K., Alsindi, N., & Li, X. (2005). Indoor geolocation distance error modeling with UWB channel measurements. In *IEEE Personal Indoor Mobile Radio Communications Conference (PIMRC)*, *1*, 418-485.

Alavi, B., & Pahlavan, K. (2006). Modeling of the TOA based Distance Measurement Error Using UWB Indoor Radio Measurements. *IEEE Communication Letters*, 10(4), 275-277.

Alavi, B., & Pahlavan, K. (2006). Studying the effect of bandwidth on performance of UWB positioning systems. In *Proceedings of the IEEE Wireless Communications and Networking Conference (WCNC)*, *2*, 884-885.

Alsindi, N., & Pahlavan, K. (2008). Cooperative localization bounds for indoor ultra wideband wireless sensor networks. *EURASIP Journal on Applied Signal Processing (ASP)*,2008. article id 852809. (pp.1-13).

Bahl, P., & Padmanabhan, V.N. (2000). RADAR: An In-Building RF-Based User Location and Tracking system. In *Proc. IEEE INFOCOM 2000*, *2*, 775-784, Tel-Aviv, Israel.

Beauregard, S. (2006). A Helmet-Mounted Pedestrian Dead Reckoning System. In *Proceedings of the 3rd International Forum on Applied Wearable Computing (IFAWC 2006).* Herzog, O., Kenn, H., Lawo, M., Lukowicz, P., & Troster, G. (Eds.), Bremen, Germany: VDE Verlag. (pp. 79–89).

Blomenhofer, H., Hein, G., Blomenhofer, E., & Werner, W., (1994). Development of a Real-Time DGPS System in the Centimeter Range. *IEEE 1994 Position, Location, and Navigation Symposium*, Las Vegas, NV, (pp. 532–539).

Bulusu, N., Heidemann, J., & Estrin, D. (2000) GPS-less Low-Cost Outdoor Localization for Very Small Devices, *IEEE Personal Communication*, 7(5), 28-34.

Cedervall, M., & Moses, R. L. (1997). Efficient maximum likelihood DOA estimation for signals with known waveforms in the presence of multipath. *IEEE Transactions on Signal Processing*, *45*(3), 808-811.

Chang, C., & Sahai, A. (2004). Estimation Bounds for Localization. *IEEE SECON.* (pp. 415-424).

Chang, C., & Sahai, A. (2006). Cramer-Rao type bounds for localization. *EURASIP Journal on Applied Signal Processing*, *2006*. article id 94287. (pp. 1-13).

Chintalapudi, K., Dhariwal, A., Govindan, R., & Sukhatme, G. (2004). Ad-hoc localization using ranging and sectoring. In *IEEE INFOCOM*, *4*, 2662-2672.

COST 231. (1991). Urban transmission loss models for mobile radio in the 900- and 1,800 MHz bands (Revision 2). COST 231 TD(90)119 Rev. 2, The Hague, The Netherlands.

Cypher, D., Chevrollier, N. Montavont, N., & Golmie, N. (2006). Prevailing over wires in healthcare environments: Benefits and challenges. *IEEE Communications Magazine*, *44*(4l), 56-63.

Dasarathy, B. V. (Ed.) (1991). Nearest Neighbor (NN) Norms: NN Pattern Classification Techniques, ISBN 0-8186-8930-7. *IEEE Computer Society.*

Davidon, W. C. (1968). Variance algorithm for minimization. *Computer Journal, 10.*

Deng, P., & Fan, P. (2000). An AOA assisted TOA positioning system. In *IEEE WCC-ICCT2000, 2,* 1501-1504.

Denis, B., Keignart, J., & Daniele, N. (2003). Impact of NLOS propagation upon ranging precision in uwb systems. In *IEEE Conference on Ultra Wideband Systems and Technologies.* (pp. 379-383).

Doherty, L., Pister, K. S. J., & Ghaoui, L. E. (2001). Convex Position Estimation in Wireless Sensor Networks, *INFOCOM'01, 3,* 1655-1663.

Duan, C. D., Orlik, P., Sahinoglu, Z., & Molisch, A. F. (2007). A Non-Coherent 802.15.4a UWB Impulse Radio. *IEEE International Conference on Ultra-Wideband.* (pp. 146-151).

Dumont, L., Fattouche, M., & Morrison, G. (1994). Super-Resolution of Multipath Channels in a Spread Spectrum Location System. *IEE Electronic Letters, 30*(19), 1583-1584.

FCC-US Federal Communications Commission. (2002). *Revision of part 15 of the commissions rules regarding ultra-wideband transmission systems.* FCC 02-48, First Report & Order.

FCC-US Federal Communications Commission. (1999). *Announcement of Commision Action, FCC Acts to Promote Competition and Public Safety in Enhanced Wireless 911 Services.* [Online]. Available: http://www.fcc.gov/Bureaus/Wireless/News_Releases/1999/nrwl9040.html

Foy, W. H. (1976). Position-location solutions by Taylor-series estimation. *IEEE Trans. Aerospace and Elect. Sys., AES-12*(2), 187-194.

Gezici, S., Tian, Z., Giannakis, G. V., Kobaysahi, H., Molisch, A. F., Poor, H. V., & Sahinoglu, Z. (2005). Localization via Ultra-Wideband Radios: A Look at Positioning Aspects for Future Sensor Networks. *IEEE Signal Processing Magazine.* ISSN: 1053-5888, *22*(4), 70-84.

Hata, M. (1980). Empirical formula for propagation loss in land mobile radio services. *IEEE Trans. Veh. Technology*, *29*, 317-325.

Hatami, A., & Pahlavan, K. (2005) A comparative performance evaluation of RSS-based positioning algorithms used in WLAN networks. In *Proceedings of the IEEE Wireless Communications and Networking Conference (WCNC '05), 4*, (pp. 2331–2337), New Orleans, La, USA.

Haykin, S. (2002). *Adaptive Filter Theory – 4th Ed.* Prentice-Hall.

Heidari, M., Akgul, F. O., & Pahlavan, K. (2007a). Identification of the Absence of Direct Path in Indoor Localization System. *IEEE PIMRC 2007.* (pp. 1-6).

Heidari, M., Akgul, F. O., Alsindi, N., & Pahlavan, K. (2007b). Neural Network Assisted Identification of the Absence of the Direct Path in Indoor Localization. *IEEE Globecom 2007.* (pp. 387-392).

Julier, S. J., & Uhlmann, J. K. (1997). A new extension of the Kalman filter to nonlinear systems. *Int. Symp. Aerospace/Defense Sensing, Simul. and Controls.*

JTC (Joint Technical Committee for PCS T1 R1P1.4). (1994). Technical Report on RF Channel Characterization and System Deployment Modeling, JTC (AIR)/94.09.23-065R6.

Kalman, R. E. (1960) A New Approach to Linear Filtering and Prediction Problems. *Transactions of the ASME - Journal of Basic Engineering, 82*, 35-45.

Kanaan, M., & Pahlavan, K. (2004a). Algorithm for TOA-based Indoor Geolocation. *IEE Electronics Letters, 40*(22).

Kanaan, M., & Pahlavan, K. (2004b). A comparison of wireless geolocation algorithms in the indoor environment. In *IEEE Wireless Communications and Networking Conference (WCNC04), 1*, 177-182.

Kanaan M., Akgul, F. O., Alavi, B., & Pahlavan, K. (2006a). A Study of the Effects of Reference Point Density on TOA-Based UWB Indoor Positioning Systems. In the *17th Annual IEEE International Symposium on Personal, Indoor and Mobile Radio Communications, PIMRC*, (pp. 1-5). Finland.

Kanaan, M., Heidari, M., Akgul, F., & Pahlavan, K. (2006b). Technical Aspects of Localization in Indoor Wireless Networks, *Bechtel Telecommunications Technical Journal, 4*(3).

Kanaan, M., Akgul, F. O., Alavi, B., Pahlavan, K. (2006c). Performance Benchmarking of TOA-Based UWB Indoor Geolocation Systems Using MSE Profiling. In *IEEE 64th Vehicular Technology Conference* (VTC-2006 Fall). (pp. 1-5).

Kim, D., & Langley, R. B. (2000). GPS Ambiguity Resolution and Validation: Methodologies,Trends and Issues. *7th GNSS Workshop-International Symposium on GPS/GNSS*, Seoul, Korea.

Knapp, C., & Carter, G. (1976). The generalized correlation method for estimation of time delay. *IEEE Transactions on Acoustics, Speech and Signal Processing, 24*(4), 320-327.

Krim, H., & Viberg, M. (1996). Two decades of array signal processing research: The parametric approach. *IEEE Signal Processing Magazine*. vol. 13, (pp. 67-94).

Kuchar, A., Tangemann, M., & Bonek, E. (2002). A Real-Time DOA-Based Smart Antenna Processor. *IEEE Transactions on Vehicular Technology, 51*(6), 1279-1293.

Kumaresan, R., & Tufts, D. W. (1983). Estimating the angles of arrival of multiple plane waves. *IEEE Trans. Aerosp. Electron. Syst., AES-19*(1), 134-139.

Larsson, E. G. (2004). Cramer-Rao bound analysis of distributed positioning in sensor networks. *IEEE Signal Processing Letters, 11*(3), 334-337.

Lee, J. Y., & Scholtz, R.A. (2002). Ranging in a dense multipath environment using an UWB radio link. *IEEE Trans. Select. Areas Commun., 20*(9), 1677–1683.

Li, J., Halder, B., & Stoica, P. (1995). Computationally efficient angle estimation for signals with known waveforms. *IEEE Transactions on Signal Processing*, 43(9), 2154-2163.

Li, X., & Pahlavan, K. (2004). Super-resolution TOA estimation with diversity for indoor geolocation. *IEEE Trans. on Wireless Communications*, *3*(1), 224-234.

Li, X. (2003). *Super-Resolution TOA Estimation with Diversity Techniques for Indoor Geolocation Applications*. PhD Thesis, WPI.

Ma, J., Min, G., Zhang, Q., Ni, L.M., & Zhu, W. (2005). Localized Low-Power Topology Control Algorithms in IEEE 802.15.4-Based Sensor Networks. *IEEE International Conference on Distributed Computing Systems*. (pp. 27-36).

Merrill, W. M., Newberg, F., Sohrabi, K., Kaiser, W., & Pottie, G. (2003). Collaborative networking requirements for unattended ground sensor systems. *IEEE Aerospace Conference, 5*, 2153-2165.

Motley, A. J. (1988). Radio Coverage in Buildings. *Proc. National Communications Forum*. Chicago. (pp. 1722-1730).

Nerguizian, C., Despins, C., & Affès, C. (2006). Geolocation in Mines With an Impulse Fingerprinting Technique and Neural Networks. *IEEE Transactions on Wireless Communications, 5*(3).

Niculescu, D., & Nath, B. (2001) Ad Hoc Positioning System (APS), *Global Telecommunications Conference, GLOBECOM '01, IEEE, 5*, 2926-2931.

Okumura Y. et al. (1968). Field Strength and its Variability in VHF and UHF Land-Mobile Radio Service. *Review of the Electrical Communication Laboratory*. Vol. 16, Numbers 9-10.

Oppermann, I., Hamalainen, M., & Linatti, J. (Eds.) (2004). *UWB Theory and Applications*. England: John Wiley and Sons.

Ottersten, B., Viberg, M., & Kailath, T. (1991). Performance analysis of the total least squares ESPRIT algorithm. *IEEE Trans. on Signal Processing, 39*, 1122-1135.

Pahlavan, K., Krishnamurthy, P., & Beneat, J. (1998). Wideband radio propagation modeling for indoor geolocation applications. *IEEE Communications Magazine*. (pp. 60–65).

Pahlavan, K., Li, X., & Makela, J. (2002). Indoor geolocation science and technology. *IEEE Commun. Mag.*, *40*(2), (pp. 112-118).

Pahlavan, K., & Levesque, A. H. (2005) *Wireless Information Networks - 2nd Edition*. Wiley – Interscience. ISBN: 0-471-72542-0, Hardcover, 722 pages.

Pahlavan, K., Akgul, F. O., Heidari, M., Hatami, A., Elwell, J. M., & Tingley, R. D. (2006). Indoor geolocation in the absence of direct path. *IEEE Wireless Communications*, *13*(6), 50-58.

Pathan, A.-S.K., Choong, S. H., Hyung-Woo, L. (2006). Smartening the environment using wireless sensor networks in a developing country. *The 8th International Conference on Advanced Communication Technology, 1*, 705-709.

Patwari, N., Ash, J. N., Kyperountas, S., Hero, A. O., Moses, R. L., & Correal, N. S. (2005) Locating the nodes: cooperative localization in wireless sensor networks, *IEEE Signal Proc. Mag.*, 22(4), 54-69.

Poling, T. C., & Zatezalo, A. (2002). Interferometric GPS ambiguity resolution, *Journal of Engineering Mathematics*, 43(2-4), 135-151(17).

Priyantha, N. B., Chakraborty, A., & Balakrishnan, H. (2000). The Cricket Location-Support System," *Proc. 6th Ann .Intl. Conf Mobile Computing and Networking (Mobicom'00)*, ACM Press, (pp. 32-43).

Qi, Y., & Kobayashi, H. (2003). On relation among time delay and signal strength based geolocation methods. In *IEEE Global Telecommunications Conference. GLOBECOM '03.*, *7*, 4079-4083.

Qi, Y., Kobayashi, H., & Suda, H. (2006). Analysis of wireless geolocation in a non-line-of-sight environment. *IEEE Transactions on Wireless Communications*, *5*(3), 672-681.

Qi, Y., Suda, H., & Kobayashi, H. (2004). On time-of arrival positioning in a multipath environment. In *Proc. IEEE 60th Vehicular Technology Conf. (VTC 2004-Fall).* Los Angeles, CA., *5*, 3540–3544.

Randell, C., Djiallis, C., & Muller, H. (2003). Personal position measurement using dead reckoning. In *Proceedings of the Seventh International Symposium on Wearable Computers, IEEE Computer Society.* (pp. 166–173).

Rao, B. D., & Hari, K. V. S. (1989). Performance analysis of root-music. *IEEE Trans. ASSP-37*(12), 1939-1949.

Roy, R., & Kailath, T. (1989). ESPRIT-Estimation of Signal Parameters via Rotational Invariance Techniques. *IEEE Transactions on Signal Processing.* 37(7), 984-995.

Röhrig, C., & Spieker, S. (2008). Tracking of Transport Vehicles for Warehouse Management Using a Wireless Sensor Network. *IEEE/RSJ International Conference on Intelligent Robots and Systems.*

Savarese, C., Rabaey, J. M., & Beutel, J. (2001). Locationing in Distributed Ad-Hoc Wireless Sensor Networks, *ICASSP.* (pp. 2037-2040).

Savvides, A., Han, C. C., & Srivastava, M. B. (2001). Dynamic Fine Grained Localization in Ad-Hoc Sensor Networks, *Proceedings of the Fifth International Conference on Mobile Computing and Networking, Mobicom.* (pp. 166-179).

Savvides, A., Garber, W. L., Moses, R. L., & Srivastava, M. B. (2005). An analysis of error inducing parameters in multihop sensor node localization. *IEEE Transactions on Mobile Computing*, *4*(6), 567-577.

Schelkunoff, S. A. (1943). A mathematical theory of linear arrays. *Bell System Technical Journal*, *22*, 80-107.

Schmidt, R. (1986). Multiple emitter location and signal parameter estimation. *IEEE Transactions on Antennas and Propagation*, *AP-34*(3), 276-280.

Schroer, R. (2003). Navigation and landing [A century of powered flight 1903-2003]. IEEE Aerospace and Electronic Systems Magazine, *18*(7), 27-36.

Sohraby, K., Minoli, D., & Znati, T. (2007). Wireless Sensor Networks – Technology, Protocols and Applications. Hoboken, New Jersey: John Wiley & Sons.

Steiner, C., Althaus, F., Troesch, F., & Wittneben, A. (2008). Ultra-Wideband Geo-Regioning: A Novel Clustering and Localization Technique. *EURASIP Journal on Advances in Signal Processing.* article id 296937.

Stoica, P., & Sharman, K. (1990). A Novel Eigenanalysis Method for Direction of Arrival Estimation. *Proc. IEE*, (pp. 19-26).

Stoleru, R., & Stankovic, J. A. (2004). Probability Grid: A location estimation scheme for wireless sensor networks. In *IEEE SECON*. (pp. 430-438).

Tewfik, A. H., & Hong, W. (1992). On the application of uniform linear array bearing estimation techniques to uniform circular arrays. *IEEE Transactions on Signal Processing, 40*, 1008-1011.

Tingley, R. (2000). *Time-Space Characteristics of Indoor Radio Channel*. PhD Thesis, WPI.

Torrieri, D. (1984). Statistical theory of passive location systems. *IEEE Trans. Aerospace and Elect. Sys. AES-20.*

Van Trees, H. L. (1968). Detection, Estimation and Modulation Theory: Part I. New York: Wiley.

Unbehaun, M. (2002). On the deployment of unlicensed wireless infrastructure. Ph.D. Thesis, Royal Institute of Technology, Department of Signals, Sensors & Systems, Stockholm.

Ward, A., Jones, A., & Hopper, A. (1997). A New Location Technique for the Active Office. In *IEEE Personal Communications, 4*(5), (pp. 42-47).

Wylie, M. P., Holtzman, J. (1996). The non-line of sight problem in mobile location estimation. *5th IEEE International Conference on Universal Personal Communications*. (pp. 827-831).

Yamasaki, R., Ogino, A., Tamaki, T., Uta, T., Matsuzawa, N., & Kato, T. (2005). TDOA location system for IEEE 802.11b WLAN. *IEEE Wireless Communications and Networking Conference,* 4, 2338-2343.

Zaidi, A. S., & Suddle, M. R. (2006). Global Navigation Satellite Systems: A Survey. In *International Conference on Advances in Space Technologies,* (pp. 84-84).

KEY TERMS

Throughout this chapter the words *anchor, reference point, transmitter* have been used interchangeably. Likewise *node, sensor, sensor node, receiver* have been used interchangeably.

ANSI	American National Standards Institute
AOA	Angle-of-Arrival
AP	Access Point
AWGN	Additive White Gaussian Noise
CDF	Cumulative Distribution Function
CEP	Circular Error Probability
CIR	Channel Impulse Response
CN-TOAG	Closest Neighbor Time-of-Arrival
CRLB	Cramer Rao Lower Bound
DDP	Detected Direct Path
DGPS	Differential Global Positioning System
DME	Distance Measurement Error

DOA	Direction-of-Arrival
DP	Direct Path
EKF	Extended Kalman Filter
ESPRIT	Estimation of Signal Parameters via Rotational Invariance Techniques
FCC	Federal Communications Commission
FDP	First Detected Path
FIM	Fisher Information Matrix
GLONASS	GLObal NAvigation Satellite System
GPS	Global Positioning System
IFT	Inverse Fourier Transform
JTC	Joint Technical Committee
LOS	Line-of-Sight
LS	Linear Squares
MAC	Medium Access Control
ML	Maximum Likelihood
MPC	Multipath Component
MSE	Mean Squared Error
MUSIC	Multiple Signal Classification
OLOS	Obstructed-Line-of-Sight
PCS	Personal Communication System
QoS	Quality-of-Service
RF	Radio Frequency
RMS	Root Mean Square
RP	Reference Point
RSS	Received Signal Strength
RT	Ray Tracing
SNR	Signal-to-Noise Ratio
SP	Strongest Path
TDOA	Time-Difference-of-Arrival
TIA	Telecommunication Industry Association
TOA	Time-of-Arrival
UDP	Undetected Direct Path
UKF	Unscented Kalman Filter
ULA	Uniform Linear Array
UWB	Ultra-Wideband
WAAS	Wide Area Augmentation System
WLAN	Wireless Local Area Network
WSN	Wireless Sensor Network

Chapter IV
RF Ranging Methods and Performance Limits for Sensor Localization

Steven Lanzisera
University of California, Berkeley, USA

Kristofer S.J. Pister
University of California, Berkeley, USA

ABSTRACT

Localization or geolocation of wireless sensors usually requires accurate estimates of the distance between nodes in the network. RF ranging techniques can provide these estimates through a variety of methods some of which are well suited to wireless sensor networks. Noise and multipath channels fundamentally limit the accuracy of range estimation, and a number of other implementation related phenomena further impact accuracy. This chapter explores these effects and selected mitigation techniques in the context of low power wireless systems.

INTRODUCTION

In this chapter we will discuss techniques for estimating the range between wireless sensor nodes using radio frequency (RF) measurements. Localization is a two part process that can roughly be divided into a phase where the relationships between nodes are estimated (range or angle) and a phase where these relationships are used to estimate locations of the devices. RF ranging, one of the options for the first phase, will be the topic of this chapter. In particular, RF time of flight methods where RF propagation time is estimated will be considered in depth. Other ranging methods (ultrasonic, sonic, light) have been proposed and tested but they are all limited from widespread adoption. Ultrasonic and sonic signals have limited range and do not pass through obstacles well when compared to RF signals. Acoustic systems also require the addition of speakers and microphones that are cumbersome for most applications. Light based systems require line of sight and are typically directional. Radios are pervasive in WSNs, and adding an accurate ranging feature would enable location aware networks in ways that are not possible using other technologies (Pahlavan, 2002).

Ranging accuracy is limited by noise, multipath channel effects, clock synchronization, clock frequency accuracy, and sampling artifacts. Fundamental performance limits exist due to these error sources, and these limits will be discussed qualitatively and mathematically. Signal bandwidth is an important factor when considering performance limits, and the impact of varying bandwidth will be shown.

Ranging methods will be discussed in the context of how well they meet application requirements for accuracy, energy consumption, latency, and useful range, and these requirements will be based on sample wireless sensor network applications. The major commercial application is asset tracking and management in factories, hospitals and other large spaces, and some commercial systems are available for these applications. Other applications including network configuration will be considered.

A number of RF based ranging systems have been proposed and implemented. The most common is the Global Positioning System (GPS), but others including cellular phone based systems are also widespread. Currently, ultra-wideband techniques are starting to be demonstrated along with more advanced narrowband techniques. The methods used and performance capabilities and limitations in selected systems will be discussed.

APPLICATION REQUIREMENTS

The requirements of a localization system are dependent on the application. This section will discuss a few applications to determine requirements on accuracy, latency, useful range, and infrastructure complexity of a ranging system. The accuracy requirement is defined to be the maximum error between true and estimated position that is acceptable for some percent of all estimates. For example, if 80% of estimates must be accurate to within 2 m, then 20% of measurements can have larger error. It is important to understand that localization is probabilistic in that the environment among other factors randomly degrades the accuracy of a measurement. Latency is the time it takes from when a request for a location update is made to when the update is presented to the user for a single device in the network. The range requirement is roughly how large of a sphere must one make around any node to find at least 4 other nodes or infrastructure points in 3D and 3 infrastructure points in 2D. Infrastructure requirements impact the cost of a network, and this impact can be considered qualitatively.

Relationship between Range Accuracy and Location Accuracy

Location accuracy requirements are in terms of difference from estimated location to true location as opposed to range accuracy. Localization algorithms and network geometries differ in how ranging accuracy translates to location accuracy, and many range based localization methods are presented in this book. In order to address the link between location and range accuracy, we apply a common method of range based location estimation: the maximum likelihood estimate (MLE) of the location based on a set of range estimates. The MLE of the location is found by calculating probability density function (PDF) of the location based on each range estimate, multiplying the

PDFs together for each range estimate, and finding the point where the resulting joint probability is maximized. Consider the case where the PDF of the location given a range estimate is given by $f(r_{est}|r_{true})$. If n independent range estimates (r_{est1}, r_{est2},...,r_{estn}) are used to find the MLE of the location, then the joint probability distribution of the location is given by the product of the individual PDFs,

$$f(\{r_{est_i}\} \mid l) = \prod_i f(r_{est_i} \mid r_{true_i})$$

where l is the location. When $f(\{r_{est_i}\} \mid l)$ is maximized, the corresponding location is the MLE. Figures 1 and 2 show the results of a random simulation of one simple 2D case when $f(r_{est}|r_{true})$ is normally distributed with parameters ($\mu=r_{est}$, σ). In Figure 1 the cumulative distribution function (CDF) of the location error normalized to the root mean square (RMS) ranging error is plotted when there are 3, 4 and 5 reference points. In Figure 2 the CDF of the location error normalized to the worst case ranging error is plotted. When more than 3 reference nodes are available, performance improves significantly especially when compared to the worst case ranging error. From this simulation two conclusions result: 1) increasing the density of nodes with known location is important for improving accuracy; 2) ranging accuracy and location accuracy are very similar. Although the location accuracy can be better or worse than the ranging accuracy depending on the conditions and localization algorithm used, we will assume that location error is equal to the RMS ranging error for simplicity.

Asset Tracking

In the hospital environment, equipment, staff and patients could all be tagged to increase the efficiency and safety of the healthcare environment. There are many cases in which hospitals own many extra pieces of equipment in hopes of ensuring that the appropriate items can be located and used quickly. Despite this preventive measure much time is often wasted searching for equipment. Because wasted time is so costly in terms of both dollars and care, this environment would benefit significantly by location aware devices. Everything must be monitored occasionally without tight latency requirements,

Figure 1. CDF of location error normalized by the RMS ranging error

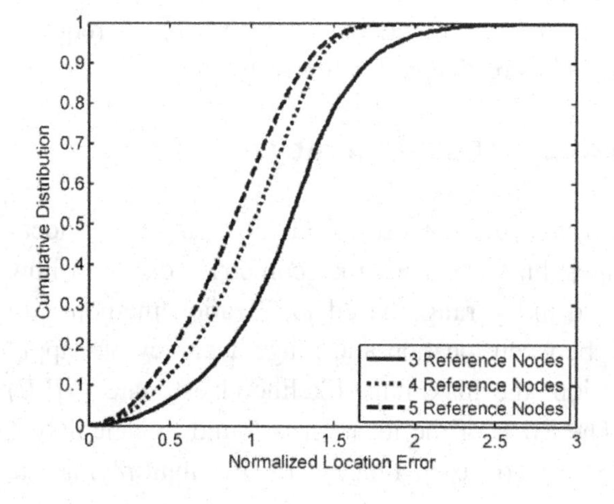

Figure 2. CDF of location error normalized by the worst case ranging error

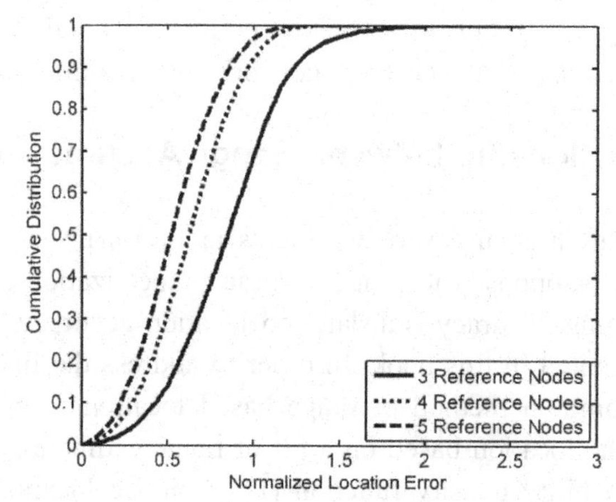

but short latency updates of specific items are required. Accuracy must be good enough to ensure that the correct room is shown almost all of the time. Given that a typical hospital room is about 4 by 7 m, accuracy of better than 1.5m ensures the correct room is indicated 80% of the time. Alarms or query targets must be localized within a few seconds, and only one or two devices may be in a room. In order to ensure enough connectivity for localization, a range of 15 m is required.

Large Data Collection Network Configuration

A primary cost of deploying a large scale wireless sensor network is the installation of nodes and recording node locations. Localization systems can reduce this cost by determining the locations of devices after deployment. Latency requirements are minimal in that it is acceptable for the initial network configuration to take hours to complete. The scale of many industrial campuses requires long ranges in the hundreds of meters but accuracy requirements depend on the location of the device. Devices outdoors can be located with less accuracy and longer range, and indoor devices are more densely populated and require the typical 1.5 m accuracy.

Security

Security systems such as radio frequency identification (RFID) operated systems are commonly used to grant privileges (e.g. room and building access), and localization systems will be able to enhance these capabilities. If the correct person or people are in the correct rooms, privileges can be granted or revoked to ensure a secure environment for sensitive information, prison populations and many other situations. Latency must be on the scale of a second, and accuracy must ensure correct room identification (Anjum, 2005).

Summary of Specifications for Ranging Systems

Location accuracy, latency, range and infrastructure complexity are quite consistent across a broad spectrum of applications. For most networks, inter-node ranges are a few tens of meters and accuracy of 1.5 m with latency of a few seconds provides a robust solution. Much higher accuracy may be required in some applications not discussed here, but it isn't all that common. Infrastructure points, or nodes, can vary in cost by orders of magnitude depending on the ranging method used, and reducing the cost of these points is important to a successful location aware wireless sensor network.

Table 1. Summary of ranging specifications for typical indoor and outdoor sensor networks

Specification	Value	Conditions
Accuracy	1.5 m	80% of estimates indoors
	5 m	80% of estimates outdoors
Range	>15 m	Indoors, through walls
	100 m	Outdoors, line of sight
Latency	< 5 s	Including data relay across network
Infrastructure Cost	Low	

ACCURACY LIMITS

The achievable accuracy of ranging systems is limited by four primary factors which are noise, time synchronization, sampling artifacts, and multipath channel effects. These factors introduce random, time and spatially varying errors into the estimate resulting in reduced accuracy. Frequency accuracy between the devices involved in the measurement can also impact ranging system accuracy significantly. Each effect can dominate the error under different circumstances, and a system must be designed so that the combination of these effects does not degrade accuracy beyond useful limits. Because the introduced errors are stochastic, the errors can never be eliminated, but it is possible that measurement techniques can be used to mitigate these effects.

Noise

Noise and interference introduce unknown errors into measurements. The effect of white noise processes such as thermal and electronic noise is well understood and can be quantified. A range measurement degraded only by noise is limited in accuracy by the signal energy to noise ratio at the receiver and the occupied bandwidth.

A ranging system suffers in low signal to noise ratio (SNR) environments because the exact time of an event cannot be resolved precisely. In a simple example "edge detection" ranging system, the ranging signal is a step function sent by the transmitter at $t = 0$ and the receiver measures the time of the rising edge it observes. When this signal is received, the edge time may be detected slightly early or slightly late due to noise added to the signal. For RF measurements radio waves move at the speed of light ($c = 3\times10^8$ m/s) meaning that a distortion of just 10ns results in 3m of measurement error. The speed of this rising edge at the receiver is proportional to the bandwidth of the communications system, and wider bandwidth typically results in better performance. Because the noise amplitude increases as the square root of bandwidth and the signal transition speed increases linearly with bandwidth, a faster rising edge is more tolerant to noise. This qualitative understanding of how SNR and bandwidth affect the noise performance of ranging is useful, but a quantitative limit of ranging accuracy in a noisy environment is needed.

The mathematical expression that links SNR and bandwidth together to give a bound on ranging performance can be derived from the Cramér-Rao lower bound (CRB). The CRB can be calculated for any unbiased estimate of an unknown parameter. Van Trees (1968) discusses ranging as a parameter estimation problem studied in the context of radar and sonar applications, and the CRB under a variety of conditions has been calculated. For the prototype "edge detection" ranging system discussed earlier, the CRB can be used to calculate a lower bound for the variance of the estimate for the range, $\hat{r}$, as

$$\sigma_{\hat{r}}^2 \geq \frac{c^2}{(2\pi B)^2 \frac{E_s}{N_0}} \left(1 + \frac{1}{\frac{E_S}{N_0}} \right) \tag{3}$$

where $\sigma_{\hat{r}}^2$ is the variance of the range estimate, c is the speed of light, B is the occupied signal bandwidth in Hertz, and E_s/N_0 is the signal energy to noise density ratio. The SNR is related to E_s/N_0 in that

$$SNR = \frac{P_s}{P_n} = \frac{E_s}{N_0 t_s B} \tag{4}$$

where P_s is the signal power, P_n is the noise power, t_s is the signal duration during which the bandwidth, B, is occupied. The concepts of occupied bandwidth and signal duration are important as illustrated by our step function example. The maximum bandwidth of the signal is set by the transmitter filter, and increasing the receiver's filter bandwidth does not increase the bandwidth used by the signal. Similarly, t_s is not simply the length of time that the signal was observed at the receiver but the length of time that the signal was observed when it was doing anything meaningful (such as changing in value). In the case of this step function, a small window of time contains nearly all of the useful information about the transition, and observing the signal for a longer period time contributes almost no additional information. In this example and in many common signals, the bandwidth and duration are tied together such that $t_s B \approx 1$. Therefore, the E_s/N_0 ratio is approximately equal to the SNR. By exchanging the locations of the factors in (4),

$$\frac{E_s}{N_0} = t_s B \cdot SNR \tag{5}$$

one advantage of having a $t_s B$ product greater than unity becomes clear. Signals with this property would exhibit better noise performance at lower SNR values. One class of signals that exhibit this property are pseudorandom number (PN) sequences that result in long duration while retaining the same bandwidth as the constituent sub-symbols. These sub-symbols are called chips to differentiate them from bits (information) and symbols (collections of bits). Taking advantage of signals with $t_s B > 1$ improves noise performance, but it comes at the cost of increased signal processing. Often there is no other way to improve noise performance (i.e. fixed transmitter output power and receiver noise floor), and the signal processing cost is acceptable. For a fixed signal energy and noise density, increasing the bandwidth provides significant improvements in noise performance. This fact is one argument for increasing the bandwidth of RF based ranging systems, but the bandwidth required to achieve reasonable noise performance is not very large (Lanzisera, 2008).

One common example can be found in GPS. The C/A (course acquisition or civilian) signal in GPS uses a PN sequence modulated with binary phase shift keying (BPSK) at 1.023×10^6 chips/s. At a receiver on the ground, the observed SNR is typically -20dB, the bandwidth occupied is about 2MHz, and there are 1023 chips per symbol (Kaplan, 2005). This is all the information required to determine the best case noise performance of GPS. First we calculate E_s/N_0 through the application of (5):

$$\frac{E_s}{N_0} = \frac{1023}{1.023\times10^6} \cdot 2\times10^6 \cdot 10^{-2} = 20$$

Applying this result to (3)

$$\sigma^2_{\hat{r}_{GPS}} \geq \frac{(3\times10^8)^2}{(2\pi \cdot 2\times10^6)^2 \cdot 20}\left(1+\frac{1}{20}\right) = (5.5m)^2$$

This accuracy is close to what GPS routinely provides, but this range estimate is updated at 1kHz in the above calculation, and the typical user uses systems that update at less than 10 Hz. This can be used to reduce the variance by a factor of 100 resulting in $\sigma^2_{\hat{r}_{GPS}} \geq (0.6m)^2$. GPS users are accustomed to accuracy of better than 5m (80% of trials) in open, flat terrain suggesting that the noise limit is not obtained or that other factors are reducing accuracy. In this case, approaching the CRB is possible because of the high value of E_s/N_0 and the signal design, but random atmospheric effects contribute the majority of the remaining error. The P (precise or military) GPS signal is broadcast at two different carrier frequencies so that these atmospheric effects can be estimated and removed which greatly enhances accuracy. It is also worth noting that $1+E_s/N_0$ term contributes very little to the CRB, and it is commonly ignored for $E_s/N_0 \gg 1$.

GPS provides a good reference for looking at other ranging systems because it is familiar and has some characteristics in common with communications systems, but it has significant differences as well. In typical wireless communications systems, the distances traveled are much less, and atmospheric effects are not significant. In addition, narrowband systems have signal SNR that is large such that, when coupled with processing gain, high values of E_s/N_0 result. These high values for E_s/N_0 allow the CRB to be nearly achieved in many systems, but the CRB is not a tight bound at low E_s/N_0 (Van Trees, 1968). If the desired error variance is not achievable directly, averages of multiple measurements will yield improved results. GPS occupies a 2MHz bandwidth which is comparable to the common IEEE 802.15.4 radios used in WSNs, but GPS signals are broadcast at a single carrier frequency. WSN radios are usually frequency agile, and information from different frequencies can be used to improve ranging performance (Lanzisera, 2006).

The CRB can also be improved through the use of additional bandwidth. Ultra wideband (UWB) technologies are being developed partially to provide accurate ranging capability to wireless systems. A UWB signal is defined to be a signal that either uses at least 500MHz or that occupies as much bandwidth as half the signal's center frequency. The use of 500MHz of bandwidth and an E_s/N_0 of -10dB yield a CRB of

$$\sigma_{\hat{r}}^2 \leq \frac{(3\times 10^8)^2(1+\frac{1}{0.1})}{(2\pi\cdot 500\times 10^6)^2\cdot 0.1} = (1m)^2$$

Although the CRB may not be achievable at this low value for E_s/N_0, small bounds are possible. This promise, along with superior performance in multipath environments (to be discussed later), has driven much interest in UWB for extremely accurate location systems.

Both bandwidth and E_s/N_0 play significant roles in determining noise limited performance. Figure 3 shows the CRB as a function of bandwidth for E_s/N_0 of 10 dB and 26 dB. Signals with t_sB products of 10 to over 1000 are commonly used enabling large E_s/N_0 in communication systems. It is interesting to note that that noise alone does not prevent 1 m accuracy for bandwidths of a few megahertz or more.

Time Synchronization

RF time of flight measurement systems must be able to estimate the time of transmission and arrival using a common time base for accurate measurements. When two wireless devices, A and B perform range estimation, the most straightforward method is for A to send a signal at $t = 0$ and for B to start a timer at $t = 0$ and stop it when it receives the signal sent by A. The value of the timer at B is equal to the time of flight (TOF). If the clocks are not perfectly time synchronized, however, and B's notion of $t = 0$ is offset in time from A's, then this offset, Δt, directly adds a bias to the measurement. Time synchronized wireless networks are typically synchronized to on the order of 1μs resulting in errors of up to 300m, but high power and expensive systems can achieve time synchronization of better than 10ns or 3m. This method is shown in Figure 4a.

Figure 3. Cramér Rao Lower Bound as a function of bandwidth for 10dB and 26dB E_s/N_0. Common radio standards used in wireless sensor networks such as IEEE 802.15.1 (Bluetooth), IEEE 802.15.4 (Zigbee and others), and wireless LAN (802.11a/b/g) are shown. Ultra-wideband (UWB) radios with more than 500MHz of bandwidth have excellent noise performance, but even a few megahertz of bandwidth can enable the 1.5m accuracy required for most applications.

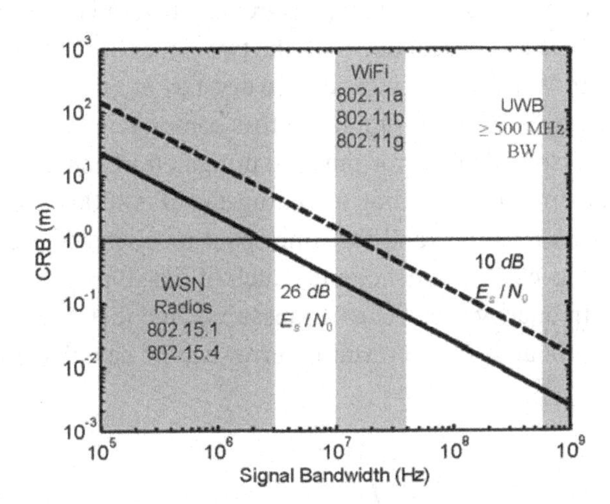

If A and B have full duplex radios, that is, they can transmit and receive at the same time, then a two way or round trip measurement can be made. A sends a signal to B at a carrier frequency f_{c1} and B translates this signal to a different carrier frequency f_{c2} and retransmits that signal in real time. The signal is received back at A at f_{c2} such that A can compare the signal it is receiving from B to the signal it is sending to B. By measuring the delay between these two signals, the round trip TOF, $\hat{\tau}_{RT}$, is estimated, and the range estimate is $c \cdot \frac{\hat{\tau}_{RT}}{2}$. This method is shown in Figure 4b.

Most WSN nodes do not have full duplex radios because they are more complicated and expensive than half duplex transceivers. Many other wireless systems are half duplex as well (e.g. wireless LAN and GSM), and the round trip method can be adapted for these systems. A round trip method known as two way time transfer (TWTT) has been developed to improve time synchronization between wireless base stations after the first communications satellites were launched, and it provides both range estimation and improved time synchronization capability (Kirchner, 1991). This method, shown in Figure 4c, allows the time offset between A and B to be ignored. Both A and B are responsible for measuring a time delay accurately using a local clock. A must measure the time that it takes for the signal it sends to return to it, and B must measure the time that the signal spends at B accurately. If the time A sends the signal is t_{sA}, the time B receives the signal from A is t_{rB}, the time B replies to A is t_{sB}, the time A receives the signal is back from B is t_{rA} such that $t_{sA} < t_{rB} < t_{sB} < t_{rA}$ then A measures $t_A = t_{rA} - t_{sA}$ and B measures $t_B = t_{sB} - t_{rB}$. By combining these two measurements together both the time of flight (τ) and clock offset ($\Delta\hat{t}$) can be estimated.

$$\Delta\hat{t} = \frac{1}{2(t_A + t_B)} \tag{6}$$

$$\hat{\tau} = \frac{1}{2(t_A - t_B)} \quad (7)$$

This or related methods are used with less accurate hardware to provide the rough time synchronization common in wireless systems.

One problem with two way ranging is that the measurement takes place over a relatively long period of time such that if the reference clock frequencies at the two nodes are not identical, an unknown bias can be added to the signal. In WSN nodes, inexpensive crystals are used where the frequency difference from crystal to crystal may be 100ppm or more across commercial temperature ranges. This clock frequency offset (also called clock drift) error must be mitigated in some fashion (Lanzisera, 2006). Consider a system where the time spent sending a ranging signal is 100μs and the time spent changing from transmit to receive mode is 200μs, and the time spent receiving the returned ranging signal is 100μs. Over this 400μs time, a clock frequency mismatch of just 10ppm would result in about 1m of estimation error. The clock frequency offset can be measured, and then the clock frequency can either be corrected to match within bounds or the resulting error can be calculated and subtracted from the

Figure 4. Three methods of performing time of flight ranging measurements: (a) time of arrival which is susceptible to clock offset Δt; (b) full duplex two way ranging; (c) half duplex two way ranging called two way time transfer

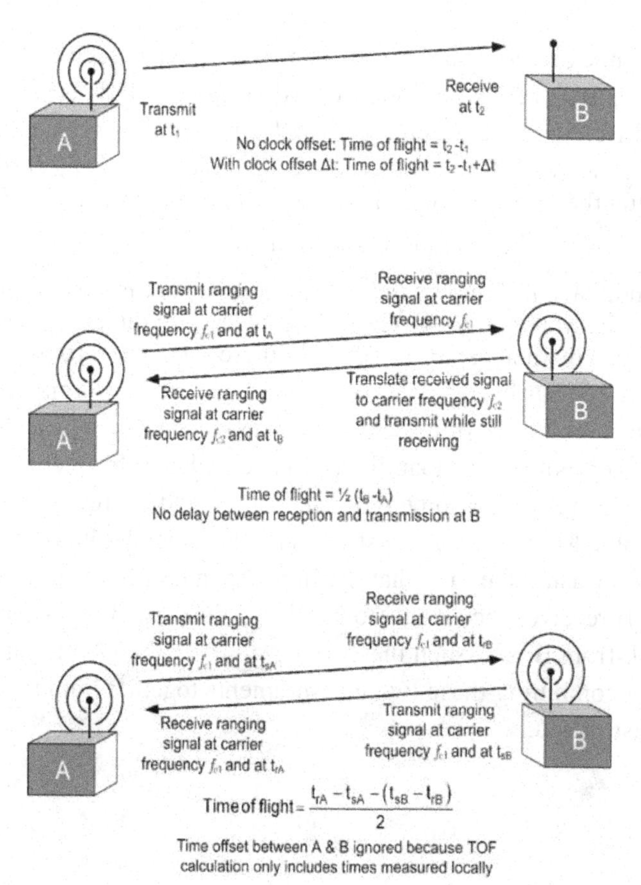

estimate later. Many methods have been used to measure frequency offsets in wireless systems, and we summarize one simple method here. This method is to run a counter over a long period of time to measure the offset. One node sends a start packet to the second node and starts a local timer, and the second node starts a local timer when it receives this packet. After waiting a sufficiently long time, the timer at the first node expires, and it sends a stop packet. The second node receives this stop packet, stops its timer, and compares the value left on the timer to the expected value (zero if the counter is counting down). This difference is a measure of the clock offset. The minimum time between packets, T_{wait} can be calculated as follows:

$$T_{wait} \geq \frac{1}{\Delta f_{xo}} \quad (8)$$

where Δ is the required matching , and f_{xo} is the crystal frequency. For a 20MHz crystal and a system requiring 10ppm accuracy, T_{wait} must be great than 5ms. This process is rather long but very simple, and other methods trade complexity for time savings.

Sampling Artifacts

Ranging systems estimate the time of arrival of a signal and compare that time with the time the signal was transmitted to calculate the time of flight and thus the range. It is commonly assumed that ranging accuracy is limited to c/f_s where f_s is the receiver sampling rate (Richards, 2005). This limit is known as range binning, and it can impact resolution if steps are not taken to mitigate its impact. A common implementation is to estimate the time of arrival using a matched filter that is sampled at the signal bandwidth resulting in time resolution of *1/B*. This sampling adds error to the estimate because the estimate space is divided up into range bins that are *c/B* wide. The error associated with this process is uniformly distributed inside the range bin. By using the variance of the uniform distribution, the impact of sampling can be found (Hoel, 1971).

$$\sigma^2_{sample} = \frac{1}{12 \cdot f_s^2} \quad (9)$$

In the case of the GPS example, with sampling at *1/B* the variance due to sampling can be calculated.

$$\sigma^2_{sample} = \frac{1}{12 \cdot (2 \times 10^6)^2} = (144ns)^2$$

This results in a range resolution of 43 m. In GPS, this coarse estimate is filtered (averaged) to improve the resolution, and a feedback loop can be used to null out the sampling error while the receiver tracks the satellites (Kaplan, 2005). Using just averaging, over 1500 measurements are required to achieve a variance of $(1\text{m})^2$. These methods are not realistic for many wireless sensor network applications where extremely low power consumption and therefore duty cycle is required. An accurate range estimate must be made in a short period of time. To reduce the sampling error, the signal can be over sampled. Figure 5 shows the CRB for a 2 MHz bandwidth signal with E_s/N_0 of 26 dB, the standard deviation of the range error due to sampling, and the combined effect of both error sources as a function of sampling frequency. This plot shows that with a 2 MHz bandwidth, the required sampling rate to ensure that the error is not dominated by sampling is over 70 MHz. It is clear that one must sample very fast to have the error dominated by the CRB rather than sampling. As the CRB improves due to increased bandwidth, the sampling speed required remains higher than twice the signal bandwidth down to E_s/N_0 of about 3 dB.

If the signal is sampled above Nyquist ($f_s>2B$), then the entire information content of the signal is captured in the sampling process (Oppenheim, 1975). Therefore, it is possible to extract better time resolution than σ_{sample}. In Figure 6, a signal is shown along with dots representing the samples of that signal that is band limited to a 2 MHz bandwidth. This signal is sampled at 10 Msps which is above the Nyquist rate of 4 Msps, but the sample rate still is far too low to achieve the CRB. The range bins are 100ns (30m) wide in this case where as the CRB from Figure 3 is only 3.5ns (1.1m) demonstrating a dramatic resolution reduction. Looking at the time of the zero crossing, it is clear that even a linear interpolation between the two adjacent samples would improve the estimate of that zero crossing location significantly. A major challenge is that many systems would need to perform this interpolation in real time increasing system complexity and power consumption beyond reasonable limits.

A round trip time of flight method known as code modulus synchronization (CMS) that takes advantage of Nyquist sampling has demonstrated its ability to approach the CRB while maintaining low sampling rates. CMS emulates a full duplex ranging system where the repeating node is retransmitting the signal that it is receiving from the first node without any delay. In CMS, however, half duplex radios such as those used in wireless sensor networks are used so the delay between reception and retransmission must be managed carefully. CMS as implemented uses a short PN code modulating an RF carrier as the ranging signal. Figure 7 shows the basic operation of CMS for a time of flight of zero. When the time of flight is greater than zero the Node B and Node A Receive lines would each be circularly shifted to the right by an amount equal to the time of flight and twice the time of flight respectively. For example, a range of 9m would have a time of flight of 30ns. The second line would be shifted by 30ns and the Node A receive line would be shifted by 60ns but the chip period may be 500ns, and it is acceptable that the shifts are much smaller than the chip period. The first node, A, generates a local code that is synchronized with a local clock called the event clock that has the same period as the PN

Figure 5. A comparison of range binning due to sampling error and the Cramér Rao bound on noise limited ranging for a 2 MHz bandwidth with a E_s/N_0 of 26 dB. The sampling rate required is much higher than required by sampling theory to achieve noise limited resolution.

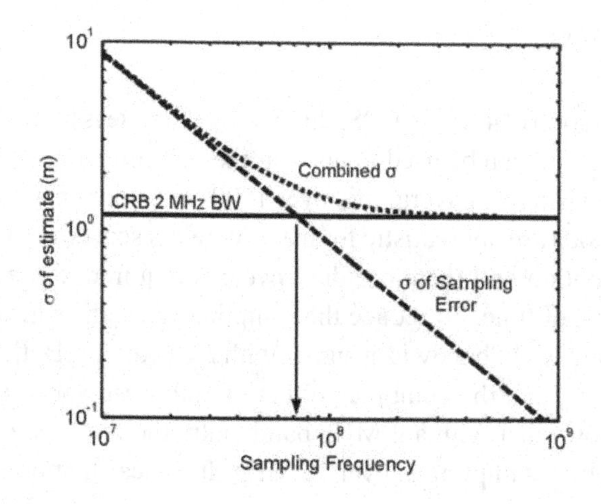

Figure 6. An above Nyquist sampled waveform is shown with the sample points marked in an example of sample based range binning. An interpolation between points enables time resolution of the zero crossing far better than 1/B and $1/f_s$ reducing the size of the range bins significantly.

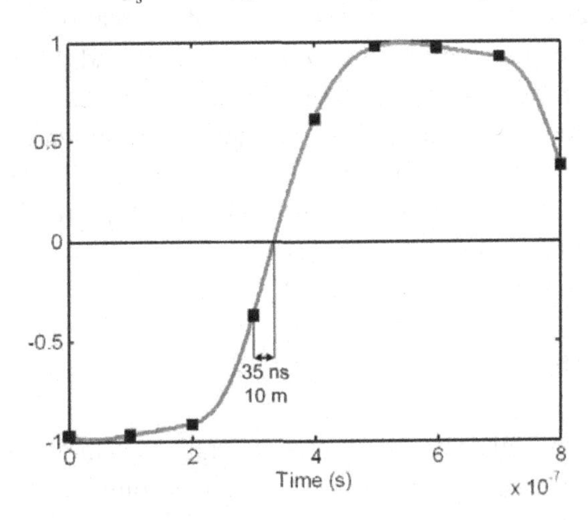

code. This code is used to modulate the carrier and is transmitted to the second node, B. B has a local event clock with the same period as at A, but the phase of the clocks are offset. As a result, B knows the length of the incoming PN code. B samples and demodulates this signal, and exactly one circularly shifted copy of the code is stored in memory. The system can accumulate multiple copies of the code in order to improve SNR, but they are all exactly one copy of the code that is circularly shifted in exactly the same way as the other received copies. At this point, B has a local copy of the code that is an average of multiple receptions and that is circularly shifted due to the event clock phase offsets between A and B. After A has sent a predetermined number of code copies and B has received some of these copies, the transceivers switch states, and B is now the source of the code. Starting on its event clock rising edge, it transmits the circularly shifted code it received back to A. On the next rising edge of its

Figure 7. Code modulus synchronization (CMS) achieves noise limited ranging performance through interpolation of data points in an non-real time time of flight (TOF) estimation phase, and this figure shows the case where the TOF is zero. Non-zero TOF would result in sub-chip width circular shifts to the signals on the node B and node A receive signals. CMS is a two way ranging technique that emulates a full duplex ranging system (Figure 4b) to eliminate clock synchronization errors.

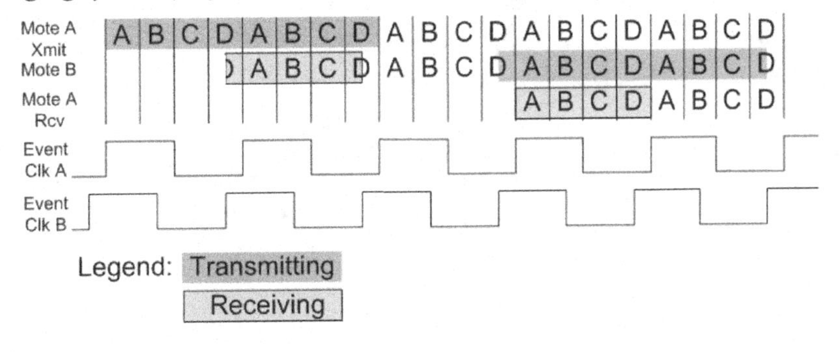

event clock, A starts to record exactly one copy of the code. Again, A can accumulate multiple copies to improve SNR. Because of the roundtrip nature of the system, the circular shift that occurred going from A to B is exactly undone going from B to A. After A has received and accumulated the desired number of code copies, the transceivers are shut off, and all of the real time processing is completed. A then computes the cross correlation between the code it recorded and the code that it sent, and zero code offset exists if the time of flight is zero. Because this system relies on sampling the signal at or above Nyquist, the received code can be interpolated to improve resolution up to the noise limit of the system. The correlation and code offset estimation are not done in real time enabling the computation to be done at any time using any method the user desires. This system can achieve the CRB in a single measurement as long as the sampling rate of the received code is above Nyquist, substantially improving over other two way ranging methods (Lanzisera, 2008).

Multipath Channel Effects

When a ranging system has been well designed, it often still fails to achieve the expected performance because the measurement is not taken in free space. In real environments the RF signals bounce off objects in the environment causing the signal to arrive at the receiving antenna through multiple paths as shown in Figure 8. In this figure, the direct path is obstructed by walls, but the other paths are not. This is common indoors, and it is likely that the non-direct paths have higher power than the direct path (Spencer, 2000). The communication environment is called the channel, and multipath channels not only vary by the type of environment (office building, residential or outdoors) but are specific to the geometry of the transmitter and receiver in that environment. The channel is often time varying resulting in a multipath environment that changes from one time to another. For narrowband radios like those common in wireless sensor networks, moving one transceiver by just a fraction of a wavelength (12cm at 2.4GHz) will cause the receiver to see what looks like an entirely new multipath environment because the paths will interfere constructively or destructively differently. The path length change is referenced to the wavelength of the RF making these small changes have large effects. The speed that the channel changes depends on how quickly objects are moving in that environment. Slower objects

Figure 8. A multipath environment that exhibits a common condition. The direct path (P_d) which is to be estimated for ranging is obstructed and heavily attenuated while the reflected paths (P_{m1}, P_{m2}) have much higher signal power.

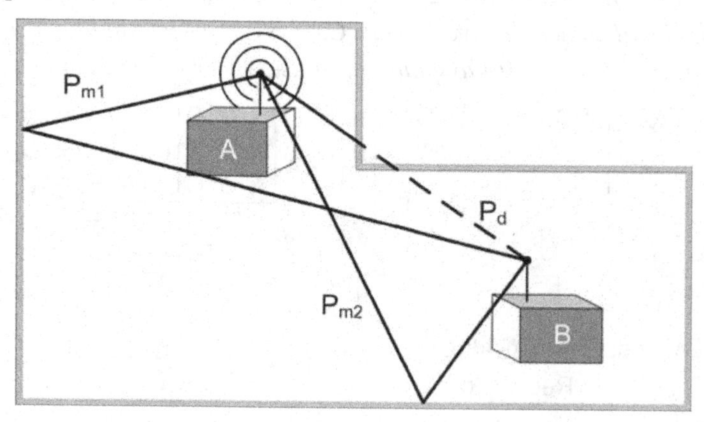

result in slower changes to the channel. This typically means that indoor channels change more slowly than outdoor channels, and the time it takes for the channel to change significantly is called the coherence time, t_c, of the channel. The value of t_c is roughly $c/(2fv)$ where c is the speed of light, f is the carrier frequency, and v is the speed of the fastest moving object in the environment. Recall that the wavelength of radio waves, λ, is c/f, and a more intuitive form of t_c is $\lambda/(2v)$ where it is clear that the time it takes to move a half wavelength corresponds to the coherence time (Tse, 2005). A series of measurements that take much less than t_c to complete can be used together as if the channel was time invariant over those measurements. This fact is useful when attempting to reduce the impact of multipath because multiple measurements taken at different frequencies can be used together. Because this interference effect is closely tied to the wavelength, changing carrier frequency even by 1% or less can dramatically affect the apparent multipath environment in narrowband systems. This can be easily observed by considering the received signal strength (RSS) profile across carrier frequency in an indoor environment as shown in Figure 9 (Werb 2005). At some carrier frequencies, the signal is in deep fade (destructive interference), while at others it has much higher signal strength (constructive interference). Without knowing the channel characteristics, knowledge of the RSS at one frequency tells you nothing about the RSS at another frequency. Wider bandwidth signals suffer less from this effect, and the bandwidth required to combat this is related to the time difference between the first and last significant path arrivals known as the delay spread, t_d. The coherence bandwidth, W_c, is approximately $1/(2\pi t_d)$ and it is the bandwidth over which the channel can be considered to be flat (either in deep fade or not, for example). If the bandwidth, B, is much larger than W_c the signal does not depend on carrier frequency to the same extent as a signal with a bandwidth less than W_c (Tse, 2005). Typical delay spreads for indoor channels are between 10ns and 100ns yielding coherence bandwidths between 1 MHz and 20 MHz. Outdoors, the delay spread can be up to microseconds, significantly reducing W_c. In RF ranging systems, the inter-path delay, $t_{\Delta p}$, is more important than the delay spread, however, because short inter-path delays can significantly impact ranging accuracy. Indoors, inter-path delays of 5ns to 10ns are very common and must be resolved if accuracy is to be better than $c \cdot t_{\Delta p}$ (Van Trees, 1968).

In a multipath environment, the receiver must somehow choose or estimate the direct path length and ignore the other paths. If a receiver can determine when only the first path arrives, then this will be the shortest distance and desired estimate. If the system is not able to resolve the individual paths, then the estimate is blurred by the multipath effects resulting in measurement error. In this case, if the receiver

Figure 9. Received signal strength verses frequency measured in a line of sight multipath channel with a 2MHz RF bandwidth. The significant changes in signal strength show that changing carrier frequency changes the apparent multipath environment significantly. Adapted from Werb (2005).

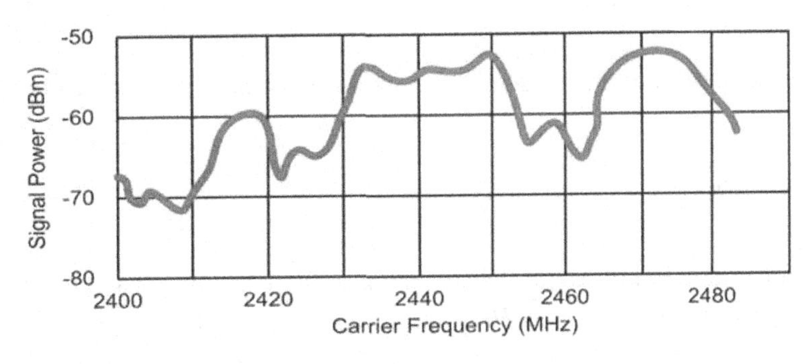

has an estimate of the channel impulse response, it can calculate the bias caused by the multipath channel and subtract the bias from its estimate. This leads to two classes of multipath mitigation methods: 1) resolving the direct path through increased bandwidth, or 2) estimating the channel response and using this information to improve or generate a range estimate.

In the first case, the ability to resolve the response of the multipath channel is directly linked to the bandwidth of the signal. Inter-path delays, $t_{\Delta p}$, separated by more than $1/B$ in time are resolvable and paths separated by less are generally not. To resolve paths that are separated by 1m or more, a bandwidth of at least 300MHz is required which shows a significant advantage of UWB systems. Using bandwidths in excess of 500MHz enables accuracy better than 1m in many cases, but this accuracy is not always achieved (Shah, 2005). Sometimes there is line of sight between the transmitter and receiver, or, in other cases, the direct path is attenuated somewhat by obstacles but still reaches the receiver with sufficient strength to be resolved, resulting in acceptable accuracy. When the direct path is too weak compared to other paths, however, a secondary path will be chosen to estimate the range resulting in an over estimate. In indoor environments, 10% to 20% of all measurements will fall into this category, but some environments are worse and a direct path is rarely available. True line of sight paths are not very common indoors, and most indoor channels will have a few strong paths spread across a few tens of nanoseconds (Spencer, 2000). Localization systems typically mitigate the severe cases of obstructed ranging by adding extra devices to "see" the obstructed areas and through localization algorithms that reject large ranging errors.

The second mitigation strategy relies on estimating the impact of the multipath environment on the range estimate and then subtracting off this error. This method is used when the signal bandwidth is too small to sufficiently resolve the multipath environment, and it is somewhat analogous to channel equalization. There are two critical steps to this method: 1) estimating the channel frequency/impulse response and 2) estimating the impact the channel has on the range estimate. Each step can be completed in different ways, and the solutions fall into the family of super-resolution algorithms. A super resolution algorithm is one that attempts to provide range resolution that is better than c/B (Dickey, 2001). If the impulse response can be estimated to include the static offset due to the time of flight, then the range can be estimated directly from the impulse response. If the impulse response is estimated with the first path always being at a delay of zero, then some other ranging method must also be used. One method to estimate the channel impulse response is to send a modulated signal that consists of a sequence of chips (Nefedov, 2000). Recall that the inter-path delay is a few nanoseconds compared to the chip duration of 100's of ns to μs, and the chip width used must typically be shorter in time than the features to be resolved. A super-resolution technique resolves features that would be too close together in time to be resolved normally. If the signal sent is x, the channel impulse response is h, and the received signal is y, then

$$y = x * h + \tilde{n}$$

Where * denotes convolution, and $\tilde{n}$ is complex noise. This can be rewritten in the frequency domain.

$$Y(\omega) = X(\omega)H(\omega) + N(\omega)$$

If the signal to noise ratio is large, and the spectrum of the transmitted signal (including the transmitter frequency response) is known, then $H(\omega)$ can be approximated.

$$H(\omega) = \frac{Y(\omega)}{X(\omega)} + \frac{N(\omega)}{X(\omega)} \approx \frac{Y(\omega)}{X(\omega)}$$

This approximation is only valid in sufficiently high SNRs, and noise causes significant problems in super-resolution estimation methods. $Y(\omega)$ is calculated by taking the Fourier transform of the received signal, and $X(\omega)$ is a system parameter known a priori. Once $H(\omega)$ has been estimated, $h(t)$ must be estimated. The inverse Fourier transform will solve this problem, but a number of substantially more complicated algorithms exist that provide better time resolution. Examples of such algorithms include Multiple Signal Classification (MUSIC) and matrix-pencil methods that have been developed for use in imaging and radar systems (Dharamdial, 2003; Song, 2004; Pahlavan, 2002). These algorithms achieve time resolution that is up to ten times better than the Fourier transform method when the SNR is high enough. Due to the narrowband nature of many radios used in WSNs (i.e. IEEE 802.15.4's 2MHz bandwidth), a resolution enhancement of even ten times may be insufficient to provide reasonable accuracy. Once the channel estimate has been made, an additional algorithm to estimate the impact of the estimated channel on a ranging measurement using TOF techniques (i.e. TWTT) can be used. Such an algorithm can include, to some degree, the effect of paths buried inside the estimated channel response, resulting in a good estimate of the range error. This error can be subtracted from the estimated range to achieve a better range estimate.

Summary of Performance Limits

In wireless sensor networks, the devices are resource and power limited, and efforts should be made to reduce the time the radio is active and reduce the amount of signal processing while preserving performance. The above discussions show that signal bandwidth is a system parameter of high importance. Increasing signal bandwidth improves noise and multipath performance linearly with bandwidth. The bandwidth required to achieve very fine resolution in a Gaussian white noise environment is far smaller than that required to achieve equivalent resolution in a typical indoor multipath environment, and the techniques to improve multipath performance are far more computationally intensive than those to combat noise. Many measurements in indoor environments will not have a resolvable direct path using any method or bandwidth, and the resulting range estimate will be highly inaccurate. Localization algorithms must deal gracefully with range measurements that are widely inaccurate some of the time. Methods to deal with other error sources such as synchronization and sampling exist and should be applied to minimize energy while maximizing performance. Although ultra-wideband systems are sure to provide high range resolution, the energy cost of data communication over an ultra-wideband radio remains very high compared to narrowband radios. Therefore, ranging methods that use small bandwidths are critical to many low power wireless networks, and methods to improve range accuracy given fixed, small bandwidths are an unsolved problem.

DEPLOYED SYSTEMS

The localization problem has seen widespread attention in the research community, and a number of RF range based methods have been proposed and implemented. Location information has significant value as represented by the E911 location requirements for cell phones (along with similar requirements around the globe) and the commercial tracking systems available. This section provides a brief survey of ranging techniques that could be applied to wireless sensor systems. Important characteristics of systems are the noise performance, suitability for indoor use, cost, and infrastructure requirements.

Received Signal Strength Range Estimation

The RF received signal strength (RSS) has been used as a surrogate for range measurement in many systems. In free space, the power of an RF signal can be calculated using the Friis transmission formula (Ulaby, 2004).

$$P_{rx} = \frac{P_{tx}}{(4\pi R/\lambda)^2}$$

The received power, P_{rx}, decreases as the range, R, squared, and there is a unique correspondence between RSS and range. Real environments have multipath channels, however, and the signal power does not behave predictably. Other models have been proposed such as the popular $1/R^{\alpha}$ model where the signal power now decreases as one over distance to the power α. Alpha is a fitting parameter that is environmentally dependent and is usually between 1 and 4. These models are unreliable because the power does not decrease monotonically with range and passing through walls causes sudden drops in power over short distances. The multipath environment causes areas of constructive and then deconstructive interference so that a user's location will be uncorrelated with signal power. This effect is frequency dependent as well, so a measurement at one carrier frequency will be uncorrelated with a measurement at another carrier frequency. Even if this multipath effect could be successfully mitigated, the effect of wall and obstacle attenuation prevents this method from providing reliable range estimates (Cheng, 2005). Figure 10 shows a plot of the received signal strength verses distance for an indoor environment. The same signal strength corresponds to more than half of the useful range of the radio, and this accuracy is typical for these systems. Range estimation error, when it can be quantified, is typically proportional to range such that short range measurements may be accurate within a few meters, and longer range measurements are less accurate. RSS measurement for ranging is often considered a near verses far technique that can provide some information regarding proximity but less about true range.

Most radios include a received signal strength indicator (RSSI), and the measurement is available to the user without any additional hardware or power costs which explains the technique's popularity. Because the RF ranging problem is challenging, RSS based techniques have received tremendous attention, and many RSS based localization systems have been proposed and implemented with varying degrees of success at turning poor range estimates into accurate location estimates.

Some RSS based systems use RSS "fingerprinting" techniques rather than RSS for ranging and achieve improved accuracy. The RSS at different carrier frequencies is recorded for many locations in the network through a site survey at the time of network deployment. In normal use the network tries to match the measured RSS of a mobile node with the fingerprint map it has stored to estimate location. Accuracy of these methods can be a few meters, but changes in the environment (an open door that was closed) can cause significant errors in location estimates (Lorincz, 2006).

Figure 10. Received signal strength plotted verses distance. A best fit does not capture the large deviation of data points showing that models with unique correspondence between range and signal power cannot provide reasonable accuracy.

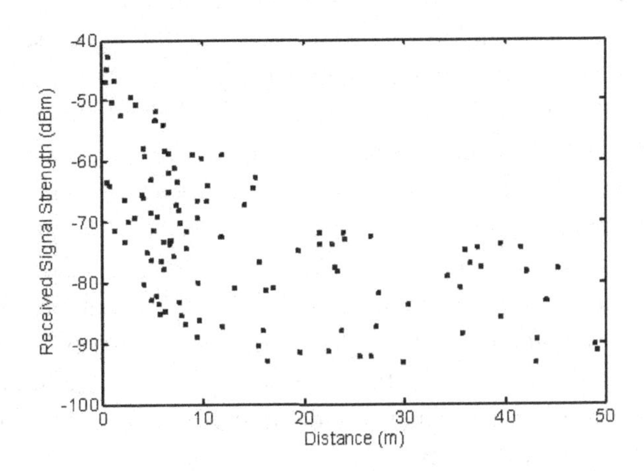

Global Positioning System

The most widespread RF localization scheme is the global positioning system (GPS). A constellation of at least 24 GPS satellites orbit the earth and constantly transmit unique signals. User receivers take four or more of these signals and use them to estimate values for position and time. GPS has a coarse/acquisition signal (C/A code) that is used for civilian uses and to aid in the synchronization to the precise (P) code used by the military. The P code is encrypted to prevent general use, and is called the P(Y) code in the encrypted state. C/A code users generally enjoy location accuracy of better than 10 m on the ground with slightly less accuracy in the vertical direction as long as they have a clear view of the sky without any significant multipath effects. Because the C/A code only occupies about 2MHz of bandwidth at a single carrier frequency, multipath can greatly degrade performance. The received signal power on the ground is extremely low making it difficult to receive the signal when a clear view of the sky is unavailable (Kaplan, 2005). GPS receivers have become much lower power in recent years, but they still consume tens of milli-Joules for the first fix. The SiRFstarIII is a low power receiver with good low received signal power performance and consumes 50 mW typically while taking 5 s to provide a location after power on. With assistance from a cellular phone network, this fix can take 1 s, but now the cellular phone radio will dominate the energy consumption (SiRF, 2008). Given that many WSN applications require location updates at much lower rates than the 1Hz typical of GPS, GPS is far from power optimized. The price of the GPS units is also high due to the complexity associated with signal acquisition and processing.

Time Difference of Arrival

Time difference of arrival (TDOA) is a powerful and commonly used technique that relies on time synchronized infrastructure to estimate the range of a mobile device. The most common scenario is where the mobile device transmits a signal that is simultaneously received by multiple base stations. These base stations estimate the time of arrival of the signal and compare the estimates among multiple stations to estimate the user location. For each receiver pair, a TDOA estimate can be made, and 3 pairs are required to determine a location. As seen in Figure 11, the time difference of arrival at the three base stations is a function only of the unknown distances. When the three measurements are made, a system of three equations with three unknowns results enabling the ranges to be calculated. The primary advantage of this system is that a mobile unit can be very simple because all of the complexity is at the base station. The disadvantage is that time synchronized infrastructure is required which increases cost and complexity of the overall network. Accuracy is linked not only to the environment but to the density of base stations thus requiring large numbers of expensive base stations to cover a network. As a rule of thumb, the density of base stations for TDOA must be four times the density required for data coverage. This technique enables the use of simple mobile devices that can periodically send a ranging signal to be detected by the always-on infrastructure providing a low energy location on schedule or on demand. This technique is not limited to a particular bandwidth or multipath mitigation scheme, and can provide highly accurate or poor performance indoors depending on the implementation. Three commercial systems will be discussed briefly below.

GSM Cellular Phone Networks

Time difference of arrival (TDOA) localization in GSM cellular phone networks has been standardized as part of GSM since 1999 (GSM 03.71 1999 and 2001). In the 1999 version, the handset sends network access packets that are received by three or more base stations which use TDOA to estimate the position of the handset. The 2001 version requires that the handset measure the time difference of arrival of signals sent from base stations. Location accuracy depends heavily on the number of base stations within communication range. In urban environments it is common to be able to communicate

Figure 11. Time difference of arrival (TDOA) uses time synchronized infrastructure nodes (B,C,D) to simultaneously measure the time of arrival of a signal transmitted by A. Because the transmission time, t_{tA}, is unknown, the time differences between arrivals, Δt, can be used to setup three equations to solve for the three unknown ranges (d_{AB}, d_{AC} and d_{AD}).

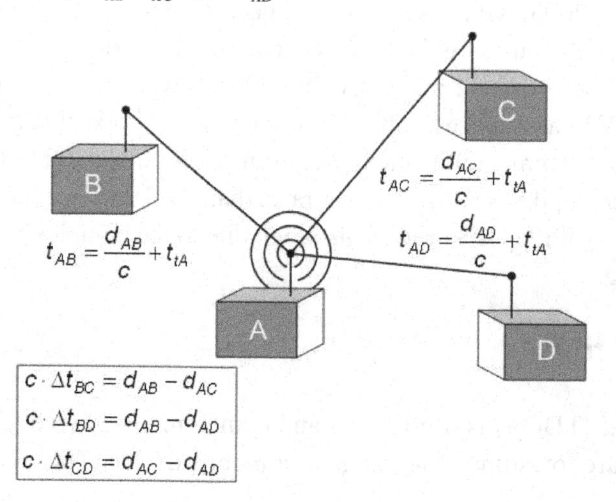

with more than three, and accuracy of better than 100m is common. When only two or less base stations are available, estimates can be many 100's of meters off resulting in an unreliable system. Due to the widespread coverage of cellular systems, cellular based localization can provide location accuracy to within a single building in areas where GPS access is denied such as indoors or in urban canyons. Room level accuracy is not possible with this technology because GSM is a very narrowband system with limited frequency diversity. Performance may improve as wider bandwidth 3G devices become more common and these methods are applied to the newer technologies. The power consumption and cost of cellular radios is very high when compared to the inexpensive and low power radios typically used in WSN (Sahi, 2002).

ANSI 371.1 RTLS (Wherenet, Inc)

ANSI 371.1 is a standard that specifies physical layer requirements and location accuracy for a real time location system (RTLS) that is based on the system developed and marketed by WhereNet, Inc. This is a 2.4 GHz direct sequence spread spectrum (DSSS) based system that consists of time synchronized base stations and low complexity tags that can be programmed to send a signal at regular intervals. The tag location is estimated using time difference of arrival, and the base stations are mounted on either the ceilings of manufacturing facilities or on tall posts for outdoor networks. The tags are programmed to send localization signals at regular intervals, and multi-year lifetimes are achievable when location updates occur every few minutes. The 60 MHz bandwidth localization signal is transmitted in the 2.4 GHz band and contains the tag's ID as well as a small payload that can be filled by user applications to transmit fault conditions or to send other brief messages. In one example deployment over a 280,000 m^2 outdoor facility, an access point was mounted to a post approximately every 90m to ensure localization accuracy of within 3m. This network was deployed just like a traditional wireless LAN starting with a site survey followed by access point installation. A full trial of the system was completed within 75 days of the start of the site survey showing both that deployments are quick on the scale of industrial automation but are also slow and expensive compared to what is expected in the WSN space. With a reported accuracy of 3m and a bandwidth of 60 MHz, it is clear that the accuracy is not limited by noise. Multipath effects common in the industrial environment cause significant accuracy degradation, and a fairly complex system with powerful, centralized base stations is required to provide reasonable performance (Wherenet, 2008)

Ubisense UWB Localization

Ubisense developed and markets an UWB based localization system that combines TDOA and angle of arrival (AoA) measurements to estimate tag location. The tags are equipped with 802.11b transceivers for data communication and proprietary UWB transmitters for localization. The tags are capable of operating at very low duty cycles to enable multi-year battery lifetimes and typically operate by sending UWB signals at regular intervals for localization. The base stations are complicated devices consisting of an array of UWB antennas that are used to estimate the angle of arrival of the UWB signal. These antennas are attached to UWB receivers that precisely estimate the time of arrival of the incoming signal before this information is passed to a central server where the location estimation occurs. In typical deployments, location accuracy of 15cm has been reported, and the UWB signal occupies 2GHz of RF bandwidth. The location accuracy is equal to c/B suggesting that the system bins incoming signals into $1/B$ bins allowing the direct (first) path to be resolved whenever there is a direct path signal. Just as with any ranging system, this excellent accuracy is not achieved when there is no direct path. The site survey process attempts to determine ideal base station positions to reduce the number of locations in which this occurs. Only two base stations must be within range of the tag due to the combination of AoA and TDOA resulting in a lower base station density than would be required otherwise. As with all systems relying on time synchronized and wired infrastructure, the installation process is protracted and expensive (Ubisense, 2008).

Radio Interferometric Positioning System

Radio interferometric positioning system (RIPS) is an idea that uses the effect of interference between RF signals that are closely spaced in frequency to estimate position. This technique is not strictly a ranging technique as discussed in this chapter because a large number of ranges between nodes are estimated simultaneously using many measurements across the network. Four nodes are needed to perform an interferometric measurement under this scheme as shown in Figure 12, and at least 6 nodes are required to achieve network localization. To take a measurement, four loosely time synchronized devices within range of one another negotiate an operation. Two of the devices transmit unmodulated carrier signals that are separated by a very small frequency offset of about 1kHz. The signals interfere at the receivers to generate a signal that has a time varying envelope at the difference frequency. This envelope can be measured by using the RSSI on the radio, and the relative phase difference, ϕ, between the envelopes at the two receivers is recorded. This phase contains information regarding the distance between the four nodes in that $\phi = 2\pi(d_{AB} - d_{BD} + d_{BC} - d_{AC}) / \lambda_{carrier}$. Once enough measurements are collected to fully define the problem, all of the ranges and locations can be calculated simultaneously. This scheme does not require precise time synchronization or significant signal processing, but it does require radios with highly precise control over the transmitted RF carrier frequency limiting its applicability to a small subset of available radios. In an open, outdoor space, RIPS has achieved accuracy of a few centimeters over ranges of many tens of meters. The primary drawback to this system is that it intrinsically relies on the carrier phase to estimate range, and the carrier phase is a strong function of multipath propagation. As a result, the system is largely unusable indoors due to poor accuracy (Maroti, 2005).

Two Way Ranging

In many WSN applications, time synchronized infrastructure is too expensive or time consuming to deploy. As discussed in the clock synchronization section, two way ranging techniques can be implemented to eliminate the effects of unknown time offset, and a few systems using these methods of two way ranging have been proposed.

Two Way Time Transfer

Two way time transfer (TWTT) was first proposed in the 1960's to provide better time synchronization between ground stations using the first communications satellite links, and the method was discussed earlier in this chapter. By performing many TWTT measurements over a period of time, time synchronization and time of flight estimates could be made accurate to within a few nanoseconds. This method has recently been proposed for use in UWB ranging systems where the E_s/N_0 values are low enough

Figure 12. The Radio Interferometric Positioning System uses the interference between two RF Signals closely spaced in frequency to generate a varying envelope that can be measured using the received signal strength indicator in a radio. The phase offset, ϕ ,between two envelopes at two nodes is a function of the unknown distances. Adapted from Maroti (2005).

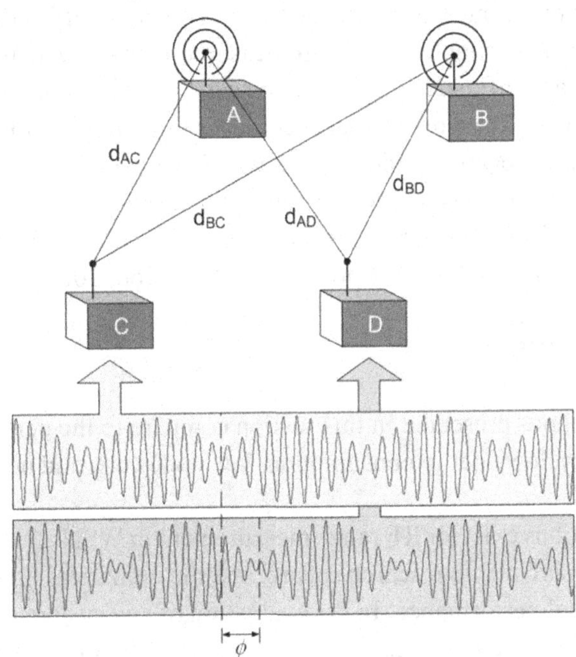

such that matched filter sampling is sufficient to achieve the CRB. Some systems have been published showing accuracy of *c/B*, but no low power systems have been demonstrated. Recent work on low power UWB transceivers have reduced power consumption compared to their high data rate counterparts, but the receivers still consume a great deal of power and/or communicate over just a meter or two of range . Although the accuracy of UWB ranging systems is quite good, the power consumption and complexity of these devices are very high. Narrowband 802.15.4 radios can turn on their radio, transmit a full packet (150B total) tens of meters, receive an acknowledgement and shut down in about 5ms. The radio consumes about 20mW during this time for a total 100μJ of energy per packet. When everything is considered, it is not clear that a viable wireless sensor network UWB communication system will be developed in the near future that can compete with this low energy consumption and reasonable range.

ISO/IEC WD 24730-5 (Nanotron Technologies)

The ranging system implemented by Nanotron Technologies and standardized under ISO/IEC WD 24730-5 uses Chip Spread Spectrum (CSS) over an 80MHz bandwidth in the 2.4GHz band. CSS is a form of linear frequency modulation that can have $t_sB>1$, and chirp pulses have been widely used in radar because they exhibit excellent spectral occupancy and correlation properties (Richards, 2005). They also propose a two way ranging method called Symmetric Double Sided-Two Way Ranging (SDS-TWR) to combat the effects of frequency reference mismatch at the two cooperating nodes. SDS-TWR measures the round trip TOF between two nodes twice. A signal is sent from node A to B back to A, and then a signal is sent from B to A and back to B. The resulting estimates are averaged, and the effect of the reference clock frequency offset is eliminated because the two measurements have the same bias in magnitude but opposite in sign. This method is simple in implementation but it increases the required signal processing because the range must be estimated twice. In order to offset the costs associated with SDS-TWR, ranges can be taken as simple round trip measurements or even as TDOA measurements depending on how the system and infrastructure is configured. Nanotron reports location accuracy for a typical indoor office building to be 2m showing that accuracy is limited by multipath. Because roundtrip measurements are possible, fixed location nodes can be added cheaply to improve location accuracy in difficult areas. Available devices are reasonably low power while providing peer to peer, infrastructure free ranging capability with sufficient accuracy for many applications. The devices are also capable of data communication for a complete RF solution for WSNs (Nanotron, 2008).

Ranging System Comparison

The ranging systems and methods presented in this section compare to the performance and specification wish list of good accuracy, low energy consumption, low node cost, and no infrastructure requirements.

Commercially, limited options exist for RF range measurement in WSNs. Table 2 summarizes available options in both the commercial and research sphere, and the numerical information is provided to give a rough estimate of typical performance. In the evaluation of any system (ranging or localization) accuracy cannot be simply expressed as a single number because estimates are impacted by random environmental parameters resulting in estimates with random error as discussed in this chapter, yet this table provides an approximate comparison of techniques. Most available systems are extremely limited in accuracy or require significant and costly infrastructure, but it is clear that progress has been made in enabling real time localization in wireless networks. Significant research is ongoing in this area to develop adequate, low power ranging systems. This comparison should provide insight into the capabilities, limitations, and challenges of RF ranging systems and show that local area RF ranging is an open problem for research.

SUMMARY

This chapter has provided an overview of the important factors that influence RF ranging system accuracy. A number of techniques and systems designed to address these factors have been presented in order to provide an understanding of issues at hand.

Fundamental Limits

The fundamental limits to performance are the result of noise and finite bandwidth in multipath environments, but accuracy is almost always limited by multipath induced error rather than noise. Indoors, most systems are limited to resolving multiple paths that are spaced by less than $1/(2B)$ in time, but super resolution and sampling aware techniques can improve accuracy to better than $c/(2B)$ in range. Predicting error in multipath environments requires knowledge of the channel impulse response as

Table 2. List of various ranging techniques and approximate values for performance

Method	Class	Outdoor accuracy	Indoor accuracy	Noise performance	Energy Consumption	Node hardware cost	Infrastructure hardware cost
RSSI – uncalibrated	RSSI	5m	10m	Poor	Low	Minimal	None
RSSI - calibrated	RSSI	3m	3m	Poor	Moderate	Minimal	High
GPS	TOA	5m	-	Good	High	High	None
GSM TDOA	TDOA	20m	>100m	Good	High	High	High
Wherenet	TDOA	1m	3m	Good	Moderate	Low	High
RIPS	Interferometric	< 10cm	-	Good	Low	Minimal	None
UWB TWTT	TWTT	10 cm	1m	Good	High	Moderate	None
Nanotron	TWR	1m	2m	Good	Low	Low	None

well as the characteristics of the transmitter and receiver making high performance ranging systems challenging to design.

Narrowband vs. Wideband

Most research on RF ranging has relied on increasing bandwidth as much as possible to obtain reasonable ranging accuracy. Wide bandwidth is a good thing to have to improve accuracy in a variety of environments, but it is not the only solution. Multipath mitigation schemes for narrowband radios have been proposed that enable good accuracy, and it is likely that both wideband and narrowband systems will see broad application. In systems requiring the best accuracy, wider bandwidth is a good choice. For general applications, however, the additional energy costs may limit the application of UWB and other wideband systems. Narrowband radios will likely remain cheaper to design (and therefore purchase), cheaper in energy consumption per packet, and widespread in wireless sensor networks.

Conclusions

Ranging systems for low power systems have just started to be developed, and the field is open for new ideas and improvements. Significant work remains to provide the ideal ranging platform for wireless sensor networks, but some systems can provide reasonable performance for a number of applications. The future promises to provide truly location aware wireless networks, and RF ranging is critical for widespread use.

REFERENCES

Aiello, G. R., & Rogerson, G. D. (2003). Ultra-wideband wireless systems. *IEEE Microwave Magazine*, *4*(2), 36-47.

Anjum, F., Pandey, S., & Agrawal, P. (2005). Secure Localization in Sensor networks using transmission range variation. *Proceedings of the IEEE Mobile Adhoc and Sensor Systems Conference*.

Carter, M., Jin, H., Saunders, M., & Ye, Y. (2006). SpaseLoc: An adaptive subproblem algorithm for scalable wireless sensor network localization. *SIAM Journal on Optimization. 17*(4), 1102-1128.

Cheng , Y., Chawathe, Y., LaMarca, A., & Krumm, J. (2005) Accuracy Characterization for Metropolitan-scale Wi-Fi Localization. *Proceedings of the Third International Conference on Mobile Systems, Applications, and Services*. (pp. 233-245).

Dharamdial, N., Adve, R., & Farha, R. (2003). Multipath Delay Estimations using Matrix Pencil. *Proceedings of the IEEE Wireless Communication and Networking Conference, 1*, 632-635.

Dickey, F., Romero, L., & Doerry, A. (2001). *Superresolution and Synthetic Aperture Radar. Sandia Report SAND2001-1532.* Retrieved March 14, 2008 from Department of Energy Scientific and Technical Information Bridge. Web site: http://www.osti.gov/bridge/purl.cover.jsp?purl=/782711-Y2uIQp/native/

Doherty, L., Pister, K. S. J., & El Ghaoui, L. (2001). Convex position estimation in wireless sensor networks. *Proceedings of the IEEE Conference on Computer Communications, 3*, 1655-1663.

Hoel, P., & Stone, J. (1971). *Introduction to Probability Theory.* Boston: Houghton Mifflin.

Kaplan, E., & Hegarty, C. (2005). *Understanding GPS: Principles and Applications*. Norwood, MA: Artech House Publishers.

Kirchner, D. (1991). Two-way time transfer via communication satellites. *Proceedings of the IEEE, 79*(7), 983-990.

Lanzisera, S., Lin, D., & Pister, K. (2006). RF Time of Flight Ranging for Wireless Sensor Network Localization. *Proceedings of the IEEE Workshop on Intelligent Solutions in Embedded Systems.*

Lanzisera, S., & Pister, K. (2008) Burst Mode Two-way Ranging with Cramér-Rao Bound Noise Performance. *Proceedings of the 2008 IEEE Global Communications Conference.*

Lorincz, K., & Welsh, M. (2006). MoteTrack: A Robust, Decentralized Approach to RF-Based Location Tracking. *Springer Personal and Ubiquitous Computing, Special Issue on Location and Context-Awareness.*

Maroti M., Kusy B., Balogh G., Volgyesi P., Molnar, Karoly, Dora S., & Ledeczi A. (2005). Radio Interferometric Positioning. *Proceedings of the ACM Conference on Embedded Networked Sensor Systems.*

Nanotron Technologies nanoLOC TRX Transceiver (NA5TR1) Datasheet Version 1.03 (2008). Retrieved March 14, 2008 from Nanotron Technologies Web site: http://www.nanotron.com/EN/docs/nanoLOC/DS_nanoLOC_TRX_NA5TR1.pdf

Nefedov, N., & Pukkila, M. (2000). Iterative channel estimation for gprs. *Proceedings of the 11th IEEE International Symposium on Personal, Indoor and Mobile Radio Communications, 2*, 999-1003.

Oppenheim, A., & Schafer, R. (1975). Digital Signal Processing. Englewood Cliffs, N.J.: Prentice-Hall.

Pahlavan, K., Xinrong L., & Makela, J. P. (2002) Indoor Geolocation Science and Technology. *IEEE Communications Magazine, 2002*(2), 112-118.

Richards, M. (2005). *Fundamentals of Radar Signal Processing.* New York: McGraw-Hill.

Sahai, P. (2002). Geolocation on Cellular Networks. In B. Sarikaya (Ed.) *Geographic location in the Internet.* (pp. 13-49). Boston: Kluwer Academic Publishers.

Shah, S., & Tewfik, A. (2005). Enhanced Position Location With UWB In Obstructed Los And NLOS Multipath Environments. *Proceedings of the XIII European Signal Processing Conference.*

SiRFStarIII Product Insert (2008). Retrieved March 14, 2008 from SiRF Technologies Web site: http://www.sirf.com/products/GSC3LPProductInsert.pdf

Song, L., Adve, R., & Hatzinakos, D. (2004). Matrix pencil positioning in wireless ad hoc sensor networks. *Proceedings of First European Workshop on Wireless Sensor Networks,* (pp. 18-27).

Spencer, Q., Jeffs, B., Jensen, M., Swindlehurst, A. (2000). Modeling the statistical time and angle of arrival characteristics of an indoor multipath channel. *IEEE Journal on Selected Areas in Communications, 18*(3), 347-360.

Tse, D., & Viswanath, P. (2005). *Fundamentals of Wireless Communication.* Cambridge, UK: Cambridge University Press.

Ubisense Limited (2008). *Ubisense System Overview.* Retrieved July 29, 2008 from Ubisense Limited Web site: http://www.ubisense.net/media/pdf/Ubisense%20System%20Overview%20V1.1.pdf

Ulaby, F. (1999). *Fundamentals of Applied Electromagnetics.* Upper Saddle River, NJ: Prentice Hall.

Werb, J., Newman, M. Berry, V., & Lamb, S. (2005). Improved Quality of Service in IEEE 802.15.4 Mesh Networks. *Proceedings of International Workshop on Wireless and Industrial Automation.*

Wherenet, (2008). *NYK Logistics Case Study.* Retrieved July 29, 2008 from Wherenet Web site: http://www.wherenet.com/NYKLogisticsCaseStudy.shtml

Chapter V
Calibration and Measurement of Signal Strength for Sensor Localization

Neal Patwari
University of Utah, USA

Piyush Agrawal
University of Utah, USA

ABSTRACT

A number of practical issues are involved in the use of measured received signal strength (RSS) for purposes of localization. This chapter focuses on device effects and modeling problems which are not well covered in the literature, such as transceiver device manufacturing variations, battery effects on transmit power, nonlinearities in RSSI circuits, and path loss model parameter estimation. The authors discuss both the negative impacts of these effects and inaccuracies, and adaptations used by particular localization algorithms to be robust to them, without discussing any algorithm in detail. The authors present measurement methodologies to characterize these effects for wireless sensor nodes, and report the results from several calibration experiments to quantify each discussed effect and modeling issue.

INTRODUCTION

Signal-strength based localization can be deceptively simple. Receivers are generally capable of measuring and reporting to higher layers information about received signal strength, so it can seem like it should be easy to take these measurements and use them directly in a localization algorithm. Significant research has developed algorithms for localization, assuming that measured signal strength has already been converted into distance estimates, and little research discusses the details of how to perform those conversions.

Multipath fading in radio channels is universally regarded to be the main degradation to RSS-based location estimates, and rightfully so – significant shadowing and small-scale fading caused by the channel is largely unavoidable and unpredictable. Beyond that, however, there can be severe degradations caused by a lack of understanding of the non-idealities of the measurement process, and inaccurate knowledge of channel parameters. If RSS-based localization is to be attempted, a designer must be able to characterize and cope with these non-idealities and imperfect knowledge.

This chapter is written to present real-world calibration and non-linearity problems in RSS measurements and how to deal with them. We follow RSS-based localization from the transmitter to the receiver, and in multiple stages in the receiver, as shown in Figure 1. The intended audience is anyone who intends to implement or has already implemented RSS-based localization algorithms which are to operate well in real-world deployments. We present our work in RSS-based localization algorithms only briefly. We have found that the experience of accurately using measured signal strength, in general, is as challenging as the localization algorithm itself.

Our chapter is organized as follows. First, in section PROPAGATION EFFECTS, we relate some of the literature on path loss models as a function of distance and the effects of shadowing and multipath fading. Then, in section DEVICE EFFECTS, we discuss a method for accurately characterizing transmit power as a function of device settings and battery voltage, and receiver RSSI values as a function of the particular device performing the measurement. The transmit power characteristics are necessary to translate measured RSS into accurate path loss values. The receiver characterization reveals the details of the nonlinearities in measured RSSI. Then, section CHANNEL EXPERIMENTS WITH POWER CONTROL describes a protocol and algorithm for transmit power control, to avoid RSSI saturation without sacrificing node connectivity. Thirdly, the section titled RANGING USING MEASURED PATH LOSS discusses the conversion of path loss, calculated using the results of the DEVICE EFFECTS section, into an estimate of range. Finally, a section called RSS-BASED LOCALIZATION ALGORITHMS discusses how the lessons discussed in this chapter apply in our RSS-based location algorithm implementation.

Figure 1. RSS-based localization requires characterization of both the transmitter and receiver, and the ability to convert measured RSSI into path loss prior to input into a localization algorithm. Path loss estimation requires knowing transmitter parameters and may require feedback to the transmitter to control its transmit power.

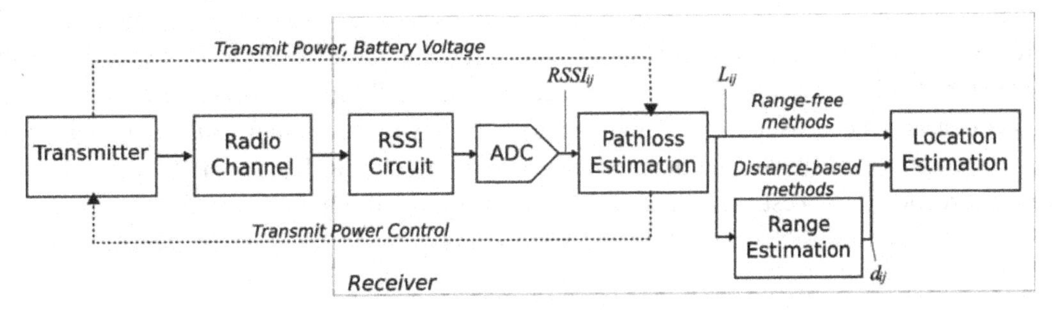

PROPAGATION EFFECTS

Multipath fading, shadowing, and antenna effects cause great variations in the measured RSS in real-world environments, degrading its ability to produce accurate distance or position estimates. This section serves to emphasize these well-reported effects in order to position the importance of studying device effects on measured RSS. As we delve deeper into device effects which cause RSS errors, we will be able to position them in context to the larger problem of RSS-based position estimation.

Path losses, on average, increase with distance – the increase is due to "large-scale" path loss (Hashemi, 1993), which are proportional to $10 n_p \log_{10} d$, where n_p is a path loss exponent, and d is the path length. But the path loss between two radios at particular positions is very much a function of the objects in the environment between them and the position and orientation of the antennas. Movement on the order of centimeters or changing channel from one frequency to another can cause dramatic path loss differences because of "multipath fading" or "small-scale fading".

Indoor environment propagation effects are particularly a problem for localization. As defined in Pahlavan, Li, and Makela (2002), when the line-of-sight path arrives with more power than any other multipath, it is called a dominant line-of-sight (DLOS) link. In indoor environments, only receivers in a small area around the transmitter are found to be likely to be DLOS. Most areas indoors have an LOS path shadowed by walls and objects; this shadowing decreases the received power, and the RSS becomes dominated mostly by multipath power from many different directions. Besides being impacted by shadowing, these situations are more strongly affected by small-scale fading because many multipath signals contribute to the received signal. When more multipath signals arrive from more directions, the statistics of small-scale fading change from Ricean to Raleigh and thus become more severe (Rappaport, 1996). Further, as we will discuss in more detail in the "Distance Estimation Equations" section, the errors in distance estimates are multiplicative; thus are more severe at longer path lengths. In short, the fading problems increase as the distance between the transmitter and receiver increases.

It is important to note that shadowing is not solely a degradation for RSS-based localization systems. RSS fingerprinting algorithms (as described in the "RF Fingerprinting Algorithms" section) take advantage of the fact that shadow fading is a feature-rich, spatially-correlated random field which is mostly stationary over time. By recording this field and using it to match measurements to a location, RF fingerprint-based algorithms exploit location-specific shadowing variations to benefit localization. In such systems, it is only small-scale fading and the change in shadow fading which degrade localization algorithms.

Errors in RSS measurements also come from the fact that real-world antennas have directionality and the orientation of a node is not known a priori. This is exacerbated by objects to which devices are attached – anything metal or mostly water (people) would block RF propogation in its direction, attenuating the signal by as much as 15 dB. King et. al. (2006) have found that these antenna orientation issues are important to measure in calibration of RSS fingerprint-based localization systems, as will be discussed in the "RF Fingerprinting Algorithms" section. Generally, multipath arrive at a receiver from many directions, and the measured RSS will depend on the powers of the multipath which arrive in the same direction as the directionality of the antenna. Algorithms such as King et. al. (2006) actually use the antenna directionality to estimate the device orientation in addition to its position.

DEVICE EFFECTS

Simple wireless devices can measure and report a quantized measurement of the received signal strength (RSS) of a received packet. This measurement of RSS is typically referred to as the received signal strength indicator (RSSI). For localization purposes, we actually require the path loss, that is, the actual dB loss experienced between transmitter and receiver antennas, which is related to distance between the two antennas. It is not trivial to convert RSSI into path loss. In this section, we show how wireless nodes can be calibrated so that RSSI measurements can be converted into path losses.

Path loss L_{ij}, in dB, on a link between transmitter j and receiver i is defined here to be the difference between the dBm transmit power P_T and the dBm received power P_R,

$$L_{ij} = P_T - P_R$$

In this section, we detail how transmit power and received power are functions of the device characteristics, parameter settings, and battery voltage.

This section does not propose that all nodes should be characterized and calibrated for purposes of system deployment. However, it is often important to calibrate one set of nodes for research and development purposes. It is critical to know how much devices vary and thus how much is lost when devices are not calibrated, even if we have no intention of deploying systems of calibrated nodes.

We intend to characterize two device characteristics that make measured RSSI vary:

- Transmit power device variations
- Transmit power battery variations
- Receiver RSSI circuit device variations

We show calibration measurements for the Crossbow Mica2 sensor in this chapter, but other wireless sensor modules can also be calibrated using this basic procedure.

Transmit Power Device Variations

Typically, the transmit power of a radio IC can be set among a discrete set of possible power levels. For wireless sensors, the need to conserve energy makes this a key requirement for the radio IC. When nodes are communicating at short range, it will conserve energy to transmit at a lower power. In addition, in interference-limited networks, reducing transmit power to the minimum required level helps reduce interference and allow higher communications rates. When nodes are transmitting at different power levels, it will be critical to know exactly what transmit power is being sent so that path loss L_{ij} (above) can be computed. And, as we will show, transmit powers may vary from device to device even for the same nominal "power level" and battery voltage, due to manufacturing variations.

Transmit power differences between nodes are to blame for many of the asymmetric links in sensor networks, that is, when node i can be received at node j, but node j cannot be received at node i. Extensive analysis is reported by Zuniga and Krishnamachari (2007). Similar to the communications case, an asymmetric link can cause difficulties for a sensor localization algorithm. Some localization algorithms would not collect data on an asymmetric links because of the lack of a bidirectional communications link – data collection and exchange protocols could fail on that link. Zuniga and Krishnamachari (2007) report a variance of transmit power of 6 $(dB)^2$. This variation includes the effects of battery variations and device variations.

To provide a controlled measurement-based quantification of transmit power variation, we present measurements of a set of Crossbow Mica2 nodes, which use the TI/ChipCon CC1000 radio IC. Our calibration experiment is shown in the block diagram in Figure 2. The calibration procedure can be described in the following steps:

- Each node is powered from a variable DC power supply set to 3.0 V so that we can study the transmit power variation separately from battery variation.
- Each node is programmed to periodically transmit data packets at a specified transmit power code.
- A node is RF shielded by placing it in an aluminum box, and its RF output port is connected via RF/coaxial cable to a spectrum analyzer (Agilent E4405B) in peak power mode, as shown in Figure 2.
- While connected, the transmit power code of the node is changed from one to the next and the measured transmit power is recorded for each.

Table 1 reports the transmit power values (in dB) for the different transmit power codes for the particular device under test. In the experimental process, we kept the transmit frequency at 903 MHz.

We have characterized multiple transmitters in order to consider the variations between devices. The transmit power calibration values for sixteen different Mica2 nodes are plotted in Figure 3 as a function of transmit power code. There is noticeably higher variance in transmitted power between devices at low transmit powers. The largest variation is at the lowest transmit power level, for which transmit power varies from -29.1 dBm to -32.7 dBm, a difference of 3.6 dB. But, over all transmit power levels, the standard deviation of the transmit power ranges from 0.15 to 0.91 dB, for an overall RMS standard deviation of 0.45 dB. On the whole, we have that the difference between the average and the actual transmit power for a given device is random with standard deviation 0.45 dB.

Transmit Power Variation with Battery Voltage

However, this is the device variation, without considering the battery voltage. As time passes, the battery power is drained, and the voltage produced by the battery reduces. At some point, when the battery voltage becomes too low, the sensor, or the radio will fail. But before failure, changes in battery voltage

Figure 2. Block diagram for transmitter calibration

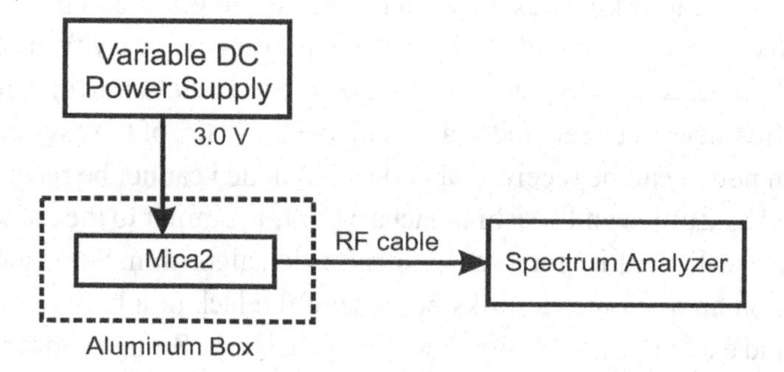

Table 1. Transmit power vs. transmit power code for Mica2 device under test

Power Code (hex)	Output Power (dBm)	Power Code (hex)	Output Power (dBm)	Power Code (hex)	Output Power (dBm)
0x02	-31.00	**0x0B**	-14.71	**0x70**	-5.23
0x03	-26.56	**0x0C**	-13.30	**0x80**	-3.94
0x04	-23.84	**0x0D**	-12.64	**0x90**	-3.09
0x05	-21.56	**0x0F**	-11.22	**0xB0**	-1.42
0x06	-19.92	**0x40**	-9.99	**0xC0**	-0.94
0x07	-18.33	**0x50**	-8.08	**0xF0**	-0.37
0x08	-16.97	**0x60**	-6.49	**0xFF**	-0.03
0x09	-15.94				

Figure 3. Transmit power in dB vs. transmit power code for 16 different Mica2 devices

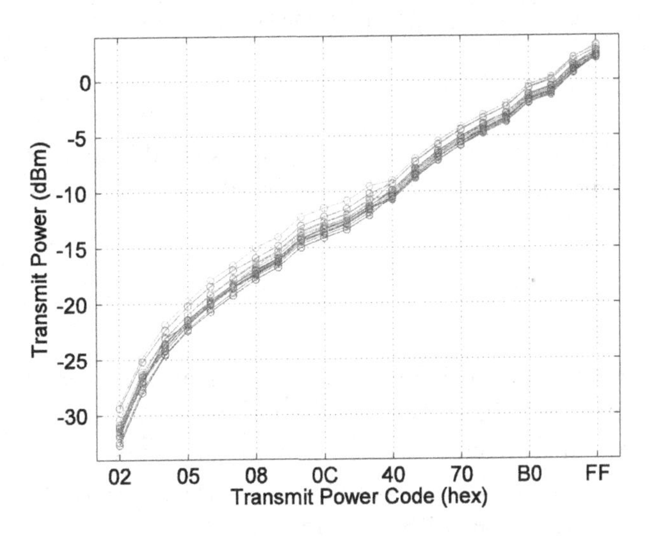

affect the transmitted power. In fact, the power that a transmit amplifier can produce is proportional to the square of the battery voltage. An amplifier has an efficiency, which relates to how much of the battery power can be converted to RF transmit power. In general, we can consider the transmit power, P_T, as a function of battery voltage V_{batt},

$$P_T(V_{batt}) = P_T(V_0) + \alpha 20 \log_{10}(V_{batt}/V_0) \quad (1)$$

where $P_T(V_0)$ is the transmit power measured at a reference voltage, V_0, and α is an efficiency constant. For our measurements, we used a reference voltage of $V_0 = 3.0$ Volts.

To characterize the transmit power as a function of battery voltage, we tested multiple Mica2 nodes using a modification of the test procedure shown Figure 2, in which we changed the DC voltage provided by the variable DC power supply in the range of about 2.4 to 3.3 Volts. At each tested DC voltage, we recorded the transmit power on the spectrum analyzer. The results are shown in Figure 4 for two of the nodes. For the linear model above, the value of α is calculated to be 0.67. This value is

Figure 4. Transmit power in dBm as a function of the battery voltage. The solid line indicates the best fit to the linear model, with α = 0.67.

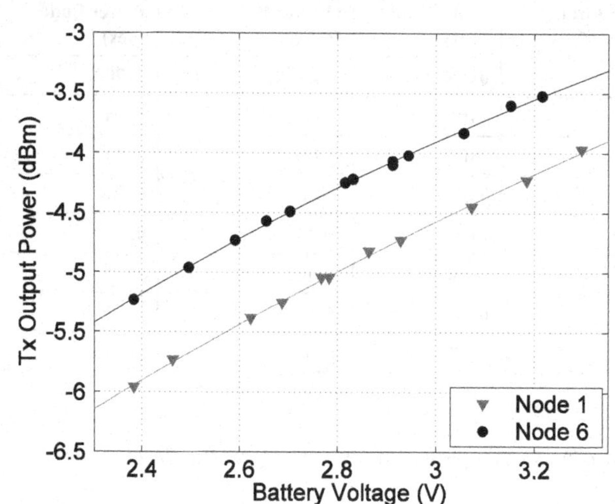

largely constant across nodes. So, while the value of $P_T(V_0)$ is not the same between nodes, the slope of the transmit power as a function of battery voltage near 3.0 Volts may be approximated as identical between nodes.

Discussion

Since devices do vary, even with identical devices, it is clear that in a large-scale sensor localization system, device transmit powers will differ from their stated, nominal, values. An important question is, how are localization algorithms affected by these random transmit power variations? In the "Non-parametric Localization Algorithms" section, we discuss one type of algorithm which can be robust to transmit power variations, due to its use of RSS differences.

We also mention existing analytical analysis reported by Patwari and Hero (2006) which reports the Bayesian Cramèr-Rao bound (CRB) on coordinate estimator variance when the transmit powers are random. The bound provided by the Bayesian CRB is useful because it is independent of algorithm; any estimator will have root mean-squared error (RMSE) greater than its bound. In this analysis, the transmit powers of each device was assumed to be random, and Gaussian, with a given standard deviation. The increase of the bound is reported to describe the performance degradation caused by not knowing the exact transmit power. From the example numerical results, for a standard deviation of transmit power of 1 dB, the bound on RMSE increases by 1-3% (depending on the setup). However, a key finding of the bound analysis is that the two bi-directional measurements on a link (both L_{ij} and L_{ji}) must both be used in the localization algorithm; it is not sufficient to average the two together, $\frac{1}{2}(L_{ij} + L_{ji})$, and use only that average. When using only the link bi-directional average, an algorithm is no longer able to adaptively estimate node transmit powers, and the RMSE bound increases much more dramatically, as much as 25% when the standard deviation of transmit power is 5 dB. In sum, the impact of transmit power variations can be severe, but only when the standard deviation of transmit power is high. Algorithms could be designed to adaptively estimate each node's deviation from its nominal transmit power and then adjust the coordinate estimate as a result. For example, use the difference values, $L_{ij} - L_{ji}$, for all j, to estimate the deviation of the transmit power of node i from its nominal value.

Receiver RSSI Circuit Device Variations

Assuming that calibrations described in past two sections are complete, we know very accurately the transmit power of a calibrated node, given its transmit power code and battery voltage. Now, we can proceed to use a fully characterized transmitter to calibrate a receiver's RSSI characteristic. This process will allow us to accurately translate the measured RSSI in Volts to the actual path loss in the channel.

We show in this section that wireless sensors report an RSSI which has a non-linear relationship with received power. For some range of RSSI, we can make a linear approximation, but we then must be careful when using signal strength measurements that the receiver is operating in a linear region of RSSI. In this section, we report characterization of the Crossbow Mica2 mote, which uses the CC1000 radio IC, but we have also observed the non-linear characteristic on the Crossbow TelosB mote (with a CC2420). We believe that, in general, inexpensive radios are not designed to achieve linear measurements of received power, because communications applications generally do not require it.

In our measurements, we input a known received power in dBm into a device and record the RSSI value in Volts. We do this using the experimental setup in Figure 5, designed to provide a known received power, and to remove the effects of RF leakage. The mote 1 in Figure 5 is placed in a RF shielded box and programmed to transmit with known transmit power code (0x0F) and known battery voltage (3.0V) from a power supply. From the results presented already, we know that this node transmits -11.22 dBm with these settings (from Table 1). We choose from among several SMA-connectorized attenuators, with a range of attenuation values, to achieve the desired link loss. Inside a second shielded box we place mote 2, which has the receiver under test, and is connected to the output of the attenuators via an SMA to MMCX cable. A third Mica2 node is programmed to receive the measured data from mote 2 and to communicate it to a laptop.

Figure 6 shows the recorded RSSI values from the experiment for a wide range of received power values. The results show two critical lessons for the use of measured signal strength:

- **At high received power (-50 dBm and above) the RSSI values saturate**, and do not have a linear relationship with the actual received power. Although it is very possible for two proximate

Figure 5. Block diagram for receiver calibration

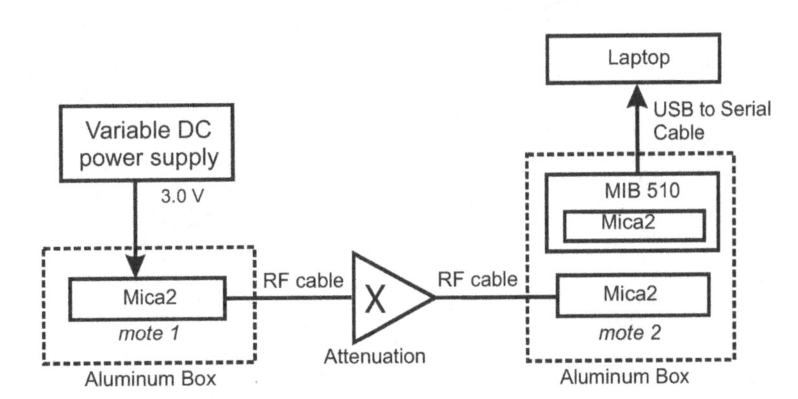

nodes to record a received power at or above -50 dBm, it is not useful for the purpose of accurately measuring link path loss.

- **At power levels below -90 dBm, packets are not correctly demodulated.** Thus path loss integers are not reported for received power levels below -90 dBm. However, we note that in the above calibration setup, if RF shielding is not done properly, and RF power leakage is allowed to circumvent the attenuators (possibly from poorly connected, or low quality cables), you would receive packets, and for this node, the RSSI would be around 0.85 V. If a second non-linear region exists at the bottom right (at low received power) of Figure 6, then it is likely that RF leakage is the problem.

For the values in the linear range (-50 to -90 dBm) we calculate a linear fit. This linear fit for this node's receiver is given by

$$P_R(RSSI) = -50.6RSSI - 44.36 \text{ [dBm]} \tag{2}$$

where RSSI is the path loss integer reported by the receiver, and P_R is the actual received power. This measured result is slightly different from the result listed on the CC1000 data sheet, which had reported that the $P_R = -51.3$ RSSI $- 49.2$ in dBm (as cited in Whitehouse, Karlof, and Culler, 2007). We do not know how the data sheet formula was determined, so our best guess regarding the 5-6 dB offset is that it is due to RF front end differences prior to the CC1000 radio IC in the tested device (the Mica2) compared to the device used in the CC1000 manufacturer calibration tests.

We note also that other research has been conducted to characterize receiver differences for communications applications. Zuniga and Krishnamachari (2007) also included characterization of receiver noise floor. The analysis uses the noise floor power value in order to determine whether or not a packet received at a given power can be demodulated or not. Such a model only requires one parameter, the noise floor, while our analysis for localization purposes must characterize the function of received

Figure 6. Measured RSSI value vs. actual received power for a tested Mica2 node, and a linear fit for the data points which do not experience saturation

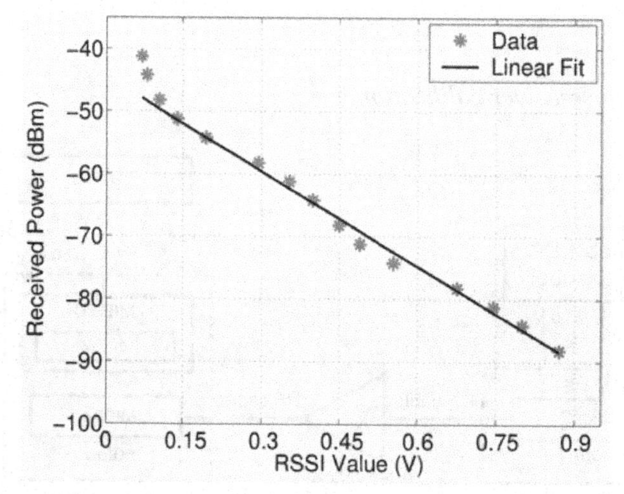

power vs. RSSI value. Zuniga and Krishnamachari (2007) find that this receiver noise floor value has a variance of 3 $(\text{dB})^2$.

In the "linear" region of the measured characteristic of Figure 6, there is some small non-linearity in the Mica2 RSSI characteristic. The actual received power in the 0.25 V < RSSI < 0.35 V range is slightly higher than the linear fit, while the actual received power in the 0.45 V < RSSI < 0.6 V range is slightly lower than the linear fit. We find this consistent across Mica2 nodes. We believe this is specific to the Mica2 implementation, but that other radio ICs will also see small non-linearities. The standard deviation of error in the linear fit, for the data recorded in the linear region, is about 1.1 dB.

Discussion

Point (1.) above deserves extra emphasis, because it indicates a key requirement for use of signal strength between nearby nodes. Intuitively, one might expect that recorded RSSI is most accurate when there are short distances between nodes. However, this is not true if the nodes are transmitting at high power levels. In fact, if nodes are to be deployed densely, nodes must turn down their transmit power! This is counter-intuitive, since lower transmit power results in lower signal to noise ratio, and shorter range. However, for purposes of path loss measurement and path loss-based ranging, very high received powers saturate the receiver and make the measured RSSI uninformative.

Nodes should not turn down their power at the expense of node connectivity. The large-scale experiments in Whitehouse, Karlof, and Culler (2007) show that localization performance degrades quickly when the connectivity degrades. By lowering the transmit power or by lowering the antennas closer to the ground (which increases path losses) the experiments reduce connectivity and show that an algorithm's ability to localize is quickly lost.

Point (2.) also is critical to understand why observed received powers are sometimes higher than predicted by a path loss model at long distances. This saturation is a "observation bias": we can only measure received power if it is above a threshold (about -90 dBm in Figure 6). Assume that at a given distance, our best propagation model says that the average received power should be below the threshold. But we can only measure received powers above the threshold. Thus when we report an average experimental received power for nodes separated by that distance, it will be above the threshold. This effect introduces bias into localization algorithms (Costa, Patwari, and Hero, 2006).

Summary

We have used calibration in two ways: to enable high accuracy path loss measurements, and to quantify the errors caused by uncalibrated nodes. Transmitter power outputs variations have a standard deviation from 0.15 to 0.91 dB at the highest and lowest transmit power level, respectively. Transmit power levels change about 1 dB between new batteries to dead batteries. Receiver RSSI can be used to indicate received power, and as long as the RSSI is in the linear region, we can expect it to contribute 1 dB of standard deviation to our measurement. In total, since variances add, the overall standard deviation of error due to device non-characterization could be as little as 1.4 to 1.7 dB. This, again, assumes that the average device characteristic is well known, and that receivers are operating in the linear region.

This standard deviation of error may have minimal impact on some common applications. For example, for ranging, the multipath fading and shadowing variances will far outweigh this non-characterization error variance. Thus we do not suggest device calibration as a necessary procedure for all applications. However, we must ensure that received powers are not in the saturation range of the receiver.

CHANNEL EXPERIMENTS WITH POWER CONTROL

One main lesson learned from the device calibration work is that accurate RSS-based ranging and localization should take place when the received power does not reach either of two extremes. The received power must be high enough to demodulate packets, but not so high power as to place them in their saturation region. In the first extreme, if the RSSI of all neighbors' packets are measured to below about .13 V, it will be very difficult to distinguish the actual received power (in particular when device variations are taken into account), and all neighbors will seem equally close to the receiver. At the other extreme, if neighbors' packets arrive with power near or below the receiver threshold, then we are missing packets from neighbors, and we will not do as good of a job in localization as we could.

To some extent, these problems can be controlled by changing the transmit power. The Mica2 node has a transmit power range of about 30 dB, and in general, wireless sensors have the ability to change their transmit power over a wide range. By lowering the transmit power to its minimum it is possible to completely avoid the first extreme, in which received powers are too high. We have run many localization experiments with a node density of 10-20 nodes per square meter with the transmit power set to its lowest power setting, and see good results from our RSS-based localization algorithm, described in Patwari, Agrawal, and Hero (2006). However, when transmit powers are set higher, measured RSSI values are nearly identical, regardless of path length. In the second extreme, with the transmit power set to its highest, we can run localization experiments with long distances of 10-20 m between nodes, as long as each node is within communication range of a few other nodes.

However, we cannot limit WSN localization to situations in which local node densities are well known prior to deployment. In this section, we introduce a simple closed-loop power control protocol and algorithm to automatically avoid the two extremes of low and high received powers.

Protocol

In order to obtain feedback regarding its transmit power, a node must learn from its neighbors. Given that node i can communicate with K nodes, $j_1, j_2, \ldots, j_K$, we denote the neighbor set of node i as $H_i = \{ j_1, j_2, \ldots, j_K \}$. In this power control protocol, node i transmits feedback to each node $j \in H_i$ regarding the RSSI node i measured during reception of packets from j. We denote this as $RSSI_{ji}$, and it is the measured RSSI at node i for the packet transmitted by node j. With this notation, the "power control packet" transmitted by node i has contents shown in Figure 7.

Each node, each round, transmits one power control packet to provide feedback to its K nearest neighbors regarding its measured RSSI values. In our experiments, we use K = 12.

Figure 7. Power control packet transmitted by node i

	1 byte	2 bytes		2 bytes			2 bytes	
Header	i	j_1	$RSSI_{j_1,i}$	j_2	$RSSI_{j_2,i}$	••••	j_K	$RSSI_{j_K,i}$

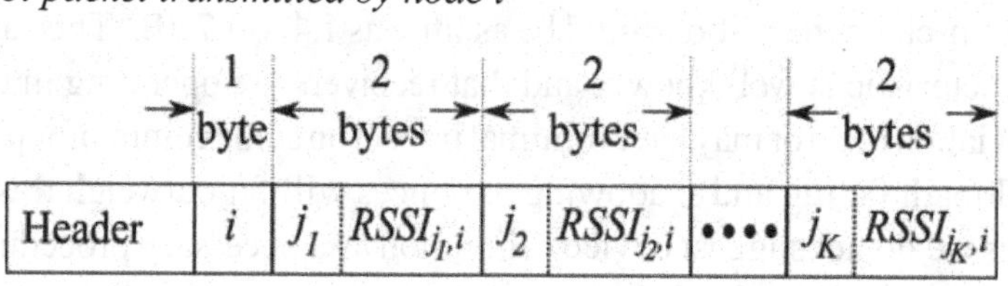

Algorithm

When node i receives any power control packet, it searches in the list of j's neighbors, $i_1, i_2, \ldots, i_K$, to find its own node number. If i is a neighbor of j, then it records $RSSI_{j,i}$ as an "incoming RSSI" from node j. It must collect each incoming RSSI value from the power control packet of one of its neighbors. To aggregate them, node i then averages all RSSI values on record to calculate its mean incoming RSSI. This value summarizes the feedback received from all of its neighbors.

The decision about the transmit power, in this algorithm, is solely a threshold test on the mean incoming RSSI. We have two rules:

- If the mean incoming RSSI is less than 0.25 V, reduce the transmit power.
- If the mean incoming RSSI is greater than 0.6 V, increase the transmit power.

The transmit power is raised or lowered by one transmit power code at a time. Note that at low or very high RSSI, we cannot predict how changing the transmit power code will impact the mean incoming RSSI, because of the nonlinear effects of either RSSI saturation or nodes being out-of-range. Thus we do not believe that an optimal transmit power can be directly computed.

Results

We implemented the above transmit power control algorithm and ran it to show that the above simple algorithm is able to adapt the transmit powers in a sensor network to an appropriate level. In an experiment, we program a group of 16 nodes to run the power control algorithm described above. This program initializes at a transmit power code 0x0F, which from Table 1, corresponds to a transmit power of -11 dBm. Then, we place the 16 nodes on a carpeted floor in a 2 m by 2 m square area. These nodes are at a high density, so many of the received powers are initially higher than -50 dBm, within the non-linear RSSI region we wish to avoid. Clearly, the transmit powers should be reduced.

We watch how the sensors respond and change their transmit power code at each iteration, or "sequence number", of the algorithm. Nodes transmit packets with a sequence number which is incremented after each transmission. Note also that nodes use a slow frequency hopping, so the frequency changes with

Figure 8. (a) Transmit power code and (b) mean incoming RSSI vs. sequence number, recorded by nodes one and two during the power control experiment

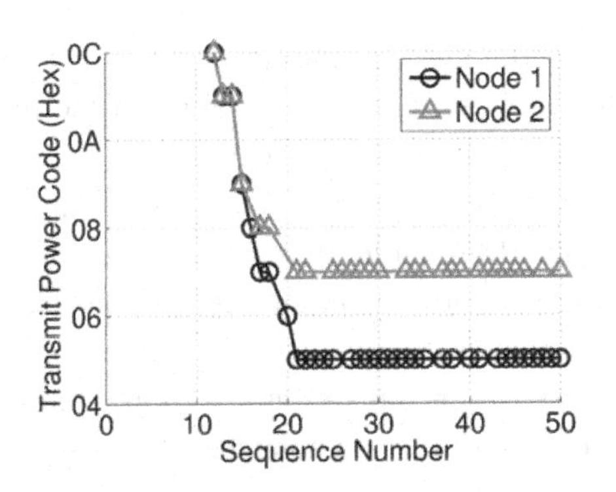

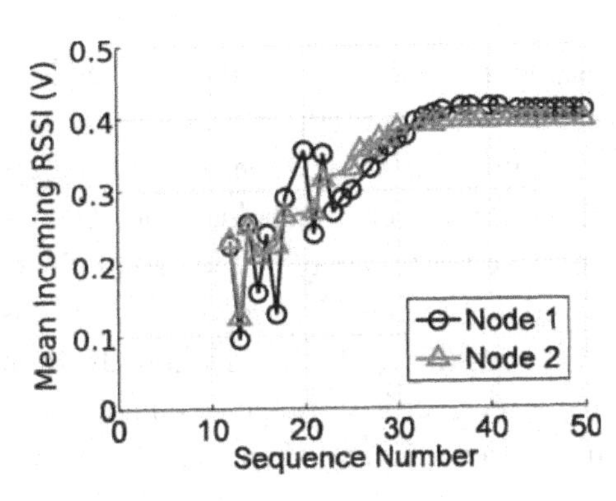

each transmission. Our base unit does not synchronize fast enough to receive packets in the first few sequence numbers. First, we observe in Figure 8(b) how the mean incoming RSSI value is initially as low as 0.1 V. While this value is below 0.25 V, the transmit power decrements each sequence number. The mean incoming RSSI is noisy during this time, because it is re-initialized each time the transmit power is changed. During each time period, it may hear from a large number of neighbors, but since packet reception rates are approximately in the 70-80% range, we will hear from a random subset each time, which leads to a large variation in the mean incoming RSSI over time. There is also a noticeable delay in the control loop, which leads to the asymptotic mean incoming RSSI approaching a value around 0.4 V, rather than the threshold of 0.25 V. This is due to the latency introduced by frequency hopping.

Summary

One major difficulty with RSS-based ranging is the nonlinear effect at close range, and the lack of connectivity at low node densities. Instead of a constant transmit power, we use an adaptive power control algorithm to set the transmit power to an acceptable value. Assuming that we have performed the transmit calibration procedure (discussed in the earlier section) on a few of the nodes, we will be able to program the nodes to compute path loss from the RSSI and the known transmit powers of its neighbors.

RANGE-FREE LOCALIZATION FROM MEASURED PATH LOSS

Up until this point, we have only talked about accurately computing measured path losses on links. Much of the sensor network localization algorithm literature deals with range-based localization algorithms, which require converting path loss measurements into a distance estimate prior to estimating the node coordinates. However, there is a significant interest in localization without such conversion; these "range-free" algorithms can provide useful characteristics compared to range-based localization.

In this section, we discuss three types of algorithms which do not require distance estimates. These three types are:

- Connectivity-based localization
- Non-parametric RSS-based localization
- RF fingerprint-based localization

Connectivity-Based Localization

Localization methods which use only whether or not two nodes can communicate are called proximity or connectivity-based methods. Connectivity is effectively a binary quantization of the received power, since digital communications receivers are largely unable to receive packets when the received power is below a receiver threshold. Thus two nodes are either "in-range" or "out-of-range". The threshold between the two may be the actual physical limits of the radio, or may be set to some other pre-determined RSSI value. Range-free algorithms are excellent low-maintenance and low-setup cost localization systems for applications where the highest accuracy is not required. In some applications, knowing that a receiver is in-range of one or more known-location transmitters is more than enough location information. Several algorithms have been proposed for use with connectivity, including Bulusu, Heidemann, and Estrin (2000), and Niculescu and Nath (2001). They range from simply finding the centroid of the coordinates of in-range nodes (Bulusu et. al., 2000) to using the shortest-path hop count as a distance metric (Niculescu and Nath, 2001).

The main problem with range-free localization is that quite a bit of information is lost by quantizing the RSSI into one bit. In our theoretical analysis, we found that we should expect at least a 50% increase in standard deviation of localization error compared to using unquantized RSSI values, even if the threshold is set optimally (Patwari and Hero, 2003).

Non-Parametric Localization Algorithms

Non-parametric RSS-based localization algorithms directly use RSSI measurements. Ecolocation (Yedavalli, Krishnamachari, Ravulla, and Srinivasan, 2005) and ROCRSSI (Liu, Wu, and He 2004) are algorithms which use the order information of the path loss values at node i, when the path loss values are sorted from smallest to largest. Then, these algorithms constrain the distance between node i and its closest neighbor to be less than the distance to its 2nd closest neighbor, which must be less than the distance to its 3rd closest neighbor, and so on. These constraints graphically imply concentric circles, for the closest to the furthest neighbor. The Ecolocation algorithm finds a region for each sensor which best meets the simultaneous constraints imposed by all the neighbor distance orderings. However, the full solution can be computationally complex because of the large number of simultaneous constraints, and is a centralized algorithm. The APIT method of He, Huang, Blum, Stankovic, and Abdelzahar (2003) similarly reduces the area of possible location, in its case, by testing each set of three nodes to see whether the device-to-be-located is within, or outside of, the triangular area formed by the three nodes. The APIT triangle test requires only a comparison of path loss measurements between neighboring nodes, but it assumes a relatively high density of anchor nodes (or anchor nodes with high transmit power) compared to the APS method of Niculescu and Nath (2001).

RF Fingerprinting Algorithms

In RF fingerprint-based algorithms, the RSS between a node and many fixed access points are recorded, and used together as a vector "fingerprint" of the location of the node (King et. al., 2006, and Bahl and Padmanabhan, 2000). This method, prior to deployment, takes thorough measurements of the fingerprint of a test node moved to every possible location in the area of deployment (e.g., in a building), and possibly facing each direction at that location. At each location and facing direction, the fingerprint of the test node is recorded and stored in a database. The localization algorithm, once deployed, simply searches for the closest measurement in the database – that measurement's location is estimated to be the current node location. This notion of "closest" is a function of the measured RSSI values (e.g., Euclidean distance between RSSI vectors in Bahl and Padmanabhan (2000)). The chosen distance metric may be optimal for certain statistical error model (e.g., the Euclidean distance metric for Gaussian errors), so an RF fingerprinting algorithm makes an implicit assumption about the statistical model for path loss measurements. The RF fingerprinting algorithm does avoid significant other modeling requirements of distance-based localization algorithms.

Clearly, RF fingerprinting methods require a huge investment of time prior to deployment, and a large fixed infrastructure – two requirements not typically present for environmental wireless sensor networks. However, for sensor systems which operate in buildings where other WiFi backbones exist,

RF fingerprinting methods can provide a high-accuracy, large scale localization system. RF fingerprinting methods have been commercialized for the active RFID / real-time location systems (RTLS) market segment, and companies such as Ekahau, Inc. and AeroScout, Inc. deploy such systems. As time passes and the arrangement of a building interior changes (e.g., every 3 months), such systems require re-measurement of the fingerprint database to keep accuracy high. Current research efforts show promise in the reduction of measurement requirements to build the database, for example, as in Fang, Lin, and Lin (2008).

Discussion

These three types of range-free algorithms require only relative notions of path loss, and thus do not require a conversion to distance. In all methods, measured RSSI is either compared to a threshold or to other RSSI measurements. When all nodes transmit at identical power levels, the RSSI can be used directly. But when transmit powers differ, either purposefully due to power control or simply due to device variations, RSSI should be converted to path loss to see the full benefits of the proposed non-parametric localization algorithms.

Non-parametric localization algorithms do eliminate the need to estimate distance from path loss. As will be discussed in the following sections, distance estimation requires additional knowledge of environmental propagation characteristics, which can be difficult to accurately provide to a general-purpose localization system.

RANGE-BASED LOCALIZATION FROM MEASURED PATH LOSS

Despite advances in "range-free" algorithms, there is often a need to use a distance-based location algorithms. When sensor networks are deployed where little infrastructure exists, and thus RF fingerprinting approaches will not work, there are often benefits to having an explicit range estimate in localization algorithms. For distributed wireless sensor network localization, there are many examples of range-based localization algorithms in the literature.

Another argument for including range estimation is that, while there is a clear delineation in the literature between range-based and range-free algorithms, practical deployed localization systems will often benefit from using a mix of both types of algorithms. For example, range-free localization algorithms may gain from the use of range estimates as a consistency check. Similarly, range-based algorithms will likely need a method for online local determination of propagation and device parameters, which at one extreme, is no different than non-parametric localization approaches.

In this section, we present a discussion of the conversion of path loss to distance between two nodes. This conversion requires models for path loss, which is not only a function of distance, but many other environmental and propagation effects. We first present these distance and noise models, and then present information about estimators of distance given measured path loss.

Exponential Decay Model

The most common model for the ensemble mean of path loss at a particular distance between transmitter and receiver is that path loss is linearly proportional to the logarithm of that distance (Rappaport, 1996, and Hashemi, 1993). This proportionality is typically written as follows:

$$E[L_{ij}] = L_0 + 10n_p \log_{10} \frac{d_{ij}}{d_0} \tag{3}$$

where L_0 is the path loss at a reference distance d_0, and n_p is called the path loss exponent, and the $E[L_{ij}]$ indicates the expected value. Here, path loss is expressed in dB, which is 10 times the log base 10 of the linear multiplicative channel loss. The values of L_0 and n_p are dependent on the environment in which the sensors are deployed.

The model of (3) assumes no site-specific knowledge. In situations where we know the positions of the two nodes and the environmental obstructions in between them (interior or exterior walls, floor losses, trees, buildings, etc.), we could estimate the mean loss much more accurately. For example, Durgin, Rappaport, and Xu (1998) models each type of obstruction in the path as an attenuator with a constant loss, and estimates these constants from a set of measurements. For the purposes of localization in WSN, site-specific information may be limited. When it is available, it introduces sharp discontinuities in the path loss as a function of the two node coordinates, making optimization-based algorithms fail. However, such models have made impact in RSS fingerprint-based localization by allowing fingerprint measurements to be interpolated, and thus requiring much less dense manual measurements (Zhu, 2006).

We also note that the model of (3) is often bifurcated into "near-field" and "far-field" cases (Feuerstein, Blackard, Rappaport, Seidel, and Xia, 1994). Within the near-field, which is defined as within the first Fresnel zone, the model of (3) is an approximation with a relatively low path loss exponent. Beyond the first Fresnel zone, there tends to be more significant destructive multipath interference and the effective path loss exponent increases. As a result, more complete path loss vs. distance models are called "piecewise linear" path loss models and include two (or possibly more) $E[L_{ij}]$ log-linear functions of (3) for use with different ranges of d_{ij}. Within RSS-based localization systems, a piecewise linear path loss model requires more parameters to be estimated, but could provide more accurate localization, in particular from very short-range or very long-range RSS measurements. For WSN with sometimes high densities, sensors can often be in each others' near-field.

Noise: Frequency-Selective Fading and Shadowing

Clearly, the measured path loss has other contributions which are not a function of distance. If measured path loss $\hat{L}_{ij} = E[L_{ij}]$ at all times on all links, we would have no problem estimating distance and we would be able to calculate it exactly. The measured $\hat{L}_{ij}$ suffers from fading and shadowing, which are commonly lumped together and called "fading error" or "noise". These errors are severe, much more significant than the 1 dB standard deviations of measurement or characterization error which we've been discussing so far. We refer to fading error on the link between i and j as Y_{ij}, and the shadowing loss on the link as X_{ij}. Then,

$$\hat{L}_{ij} = E[L_{ij}] + X_{ij} + Y_{ij} \tag{4}$$

where $E[L_{ij}]$ is given in Equation (3).

There are several key points about the model in (4) that allow us to analyze RSS-based localization.

- The shadowing loss X_{ij} can be well approximated as Gaussian, since it is expressed in dB (Rappaport, 1996). Shadowing losses in linear terms are multiplicative, but in dB are additive. Thus, by a central limit theorem argument, the dB total loss after interaction with multiple attenuators can be considered to be Gaussian. In linear terms, this distribution is called log-normal.
- Fading errors are due predominantly to frequency-selective fading. The arriving multipath components add together, each with a phase and amplitude, and a phase that is a function of the frequency. At some frequencies, the phases are opposite and cancel, and we experience a "frequency null". At other frequencies, the phases are such that amplitudes add constructively. At one frequency, the fading error Y_{ij} has a non-Gaussian distribution such as a Raleigh or Ricean (when loss expressed in linear terms).

We note that frequency selective fading can be reduced by performing wideband measurements. With narrowband radios, this translates into frequency hopping in order to make measurements at multiple different frequencies. When these frequencies are separated by more than the correlation bandwidth of the channel, we can expect the experienced fading losses to be uncorrelated (Rappaport, 1996). For example, both Mica2 and 802.15.4-compatible radios can be designed to switch frequencies. We suggest operating a slow frequency hopping protocol when using narrowband radios, in which sensors measure RSSI with their neighbors sequentially across a list of different frequencies. In our implementation (Patwari and Agrawal, 2006), we hop over 16 frequencies over the range of 900-928 MHz, and nodes save only an averaged path loss measurement. We find significant improvement in measured path loss when averaging, compared to without averaging. In general, for the average of N uncorrelated frequency measurements, we expect the standard deviation of Y_{ij} to reduce by a factor of $\sqrt{N}$. Note that this average does little to reduce the variance of X_{ij}; small percentage frequency changes do not significantly change the attenuation experienced due to obstructions.

The distribution of L_{ij}, due to these two contributions, has both similarities and differences from the Gaussian distribution. Although X_{ij} is approximately Gaussian, Y_{ij} may or may not be Gaussian. When averaging over many different frequencies, it can be argued that Y_{ij} will also be approximately Gaussian as well by another central limit argument. However, a Ricean or Rayleigh random variable converted to dB units will have noticeably heavier tails than the Gaussian distribution.

We can quantify from experiments how well the Gaussian assumption holds. For example, measurements of RSS from (Patwari, Wang, and O'Dea 2002) are used to test the Gaussian assumption in Figure 9. These measurements were of path loss in a narrow-band channel, and did not do any frequency averaging. First, we computed the fading error by subtracting $\bar{L}_{ij}$ from the path loss measurements L_{ij}. This fading error is then compared to the a unit-variance Gaussian distribution in a quantile-quantile plot, shown in Figure 9. If the data were exactly Gaussian, the data points would lie in a diagonal line. Within the -2 to +2 quantiles, the data lie very close to a Gaussian distribution. But, the extreme values of the measured fading errors show heavier tails, indicating the non-Gaussian nature of fading errors.

Another measurement set reported in (Patwari, Hero, Perkins, Correal, and O'Dea, 2003) showed similar results for RSS fading errors. In fact, it was reported that the RSS modeling error could be more accurately described as a Gaussian mixture, that is, one in which a large majority of data is described as Gaussian with one (smaller) variance, and a small fraction of the data is described as Gaussian with a different (larger) variance. Such a mixture distribution would explain the heavier tails seen in RSS fading errors. The work of Whitehouse, Karlof, Woo, Jiang, and Culler (2005), advocate using empirical cumulative distribution functions (CDFs) of measurement data directly as the distribution of

Figure 9. Fading and shadowing losses on RSS measurements reported in (Patwari, Wang, and O'Dea, 2002) compared to a zero-mean, unit variance Gaussian distribution in a quantile-quantile plot. Data shows agreement between -2 to +2 quantiles, but somewhat heavier tails.

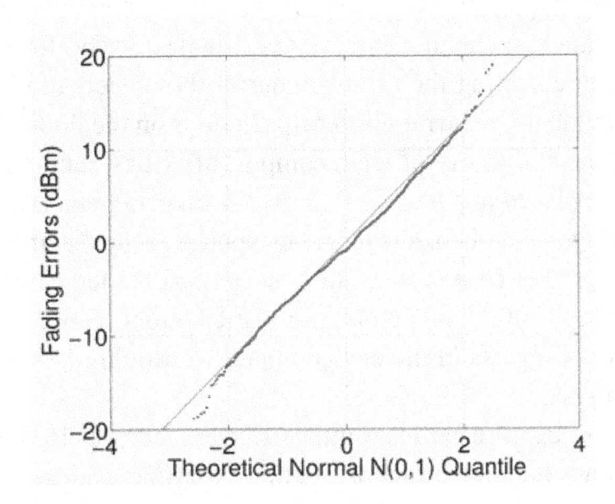

fading errors. Such CDFs could improve the accuracy of studies which use simulation to quantify the performance of RSS localization.

Mixture and empirical heavy-tailed distributions are analytically more difficult to deal with, but they motivate "robust estimation" of localization. Robust estimators down-weight or eliminate data points which seem inconsistent with other data. One example is to eliminate triplets of ranges which do not meet the triangle inequality, or distances among four nodes which do not meet the "robust quadrilateral" requirement (Moore, Leonard, Rus, and Teller, 2004). Such algorithms can eliminate some of the worst distance estimates.

We also note that in simulation, that shadowing errors experienced on different links (i,j) are taken to be independent and identically distributed, even though this is not a realistic assumption (Patwari, Wang, and O'Dea, 2002). In real life, the shadowing on different links may be correlated if the links are shadowed by the same obstructions. For example, two links passing through the same concrete wall would experience similar shadowing. In general, two links which cover similar ground may experience correlated shadowing (Agrawal and Patwari, 2008).

Parameter Estimation

While we know that fading and shadowing will cause range errors when using a general path loss model such as (3), we also must have an accurate estimate for the parameters of the model in order to estimate distance. For (3), there are two parameters, L_0 and n_p, which are functions of the environment of deployment. One main difficulty in range estimation is determining these two parameters. In our implementations, we conduct measurements in the environment of interest, which allow us to estimate these parameters. This is a significant task. Current research is evaluating algorithms to adaptively estimate the two parameters, which would be a huge benefit to quick deployment of RSS-based localization systems, and would be essential for mobile sensor networks which may periodically change environment. One such adaptive approach is presented in (Li, 2006), which simultaneously estimates coordinates and the path loss exponent.

Without adaptive parameter estimation, we can, and often do, perform deployment experiments to determine path loss parameters. Typically, we would perform one deployment experiment in the environment of interest, which would then allow us to estimate the values of L_0 and n_p.

We must be aware that such experiments will also result in parameter estimates which are noisy. In order to demonstrate the variation in channel parameter estimates, we perform 15 deployment experiments in one single area, only changing the arrangement of the objects in that area between experiments. In this series of experiments, we arrange 16 Mica2 nodes on the floor in an empty classroom in an Engineering building on the University of Utah campus, after first removing all existing furniture from the room. The nodes are placed in a four by four grid, with grid points each four feet (1.2 meters). We generate a "random" arrangement of objects for each experiment by placing obstructions as dictated by a randomized Matlab script. For ease of use, our obstructions are ten cardboard boxes wrapped in aluminum foil. These are significant RF reflectors, if not attenuators. Since each experiment ran with identical quantity and quality of objects in the environment, we would intuitively expect the path loss model parameters to be identical.

During each experiment, we capture path loss integers, averaged over 16 frequencies (to which sensors hop during a slow frequency hopping protocol). In this experiment, all sensors use the same known transmit power code. Since we know the actual coordinates, we can plot the measured RSSI vs. actual distance. Then, we use linear regression to find the best linear fit between RSSI and distance. These linear fits are plotted in Figure 10. Finally, over all 15 experiments, we find the average linear fit, which is also plotted in Figure 10. We could have equivalently plotted a linear fit with path loss in dB using the known transmit power, reported battery voltages, and receiver characteristics of the nodes involved in the measurements, but have chosen to plot the direct RSSI measurement for simplicity.

Discussion

At short range, there can be significant variations in the estimated path loss parameters – at 1 meter (0 dB meters), the standard deviation in RSSI is 4, which corresponds to approximately 1 dB. At longer distances, the standard deviation is in the range of 0.01 to 0.02 V RSSI, or 0.5 to 1 dB path loss. Al-

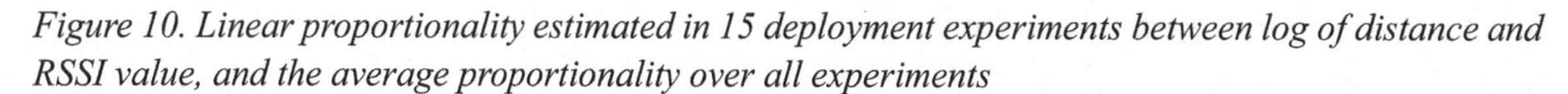

Figure 10. Linear proportionality estimated in 15 deployment experiments between log of distance and RSSI value, and the average proportionality over all experiments

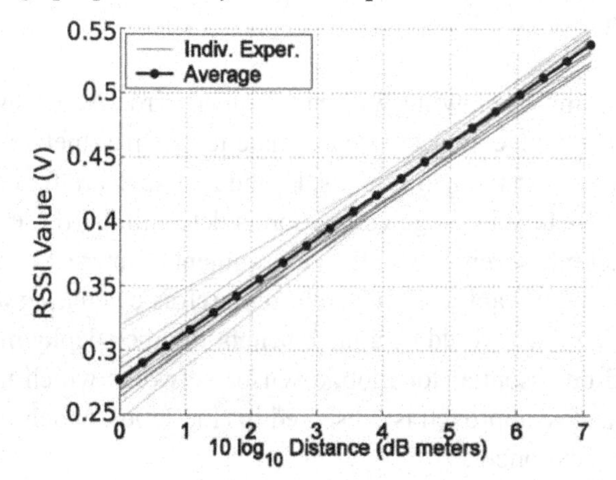

though these numbers are not severe, it is possible that the particular environment measured is not going to characterize that environment for all time. We would recommend that more than one deployment experiment be performed in order to robustly characterize an environment. In general, as we mentioned at the start of the "Parameter Estimation" section, it may be critical for large-scale deployments to use an algorithm which estimates model parameters and coordinates simultaneously, so that the specific environment in which deployed sensors is modeled appropriately.

Distance Estimation Equations

Assuming that path loss parameters have been estimated for the environment of interest, we can estimate distance from path loss. Solving (4) for d_{ij},

$$\hat{d}_{ij} = d_0 10^{(L_{ij} - L_0)/(10n_p)} \qquad (5)$$

As it turns out, this equation for $\hat{d}_{ij}$ is also the maximum likelihood estimator (MLE) of distance given measured path loss L_{ij} and measured parameters L_0 and n_p. This means that as the noise reduces towards zero (as fading errors lessen), this estimator is efficient, that is, its variance approaches the lower bound on the variance of an unbiased estimator. Clearly, the MLE has favorable features. However, the MLE is a biased estimator of distance in this case. As pointed out in Patwari, Hero, Perkins, Correal, and O'Dea (2003), given the discussed Gaussian distribution for L_{ij},

$$E[\hat{d}_{ij}] = C d_{ij}, \quad \text{where} \quad C = \exp\left\{\frac{1}{2}\left[\frac{\sigma_{dB} \log 10}{10 n_p}\right]^2\right\} \qquad (6)$$

For typical channel parameters (Rappaport 1996), C is between 1.08 and 1.2, adding 8-20% bias to the range estimate. Motivated by (6), we might also estimate distance d_{ij} as,

$$\hat{d}_{ij}(BR) = \frac{d_0}{C} 10^{(L_{ij} - L_0)/(10n_p)} \qquad (7)$$

in order to remove the bias in the distance estimate. We denote this "bias-removed" estimator as $\hat{d}_{ij}(BR)$. The choice between (5) and (7) is left to the system designer – it is often the case that the localization algorithm introduces other biases, which may counteract the bias in Equation (5). However, if it does not, the use of (7) may reduce the localization errors in the algorithm.

Moreover, it is critical to note that when using RSS-based range estimates, the nodes separated by the shortest distances are going to produce the most accurate range estimates. Equivalently, longer distances will measure distance estimates which have higher variance. Intuitively, we can see that at longer range, the same error in measured RSSI will correspond to a larger distance error. For example, consider Figure 10. The conversion between RSSI and log distance is linear, but if log distance is high, then a change in log distance results in more linear distance change than at a low log distance. From another perspective, we can look at the variance of the MLE estimate of distance from (5),

$$\text{Var}[\hat{d}_{ij}] = (C^4 - C) d_{ij}^2 \qquad (8)$$

This shows that the standard deviation of the distance estimate is directly proportional to the actual distance. For example, if C = 1.15 and the actual distance is 1 meter, using (6) we would observe a standard deviation of the MLE range estimate of 0.77 meters -- but for an actual distance of 10 meters, we would expect a standard deviation of 7.7 meters. Clearly while 77 cm error may be acceptable, a range error of 7.7 meters would be very severe.

A good localization algorithm would use this relationship to strongly down-weight range estimates to distant neighbors and instead emphasize range estimates between nearby neighbors. Further, it quantifies the notion that RSS will be most accurate at short range, as we expect from the discussion in the "Propagation Effects" section.

DISCUSSION

There are clearly many issues that negatively impact RSS-based localization, and it is easy to criticize or neglect the capabilities of RSS in localization without considering its capabilities and successes. In terms of commercial application, we have mentioned deployments of indoor localization systems using RF fingerprint-based algorithms, and there have also been successful commercial deployments by AwarePoint, Inc., which use a WSN-like mesh network of nodes and use distance-based algorithms to estimate node locations.

In terms of research literature, there have been a number of algorithms proposed and studied which consider the effects of ranging error and device inaccuracies. We have mentioned many such references in the text, and many good reviews of RSS-based localization algorithms exist in the literature, for example, Elnahraway, Li, and Pahlavan (2004).

This chapter is not intended to discuss the wide literature on localization algorithms, which may be the subject of many chapters of this book. Instead, we reiterate major lessons learned during device calibration and path loss model parameter estimation to recommend several simple adaptations which RSS-based location algorithms can take in order to improve their robustness and accuracy:

- Use of slow frequency hopping and averaging RSSI over frequency to reduce frequency-selective fading effects,
- Use of transmit power control to avoid the two opposite extreme problems of RSSI saturation and lack of connectivity,
- If transmit powers are known to have high variation, do not average bi-directional path losses L_{ij} and L_{ji} and instead use their differences to estimate the transmit power of each node.
- Down-weighting of the range estimates to the furthest neighbors (or equivalently emphasis given to nearest neighbors' range estimates),
- Adaptive estimation or learning of path loss parameters in the local environment of the deployment when using distance-based algorithms,
- Non-parametric approaches which compare signal strength measurements to each other, rather than to estimate distances directly.

For each of these bullet points, some existing literature has been discussed in this chapter. Future research in algorithms will be critical to achieve the best possible RSS-based localization performance without large deployment expense. Researchers and developers with experience in the measurement and calibration of received signal strength will be best able to understand the interactions between real-world measurements of RSS and localization algorithms.

CONCLUSION

While signal strength may be a highly desirable measurement for sensor self-localization algorithms, it is not necessarily straight-forward to obtain measurements which can be used to estimate sensor location to the best degree possible. We have presented a calibration procedure which can be used with any wireless sensor, and shown the results of one such set of calibration experiments. Critically, deployed sensor location systems must localize based on path loss rather than RSSI or received power, since the transmit power of sensors may vary. We have motivated key location system adaptations which may be used to reduce measurement and thus localization errors, for example, transmit power control to reduce the problem of RSSI saturation; and frequency hopping to reduce multipath fading error. We discuss both distance-based estimation and range-free estimation, in particular adaptations which eliminate modeling requirements or transmit power variation effects. Using these methods, we can maximize the ability of signal strength measurements to provide accurate inputs into localization algorithms, such as those described within this book.

ACKNOWLEDGMENT

This material is based in part upon work supported by the National Science Foundation under CAREER grant ECCS-0748206 and CyberTrust grant CNS-0831490. Any opinions, findings, and conclusions or recommendations expressed in this material are those of the author(s) and do not necessarily reflect the views of the National Science Foundation.

REFERENCES

Agrawal, P., & Patwari, N. (2008). *Correlated Link Shadow Fading in Multi-hop Wireless Networks*, (Tech Report arXiv:0804.2708v2), arXiv.org. Retrieved 18 Apr 2008 from http://arxiv.org/abs/ 0804.2708v2

Bahl, P., & Padmanabhan, V. N. (2000). RADAR: An In-Building RF-Based User Location and Tracking System. *In Proc. 19th International Conference on Computer Communications (Infocom)*, 2, 775–784.

Bulusu, N., Heidemann, J., & Estrin, D. (2000). GPS-less low-cost outdoor localization for very small devices. *IEEE Personal Communications, 7*(5), 28-34.

Costa, J., Patwari, N., & Hero, A. O. (2006). Distributed weighted-multidimensional scaling for node localization in sensor networks. *ACM Trans. Sensor Networks, 2*(1), 39-64.

Durgin, G., Rappaport, T. S., & Xu, H. (1998). Measurements and models for radio path loss and penetration loss in and around homes and trees at 5.85 GHz. *IEEE Trans. Communications, 46*(11), 1484-1496.

Elnahraway, E., Li, X., & Martin, R. P. (2004). The limits of localization using RSS. *In Proceedings of the 2nd Intl. Conf. on Embedded Networked Sensor Systems* (pp. 283-284), Baltimore, MD.

Fang, S.-H., Lin, T.-N., & Lin, P.-C. (2008), Location Fingerprinting In A Decorrelated Space. *IEEE Trans. Knowledge and Data Engineering, 20*(5), 685-691.

Feuerstein, M. J., Blackard, K. L., Rappaport, T. S., Seidel, S. Y., & Xia, H. H. (1994). Path loss, delay spread, and outage models as functions of antenna height for microcellular system design. *IEEE Trans. Vehicular Technology, 43*(3), 487-498.

Hashemi, H. (1993). The indoor radio propagation channel. *Proc. IEEE, 81*(7), 943–968.

He, T., Huang, C., Blum, B. M., Stankovic, J. A., & Abdelzaher, T. (2003). Range-free localization schemes for large scale sensor networks. *In Proc. 9th Intl. Conf. on Mobile Computing and Networking (Mobicom'03)*, (pp. 81-95), San Diego, CA.

King, T., Kopf, S., Haenselmann, T., Lubberger, C., & Effelsberg, C. W. (2006). COMPASS: A Probabilistic Indoor Positioning System Based on 802.11 and Digital Compasses. *In Proc. 1st ACM Intl. Workshop on Wireless Network Testbeds, Experimental Evaluation & Characterization (WiNTECH)*, (pp. 34-40), Los Angeles, USA.

Krishnamachari, B. (2006). *Networking Wireless Sensors.*Cambridge University Press.

Li, X. (2006). RSS-Based Location Estimation with Unknown Pathloss Model. *IEEE Trans. Wireless Communications, 5*(12), 3626-3633.

Liu, C., Wu, K., & He, T. (2004). Sensor localization with ring overlapping based on comparison of received signal strength indicator. *In Proc. IEEE Mobile Ad-hoc and Sensor Systems (MASS)*, (pp. 516–518).

Moore, D., Leonard, J., Rus, D., & Teller, S. (2004). Robust distributed network localization with noisy range measurements. *In Proc. 2nd Intl Conf. Embedded Networked Sensor Systems*, (pp. 50-61), Baltimore, MD.

Niculescu, D., & Nath, B. (2001). Ad Hoc Positioning System (APS). *In Proc. IEEE Global Communications Conference (GLOBECOM '01), 3*, 1734- 1743.

Pahlavan, K., Li, X., & Makela, J. P. (2002) Indoor geolocation science and technology. *IEEE Communications Magazine, 40*(2), 112-118.

Patwari, N., Wang, Y., & O'Dea, R. J. (2002). The Importance of the Multipoint-to-Multipoint Indoor Radio Channel in Ad Hoc Networks. *In Proceedings of the IEEE Wireless Communication and Networking Conference (WCNC'02), 2*, 608-612, Orlando FL.

Patwari, N., Hero, A. O., Perkins, M., Correal, N. S., & O'Dea, R. J. (2003). Relative Location Estimation in Wireless Sensor Networks. *IEEE Trans. Signal Processing, 51*(8), 2137-2148.

Patwari, N., & Hero, A. O. (2003). Using proximity and quantized RSS for sensor localization in wireless networks. *In Proc. 2nd ACM Intl. Conf. Wireless Sensor Networks and Applications (WSNA '03)* (pp. 20-29), San Diego, CA.

Patwari, N., Agrawal, P., & Hero, A. O. (2006). Demonstrating Distributed Signal Strength Location Estimation. *In Proc. Fourth Intl. Conf. Embedded Networked Sensor Systems (SenSys'06)*, (pp. 353-354), Boulder, CO.

Patwari, N., & Hero, A.O. (2006). Signal strength localization bounds in ad hoc & sensor networks when transmit powers are random. *In Proceedings of the Fourth IEEE Workshop on Sensor Array and Multichannel Processing* (SAM-2006) (pp. 299-303), July 12-14, 2006, Waltham, MA.

Rappaport, T. S. (1996). *Wireless Communications: Principles and Practice.* Englewood Cliffs, NJ: Prentice-Hall.

Whitehouse, K., Karlof, C., Woo, A., Jiang, F., & Culler, D. (2005). The effects of ranging noise on multihop localization: an empirical study. *In Proc. 4th Intl. Symp. Information Processing in Sensor Networks (IPSN'05)* (pp. 73-80), April 24 - 27, 2005 Los Angeles, California.

Whitehouse, K., Karlof, C., & Culler, D. (2007). A practical evaluation of radio signal strength for ranging-based localization. SIGMOBILE Mob. *Comput. Commun. Rev. 11*(1), 41-52.

Yedavalli, K., Krishnamachari, B., Ravula, S., & Srinivasan, B. (2005). Ecolocation: a sequence based technique for RF localization in wireless sensor networks. *In Proc. 4th Intl. Symp. Information Processing in Sensor Networks*, (pp. 285-292), Los Angeles, CA.

Zhu, J. (2006). *Indoor/Outdoor Location of Cellular Handsets Based on Received Signal Strength.* Doctoral dissertation, Georgia Tech, Atlanta. Retrieved Aug 12, 2008, from http:// etd.gatech.edu/theses/available/etd-05182006-154920/

Zuniga, M. Z., & Krishnamachari, B. (2007). An analysis of unreliability and asymmetry in low-power wireless links. *ACM Trans. Sensor Networks, 3*(2), 1-7.

Chapter VI
Graph Theoretic Techniques in the Analysis of Uniquely Localizable Sensor Networks

Bill Jackson
University of London, UK

Tibor Jordán
Eötvös University, Hungary

ABSTRACT

In the network localization problem the goal is to determine the location of all nodes by using only partial information on the pairwise distances (and by computing the exact location of some nodes, called anchors). The network is said to be uniquely localizable if there is a unique set of locations consistent with the given data. Recent results from graph theory and combinatorial rigidity made it possible to characterize uniquely localizable networks in two dimensions. Based on these developments, extensions, related optimization problems, algorithms, and constructions also became tractable. This chapter gives a detailed survey of these new results from the graph theorist's viewpoint.

INTRODUCTION

In the network localization problem the locations of some nodes (called anchors) of a network as well as the distances between some pairs of nodes are known, and the goal is to determine the location of all nodes. This is one of the fundamental algorithmic problems in the theory of wireless sensor networks and has been the focus of a number of recent research articles and survey papers, see for example (Aspnes et al., 2007; Eren et al., 2004; Mao et al., 2007; So & Ye, 2007).

A natural additional question is whether a solution to the localization problem is unique. The network, with the given locations and distances, is said to be *uniquely localizable* if there is a unique set of locations consistent with the given data. The unique localizability of a two-dimensional network, whose nodes are 'in generic position', can be characterized by using results from graph rigidity theory. In this case unique localizability depends only on the combinatorial properties of the network: it is determined completely by the *distance graph* of the network and the set of anchors, or equivalently, by the *grounded graph* of the network and the number of anchors. The vertices of the distance and grounded graph correspond to the nodes of the network. In both graphs two vertices are connected by an edge if the corresponding distance is explicitly known. In the grounded graph we have additional edges: all pairs of vertices corresponding to anchor nodes are adjacent. The grounded graph represents all known distances, since the distance between two anchors can be obtained from their locations. Before stating the basic observation about unique localizability we need some additional terminology. It is convenient to investigate localization problems with distance information by using frameworks, the central objects of rigidity theory.

A d-dimensional *framework* (also called *geometric graph* or *formation*) is a pair (G, p), where $G = (V,E)$ is a graph and p is a map from V to $\mathbb{R}^d$. We consider the framework to be a straight line realization of G in $\mathbb{R}^d$. Two frameworks (G, p) and (G, q) are *equivalent* if corresponding edges have the same lengths, that is, if $\|p(u) - p(v)\| = \|q(u) - q(v)\|$ holds for all pairs u, v with $uv \in E$, where $\|.\|$ denotes the Euclidean norm in $\mathbb{R}^d$. Frameworks (G, p), (G, q) are *congruent* if $\|p(u) - p(v)\| = \|q(u) - q(v)\|$ holds for all pairs u, v with $u, v \in V$. This is the same as saying that (G, q) can be obtained from (G, p) by an isometry. We shall say that (G, p) is *globally rigid*, or that (G, p) is a *unique realization* of G, if every framework which is equivalent to (G, p) is congruent to (G, p), see Figure 1.

The next observation shows that the theory of globally rigid frameworks is the mathematical background which is needed to investigate the unique localizability of networks.

Theorem 1. (Aspnes et al., 2006; So & Ye, 2007) Let N be a network in $\mathbb{R}^d$ consisting of m anchors located at positions $p_1, ..., p_m$ and $n-m$ ordinary nodes located at $p_{m+1}, ..., p_n$. Suppose that there are at least $d+1$ anchors in general position. Let G be the grounded graph of N and let $p = (p_1, ..., p_n)$. Then the network is uniquely localizable if and only if (G, p) is globally rigid.

We shall give a survey of the current status of the theory of globally rigid graphs and frameworks, focusing on the most relevant cases of two and three-dimensional frameworks, but stating results for higher dimensions, wherever possible. We will assume that the reader is familiar with the basic terms of graph theory. Readers who are not can find them in the Appendix.

Figure 1. Two realizations of the same graph G in R^2: F_1 is globally rigid; F_2 is not since we can obtain a realization of G which is equivalent but not congruent to F_2 by reflecting p_2 in the line through p_1, p_5, p_3.

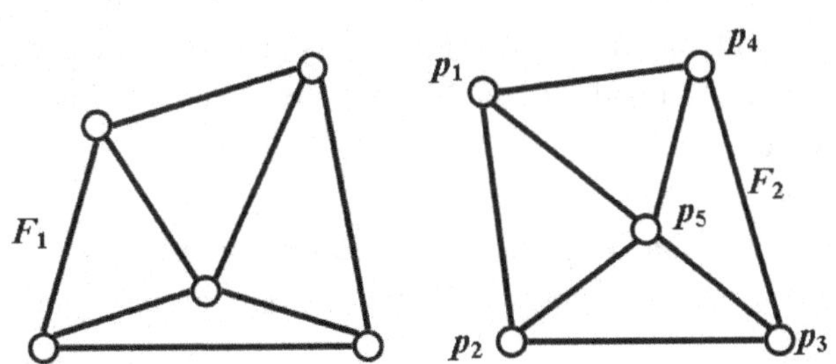

Generic Frameworks

It is a hard problem to decide if a given framework is globally rigid. Indeed Saxe (1979) has shown that this problem is NP-hard even for 1-dimensional frameworks. Further hardness results can be found in (Aspnes et al., 2004), see also (Aspnes et al., 2006; Yemini, 1979). The problem becomes more tractable, however, if we assume that there are no algebraic dependencies between the coordinates of the points of the framework.

A framework (G, p) is said to be *generic* if the set containing the coordinates of all its points is algebraically independent over the rationals. (A set $\{\alpha_1, \alpha_2, ..., \alpha_t\}$ of real numbers is *algebraically independent* over the rationals if, for all non-zero polynomials with rational coefficients $p(x_1, x_2, \ldots, x_t)$, we have $p(\alpha_1, \alpha_2, ..., \alpha_t) \neq 0$.) Restricting to *generic* frameworks gives us two important 'stability properties'. The first is that, if (G, p) is a globally rigid d-dimensional generic framework then there exists an $\varepsilon > 0$ such that all frameworks (G, q) which satisfy $\|p(v)-q(v)\| < \varepsilon$ for all $v \in V$ are also globally rigid, see for example (Cheung & Whiteley, 2008). The second, which follows from a recent result of Gortler et al. (2007), is that if some d-dimensional generic realization of a graph G is globally rigid, then all d-dimensional generic realizations of G are globally rigid. We will return to this in the last section.

RIGIDITY AND GLOBAL RIGIDITY OF GRAPHS

Rigidity, which is a weaker property of frameworks than global rigidity, plays an important role in the exploration of the structural results of global rigidity as well as in the corresponding algorithmic problems. Intuitively, we can think of a d-dimensional framework (G, p) as a collection of bars and joints where vertices correspond to joints and each edge to a rigid bar joining its end-points. The framework is rigid if it has no continuous deformations. Equivalently, and more formally, a framework (G, p) is *rigid* if there exists an $\varepsilon > 0$ such that, if (G, q) is equivalent to (G, p) and $\|p(v) - q(v)\| < \varepsilon$ for all $v \in V$, then (G, q) is congruent to (G, p).

Rigidity, like global rigidity, is a generic property of frameworks, that is, the rigidity of a generic realization of a graph G depends only on the graph G and not the particular realization. We say that the graph G is *rigid*, respectively *globally rigid* or *uniquely realizable*, in $\mathbb{R}^d$ if every (or equivalently, if some) generic realization of G in $\mathbb{R}^d$ is rigid, respectively globally rigid.

The problem of characterizing when a graph is rigid in $\mathbb{R}^d$ has been solved for $d = 1, 2$. A graph is rigid in $\mathbb{R}$ if and only if it is connected. The characterization of rigid graphs in $\mathbb{R}^2$ is a result of Lovász and Yemini (1982), which we will return to in the fourth section. We refer the reader to (Graver et al., 2003; Whiteley, 1996) for a detailed survey of the rigidity of d-dimensional frameworks.

A similar situation holds for global rigidity: the problem of characterizing when a generic framework is globally rigid in $\mathbb{R}^d$ has also been solved for $d = 1, 2$. A generic framework (G, p) is globally rigid in $\mathbb{R}$ if and only if either G is the complete graph on two vertices or G is 2-connected. The characterization for $d = 2$ uses the following general result of Hendrickson. We say that G is *redundantly rigid* in $\mathbb{R}^d$ if $G - e$ is rigid in $\mathbb{R}^d$ for all edges e of G.

Theorem 2. (Hendrickson, 1992) Let (G, p) be a generic framework in $\mathbb{R}^d$. If (G, p) is globally rigid then either G is a complete graph with at most $d + 1$ vertices, or G is $(d + 1)$-connected and redundantly rigid in $\mathbb{R}^d$.

Hendrickson conjectured that the necessary conditions for global rigidity of generic frameworks given in Theorem 2 are also sufficient. When $d = 1$, this follows from the above mentioned characterizations of rigidity and global rigidity. Counterexamples to Hendrickson's conjecture were constructed by Connelly (1991) for all $d \geq 3$. The remaining open case, $d = 2$, was settled by the following result, which incorporated earlier results from (Connelly, 2005; Hendrickson, 1992) and a new inductive construction for the family of 3-connected redundantly rigid graphs (see in the section on inductive constructions).

Theorem 3. (Jackson & Jordán, 2005) Let (G, p) be a 2-dimensional generic framework. Then (G, p) is globally rigid if and only if either G is a complete graph on two or three vertices, or G is 3-connected and redundantly rigid in $\mathbb{R}^2$.

Note that the characterizations of globally rigid generic frameworks for $d = 1, 2$ depend only on the structure of the underlying graph and hence imply the above mentioned result that global rigidity is a generic property in $\mathbb{R}^d$, for the special cases when $d = 1, 2$.

MATROIDS

A matroid is an abstract structure which extends the notion of linear independence of vectors in a vector space. We will see that many of the rigidity properties of a generic framework (G, p) are determined by an associated matroid defined on the edge set of G. We first need some basic definitions. We refer the reader to the books (Recski, 1989; Schrijver, 2003) for more information on matroids.

A *matroid* is an ordered pair $\mathcal{M} = (E, \mathcal{I})$ where E is a finite set, and $\mathcal{I}$ is a family of subsets of E, called *independent sets*, which satisfy the following three axioms.

(M1) $\emptyset \in \mathcal{I}$,
(M2) if $I \in \mathcal{I}$ and $D \subseteq I$ then $D \in \mathcal{I}$,
(M3) for all $F \subseteq E$, the maximal independent subsets of F have the same cardinality.

The fundamental example of a matroid is obtained by taking E to be a set of vectors in a vector space and $\mathcal{I}$ to be the family of all linearly independent subsets of E.

Given a matroid $\mathcal{M} = (E, \mathcal{I})$, the cardinality of a maximum independent subset of a set $F \subseteq E$ is defined to be the *rank* of F and denoted by $r(F)$. The rank of E is referred to as the rank of $\mathcal{M}$. A *base* of $\mathcal{M}$ is a maximum independent subset of E. A subset of E which is not independent is said to be *dependent.* A *circuit* of $\mathcal{M}$ is a minimal dependent subset of E. The matroid $\mathcal{M}$ is said to be *connected* if every pair of elements of E are contained in a circuit.

Given a graph $G = (V, E)$, we may define a matroid $\mathcal{M} = (E, \mathcal{I})$ by letting $\mathcal{I}$ be the family of all edge sets of forests in G. The rank of a set $F \subseteq E$ is given by $r(F) = |V| - k(F)$, where $k(F)$ denotes the number of connected components in the graph (V, F). A base of $\mathcal{M}$ is the edge set of a forest which has the same number of components as G. A circuit of $\mathcal{M}$ is the edge set of a cycle of G, and $\mathcal{M}$ is connected if and only if G is 2-connected. This matroid is called the *cycle matroid* of G.

Rigidity Matrices and Matroids

Let (G, p) be a d-dimensional realization of a graph $G = (V,E)$. The *rigidity matrix* of the framework (G, p) is the matrix $R(G, p)$ of size $|E|\times d|V|$, where, for each edge $e = v_i v_j \in E$, in the row corresponding to e, the entries in the two columns corresponding to vertices i and j contain the d coordinates of $(p(v_i)-p(v_j))$ and $(p(v_j)-p(v_i))$, respectively, and the remaining entries are zeros. See (Graver et al., 2003; Whiteley, 1996) for more details. The rigidity matrix of (G, p) defines the rigidity matroid of (G, p) on the ground set E where a set of edges $F \subseteq E$ is *independent* if and only if the rows of the rigidity matrix indexed by F are linearly independent. Any two generic d-dimensional frameworks (G, p) and (G, q) have the same rigidity matroid. We call this the *d-dimensional rigidity matroid* $R_d(G)$ of the graph G. We denote the rank of $R_d(G)$ by $r_d(G)$.

As an example, consider a 1-dimensional framework (G, p). In this case, the rows of $R(G, p)$ are just scalar multiples of a directed incidence matrix of G. It is well known that a set of rows in this matrix is independent if and only if the corresponding edges induce a forest in G. Thus $R_1(G)$ is the cycle matroid of G.

Gluck (1975) characterized rigid graphs in terms of their rank.

Theorem 4. (Gluck, 1975) Let $G = (V,E)$ be a graph. Then G is rigid in $\mathbb{R}^d$ if and only if either $|V| \leq d + 1$ and G is complete, or $|V| \geq d + 2$ and $r_d(G) = d|V| - (d + 1)$.
2

This characterization does not give rise to a polynomial algorithm for deciding whether a graph is rigid in $\mathbb{R}^d$. The problem is that to compute $r_d(G)$ we need to determine the rank of the rigidity matrix of a generic realization of G in $\mathbb{R}^d$. There is no known polynomial algorithm for calculating the rank of a matrix in which the entries are linear functions of algebraically independent numbers.

We say that a graph $G = (V,E)$ is *M-independent* in $\mathbb{R}^d$ if E is independent in $R_d(G)$. Knowing when subgraphs of G are M-independent allows us to determine the rank of G (and hence determine whether G is rigid), since we can construct a base for $R_d(G)$ by greedily constructing a maximal independent set of $R_d(G)$. This follows from axiom (M3) which guarantees that an independent set which is maximal with respect to inclusion is also an independent set of maximum cardinality. For example, when $d = 1$, we have seen that a subgraph is independent if and only if it is a forest. Thus we can determine the rank of G by greedily growing a maximal forest F in G. By Theorem 4, G is rigid if and only if F has $|V|-1$ edges, i.e. F is a spanning tree of G.

THE 2-DIMENSIONAL RIGIDITY MATROID

Subsequent sections of this chapter will mainly be concerned with the case when $d = 2$. We will assume that this is the case unless specifically stated otherwise, and suppress the subscript d accordingly.

We first describe the characterization of M-independent graphs due to Laman (1970). For $X \subseteq V$ let $i_G(X)$ denote the number of edges in $G[X]$, that is, in the subgraph induced by X in G.

Theorem 5. (Laman, 1970) A graph $G = (V,E)$ is M-independent if and only if $i_G(X) \leq 2|X| - 3$ for all $X \subseteq V$ with $|X| \geq 2$.

The following characterization of rigid graphs due to Lovász and Yemini, which can be deduced from Theorems 4 and 5, is a slight reformulation of Corollary 4 in (Lovász & Yemini, 1982), see also Corollary 2.5 in (Jackson & Jordán, 2005). A *cover* of $G = (V,E)$ is a collection $\chi = \{X_1, X_2, ..., X_t\}$ of subsets of V such that $\{E(X_1), E(X_2), ..., E(X_t)\}$partitions E, where $E(X)$ denotes the set of edges in $G[X]$.

Theorem 6. (Lovász & Yemini, 1982) Let $G = (V,E)$ be a graph. Then G is rigid if and only if for all covers χ of G we have $\sum_{X \in \chi} (2|X| - 3) \geq 2|V| - 3$.

Theorem 6 is illustrated in Figure 2.

A graph $G = (V,E)$ is *minimally rigid*, or *isostatic*, if G is rigid, but $G - e$ is not rigid for all $e \in E$. Theorems 4 and 5 imply that G is minimally rigid if and only if $i_G(X) \leq 2|X| - 3$ for all $X \subseteq V$ with $|X| \geq 2$ and $|E| = 2|V| - 3$. Other characterizations for minimally rigid graphs have been given by Lovász and Yemini (1982) (for each $e \in E$, the graph obtained from G by adding a new edge parallel to e is the union of two edge disjoint spanning trees), and by Crapo (1990) (G contains three trees such that their edge sets partition E, each vertex in V is incident to exactly two of the trees, and the vertex sets of any two non-tivial subtrees that belong to different trees are different.) Note that if G is rigid, then the edge sets of the minimally rigid spanning subgraphs of G form the bases in the rigidity matroid of G.

Given a graph $G = (V,E)$, a subgraph $H = (W,C)$ is said to be an *M-circuit* (also called *rigidity circuit* or *generic cycle*) in G if C is a circuit (i.e. a minimal dependent set) in $R(G)$. In particular, G is an *M-circuit* if E is a circuit in $R(G)$. For example, K_4, $K_{3,3}$ plus an edge, and $K_{3,4}$ are all *M*-circuits, see Figure 3. Using Theorem 5 we may deduce:

Lemma 7. Let $G = (V,E)$ be a graph. The following statements are equivalent.

(a) G is an *M-circuit.*

(b) $|E| = 2|V| - 2$ and $G - e$ is minimally rigid for all $e \in E$.

(c) $|E| = 2|V| - 2$ and $i_G(X) \leq 2|X| - 3$ for all $X \subseteq V$ with $2 \leq |X| \leq |V| - 1$.

Figure 2. Let $X_1 = \{v_1, v_2, v_5, v_6\}$, $X_2 = \{v_3, v_4, v_7, v_8\}$, $X_3 = \{v_6, v_7, v_9, v_{10}\}$, $X_4 = \{v_2, v_3\}$, and $\chi = \{X_1, X_2, X_3, X_4\}$. Then χ is a cover of G. Furthermore $\sum X \in \chi\ (2|X| - 3) = 16 < 17 = 2|V| - 3$ so G is not rigid by Theorem 6.

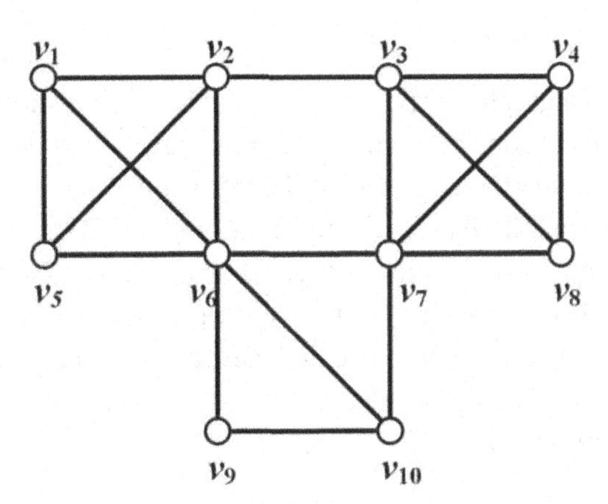

Figure 3. Three examples of M-circuits: G_1 is K_4 and G_3 is K_3,3 plus an edge

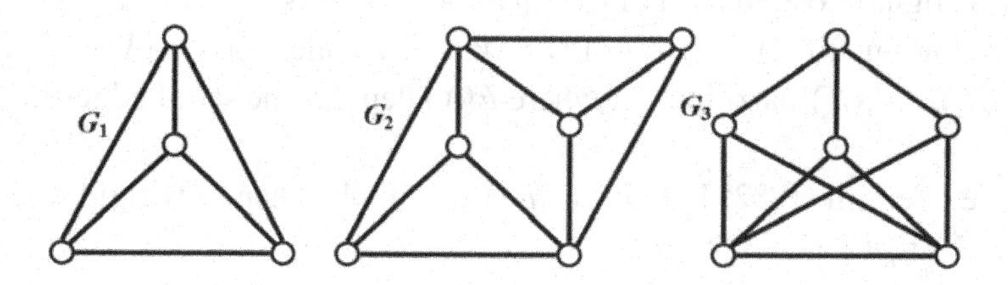

Recall that a graph $G = (V,E)$ is redundantly rigid if $G - e$ is rigid for all $e \in E$. Equivalently, a graph G is redundantly rigid if and only if G is rigid and each edge of G belongs to an *M*-circuit of G. If G is redundantly rigid then $|E| \geq 2|V| - 2$ with equality only if G is an *M*-circuit.

We say that a graph G is *M-connected* if $R(G)$ is connected i.e. every pair of edges of G belongs to an *M*-circuit. We will see that this property has important implications for global rigidity. The graph $K_{3,m}$, for $m \geq 4$, is an example of an *M*-connected graph. It is minimally *M*-connected in the sense that deleting any of its edges results in a graph which is no longer *M*-connected. Note that $K_{3,m}$ is an *M*-circuit only when $m = 4$.

The facts that *M*-circuits are rigid and the union of two rigid graphs with at least two vertices in common is rigid imply that *M*-connected graphs are rigid. Since every edge of an *M*-connected graph belongs to an *M*-circuit, we have:

Lemma 8. Every *M*-connected graph is redundantly rigid.

On the other hand, sufficiently connected redundantly rigid graphs are *M*-connected.

Theorem 9. Every 3-connected redundantly rigid graph is *M*-connected (Jackson & Jordan, 2005).

In fact *M*-connected graphs can be characterized as redundantly rigid graphs which have no vertex cut sets of a certain type, see Theorem 3.7 in (Jackson & Jordán, 2005;). Note that Theorems 3, 9, and Lemma 8 imply that a graph is globally rigid if and only if it is either a complete graph on at most three vertices or is both 3-connected and *M*-connected.

Decompositions

We define a *rigid component* of a graph $G = (V,E)$ to be a maximal rigid subgraph of G. It is known (see for example Corollary 2.14 in (Jackson & Jordán, 2005)), that any two rigid components of G intersect in at most one vertex and hence that the edge sets of the rigid components of G partition E.

It is also known that any two maximal redundantly rigid subgraphs of a graph G can have at most one vertex in common, and hence are edge disjoint, see (Jackson & Jordán, 2005). Defining a *redundantly rigid component* of G to be either a maximal redundantly rigid subgraph of G, or a subgraph induced by an edge which belongs to no *M*-circuit of G, we deduce that the redundantly rigid components of G partition E. Since each redundantly rigid component is rigid, this partition is a refinement of the partition of E given by the rigid components of G.

We may further define an *M-component* of G to be either a maximal M-connected subgraph of G, or a subgraph induced by an edge which belongs to no M-circuit of G. We again have the property that any two M-components of G can have at most one vertex in common, and hence are edge-disjoint, see (Jackson & Jordán, 2005). Since the M-components of G are redundantly rigid by Lemma 8, the partition of E given by the M-components is a refinement of the partition given by the redundantly rigid components and hence a further refinement of the partition given by the rigid components, see Figure 4.

The partitions of E described above have a stronger matroid property. We say that a matroid $\mathcal{M} = (E, \mathcal{I})$ is the *direct sum* of two matroids $\mathcal{M}_1 = (E_1, \mathcal{I}_1)$ and $\mathcal{M}_2 = (E_2, \mathcal{I}_2)$ if E is the disjoint union of E_1 and E_2 and

$$\mathcal{I} = \{I_1 \cup I_2 : I_1 \in \mathcal{I}_1, I_2 \in \mathcal{I}_2\}.$$

The rigidity matroid of a graph G is the direct sum of the rigidity matroids of either the rigid components of G, the redundantly rigid components of G, or the M-components of G. Furthermore, the vertex sets of the components in each of the above decompositions form a cover of G which minimizes the sum in Theorem 6. This minimum value is equal to the rank of $R_2(G)$.

It is an open problem to determine the maximal 'globally rigid components' of vertices of a graph. We will return to this problem in the section on globally linked pairs.

INDUCTIVE CONSTRUCTIONS

One of the most useful tools in the analysis of (global) rigidity properties of a family of graphs is an inductive construction. In this section we will describe such constructions for rigid graphs and globally rigid graphs.

Let H be a graph. The operation 0-*extension* (or *vertex addition*, or *Henneberg operation of type I*) adds a new vertex v to G and two edges vu, vw with $u \neq w$. The operation *1-extension* (or *edge-split*, or

Figure 4. This graph G is rigid so has exactly one rigid component. It has three redundantly rigid components, consisting of $G - z$ and the remaining two copies of K_2. It has five M-components: each of the three copies of K_4, and the remaining two copies of K_2.

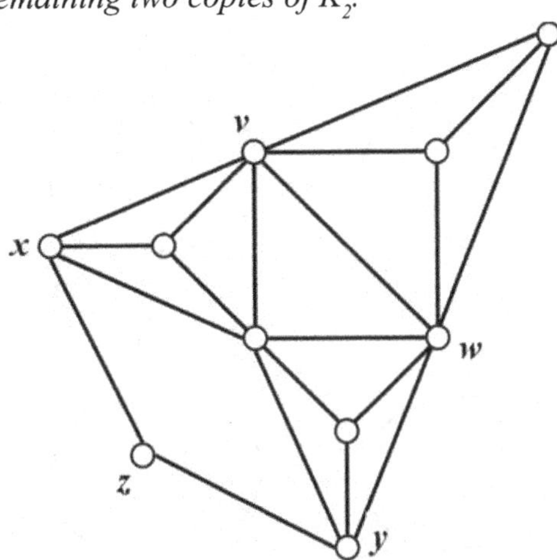

Henneberg operation of type II) subdivides an edge uw of G by a new vertex v and adds a new edge vz for some $z \neq u,w$. An *extension* is either a 0-extension or a 1-extension, see Figure 5. The next lemma follows easily from Theorem 5.

Lemma 10. Let H be a minimally rigid graph and let G be obtained from H by an extension. Then G is minimally rigid.

The following result gives a converse.

Theorem 11. (Jackson & Jordán, 2005) Let G be minimally rigid and let H be a minimally rigid subgraph of G. Then G can be obtained from H by a sequence of extensions.

By choosing H to be the subgraph induced by an arbitrary edge of G we obtain the following constructive characterization of minimally rigid graphs (called the *Henneberg construction*, see (Laman, 1970; Tay & Whiteley, 1985)).

Theorem 12. (Tay & Whiteley, 1985) A graph is minimally rigid if and only if it can be obtained from K_2 by a sequence of extensions.

As an immediate corollary we deduce that a graph is rigid if and only if it can be obtained from K_2 by extensions and edge additions.

The analogue of Lemma 10 for global rigidity is as follows.

Theorem 13. (Jackson et al., 2006) Let H be a globally rigid graph with at least four vertices and let G be obtained from H by a 1-extension. Then G is globally rigid.

A slightly weaker result was previously obtained by Connelly (2005), who showed that if G can be obtained from K_4 by a sequence of 1-extensions then G is globally rigid. His result was a key step in

Figure 5. The extension operations

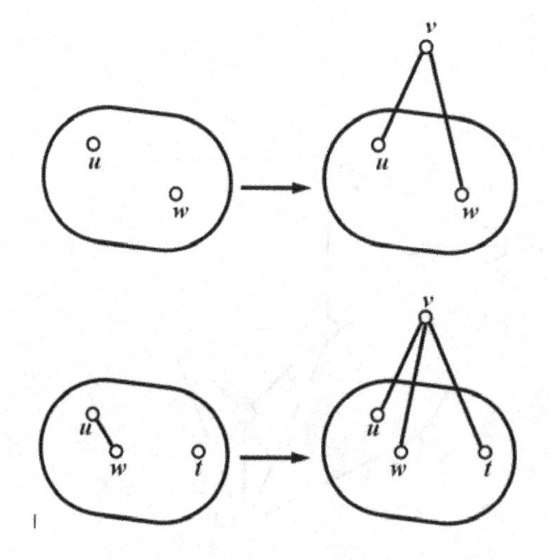

the characterization of globally rigid graphs. The other key ingredient was an inductive construction of 3-connected *M*-connected graphs from (Jackson & Jordán, 2005). Combining both results we obtain

Theorem 14. (Jackson & Jordán, 2005) A graph with at least four vertices is globally rigid if and only if it can be obtained from K_4 by a sequence of 1-extensions and edge additions.

Given a graph $G = (V,E)$, an edge $uv \in E$, and a bipartition F_1, F_2 of the edges incident to v in $G - uv$, the *vertex splitting* operation *on edge uv at vertex v* replaces the vertex v by two new vertices v_1 and v_2, replaces the edge uv by three new edges uv_1, uv_2, v_1v_2, and replaces each edge $wv \in F_i$ by an edge wv_i, $i = 1, 2$, see Figure 6. The vertex splitting operation is said to be *non-trivial* if F_1, F_2 are both non-empty, or equivalently, if each of the split vertices v_1, v_2 has degree at least three.

It is known that vertex splitting preserves rigidity (Whiteley, 1990; Whiteley, 1996). Theorem 3 can be used to show that it also preserves global rigidity.

Theorem 15. (Jordán & Szabadka, in press) Let H be a globally rigid graph and let G be obtained from H by a non-trivial vertex splitting. Then G is also globally rigid.

Inductive constructions can also be used in the problem where a graph G is given and the goal is to construct (a non-generic) globally rigid realization of G. Given a graph $G = (V,E)$ we say that a 1-extension on the edge uw and vertex t is a *triangle-split* if $\{ut,wt\} \subseteq E$ (that is, if u,w, t induce a triangle of G). A graph will be called *triangle-reducible* if it can be obtained from K_4 by a sequence of triangle-splits. It is easy to check that triangle-reducible graphs are 3-connected *M*-circuits. A polynomial time construction for a globally rigid realization of a triangle-reducible graph is given in (Jordán & Szabadka, in press).

The only other known result on the construction of globally rigid realizations is the following. A *d-dimensional trilateration ordering* of a graph $G = (V,E)$ is an ordering $(v_1, v_2, ..., v_n)$ of V for which the first $d + 1$ vertices are pairwise adjacent and at least $d + 1$ edges connect each vertex v_j, $d + 2 \leq j \leq n$, to the set of the first $j - 1$ vertices. The graph G is a *d-dimensional trilateration graph* if it has a *d*-dimensional trilateration ordering. It is shown in (Aspnes et al., 2006) that a *d*-dimensional trilateration graph G is globally rigid in $\mathbb{R}^d$, and a construction for a globally rigid realization is given in (Eren

Figure 6. The vertex splitting operation on edge uv at vertex v

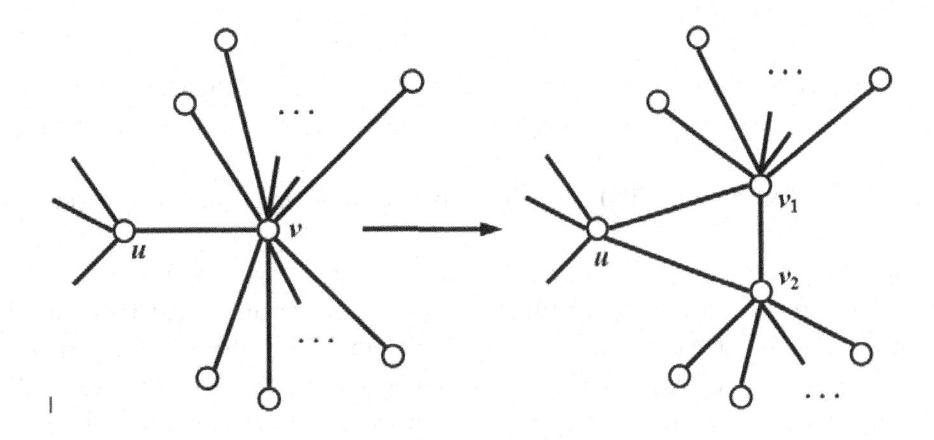

et al., 2004). Note that 2-dimensional trilateration graphs satisfy $|E| \geq 3|V| - 6$, while triangle-reducible graphs are sparser: they have $2|V|-2$ edges, which is the smallest possible number in a globally rigid graph in $\mathbb{R}^2$.

SPECIAL FAMILIES OF GRAPHS

Graphs of Large Minimum Degree

We may obtain a sufficient condition for global rigidity in terms of the minimum degree of G. The lower bound on the minimum degree in the next theorem is best possible, as shown by two complete graphs of equal size with two vertices in common.

Theorem 16. Let $G = (V,E)$ be a graph on $n \geq 4$ vertices with $\delta(G) \geq \frac{n+1}{2}$, where $\delta(G)$ denotes the minimum degree in G. Then G is globally rigid.

We sketch the proof of Theorem 16. By Theorem 3 it suffices to show that G is 3-connected and redundantly rigid. If $n \leq 4$ then G is complete, so we may suppose that $n \geq 5$. The hypothesis that $\delta(G) \geq (n + 1)/2$ implies that G cannot have a vertex cut of size less than three and hence G is 3-connected. For a contradiction suppose that $H = G - e$ is not rigid for some $e \in E$. Let C be a rigid component of H with as few vertices as possible. Put $D = H - V(C)$. The facts that distinct rigid components of H can share at most one vertex and $\delta(G) \geq \frac{n+1}{2}$, imply that $|V(D)| \geq 4$ and $|V(C)| \leq \frac{n-1}{2}$. Since C is a rigid component of H, each vertex of D is adjacent to at most one vertex of C in H by Lemma 10. Since $\delta(G) \geq \frac{n+1}{2}$, this implies that $\delta(D) \geq \frac{n-3}{2}$ and all but at most two non-adjacent vertices of D have degree at least $\frac{n-1}{2}$. Hence we may construct a graph D^* with $\delta(D^*) \geq \frac{n-1}{2}$ by adding at most one edge to D. Since $|V(C)| \geq 2$, we have $|V(D)| \leq n-2$. We may now use induction on n to deduce that D^* is globally rigid. Since $|V(D)| \geq 4$, D^* is redundantly rigid and hence D is rigid. Since $|V(C)| \leq \frac{n-1}{2}$ and $\delta(G) \geq \frac{n+1}{2}$, each vertex of C is adjacent to at least one vertex of D in H, and all but at most two non-adjacent vertices of C are adjacent to at least two vertices of D. We may now use Lemma 10 to deduce that H is rigid, a contradiction.

Three-dimensional analogues of Theorem 16 are given in (Berger et al., 1999).

Highly Connected Graphs

Lovász and Yemini (1982) proved that 6-connected graphs are redundantly rigid. Combining this with Theorem 3, we may deduce that the same degree of connectivity suffices to give global rigidity.

Theorem 17. (Jackson & Jordán, 2005) Let G be 6-connected. Then G is globally rigid.

An infinite family of 5-connected non-rigid graphs given in (Lovász & Yemini, 1982) shows that the hypothesis on vertex connectivity in both the Lovász-Yemini theorem and Theorem 17 cannot be reduced from six to five. On the other hand, Jackson et al. (2007) show that the connectivity hypothesis can be replaced by a slightly weaker hypothesis of 'essential 6-vertex-connectivity' which allows vertex cuts of size four or five as long as they only separate one or at most three vertices, respectively, from the rest of the graph.

Jackson and Jordán (2008/a) show that the connectivity hypothesis can be weakened in a more substantial way and still guarantee the rigidity and global rigidity of the graph. To this end we define the following form of mixed vertex and edge connectivity. Let $G = (V,E)$ be a graph. A pair (U,D) with $U \subseteq V$ and $D \subseteq E$ is a *mixed cut* in G if $G - U - D$ is not connected. We say that G is *6-mixed-connected* if $2|U|+|D| \geq 6$ for all mixed cuts (U,D) in G. Equivalently, G is 6-mixed-connected if G is 6-edge-connected, $G - v$ is 4-edge-connected for all $v \in V$, and $G-\{u, v\}$ is 2-edge-connected for all pairs $u, v \in V$. It follows that 6-vertex-connected graphs are 6-mixed-connected and 6-mixed-connected graphs are 3-vertex-connected.

Theorem 18. (Jackson & Jordán, 2008/a) Let $G = (V,E)$ be a 6-mixed-connected graph. Then $G - e$ is globally rigid for all $e \in E$.

The final result of this subsection observes that an even weaker connectivity condition is sufficient to imply that 4-regular graphs are globally rigid. A graph $G = (V,E)$ is said to be *cyclically k-edge-connected* if, for all $X \subseteq V$ such that $G[X]$ and $G[V-X]$ both contain cycles, we have at least k edges from X to $V - X$.

Theorem 19. (Jackson et al., 2007) Let $G = (V,E)$ be a cyclically 5-edge-connected 4-regular graph. Then G is globally rigid.

Examples of 4-regular 4-connected graphs and 5-regular 5-connected graphs which are not globally rigid are given in Theorem 20 (c),(d) below.

Vertex Transitive Graphs

Vertex transitive graphs which are rigid or globally rigid were characterized by Jackson, Servatius, and Servatius.

Theorem 20. (Jackson et al., 2007) Let $G = (V,E)$ be a connected k-regular vertex transitive graph on n vertices. Then G is not globally rigid if and only if one of the following holds:
(a) $k = 2$ and $n \geq 4$.
(b) $k = 3$ and $n \geq 6$.
(c) $k = 4$ and G has a 3-factor F consisting of s disjoint copies of K_4 where $s \geq 3$.
(d) $k = 5$ and G has a 4-factor F consisting of s disjoint copies of K_5 where $s \geq 6$.

As a corollary they determine all vertex transitive graphs which are rigid but not globally rigid.

Corollary 21. (Jackson et al., 2007) There are exactly four vertex transitive graphs which are rigid but not globally rigid. These are $K_{3,3}$, the triangular prism, the graph obtained from $2C_4$ by replacing each vertex by a copy of K_4, and the graph obtained from K_6 by replacing each vertex by a copy of K_5.

(The triangular prism is the 3-regular graph consisting of two disjoint triangles joined by three edges. The graph $2C_4$ is the 4-regular graph obtained from a cycle on four vertices by replacing each edge by two parallel edges.)

Random Graphs

We consider three different models of random graphs. Throughout this subsection, we assume that all logarithms are natural.

Our first model is the Erdős-Rényi model of random graphs. Let $G(n, p)$ denote the probability space of all graphs on n vertices in which each pair of vertices is joined by an edge with independent probability p, see (Bollobás, 1985). A sequence of graph properties A_n holds asymptotically almost surely, or a.a.s. for short, in $G(n, p)$ if $\lim_{n\to\infty} \Pr_{G(n,p)}(A_n) = 1$.

Theorem 22. (Jackson et al., 2007) Let $G \in G(n, p)$, where $p = (\log n{+}k \log\log n{+}w(n))/n$, and $\lim_{n\to\infty} w(n) = \infty$.

(a) If $k = 2$ then G is a.a.s. rigid.
(b) If $k = 3$ then G is a.a.s. globally rigid.

The bounds on p given in Theorem 22 are best possible since if $G \in G(n, p)$ and $p = (\log n + k \log\log n + c)/n$ for any constant c, then G a.a.s. does not have minimum degree at least k, see (Bollobás, 1985).

Our second model is of random regular graphs. Let $G_{n,d}$ denote the probability space of all d-regular graphs on n vertices chosen with the uniform probability distribution. (We refer the reader to (Bollobás, 1985) for a mathematical procedure for generating the graphs in $G_{n,d}$.) Since globally rigid graphs on at least four vertices are redundantly rigid, the only globally rigid graphs in $G(n, d)$ for $d \leq 3$ are K_2, K_3, and K_4. The situation changes drastically for $d \geq 4$.

Theorem 23. (Jackson et al., 2007) If $G \in G_{n,d}$ and $d \geq 4$ then G is a.a.s. globally rigid.

Our third model is of geometric random graphs. Let $Geom(n, r)$ denote the probability space of all graphs on n vertices in which the vertices are distributed uniformly at random in the unit square and all pairs of vertices of distance at most r are joined by an edge. Suppose $G \in Geom(n, r)$. Li et al. (2003) have shown that if $n\pi r^2 = \log n + (2k-3) \log\log n + w(n)$ for $k \geq 2$ a fixed integer and $\lim_{n\to\infty} w(n) = \infty$, then G is a.a.s. k-connected. As noted by Eren et al. (2004), this result can be combined with Theorem 17 to deduce that if $n\pi r^2 = \log n + 9 \log\log n + w(n)$ then G is a.a.s. globally rigid. We do not know if this result is best possible. However, it is also shown in (Li et al., 2003) that if $n\pi r^2 = \log n + (k-1) \log\log n + c$ for any constant c, then G is not a.a.s. k-connected. Thus, if $n\pi r^2 = \log n + 2 \log\log n + c$ for any constant c, then G is not a.a.s. 3-connected, and hence is not a.a.s. globally rigid.

Unit Disk Graphs

A *unit disk framework* (with radius R) is a framework (G,p) for which $uv \in E(G)$ if and only if $\|p(u)-p(v)\| \leq R$. A graph G is called a *unit disk graph* if there is a unit disk realization of G. The family of unit disk graphs is a natural model for sensor networks in which the distance between two nodes is known if and only if this distance is at most the sensing radius R of the nodes. It is NP-hard to test whether a graph is a unit disk graph (Breu & Kirkpatrick, 1998), and it is also NP-hard to test whether a unit disk framework is globally rigid (Aspnes et al., 2006). However, it may be possible to use the unit disk property of a unit disk framework (G, p) and bounds on the radius to deduce necessary or sufficient conditions which imply that (G, p) is globally rigid in the sense that it is a unique realization of G, with the given edge lengths, as a unit disk framework. This is a largely unexplored area of research.

Squares of Graphs

The square G^2 of a graph G is obtained from G by adding a new edge uv for each pair $u, v \in V(G)$ of distance two in G. For example, if we double the sensing radius of a unit disk framework with distance graph G, the augmented distance graph will contain its square G^2. This operation makes a network, whose underlying graph is 2-edge-connected, uniquely localizable.

Theorem 24. (Anderson et al., 2007) Let G be a 2-edge-connected graph. Then G^2 is globally rigid.

GLOBALLY LINKED PAIRS AND UNIQUELY LOCALIZABLE NODES

Even if a network is not uniquely localizable, the location of some of its vertices, or the distance between some additional pairs of vertices, may be uniquely determined by the distance graph of the network and the set of anchors. This can be modelled as follows.

A pair of vertices $\{u, v\}$ in a d-dimensional framework (G, p) is *globally linked* in (G, p) if, in all equivalent frameworks (G, q), we have $\|p(u) - p(v)\| = \|q(u) - q(v)\|$. Thus (G, p) is globally rigid if and only if all pairs of vertices of G are globally linked in (G, p). When $d = 1$, it can be seen that a pair of vertices $\{u, v\}$ is globally linked in a generic framework (G, p) if and only if there are two openly disjoint uv-paths in G. Thus global linkedness is a generic property for 1-dimensional frameworks. Unlike global rigidity, however, 'global linkedness' is not a generic property in $\mathbb{R}^d$ when $d \geq 2$. Figures 7 and 8 give an example of a pair of vertices in a rigid graph G which is globally linked in one 2-dimensional generic realization, but not in another: the global linkedness of $\{u, v\}$ depends on the lengths of the edges incident with vertex w. We say that a pair of vertices $\{u, v\}$ is *globally linked in a graph G in $\mathbb{R}^d$* if it is globally linked in *all* generic d-dimensional frameworks (G, p).

We will assume henceforth in this section that $d = 2$ and supress specific reference to this value of d. We first give a necessary condition for two vertices in a framework to be globally linked. Given two vertices x, y in a graph G, let $\kappa_G(x, y)$ denote the maximum number of pairwise openly disjoint xy-paths in G.

Lemma 25. (Jackson et al., 2006) Let (G, p) be a generic framework, $x, y \in V(G)$, $xy \notin E(G)$, and suppose that $\kappa_G(x, y) \leq 2$. Then $\{x, y\}$ is not globally linked in (G, p).

This necessary condition is also sufficient when G is an M-connected graph.

Theorem 26. (Jackson et al., 2006) Let $G = (V, E)$ be an M-connected graph and $x, y \in V$. Then $\{x, y\}$ is globally linked in G if and only if $\kappa_G(x, y) \geq 3$.

Theorem 26 has the following immediate corollary for graphs which are not necessarily M-connected.

Figure 7. Two equivalent realizations of a rigid graph G which show that the pair {u, v} is not globally linked in G

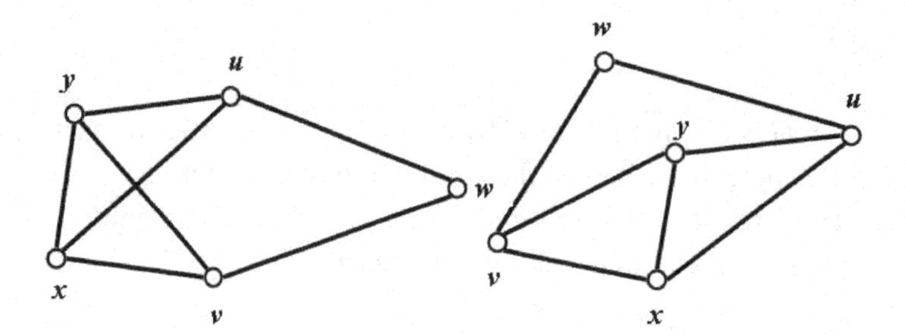

Corollary 27. (Jackson et al., 2006) Let $G = (V,E)$ be a graph and $x, y \in V$. If either $xy \in E$, or there is an M-component H of G with $\{x, y\} \subseteq V(H)$ and $\kappa_H(x, y) \geq 3$, then $\{x, y\}$ is globally linked in G.

We conjecture that the converse is also true.

Conjecture 28. (Jackson et al., 2006) Let $G = (V,E)$ be a graph and $x, y \in V$. Then $\{x, y\}$ is globally linked in G if and only if either $xy \in E$, or there is an M-component H of G with $\{x, y\} \subseteq V(H)$ and $\kappa_H(x, y) \geq 3$.

Corollary 27 can be used to identify large 'globally linked sets of vertices' in a graph G. A *globally rigid cluster* of G is a maximal subset of V in which all pairs of vertices are globally linked in G. By Corollary 27, the vertex sets of the 3-connected '*cleavage units*', (sometimes called 3-*connected components* or 3-*blocks*), see in Section 3 of Jackson and Jordán (2005), of the M-components of G are globally linked sets in G. The truth of Conjecture 28 would imply that the vertex sets of these cleavage units (and the pairs of adjacent vertices not included in these units) are precisely the globally rigid clusters of G. Note that the vertices of a globally rigid cluster of G need not induce a globally rigid subgraph in G. For example, the maximal globally rigid subgraphs of the graph G in Figure 9 are the six copies of K_3 and the remaining four copies of K_2.

Figure 8. Another realization (G, p) of the rigid graph G of Figure 7. The pair {u, v} is globally linked in (G, p) since the lengths of the edges uw, vw preclude the flipping of vertex v about the line containing vertices x and y.

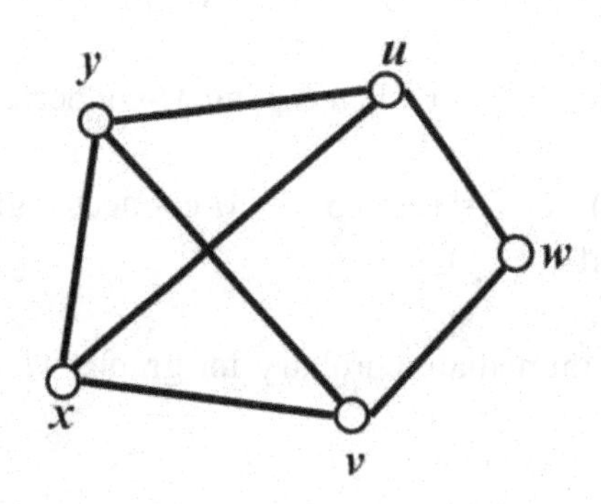

On the other hand, G has three cleavage units, a copy of the wheel on six vertices and two copies of K_4. The globally rigid clusters of G are precisely the vertex sets of these three cleavage units.

We next consider another generalization of global rigidity, unique localizability, which also has direct applications in sensor network localization, see (Goldenberg et al., 2005). Let (G, p) be a generic framework with a designated set $P \subseteq V(G)$ of vertices. We say that a vertex $v \in V(G)$ is *uniquely localizable* in (G, p) *with respect to* P if whenever (G, q) is equivalent to (G, p) and $p(b) = q(b)$ for all vertices $b \in P$, then we also have $p(v) = q(v)$. We can think of P as the set of *pinned vertices* (or *anchor nodes* in a sensor network). Vertices in P are, by definition, uniquely localizable. It is easy to observe that if $v \in V - P$ is uniquely localizable then $|P| \geq 3$ and there exist three openly disjoint paths from v to P (c.f. Lemma 25). As was the case for global linkedness, unique localizablity is not a generic property. Consider the graph given in Figures 7 and 8. If we pin the set $P = \{u, x, y\}$ in the framework of Figure 8, then v is uniquely localizable with respect to P. This is not the case if we pin the same set in Figure 7. Thus the unique localizablity of v with respect to P depends on the lengths of the edges incident with w.

We say that v is *uniquely localizable in the graph G with respect to P*, if v is uniquely localizable with respect to P in all generic frameworks (G, p). Let $G+K(P)$ denote the graph obtained from G by adding all edges bb' for which $bb' \notin E$ and $b, b' \in P$. The following lemma is easy to prove.

Lemma 29. (Jackson et al., 2006) Let $G = (V,E)$ be a graph, $P \subseteq V$ and $v \in V - P$. Then v is uniquely localizable in G with respect to P if and only if $|P| \geq 3$ and $\{v, b\}$ is globally linked in $G + K(P)$ for all (or equivalently, for at least three) vertices $b \in P$.

Lemma 29 and Theorem 26 imply the following characterization of uniquely localizable vertices when $G + K(P)$ is M-connected.

Theorem 30. (Jackson et al., 2006) Let $G = (V,E)$ be a graph, $P \subseteq V$ and $v \in V - P$. Suppose that $G+K(P)$ is M-connected. Then v is uniquely localizable in G with respect to P if and only if $|P| \geq 3$ and $\kappa(v, b) \geq 3$ for all $b \in P$.

Similarly, Lemma 29 and Conjecture 28 would imply the following characterization of uniquely localizable vertices in an arbitrary graph.

Figure 9. An M-connected graph G with three 'cleavage units'. The pairs {u, v} and {x, y} are globally linked

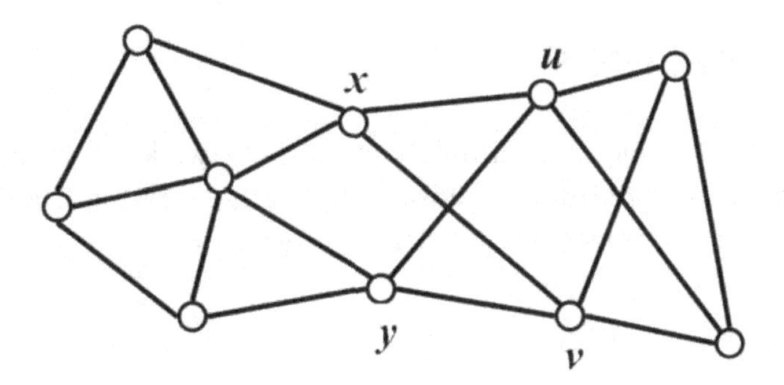

Conjecture 31. (Jackson et al., 2006) Let $G = (V,E)$ be a graph, $P \subseteq V$ and $v \in V - P$. Then v is uniquely localizable in G with respect to P if and only if $|P| \geq 3$ and there is an M-component H of $G + K(P)$ with $P + v \subseteq V(H)$ and $\kappa_H(v, b) \geq 3$ for all $b \in P$.

Theorems 26 and 30 imply that the sets of globally linked pairs and uniquely localizable vertices can be determined for M-connected graphs. Conjectures 28 and 31 would extend this to all graphs.

We close this section by noting that if (G, p) is a generic realization of an M-connected graph G, then we can obtain a representative of each distinct congruence class of frameworks which are equivalent to (G, p) by iteratively applying the following operation to (G, p): choose a 2-vertex-cut $\{u, v\}$ of G and reflect some, but not all, of the components of $G - \{u, v\}$ in the line through the points $p(u)$ and $p(v)$, see (Jackson et al., 2006). Thus, even though a network with an M-connected grounded graph may not be uniquely localizable, all possible sets of locations can be obtained from one set of feasible locations in a simple manner.

OPTIMAL SELECTION OF ANCHORS

Throughout this section, we will again restrict our attention to the 2-dimensional case. Consider the following optimization problem:

Given the set of known distances in a network and a cost function on the nodes, make the network uniquely localizable by designating a minimum cost set of anchor nodes.

Theorems 1 and 3 imply that, for generic networks, we may reformulate the above problem in the following purely combinatorial form:

Given a graph $G = (V,E)$ and a function $c : V \rightarrow \mathbb{R}_+$, find a set $P \subseteq V$, $|P| \geq 3$, for which $G + K(P)$ is globally rigid, and $c(P) = \Sigma_{v \in P}\, c(v)$ is minimum.

For example, suppose that G is the graph obtained from a complete graph on four vertices $\{a, b, c, d\}$ by adding two vertices $\{u, v\}$ and two edges $\{du, uv\}$, and let the cost function be constant. Then an optimal anchor set has cardinality four, and must contain the vertices $\{u, v\}$ as well as two vertices from the set $\{a, b, c\}$. Efficient approximation algorithms for solving this problem were obtained by Fekete and Jordán, using techniques from matroid optimization.

Theorem 32. (Fekete & Jordán, 2006; Fekete & Jordán, 2008) There is a polynomial time 5/2-approximation algorithm for the problem of finding a minimum cost anchor set which makes a generic framework globally rigid.

DISTANCES AND DIRECTIONS

In this section we consider the unique localizability of sensor networks in which some, or all, of the constraints concern the direction, or bearing, between pairs of nodes rather than the distance. We first look at the simpler case when all constraints are direction constraints.

Direction Constraints

Let (G, p) and (G, q) be two d-dimensional realizations of a graph G. We say that (G, p) and (G, q) are *direction equivalent* (also called *parallel*) if $p(u) - p(v)$ is a scalar multiple of $q(u) - q(v)$ for all $uv \in E$. We say that (G, p) and (G, q) are *direction congruent* if there exists a scalar λ and a vector t such that $q(v) = \lambda p(v)+t$ for all $v \in V$. (This is equivalent to saying that (G, q) can be obtained from (G, p) by a translation and dilation.) We call (G, p) *globally direction rigid* (or *tight*) if every framework which is direction equivalent to (G, p) is direction congruent to (G, p).

The fact that the direction constraint is a linear constraint enabled Whiteley to characterize globally direction rigid d-dimensional frameworks in terms of the rank of their `direction rigidity matrix'. He then used this result to obtain a combinatorial characterization for the generic case. The following result can be derived from Theorem 8.2.2 in (Whiteley, 1996).

Theorem 33. (Whiteley, 1996) Let (G,p) be a d-dimensional generic framework. Then (G,p) is globally direction rigid if and only if

$$\sum_{X\in\chi}\left(d|X| - d - 1\right) \geq d|V| - d - 1$$

for all covers χ of G.

Note that the characterization of 2-dimensional globally direction rigid generic frameworks given by Theorem 33 is identical to the characterization of 2-dimensional rigid generic frameworks given in Theorem 6. On the other hand, there are three important differences between the cases when all constraints are directions and all constraints are lengths. The direction problem can be solved for all d, whereas the length problem has been solved only when $d = 1, 2$. Furthermore, there is no need to assume that the framework is generic to solve the direction problem and there is no distinction between 'local behaviour' and 'global behaviour' in the direction problem. Theorem 33 implies that a d-dimensional generic sensor network in which positions of some anchor nodes and directions between some pairs of nodes are known is uniquely localizable if and only if there are at least two anchor nodes and the underlying 'direction graph' of the network satisfies the condition on covers given in Theorem 33. Further results on sensor networks with direction constraints can be found in (Eren, 2007; Katz et al., 2007).

Mixed Constraints

We consider sensor networks in which either distances, directions, or both, are known for some pairs of vertices. The unique localization problem for such networks seems to be at least as hard as that for networks with only distance constraints so we will restrict our attention to the 2-dimensional case (direction constraints are meaningless in one dimension).

A *mixed graph* is a graph together with a bipartition $D \cup L$ of its edge set. We refer to edges in D as *direction edges* and edges in L as *length edges*. A *mixed framework* (G, p) is a mixed graph $G = (V; D, L)$ together with a map $p : V \rightarrow \mathbb{R}^2$. Two mixed frameworks (G, p) and (G, q) are equivalent if $p(u)-p(v)$ is a scalar multiple of $q(u)-q(v)$ for all $uv \in D$ and $\|p(u) - p(v)\| = \|q(u) - q(v)\|$ for all $uv \in L$. The mixed frameworks (G, p) and (G, q) are *congruent* if there exists a vector $t \in \mathbb{R}^2$ and $\lambda \in \{-1, 1\}$

such that $q(v) = \lambda p(v) + t$ for all $v \in V$. This is equivalent to saying that (G, q) can be obtained from (G, p) by a rotation by 0 or 180 degrees and a translation. The mixed framework (G, p) is *globally rigid* if every framework which is equivalent to (G, p) is congruent to (G, p). It is *rigid* if there exists an $\varepsilon > 0$ such that every framework (G, q) which is equivalent to (G, p) and satisfies $\|p(v) - q(v)\| < \varepsilon$ for all $v \in V$, is congruent to (G, p).

Servatius and Whiteley (1999) developed a rigidity theory for mixed frameworks analogous to that for 'distance constrained' frameworks. One may construct a $|D \cup L| \times 2|V|$ mixed rigidity matrix for a mixed framework (G, p) and use its rows to define the *mixed rigidity matroid* of (G, p). A generic mixed framework is rigid if and only if its rigidity matrix, or matroid, has rank $2|V| - 2$. It follows that the rigidity of mixed frameworks is a generic property and we may define a mixed graph G to be *rigid* if every, or equivalently, if some, generic realization of G is rigid. All generic realizations of a mixed graph G have the same mixed rigidity matroid and this matroid is defined to be the *mixed rigidity matroid* of G.

Minimally rigid mixed graphs were characterized by Servatius and Whiteley (1999). For a set X of vertices in a mixed graph G we use $D(X)$ and $L(X)$ to denote the set of direction, resp. length edges in $G[X]$.

Theorem 34. (Servatius & Whiteley, 1999) Let $G = (V; D, L)$ be a mixed graph with $|D \cup L| = 2|V| - 2$. Then G is rigid if and only if for all $X \subseteq V$ with $|X| \geq 2$
$i(X) \leq 2|X| - 2$ when $D(X) \neq \emptyset \neq L(X)$,
and
$i(X) \leq 2|X| - 3$ otherwise.

Their result implies the following characterization of rigid mixed graphs.

Theorem 35. Let $G = (V; D, L)$ be a mixed graph. Then G is rigid if and only if for all covers χ of G we have $\sum_{X \in \chi} f(X) \geq 2|V| - 2$, where $f(X) = 2|X| - 2$ if $D(X) \neq \emptyset \neq L(X)$, and $f(X) = 2|X| - 3$ otherwise.

The problem of characterizing when a generic mixed framework (G, p) is globally rigid is still an open problem. We have, however, been able to obtain some partial results. In order to state these in terms of graphs we define a mixed graph G to be *globally rigid* if all generic realizations of G are globally rigid. (It is not known whether global rigidity of mixed frameworks is a generic property. Thus there may exist mixed graphs which have both a globally rigid generic realization and a non-globally rigid generic realization.)

Figure 10. Two equivalent realizations of the same mixed graph. Solid (dashed) edges indicate length (resp. direction) constraints.

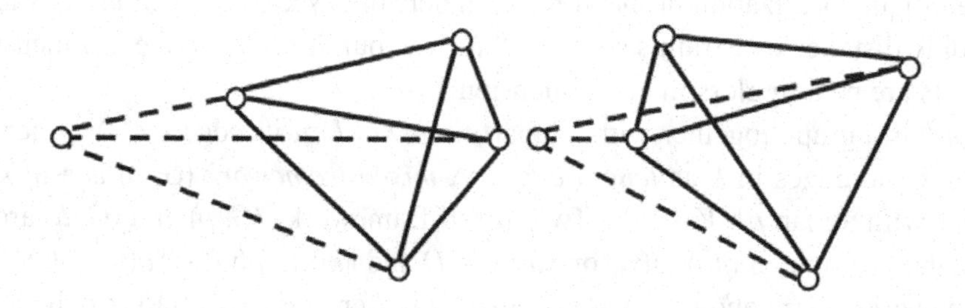

We first give a necessary condition for global rigidity, which is analogous to the '3-connectedness condition' of Theorem 3. It uses the following concept. Let G be a 2-connected mixed graph. A *2-separation* of G is a pair of subgraphs G_1, G_2 such that $G = G_1 \cup G_2$, $|V(G_1) \cap V(G_2)| = 2$ and $V(G_1) - V(G_2) \neq \emptyset \neq V(G_2) - V(G_1)$. The 2-separation is *direction balanced* if both G_1 and G_2 contain a direction edge. We say that G is *direction balanced* if all 2-separations of G are direction balanced.

Theorem 36. (Jackson & Jordán, 2008/b) Every globally rigid mixed graph is 2-connected and direction balanced.

Rigidity is also a necessary condition for global rigidity. Redundant rigidity, however, is no longer necessary. To see this consider a minimally rigid mixed graph G with exactly one length edge e. Then G satisfies the hypotheses of Theorem 34, and hence $G - e$ satisfies the hypotheses of Theorem 5. Theorem 33 now implies that $G - e$ is globally direction rigid, which in turn implies that the mixed graph G is globally rigid. On the other hand, $G - f$ is not rigid for all edges f of G.

We next give a result on 1-extensions which is analogous to Theorem 13. The operation *1-extension* (on edge uw and vertex z) for a mixed graph G deletes an edge uw and adds a new vertex v and new edges vu, vw, vz for some vertex $z \in V(G)$, with the provisos that at least one of the new edges has the same type as the deleted edge and, if $z = u$, then the two edges from z to u are of different type.

Theorem 37. (Jackson & Jordán, 2008/c) Let H be a globally rigid mixed graph with at least three vertices and let G be obtained from H by a 1-extension on an edge uw. If $H - uw$ is rigid, then G is globally rigid.

In mixed graphs, a special kind of 0-extension also preserves global rigidity.

Theorem 38. (Jackson & Jordán, 2008/c) Let G and H be mixed graphs with $|V(H)| \geq 2$. Suppose that G can be obtained from H by a 0-extension which adds a vertex v incident to two direction edges. Then G is globally rigid if and only if H is globally rigid.

Note that if G is obtained by a 0-extension then G is not redundantly rigid.

We can use Theorems 37 and 38 to show that a special family of generic mixed frameworks are globally rigid. We say that a mixed graph $G = (V; D, L)$ is a *mixed M-circuit* if $D \neq \emptyset \neq L$ and $D \cup L$ is a circuit in the mixed rigidity matroid of G. Theorem 34 can be used to characterize mixed M-circuits and, in particular, show that mixed M-circuits are 2-connected. We recently showed that the other necessary condition for global rigidity given in Theorem 36 is also sufficient to imply that mixed M-circuits are globally rigid.

Theorem 39. (Jackson & Jordán, 2008/b) Let G be a mixed M-circuit. Then G is globally rigid if and only if G is direction balanced.

We close this section by noting that we can characterize global rigidity in all dimensions for mixed graphs in which every pair of adjacent vertices is connected by both a length and a direction edge.

Theorem 40. (Jackson & Jordán, 2008/b) Let G be a mixed graph in which every pair of adjacent vertices is connected by both a length and a direction edge, and (G, p) be a generic realization of G in $\mathbb{R}^d$. Then (G, p) is globally rigid if and only if G is 2-connected.

ALGORITHMS

The structural results presented in this chapter give rise to efficient combinatorial algorithms for testing different localizability properties of generic networks and for solving a number of related algorithmic problems in the plane. Here we simply list the basic questions and refer the reader to the references below. The combinatorial characterizations lead us to two kinds of problems. How can we decide whether a given graph $G = (V,E)$ is rigid, redundantly rigid, or M-connected? How can we test if G is 3-connected, identify its 'cleavage units', or find its 2-vertex-cuts?

The solution for the questions in the first group boils down to the existence of an efficient subroutine for checking if a set of edges satisfies the count in Theorem 5, that is, whether it is M-independent in the rigidity matroid (c.f. last paragraph of the section on matroids). This subroutine can be implemented in $O(|V|^2)$ time by using various alternating path algorithms: methods from matching theory (Hendrickson, 1992), network flows (Imai, 1985), matroid optimization (Gabow & Westermann, 1992), and graph orientations (Berg & Jordán, 2003/b; Hendrickson & Jacobs, 1997) have been used for this job. By using additional algorithmic techniques, each of these properties can be tested in $O(|V|^2)$ time.

The connectivity related questions of the second group can be solved in $O(|V| + |E|)$ time (Hopcroft & Tarjan, 1973).

HIGHER DIMENSIONAL RESULTS

Although most of the known results on globally rigid graphs are concerned with the two dimensional case, some of them extend to higher dimensions, leading to partial results on the unique localizability of 3-dimensional networks. The main tool for working with higher dimensional global rigidity is the 'stress matrix', which was introduced by Connelly (1982) and plays a similar role for global rigidity as the rigidity matrix does for rigidity. Indeed, we will see that the global rigidity of a generic framework can be characterized in terms of the rank of a stress matrix in much the same way as Theorem 4 characterizes rigidity in terms of the rank of the rigidity matrix. Although the 3-dimensional case is of most

Figure 11. A direction balanced mixed M-circuit

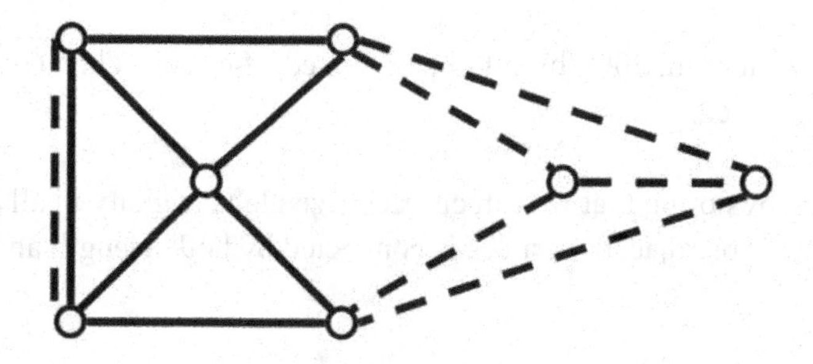

practical interest, all the results we know for the 3-dimensional case extend to d-dimensions, and we will present them in this context.

Let $G = (V,E)$ be a graph and (G, p) be a realization of G in $\mathbb{R}^d$. For each $v \in V$, let $E(v)$ be the set of edges of G which are incident to v. An *equilibrium stress*, or *self-stress*, for (G, p) is a map $\omega : E \to \mathbb{R}$ such that, for each $v \in V$, $\sum_{e=uv\in E(v)} \omega(e)(p(u) - p(v)) = 0$. We associate a symmetric $|V|\times|V|$ *stress matrix* Ω with each equilibrium stress ω for (G, p) as follows. For each distinct $u, v \in V$, the entry in row u and column v of Ω is $-\omega(e)$ if $e = uv \in E$ and zero otherwise. The diagonal entries of Ω are then chosen so that its row and column sums are equal to zero. Connelly (1982) proved that the rank of Ω is at most $|V| - d - 1$ and that having a stress matrix with this maximum possible rank is a sufficient condition for the global rigidity of a generic framework.

Theorem 41. (Connelly, 2005) Let $G = (V,E)$ be a graph and (G, p) be a generic realization of G in $\mathbb{R}^d$. If (G, p) has an equilibrium stress ω such that the associated stress matrix Ω has rank $|V|-d-1$, then (G, p) is globally rigid.

Connelly (2005) conjectured that this condition is also a necessary condition for the global rigidity of a generic framework (G, p) when G is not complete. His conjecture was recently verified by Gortler, Healy and Thurston.

Theorem 42. (Gortler et al., 2007) Let $G = (V,E)$ be a graph and (G, p) be a globally rigid generic realization of G in $\mathbb{R}^d$. Then either G is a complete graph on at most $d + 1$ vertices, or (G, p) has an equilibrium stress ω such that the associated stress matrix Ω has rank $|V| - d - 1$.

Theorem 42 implies that global rigidity in $\mathbb{R}^d$ is a generic property. It also gives rise to a randomized algorithm for checking whether a graph is globally rigid in $\mathbb{R}^d$, see (Gortler et al., 2007) for more details.

Theorem 41 implies the result of Connelly mentioned in the section on inductive constructions: if a graph can be obtained from K_4 by a sequence of 1-extensions, then it is globally rigid in $\mathbb{R}^2$. In fact, Connelly used Theorem 41 to obtain a d-dimensional version of this result. A *d-dimensional 1-extension* of a graph G subdivides an edge uw with a new vertex v and then adds $d-1$ new edges vz_i, $1 \le i \le d - 1$, with $u \ne z_i \ne w$ and $z_i \ne z_j$ for $1 \le i < j \le d - 1$. Connelly (2005) showed that the property that a d-dimensional generic framework has a stress matrix of maximum rank is preserved by a d-dimensional 1-extension of the underlying graph. Combining this with Theorem 42 we obtain the following d-dimensional version of Theorem 13.

Theorem 43. Let H be a graph with at least $d + 2$ vertices and let G be obtained from H by a d-dimensional 1-extension. If H is globally rigid in $\mathbb{R}^d$ then G is globally rigid in $\mathbb{R}^d$.

As a special case, we obtain the result of Connelly (2005) that if a graph can be obtained from K_{d+2} by a sequence of d-dimensional 1-extensions, then it is globally rigid in $\mathbb{R}^d$.

Cheung and Whiteley conjecture that a d-dimensional version of the vertex splitting operation also preserves global rigidity in d-dimensions. Given a graph G, edges $u_iv \in E$ for $1 \le i \le d - 1$, and a bipartition F_1, F_2 of the edges incident to v in $G - \{u_1v, u_2v, \ldots, u_{d-1}v\}$, the d-dimensional vertex splitting operation replaces the vertex v by two new vertices v_1 and v_2 and a new edge v_1v_2, replaces each edge

$u_i v$ by two new edges $u_i v_1$, $u_i v_2$ for $1 \leq i \leq d-1$, and replaces each edge $wv \in F_i$ by an edge wv_i, for i = 1, 2. The d-dimensional vertex splitting operation is said to be *non-trivial* if F_1, F_2 are both non-empty, or equivalently, if each of the split vertices v_1, v_2 has degree at least $d + 1$.

Conjecture 44. (Cheung & Whiteley, 2008) If H is a globally rigid graph in $\mathbb{R}^d$ and G is obtained from H by a non-trivial d-dimensional vertex splitting operation, then G is globally rigid in $\mathbb{R}^d$.

Theorem 15 verifies the 2-dimensional version of Conjecture 44. It is open for $d \geq 3$. Cheung and Whiteley (2008) discuss several additional conjectures on globally rigid graphs in higher dimensions.

ACKNOWLEDGMENT

This work was supported by the Mobile Innovation Centre, funded by the Hungarian National Office for Research and Technology.

REFERENCES

Anderson, B.D.O., Belhumeur, P.N., Eren, T., Goldenberg, D.K., Morse, A.S., Whiteley, W. & Yang, Y.R. (2007) *Graphical properties of easily localizable sensor networks*, Wireless Netw., to appear.

Aspnes, J., Eren, T., Goldenberg, D.K., Morse, A.S., Whiteley, W., Yang, Y.R., Anderson, B.D.O. & Belhumeur, P.N. (2006) A theory of network localization, IEEE Trans. on *Mobile Computing*, *5*(12),1663-1678.

Aspnes, J., Goldenberg, D.K. & Yang, Y.R. (2004) On the computational complexity of sensor network localization, Springer Lecture Notes in Computer Science 3121, *Algorithmic Aspects of Wireless Sensor Networks*, (pp. 32-44). Berlin, Springer.

Berg, A. & Jordán, T. (2003/a) A proof of Connelly's conjecture on 3-connected circuits of the rigidity matroid, *J. Combinatorial Theory*, Ser. B. 88, 77-97.

Berg, A. & Jordán, T. (2003/b) Algorithms for graph rigidity and scene analysis, In G. Di Battista & U. Zwick, (Ed), *11th Annual European Symposium on Algorithms (ESA)* (pp.78-89), Springer Lecture Notes in Computer Science 2832. Berlin, Springer.

Berger, B., Kleinberg, J. & Leighton, T. (1999) Reconstructing a three-dimensional model with arbitrary errors, *Journal of the ACM*, 46 (2), 212-235.

Bollobás, B. (1985) *Random graphs*, New York, Academic Press.

Bondy, J.A. & Murty, U.S.R. (2008) *Graph theory*, Springer.

Breu, H. & Kirkpatrick, D.G. (1998) Unit disk graph recognition is NP-hard, *Comput. Geom. 9* (1-2), 3-24.

Cheung, M. & Whiteley, W. (2008) *Transfer of global rigidity results among dimensions: graph powers and coning*, preprint, York University.

Connelly, R. (1982) Rigidity and energy, *Invent. Math.*, 66(1), 11-33.

Connelly, R. (1991) On generic global rigidity, Applied geometry and discrete mathematics, 147–155, *DIMACS Ser. Discrete Math. Theoret. Comput. Sci., 4*, Amer. Math. Soc., Providence, RI.

Connelly, R. (2005) Generic global rigidity, *Discrete Comput. Geom.* 33, 549-563.

Crapo, H. (1990) *On the generic rigidity of plane frameworks*, INRIA research report No. 1278.

Eren, T., Goldenberg, D., Whiteley, W., Yang, Y.R., Morse, A.S., Anderson, B.D.O. & Belhumeur, P.N. (2004) Rigidity, Computation, and Randomization in Network Localization, In *IEEE INFOCOM Conference*, (pp.2673-2684) Hong Kong.

Eren, T., Whiteley, W., Morse, A.S., Belhumeur, P.N. & Anderson, B:D.O. (2003) Sensor and network topologies of formations with direction, bearing and angle information between agents, In *42nd IEEE Conference on Decision and Control*, (pp. 3064-3069).

Eren, T. Using angle of arrival (bearing) information for localization in robot networks, *Turk J Elec Engin*, Vol *15*(2), 169-186.

Fekete, Z. (2006) Source location with rigidity and tree packing requirements, *Operations Research Letters 34*(6), 607-612.

Fekete, Z. & Jordán, T. (2006) Uniquely localizable networks with few anchors, In S. Nikoletseas, S. & Rolim, J.D.P. (Ed.) *Algosensors 2006*, (pp. 176-183). Springer Lecture Notes in Computer Science 4240.

Fekete, Z. & Jordán, T. (2008) Algorithms for minimum cost anchor sets in uniquely localizable networks, preprint.

Gabow, H.N. & Westermann, H.H. (1992) Forests, frames and games: Algorithms for matroid sums and applications, *Algorithmica* 7, 465-497.

Gluck, H. (1975) Almost all simply connected closed surfaces are rigid, Geometric topology (Proc. Conf., Park City, Utah, 1974), pp. 225–239. *Lecture Notes in Math., 438*, Springer, Berlin.

Goldenberg, D.K., Krishnamurthy, A., Maness, W.C., Yang, Y.R., Young, A., Morse, A.S., Savvides, A. & Anderson, B.D.O. (2005) Network localization in partially localizable networks, In *IEEE INFOCOM Conference*, (pp.313-326), Miami.

Gortler, S.J., Healy, A.D. & Thurston, D.P. (2007) Characterizing generic global rigidity, arXiv:0710.0926v3.

Graver, J., Servatius, B. & Servatius, H. (2003) Combinatorial Rigidity, *AMS Graduate Studies in Mathematics* 2.

Hendrickson, B. (1992) Conditions for unique graph realizations, *SIAM J. Comput.*21 (1), 65-84.

Hendrickson, B. & Jacobs, D. (1997) An algorithm for two-dimensional rigidity percolation: the pebble game, *J. Computational Physics* 137, 346-365.

Henneberg, L. (1911) *Die graphische Statik der starren Systeme*, Leipzig.

Hopcroft J.E. & Tarjan, R.E. (1973) Dividing a graph into triconnected components, *SIAM J. Comput.* 2, 135–158.

Imai, H. (1985) On combinatorial structures of line drawings of polyhedra, *Discrete Appl. Math.* 10, 79-92.

Jackson, B. & Jordán, T. (2005) Connected rigidity matroids and unique realizations of graphs, *J. Combinatorial Theory, Ser. B.*, 94, 1-29.

Jackson, B., Jordán, T. & Szabadka, Z. (2006) Globally linked pairs of vertices in equivalent realizations of graphs, *Discrete and Computational Geometry*, 35, 493-512.

Jackson, B. & Jordán, T. (2008a) A sufficient connectivity condition for generic rigidity in the plane, Discrete Applied Math, in press.

Jackson, B & Jordán, T. (2008b) Globally rigid circuits of the two-dimensional direction-length rigidity matroid, J. Combinatorial Theory, Ser. B., to appear.

Jackson, B & Jordán, T. (2008c) Operations preserving global rigidity of generic direction-length frameworks, Egerváry Research Group TR-2008-08, submitted.

Jackson, B., Servatius, B. & Servatius, H. (2007) The 2-dimensional rigidity of certain families of graphs, *J. Graph Theory* 54 (2), 154–166.

Jordán, T. & Szabadka, Z. (in press) Operations preserving the global rigidity of graphs and frameworks in the plane, *Comput.Geom.*.

Katz, B., Gaertler, M. & D. Wagner, (2007) Maximum rigid components as means for direction-based localization in sensor networks, In Jan van Leeuwen et al. (Ed) *SOFSEM 2007, LNCS 4362*, (pp. 330-341), Springer, Berlin.

Laman, G. (1970) On graphs and rigidity of plane skeletal structures, *J. Engineering Math.* 4, 331-340.

Li, X-Y., Wan, P-J., Wang, Y. & Yi, C-W. (2003) Fault tolerant deployment and topology control in wireless networks. In *ACM Symposium on Mobile Ad Hoc Networking and Computing (MobiHoc)* (pp. 117-128), Annapolis, MD.

Lovász, L. & Yemini, Y. (1982) On generic rigidity in the plane, *SIAM J. Algebraic Discrete Methods* 3 (1), 91–98.

Mao, G., Fidan, B. & Anderson, B.D.O. (2007) Localisation, In N.P. Mahalik (Ed.) *Sensor networks and configuration* (pp. 281-315), Springer, Berlin.

Recski, A. (1989) *Matroid theory and its applications in electric network theory and in statics*, Budapest, Akadémiai Kiadó.

Saxe, J.B. (1979) *Embeddability of weighted graphs in k-space is strongly NP-hard*, (Tech. Report), Computer Science Department, Carnegie-Mellon University, Pittsburgh, PA.

Schrijver, A. (2003) *Combinatorial Optimization*, Berlin, Springer.

Servatius B. & Whiteley, W. (1999) Constraining plane configurations in CAD: Combinatorics of directions and lengths, *SIAM J. Discrete Math.*, 12, 136–153.

So, A.M. & Ye, Y. (2007) Theory of semidefinite programming for sensor network localization, *Math. Program.* 109 (3) Ser. B, 367–384.

Tay, T.S. & Whiteley, W. (1985) Generating isostatic frameworks, *Structural Topology* 11, pp. 21-69.

Whiteley, W. (1990) Vertex splitting in isostatic frameworks, *Structural Topology* 16, 23–30.

Whiteley, W. (1996) Some matroids from discrete applied geometry. Matroid theory (Seattle, WA, 1995), 171–311, *Contemp. Math.,* 197, Amer. Math. Soc., Providence, RI.

Whiteley, W. (2004) Rigidity and scene analysis, In J. E. Goodman and J. O'Rourke (Eds.) *Handbook of Discrete and Computational Geometry*, (pp. 1327-1354), CRC Press, Second Edition.

Yemini, Y. (1979) Some theoretical aspects of position-location problems, In *20th Annual IEEE Symposium on Foundations of Computer Science*, (pp. 1–8).

APPENDIX

In what follows we introduce the basic graph theoretical notions that are used in this chapter. For more details see for example (Bondy & Murty, 2008).

A *graph* $G = (V,E)$ consists of two sets V and E. The elements of V are called *vertices* (or *nodes*). The elements of E are called *edges*. Each edge $e \in E$ joins two vertices from V, which are called the *endvertices* of e. The notations $V(G)$ and $E(G)$ are also used for the vertex- and edge-sets of a graph G. If vertex v is an endvertex of edge e then v is said to be *incident* with e and e is incident with v. A vertex v is *adjacent* to vertex u if they are joined by an edge. A graph is *simple* if the pairs of endvertices of its edges are pairwise distinct. With the exception of the mixed graphs in the section on distance and direction constraints, all graphs considered in this chapter will be simple.

The *degree* of a vertex v in a graph G, denoted by $d_G(v)$, is the number of edges incident with v. A graph is *regular* if every vertex is of the same degree. It is *k-regular* if every vertex is of degree k.

A *path* in a graph G from vertex u to vertex v is an alternating sequence of vertices and edges, which starts and ends with u and v (which are its initial and final vertices, respectively), and for which consecutive elements are incident with each other and no internal vertex is repeated. A *cycle* is a path which contains at least one edge and for which the initial vertex is also the final vertex. A graph is *connected* if between every pair of vertices there is a path.

A *subgraph* of a graph G is a graph H with $V(H) \subseteq V(G)$ and $E(H) \subseteq E(G)$. In a graph G the *induced subgraph* on a set X of vertices, denoted by $G[X]$, has X as its vertex set and it contains every edge of G whose endvertices are in X. A subgraph H is a *spanning subgraph* if $V(H) = V(G)$. A *component* of a graph G is a maximal connected subgraph. A *k-factor* of a graph G is a k-regular spanning subgraph.

The operation of *deleting a vertex set* $X \subseteq V(G)$ from a graph G removes the vertices in X from $V(G)$ and also removes every edge which has an endvertex in X from $E(G)$. The resulting graph is denoted by $G - X$ (or $G - x$, if $X = \{x\}$ is a single vertex). The operation of *deleting an edge* set $F \subseteq E(G)$ from a graph G removes the edges in F from $E(G)$. The resulting graph is denoted by $G-F$ (or $G-f$, if $F = \{f\}$ is a single edge).

A *forest* is a graph without cycles and a *tree* is a connected forest. A *spanning tree* of a graph G is a spanning subgraph which is a tree.

A graph is a *complete graph* if each pair of its vertices is joined by an edge. A complete graph on n vertices is denoted by K_n. A graph is *bipartite* if its vertices can be partitioned into two sets in such a way that no edge joins two vertices in the same set. A *complete bipartite graph* is a bipartite graph in which each vertex in one partite set is adjacent to all vertices in the other partite set. If the two partite sets have cardinalitites m and n, then this graph is denoted by $K_{m,n}$. A graph G on n vertices is a *wheel*, denoted by W_n, if it has an induced subgraph which is a cycle on $n - 1$ vertices and the remaining vertex is joined to all vertices of this cycle.

A *k-vertex-cut* in a graph G is a set $X \subseteq V(G)$ of k vertices for which $G - X$ is not connected. A *k-edge-cut* is a set $F \subseteq E(G)$ of k edges for which $G - F$ is not connected. A graph is called *k-vertex-connected* (or *k-connected*) if it has at least $k + 1$ vertices and contains no l-vertex-cut for $l \leq k - 1$. A graph is *k-edge-connected* if it contains no l-edge-cuts for $l \leq k - 1$.

Two paths are called *openly disjoint* if they have no common internal vertex. They are called *edge disjoint* if they have no common edge. A fundamental theorem of Menger states that if u and v are non-adjacent vertices in graph G then the smallest integer k for which there is a k-vertex-cut X in G such that u and v are in different components of $G - X$ is equal to the maximum number of pairwise openly disjoint paths from u to v. The edge disjoint version of Menger's theorem is as follows. For any pair of vertices u, v in G the smallest integer k for which there is a k-edge-cut F in G such that u and v are in different components of $G-F$ is equal to the maximum number of pairwise edge disjoint paths from u to v.

An *isomorphism* between two graphs G and H is a vertex bijection $\phi : V(G) \rightarrow V(H)$ such that $uv \in E(G)$ if and only if $\phi(u)\, \phi(v) \in E(H)$. A *graph automorphism* is an isomorphism of the graph to itself. The *orbit of a vertex u* of a graph G is the set of all vertices $v \in V(G)$ such that there is an automorphism ϕ such that $\phi(u) = v$. A graph is *vertex-transitive* if all the vertices are in the same orbit.

The *incidence matrix* of a graph $G = (V,E)$ is an $|E| \times |V|$ matrix I where the entry in the row of edge e and vertex v is equal to 1 if e is incident with v, and 0 otherwise. The *directed incidence matrix* of G is obtained from I by replacing exactly one of the two 1's in each row of I by -1.

Chapter VII
Sequential Localization with Inaccurate Measurements

Jia Fang
Yale University, USA

Dominique Duncan
Yale University, USA

A. Stephen Morse
Yale University, USA

ABSTRACT

The sensor network localization problem with distance information is to determine the positions of all sensors in a network given the positions of some sensors and the distances between some pairs of sensors. In this chapter the authors present a sequential algorithm for estimating sensor positions when only inaccurate distance measurements are available, and they use experimental evaluation to demonstrate network instances on which the algorithm is effective.

INTRODUCTION

In many situations where wireless sensor networks are used, only the positions of some of the sensors are known, and the positions of the remaining sensors must be inferred from the known locations and available inter-sensor distance measurements. More formally, consider n sensors in the plane labelled 1 through n, where the positions of some sensors are known, and the measured distances between some pairs of sensors are known. The sensors with known positions are called *anchors*. Since ranging devices are never exact, we consider the following model for the type of distance measurements obtained. For each inter-sensor distance measurement $\tilde{d}$, we assume that an *accuracy guarantee* denoted by $\varepsilon > 0$ of $\tilde{d}$ is given such that the actual inter-sensor distance is within ε of the measured distance $\tilde{d}$. Obviously, sensor positions are generally not uniquely determined by inaccurate distance measurements. In Moore et al.(2004) and Priyantha et al. (2003), it is pointed out that it is more important to obtain position estimates which reflect the general layout of the actual sensor positions, rather than simply position estimates which induce inter-sensor distances within some desired tolerance of the given inter-sensor distance measurements. In Priyantha et al. (2003), a modified spring based relaxation method is used to obtain sensor estimates, and it is shown via experimental evaluations that the estimated positions reflect the general layout of the actual sensor positions. Our work is most closely related to Moore et al. (2004) where the aim was to compute position estimates for subnetworks called "robust quadrilaterals" with correctness guarantees. More specifically, an algorithm is given which assigns position estimates to the sensors in a robust quadrilateral only if the position estimates can be guaranteed to be free of "flip ambiguities" with high probability Moore et al. (2004).

Roughly speaking, position estimates are desired so that the configuration of the estimated positions are approximately congruent to that of the actual positions. We capture this notion using the concept of "correctly oriented" position estimates which we define as follows. For two points p and q in R^2, let $l(p,q)$ denote the line segment with endpoints p and q. For $m \geq 4$ sensors labelled 1 through m and $i \in \{1,...,m\}$ let p_i denote the position of sensor i, and let q_i denote the estimated position of sensor i. The estimated positions $q_1,..., q_m$ are said to be *correctly oriented* if for all distinct $i,j,k,l \in \{1,...,m\}$, the line segments $l(p_i, p_j)$ and $l(p_k, p_l)$ intersect if and only if the line segments $l(q_i, q_j)$ and $l(q_k, q_l)$ intersect. As an illustration, suppose sensors 1, 2, 3 and 4 are positioned at p_1, p_2, p_3 and p_4 respectively as shown in Figure 1(a) and 1(b). For each sensor i, let q_i and q'_i denote two estimated positions of sensor i. Suppose q_i for i=1,2,3,4 are as shown in Figure 1(a), and q'_i are as shown in Figure 1(b). It is easy to see that $\{q_1, q_2, q_3, q_4\}$ are correctly oriented while $\{q'_1, q'_2, q'_3, q'_4\}$ are not correctly oriented.

In Section Correctly Oriented Position Estimates, we will demonstrate that correctly oriented position estimates can be used to deduce important geometric properties of the configuration of the actual sensor positions. The key difficulty however lies in determining if a set of position estimates are correctly oriented without knowing the corresponding actual sensor positions. In this work we will propose a position estimation algorithm for computing position estimates with *error bounds*, and give a sufficient condition on the position estimates and the corresponding error bounds for the position estimates to be correctly oriented.

In Anderson et al. (2007) a sequential localization algorithm for exact distance measurements was proposed in which the sensors of the network are processed one by one in a pre-determined order. That

Figure 1. (a) Correctly oriented position estimates, (b) Position estimates which are not correctly oriented

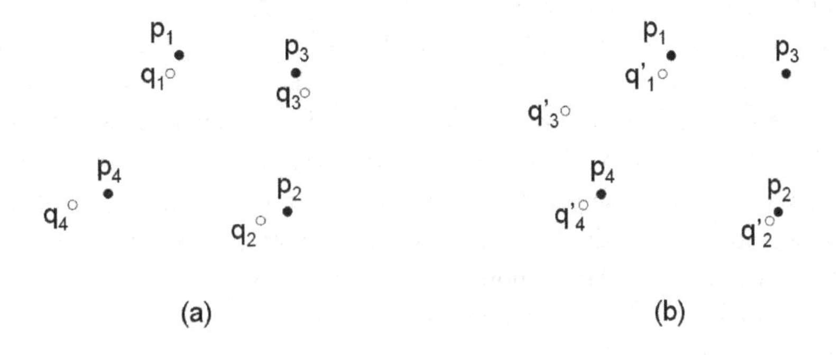

work was extended in Fang et al. (2006), Goldenberg et al. (2006) and Fang et al. (2008) to a sequential localization algorithm called Sweeps again for the case of *exact* distance measurements. In this chapter, we present an algorithm based on Sweeps, which we call modified Sweeps, for estimating sensor positions of a network when only inaccurate distance measurements can be obtained. In modified Sweeps, we aim to give correctness guarantees on estimated positions and characterize position estimates which are oriented correctly. More specifically, for each position estimate p computed by modified Sweeps, an error bound $e(p)$ is also computed such that the maximum distance between the position estimate p and the actual position is at most $e(p)$. The error bounds can be used by the final application to determine which sensor estimates are precise enough to be useful. More importantly, we will use the error bounds to give a sufficient condition for the position estimates to be correctly oriented, for as will be shown in Section Correctly Oriented Position Estimates, position estimates which are correctly oriented can be used to deduce geometric properties of the configuration of actual sensor positions. We note that however not all sensors are assigned position estimates even when the average degree of the network exceeds six. Like the algorithms proposed in Moore et al. (2004) and Anderson et al. (2007), modified Sweeps is a sequential algorithm in the sense that the sensors are processed one by one in some order. As was shown in Moore et al. (2004) and Anderson et al. (2007), the notion of processing sensors in a particular order can be used to gain important insight into the graphical characterization of networks whose positions or position estimates can be computed *efficiently*. In Section Efficiently Localizable Networks, we will give the graphical characterization of some networks for which modified Sweeps can be used to efficiently compute position estimates.

In Section Experimental Evaluations, we discuss the performance of modified Sweeps on a network of 100 sensors randomly deployed in a unit square. A noisy distance measurement is generated for each pair of sensors within a specified sensing radius. Roughly speaking, a distance measurement with a guaranteed accuracy of ε is "noisier" than a distance measurement with a guaranteed accuracy δ when $\delta < \varepsilon$. In general, we note that the number of sensors for which a position estimate is obtained increases with the average degree of the network, and as expected, the error bound associated with each position estimate increases as distance measurements become noisier. However, even for the case where the guaranteed accuracy of each distance measurement d is 8% of d, the respective error bounds of the position estimates is on average less than 1/6 of the sensing radius. The trade-off however is that only ~55% of the sensors are assigned position estimates.

HIGH LEVEL DESCRIPTION OF MODIFIED SWEEPS

In this section, we give a high level description of modified Sweeps and illustrate some key aspects of the algorithm via examples. A *candidate regions set* of a sensor is a set consisting of a finite number of regions in the plane with the property that the actual position of the sensor is in one of the regions, and the regions are either all polygons or all disks. We call each region in a candidate regions set of a sensor a *candidate region* of the sensor. We say that a candidate region of a sensor is *false* if the region does not contain the sensor's position. When the candidate regions set of a sensor consists of only one region, say with diameter d, then the centroid of the region must be within at most distance $d/2$ from the sensor's actual position. Obviously, it is most desirable to obtain for each sensor a candidate regions set which consists of just one candidate region with a "small" diameter.

Roughly speaking, modified Sweeps first computes a candidate regions set for each sensor by processing the sensors one by one in some order, and then refines each candidate regions set by processing the sensors in an ordering different from the previous ordering. More specifically, to determine a candidate regions set for each sensor, an ordering of the sensors is first determined so that the anchors precede all other sensors in the ordering, and each non-anchor sensor has at least one "predecessor" in the ordering. A *predecessor* of a sensor is any other sensor preceding it in the ordering such that the measured distance between the two sensors is obtained. Assuming such an ordering exists, the algorithm "sweeps" through the network by processing the sensors sequentially according to the ordering, beginning with the first non-anchor sensor in the ordering. For each non-anchor sensor, a candidate regions set is computed for the sensor using the measured distances between the sensor and its predecessors, and the candidate regions sets, or known positions, of its predecessors. Once the last sensor in the ordering is processed, a candidate regions set will have been computed for each sensor. We call this first "sweep" a *candidate regions set generating sweep.* Once a candidate regions set has been computed for each sensor, subsequent "refining" sweeps are performed to remove, if possible, regions from each candidate regions set so as to obtain a candidate regions set of fewer elements. To perform a refining sweep, an ordering distinct from the one used to perform the previous sweep is determined, and the sensors are again processed sequentially according to the new ordering. In the following, we will use *singleton* to refer to a set consisting of exactly one element. For each non-anchor sensor v with a non-singleton candidate regions set, the candidate regions sets of v's predecessors, and the measured distances between v and its predecessors, are used to identify, if possible, those candidate regions of sensor v which do not contain v's actual position. More specifically, for each sensor, the minimum and maximum distances between the sensor's candidate regions and the candidate regions of its predecessors in the new ordering are used as a means of identifying false candidate regions of the sensor. The false candidate regions are then removed from the candidate regions set of the sensor to obtain a candidate regions set of fewer elements. Note that when two candidate regions are both disks or both polygons, the minimum and maximum distances between them are efficient to compute.

A sensor is localized if its candidate regions set consists of a point, i.e. a region with diameter zero. As noted previously, exact positions in general cannot be computed when distance measurements are inaccurate. Hence, the desired outcome for each sensor is that after a finite number of sweeps through the network, the candidate regions set of the sensor contains just one candidate region with a "small" diameter. Whether this will be the case will depend on the geometry of the configuration of the actual sensor positions. In Section Experimental Evaluations, we will show that for randomly deployed networks of 100 sensors, it is possible to obtain singleton candidate regions sets with error bounds less than 1/6 of the sensing range. In the following two sections, we will give examples of using modified Sweeps to estimate sensor positions in two simple networks. Although the networks are simple, they illustrate the two key aspects of modified Sweeps, namely generating and refining candidate regions sets.

Generating Candidate Regions Sets

For a pair of sensors i and j in a network, let d_{ij} denote the actual distance between the sensors, and if the measured distance between them is obtained, let $\tilde{d}_{ij}$ denote the measured distance, and let ε_{ij} denote the guaranteed accuracy of $\tilde{d}_{ij}$. By definition, if the measured distance $\tilde{d}_{ij}$ is obtained for sensors i and j, then the actual distance d_{ij} between i and j must be within ε_{ij} of the measured distance:

$$d_{ij} \in [\tilde{d}_{ij} - \varepsilon_{ij}, \tilde{d}_{ij} + \varepsilon_{ij}] \quad (1)$$

Consider the network of four sensors labelled *a,b,c,v* and positioned at points π(a), π(b), π (c) and π (v) respectively, so that no three of the sensor positions are collinear. The sensors labelled *a, b* and *c* are anchors, and suppose distance measurements $\tilde{d}_{av}$, $\tilde{d}_{bv}$ and $\tilde{d}_{cv}$ are obtained. Using the guaranteed accuracies of the distance measurements, we get the following:

$$d_{av} \in [\tilde{d}_{av} - \varepsilon_{av}, \tilde{d}_{av} + \varepsilon_{av}],\ d_{bv} \in [\tilde{d}_{bv} - \varepsilon_{bv}, \tilde{d}_{bv} + \varepsilon_{bv}],\ d_{cv} \in [\tilde{d}_{cv} - \varepsilon_{cv}, \tilde{d}_{cv} + \varepsilon_{cv}] \quad (2)$$

The network is denoted by the graph shown in Figure 2 where each vertex corresponds to the sensor of the same label, and two vertices are adjacent if either the corresponding sensors are both anchors, or the distance measurement between the sensors is obtained.

An ordering of the sensors is first determined so that the anchors precede all other sensors, and each non-anchor sensor has at least one predecessor. One such ordering is *a,b,c,v*, and the predecessors of sensor *v* are *a*, *b* and *c* since the distance measurements between sensor *v* and anchors *a,b,c* are given. For $x \in \{a,b,c\}$, let A_x denote the ring centered at $\pi(x)$ with inner radius $\tilde{d}_{xv} - \varepsilon_{xv}$ and outer radius $\tilde{d}_{xv} + \varepsilon_{xv}$. From (2) it follows that $\pi(v) \in A_a \cap A_b \cap A_c$. Suppose the three rings intersect as shown in Figure 3a.

The rings' intersection can be approximated by a disk D which contains $A_a \cap A_b \cap A_c$ as shown in Figure 3b. Clearly, $\pi(v) \in$ D, so the singleton set containing just D is a candidate regions set for sensor *v*. Ideally, D would be a disk with the least diameter that contains the common intersection of the three rings. However, modified Sweeps does not require this. In our implementation of modified Sweeps, we use an efficient algorithm to compute an approximating polygon which contains the ring intersection. Although the computed polygon is by no means the polygon of the least diameter which contains the ring intersection, experimental evaluations indicate that it is adequate in the sense that it yields position estimates for a non-trivial number of non-anchor sensors with reasonable error bounds. We discuss implementation details in Section Experimental Evaluations.

In the network above, it may have seemed redundant to approximate the common intersection of the three rings by a disk. In the next section we consider a network of five sensors, three of which are anchors, for which no ordering exists so that each of the non-anchor sensors has three anchors as predecessors. We use such a network to illustrate a refining sweep and in the process justify the assumption that candidate regions are constrained to be either disks or polygons.

Figure 2. Sensors a, b and c are anchors

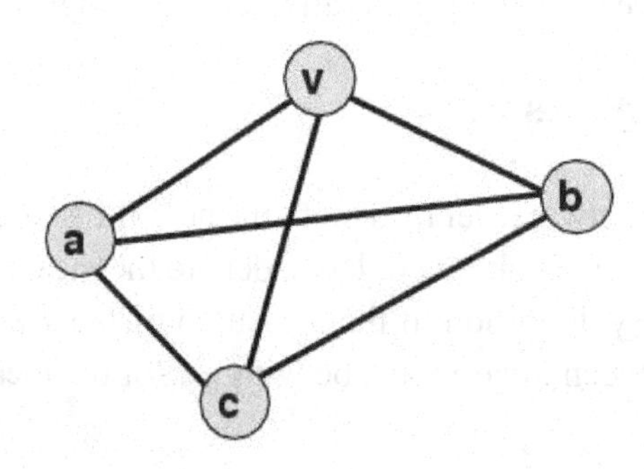

Figure 3. Rings intersection

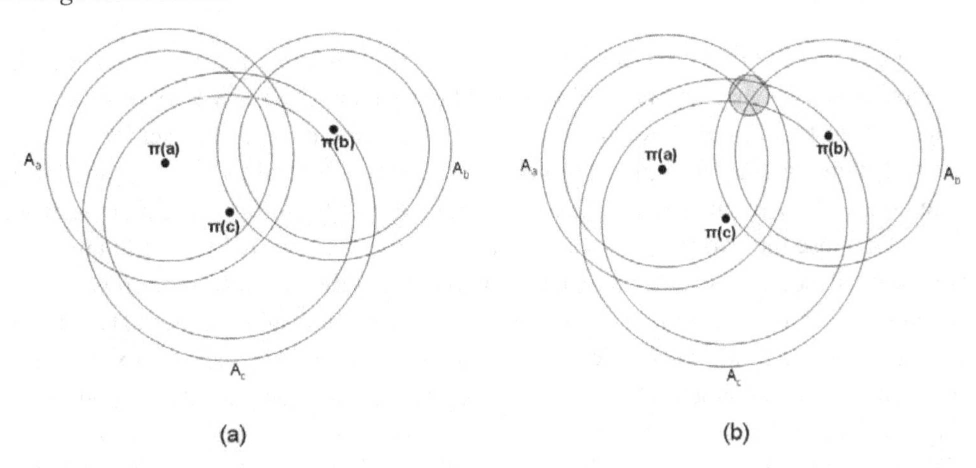

Refining Sweep

Consider the network which consists of five sensors labelled a,b,c,u,v and positioned at points $\pi(a)$, $\pi(b)$, $\pi(c)$, $\pi(u)$ and $\pi(v)$ respectively, so that no three of the sensor positions are collinear. Sensors labelled a, b and c are anchors, and measurements $\tilde{d}_{au}$, $\tilde{d}_{bu}$, $\tilde{d}_{av}$, $\tilde{d}_{cv}$ and $\tilde{d}_{uv}$ are obtained. The network is denoted by the graph in Figure 4. Note that even when the distance measurements are exact, neither the positions of sensor u nor v are uniquely determined by their distances to the anchors alone. Let ε_{au}, ε_{bu}, ε_{av}, ε_{cv} and ε_{uv} denote the guaranteed accuracies of $\tilde{d}_{au}$, $\tilde{d}_{bu}$, $\tilde{d}_{av}$, $\tilde{d}_{cv}$ and $\tilde{d}_{uv}$ respectively.

Using the measured distances and the corresponding guaranteed accuracies, we get:

$$d_{xu} \in [\tilde{d}_{xu} - \varepsilon_{xu}, \tilde{d}_{xu} + \varepsilon_{xu}], \text{ for all } x \in \{a, b\} \tag{3}$$

$$d_{xv} \in [\tilde{d}_{xv} - \varepsilon_{xv}, \tilde{d}_{xv} + \varepsilon_{xv}], \text{ for all } x \in \{a, c\} \tag{4}$$

Figure 4. Sensors a, b and c are anchors

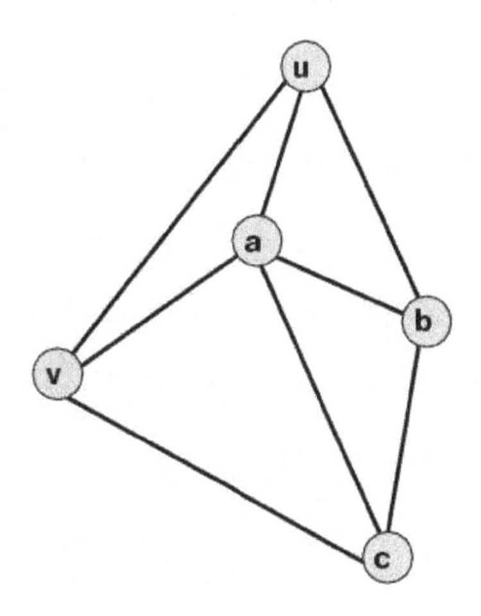

$$d_{uv} \in [\tilde{d}_{uv} - \varepsilon_{uv}, \tilde{d}_{uv} + \varepsilon_{uv}] \tag{5}$$

An ordering of the sensors is first determined so that the anchors precede all other sensors, and each non-anchor sensor has at least one predecessor. One such ordering is *a,b,c,u,v*. Since *u* is the first non-anchor sensor, the first "sweep" begins by determining a candidate regions set for sensor *u*. Let A_a denote the ring centered at $\pi(a)$ with inner radius $\tilde{d}_{au} - \varepsilon_{au}$ and outer radius $\tilde{d}_{au} + \varepsilon_{au}$, and let A_b denote the ring centered at $\pi(b)$ with inner radius $\tilde{d}_{bu} - \varepsilon_{bu}$ and outer radius $\tilde{d}_{bu} + \varepsilon_{bu}$. It follows from (3) that $\pi(u) \in A_a \cap A_b$. Suppose the two rings intersect in two disjoint regions as shown in Figure 5a.

For reasons we will soon make clear, a disk approximation of the ring intersection is computed. The disk approximation is required to contain all the regions of intersection, and consists of two disjoint disks, each containing one of the contiguous regions of intersection. Ideally, the disk approximation would be two disks with the smallest diameters whose union contains the regions of intersection; however, the modified Sweeps algorithm does not require this. A valid disk approximation is shown in Figure 5b, and let D_u and D'_u denote the two disks. Note that D_u and D'_u are disjoint, and each of D_u and D'_u contains one of the contiguous regions of intersection. Obviously, $\pi(u) \in D_u \cup D'_u$, and the set consisting of D_u and D'_u is a candidate regions set for sensor *u*. In the limit case of exact distance measurements, the two rings will actually be circles and intersect in at most two points, in which case the disk approximation of the intersection region should simply be the set of intersection points.

The candidate regions set for sensor *v* is computed in the same way as that for sensor *u*. Let A'_a denote the ring centered at $\pi(a)$ with inner radius $\tilde{d}_{av} - \varepsilon_{av}$ and outer radius $\tilde{d}_{av} + \varepsilon_{av}$, and let A'_c denote the ring centered at $\pi(c)$ with inner radius $\tilde{d}_{cv} - \varepsilon_{cv}$ and outer radius $\tilde{d}_{cv} + \varepsilon_{cv}$. It follows from (4) that $\pi(v) \in A'_a \cap A'_c$. Suppose the two rings intersect at two disjoint regions as shown in Figure 6a. As shown in Figure 6b, let D_v and D'_v denote the two disks in the disk approximation. Clearly, the set consisting of D_v and D'_v is a candidate regions set for sensor *v*. In the actual modified Sweeps algorithm, the candidate regions set of sensor *u* is also used in the computation of the candidate regions set for sensor *v* since *u* is a predecessor of *v* in the chosen ordering. However, we skip this step in an effort to keep this example simple.

Since both computed candidate regions sets consist of more than one disjoint regions, a refining sweep will be performed to identify false candidate regions in each of the candidate region sets. We

Figure 5. Intersection of two rings, and its disk approximation

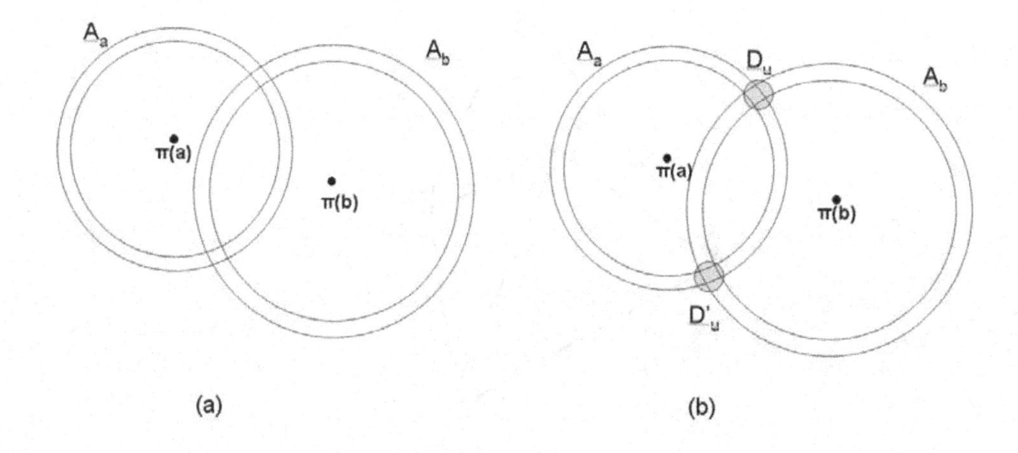

Figure 6. Intersection of two rings, and its disk approximation

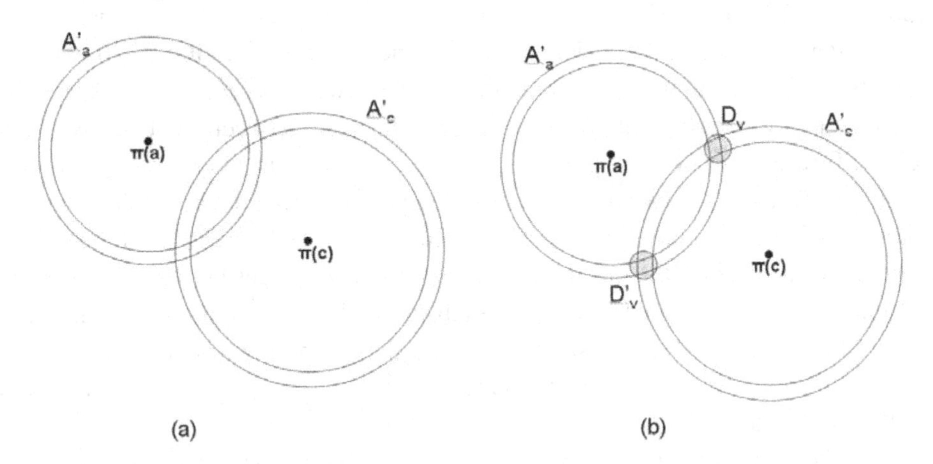

first describe the underlying idea behind identifying false candidate regions of a sensor. Let D_1 and D_2 be either two disks or two polygons in the plane, and let $d_{min}(D_1, D_2)$ denote the minimum among all distances between a point in D_1 and a point in D_2:

$$d_{min}(D_1, D_2) = min_{p1 \in D1, p2 \in D2} \| p_1 - p_2 \| \tag{6}$$

Let $d_{max}(D_1, D_2)$ denote the maximum among all distances between a point in D_1 and a point in D_2:

$$d_{max}(D_1, D_2) = max_{p1 \in D1, p2 \in D2} \| p_1 - p_2 \| \tag{7}$$

Clearly, if $\pi(u) \in D_1$ and $\pi(v) \in D_2$, then it must be the case that $d_{uv} \in [d_{min}(D_1, D_2), d_{max}(D_1, D_2)]$. Moreover, from (5), we have that $d_{uv} \in [\tilde{d}_{uv} - \varepsilon_{uv}, \tilde{d}_{uv} + \varepsilon_{uv}]$, which implies $[d_{min}(D_1, D_2), d_{max}(D_1, D_2)]$ and $[\tilde{d}_{uv} - \varepsilon_{uv}, \tilde{d}_{uv} + \varepsilon_{uv}]$ cannot be disjoint when $\pi(u) \in D_1$ and $\pi(v) \in D_2$.

Without loss of generality, suppose $\pi(v) \in D_v$ and $\pi(u) \in D_u$. The crux of the modified Sweeps algorithm is based on the simple observation that if $\pi(v) \in D^*$, where D^* is a candidate region in the candidate regions set of sensor v, then there must be at least one candidate region D in the candidate regions set of sensor u, namely the candidate region which contains $\pi(u)$, such that:

$$[d_{min}(D^*, D), d_{max}(D^*, D)] \cap [\tilde{d}_{uv} - \varepsilon_{uv}, \tilde{d}_{uv} + \varepsilon_{uv}] \neq \emptyset \tag{8}$$

In other words, if for some candidate region D^* in the candidate regions set of v, we have that

$$[d_{min}(D^*, D), d_{max}(D^*, D)] \cap [\tilde{d}_{uv} - \varepsilon_{uv}, \tilde{d}_{uv} + \varepsilon_{uv}] = \emptyset \tag{9}$$

for *all* disks D in the candidate regions set of u, then it cannot be the case that $\pi(v) \in D^*$. In this case D^* can be removed from the candidate regions set of v, and the resulting set is again guaranteed to be a candidate regions set of sensor v. Note also that the minimum (maximum) among all distances between a point in one region in the plane and another region in the plane is particularly easy to compute when the regions are either both disks or both polygons. In the case of two disjoint disks, the minimum and maximum distances between the disks can be obtained by first drawing the line containing the centers of both disks. The line intersects the boundaries of the two disks at four points, and the distance between the closest pair of those points is the minimum distance between the two disks, and the distance between the furthest pair of those points is the maximum distance between the two disks. In the case of two disjoint polygons, the two points, each belonging to one of the polygons, which induce the minimum (maximum) distance between the polygons must lie on the boundaries of the polygons which are straight line segments. Hence, the minimum and maximum distances between two polygons can be obtained by computing the minimum and maximum distances between pairs of straight line segments. When the regions are not confined to be disks or polygons, the computation is more complicated. Hence, in order to keep the computations simple and efficient, candidate regions sets are required to consist of either all disks or all polygons.

Recall that $\{D_u, D'_u\}$ and $\{D_v, D'_v\}$ are the candidate regions sets computed for *u* and *v* respectively in the first sweep. To refine the candidate regions set computed for sensor *v*, we process the sensors in the ordering *u,v,a,b,c*. Only sensor *v* is processed in this "refining sweep" since it is the only non-anchor sensor with a predecessor, namely sensor *u*, in the new ordering. Suppose the disks D_u, D'_u, D_v, and D'_v are positioned in the plane as shown in Figure 7.

If ε_{uv} is not too "large," then it is easy to see that $[d_{min}(D'_v, D_u), d_{max}(D'_v, D_u)]$ and $[d_{min}(D'_v, D'_u), d_{max}(D'_v, D'_u)]$ are both disjoint from the interval $[\tilde{d}_{uv} - \varepsilon_{uv}, \tilde{d}_{uv} + \varepsilon_{uv}]$. As shown above, the previous imply D'_v cannot contain *π(v)* and so can be removed from the candidate regions set of *v* to obtain the new candidate regions set $\{ D_v \}$ of *v*. Removing D'_v from the candidate regions set of *v* provides a refinement of the candidate regions set of *v* in that the set of points in its candidate regions set is strictly reduced while still retaining the property that the remaining disks contain the sensor's actual position. A set consisting of a finite number of elements from R is said to be *algebraically independent over the rationals* if its elements do not satisfy a non-zero polynomial equation with rational coefficients. In other words, a set

Figure 7. Candidate regions sets of sensors u and v

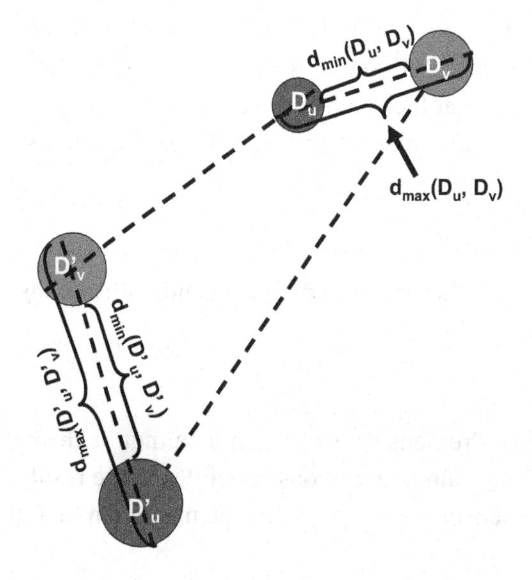

consisting of i real numbers $x_1,...,x_i$ is algebraically independent over the rationals if there does not exist a non-zero polynomial equation (with rational coefficients) in i variables which is satisfied by $x_1,...,x_i$. The positions of a network's sensors are said to be *generic* if the set consisting of the position coordinates is algebraically independent over the rationals. When distance measurements are exact and the sensor positions are generic, it can be shown that $[d_{min}(D'_v, D_u), d_{max}(D'_v, D_u)]$ and $[d_{min}(D'_v, D'_u), d_{max}(D'_v, D'_u)]$ are always disjoint from the interval $[\tilde{d}_{uv} - \varepsilon_{uv}, \tilde{d}_{uv} + \varepsilon_{uv}]$, and so D'_v can be removed to obtain a smaller candidate regions set for v (Fang et al. (2006)).

After the refining sweep, $\{D_u, D'_u\}$ and $\{D_v\}$ are the new candidate regions sets of u and v respectively. A different ordering is now chosen so as to refine the candidate regions set of sensor u. Let the ordering be v, u, a, b, c. Again, if ε_{uv} is not too large, then it will be the case that $[d_{min}(D'_u, D_v), d_{max}(D'_u, D_v)]$ is disjoint from the interval $[\tilde{d}_{uv} - \varepsilon_{uv}, \tilde{d}_{uv} + \varepsilon_{uv}]$. If this holds, then D'_u cannot contain $\pi(u)$, so D'_u can be removed from the candidate regions set of u. The sum of the diameters of the disks in the resulting set is strictly reduced while retaining the property that at least one of the disks contains $\pi(u)$, thus refining the position estimate of sensor u.

Whether smaller candidate region sets of sensors u and v can be obtained depends on the configuration of the actual sensor positions, the distance measurements, and the guaranteed accuracies of the distance measurements. Intuitively, and as have been confirmed by experimental evaluations, the more accurate the distance measurements are, i.e. as $\varepsilon_{ij} \rightarrow 0$ for each measured distance $\tilde{d}_{ij}$, the more likely it is that the candidate regions sets can be refined to be a singleton. In the limit case where distance measurements are exact, and the sensor positions are generic, it is easy to see that the candidate region sets computed for sensors u and v in the first sweep must each consist of two points. Furthermore, each of the candidate region sets can be refined to be a singleton candidate region set in the subsequent refining sweeps. In other words, if distance measurements are exact, then the two refining sweeps will remove all but the actual position from the candidate regions set of each sensor, and thus localize the network.

GLOBAL RIGIDITY AND LOCALIZABILITY

A *multi-point* $p=\{p_1,...,p_n\}$ in d-dimensional space is a set of n points in R^d labelled $p_1,...,p_n$. A multi-point is said to be *generic* if the set consisting of the coordinates of its points is algebraically independent over the rationals. Because we are only concerned with networks in the plane, we will henceforth restrict our attention to the case of $d=2$. A graph with vertex set V and edge set E is denoted by (V, E). A *point formation* in R^2 of n points at a multi-point $p=\{p_1,...,p_n\}$ consists of p and a simple undirected graph G with vertex set $V=\{1,...,n\}$, and is denoted by (G, p). If (i, j) is an edge in G, then the *length of edge* (i, j) in the point formation (G, p) is the distance between p_i and p_j. Two point formations with the same graph have the same edge lengths if the length of each edge in the graph is the same in both point formations. Two point formations with the same graph are *congruent* if all inter-vertex distances are the same.

A point formation (G,p) in R^2 is *globally rigid* in R^2 if *every* other point formation in the plane with the same graph and edge lengths is congruent to (G,p). For any multi-point $p=\{p_1,...,p_n\}$ in R^2 and $\varepsilon > 0$, let $B_p(\varepsilon)$ denote the set of all multi-points $q=\{q_1,...,q_n\}$ in R^2 where $\| p_i - q_i \| < \varepsilon$ for all $i \in \{1,...,n\}$. A graph G is said to be *globally rigid* in R^2 if there exist multi-point p in R^2 and $\varepsilon > 0$ such that (G, q) is globally rigid in R^2 for all $q \in B_p(\varepsilon)$. It is known that if a multi-point p in R^2 is generic, then the point formation (G,p) is globally rigid in R^2 if and only if G is globally rigid in R^2 [6,7]. A number of efficient algorithms, such as Pebble Game, can be used for determining if a graph is globally rigid in R^2 [6,7,8].

A network with n sensors is modelled by a point formation (G,p) where each sensor corresponds to exactly one vertex of G, and vice versa, with (i,j) being an edge of G if the sensors corresponding to i and j are both anchors or if i and j are distinct and the distance measurement between the corresponding sensors is obtained, and $p=\{p_1,...,p_n\}$ where p_i is the position of the sensor corresponding to vertex i. We say that G is the graph of the network, and p is the multi-point of the network. We will restrict our attention to those networks with generic multi-points. In particular, this implies no two sensors occupy the same point and no three sensors are collinear in the networks we consider.

A network in which all distance measurements are exact is *localizable* if there corresponds exactly one position to each non-anchor sensor so that the given inter-sensor distances are satisfied. We consider the natural extension of this definition to networks with inaccurate distance measurements where computing exact sensor positions is (in general) impossible. We say that a network of n sensors positioned at $\pi(1),..., \pi(n)$ respectively is *localizable with precision* ρ if for all points $p_1,...,p_n \in R^2$ where $p_i = \pi(i)$ for all anchors i and $\| p_i - p_j \| \in (\tilde{d}_{ij} - \varepsilon_{ij}, \tilde{d}_{ij} + \varepsilon_{ij})$ for each distance measurement $\tilde{d}_{ij}$, we have that $\| p_i - \pi(i) \| \leq \rho$ for all $i \in \{1,...,n\}$. When distance measurements are exact, a network is localizable if and only if the network has three anchors and the graph of the network is globally rigid in R^2. When the distance measurements in a network are inaccurate, it is straightforward to show from the definitions that:

Lemma 1. If the graph of a network is not globally rigid in R^2, then there exists $\rho > 0$ such that the network is not localizable with precision ρ.

MODIFIED SWEEPS

We first give the terms and definitions to be used in defining the modified Sweeps algorithm. In the following, let N be a network of n sensors labelled 1 through n where each sensor i is positioned at $\pi(i)$. Let G=(V,E) denote the graph of N where $V=\{1,...,n\}$, and each vertex i corresponds to sensor i. For each vertex v, let $N(v)$ denote the set of vertices adjacent to v in G. We assume there are at least three anchors and that G is connected. For each $(i,j) \in E$ where at least one of i or j is a non-anchor sensor, let $\tilde{d}_{ij}$ denote the distance measurement obtained between sensors i and j, and let d_{ij} denote the actual distance between sensors i and j, i.e. $d_{ij} = \| \pi(i) - \pi(j) \|$. For each measured distance $\tilde{d}_{ij}$, let ε_{ij} denote the guaranteed accuracy of the measured distance. This implies that for each $(i,j) \in E$, $d_{ij} \in [\tilde{d}_{ij} - \varepsilon, \tilde{d}_{ij} + \varepsilon]$. To avoid degenerate cases, we will assume that $\varepsilon_{ij} < \tilde{d}_{ij}$ for all measured distance $\tilde{d}_{ij}$.

A *mapping* α has as its domain a non-empty subset U of V, and for each element u in U, $\alpha(u)$ is either a disk or polygon in the plane. Given any mapping α with domain U, α is called a *disk mapping* if $\alpha(u)$ is a disk for all $u \in U$, and a *polygon mapping* if $\alpha(u)$ is a polygon for all $u \in U$. For mapping α, let $\Delta(\alpha)$ denote the domain of α. Two mappings α and β are said to be *consistent with each other*, and we write $\alpha \sim \beta$, if for all $u \in \Delta(\alpha) \cap \Delta(\beta)$, α and β map u to the same region in the plane. Two mappings are always consistent if their domains are disjoint. For any positive integer k, consider a collection of k pairwise consistent mappings $\alpha_1,..., \alpha_k$ which are either all disk mappings or all polygon mappings. Let $u_k(\alpha_1,..., \alpha_k)$ denote the mapping with domain $\bigcup_{i \in \{1,...,k\}} \Delta(\alpha_i)$ whose restriction to $\Delta(\alpha_i)$ is equal to α_i for each $i \in \{1,...,k\}$. For a disk or polygon P in the plane, let *centroid(P)* denote the centroid of the convex hull of the points in P, and let *radius(P)* denote the maximum distance between *centroid(P)* and the boundary of P. Note that *centroid(P)* and *radius(P)* are easy to compute since P is constrained to be either a disk or polygon. For positive reals d and ε, let A(P, d, ε) denote the ring centered at *centroid(P)* with outer radius $d+\varepsilon+radius(P)$, and inner radius $d-\varepsilon-radius(P)$ if $d-\varepsilon-radius(P)>0$, and zero otherwise.

Consider sensor v and suppose that the measured distances between v and sensors $u_1,\ldots, u_m$ are known. For $i\in\{1,\ldots,m\}$, let d_i denote the measured distance between sensor v and sensor u_i, and let ε_i denote the accuracy guarantee of d_i. Now suppose $\alpha_1,\ldots, \alpha_m$ are m pairwise consistent mappings such that $u_i\in\Delta(\alpha_i)$ and $v\notin \Delta(\alpha_i)$ for each $i\in\{1,\ldots,m\}$. The mappings $\alpha_1,\ldots, \alpha_m$ are also required to be either all disk mappings or all polygon mappings. Assuming $\bigcap_{i\in\{1,\ldots,m\}} A(\alpha_i (u_i),d_i, \varepsilon_i)\neq \emptyset$, any collection of disks or polygons in the plane containing $\bigcap_{i\in\{1,\ldots,m\}} A(\alpha_i (u_i), d_i, \varepsilon_i)$ can be considered a candidate regions set of sensor v *under the assumption* that region $\alpha_i (u_i)$ contains the position of u_i for each $i\in\{1,\ldots,m\}$. We define the set $M(\alpha_1,\ldots, \alpha_m,v, u_1,\ldots, u_m)$ with the goal of keeping track of the candidate regions of sensor v assuming the position of each u_i is contained in the region $\alpha_i (u_i)$. If in fact each $\alpha_i (u_i)$ does contain the position of u_i , then $\bigcap_{i\in\{1,\ldots,m\}} A(\alpha_i (u_i),d_i, \varepsilon_i)$ is non-empty. However, if some $\alpha_i (u_i)$, $i\in\{1,\ldots,m\}$, does not contain the position of u_i , then $\bigcap_{i\in\{1,\ldots,m\}} A(\alpha_i (u_i),d_i, \varepsilon_i)$ may be the empty set.

Clearly $\bigcap_{i\in\{1,\ldots,m\}} A(\alpha_i (u_i), d_i, \varepsilon_i)$ consists of either zero, one or more contiguous regions. If $\bigcap_{i\in\{1,\ldots,m\}} A(\alpha_i (u_i), d_i, \varepsilon_i) = \emptyset$, then $M(\alpha_1,\ldots, \alpha_m,v, u_1,\ldots, u_m)$ is defined to be $\emptyset$:

$$M(\alpha_1,\ldots, \alpha_m,v, u_1,\ldots, u_m) = \emptyset \tag{10}$$

If $\bigcap_{i\in\{1,\ldots,m\}} A(\alpha_i (u_i), d_i, \varepsilon_i)$ consists of one contiguous region, then let P_v be any region in the plane which contains $\bigcap_{i\in\{1,\ldots,m\}} A(\alpha_i (u_i), d_i, \varepsilon_i)$. If $\bigcap_{i\in\{1,\ldots,m\}} A(\alpha_i (u_i), d_i, \varepsilon_i)$ is a point, then P_v is required to be that point. If $\alpha_1,\ldots, \alpha_m$ are all disk (polygon) mappings, then we require that P_v be a disk (polygon). When $\bigcap_{i\in\{1,\ldots,m\}} A(\alpha_i (u_i), d_i, \varepsilon_i)$ is not a point, then in keeping with the desire to compute "small" candidate regions for each sensor, P_v would ideally be the disk or polygon with the smallest diameter which contains $\bigcap_{i\in\{1,\ldots,m\}} A(\alpha_i (u_i), d_i, \varepsilon_i)$. However, the modified Sweeps algorithm does not require this to be so. In the instance of the modified Sweeps algorithm we implemented, we used a simple algorithm to determine a polygon containing $\bigcap_{i\in\{1,\ldots,m\}} A(\alpha_i (u_i), d_i, \varepsilon_i)$ which has been shown to be both computationally efficient and adequate in our experimental evaluations on randomly deployed networks. Let β denote the mapping with domain $\{v\}\cup\bigcup_{i\in\{1,\ldots,m\}} \Delta(\alpha_i)$. Let $\beta(v)$ be P_v, and for $i\in\{1,\ldots,m\}$ and each $u\in \Delta(\alpha_i)$, define $\beta(u)= \alpha_i (u)$. Note that β is well defined since α_i, $i\in\{1,\ldots,m\}$, are pairwise consistent, and β, $\alpha_1,\ldots, \alpha_m$ are either all disk mappings or all polygon mappings. Let $M(\alpha_1,\ldots, \alpha_m, v, u_1,\ldots, u_m)$ be:

$$M(\alpha_1,\ldots, \alpha_m,v, u_1,\ldots, u_m)=\{\beta\} \tag{11}$$

Note that m being greater than or equal to three will not guarantee that $\bigcap_{i\in\{1,\ldots,m\}} A(\alpha_i (u_i), d_i, \varepsilon_i)$ consists of one contiguous region. In Figure 8 below, the ring with the dotted boundary and the ring with the solid line boundary intersect in two disjoint regions, and the ring with the solid interior has a non-empty intersection with both of those regions. Hence, the three rings intersect in two disjoint contiguous regions.

If $\bigcap_{i\in\{1,\ldots,m\}} A(\alpha_i (u_i), d_i, \varepsilon_i)$ consists of two or more disjoint regions, then let P_{v1} and P_{v2} be any two regions whose union contains $\bigcap_{i\in\{1,\ldots,m\}} A(\alpha_i (u_i), d_i, \varepsilon_i)$. If $\alpha_1,\ldots, \alpha_m$ are all disk (polygon) mappings, then we require that P_{v1} and P_{v2} both be disks (polygons). In keeping with our desire to obtain "small" candidate regions for each sensor, P_{v1} and P_{v2} would ideally be disjoint regions with the smallest possible diameters whose union contains the regions of intersection. However, the modified Sweeps algorithm does not require this to be so. If $\bigcap_{i\in\{1,\ldots,m\}} A(\alpha_i (u_i), d_i, \varepsilon_i)$ consists of two points, then we require that P_{v1} and P_{v2} be the two points. Let β_1 and β_2 denote mappings both with domain $\{v\}\cup\bigcup_{i\in\{1,\ldots,m\}} \Delta(\alpha_i)$ defined as follows. Let β_1 (v) be P_{v1}, β_2 (v) be P_{v2}, and for $i\in\{1,\ldots,m\}$ and each $u\in \Delta(\alpha_i)$, let $\beta_1 (u)= \alpha_i (u)$ and β_2

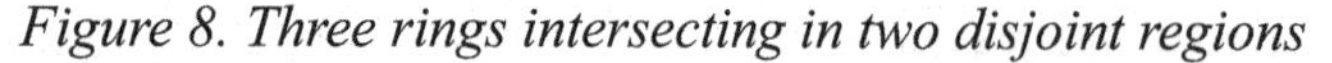

Figure 8. Three rings intersecting in two disjoint regions

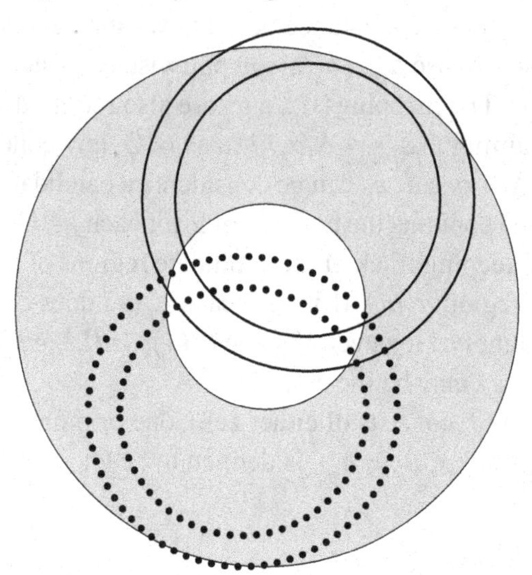

$(u) = \alpha_i\,(u)$. Note that β_1 and β_2 are well defined since α_i, $i \in \{1,...,m\}$, are pairwise consistent, and β_1, β_2, α_1,..., α_m are either all disk mappings or all polygon mappings. Let $M(\alpha_1,..., \alpha_m, v, u_1,..., u_m)$ be:

$$M(\alpha_1,..., \alpha_m, v, u_1,..., u_m) = \{\beta_1, \beta_2\} \tag{12}$$

Modified Sweeps will use M to compute candidate regions sets of sensors.

Let P_1 and P_2 be either two disks or two polygons in the plane. Let $min(P_1, P_2)$ and $max(P_1, P_2)$ denote the minimum and maximum distance, respectively, between P_1 and P_2. Note that it is computationally efficient to determine $min(P_1, P_2)$ and $max(P_1, P_2)$ since P_1 and P_2 are either both disks or both polygons. Let $I(P_1, P_2)$ denote the real line interval with left endpoint $min(P_1, P_2)$ and right endpoint $max(P_1, P_2)$.

Algorithm

In the following, we give an algorithm which computes for each sensor v a sequence of sets $S(v, i)$, $i = 1, 2, ...$ such that each $S(v,i)$ consists of a finite number of mappings and $\{\alpha\,(v) \mid \alpha \in S\,(v, i)\}$ is a candidate regions set for sensor v.

Let $[v] = v_1, v_2, v_3,..., v_n$ be an ordering of the vertices of G where v_1, v_2, v_3 are anchors. For v_i, $i \in \{1,2,3\}$, let α_i be the mapping with domain $\{v_i\}$ where $\alpha_i\,(v_i)$ is the known position of v_i. We require that α_i, $i \in \{1,2,3\}$, be either all disk mappings, or all polygon mappings. A point in the plane is considered a degenerate disk (polygon) with diameter zero. For $i \in \{1,2,3\}$, let $S(v_i, 1)$ be defined as:

$$S(v_i, 1) = \{\,\alpha_i\,\}, \quad i \in \{1,2,3\} \tag{13}$$

The sets $S(v_i, 1)$, $i > 3$, are computed iteratively as follows. For v_i, $i > 3$,, let u_1,...,u_m denote the vertices adjacent to v_i in G and which precedes v_i in the ordering v_1,..., v_n: $N(v_i) \cap \{\,v_1,..., v_{i-1}\,\} = \{\,u_1,...,u_m\}$. Define $S(v_i, 1)$ using M as :

$$S(v_i,1)=\bigcup_{\alpha_j\in S(u_j,1)\ \ j\in\{1,\dots,m\},\ \ and\,\alpha_j\sim\alpha_k\ \forall\ j,k\in\{1,\dots,m\}} M(\alpha_1,\alpha_2,\dots,\alpha_m,v_i,u_1,u_2,\dots,u_m) \quad (14)$$

Roughly speaking, $S(v_i, 1)$ is the set of mappings "storing" the candidate regions of sensor v_i corresponding to each combination of candidate regions of v_i's predecessors in the ordering. It is easy to see that each $S(v, 1)$ consists of a finite number of disk (polygon) mappings, and for each sensor v, the set $\{\alpha(v) \mid \alpha \in S(v, 1)\}$ is a candidate regions set for sensor v. We call $S(v, 1)$ a *candidate mapping set* of sensor v and we call $\{\alpha(v) \mid \alpha \in S(v, 1)\}$ the candidate regions set of sensor v *obtained by the first sweep*.

Suppose for some $k \geq 1$ that $S(v, k)$, $v \in V$, have been computed, and that for each sensor v, the set $\{\alpha(v) \mid \alpha \in S(v, k)\}$ is a finite candidate regions set for sensor v. Let $u_1,\dots, u_n$ be any ordering of the vertices of G such that at least one vertex u_i is adjacent to some vertex u_j where $j<i$, and all vertices v where $S(v, k)$ is a singleton precede all vertices u where $S(u, k)$ is not a singleton. For each vertex u_i, let $P(u_i)$ denote the set of vertices adjacent to u_i in G and which precede u_i in the ordering $u_1,\dots, u_n$: $P(u_i)= N(u_i) \cap \{u_1,\dots, u_{i-1}\}$. Let s denote the number of vertices v for which $S(v, k)$ is a singleton. For $i\in\{1,\dots,s\}$, define:

$$S(u_i, k+1) = S(u_i, k) \quad (15)$$

For $i\in\{s+1,\dots,n\}$, if $P(u_i)=\emptyset$, then define:

$$S(u_i, k+1) = S(u_i, k) \quad (16)$$

Now suppose $P(u_i)$ is non-empty. Recall that for each $w\in P(u_i)$, $\tilde{d}_{wui}$ is the measured distance between sensors u_i and w, and ε_{wui} is the guaranteed accuracy of $\tilde{d}_{wui}$. The underlying idea behind obtaining $S(u_i, k+1)$ from $S(u_i, k)$ and $S(w, k+1)$, $w\in P(u_i)$, is as follows. By assumption, the set $\{\alpha(u_i) \mid \alpha \in S(u_i, k)\}$ is a candidate regions set of sensor u_i. Let P be any candidate region of sensor u_i from the set. Suppose that for all $w\in P(u_i)$, $\{\alpha(w) \mid \alpha \in S(w,k+1)\}$, is a candidate regions set of w. This implies that for all $w\in P(u_i)$, there is a region P_w^* in $\{\alpha(w) \mid \alpha\in S(w, k+1)\}$ which contains the position of w. Suppose that for some sensor $w\in P(u_i)$ that $I(P,P')$ is disjoint from $[\tilde{d}_{wui}-\varepsilon_{wui}, \tilde{d}_{wui}+\varepsilon_{wui}]$ for all $P'\in \{\alpha(w) \mid \alpha \in S(w,k+1)\}$. This implies P cannot contain the position of sensor u_i, for if P did contain the position of sensor u_i, then $I(P, P_w^*)$ is not disjoint from $[\tilde{d}_{wui}-\varepsilon_{wui}, \tilde{d}_{wui}+\varepsilon_{wui}]$. In this case we say that P is an *identified* false candidate region of sensor u_i. To obtain the set $S(u_i, k+1)$, we remove all mappings β from $S(u_i, k)$ where $\beta(u_i)$ is an identified false candidate region of sensor u_i. Since only mappings β where $\beta(u_i)$ is a false candidate region of u_i is removed from $S(u_i, k)$ to obtain $S(u_i, k+1)$, it follows that $\{\alpha(u_i) \mid \alpha \in S(u_i, k+1)\}$ must still be a candidate regions set for sensor u_i. In the following, for notational convenience, let $w_1,\dots,w_m$ be the elements of $P(u_i)$, and define $S(u_i, k+1)$ as:

$$S(u_i,k+1)=\{u_{m+1}(\alpha,\alpha_1,\dots,\alpha_m) \mid \alpha\in S(u_i,k),\alpha_j\in S(w_j,k+1)\ \forall j\in\{1,\dots,m\},$$

$$\beta\sim\gamma\ \forall\beta,\gamma\in\{\alpha,\alpha_1,\dots,\alpha_m\},$$

$$I(\alpha(u_i),\alpha_j(w_j))\cap[\tilde{d}_{uiwj}-\varepsilon_{uiwj},\tilde{d}_{uiwj}+\varepsilon_{uiwj}]\neq\emptyset\ \forall j\in\{1,\dots,m\}\} \quad (17)$$

Since each $S(v, k)$, $v\in V$, consists of a finite number of elements, it follows from equation 17 that $S(v, k+1)$ must also consist of a finite number of elements. Furthermore, for each sensor v the set $\{\alpha(v) \mid \alpha\in$

$S(v, k+1)\}$ is a candidate regions set for sensor v. We call $\{\alpha(v) \mid \alpha \in S(v, k+1)\}$ the candidate regions set of sensor v obtained by the $(k+1)$th sweep, and we call $S(v, k+1)$ a *candidate mapping set* of sensor v. It is easy to see that for each sensor v, the candidate regions set of sensor v obtained by the $(k+1)$th sweep is a subset, not necessarily proper, of the candidate regions set of sensor v obtained by the kth sweep.

Suppose that for sensor v and some $k \geq 1$ that $S(v, k)$ has been computed, and $S(v, k)$ is a singleton. Let α be the mapping in $S(v, k)$. Furthermore, suppose there exist $u,w \in \Delta(\alpha)$ where distance measurements $\tilde{d}_{uv}$ and $\tilde{d}_{wv}$ are obtained, i.e. $u,w \in N(v)$, and $A(\alpha(u), \tilde{d}_{uv}, \varepsilon_{uv}) \cap A(\alpha(w), \tilde{d}_{wv}, \varepsilon_{wv})$ consists of two disjoint regions such that $\alpha(v)$ contains just one of those regions. When the previous hold, the position estimate of sensor v is taken to be the centroid of $\alpha(v)$ and the error bound is taken to be radius($\alpha(v)$).

The order in which the sensors are processed in the first sweep can be a determining factor in the diameter of the candidate regions in each of the generated candidate region sets, and the number of regions in the set. This is because each sensor's candidate region sets is computed using the candidate regions sets of its neighbors preceding it in the orderings, and the distance measurements between itself and those neighbors. Intuitively, candidate regions with small diameters can be obtained for a sensor if the sensor has at least two neighbors preceding it in the ordering such that the distance measurements between the sensor and those neighbors have small accuracy guarantees and the candidate regions of those neighbors are also small. An ordering can also be chosen "on the go." For example, suppose we have sensors a,b,c,u,v,w where a,b,c are anchors. The anchors always comprise the first three sensors in the ordering. Suppose the graph of the network is as shown in Figure 9 below. So anchors a and c are neighbors of sensor u, anchors a and b are neighbors of sensor v, and anchor a and sensors u and v are neighbors of sensor w. Furthermore, suppose ε_{av}, ε_{bv}, ε_{aw} and ε_{vw} are "small" as compared to ε_{uc}. One way to order the sensors on the go is as follows. Since sensor v has distance measurements with small accuracy guarantees to two sensors with known positions, sensor v can be ordered as the first sensor to follow the anchors. If the candidate regions set computed for sensor v consists of candidate regions with "small" diameters, then sensor w has two distance measurements with small accuracy guarantees to two sensors with either exact positions or candidate regions with small diameters. Hence, let sensor w be the next sensor in the ordering following sensor v. Furthermore, if ε_{uw} is small, then there is a clear advantage in placing w in front of u in the first ordering.

In the following, we give a heuristic for choosing an ordering for generating candidate regions sets based on the factors discussed above. Given the distance measurement $\tilde{d}_{ij}$ between sensors i and j with accuracy guarantee ε_{ij}, let $p_{ij} = \varepsilon_{ij}/\tilde{d}_{ij}$. Let the anchors be the sensors which precede all others in the ordering. Suppose the first ith sensors in the ordering has already been chosen, where $i \geq 3$, and denote the ordering thus far by $v_1,\dots,v_i$. Let W denote the set of all sensors which are not in the ordering thus far. To choose the $i+1$th sensor, let U denote the set of all sensors in W which are adjacent to two or more sensors in $\{v_1,\dots,v_i\}$. If U is empty, then let v_{i+1} be the sensor $v \in W$ such that v is a neighbor of some v_j, $j \leq i$, and p_{vvj} is less than or equal to all other p_{xy} where $x \in W$ and $y = v_k$ for some $k \leq i$. Otherwise, if U is not empty, then let v_{i+1} be a sensor $u \in U$ such that the average of p_{uvj}, $v_j \in \{v_1,\dots,v_i\} \cap N(u)$, is less than or equal to the average of p_{wvj}, $v_j \in \{v_1,\dots,v_i\} \cap N(w)$, for all $w \in U$.

CORRECTLY ORIENTED POSITION ESTIMATES

In this section, we introduce the concept of "correctly oriented" position estimates, and demonstrate that a set of correctly oriented position estimates can be used to deduce geometric properties of the configu-

Figure 9. a, b and c are anchors

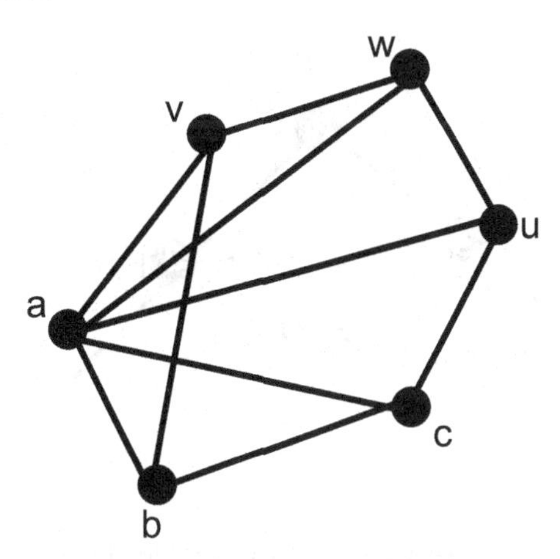

ration of the corresponding actual sensor positions. We then give a sufficient condition on the position estimates and their corresponding error bounds for the position estimates to be correctly oriented.

Consider in the plane a set of four sensors labelled 1,2,3,4 such that no three sensors are collinear. For $i \in \{1,2,3,4\}$, let p_i and q_i denote the actual and estimated position of sensor *i* respectively, and suppose the estimated positions are correctly oriented and no three of the estimated positions are collinear. Let $H(1)$ denote the region of the plane which does not contain p_1 and is bounded by the line segment $l(p_2, p_3)$ and the two half-lines both with origin at p_1, and containing the points p_2 and p_3, respectively. The regions $H(2)$, $H(3)$ are defined analogously. See Figure 10 for an illustration of $H(1)$ which is the region of the plane bounded by the dotted lines.

Suppose the actual position of sensor 4 lies in $H(1)$. Clearly, p_4 lies in $H(1)$ if and only if the line segments $l(p_1, p_4)$ and $l(p_2, p_3)$ intersect. Since q_1, q_2, q_3, q_4 are correctly oriented, it follows that $l(p_1, p_4)$ intersects $l(p_2, p_3)$ if and only if $l(q_1, q_4)$ intersects $l(q_2, q_3)$. Therefore, using the estimated positions, it can be determined if the actual position of sensor 4 lies in $H(1)$. By similar reasoning, the estimated positions can be used to determine if p_4 lies in $H(2)$ and $H(3)$. If the actual position of sensor 4 does not lie in any of the $H(i)$, $i \in \{1,2,3\}$, then the actual position of one of the sensors must lie in the convex hull determined by the actual positions of the other three sensors. Hence, a set of correctly oriented sensor position estimates can be used to deduce certain geometric properties of the configuration of actual sensor positions. However, the difficulty lies in determining if a set of position estimates are correctly oriented without knowing the corresponding actual sensor positions.

In the following, we will give a sufficient condition for a set of position estimates to be correctly oriented. We begin by defining a quadrilateral given two circles in the plane. Let C_i and C_j be two circles in the plane centered at q_i and q_j respectively, and let $Q(C_i, C_j)$ denote the quadrilateral defined as follows. Let l denote the line segment obtained by extending the line segment $l(q_i, q_j)$ to C_i and C_j as shown in Figure 11a. So l is the line passing through q_i and q_j and whose endpoints are on C_i and C_j. Let T_i denote the line tangent to C_i at the endpoint of l on C_i, and define T_j similarly. Lines T_i and T_j are denoted in bold in Figure 11a. Let l_i denote the line passing through q_i and which is also perpendicular to l. Clearly, l_i intersects C_i at exactly two points, which we denote by s_{i1} and s_{i2}. Similarly, if l_j denotes the line passing through q_j which is perpendicular to l, then l_j must intersect C_j at exactly two points, which we denote by s_{j1} and s_{j2}. Without loss of generality, suppose s_{i1} and s_{j1} lie on the same side of line

Figure 10. The region H(1)

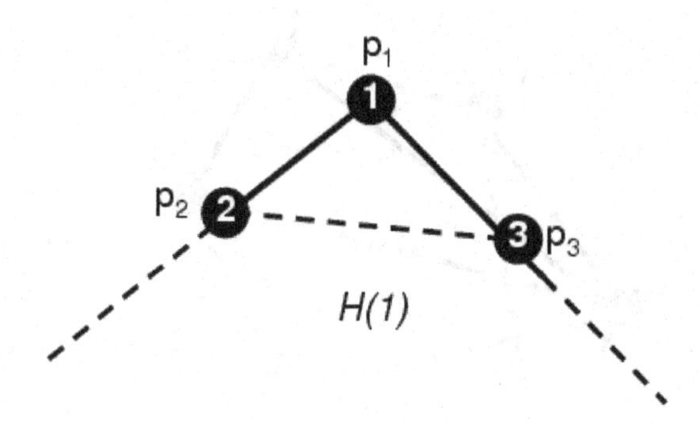

segment l. Let T_L denote the line passing through s_{i1} and s_{j1}, and let T_R denote the line passing through s_{i2} and s_{j2}. Lines T_L and T_R are denoted in bold in Figure 11a. It is easy to see that T_i and T_j are parallel, but that neither T_i and T_j are parallel to the two lines T_L and T_R. So let c_1 and c_2 be the two points where T_L intersects with T_i and T_j, and let c_3 and c_4 be the two points where T_R intersects with T_i and T_j. Let $Q(C_i, C_j)$ denote the convex hull of the four intersection points c_1, c_2, c_3, c_4 as shown in Figure 11b.

Consider $t > 3$ sensors $u_1,..., u_t$ with estimated positions $q_1,..., q_t$ respectively. For $i \in \{1,...,t\}$, let e_i denote the error bound of q_i, and let D_i and C_i denote respectively the disk and circle in the plane centered at q_i with radius e_i. Consider the following condition on the geometry of $C_1,..., C_t$:

Condition 1. For all distinct $i, j, k, l \in \{1,...,t\}$, the quadrilaterals $Q(C_i, C_j)$ and $Q(C_k, C_l)$ are either disjoint, or $Q(C_i, C_j)$ and $Q(C_k, C_l)$ intersect in a quadrilateral Q such that Q is disjoint from each of the circles C_i, C_j, C_k, C_l, and if e and e' are opposite edges of Q, then both edges are contained in the interior of $Q(C_i, C_j)$ or $Q(C_k, C_l)$.

Figure 12 shows the relative positions of four circles C_i, C_j, C_k and C_l in the plane which satisfy Condition 1. If Condition 1 is satisfied by C_i, $i \in \{1,...,t\}$, then the position estimates $q_1,..., q_t$ must be correctly oriented. The proof of this is straightforward and relies upon the observation that if quadrilaterals $Q(C_i, C_j)$ and $Q(C_k, C_l)$ are disjoint, then no line segment with endpoints in C_i and C_j can intersect any line segment with endpoints in C_k and C_l. If $Q(C_i, C_j)$ and $Q(C_k, C_l)$ are not disjoint, and they satisfy Condition 1, then the opposite is true, i.e. given a line segment with endpoints in C_k and C_l, and a line segment with endpoints in C_i and C_j, the two line segments must intersect. Assuming Condition 1 holds, then for all distinct $i, j, k, l \in \{1,...,t\}$, if $q'_i \in D_i$, $q'_j \in D_j$, $q'_k \in D_k$ and $q'_l \in D_l$, then $l(q'_i, q'_j)$ intersects $l(q'_k, q'_l)$ if and only if $l(q_i, q_j)$ intersects $l(q_k, q_l)$. The previous implies $q_1,..., q_t$ must be correctly oriented since for each $i \in \{1,...,t\}$, the actual position of sensor u_i is contained in D_i.

EFFICIENTLY LOCALIZABLE NETWORKS

As we have noted previously, whether a position estimate of some desired precision can be computed by modified Sweeps for a sensor depends on the geometry of the configuration of actual sensor positions, the accuracy of the distance measurements, and the graph of the network. This observation applies to any

Figure 11. $Q(C_i, C_j)$

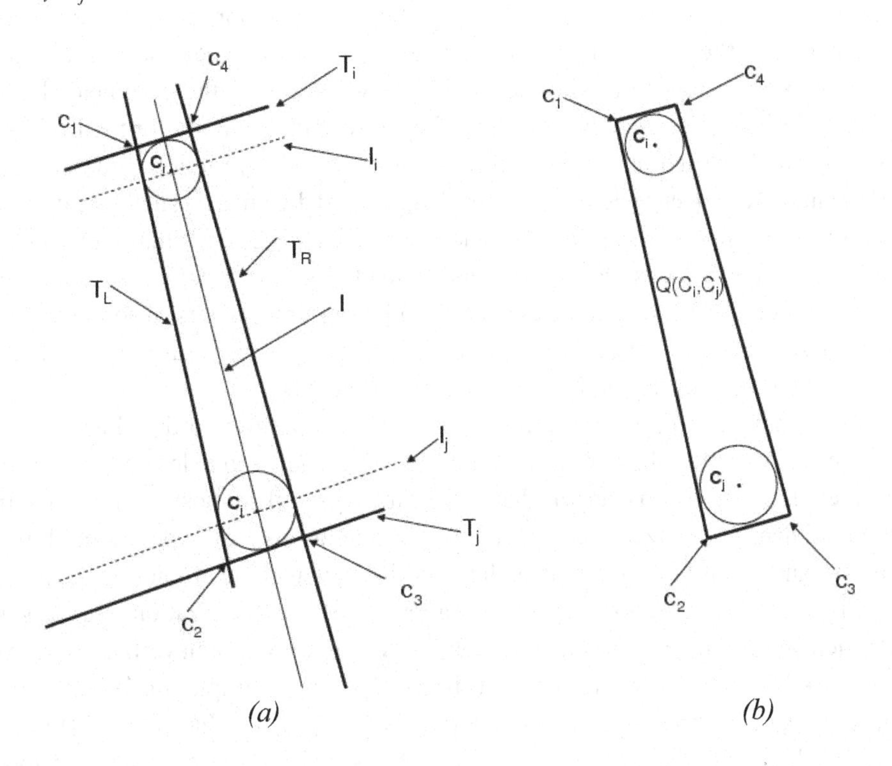

Figure 12. Four circles satisfying Condition 1

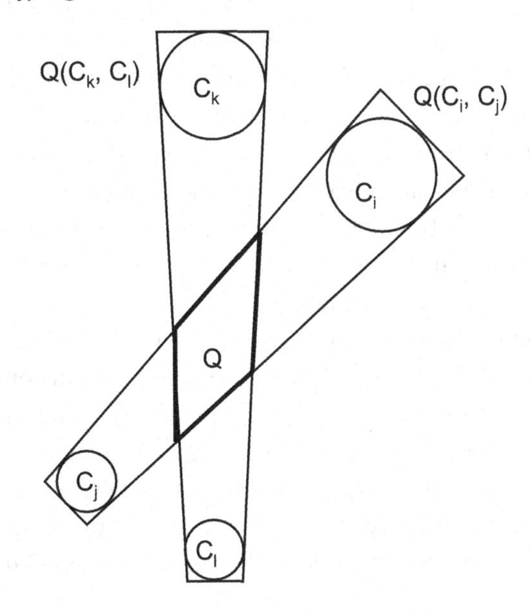

localization algorithm when distance measurements are inaccurate. For a general network, it is difficult to characterize the conditions under which a sufficiently accurate position estimate can be computed for each sensor. Furthermore, the characterization of efficiently localizable networks is far from complete even for the case of exact distance measurements. Hence, we will give the graphical characterization of some networks whose positions can be determined exactly and efficiently for the limit case of exact distance measurements. For such networks, it follows by a simple continuity argument that there are sufficiently small guaranteed accuracies $\varepsilon_{ij} > 0$ for each measured distance $\tilde{d}_{ij}$ such that modified Sweeps can be used to compute a position estimate for each sensor which has a "small" error bound. A key component of the modified Sweeps algorithm is its sequential nature whereby sensors are processed in some order. As will be illustrated below, the notion of processing sensors in some order enables the graphical characterization of networks for which modified Sweeps can be used to *efficiently* compute position estimates in the limit as $\varepsilon_{ij} \rightarrow 0$ for each measured distance $\tilde{d}_{ij}$.

A graph is said to have a *trilateration ordering* if its vertices can be ordered as $u_1, u_2, u_3, \ldots, u_n$ so that u_1, u_2, u_3 induce a complete subgraph, and each u_i, $i>3$, is adjacent to at least three vertices u_j where $j<i$. In Anderson et al. (2007) it was shown that localizable networks whose graphs have trilateration orderings can be efficiently localized, i.e. in a number of operations that is polynomial in the number of sensors. It is straightforward to show that such networks can also be efficiently localized by modified Sweeps in one sweep. In the following, we give a graphical characterization of a class of networks which can be efficiently localized but whose graphs do not necessarily have trilateration orderings. Suppose the network N has $n>4$ sensors and that the vertices of its graph G can be ordered as $v_1, v_2, v_3, \ldots, v_n$ so that v_1, v_2, v_3 correspond to anchors, and that each v_i, $i>3$, is adjacent to vertices v_{i-1} and v_{i-2}. Furthermore, for each v_i, where either $i=n$ or $3<i<n$ and i is odd, v_i is also adjacent to some v_j where $j \neq i-3$ and j is either odd or $j \leq 3$. We call such an ordering an *augmented triangulation* ordering. If G has an augmented triangulation ordering, then G must be globally rigid. This implies any network with a generic multi-point, three anchors, and whose graph has an augmented triangulation ordering must be localizable. Furthermore, G can have an augmented triangulation ordering without possessing any trilateration orderings. To see this note that if G has a trilateration ordering then G must have at least $3(n\text{-}3)+3$ edges since each vertex following the first three vertices in the ordering must be adjacent to at least three vertices preceding it. Suppose G has an augmented triangulation ordering where each v_i, where $i<n$ and i is even, is adjacent to *exactly* two vertices preceding it, and each v_j where j is either equal to n or j is odd, is adjacent to *exactly* three vertices preceding it. In this case G has $(5(n\text{-}3)/2)+3$ edges when n is odd and $(5(n\text{-}4)/2)+6$ when n is even. In either case, G would have less than the minimum number of edges required for a trilateration ordering.

Suppose an augmented triangulation ordering $v_1, v_2, v_3, \ldots, v_n$ of G is chosen to compute the first sweep in modified Sweeps. In other words, $v_1, v_2, v_3, \ldots, v_n$ is the ordering used in computing $S(v,1)$, $v \in V$. In this case, each $S(v,1)$, $v \in V$, contains at most two mappings, and $S(v_i,1)$ where i is odd must be a singleton. This implies N can be efficiently localized by modified Sweeps since the computational complexity of modified Sweeps is entirely dependent on the size of the sets generated during each sweep. Since an augmented triangulation ordering is not necessarily a trilateration ordering, the previous also implies that N can be efficiently localized by modified Sweeps even if its graph G does not have any trilateration orderings.

EXPERIMENTAL EVALUATIONS

In our experimental evaluations, we used Matlab to first generate a random network of 100 nodes using three input parameters R, m and η, where R is the sensing range, m is the number of anchors, and η is the noise factor. Sensor positions are randomly generated from the distribution that is uniformly distributed on the 1-by-1 two dimensional space, and m of those sensors are randomly chosen to be anchors. For each pair of sensors within sensing range R a noisy distance measurement is generated using the input parameter noise factor η. More specifically, η is specified to be between zero and one, and for each pair of sensors within sensing range, a distance measurement $\tilde{d}$ is generated such that the actual inter-sensor distance d is within $\eta \cdot \tilde{d}$ of the generated distance measurement, i.e. $| \tilde{d} - d | \leq \eta \cdot \tilde{d}$. The limit case of exact distance measurements corresponds to when noise factor η is zero: $| \tilde{d} - d | = 0$. In other words, the guaranteed accuracy of each generated distance measurement is $\eta \cdot 100$ percent of the distance measurement. Roughly speaking this corresponds to the notion that the distance measurement between two sensors become less accurate as the distance between the two sensors increase. For reasons we will specify below, we will also consider a class of networks such that distance measurements between particular pairs of sensors within sensing range are *not* generated.

For ease of implementation, we used convex polygons, as opposed to just general polygons, to approximate ring intersection regions since convex polygons are particularly easy to manipulate. We compute the convex polygon approximation of ring intersections by means of a simple algorithm using tangent lines and convex hulls. More specifically, consider the intersection region of two rings that intersect in two disjoint regions as shown in Figure 13. Each intersection region is comprised of four arcs which we denote by a_1, a_2, a_3, and a_4. Without loss of generality, suppose a_i is disjoint from a_{i+2} for $i=1,2$. Note that each arc lies on either the inner or outer circle of one of the two rings. If arc a_i lies on the outer circle of one of the rings, then let T_i denote a line that is tangent to the arc at roughly the midpoint of the arc. If arc a_i lies on the inner circle of one of the rings, then let T_i denote the line passing through the endpoints of a_i. See Figure 13 below for an illustration. For $i=1,2,3$, let t_i denote the point where T_i and T_{i+1} intersects, and let t_4 denote the point where T_4 and T_1 intersects. We obtain a convex polygon which contains one of the contiguous regions of the ring intersection by taking the convex hull of t_1, t_2, t_3, and t_4.

We evaluated modified Sweeps on networks whose graphs have "augmented bilateration" orderings. A graph is said to have a *bilateration ordering* if its vertices can be ordered so that v_1, v_2, v_3 induce a complete subgraph, and each v_i, $i>3$, is adjacent to at least two vertices v_j where $j<i$. A network's graph is said to have an *augmented* bilateration ordering if it has a bilateration ordering $v_1,\ldots, v_n$ where $v_1,\ldots, v_m$ are the $m\geq 3$ anchors and if v_i, $i>m$, is only adjacent to two vertices v_j where $j<i$, then v_i must also be adjacent to v_{i+1} where v_{i+1} is adjacent to at least two vertices v_j where $j<i$ and at least one of which is not adjacent to v_i. If a graph has at least one augmented bilateration ordering, but no such ordering is also a trilateration ordering, then we say that the graph's augmented bilateration orderings are *untrilaterable*. In our evaluations, we consider both networks whose graphs have augmented bilateration orderings, and untrilaterable augmented bilateration orderings. The ordering chosen for the first sweep is an augmented bilateration ordering of the network, and we use a simple connectivity based procedure for obtaining such an ordering. We begin by labelling the $m\geq 3$ anchors to be $v_1,\ldots, v_m$. Assuming the ordering determined thus far is $v_1,\ldots, v_i$ where $i\geq m$, the $i+1$th sensor in the ordering is determined as follows. If v is any sensor that is not in the ordering and v is the neighbor of three or more sensors v_j where $j<i+1$, then we let v be the $i+1$th sensor in the ordering. Otherwise, if there is a pair of sensors u and v

Figure 13. Convex polygon approximation of ring intersection

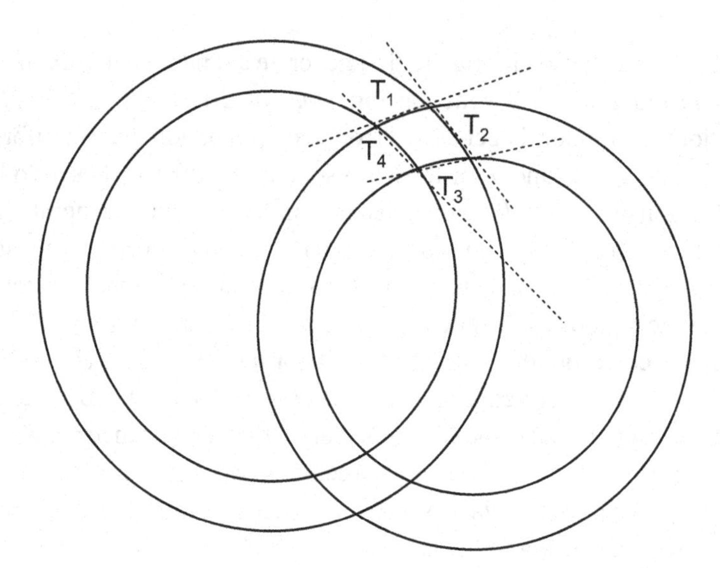

such that u and v are neighbors and each of u and v is adjacent to two sensors v_j where j <i+1, and v is adjacent to at least one sensor $v\_j, j < i+1$, which is not adjacent to v_i, then we let u and v be the $i+1$th and $i+2$th sensors in the ordering. Let $P(v,1)$, $v \in V$, denote the set of sensors adjacent to v and preceding v in the ordering of the first sweep. We implemented a slightly altered version of modified Sweeps in that if $S(v,1)$ is a singleton for some $v \in V$, then each $S(u,1)$, $u \in P(v,1)$, is refined using $S(v,1)$ before proceeding with the first sweep. More specifically, if α is the mapping in the singleton $S(v,1)$, then for each $u \in P(v,1)$, $S(u,1)$ is refined using $S(v,1)$ by removing all mappings β from $S(u,1)$ where $\beta(u) \neq \alpha(u)$. After $S(u,1)$, $u \in P(v,1)$, have each been refined by $S(v,1)$, each $\{\gamma(u) \mid \gamma \in S(u,1)\}$ remains a candidate region set for sensor u. Using this slightly altered version of modified Sweeps, we sweep through the network only once, i.e. by only computing $S(v,1)$, $v \in V$. Theoretically, similar results can be obtained by sweeping through the network multiple times using the original modified Sweeps algorithm. However, for evaluation purposes, we have found the altered implementation to be more computationally efficient and the computed error bounds on the position estimates were reasonable.

We now discuss two scenarios in detail. First, networks of 100 sensors are generated whose graphs have an augmented bilateration ordering for which the input parameters are as follows: sensing range R=0.2, number of anchors m=15, and noise factor $\eta = 0.08$. We averaged the results of modified Sweeps over 100 randomly deployed instances of such networks with the aforementioned parameters. Since there are 15 anchors, there are a total of 85 non-anchor sensors. We found that on average 47 of the 85 non-anchor sensors are assigned a position estimate, and the average error bound was 0.0303, which is less than 1/6 of the sensing range. Hence, on average, the actual position of a sensor can be guaranteed to be within 0.0303 of its estimated position. The average of the distance between each position estimate and the actual sensor position, did not exceed 0.02, and was in general far less than the average error bound. As expected, when the noise factor is decreased to 0.05, more sensors on average are assigned position estimates, and the corresponding error bounds are also lower.

We next considered networks of 100 sensors whose graphs have only untrilaterable augmented bilateration orderings. The input parameters used to generate the networks are identical as those used to generate networks in the previous scenario: sensing range R=0.2, number of anchors m=15, and a noise factor of $\eta = 0.08$. However, after the network is generated, the sensors which are adjacent to more than two anchors are identified, and for each such sensor, distance measurements are generated only between the sensor and two of the anchors the sensor is adjacent to. If the graph of the resulting network contains an augmented bilateration ordering, then the ordering can be guaranteed to be untrilaterable. The results are averaged over 100 instances of randomly deployed networks with the aforementioned parameters and whose graphs have only untrilaterable augmented bilateration orderings. For these networks, less sensors on total are assigned position estimates by Sweeps as compared to the previous scenario. More specifically, 40 of the sensors, as opposed to the previous 47, are assigned position estimates. The average error bound of the position estimates was 0.0366, which is just slightly higher than the previous scenario.

Generally speaking, we found that as the sensing range or the number of anchors of the network increased, the number of sensors for which a position estimate was computed increased. To illustrate this trend, we considered networks of 100 sensors whose graphs have an augmented bilateration ordering for which the input parameters are: sensing range R=0.2, and noise factor $\eta = 0.05$. By varying the number of anchors, we see what effect this has on the number of sensors for which a position estimate was computed and on the average error bound of the position estimates. We found the most meaningful results to occur when 10-25 anchors were used. There is a significant amount of variation in the case when there are less than 10 anchors, due to the fact that the sample size is too small. Twenty simulations were run for each of the following cases: 10 anchors, 15 anchors, 20 anchors and 25 anchors, and averaged in the Monte Carlo method. In Figure 14, we see an increase in the number of sensors for which a position estimate is computed as the number of anchors increases from 10 to 25. In Figure 15, we see that the mean error decreases from 0.0054 to 0.004 as the number of anchors increases from 10 to 25, which is what we expect.

CONCLUSION

In this chapter we presented a sequential algorithm called modified Sweeps for estimating sensor positions of a network when only inaccurate distance measurements and some anchor positions are available. If a position estimate p is computed for a sensor, then an error bound $e(p)$ is also computed so that the actual position of the sensor can be guaranteed to be within distance $e(p)$ of the estimated sensor position. We define the concept of correctly oriented position estimates, and show by example that a set of correctly oriented estimated sensor positions can be used to deduce certain geometric properties of the configuration of actual sensor positions. We also give a sufficient condition on the estimated positions and the corresponding error bounds in order to guarantee that the estimated positions are correctly oriented. We show by experimental evaluations that for randomly deployed networks whose graphs have an augmented bilateration ordering, modified Sweeps is able to assign positions to more than half of the non-anchor sensors, and furthermore, the error bounds of the estimated positions are small as compared to the sensing range. As we noted in the experimental evaluations section, the orderings in which sensors are processed are determined by a simple method using only connectivity. For future work, we will evaluate the algorithm using different techniques for determining orderings. In particular, we are

Figure 14. Anchors vs. number of sensors with estimated positions

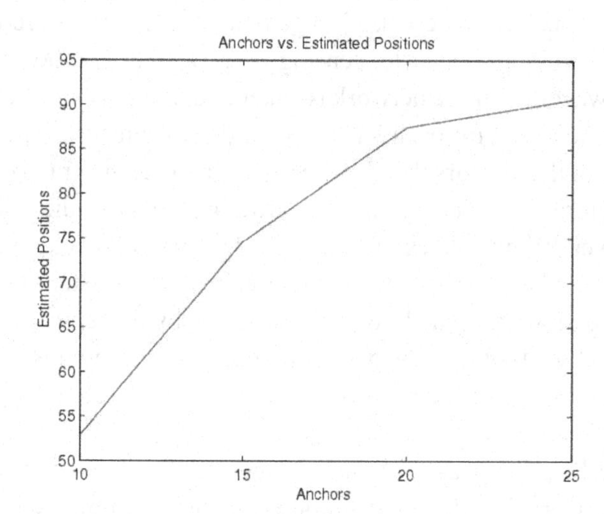

Figure 15. Anchors vs. mean error bound

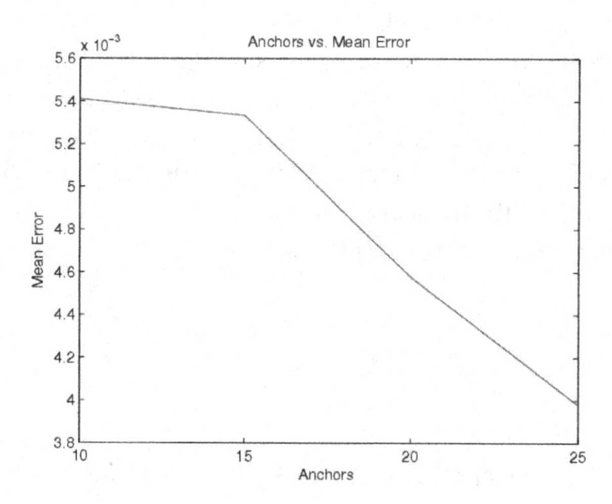

interested in those methods that take into account the guaranteed accuracies of distance measurements. We also aim to extend the proposed algorithm to a decentralized setting and carry out more extensive experimental evaluations using actual sensor data.

REFERENCES

Anderson, B. D. O., Belhumeur, P. N., Eren, T., Goldenberg, D. K., Morse, A. S., Whiteley, W., & Yang, Y. R. (2007). Graphical properties of easily localizable sensor networks. *Wireless Networks*. Thre Netherlands: Springer.

Fang, J., Cao, M., Morse, A. S., & Anderson, B. D. O. (2006). Localization of sensor networks using sweeps. In *Proceedings of CDC*, (pp. 4645–4650), San Diego, CA.

Fang, J., Cao, M., Morse, A. S., & Anderson. B. D. O. (2008). Sequential Localization of Sensor Networks. *SIAM J. on Control and Optimization*, to appear.

Goldenberg, D. K., Bihler, P., Cao, M., Fang, J., Anderson, B. D. O., Morse, A. S., & Yang, Y. R. (2006). Localization in sparse networks using sweeps. *Proceedings of Mobicom* (pp. 110–121).

Hendrickson, B. (1992). Conditions for unique graph realizations. *SIAM J. Comput.*, *21*(1), 65–84.

Jackson, B., & Jordan, T. (2005). Connected rigidity matroids and unique realizations of graphs. *J. Comb. Theory Ser. B*, *94*(1), 1–29.

Jacobs, D., & Hendrickson, B. (1997). An algorithm for two-dimensional rigidity percolation: the pebble game. *J. Comput. Phys.*, *137*(2), 346–365.

Moore, D., Leonard, J., Rus, D., & Teller, S. (2004). Robust distributed network localization with noisy range measurements. In *Proc. 2nd ACM SenSys*, (pp. 50–61), Baltimore, MD.

Priyantha, N. B., Demaine, E., & Teller S. (2003) Poster abstract: anchor-free distributed localization in sensor networks. In *SenSys '03: Proceedings of the 1st international conference on Embedded networked sensor systems*, (pp. 340–341,) New York, NY, USA. ACM.

Chapter VIII
MDS-Based Localization

Ahmed A. Ahmed
Texas State University–San Marcos, USA

Xiaoli Li
University of Missouri–Columbia, USA

Yi Shang
University of Missouri–Columbia, USA

Hongchi Shi
Texas State University–San Marcos, USA

ABSTRACT

The authors present several network node localization methods that are based on multidimensional scaling (MDS) techniques. Four algorithms are introduced: MDS-MAP(C), MDS-MAP(P), MDS-Hybrid, and RangeQ-MDS. MDS-MAP(C) is a centralized algorithm that simply applies MDS to estimate node positions. In MDS-MAP(P), a local map is built at each node of the immediate vicinity, then these maps are merged together to form a global map. MDS-Hybrid uses MDS-MAP(C) to relatively localize N_r reference nodes. Then, an absolute localization method uses these N_r nodes as anchors to localize the rest of the network. Finally, RangeQ-MDS assumes the absence of an RSSI-distance mapping function. It uses a quantized RSSI-based distance estimation technique (called RangeQ) to achieve more precise hop distances than other range-free approaches do. While MDS-MAP(C), MDS-MAP(P), and MDS-Hybrid can be range-aware or range-free, RangeQ-MDS is partially range-aware.

INTRODUCTION

The multidimensional scaling (MDS) technique has its origins in psychometrics and psychophysics. It is often used as part of exploratory data analysis or information visualization. It is related to principal component analysis, factor analysis, and cluster analysis. MDS has been applied in many fields, such as machine learning (Tenenbaum *et al.*, 2000) and computational chemistry (Glunt *et al.*, 1993). When used for localization, MDS takes full advantage of connectivity or distance information between nodes that have yet to be localized, unlike previous approaches.

In this chapter, we present four MDS-based localization methods for wireless sensor networks. The first method, called MDS-MAP(C) (Shang *et al.*, 2004), is a simple centralized approach that builds a global map using classical MDS. Like many existing methods, MDS-MAP(C) works well on networks with relatively uniform node density, but less well on more irregular networks, where the shortest-path distance between two nodes does not correspond well to their Euclidean distance.

To tackle this difficult problem, we present MDS-MAP(P) (Shang *et al.*, 2004). It is more complicated than MDS-MAP(C) because it builds for each node a local map of the small sub-network in the node's vicinity and then merges (patches) the local maps together to form a global map. This method avoids using shortest-path distances between far away nodes, and thus the smaller local maps constructed using local information are more accurate. Another advantage of the method is that it can be easily performed in a distributed fashion, which makes it appropriate for large-scale networks. A refinement step that uses minimum-squares minimization can be added to either MDS-MAP(C) or MDS-MAP(P) to improve the solution computed by MDS. We call the resulting methods MDS-MAP(C,R) and MDS-MAP(P,R), respectively.

In relative localization, nodes use the distance measurements to estimate their positions relative to some coordinate system. In absolute localization, a few nodes, called *anchors* (Savarese *et al.*, 2001), need to know their absolute positions, and all the other nodes are absolutely localized in the anchors' coordinate system. Due to the relatively high computational cost of MDS-MAP(C) and MDS-MAP(P), MDS-Hybrid is a relative localization algorithm that looks for a higher degree of granularity in terms of the performance-cost tradeoff than other localization algorithms. MDS-Hybrid, which is based on SHARP proposed by Ahmed *et al.* (2005), selects a number N_r of *reference nodes*. Then, MDS-MAP(C) relatively localizes these reference nodes using the distance information between them. Finally, an absolute localization method (M) localizes the rest of the network relative to the coordinate system of the reference nodes. Choosing N_r and M depends on both the application and the interest, either good performance, low cost, or somewhere in between.

We consider the node localization problem under two different scenarios. In the first, only connectivity (or proximity) information is available. Each node only knows what nodes are nearby, but not how far away these neighbors are or in what direction they lie. We call this scenario *range-free* localization. In the second scenario, the proximity information is enhanced by knowing the distances, perhaps with limited accuracy, between neighboring nodes. This is called *range-aware* localization. While MDS-MAP(C), MDS-MAP(P), and MDS-Hybrid can be range-free or range-aware, the RangeQ-MDS algorithm uses a sorted RSSI Partial Range Information (*PRI*)-based quantization scheme, called RangeQ (Li *et al.*, 2006), to improve the range estimation accuracy when distance information is not available.

Through simulation studies on regular as well as irregular networks, we show the improvement in localization performance the four methods presented can achieve.

PROBLEM FORMULATION

The network is represented as a connected undirected graph $G = (V, E)$, where V is the set of sensor nodes, of which there exist $A \subset V$ special nodes (called anchors) with known positions, and E is the set of edges connecting neighboring nodes. For the range-free case, the edges in the graph correspond to the connectivity information. For the range-aware case, the edges are associated with values corresponding to the estimated distances.

Given a network graph of n nodes and estimated distances $\mathbf{P}$ between some pairs of nodes (let p_{ij} represent the estimated distance between nodes i and j), the localization problem is to find the coordinates $\mathbf{X} = (\mathbf{X}_1, \mathbf{X}_2, \ldots, \mathbf{X}_n)$ of the nodes such that the Euclidean distances between the estimated positions of the nodes equal $\mathbf{P}$, i.e., $d_{ij} = p_{ij}$ for available p_{ij}, where $d_{ij} = \left\|\mathbf{X}_i - \mathbf{X}_j\right\|_2$. When the estimates p_{ij} are just the connectivity or inaccurate local distance measurements, usually there is no exact solution to the over-determined system of equations. Thus the localization problem is often formulated as an optimization problem that minimizes the sum of squared errors. This optimization problem is generally non-convex with many local minima. Traditional local optimization techniques, such as the Levenberg-Marquardt method, require good initial points in order to produce good solutions. Global search methods such as simulated annealing or genetic algorithms are generally too slow.

There are two possible outputs when solving the localization problem. One is a relative map and the other is an absolute map. The task of finding a relative map is to find an embedding of the nodes into either two- or three-dimensional space that results in the same neighbor relationships as the underlying network. Such a relative map can provide correct and useful information even though it does not necessarily include accurate absolute coordinates for each node. Relative information may be all that is obtainable in situations in which powerful sensors or expensive infrastructure cannot be installed, or when there are not enough anchors present to uniquely determine the absolute positions of the nodes. Furthermore, some applications only require relative positions of nodes, such as some direction-based routing algorithms (Royer and Toh, 1999; Yu *et al.*, 2001). Sometimes, however, an absolute map is required. The task of finding an absolute map is to determine the absolute geographic coordinates of all the nodes. This is needed in applications such as geographic routing and target discovering and tracking (Chu *et al.*, 2002; Intanagonwiwat *et al.*, 2000; Johnson and Maltz, 1996; Karp and Kung, 2000).

Before we describe the details of the methods presented, we first introduce MDS, and then describe the simulation setup for our experiments.

MULTIDIMENSIONAL SCALING (MDS)

Multidimensional scaling (MDS) is a method for visualizing dissimilarity data. For example, instead of knowing the latitude and longitude of a set of cities, we may only know their inter-city distances. The typical goal of MDS is to create a configuration of points in one, two, or three dimensions, whose inter-point distances are "close" to the original dissimilarities. The different variants of MDS use different criteria to define "close". These points represent the set of objects, and so a plot of the points can be used as a visual representation of their dissimilarities. Recently, MDS has been successfully applied to the problem of node localization in wireless sensor networks.

Basics of MDS Models

MDS models are defined by specifying how the given similarity data p_{ij} between two objects i and j are mapped into distances d_{ij} of an m-dimensional MDS configuration $\mathbf{X}$ consisting of all objects. The mapping is specified by a representation function, $f: p_{ij} \rightarrow d_{ij}(\mathbf{X})$, which specifies how the similarity data should be related to the distances. In practice, one usually does not attempt to strictly satisfy *f*. Rather, what is sought is a configuration whose distances satisfy *f* as closely as possible. The condition "as closely as" is quantified by a badness-of-fit measure or loss function. The loss function is a mathematical expression that aggregates the representation errors, $e_{ij} = f(p_{ij}) - d_{ij}(\mathbf{X})$, over all pairs (i, j). A normalized sum-of-squares of these errors define stress, the most common loss function in MDS.

Assume that measures of similarity, for which we use the general term proximity, p_{ij}, are given for each pair (i, j) of n objects. MDS attempts to represent proximities by distances among the points (representing the objects) of an m-dimensional configuration $\mathbf{X}$, the MDS space. Given a Cartesian space, one can compute the distance between any two points i and j. The Euclidean distance between points i and j in a two-dimensional configuration $\mathbf{X}$ is computed by the following formula:

$$d_{ij}(\mathbf{X}) = \sqrt{(x_{i1} - x_{j1})^2 + (x_{i2} - x_{j2})^2},$$

which can be written as

$$d_{ij}(\mathbf{X}) = \left[\sum_{a=1}^{2} (x_{ia} - x_{ja})^2 \right]^{1/2}.$$

MDS maps proximities p_{ij} into the corresponding distances $d_{ij}(\mathbf{X})$ of an MDS space $\mathbf{X}$. That is, f: $p_{ij} \rightarrow d_{ij}(\mathbf{X})$. The distances $d_{ij}(\mathbf{X})$ are unknowns, and MDS finds a configuration $\mathbf{X}$ of a predetermined dimensionality m on which the distances are computed. The function f, on the other hand, can be either completely specified or restricted to come from a particular class of functions.

Empirical proximities always contain noise due to measurement imprecision. Hence, one should not insist, in practice, that $f(p_{ij}) = d_{ij}(\mathbf{X})$, but rather that $f(p_{ij}) \approx d_{ij}(\mathbf{X})$, where "≈" can be read as "as equal as possible". Computerized procedures for finding an MDS representation usually start with some initial configuration and improve it by moving around its points in small steps (iteratively) to approximate the ideal model relation $f(p_{ij}) = d_{ij}(\mathbf{X})$ more and more closely. A squared error of representation is defined by

$$e_{ij}^2 = [f(p_{ij}) - d_{ij}(\mathbf{X})]^2.$$

Summing e_{ij}^2 over all pairs (i, j) yields a badness-of-fit measure for the entire MDS representation, *raw stress* σ_r,

$$\sigma_r(X) = \sum_{i,j} [f(p_{ij}) - d_{ij}(X)]^2.$$

To avoid scale dependency, σ_r can be normalized as follows,

$$\sigma_1^2(\mathbf{X}) = \frac{\sigma_r(\mathbf{X})}{\sum d_{ij}^2(\mathbf{X})} = \frac{\sum [f(p_{ij}) - d_{ij}(\mathbf{X})]^2}{\sum d_{ij}^2(\mathbf{X})}.$$

Taking the square root yields a value known as *Stress-1.* Thus

$$Strtess - 1 = \sigma_1(\mathbf{X}) = \sqrt{\frac{\sum [f(p_{ij}) - d_{ij}(\mathbf{X})]^2}{\sum d_{ij}^2(\mathbf{X})}}.$$

In weighted MDS, positive weights w_{ij} are added into the stress function as follows:

$$\sigma_r(\mathbf{X}) = \sum_{i>j} w_{ij}[f(p_{ij}) - d_{ij}(\mathbf{X})]^2.$$

When working with missing data, w_{ij} is set to 1 if p_{ij} is known and 0 if p_{ij} is missing.

Classical MDS

The basic idea of classical MDS (Gower, 1966 and Torgerson, 1952) is to assume that the dissimilarities are distances and then find coordinates that explain them. Moreover, a linear transformation model is assumed, i.e., $d_{ij} = a + bp_{ij}$. The distances **D** are determined so that they are as close to the proximities **P** as possible. There are a variety of ways to define "close". A common one is a least-squares definition, which is used by classical metric MDS. In this case, we define **I(P)** = **D** + **E**, where **I(P)** is a linear transformation of the proximities, and **E** is a matrix of errors (residuals). Since **D** is a function of the coordinates **X**, the goal of classical MDS is to calculate **X** such that the sum of squares of **E** is minimized. In classical MDS, the coordinates **X** can be computed from **P** through singular value decomposition (SVD) on the double centered squared **P**. Double centering a matrix is subtracting the row and column means of the matrix from its elements, adding the grand mean and multiplying by –1/2.

Let $\mathbf{X}_{n\times m}$ be the matrix of coordinates of the points. Each row *i* of **X** gives the coordinates of point *i* on *m* dimensions, i.e., $x_{i1}, x_{i2}, \ldots, x_{im}$. The squared Euclidean distance is defined by

$$d_{ij}^2(\mathbf{X}) = d_{ij}^2 = \sum_{a=1}^{m}(x_{ia} - x_{ja})^2 = \sum_{a=1}^{m}(x_{ia}^2 + x_{ja}^2 - 2x_{ia}x_{ja}).$$

Let $\mathbf{D}^{(2)}(\mathbf{X})$ denote the matrix of squared distances. For example, when **X** contains the coordinates of three points in two dimensions, $\mathbf{D}^{(2)}(\mathbf{X})$ can be represented as

$$\mathbf{D}^{(2)}(\mathbf{X}) = \begin{bmatrix} 0 & d_{12}^2 & d_{13}^2 \\ d_{12}^2 & 0 & d_{23}^2 \\ d_{13}^2 & d_{23}^2 & 0 \end{bmatrix}$$

$$= \sum_{a=1}^{2}\begin{bmatrix} x_{1a}^2 & x_{1a}^2 & x_{1a}^2 \\ x_{2a}^2 & x_{2a}^2 & x_{2a}^2 \\ x_{3a}^2 & x_{3a}^2 & x_{3a}^2 \end{bmatrix} + \sum_{a=1}^{2}\begin{bmatrix} x_{1a}^2 & x_{2a}^2 & x_{3a}^2 \\ x_{1a}^2 & x_{2a}^2 & x_{3a}^2 \\ x_{1a}^2 & x_{2a}^2 & x_{3a}^2 \end{bmatrix} - 2\sum_{a=1}^{2}\begin{bmatrix} x_{1a}x_{1a} & x_{1a}x_{2a} & x_{1a}x_{3a} \\ x_{2a}x_{1a} & x_{2a}x_{2a} & x_{2a}x_{3a} \\ x_{3a}x_{1a} & x_{3a}x_{2a} & x_{3a}x_{3a} \end{bmatrix}$$

$$= \mathbf{c1}' + \mathbf{1c}' - 2\sum_{a=1}^{2}\mathbf{x}_a\mathbf{x}_a'$$

In general, we have

$$\mathbf{D}^{(2)}(\mathbf{X}) = \mathbf{c1}' + \mathbf{1c}' - 2\sum_{a=1}^{2}\mathbf{x}_a\mathbf{x}_a' = \mathbf{c1}' + \mathbf{1c}' - 2\mathbf{XX}' \quad (1)$$

where $\mathbf{x}_a$ is column a of matrix $\mathbf{X}$ and $\mathbf{c}$ is a vector that has elements $\sum_{a=1}^{m} x_{ia}^2$, the diagonal elements of $\mathbf{XX}'$. The matrix $\mathbf{B} = \mathbf{XX}'$ is called a scalar product matrix.

The problem here is to arrive at a scalar product matrix $\mathbf{B}$ given a matrix of squared distances $\mathbf{D}^{(2)}$. Since distances do not change under translations, we assume that $\mathbf{X}$ has column means equal to 0. Multiplying the left and the right sides of Eq. (1) by the centering matrix $\mathbf{J} = \mathbf{I} - n^{-1}\,\mathbf{11}'$ and by the factor $-\frac{1}{2}$ gives

$$\begin{aligned} -\frac{1}{2}\mathbf{J}\mathbf{D}^{(2)}\mathbf{J} &= -\frac{1}{2}\mathbf{J}(\mathbf{c1}' + \mathbf{1c}' - 2\mathbf{XX}')\mathbf{J} = -\frac{1}{2}\mathbf{Jc1}'\mathbf{J} - \frac{1}{2}\mathbf{J1c}'\mathbf{J} + \frac{1}{2}\mathbf{J}(2\mathbf{B})\mathbf{J} \\ &= -\frac{1}{2}\mathbf{Jc0}' - \frac{1}{2}\mathbf{0c}'\mathbf{J} + \mathbf{JBJ} = \mathbf{B} \end{aligned} \qquad (2)$$

The first two terms are zero because centering a vector of ones yields a vector of zeros ($\mathbf{1}'\mathbf{J} = 0$). The centering around $\mathbf{B}$ can be removed because $\mathbf{X}$ is column centered, and hence so is $\mathbf{B}$. The operation in Eq. (2) is called double centering. To find the MDS coordinates from $\mathbf{B}$, we factor

$\mathbf{B}$ by eigendecomposition, $\mathbf{Q\Lambda Q}' = (\mathbf{Q\Lambda}^{1/2})(\mathbf{Q\Lambda}^{1/2})' = \mathbf{XX}'$. In classical scaling, the $\mathbf{D}^{(2)}$ matrix is replaced by the squared dissimilarities $\Delta^{(2)}$.

The procedure for classical scaling is summarized as follows:

- Compute the matrix of squared dissimilarities $\Delta^{(2)}$.
- Apply double centering to this matrix:

$$\mathbf{B}_\Delta = -\frac{1}{2}\mathbf{J}\Delta^{(2)}\mathbf{J}.$$

- Compute the eigendecomposition of $\mathbf{B}_\Delta = \mathbf{Q\Lambda Q}'$
- Let m be the dimensionality of the solution, $\mathbf{\Lambda}_+$ the matrix of the first m eigenvalues greater than zero, and $\mathbf{Q}_+$ the first m columns of $\mathbf{Q}$. Then, the coordinate matrix of classical scaling is given by $\mathbf{X} = \mathbf{Q}_+\mathbf{\Lambda}_+^{1/2}$.

Classical scaling minimizes the loss function (strain),

$$L(\mathbf{X}) = \left\| -\frac{1}{2}\mathbf{J}[\mathbf{D}^{(2)}(\mathbf{X}) - \Delta^{(2)}]\mathbf{J} \right\|^2 = \left\| \mathbf{XX}' + \frac{1}{2}\mathbf{J}\Delta^{(2)}\mathbf{J} \right\|^2 = \left\| \mathbf{XX}' - \mathbf{B}_\Delta \right\|^2.$$

Localization Using MDS

Imagine a small cloud of colored beads suspended in mid-air. To characterize the arrangement, one could measure the straight-line distance between each pair of beads. If the cloud were shattered and the beads fell to the floor, one could imagine trying to recreate the arrangement based on the recorded inter-point distances. One would try to determine a location for each bead such that the distances in the new arrangement matched the desired distances. This recreation process is exactly the problem that MDS solves. Intuitively, it is clear that while the $O(n^2)$ distances will be more than enough to determine $O(n)$ coordinates, the result of MDS will be an arbitrarily rotated and flipped version of the true original layout because the inter-point distances make no reference to any absolute coordinates.

MDS can be seen as a set of data analysis techniques that display the structure of distance-like data as a geometrical picture (Borg and Groenen, 1997). As shown formally in the previous subsection, MDS starts with one or more distance (or similarity) matrices that are presumed to have been derived from points in a multidimensional space. It is usually used to find a placement of the points in a low-dimensional space, usually two- or three-dimensional, where the distances between points resemble the original similarities. By visualizing objects as points in a low-dimensional space, the complexity in the original data matrix can often be reduced while preserving the essential information.

There are many types of MDS techniques. They can be classified according to whether the similarity data is qualitative (non-metric MDS) or quantitative (metric MDS). They can also be classified according to the number of similarity matrices and the nature of the MDS model. Classical MDS uses one matrix. Replicated MDS uses several matrices, representing distance measurements taken from several subjects or under different conditions. Weighted MDS uses a distance model which assigns a different weight to each dimension. Finally, there is a distinction between deterministic and probabilistic MDS. In deterministic MDS, each object is represented as a single point in a multidimensional space, whereas in probabilistic MDS each object is represented as a probability distribution over the entire space.

We focus on classical metric MDS in this chapter. Classical metric MDS is the simplest case of MDS: the data is quantitative and the proximities of objects are treated as distances in a Euclidean space (Torgerson, 1965). The goal of metric MDS is to find a configuration of points in a multidimensional space such that the inter-point distances are related to the provided proximities by some transformation (e.g., a linear transformation). If the proximity data were measured without error in a Euclidean space, then classical metric MDS would exactly recreate the configuration of points. In practice, the technique tolerates error gracefully, due to the overdetermined nature of the solution. This is very helpful when we apply it to localization, as our distance estimates can be very rough indeed. Because classical metric MDS has an analytical solution, it can be performed efficiently on large matrices.

In non-metric (also called ordinal) MDS (Shepard, 1962), the goal is to establish a monotonic relationship between inter-point distances and the desired distances. Instead of trying to directly match the given distances, one is satisfied if the distances between the points in the solution fall in the same ranked order as the corresponding distances in the input matrix. The advantage of non-metric MDS is that no assumptions need to be made about the underlying transformation function. The only assumption is that the data is measured at the ordinal level. Just as classical MDS, non-metric MDS can also be applied to the localization problem. By adopting a more flexible model, the effects of a few highly incorrect measurements might be more easily tolerated.

A thorough analysis of the localization error bounds has been done (Shang *et al*., 2004). The localization problem has been treated as an estimation problem. The Cramér-Rao error bounds have been derived when the distances between all nodes were used, which was the case for MDS. However, a detailed analysis of error bounds is beyond the scope of this chapter.

SIMULATION SETUP

In the experiments reported, we assess the average-case performance of the localization methods presented by simulation on Matlab 7.0 on 2-dimensional networks of at least 100 nodes deployed inside a $10r \times 10r$ square field, where r is the placement unit length. Two example scenarios are shown in Figure 1: (a) regular networks – 200 nodes are randomly placed in a $10r \times 10r$ square and (b) irregular network

– 160 nodes are randomly placed in an area of C shape within a $10r \times 10r$ square. Circles represent sensor nodes and lines represent connections between nodes that are within communication range of each other. The radio range is $1.5r$, where r is the placement unit length. The average node degrees of the two problems are 12.1 and 11.5, respectively. For each type of networks, the algorithms are run on at least 30 randomly-generated network instances.

All nodes are assumed to have the same radio range R, which is modeled as a circle with a predefined radius. We do not consider models of non-uniform radio propagation or widely varying ranging errors. Hence, all communications are assumed to be bidirectional, i.e., if node i can communicate with node j, then node j can communicate with node i. The average node degree (average number of neighbors) is controlled by specifying R. The errors of position estimates are normalized to R (i.e., 50% position error means half of the range of the radio). The resulting average node degree is a function of both R, network type (topology), and number of nodes. For example, a 200-node regular network with R equal to $1.5r$ has an average node degree of 12.1. A 160-node irregular network with R equal to $1.5r$ has an average node degree of 11.5. The distance measure is modeled as the true distance blurred with Gaussian noise. Assume the true distance is d and the standard deviation of the range error is e_r, then the measured distance $\hat{d}$ is a random value drawn from the normal distribution $d(1+N(0,e_r))$. For simplicity, we will refer to e_r by just "range error" throughout this chapter.

MDS-MAP(C) AND MDS-MAP(C,R)

The simplest MDS-based localization method is MDS-MAP(C). It builds a global map using a single application of classical MDS. The parameter "C" refers to "centralized", as the connectivity information of the network is sent to a central location where the computation is carried out. The method with additional refinement to MDS-MAP(C) is called MDS-MAP(C,R), where the parameter "R" is for "refinement".

Figure 1. Two example problems similar to those in simulation

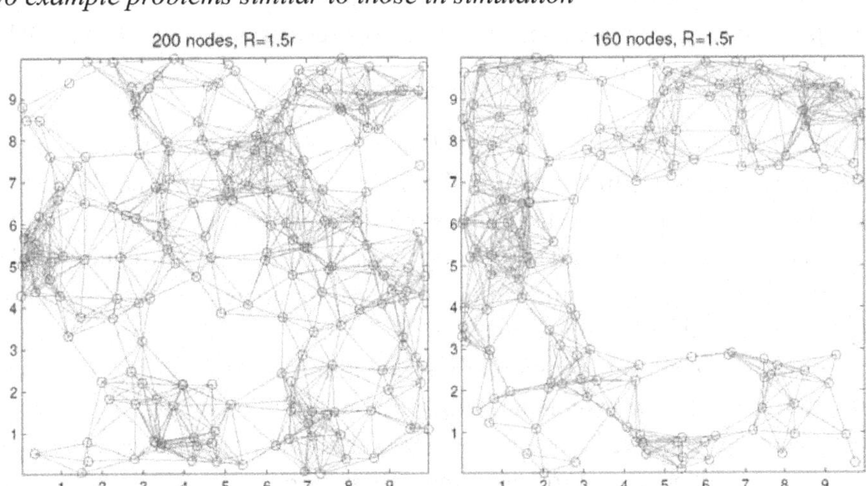

Algorithm

MDS-MAP(C) consists of three steps as follows.

- Compute the shortest paths between all pairs of nodes in the region of consideration. The shortest path distances are used to construct the distance matrix **D** for MDS.
- Apply MDS to **D**, retaining the first 2 (or 3) largest eigenvalues and eigenvectors to construct a 2-D (or 3-D) relative map.
- Given sufficient anchor nodes (3 or more for 2-D, 4 or more for 3-D), transform the relative map to an absolute map based on the absolute positions of anchors.

In step 1, we first assign distances to the edges in the graph. When the distance of a pair of neighbor nodes is known, the value of the corresponding edge is the measured distance. When we only have connectivity information, a simple approximation is to assign to all edges the value 1 multiplied by the transmission range. Then, a shortest-path algorithm, such as Dijkstra's or Floyd's, can be applied to find the shortest path between all pairs of nodes. The time complexity is $O(n^3)$, where n is the number of nodes.

In step 2, classical MDS is applied directly to the distance matrix. The core of classical MDS is singular value decomposition, which has complexity of $O(n^3)$. The result of MDS is a relative map that gives a location to each node. Although these locations may be accurate relative to one another, the entire map will be arbitrarily rotated, translated, and flipped relative to the true node positions.

In step 3, the relative map is transformed through a linear transformation, which may include scaling, rotation, and reflection. The goal is to minimize the sum of the squares of the errors between the true positions of the anchors and their transformed positions in the MDS map. Computing the transformation parameters takes $O(m^3)$ time, where m is the number of anchors. Applying the transformation to the whole relative map takes $O(n)$ time.

In MDS-MAP(C,R), the following refinement step is added between steps 2 and 3 of MDS-MAP(C) to improve the relative map.

- Using the position estimates of nodes in the MDS solution as an initial solution, apply least-squares minimization to improve the match between the measured distances between neighboring nodes and their distances in the solution.

Our formation of the refinement is more general than previous methods (Savarese *et al.*, 2002; Savvides *et al.*, 2002) in two ways: (1) In addition to the information between 1-hop neighbors, information between multihop neighbors is also used, but with different weights. (2) Instead of refining the coordinates of one node at a time while all other nodes remain fixed, the coordinates of all nodes in the relative map are variables in a single optimization. We use a refinement range R_{ref}, defined based on hops, to specify how much information is considered. $R_{ref} = 1$ means only information between 1-hop neighbors are used, $R_{ref} = 2$ means information of nodes within two hops is used, and so on. Different values of R_{ref} offer a trade-off between computational cost and solution quality.

An important advantage of our refinement approach is that MDS can provide better starting points for the least-squares minimization than other triangulation-based or heuristic methods (Savvides *et al.*, 2002). The least-squares minimization problem is high-dimensional and has lots of local minima. Random starting points usually lead to very bad solutions. MDS is good at finding the right general topology of a network, which corresponds to a starting point in the basin of attraction of an optimal or near-optimal solution.

More formally, let (x_i, y_i), $i = 1, \ldots, N$, represent the coordinates of the N nodes in a 2-D local map, d_{ij} the Euclidean distance between two nodes i and j in a candidate solution, and p_{ij} the measured proximity between nodes i and j. When only proximity information is available, $p_{ij} = 1$ if i and j are 1-hop neighbors. When distance measurements between 1-hop neighbors are available, p_{ij} is the distance between i and j if they are 1-hop neighbors or the shortest path distance if i and j are further apart. The objective of the refinement step is

$$\min_{x_k, y_k} \sum_{i,j} w_{ij} (d_{ij} - p_{ij})^2, \text{ for } k = 1, \cdots, N \tag{3}$$

where w_{ij} are the weights.

For a 2-D n-node network, the problem has $2n$ variables and no constraints. The Jacobian can be computed analytically. In our experiments, we use the Levenberg-Marquardt method ("lsqnonlin" in Matlab's optimization toolbox) to solve the problem. Usually only the first few iterations of "lsqnonlin" give significant improvement. Thus, the maximum number of iterations is set to a small number, such as 20. Although this local optimization algorithm is fast, it is considerably slower than classical MDS. For 100-node networks, it is more than an order of magnitude slower. For larger networks, the time difference becomes larger.

Illustrative Examples

Figure 2 shows the results of the range-free versions of MDS-MAP(C) and MDS-MAP(C,R) on a regular network example. Four random anchor nodes, denoted by asterisks, are used to estimate the transformation to absolute coordinates. The circles represent the true locations of the nodes and the solid lines represent the errors of the estimated positions from the true positions. The longer the line, the larger the error is. The average errors of MDS-MAP(C) and MDS-MAP(C,R) are $0.67r$ and $0.35r$, respectively, where the field in which the nodes are placed measures $10r$ by $10r$.

When distances between one-hop neighbors are known, the result of MDS-MAP(C) can be improved. Figure 3 shows results on the same network, but when distances between one-hop neighbors are known with 5% range error. The estimates of MDS-MAP(C) based on the same 4 anchor nodes have an average error of $0.25r$, much better than the result when using connectivity only ($0.67r$). The result after refinement in MDS-MAP(C,R) is excellent. The average error is reduced to $0.06r$, where r is the placement unit length and is set to 1 in the experiments.

Irregular topologies are much harder than uniform topologies. Figure 4 shows sample results on the irregular network example. Again, there are four random anchor nodes. The result of MDS-MAP(C) is poor. Although the result of MDS-MAP(C,R) is better than that of MDS-MAP(C), it is much worse than its result on the uniform example. MDS-MAP(C) does not work well because the shortest-path distance between two nodes in different wings of the network is much larger than their actual Euclidean distance. The error of MDS-MAP(C) using connectivity information is very large, $2.4r$. The refinement in MDS-MAP(C,R) is useful and reduces the error to $0.55r$.

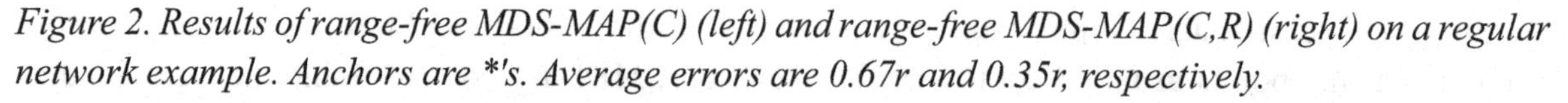

*Figure 2. Results of range-free MDS-MAP(C) (left) and range-free MDS-MAP(C,R) (right) on a regular network example. Anchors are *'s. Average errors are 0.67r and 0.35r, respectively.*

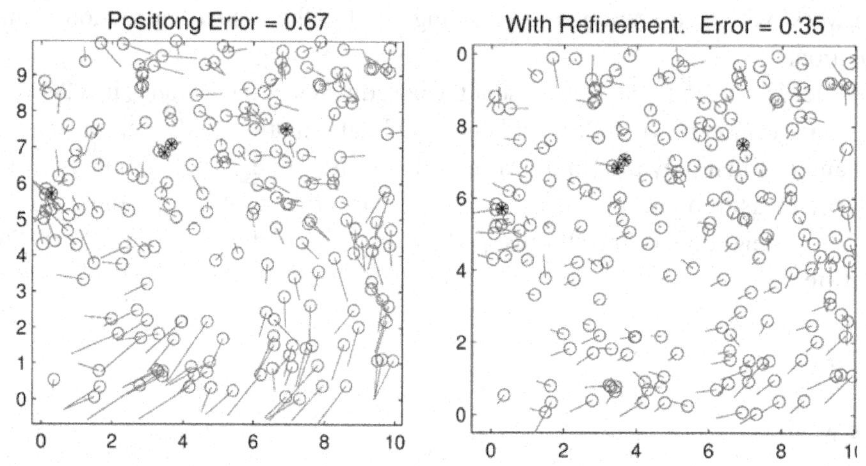

*Figure 3. Results of range-aware MDS-MAP(C) (left) and range-aware MDS-MAP(C,R) (right) on a regular network example. Anchors are *'s. Average errors are 0.25r and 0.06r, respectively.*

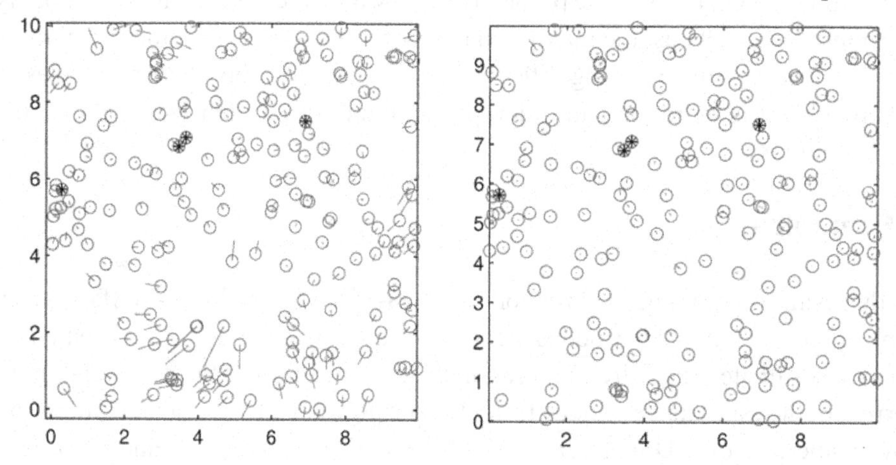

Experimental Results

In the experiments reported here, 200-node regular networks and 160-node irregular networks are used. Moreover, we set the refinement range R_{ref} to 2 and the weights w_{ij} to 1 when i and j are one hop apart (see Eq. (5)). When the two nodes i and j are two hops apart, we tried several values for the weights w_{ij}, and the value 1/4 worked the best.

For regular networks, Figure 5 shows the performance of MDS-MAP(C) as a function of average node degree and number of anchors, using 4, 6, or 10 random anchors. Position estimates by MDS-MAP(C) have an average error under 100%R in scenarios with just 4 anchor nodes and an average node degree level of 8.9 or greater. On the other hand, when the average node degree is low, e.g., 5.9, the errors can be large. Having good estimates of the distances between neighbors leads to much better solutions when the average node degree is high. When the average node degree is 12.2 or greater, the errors are about half of those by the range-free version. On the other hand, when the average node degree is low, e.g.,

Figure 4. Results of range-free MDS-MAP(C) (left) and range-free MDS-MAP(C,R) (right) on an irregular network example. Average errors are 2.4r and 0.55r, respectively.

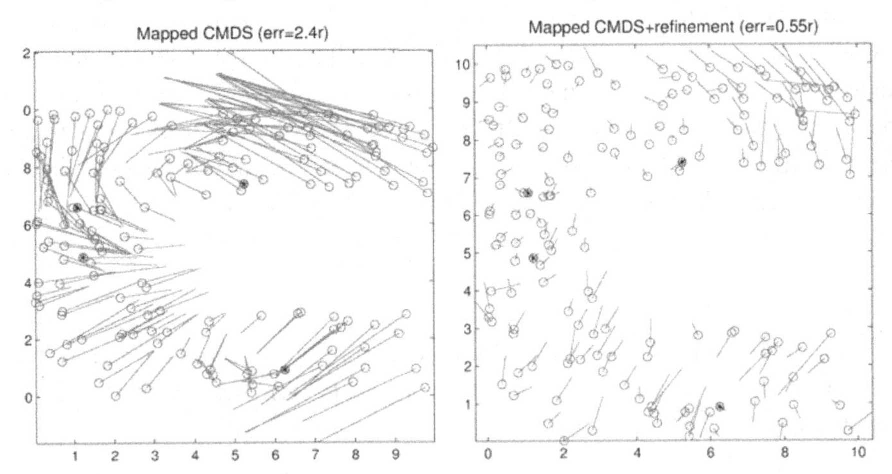

5.9, knowing the local distance does not help much. More experimental results will be shown compared to the performance of MDS-MAP(P) in the next section.

The refinement in MDS-MAP(C,R) improves the solutions significantly. The result can be misleading because it seems that the refinement is the most important and does all the work. This is not the case. From a random starting point, the refinement usually doesn't do much and just returns a bad solution, because there are many local minima. Thanks to MDS, the relative map often has the right topology, which corresponds to a good starting point in the same basin as the optimal or near-optimal solution. This is why the refinement performs so well.

MDS-MAP(P) AND MDS-MAP(P,R)

MDS-MAP(C) and MDS-MAP(C,R) do not work well on irregular networks because they rely on shortest-path distance estimation, which can have large errors for remote nodes. Another problem with

Figure 5. Results of MDS-MAP(C): range-free (left) and range-aware (right) on 200-node regular networks.

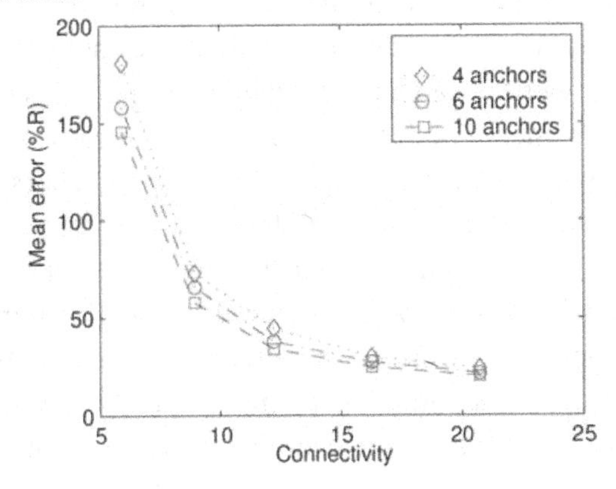

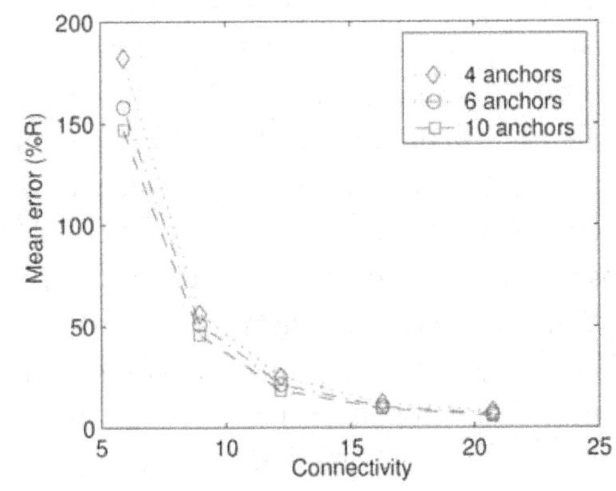

these centralized methods is that they can not be applied easily to large networks, for which reading out the connectivity and distance information is potentially prohibitive. In such cases, in-network computation of coordinates would be much more attractive. MDS-MAP(P) addresses both of these problems.

MDS-MAP(P) is more complicated than MDS-MAP(C). It builds many local maps and then patches them together to form a global map. This method relies on local information and avoids using the distance estimation between remote nodes. As we will show, it achieves better results on irregular networks. Another benefit of MDS-MAP(P) is that it can be easily executed in a distributed fashion. When we add refinement to improve the global map, we call the method MDS-MAP(P,R).

Algorithm

In MDS-MAP(P), individual nodes simultaneously compute their own local maps using their local information. Then these maps can be incrementally merged to form a global map. The steps of MDS-MAP(P) are as follows.

- Set the range for local maps, R_{lm}. For each node, neighbors within R_{lm} hops are involved in building its local map. We use $R_{lm} = 2$ in our experiments.
- For each node, apply MDS-MAP(C,R) to the nodes within range R_{lm} to generate its local map.
- Merge local maps. Local maps can be merged in various ways. We use a simple strategy: first randomly pick a node and start with its local map; then merge in the maps of neighboring nodes one by one. Each time, we choose the neighbor to merge whose local map shares the most nodes with the current map. Thus, the initial local map grows by incorporating other local maps and can eventually cover the entire network.
- Given sufficient anchor nodes (3 or more for 2-D, 4 or more for 3-D), transform the relative map to an absolute map based on the absolute positions of anchors.

Two maps are merged together based on the coordinates of their common nodes. The best linear transformation (minimizing discrepancy errors) is computed to transform the coordinates of the common nodes in one map to those in the other map. Given the coordinates of common nodes in maps A and B as matrices X_A and X_B, a linear transformation (translation, reflection, orthogonal rotation, and scaling) of X_B to best conform to X_A is determined. The "goodness-of-fit" criterion is the sum of squared errors, i.e., $\min_T \|T(X_B) - X_A\|_2$, where $T(\cdot)$ is the linear transformation.

This method allows for parallel and distributed implementations in several ways. First, the computation of local maps can be done locally at each node in parallel with the others. Second, the local maps can be merged in parallel in different parts of the network. Because the method does not require anchor nodes in order to build a relative map of a sub-network, it can be applied to many sub-networks in parallel. Third, the computation of absolute maps from anchor nodes could be applied to relative local maps and thus also be distributed in the network. For example, as soon as three or more anchors are present in a sub-network, an absolute map could be computed. Furthermore, all local maps bordering on this absolute map could be absorbed in parallel into that map using the merger step. For large networks and a sufficient number of anchor nodes, it should never be necessary to compute a single global map anywhere. Distributed map merging has a number of benefits, including more balanced computation and communication among the nodes, faster construction of the global map, and distribution of map information in the network at multiple levels of granularity, giving the opportunity for better flexibility and robustness.

The amount of error generated when two maps are merged depends on several factors, including the accuracy of the two maps and the number of common nodes. The error will propagate when a linear sequence of maps are merged. In dense networks, the adjacent local maps usually have many common nodes, and thus the error introduced in merging is small. In MDS-MAP(P,R), a refinement step is added between steps 3 and 4 of to improve the global relative map.

Illustrative Examples

Using the two example problems from Figure 2, we illustrate the performance of MDS-MAP(P) and MDS-MAP(P,R). Figures 6 and 7 show the results on a regular network example for the range-free and range-ware scenarios. Using connectivity information only (range-free), the average error of MDS-MAP(P) is 0.40*r*, about 60% of the error of MDS-MAP(C) in Figure 2, and slightly worse than MDS-MAP(C,R). After refinement, the error of MDS-MAP(P,R) is 0.31*r*, better than that of MDSMAP(C,R). Using local distances, MDS-MAP(P) and MDS-MAP(P,R) obtain much better results. The error of MDS-MAP(P) is 0.16*r*, better than the 0.25*r* error of MDS-MAP(C) in Figure 3. After refinement, the error of MDS-MAP(P,R) is 0.06*r*, at the level of the distance estimation errors.

Figure 8 shows sample results on the irregular placement example for the range-aware case. The solution of MDS-MAP(P) (error 0.72r) is quite reasonable. The solution of MDS-MAP(P,R) is even better (error 0.29*r*).

Experimental Results

Similar to the experiments done with MDS-MAP(C), 200-node regular networks and 160-node irregular networks are used. Figures 9 and 10 show the performance of MDS-MAP(P) and MDS-MAP(P,R) compared to MDS-MAP(C) and MDS-MAP(C,R) for regular networks. The errors are plotted against the average node degree. The radio range (*R*) goes from 1.25*r* to 2.5*r*, in increments of 0.25*r*, which leads to average node degrees of 8.9, 12.2, 16.4, 20.9, 25.9, and 31.1. Three or ten random anchors are used.

Figure 6. Results of range-free MDS-MAP(P) (left) and range-free MDS-MAP(P,R) (right) on a regular network example. Average errors are 0.40r and 0.31r, respectively

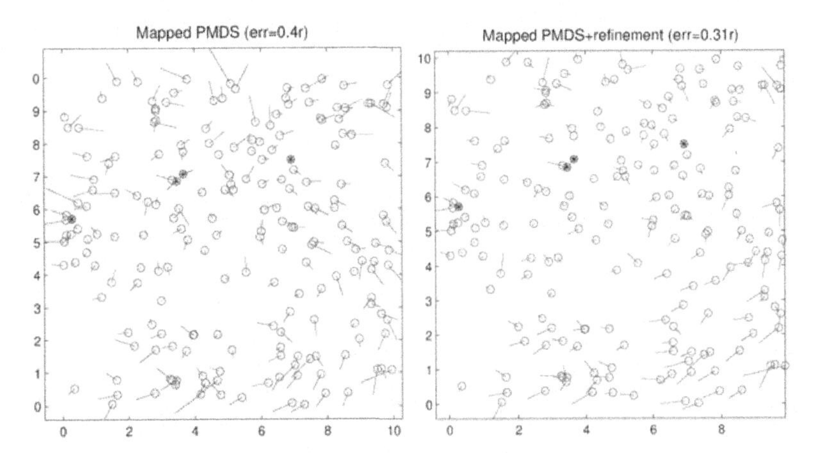

Figure 7. Results of range-aware MDS-MAP(P) (left) and range-aware MDS-MAP(P,R) (right) on a regular network example. Average errors are 0.16r and 0.06r, respectively

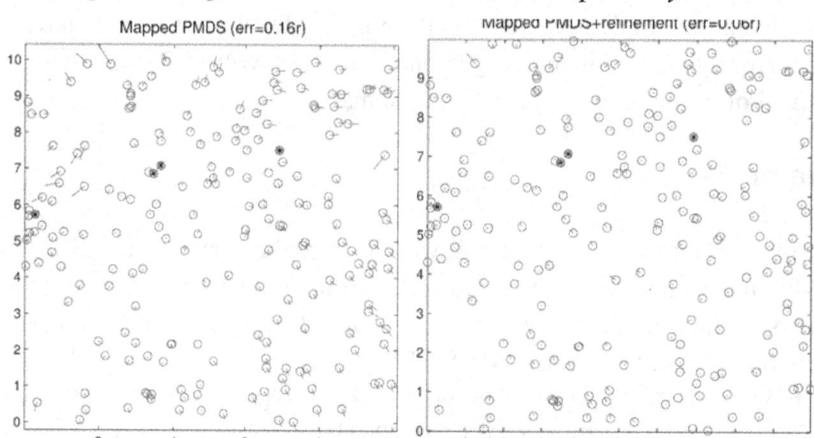

When using only connectivity information, MDS-MAP(P) is consistently better than MDS-MAP(C), more than 10%R better when the average node degree is low. MDS-MAP(C,R) and MDS-MAP(P,R) have comparable results and are better than MDS-MAP(P). Although more anchors lead to better results, the improvement with more than 6 anchors is small. For the range-free scenario, MDS-MAP algorithms are much better than the convex optimization approach in (Doherty *et al.*, 2001) when the number of anchor nodes is low. For example, with 4 to 10 anchors in a 200-node random network, the convex optimization approach has an average estimation error of more than twice the radio range when the average node degree is 8.9 and above. The results are also better than Hop-TERRAIN (Savarese *et al.*, 2002), especially when the number of anchors is small. For example, with 4 anchors (2% of the network) and an average node degree 12.2, MDS-MAP(P) using connectivity information only has an average error of about 27%*R*, whereas Hop-TERRAIN has an average error of about 90%*R*.

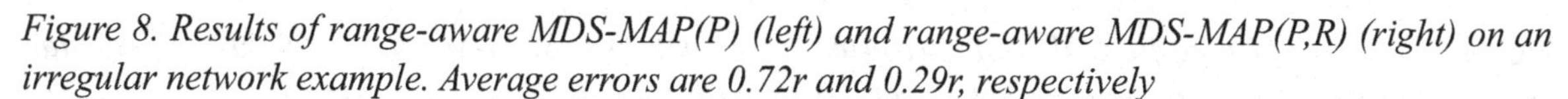
Figure 8. Results of range-aware MDS-MAP(P) (left) and range-aware MDS-MAP(P,R) (right) on an irregular network example. Average errors are 0.72r and 0.29r, respectively

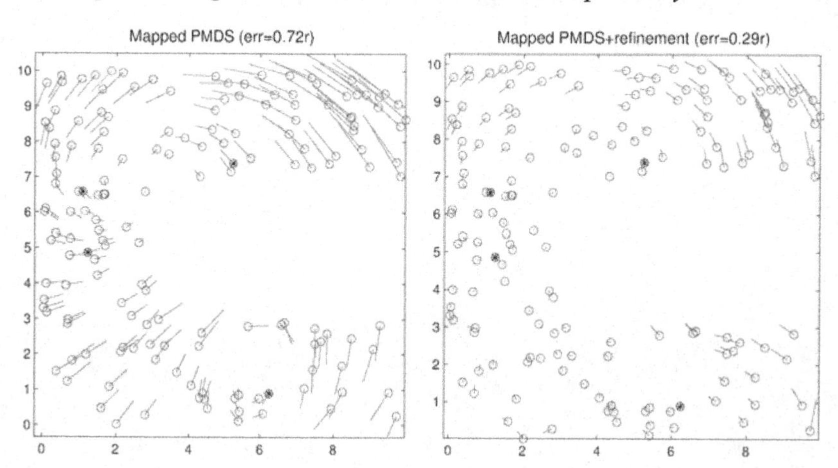

Figure 9. Performance of range-free MDS-MAP methods for regular networks with 3 (left) and 10 (right) anchors for different values of average node degree (horizontal axis)

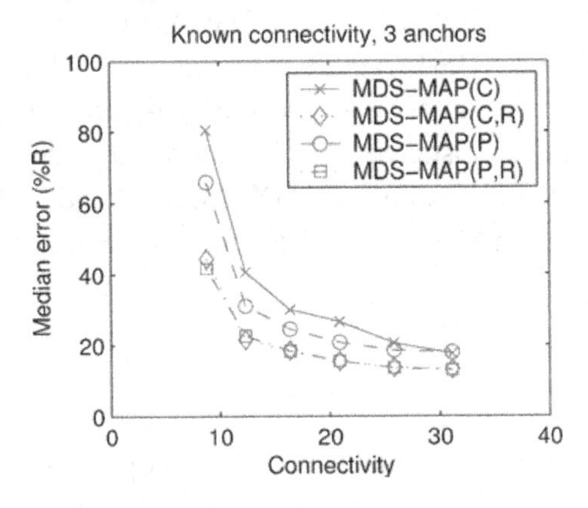

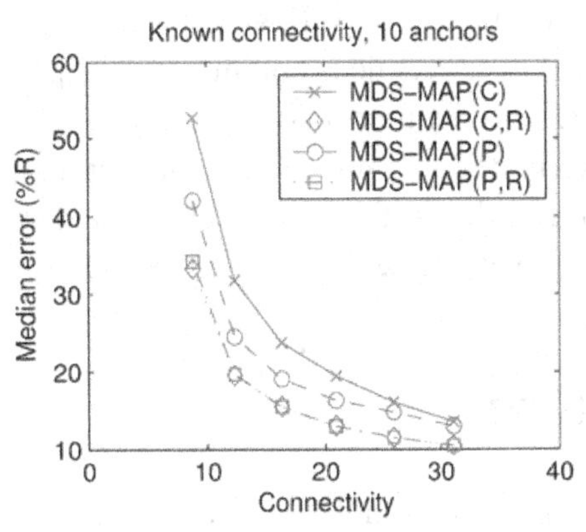

Figure 10. Performance of range-aware MDS-MAP methods for regular networks with 3 (left) and 10 (right) anchors for different values of average node degree (horizontal axis)

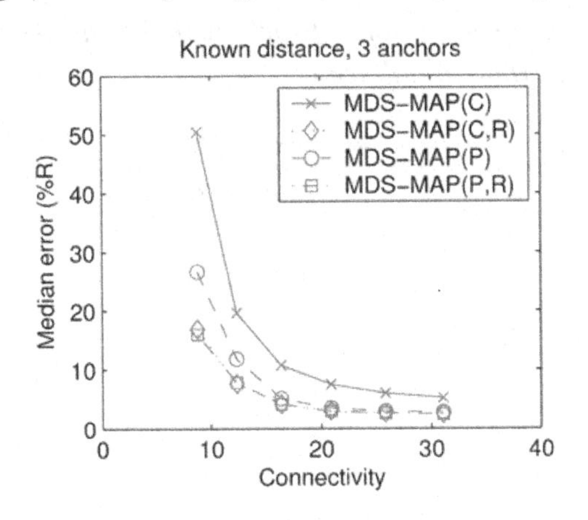

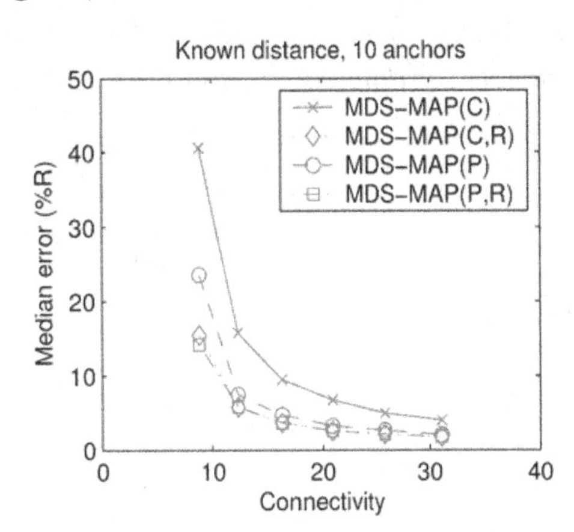

Using local distances with 5% error improves the performance of the MDS-MAP algorithms. Their errors are about half of those obtained using only proximity information. MDS-MAP(P) is comparable to MDS-MAP(C,R) and MDS-MAP(P,R) when the average node degree is 12.2 and above. MDS-MAP succeeds in localizing a higher fraction of the nodes in a network than most previous methods. MDS-MAP localizes all nodes in a connected network. In our experiments, when the average node degree is 12.2 or more, the network is usually a connected graph and all nodes are located.

Irregular topologies are much harder than uniform topologies and previous methods reported very poor results on them (Niculescu and Nath, 2001). Figures 11 and 12 show the performance of the MDS-MAP algorithms on the C-shaped irregular networks. The radio ranges (*R*) are from 1.25*r* to 2.5*r*, in increments of 0.25r, which leads to average node degrees of 8.8, 12.0, 15.4, 19.2, 23.1, and 27.1. MDS-MAP(P) performs very well on these irregular networks, especially when the average node degree is 12.0 or more, finding solutions just slightly worse than those returned by MDS-MAP(C,R) and MDS-MAP(P,R). The results of MDS-MAP(P,R) are slightly better than those of MDS-MAP(C,R).

On networks with similar average node degrees, the results of MDS-MAP(C) on the irregular networks are worse than those on the uniform networks. In contrast, MDS-MAP(C,R), MDS-MAP(P), and MDS-MAP(P,R) perform well when the average node degree is relatively high. Having accurate estimates of local distances does not improve the performance of MDS-MAP(C), but helps MDS-MAP(P) and MDS-MAP(P,R) tremendously. As pointed out earlier, this is because the shortest-path distance between two nodes does not correspond well to their Euclidean distance. This has a greater impact on the centralized solution than on the distributed (patched) one. The results of MDS-MAP(P) and MDS-MAP(P,R) are very close, indicating that the refinement step in MDS-MAP(P,R) does not do much.

MDS-HYBRID

Different relative and absolute localization methods have different performance and cost. The performance may be expressed by the localization error, which is either the average distance between the estimated and the true positions of a node in case of absolute localization or the average difference between the

Figure 11. Performance of range-free MDS-MAP methods for irregular networks with 3 (left) and 10 (right) anchors for different values of average node degree (horizontal axis)

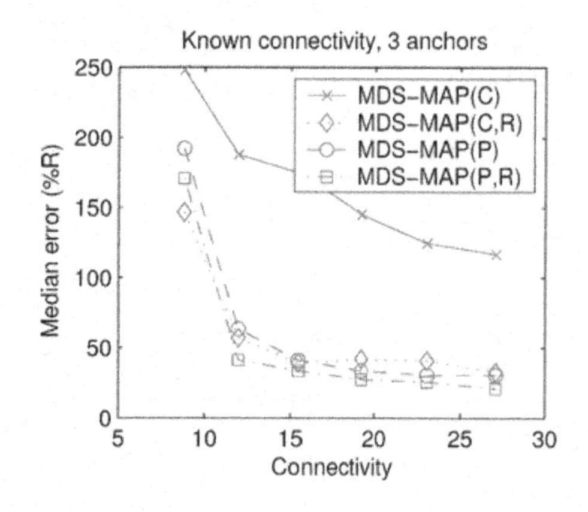

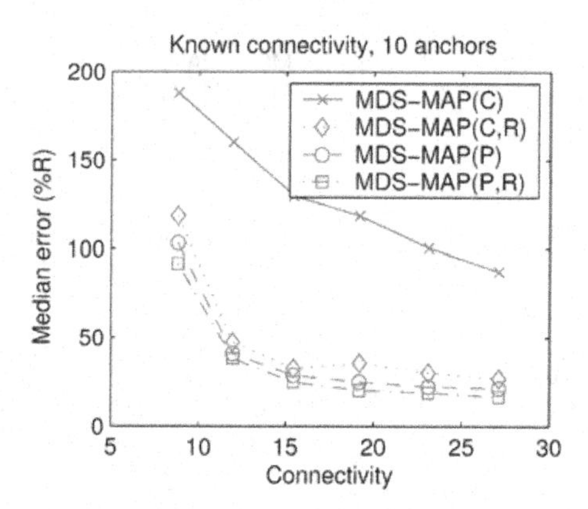

Figure 12. Performance of range-aware MDS-MAP methods for irregular networks with 3 (left) and 10 (right) anchors for different values of average node degree (horizontal axis)

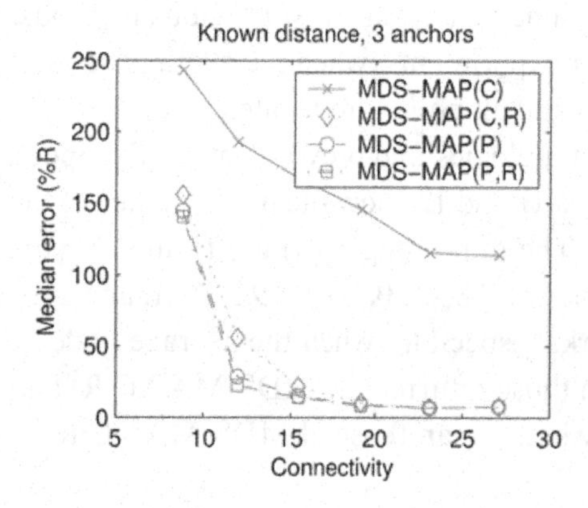

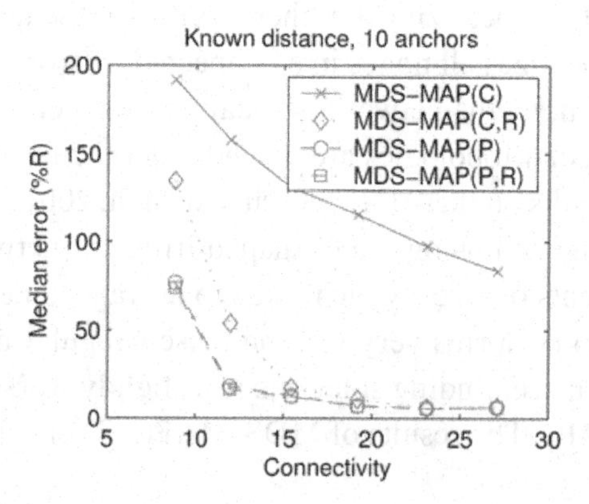

true and the estimated distances between nodes in relative localization. The cost may be measured by the localization delay or by the energy consumption. It is important when part of, or all, the nodes are moving. In this case, the localization process needs to be performed periodically or, at least, more than once. This makes the cost of localization, along with the performance, an important factor in choosing which method to use.

Performance-Cost Metric

MDS-Hybrid uses the weighted sum of localization accuracy and cost as a metric to evaluate localization. This metric is called *Performance-Cost Metric* (*PCM*). For absolute localization, the localization error may be computed as the average distance between the estimated and the true locations of all nodes. For relative localization, the resulting network is subject to translation, rotation, and reflection. Therefore, instead of using the same definition, the average estimated distances between all pairs of nodes are computed and compared to the corresponding true distances. Priyantha *et al.* (2003) proposed a performance metric that they refer to as the *Global Energy Ratio* (*GER*). *GER* is defined as:

$$GER = \frac{\sqrt{\sum_{i,j:i<j} \hat{e}_{ij}^2}}{N(N-1)/2}$$

where e_{ij} is the difference between the true distance d_{ij} and the distance in the algorithm's result $\hat{d}_{ij}$, $\hat{e}_{ij} = (\hat{d}_{ij} - d_{ij}) / d_{ij}$, and N is the number of nodes. We believe that taking the square root of the whole fraction represents the root-mean-square error in a better way. In addition, we normalize this error by the radio range (R) of a node. Therefore, to measure the localization error, we introduce the *Global Distance Error* (*GDE*) defined as:

$$GDE = \frac{1}{R}\sqrt{\frac{\sum_{i,j:i<j} \hat{e}_{ij}^2}{N(N-1)/2}}$$

Thus, the Performance-Cost Metric (*PCM*) may be defined as:

$$PCM = \alpha \cdot GDE + (1-\alpha)C \tag{4}$$

where $0 \leq \alpha \leq 1$ represents the degree of interest in the localization error as an evaluation criterion, and C is the localization cost defined as follows.

$$C = N_r c_1 + (N - N_r)c_2 \tag{5}$$

where N_r is the number of reference nodes localized with MDS-MAP(C), and c_1 and c_2 are the average costs of localizing one node using MDS-MAP(C) and method M, respectively, normalized by the initial energy at every node. Assuming that $c_1 = kc_2$ for some real $k > 0$, Eq. (5) yields

$$C = c_2[N + (k-1)N_r] \tag{6}$$

Substituting (6) in (4) gives

$$PCM = \alpha \cdot GDE + c_2(1-\alpha)[N+(k-1)N_r] \quad (7)$$

Algorithm

MDS-Hybrid consists of three phases: selecting reference nodes, localizing reference nodes, and localizing non-reference nodes. Three design parameters are associated with this method. First, how many reference nodes to use. Second, how to select them. Finally, whether to use as anchors all reference nodes or only the nearest *n* reference nodes. Determining these parameters to reach the required point of operation depends on the performance and the cost of both MDS-MAP(C) and *M*.

Phase 1: Selecting Reference Nodes

Since the performance of some absolute localization algorithms depends on the placement of the anchors (Shang *e. al.*, 2004), we consider two approaches of selecting reference nodes: random and outer (along the outer perimeter of the network). The random selection can be done in a distributed way using nodes' *IDs*. For example, N_r nodes with the smallest *IDs* can be selected using distributed algorithms. For the outer selection, we extend the algorithm used by Priyantha *et al.* (2003), keeping it distributed. First, four nodes 1, 2, 3, and 4 are selected roughly at the corners of the network using only the distance measurements. Then, the algorithm proceeds iteratively, doubling the number of selected nodes with every iteration, till all the N_r nodes are selected. The algorithm is described below. For a simple description of the algorithm, we assume that $N_r = 2^m$, for some integer $2 \le m \lfloor \lg N \rfloor$. Assume that d_{ij} is the shortest-path distance between nodes *i* and *j*.

1. Initialize two vectors *S* and *S'* of size N_r each to be empty.
2. Select a random node 0. This can be achieved by selecting the node with the smallest *ID* due to the random deployment.
3. Select reference node 1 such that d_{01} is maximized. $S[1] \leftarrow 1$.
4. Select reference node 2 such that d_{12} is maximized. $S[3] \leftarrow 2$.
5. Select reference node 3 such that $(d_{13}+d_{23})$ is maximized. $S[2] \leftarrow 3$.
6. Select reference node 4 such that d_{34} is maximized. $S[4] \leftarrow 4$..
7. $l \leftarrow 4$.
8. Repeat until size of $S = N_r$.
    - Repeat for i=1 to $(l-1)$
        - Select reference node $k \notin S'$ such that $(d_{ki}+d_{k(i+1)})$ is maximized.
        - $S'[2i-1] \leftarrow S[i]$, $S'[2i] \leftarrow k$
    - End
    - Select reference node $k \notin S'$ such that $(d_{k1}+d_{kl})$ is maximized.
    - $S'[2l-1] \leftarrow S[1]$, $S'[2l] \leftarrow k$.
    - $S \leftarrow S'$, $l \leftarrow 2l$

   End
9. Return *S*

Phase 2: Localizing Reference Nodes

MDS-MAP(C) is used to relatively localize the reference nodes selected in phase 1. Although it can give good results, it suffers from the high cost of computation to achieve a good solution. A relative map of the reference nodes is constructed. The work done by Rao *et al.* (2003) shows how the distance information can be exchanged between nodes in the case of outer reference nodes. The performance of MDS-MAP(C) is presented in section 4 of this chapter. Figure 13 illustrates the performance of MDS-MAP(C), expressed by *GDE*, under different conditions of ranging errors, average node degrees, and network topologies. In general, the localization error decreases with increasing average node degree. This is the case until the range error reaches some point beyond which the error may decrease if the average node degree is increased. The reason for that, as explained by Langendoen and Reijers (2003), is that a node will have more neighbors from which it can select the next hop on the shortest path. For large ranging errors, a node will prefer the shortest *measured* distance; this will underestimate the Euclidean distance, resulting in large localization errors for large average node degrees when the ranging error is beyond some value. For the isotropic networks, the localization error increases with larger ranging error, which is intuitive. On the other hand, it is hard to correlate the localization error and the ranging error in case of anisotropic networks, where MDS performs poorly. This is simply because the shortest-path distance becomes a bad approximation to the Euclidean distance.

Phase 3: Localizing Non-Reference Nodes

The result of phases 1 and 2 is a set of nodes with known coordinates in some coordinate system. In this phase, an absolute localization method M is used to localize the rest of the network using the reference nodes as anchors.

In our simulation of this phase, we used the DV-distance version of the APS method developed by Niculescu and Nath (2001). Each node uses the shortest-path distance information to estimate its distances to anchors. Then, it performs multilateration to estimate its position. APS has the advantages of simplicity and low cost. However, MDS-MAP(C) outperforms APS for most of the network conditions (Shang *et al.*, 2004). The APS method has the following steps:

1. Reference nodes broadcast their positions throughout the network to all nodes.
2. Each reference node k receives the positions (a_j, b_j), $j=1, ..., N_r$, of all reference nodes and also computes the shortest-path distance p_{kj} to each reference node.
3. Each reference node k computes its distance correction value, c_k.

$$c_k = \frac{\sum_{j=1}^{N_r-1} d_{kj}}{\sum_{j=1}^{N_r-1} p_{kj}},$$

   where $d_{kj} = \sqrt{(a_k - a_j)^2 + (b_k - b_j)^2}$ is the Euclidean distance between reference nodes k and j.
4. For each unknown node i, compute the shortest-path distance p_{ij}, $j=1, ..., N_r$, to all reference nodes. To estimate the position of node i, perform multilateration based on all reference nodes as follows.

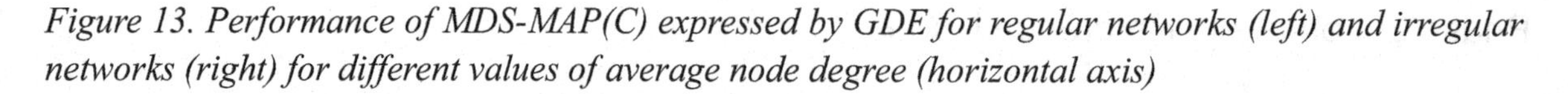

Figure 13. Performance of MDS-MAP(C) expressed by GDE for regular networks (left) and irregular networks (right) for different values of average node degree (horizontal axis)

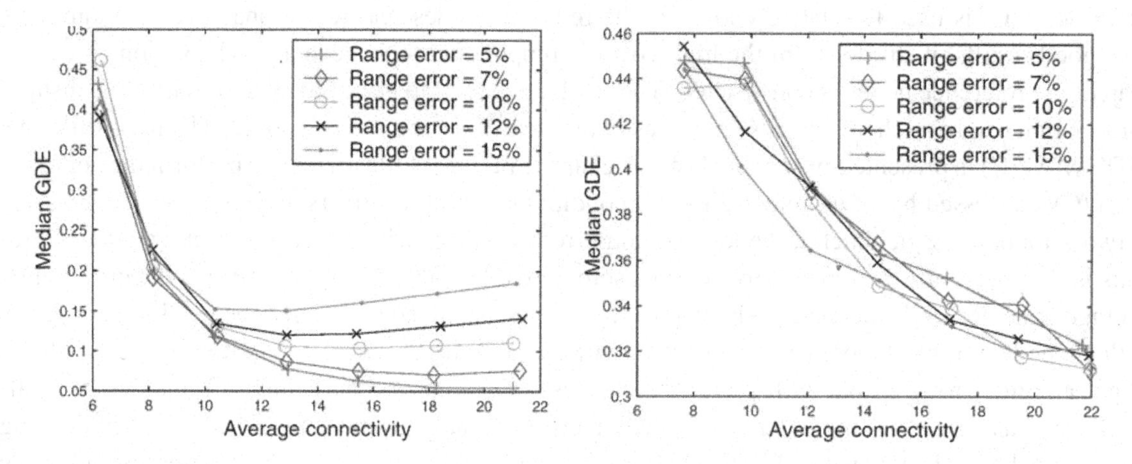

Figure 14. MDS-Hybrid phase 1: selecting reference nodes. This could be randomly (left) or along the outer perimeter of the network (right). Circles represent reference nodes.

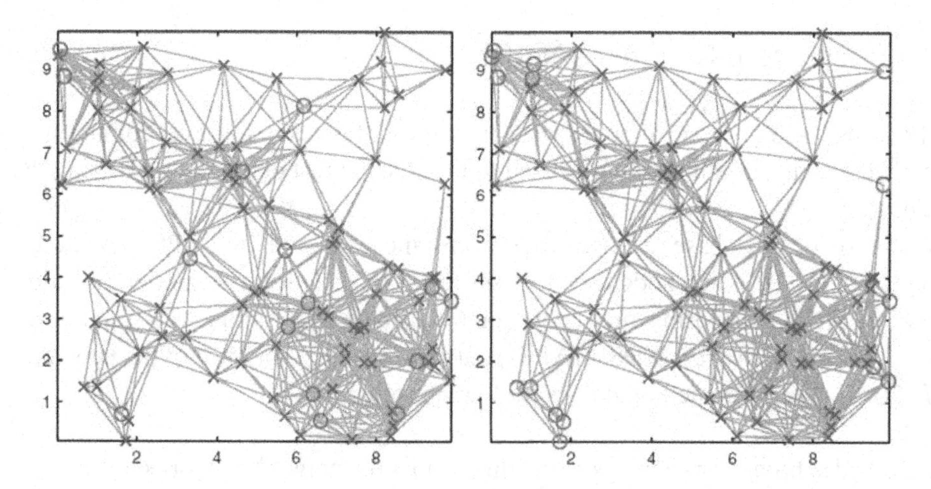

a. A system of quadratic equations of two variables is formed.

$$(x_i - a_j)^2 + (y_i - b_j)^2 = (c_s p_{ij})^2 \tag{8}$$

where s is the closest reference node to node i, i.e., $p_{is} \le p_{ij}$ for $j = 1, ..., N_r$.

b. The system of equations (8) is linearized by subtracting one equation, e.g., the first one, from the rest.

$$2(a_1 - x_j)x_i + 2(b_1 - b_j)y_i + a_j^2 + b_j^2 - a_1^2 - b_1^2 = (c_s p_{ij})^2 - (c_s p_{i1})^2 \tag{9}$$

for $j = 2, ..., N_r$.

c. The linear system (9) is solved. Then, using the solution as the initial point, the nonlinear system in (8) is solved using least-squares minimization.

Figure 15. Result of MDS-Hybrid, phase 2. GDE is 0.0944 for random selection (left) and 0.2154 for outer selection (right).

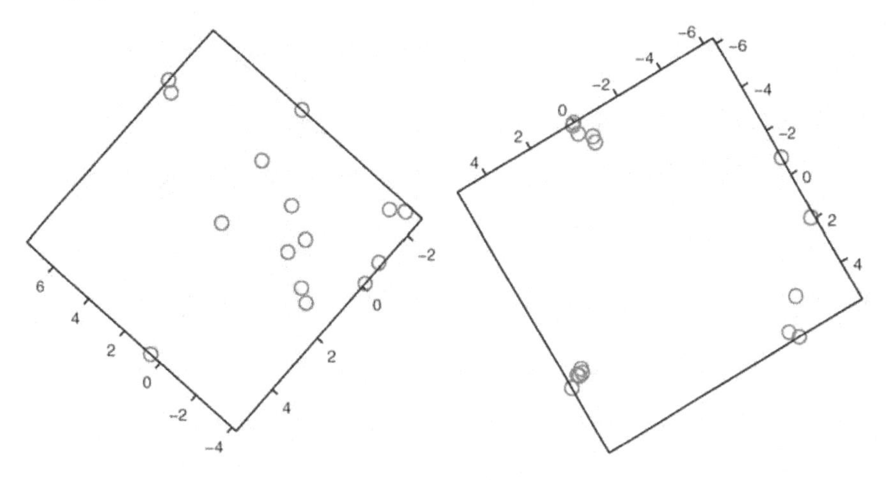

Illustrative Examples

We illustrate the operation of MDS-Hybrid by two concrete phase-by-phase examples for regular networks: one for random selection of reference nodes and one for outer selection. Figure 14 shows the result of phase 1 using 15 reference nodes selected randomly (left) and along the outer perimeter of the network (right). Circles represent reference nodes selected, and ×'s represent regular nodes. In phase 2, MDS-MAP(C) is applied to estimate the positions of reference nodes. Figure 15 shows the result of this phase, during which a coordinate system is created. The estimated networks were reflected and/or rotated to be easily compared to the true networks in Figure 14. Range error = 5%.

In the final phase, APS is used to localize the rest of the networks using the reference nodes as anchors. Figure 16 shows the result of this phase. Again, the estimated networks were reflected and/or rotated to be easily compared to the true networks in Figure 14. APS uses all reference nodes. Range error = 5%, $\alpha = 0.8$, $c_2 = 0{:}005$, and $k = 1.5$, as in Eq. (7).

Experimental Results

Experiments have been done based on the simulation setup explained earlier in this chapter with 100-node networks, and the results reported are the median of 100 runs. Figure 17 gives sample results for regular networks where only the nearest 4 reference nodes are used and for both random and outer selection of reference nodes. It is for a 5% range error, k=1.5, $c_2 = 0.005$, and $\alpha = 0.8$ (see Eq. (7)). The point of the minimum *PCM* is the required point of operation if we are 80% interested in a good accuracy and 20% in a low cost ($\alpha = 0.8$). Table 1 summarizes the values of the three design parameters associated with MDS-Hybrid that give the minimum *PCM* using APS for method *M*. An entry (*F*,*S*,*H*) is an ordered triple that represents *F* reference nodes, *S* of them were used by APS as anchors (*A* for all and *T* for nearest 4), and they were selected using the *H* method (*L* for random and *O* for outer). We note the following:

Figure 16. Result of MDS-Hybrid, phase 3. For random reference nodes (left), overall GDE = 0.1439. For outer reference nodes (right), overall GDE = 0.1634.

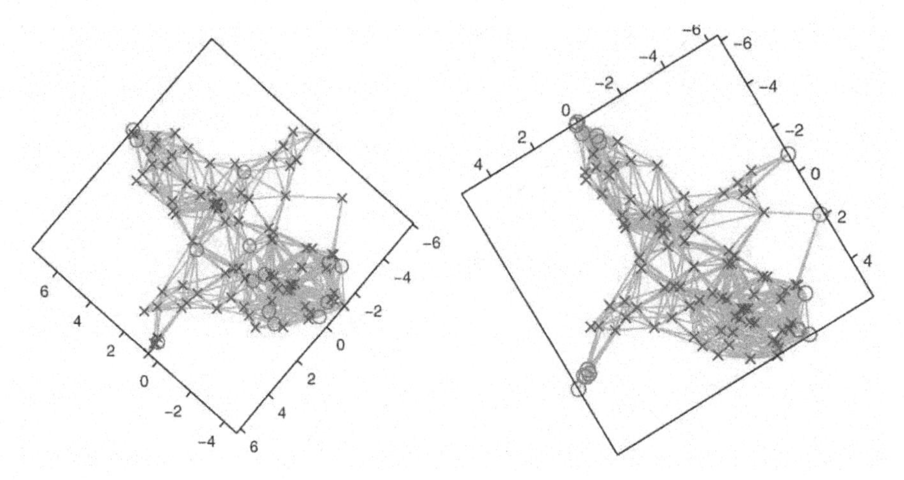

1. There is no entry that has the value 100 for the number of reference nodes. Thus, MDS-Hybrid is better than MDS-MAP(C).
2. For irregular networks, the results are the best for the smallest number of reference nodes (5 and 10). This is because MDS-MAP(C) performs poorly for irregular networks, as shown in Figure 18.
3. For regular networks, using all reference nodes is preferred to using only the nearest 4. This is due to the nature of the topology and the fact that using more anchors by APS gives more accurate position estimates.
4. For irregular networks with high average node degrees, using the nearest 4 reference nodes is better than using all of them because of the shortest-path problem discussed earlier. For low average node degrees, both approaches are close, and using all reference nodes might be better in order to tolerate the poor shortest-path distances.

Figure 17. Performance of MDS-Hybrid for regular networks with random (left) and outer (right) selections of reference nodes. Only the nearest 4 reference nodes are used. Range error = 5%, α = 0.8, c_2 = 0.005, and k = 1.5, as in Eq. (5).

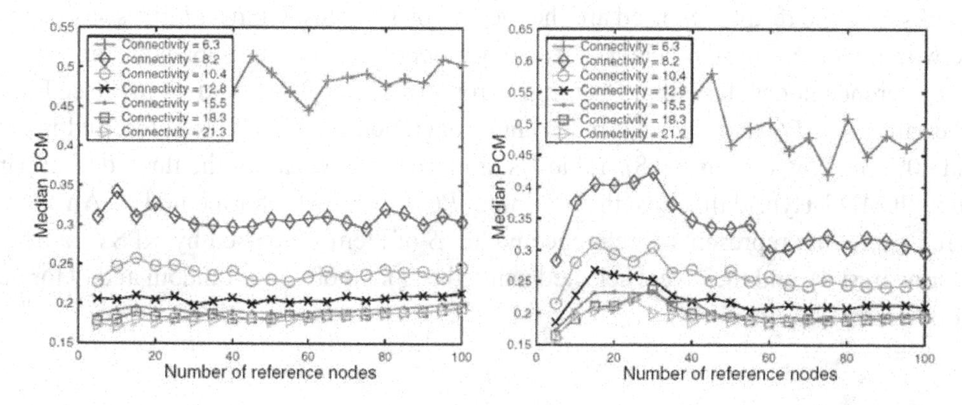

Figure 18. Performance of MDS-Hybrid for irregular networks with random (left) and outer (right) selections of reference nodes. All reference nodes are used. Range error = 5%, $\alpha = 0.8$, $c_2 = 0.005$, and $k = 1.5$, as in Eq. (5).

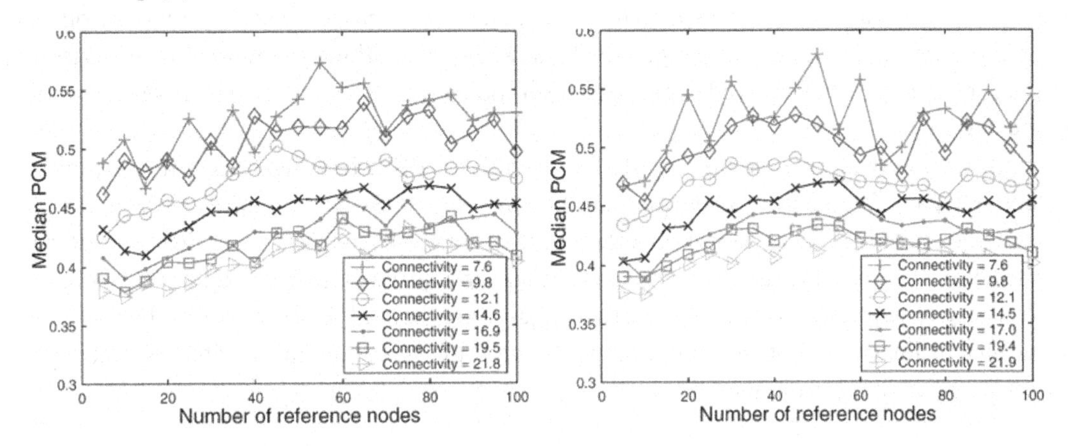

5. APS performs slightly better for outer placement of reference nodes. However, this performance might be degraded by the poorly estimated long distances to reference nodes. Therefore, we see both the random and the outer selection of reference nodes in case of regular networks. For the irregular networks, the outer selection is usually better because of the small number of reference nodes that would give a lower *PCM*.

RANGEQ-MDS

MDS-MAP(C), MDS-MAP(P), and MDS-Hybrid use the shortest-path distances for range-aware localization and the shortest-hop count multiplied by the range for range-free localization. RangeQ-MDS still uses the shortest-hop count, but each hop distance is obtained by the range quantization process. RangeQ-MDS is a *partially* range-aware localization algorithm that is based on Partial Range Information (*PRI*) presented by Li *et al.* (2004). *PRI* is defined as any type of measurement which is

Table 1. Design parameters for MDS-Hybrid's minimum PCM for both regular and irregular networks

	Average Node Degree	Range Error (%)		
		5	10	15
Regular	6.3	(15,*A*,*O*)	(15,*A*,*O*)	(15,*A*,*O*)
	10.4	(20,*A*,*L*)	(10,*A*,*L*)	(15,*A*,*L*)
	15.5	(15,*A*,*L*)	(5,*A*,*O*)	(5,*A*,*L*)
	21.3	(20,*A*,*L*)	(10,*A*,*O*)	(5,*T*,*O*)
Irregular	7.6	(5,*A*,*O*)	(10,*A*,*L*)	(5,*A*,*O*)
	12.1	(5,*T*,*O*)	(5,*T*,*L*)	(5,*A*,*O*)
	17.0	(5,*T*,*O*)	(5,*T*,*O*)	(5,*T*,*O*)
	21.8	(5,*T*,*O*)	(5,*T*,*O*)	(5,*T*,*O*)

monotonically increasing or decreasing and has an unknown or environment-dependent one-to-one relationship with the range measurement. It is called partial range information because these types of measurement can not be easily converted to accurate distance measurement due to the unknown exact mapping, yet the *PRI* values can correspond to the distance values based on their monotonic one-to-one relationship. It can be utilized in any range-free localization algorithms to improve their performance. One of the *PRI* examples is Received Signal Strength Indication (RSSI), which is a measure of the RF energy received.

Although wireless sensor systems usually have available RSSI readings, this information has not been effectively used for localization purposes. The RangeQ-MDS algorithm uses a sorted RSSI quantization scheme, called RangeQ, to improve the range estimation accuracy when distance information is not available. The output of this scheme is the distance matrix *D*, using which MDS is applied to obtain position estimates, resulting in a partially-range-aware method (Li *et al.*, 2004). The performance of RangeQ-MDS for various sensor networks is shown with experimental results from our extensive simulation with a realistic radio model.

Sorted RSSI Quantization

Received signal strength indication (RSSI) is a measure of the RF energy received and is closely related to the range. RSSI is supported by sensor node hardware, such as the Berkeley motes (Whitehouse and Culler, 2002). For localization purposes, the information provided by RSSI or similar types of measurements can be used to improve the accuracy of range-free localization algorithms.

The concept of sorted RSSI quantization is similar to that of image quantization in image processing, except that the quantization is not from continuous RSSI to discrete RSSI. The process of sorted RSSI quantization starts with sorting RSSI readings to obtain a sorted range list. It then applies a quantizer on the list to generate a range estimation. In the quantization process, range level represents the number of measurable range units in a hop, which is similar to the gray level representing the intensity of a pixel in a gray-level image. The number of range levels in range quantization is referred to as range-level resolution. For example, the range-level resolution in range-free localization algorithms is $s = 1$ since each 1-hop connection has one range level with the same range-level value. After obtaining the RSSI values, the sorted RSSI quantization algorithm follows two steps to assign a range value for each 1-hop connection as follows.

Step 1: Sorting RSSI Values

Let N_i be the set of all 1-hop neighboring nodes to node i in a randomly-deployed sensor network and p_{ij} the RSSI value between node i and node $j \in N_i$. In this step, all p_{ij} values are sorted in an ascending order by their values, and all nodes in N_i are rearranged accordingly. The result is an ordered node list L_i of all neighboring nodes to node i.

Step 2: Quantization

The goal of this step is to estimate the distance of each 1-hop connection. The range-free versions of MDS-MAP(C) and MDS-MAP(P) set all the 1-hop distances to the same value, i.e., the radio range R or a distance correction obtained by dividing the summation of the true shortest-path distance by the summation of the shortest-hop count of a number of anchors. In the sorted RSSI quantization scheme, and for node i, a hop is divided into s_i sub-unit hops of size u_i each, where $u_i = R/s_i$. The number of sub-unit hops s_i is called *range-level resolution*. Each 1-hop connection to node i is assigned a *range-level value* of u_i, $2u_i$, $\cdots$, or $s_i u_i$. If the assignment is correct, it obtains a range-level value closer to the true distance for each 1-hop connection than R as in MDS-MAP(C) and MDS-MAP(P). RangeQ-MDS assumes there is no available mapping function between the received RSSI and the corresponding distance to a neighboring node. Therefore, the problem left unsolved is to find an effective distance distribution model so that the number of nodes falling into each range level R_{ij} of node i, ($j = 1, 2, \ldots, s_i$) can be estimated with the help of the range order obtained from Step 1. For each node i, let the size of its neighboring node ordered list L_i be n_i. The quantization process involves dividing the maximum 1-hop range (R) into s_i smaller quantities of size u_i each and then assigning an appropriate quantized range-level value, R_{ij}, to each neighbor in L_i. With range-level resolution s_i, L_i is divided into s_i clusters with m_{ij} nodes in the j-th cluster C_{ij} for node i, where $\sum_{j=1}^{s_i} m_{ij} = n_i$. The range-level value for each node $k \in C_{ij}$ is set to $R_{ij} = ju_i$.

We use an area-proportional model shown in Figure 19 to assign a certain number of nodes from the sorted list into each range level. The area-proportional model estimates the distribution of nodes in the neighborhood. Assuming that the nodes are randomly distributed, we know that the nodes are equally likely to fall into any spot in a circle with radius R. For node i, we cut the whole circle into s_i annuli of equal width. The expected number of nodes falling into the j-th annulus is

$$m_{ij} = \frac{\pi\,(2j-1)n_i u_i^2}{\pi\,(s_i u_i)^2} = \frac{(2j-1)n_i}{s_i^2}$$

Therefore, the expected numbers of nodes falling into the s_i-th's annuli are

$$\frac{n_i}{s_i^2}, \frac{3n_i}{s_i^2}, \ldots, \frac{(2j-1)n_i}{s_i^2}, \ldots, \frac{(2s_i-1)n_i}{s_i^2}$$

Algorithm and Experimental Results

The pair-wise distance obtained by RangeQ is more like an estimated shortest-path distance except that the 1-hop range estimation itself is not provided by the hardware but provided by RangeQ instead. Simulation results show that the performance of RangeQ-MDS is between those of range-free and range-aware algorithms. Compared to MDS-MAP(C) and MDS-MAP(P), RangeQ-MDS is better than the range-free version of MDS when the range error is less than 35% of the radio range R, and better than the range-aware version of MDS when the error is more than about 16% of the radio range, as shown in Figure 20, where R is set to $1.75r$. RangeQ range estimation is more accurate than both range-free and range-aware when the range error is between 15% and 35% of radius. Figure 21 shows that the accuracy performance ranking of all the 6 listed algorithms from best to worst follows MDS-Range, MDS-RamgeQ, APS-Range, APS-RangeQ, APS-Hop and MDS-Hop.

Figure 19. Area-proportional model for the RangeQ-MDS algorithm

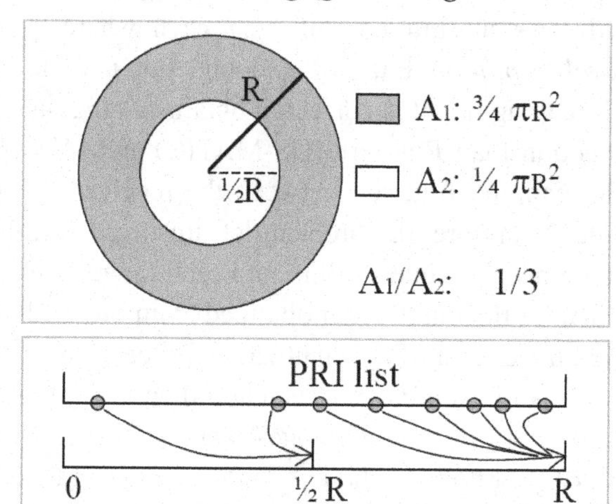

Error Analysis

In addition to the localization accuracy comparison, we also use the Cramér-Rao Lower Bound (CRLB) model to formulate the error of RangeQ range estimation technique (Shi *et al.*, 2005). Let d_{ij} be the true distance between sensor nodes *i* and *j*. That is,

$$d_{ij} = \sqrt{(x_i - x_j)^2 + (y_i - y_j)^2}$$

The power P_j (dBm) transmitted by device *i* and received at device *j* can be formulated as

$$P_i = P_\circ - 10 n_p \log_{10} \frac{d_{ij}}{\Delta_\circ}$$

Figure 20. Performance of RangeQ-MDS

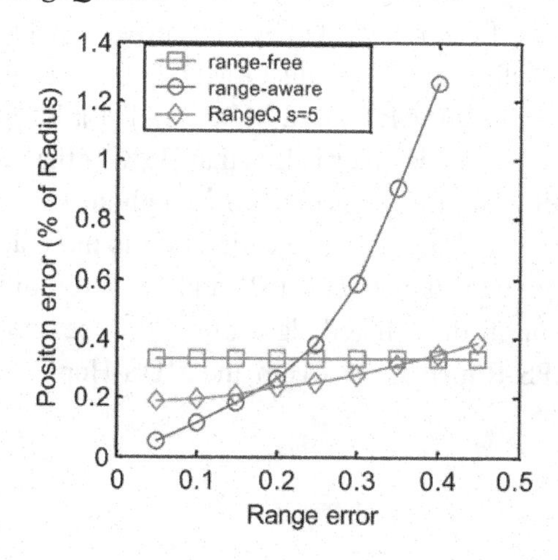

where $P_\circ$ (dBm) is the received power at the reference distance $\Delta_\circ$. Typically, $\Delta_\circ = 1$ meter, and $P_\circ$ is calculated from the free space loss formula (Rappaport, 2002). The path-loss exponent n_p depends on the environment, with typical value between 2 and 4.

Patwari and Hero III (2003) claim that the power loss has a log-normal distribution with if n_p is a fixed constant. The experimental results supporting their claim are 946 pair-wise RSSI measurements with a DS-SS transmitter and receiver in a network of 44 device locations. In the quantile-quantile plot of comparing the distribution of

$$P_i - (P_\circ - 10n_p \log_{10} \frac{d_{ij}}{\Delta_\circ})$$

(i.e., the attenuation of the channel) to the Gaussian distribution, they match well in the middle part. At both ends, however, they do not match well. A more realistic assumption is that both the environment variance and the measurement errors affect the received power. We assume that n_p is a Gaussian random variable, i.e., $n_p \approx N(\alpha, \delta_{n_p}^2)$ and the measurement error is Gaussian noise, $N(0, \delta_{n_p}^2)$. The assumption that n_p can be modeled as a Gaussian random variable is supported by a few researchers' work. Based on the data obtained by Seidel and Rappaport (1992), the mean of n_p is 3. Ghassemzadeh *et al.* (2002) find that the path loss exponent n_p follows normal distribution based on 300,000 frequency response profiles measured in 23 homes. For an outdoor environment, the work of Walden and Rowsell (2005) also shows that the distribution of n_p over the range 100 m to 2 km appears to be Gaussian in shape.

Figure 22 is a comparison of localization variance of six localization algorithms: range-free MDS-MAP(C) (MDS-HOP in the figure's legend), range-aware MDS-MAP(C) (MDS in the figure's legend), range-free APS (APS-HOP), range-aware APS, RangeQ-MDS, and RangeQ-APS. The last one uses APS for localization along with the RangeQ scheme. The Cramer-Rao Lower Bound of the localization variance is also plotted. The figure shows that range-aware MDS-MAP(C) gives the best localization precision among the six algorithms, and RangeQ-MDS performs better than the range-free versions of APS and MDS-MAP(C). MDS-MAP(C) is also very close to CRB.

CONCLUSION

We presented four MDS-based approaches for localization in wireless sensor networks, namely: MD-MAP(C), MDS-MAP(P), MDS-Hybrid, and RangeQ-MDS. We considered the localization problem under two difference scenariso: range-free and range-aware. In the first one, only proximity information is available to a sensor node, i.e., neighboring nodes. In the second scenario, distance measurements is assumed to be available between sensor nodes. MDS-MAP(C) and MDS-MAP(P) work well with mere connectivity information. It can also incorporate distance information when it is available. The strength of the MDS-MAP methods is that they can be used when there are few or no anchor nodes. Previous methods often require well-placed anchors to work well. Extensive simulations using various network topologies and different levels of ranging error show that the MDS-MAP methods are effective and surpasses previous methods.

Because MDS-MAP methods are expensive in terms of computational cost, we proposed an approach to relative localization referred to as MDS-Hybrid. This approach tries to combine the advantages of absolute and relative localization methods. It starts by selecting a number of reference nodes in the network based on some criterion. Then, MDS-MAP(C) is used to relatively localize the reference nodes.

Figure 21. Performance comparison of RangeQ-MDS, MDS, and APS

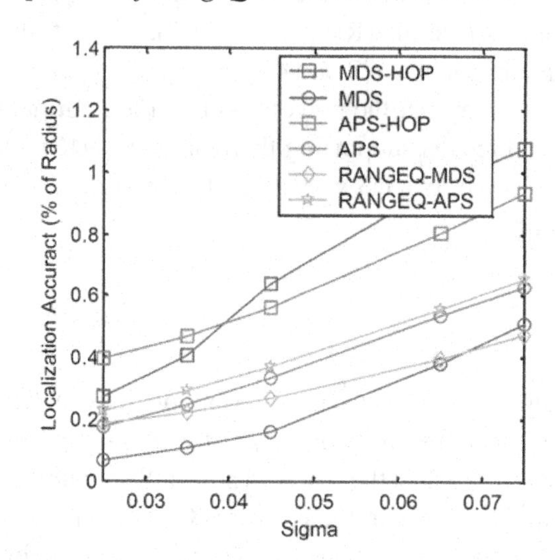

Figure 22. Comparison between 6 localization algorithms along with the Cramer-Rao lower bound on the localization variance

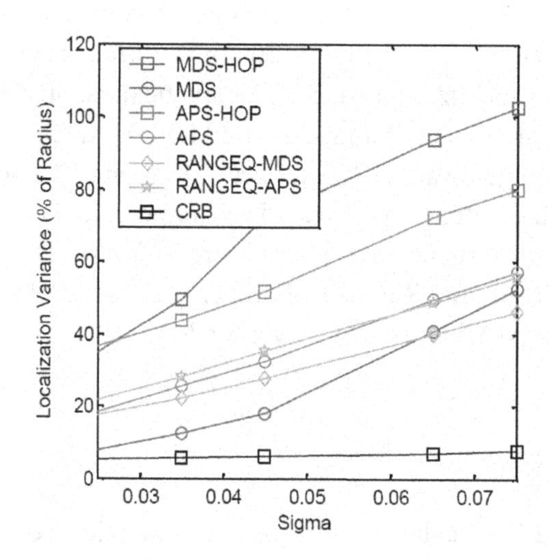

These nodes are used as anchors for an absolute localization algorithm to localize the rest of the network. We select the APS method for the absolute localization part. Simulation results show that MDS-Hybrid helps in choosing the required point of accuracy and cost. Moreover, it achieve a better performance than using MDS-MAP(C) in order to localize the whole network.

We presented the RangeQ-MDS localization algorithm. It is based on the RangeQ method used for RSSI quantization. While requiring only local PRI measurements, the partially range-aware RangeQ-MDS localization algorithm is found to be effective. In addition to being distirbuted, it can improve localization accuracy of previous range-free methods by up to 50%. It performs better than both the previous range-free and range-aware algorithms when the range error is between 15% and 35% of the radio range. Simulation results show the effectiveness of RangeQ-MDS on range estimation and localization. We analyzed the variance of localization errors using Cramer-Rao Lower Bound and have studied the effects of network parameters on localization accuracy with considering errors caused by the variance of environment factor n_p.

REFERENCES

Ahmed, A., Shi, H., & Shang, Y. (2005). SHARP: A new approach to relative localization in wireless sensor networks. In IEEE, *Proc. of the Distributed Computing Systems Workshops* (pp. 892-898). Columbus, OH.

Borg, I., & Groenen, P. (1997). *Modern Multidimensional Scaling, Theory and Applications.* New York, NY: Springer-Verlag.

Chu, M., Haussecker, H., & Zhao, F. (2002). Scalable information-driven sensor querying and routing for ad hoc heterogeneous sensor networks. *Int. Journal on High Performance Computing Applications, 16*(3), 90-110.

Doherty, L., El Ghaoui, L., & Pister, K. (2001). Convex position estimation in wireless sensor networks. In IEEE, *Proc. of INFOCOM* (pp. 1655-1663). Anchorage, AK.

Ghassemzadeh, S.S., Jana, R., Rice, C.W., Turin, W., & Tarokh, V. (2002). A statistical path loss model for in-home UWB channels. In IEEE, *Proc. of the Conference on Ultra Wideband Systems and Technologies* (pp. 59–64).

Glunt, W., Hayden, T. L., & Raydan, M. (1993). Molecular conformation from distance matrices. *J. Computational Chemistry, 14*, 114-120.

Gower, J. (1966). Some distance properties of latent root and vector methods used in multivariate analysis. *Biometrika, 53*(3,4), 325-338.

Intanagonwiwat, C., Govindan, R., & Estrin, D. (2000). Directed diffusion: A scalable and robust communication paradigm forsensor networks. In ACM, *Proc. of the 6th Int'l Conf. on Mobile Computing and Networks (MobiCom)* (pp. 56-67). Boston, MA.

Johnson, D. B. & Maltz, D. B. (1996). Dynamic source routing in ad hoc wireless networks. In T. Imielinski and H. Korth (Ed.), *Mobile Computing* (pp. 153-181). The Netherlands: Kluwer Academic Publishers.

Karp, B. & Kung, H. T. (2000). GPSR: Greedy perimeter stateless routing for wireless networks. In ACM, *Proc. of the 6th Int'l Conf. on Mobile Computing and Networks (MobiCom)* (pp. 243-254). Boston, MA.

Langendoen, K., & Reijers, N. (2003). Distributed localization in wireless sensor networks: a quantitative comparison. *Networks: The International Journal of Computer and Telecommunications Networking, 43*(4), 499-518.

Li, X., Shi, H., & Shang, Y. (2004). A Partial-Range-Aware Localization Algorithm for Ad-hoc Wireless Sensor Networks. In IEEE, *Proc. of the 29th Int. Conf. on Local Computer Networks (LCN)* (pp. 77-83). Tampa, FL.

Li, X., Shi, H., & Shang, Y. (2006). Sensor network localisation based on sorted RSSI quantisation. *Int. Journal of Ad Hoc and Ubiquitous Computing (IJAHUC), 1*(4), 222-229.

Niculescu, D. & Nath, B. (2001). Ad-hoc positioning system. In IEEE, *Proc. of Global Telecommunications Conference (Globecom)* (pp. 2926-2931). San Antonio, TX.

Patwari, N., & Hero III, A.O. (2003). Using proximity and quantized RSS for sensor localization in wireless networks. In IEEE/ACM, *Proc. of the 2nd International Workshop on Wireless Sensor Networks and Applications (WSNA)* (pp. 20-29). San Diego, CA.

Priyantha, N. B., Balakrishnan, H., Demaine, E., & Teller, S. (2003). *Anchor-free distributed localization in sensor networks* (Tech. rep. No. 892). MIT Laboratory for Computer Science.

Rao, A., Ratnasamy, S., Papadimitriou, C., Shenker, S., & Stoica, I. (2003). Geographical routing without location information. In ACM, *Proc. of the 9th Int. Conf. on Mobil Computing and Networking (MobiCom)* (pp. 96-108). San Diego, CA.

Rappaport, T. S. (2002). *Wireless Communications: Principles and Practice.* NJ: Prentice Hall.

Royer, E. & Toh, C. (1999). A review of current routing protocols for ad hoc mobile wireless networks. *IEEE Personal Communications, 6*(2), 46-55.

Savarese, C., Rabaey, J., & Beutel, J. (2001). Locationing in distributed ad-hoc wireless sensor networks. In IEEE, *Proceedings of ICASSP* (pp. 2037–2040). Salt Lake city, UT.

Savarese, C., Rabaey, J., & Langendoen, K. (2002). Robust positioning algorithm for distributed ad-hoc wireless sensor networks. In *USENIX Technical Annual Conf.* (pp. 317-328). Monterey, CA.

Savvides, A., Park, H., & Srivastava, M. (2002). The bits and flops of the n-hop multilateration primitive for node localization problems. In ACM, *1st Int'l Workshop on Wireless Sensor Networks and Applications (WSNA'02)* (pp. 112–121). Atlanta, GA.

Seidel, S.Y., & Rappaport, T. S. (1992). 914 MHz path loss prediction models for indoor wireless communications in multifloored buildings , *IEEE Transactions on Antennas and Propagation, 40*(2), 207-217.

Shang, Y., Ruml, W., Zhang, Y., & Fromherz, M. (2004). Localization from connectivity in sensor networks. *IEEE Transactions on Parallel and Distributed Systems, 15*(11), 961-974.

Shang, Y., Shi, H., & Ahmed, A. (2004). Performance study of localization methods for ad-hoc sensor networks. In IEEE, *Proceedings of the 1st IEEE International Conference on Mobile Ad-hoc and Sensor Systems (MASS'04)* (pp. 184-193). Fort Lauderdale, FL.

Shepard, R. N. (1962). Analysis of proximities: Multidimensional scaling with an unknown distance function i & ii. *Psychometrika, 27,* 125–140, 219–246.

Shi, H., Li, X., Shang, Y., & Ma, D. (2005). Cramer-Rao Bound Analysis of Quantized RSSI Based Localization in Wireless Sensor Networks. In IEEE, *Proc. of the 11th Int. Conf. on Parallel and Distributed Systems - Workshops (ICPADS'05)* (pp. 32-36). Fukuoka, Japan.

Tenenbaum, J., de Silva, V., & Langford, J. (2000). A global geometric framework for nonlinear dimensionality reduction. *Science, 290*(5500), 2319-2323.

Torgerson, W. S. (1965). Multidimensional scaling of similarity. *Psychometrika, 30*, 379–393.

Torgerson, W. S. (1952). Multidimensional scaling: I. Theory and method. *Psychometrika, 17*, 401–419.

Walden, M. C., & Rowsell, F. J. (2005). Urban propagation measurements and statistical path loss model at 3.5 GHz. In IEEE, *Proc. of the Antennas and Propagation Society International Symposium* (pp. 363-366).

Whitehouse, K., & Culler, D. (2002). Calibration as parameter estimation in sensor networks. In ACM, *Proc. of the Int'l Workshop on Wireless Sensor Networks and Applications* (pp. 59-67). Atlanta, GA.

Yu, Y., Govindan, R., & Estrin, D. (2001). *Geographical and energy aware routing: a recursive data dissemination protocol for wireless sensor networks* (Tech. Rep. No. ucla/csd-tr-01-0023). Los Angeles, CA: University of California-Los Angeles.

Chapter IX
Statistical Location Detection

Saikat Ray
University of Bridgeport, USA

Wei Lai
Boston University, USA

Dong Guo
Boston University, USA

Ioannis Ch. Paschalidis
Boston University, USA

ABSTRACT

The authors present a unified stochastic localization approach that allows a wireless sensor network to determine the physical locations of its nodes with moderate resolution, especially indoors. The area covered by the wireless sensor network is partitioned into regions; the localization algorithm identifies the region where a given sensor resides. The localization is performed using an infrastructure of stationary clusterheads that receive beacon packets periodically transmitted by the given sensor. The localization algorithm exploits the statistical characteristics of the beacon signal and treats the localization problem as a multi-hypothesis testing problem. The authors provide an asymptotic performance guarantee for the system and use this metric to determine the optimal placement of the infrastructure nodes. The placement problem is NP-hard and they leverage special-purpose algorithms from the theory of discrete facility location to solve large problem instances efficiently. They also show that localization decisions can be taken in a distributed manner by appropriate collaboration of the clusterheads. The approach is validated in a Boston University testbed.

INTRODUCTION

Localization – determining the approximate physical position of a user/device on a site – can be seen as an important enabling service in *Wireless Sensor Networks (WSNs).* The Global Positioning System (GPS) (Hofmann-Wellenhof et al. 1997) provides an effective localization technology outdoors and its popularity and the host of location-based services it has spawned is a testament to the importance of location information. The GPS technology though is unreliable in downtown urban areas and not functional indoors. Moreover, GPS receivers are expensive and power-hungry making them inappropriate for many WSN applications that emphasize very low-cost low-power sensor nodes.

A reliable indoor localization service would be extremely useful and would give rise to a plethora of innovative applications including: asset and personnel tracking in hospitals, warehouses, and other large complexes; locating faulty sensors in building automation applications; intelligent audio players in self-guided museum tours; intelligent maps for large malls and offices, smart homes (Hodes et al. 1997, Priyantha et al. 2000); as well as surveillance, military and homeland security related applications. Moreover, a location detection service is an invaluable tool for counter-action and rescue (Meissner et al. 2002) in disaster situations.

For these reasons, localization has received widespread attention in the literature and many approaches have been developed. A large class of localization systems uses special hardware (e.g., infrared sensors, ultrasound) which necessitates the deployment of a special-purpose WSN just for this purpose. Several related works are described in the following section. We are instead interested in a localization approach that can use WSN features found in virtually all existing platforms. Specifically, all WSN nodes carry a radio to communicate with each other. That radio is often rather rudimentary and the only information on the received RF signals one can obtain is signal strength. Received signal strength depends on the location of the transmitting sensor and the objective is to exploit this information to reveal the transmitter's location. At the same time, we are interested in an approach that is general enough to exploit additional RF or other information that could be obtained with more sophisticated hardware, for instance signal angle-of-arrival and signal time-of-flight. As we will see, we are able to deal with any vector of available *observations* about the transmitting sensor.

The approach we develop in this work starts with a "discretization" of the localization problem by splitting the coverage area into a set of regions. The problem is to determine the region where a sensor node we seek resides. Quantities like signal strength are highly variable indoors due to the dynamic character of the environment leading to multipath and fading in the propagation of RF signals. For example, the propagation environment inside a building is highly complex and dynamic as there are multiple reflections, doors that may be open or shut, and people (acting as RF energy absorbers and reflectors) that are constantly moving.

To accommodate this level of variability it is critical that we use a *stochastic characterization* of signal strength or other RF characteristics the localization system may rely upon. To that end, we will associate a probabilistic description of observations to every region-clusterhead pair. In some cases, e.g., for small regions, a single probability density function (pdf) of observations may suffice to characterize observations about a sensor in a certain region as received at a specific clusterhead. In other cases, especially when the region is large, a *family* of pdfs may be necessary to accurately represent observations anywhere in the region. A pdf family is intended to provide robustness with respect to the position of the sensor within a region and can be constructed from measurements taken from locations within the region. This flexible stochastic characterization of observations, ranging from a single pdf to a sufficiently rich family of pdfs combines earlier localization approaches we have developed (Ray et al. 2006, Paschalidis and Guo 2007). Once we have these probabilistic descriptors we can think of the localization problem as a hypothesis testing problem where we have to match observations to a single pdf or a pdf family. In the latter case, the problem is known as a composite hypothesis testing problem. To make decisions we will rely on likelihood ratio tests which we show to be optimal in a certain asymptotic sense. Our optimal approach increases the accuracy more than 3 times over an alternative approach; cf. Section "TESTBED AND EXPERIMENTAL RESULTS". Further, note that we are interested in locating the current position of a node; we do not assume any mobility model and as such do not attempt to predict/estimate any mobility pattern.

An advantage of the approach we advance is that we are able to characterize the performance of the localization system, quantified by the probability of error. In particular, we obtain the dominant exponent of this probability as the number of observations grows large. Having a meaningful performance metric enables us to pose the following design question: How should clusterheads be placed to minimize the probability of error? We study this optimal deployment/WSN-design problem which turns out to be NP-complete. However, we leverage results from the theory of discrete facility location and present an efficient algorithm that can solve reasonably large instances.

An important consideration in WSNs is whether the localization algorithm can run in a distributed manner by appropriate in-network processing. We demonstrate that one can organize the necessary computations so that clusterheads make observations and take local decisions which get processed as they propagate through the network of clusterheads. The final decision reaches the gateway and, as we show, there is no performance cost compared to a centralized approach. We have implemented our approach in a testbed installed at a Boston University (BU) building, and our experimental results establish that we can achieve accuracy that is, roughly, on the same order of magnitude as the radius of our regions. This is, in fact, the best possible accuracy one can expect from a discretized system that turns localization into the problem of identifying the sensor's region. We report experimental results from our testbed showing great promise in using an approach of this type in a practical setting.

The rest of this chapter is organized as follows. We first discuss related work. In PROBLEM FORMULATION, we introduce our system model. In MATHEMATICAL FOUNDATION we introduce the mathematical underpinnings of hypothesis testing. In the same section we review the standard hypothesis testing problem, study the composite binary hypothesis testing problem, establish an optimality condition for the test we propose, and obtain bounds on the error exponents which allow us to optimize performance. In OPTIMUM CLUSTERHEAD PLACEMENT we consider the WSN design problem and present a fast algorithm for solving it in an efficient manner. In LOCALIZATION DECISIONS we develop the distributed decision approach and compare it to a centralized one. Results from an implementation of our approach in the testbed are reported in TESTBED AND EXPERIMENTAL RESULTS. Final remarks are in CONCLUSIONS.

RELATED WORK

Several non-GPS location detection systems have been proposed in the literature. One class of localization systems is "deterministic" and as in (Bahl et al. 2000) compares the mean signal strength from a sensor to a pre-computed signal-strength map of the coverage area. This approach though, may be unreliable indoors due to the significant variability of the RF signal landscape (due to multipath, fading, etc.). A similar system is SpotOn (Hightower et al. 2000). The *Nibble* system improves up on *RADAR* by taking the probabilistic nature of the problem into account (Castro et al. 2001). A similar approach is also found in (Yong Wu et al. 2007). Another class of systems uses trilateration or stochastic trilateration techniques as in (Patwari et al. 2003) where signal strength measurements are used to estimate the distance and location. These techniques assume a model describing how signal strength reduces with distance (path loss formula) and the modeling error can lead to inaccuracies. In the experimental results we report in this chapter, our approach is shown to significantly reduce the mean error distance compared to stochastic trilateration techniques. In (Patwari et al. 2008), a more accurate path loss model including correlated shadow loss and non-shadow loss was introduced. In (Battiti et al. 2003), the location detection problem is cast in a statistical learning framework to enhance the models. In (Lasse Klingbeil et al. 2008), a sequential Monte Carlo simulation technique was introduced to estimate the location and motion of a mobile sensor. The Monte Carlo techniques require some information about the mobility model or probability of a node's location at a given time; we assume neither. Performance trade-off and deployment issues are explored in (Prasithsangaree et al. 2002). References to many other systems can be found in the homepage of (Youssef 2008).

In addition to the related work, the works presented in this chapter not only describe and evaluate a localization system, but also characterize the performance, outline optimization approaches and propose a distributed decision making algorithm.

PROBLEM FORMULATION

In this section we introduce our system model. Consider a WSN deployed in a site for localization purposes. The reader may assume the site to be the interior of a building. We divide the site into N regions denoted by an index set $L=\{L_1,\ldots,L_N\}$. There are M distinct positions $B=\{B_1,\ldots,B_M\}$ at which we can place the fixed infrastructure nodes we call clusterheads.

Let a sensor be located in region $l \in L$. A series of packets broadcast by the sensor are received by the clusterheads (not necessarily all of them) which observe certain physical quantities associated with each packet. In most existing WSN platforms the observed physical quantities are just the received signal strength indicator (RSSI), which is related to the voltage observed at the receiver's antenna circuit.

Let $y^{(i)}$ denote the vector of observations by a clusterhead at position $B^{(i)}$ corresponding to a packet broadcasted by the sensor. These observations are assumed to be random. To simplify the analysis we will assume that the observations take values from a finite alphabet $\Sigma=\{\sigma_1,\ldots,\sigma_{|\Sigma|}\}$, where $|\Sigma|$ denotes the cardinality of Σ. In practice, this is indeed the case since WSN nodes report quantized RSSI measurements. A series of n consecutive observations are denoted by $y_1^{(i)},\ldots,y_n^{(i)}$ and are assumed independent and identically distributed (i.i.d.) conditioned on the region the sensor node resides. This assumption is well justified for fairly dynamic sites where the various radio-paths between the receiver and the transmitter change rapidly; for typical indoor sensor networks observations separated by a few seconds can be i.i.d. Observations made by different clusterheads at about the same time need not be independent.

With every clusterhead-region pair (B_i, L_j) we associate a family of pdfs $p_{Y^{(i)}|\theta_j}(y)$ where $Y^{(i)}$ denotes the random variable corresponding to observations $y^{(i)}$ at clusterhead B_i when the transmitting sensor is in some location within L_j. Here, $\theta_j \in \Omega_j$ is a vector in some space Ω_j parameterizing the pdf family. We allow the possibility that for some (typically small) regions the pdf family degenerates into a single pdf. We will be writing $p_{Y^{(i)}|L_j}(y)$ in that case. Such a pdf can be obtained by using measurements at a single position within the region and constructing an empirical distribution. For larger regions, a single pdf may not be an appropriate representation for all positions within the region. As we will see, we will use measurements at a few locations (or even a single one) within L_j but we will associate to these measurements a family of pdfs parametrized by θ_j. For example, one could obtain an empirical pdf from the measurements and associate with L_j pdfs with the same shape as the empirical pdf and a mean lying in some interval centered at the empirical mean.

Given a family of pdfs for every pair (B_i, L_j) we are interested in placing $K \leq M$ clusterheads at positions in B and use observations by them to determine the region in which a sensor node resides. To that end, we will (i) characterize the performance of the localization system in terms of the probability of error, (ii) develop an algorithm for placing clusterheads that provides guarantees for the probability of error, and (iii) develop approaches for determining the sensor location in a distributed manner.

MATHEMATICAL FOUNDATION

In this section we take the clusterhead locations as given and formulate the hypothesis testing problem for determining the location of sensors. First we describe the simpler case where there is only one pdf associated with each location. Then we expand on the case of composite hypothesis testing where there are more than one pdf per region.

Binary Hypothesis Testing

Suppose we place clusterheads in K out of the M available positions in B. Without loss of generality let these positions be $B_1,\ldots,B_K$. Suppose also that a sensor is in some location $l \in L$ and transmitting packets. As before, let $y^{(i)}$ be the vector of observations at each clusterhead $i = 1, \ldots, K$; we write $y = (y^{(1)}, \ldots, y^{(K)})$ for the vector of observations at all K clusterheads. These observations are random; let Y denote the random variable corresponding to y and $p_{Y|L_j}(y)$ the pdf of Y conditional on the sensor being in location $L_j \in L$. Observations $y^{(i)}$ and $y^{(j)}$ made at the same instant need not be independent. If they are, however, it follows that $p_{Y|L_j}(y) = p_{Y^{(1)}|L_j}(y^{(1)}) \cdots p_{Y^{(K)}|L_j}(y^{(K)})$. Suppose that the clusterheads make n consecutive observations $y_1, \ldots, y_n$, which are assumed i.i.d. Based on these observations we want to determine the location l of the sensor.

The problem at hand is a standard N-ary hypothesis testing problem. It is known that the maximum *a posteriori* probability (MAP) rule is *optimal* in the sense of minimizing the probability of error. More specifically, we declare $l = L_j$ if

$$j = \arg\max_{i=1,\ldots,N}[\pi_i p_{Y|L_i}(y_1) \cdots p_{Y|L_i}(y_n)] \tag{1}$$

(ties are broken arbitrarily), where π_i denotes the prior probability that the sensor is in location L_i.

Next we turn our attention to binary hypothesis testing for which tight asymptotic results on the probability of error are available. These results will be useful in establishing performance guarantees for our proposed clusterhead placement later on.

Suppose that the sensor's position is either L_i or L_j. A clusterhead located at B_k makes n i.i.d. observations $y_1^{(k)},\ldots,y_n^{(k)}$. Let

$$X_{ijk}(y) = \log[p_{Y^{(k)}|L_i}(y) / p_{Y^{(k)}|L_j}(y)]$$

be the log-likelihood ratio. Define

$$\Lambda_{ijk}(\lambda) = \log \mathrm{E}_{L_j}[e^{\lambda X_{ijk}(y)}] \quad (2)$$

The expectation is taken with respect to the density $p_{Y^{(k)}|L_j}(y)$. It follows that

$$\Lambda_{ijk}(\lambda) = \log \int_{-\infty}^{\infty} p_{Y^{(k)}|L_i}^{\lambda}(y) p_{Y^{(k)}|L_j}^{1-\lambda}(y) \mathrm{d}y \quad (3)$$

The function $\Lambda_{ijk}(\lambda)$ is the log-moment generating function of the random variable X_{ijk}, hence convex (see Dembo and Zeitouni 1998, Lemma 2.2.5 for a proof). Let d_{ijk} be the Fenchel-Legendre transform (or convex dual) of $\Lambda_{ijk}(\lambda)$ evaluated at zero, i.e.,

$$d_{ijk} = \sup_{\lambda \in [0,1]} [-\Lambda_{ijk}(\lambda)]. \quad (4)$$

d_{ijk} is the so called Chernoff information or distance (it is nonnegative, symmetric, but it does not satisfy the triangle inequality) between the densities $p_{Y^{(k)}|L_i}(y)$ and $p_{Y^{(k)}|L_j}(y)$ (Dembo and Zeitouni 1998 § 3.4, Chernoff 1952).

Consider next the probability of error in this binary hypothesis testing problem when we only use the observations made by clusterhead B_k. Suppose we make decisions optimally and let S^n denote the optimal decision rule (i.e., a mapping of $y_1^{(k)}, \ldots, y_n^{(k)}$ onto either "accept L_i" or "accept L_j"). We have two types of errors with probabilities

$$\alpha_n = \mathrm{P}_{L_j}[S^n \text{rejects } L_j], \quad \beta_n = \mathrm{P}_{L_i}[S^n \text{ rejects } L_i] \quad (5)$$

The first probability is evaluated under $p_{Y^{(k)}|L_j}(y)$ and the second under $p_{Y^{(k)}|L_i}(y)$. The probability of error, $P_n^{(e)}$, of the rule S^n is simply $P_n^{(e)} = \pi_i \alpha_n + \pi_j \beta_n$. Large deviations asymptotics for the probability of error under the optimal rule S^n have been established by (Chernoff 1952, Dembo and Zeitouni 1998 Corollary 3.4.6) and are summarized in the following theorem.

Theorem IV.1 (Chernoff's bound) *If* $0 < \pi_i < 1$, *then*

$$\lim_{n\to\infty} \frac{1}{n} \log \alpha_n = \lim_{n\to\infty} \frac{1}{n} \log \beta_n = \lim_{n\to\infty} \frac{1}{n} \log P_n^{(e)} = -d_{ijk}$$

In other words, all these probabilities approach zero exponentially fast as n grows and the exponential decay rate equals the Chernoff distance d_{ijk}. Intuitively, these probabilities behave as $f(n)e^{-nd_{ijk}}$ for sufficiently large n, where $f(n)$ is a slowly growing function in the sense that $\lim_{n\to\infty}(\log f(n))/n = 0$. In the sequel, we will often consider the asymptotic rate according to which probabilities approach zero as $n \to \infty$. We will use the term *exponent* to refer to the quantity $\lim_{n\to\infty}(1/n)\log P[\cdot]$ for some probability P[·]; if the exponent is d then the probability approaches zero as e^{-nd}. When the Maximum Likelihood (ML) rule is optimal (i.e., prior probabilities of the hypotheses are equal), we have

Theorem IV.2 *Suppose* $\pi_i = 1/N \ \forall i$. *Then* $P_n^{(e)} \le e^{-nd_{ijk}}$ *for all n.*

The proof is given in Appendix A. Note the interesting fact that Chernoff distances, and thus the exponents of the probability of errors do not depend on the priors π_i. This observation is important since prior probabilities are not within the control of the system designer, but conditional pdfs are.

The Chernoff distances between the joint densities of the data observed by all the clusterheads can be defined similarly by replacing $p_{Y^{(k)}|L_i}(y)$ and $p_{Y^{(k)}|L_j}(y)$ by $p_{Y|L_i}(y)$ and $p_{Y|L_j}(y)$, respectively. The clusterhead placement problem consists of choosing the placement so as to maximize these Chernoff distances. However, the optimum clusterhead placement that maximizes distances between the joint densities turns out to be a nonlinear problem with integral constraints. It quickly becomes intractable with increasing problem size and the optimum clusterhead placement for realistic sites cannot be computed using such a formulation. Optimization of the clusterhead placement in our formulation, on the other hand, reduces to a linear optimization problem (although still with integral constraints), for which large problem instances can be solved within reasonable time. For the resultant placement, we derive bounds on the probability of error of the decision rule that uses joint distributions.

Binary Composite Hypothesis Testing

We now consider the case where regions *i* and *j* have an associated pdf family $p_{Y^{(k)}|\theta_i}(y)$ and $p_{Y^{(k)}|\theta_j}(y)$, respectively, corresponding to observations at clusterhead B_k (θ_j and θ_i depend on *k* as well but we elect to suppress this dependence in the notation for simplicity). The clusterhead makes *n* i.i.d. observations $y^{(k),n} = (y_1^{(k)}, \ldots, y_n^{(k)})$ from which we need to determine the region L_i vs. L_j. We will be using the notation $p_{Y^{(k)}|\theta_i}(y^{(k),n}) = \prod_{l=1}^{n} p_{Y^{(k)}|\theta_i}(y_l^{(k)})$.

The problem at hand is a binary composite hypothesis testing problem for which the so called *Generalized Likelihood Ratio Test (GLRT)* is commonly used. The GLRT compares the normalized generalized log-likelihood ratio

$$X_{ijk}(y^{(k),n}) = \frac{1}{n} \log \frac{\sup_{\theta_i \in \Omega_i} p_{Y^{(k)}|\theta_i}(y^{(k),n})}{\sup_{\theta_j \in \Omega_j} p_{Y^{(k)}|\theta_j}(y^{(k),n})}$$

to a threshold λ and declares L_i whenever

$$y^{(k),n} \in S_{ijk,n}^{GLRT} = \{ y^{(k),n} \mid X_{ijk}(y^{(k),n}) \ge \lambda \}$$

and L_j otherwise. Note that in the case one of the pdf families, say the one corresponding to the (B_k, L_i) clusterhead-region pair, is a singleton $p_{Y^{(k)}|L_i}(y)$ the supremum in the numerator is moot. When both pdf families are singletons GLRT becomes the standard LRT and a threshold $\lambda = 0$ should be used. There are two types of error (referred to as type I and type II, respectively) associated with a decision with probabilities

$$\alpha_{ijk,n}^{GLRT} = P_{\theta_j}[y^{(k),n} \in S_{ijk,n}^{GLRT}], \quad \beta_{ijk,n}^{GLRT} = P_{\theta_i}[y^{(k),n} \notin S_{ijk,n}^{GLRT}]$$

where $P_{\theta_j}[\cdot]$ (resp. $P_{\theta_i}[\cdot]$) is a probability evaluated assuming that $y^{(k),n}$ is drawn from $p_{Y^{(k)}|\theta_j}(y)$ (resp. $p_{Y^{(k)}|\theta_i}(y)$). We use a similar notation and write $\alpha^{S}_{ijk,n}$ and $\beta^{S}_{ijk,n}$ for the error probabilities of any other test that declares L_i whenever $y^{(k),n}$ is in some set $S_{ijk,n}$.

Since we have two probabilities of error we can not minimize both at the same time. A natural objective is to minimize one (type II) subject to a constraint on the other (type I). This is known as the *generalized Neyman-Pearson* optimality criterion and is given below.

Definition 1

Generalized Neyman-Pearson (GNP) Criterion: We will say that the decision rule $\{S_{ijk,n}\}$ is optimal if it satisfies

$$\limsup_{n\to\infty} \frac{1}{n} \log \alpha^{S}_{ijk,n}(\theta_j) < -\lambda, \quad \forall\, \theta_j \in \Omega_j \tag{6}$$

and maximizes $-\limsup_{n\to\infty} \frac{1}{n} \log \beta^{S}_{ijk,n}(\theta_i)$ *uniformly for all* $\theta_i \in \Omega_i$.

Zeitouni et al. have established conditions for the optimality of the GLRT in a Neyman-Pearson sense for general Markov sources. The analysis in (Zeitouni et al. 1992) is carried out for the case where one hypothesis corresponds to a single pdf and the other to a pdf family. We provide a generalization to the situation of interest where both hypotheses correspond to a family of pdfs. We will establish a necessary and sufficient condition for the GLRT to satisfy the GNP criterion.

Let us introduce some additional notation, which is common in information theory. For any sequence of observations $y^n=(y_1,\ldots,y_n)$, the empirical measure (or type) is given by $L_{y^n} = (L_{y^n}(\sigma_1),\ldots,L_{y^n}(\sigma_{|\Sigma|}))$, where

$$L_{y^n}(\sigma_i) = \frac{1}{n}\sum_{j=1}^{n} 1\{y_j = \sigma_i\}, \qquad i = 1,\ldots,|\Sigma|,$$

and $1\{\cdot\}$ denotes the indicator function. We will denote the set of all possible types of sequences of length n by $L_n = \{\nu \mid \nu = L_{y^n}$ for some $y^n\}$ and the type class of a probability law ν by $T_n(\nu) = \{y^n \in \Sigma^n \mid L_{y^n} = \nu\}$, where Σ^n denotes the cartesian product of Σ with itself n times. Let

$$H(\nu) = -\sum_{i=1}^{|\Sigma|} \nu(\sigma_i)\log\nu(\sigma_i)$$

be the entropy of the probability vector ν and

$$D(\nu \,\|\, \mu) = \sum_{i=1}^{|\Sigma|} \nu(\sigma_i)\log\frac{\nu(\sigma_i)}{\mu(\sigma_i)},$$

the divergence or relative entropy of ν with respect to another probability vector μ.

Lemma 3.5.3 in (Dembo and Zeitouni 1998) states that it suffices to consider functions of the empirical measure when trying to construct an optimal test (i.e., the empirical measure is a sufficient statistic). Let P_{θ_j} denote the probability law induced by $p_{Y^{(k)}|\theta_j}(\cdot)$. Considering hereafter tests that depend only on L_{y^n}, the so called generalized Hoeffding test (Hoeffding 1965) that accepts L_i when $y^{(k),n}$ is in the set

$$S^{*}_{ijk,n} = \{y^n \mid \inf_{\theta_j} D(L_{y^n} \,\|\, P_{\theta_j}) \geq \lambda\},$$

and accepts L_j otherwise, is optimal according to the GNP criterion. The following lemma generalizes Hoeffding's result and a similar result in (Zeitouni et al. 1992); the proof is in Appendix B.

Lemma IV.3 *The generalized Hoeffding test satisfies the GNP criterion.*

Next, we will determine the exponent of $\beta_{ijk,n}^{S^*}(\theta_i)$. Define the set

$$A_{ijk} = \{Q \mid \inf_{\theta_j} D(Q \| P_{\theta_j}) < \lambda\}.$$

We have

$$\beta_{ijk,n}^{S^*}(\theta_i) = P_{\theta_i}[y^{(k),n} \notin S_{ijk,n}^*] = P_{\theta_i}[L_{y^{(k),n}} \in A_{ijk} \cap L_n].$$

Due to Sanov's theorem (Dembo and Zeitouni 1998, Chap. 2)

$$\inf_{Q \in A_{ijk}} D(Q \| P_{\theta_i}) \leq -\limsup_{n \to \infty} \frac{1}{n} \log \beta_{ijk,n}^{S^*}(\theta_i) \leq -\liminf_{n \to \infty} \frac{1}{n} \log \beta_{ijk,n}^{S^*}(\theta_i) \leq \inf_{Q \in A_{ijk}^o} D(Q \| P_{\theta_i}) \quad (7)$$

where A_{ijk}^o denotes the interior of A_{ijk}. Since A_{ijk} is an open set the upper and lower bounds match and $\inf_{Q \in A_{ijk}} D(Q \| P_{\theta_i})$ is the exponent of $\beta_{ijk,n}^{S^*}(\theta_i)$.

The following theorem establishes a necessary and sufficient condition for the optimality of GLRT under the GNP criterion. The proof is omitted; we refer the interested reader to (Paschalidis and Guo 2007).

Theorem IV.4 *The GLRT with a threshold λ is asymptotically optimal under the GNP criterion, if and only if*

$$\inf_{Q \in C_{ijk}} D(Q \| P_{\theta_i}) \geq \inf_{Q \in A_{ijk}} D(Q \| P_{\theta_i}) \quad (8)$$

for all θ_i, where

$$C_{ijk} = \{Q \mid \inf_{\theta_j} D(Q \| P_{\theta_j}) - \inf_{\theta_i} D(Q \| P_{\theta_i})\} < \lambda \leq \inf_{\theta_j} D(Q \| P_{\theta_j})\}.$$

Furthermore, assuming that (8) is in effect

$$-\liminf_{n \to \infty} \frac{1}{n} \log \alpha_{ijk,n}^{GLRT}(\theta_j) \leq -\lambda, \quad \forall\, \theta_j \in \Omega_j \quad (9)$$

$$-\liminf_{n \to \infty} \frac{1}{n} \log \beta_{ijk,n}^{GLRT}(\theta_i) \leq -\inf_{Q \in A_{ijk}} D(Q \| P_{\theta_i}), \quad \forall \theta_i \in \Omega_i. \quad (10)$$

Although, Thm. IV.4 is an interesting theoretical result, in practice it is not trivial to verify whether condition (8) is satisfied or not. To that end, the following theorem derives bounds on the type I and type II error probability exponents in the absence of condition (8).

Theorem IV.5 *The GLRT with a threshold λ satisfies*

$$-\liminf_{n\to\infty} \frac{1}{n} \log \alpha_{ijk,n}^{GLRT}(\theta_j) \le -\lambda, \quad \forall \theta_j \in \Omega_j \tag{11}$$

$$-\liminf_{n\to\infty} \frac{1}{n} \log \beta_{ijk,n}^{GLRT}(\theta_i) \le -\inf_{Q\in D_{ijk}} D(Q \| P_{\theta_i}), \quad \forall \theta_i \in \Omega_i. \tag{12}$$

where

$$D_{ijk} = \{Q \mid \inf_{\theta_j} D(Q\|P_{\theta_j}) - \inf_{\theta_i} D(Q\|P_{\theta_i})\} < \lambda\}.$$

Proof: For $y^{(k),n} \in S_{ijk,n}^{GLRT}$

$$\begin{aligned}\lambda &\le \frac{1}{n}\log \sup\nolimits_{\theta_i\in\Omega_i} p_{Y^{(k)}|\theta_i}(y^{(k),n}) - \frac{1}{n}\log \sup\nolimits_{\theta_j\in\Omega_j} p_{Y^{(k)}|\theta_j}(y^{(k),n}) \\ &= \sup_{\theta_i}[-H(L_{y^{(k),n}}) - D(L_{y^{(k),n}}\|P_{\theta_i})] - \sup_{\theta_j}[-H(L_{y^{(k),n}}) - D(L_{y^{(k),n}}\|P_{\theta_j})] \\ &= -\inf_{\theta_i} D(L_{y^{(k),n}}\|P_{\theta_i}) + \inf_{\theta_j} D(L_{y^{(k),n}}\|P_{\theta_j})\end{aligned} \tag{13}$$

This implies that $y^{(k),n} \in S_{ijk,n}^{GLRT}$ is equivalent to $y^{(k),n} \notin D_{ijk}$.

Next note that the right hand side of (13) is upper bounded by $\inf_{\theta_j} D(L_{y^{(k),n}}\|P_{\theta_j})$ which implies that $y^{(k),n} \in S_{ijk,n}^{*}$ as well. It follows that $\alpha_{ijk,n}^{GLRT}(\theta_j) \le \alpha_{ijk,n}^{*}(\theta_j)$ which establishes that the GLRT satisfies (11) due to Lemma IV.3.

To compute the type II exponent note that

$$\beta_{ijk,n}^{GLRT}(\theta_i) = P_{\theta_i}[y^{(k),n} \notin S_{ijk,n}^{GLRT}] = P_{\theta_i}[y^{(k),n} \in D_{ijk}]$$

An immediate application of Sanov's theorem (Dembo and Zeitouni 1998, Chap. 2) yields (12). ■

Determining the Optimal Threshold

It can be seen from (11) and (12) that the exponent of the type I error probability is increasing with λ but the exponent of the type II error probability is nonincreasing with λ. We have no preference on the type of error we make, thus, we would like to balance the two exponents and determine the value of λ at which they become equal. In this subsection we detail how this can be done and obtain a λ_{ijk}^{*} that bounds the worst case (over Ω_j and Ω_i) exponents of the type I and type II error probabilities. To simplify the exposition we will be assuming that Ω_j and Ω_i are discrete sets; this is also the case in the experimental setup we describe later on.

Let us consider the exponent of the type II GLRT error probability (cf. (12)):

$$Z_{ijk}(\lambda, \theta_i) = \min\nolimits_Q D(Q\|P_{\theta_i})$$

$$\text{s.t. } \min\nolimits_{\theta_j} D(Q\|P_{\theta_j}) - \min\nolimits_{\theta_i} D(Q\|P_{\theta_i}) \le \lambda,$$

which is equivalent to

$$Z_{ijk}(\lambda,\theta_i) = \min_Q D(Q \| P_{\theta_i})$$
$$\text{s.t. } \min_{\theta_j} D(Q \| P_{\theta_j}) - D(Q \| P_{\theta_i}) \le \lambda \quad \forall \theta_i \tag{14}$$

The worst case exponent over $\theta_i \in \Omega_i$ is given by

$$Z_{ijk}(\lambda) = \min_{\theta_i} Z_{ijk}(\lambda, \theta_i).$$

Note that $Z_{ijk}(\lambda)$ is nonincreasing in λ, and $\lim_{\lambda\to\infty} Z_{ijk}(\lambda) = 0$. Assuming that $Z_{ijk}(0) > 0$, there exists a $\lambda^*_{ijk} > 0$ such that $Z_{ijk}(\lambda^*_{ijk}) = \lambda^*_{ijk}$. Furthermore, both error probability exponents in (11) and (12) are no smaller than λ^*_{ijk}.

Now consider the clusterhead at B_k observing $y^{(k),n}$ and seeking to distinguish between L_i and L_j. The clusterhead has the option of using the GLRT by comparing $X_{ijk}(y^{(k),n})$ to the threshold λ^*_{ijk}, or comparing $X_{jik}(y^{(k),n})$ to a threshold λ^*_{jik} that can be obtained in exactly the same way as λ^*_{ijk}. Let

$$\hat{d}_{ijk} = \max\{\lambda^*_{ijk}, \lambda^*_{jik}\} \tag{15}$$

and set $(\bar{i},\bar{j}) = (i,j)$ if λ^*_{jik} is the maximizer above; otherwise set $(\bar{i},\bar{j}) = (j,i)$. Define the maximum probability of error as

$$P^{(e)}_{ijk,n} = \max\{\max_{\theta_{\bar{j}}} \alpha^{GLRT}_{\bar{i}\bar{j}kn}(\theta_{\bar{j}}), \max_{\theta_{\bar{i}}} \beta^{GLRT}_{\bar{i}\bar{j}kn}(\theta_{\bar{i}})\}.$$

The discussion above leads to the following proposition.

Proposition IV.6 *Suppose that the GLRT at clusterhead B_k compares $X_{\bar{i}\bar{j}k}(y^{(k),n})$ to $\hat{d}_{ijk}$. The maximum probability of error satisfies*

$$\limsup_{n\to\infty} \frac{1}{n} P^{(e)}_{ijk,n} \le -\hat{d}_{ijk}.$$

One of the challenges computing $\hat{d}_{ijk}$ is that the problem in (14) is nonconvex. This may not be an issue when there are relatively few possible values of θ_j and θ_i but for large sets Ω_j and Ω_i computing $\hat{d}_{ijk}$ becomes expensive. To address this issue, we will next develop a lower bound to $Z_{ijk}(\lambda, \theta_i)$ using nonlinear duality.

Let $\bar{Z}_{ijk}(\lambda,\theta_i)$ be the optimal value of the dual of (14); by weak duality it follows that $\bar{Z}_{ijk}(\lambda,\theta_i) \ge Z_{ijk}(\lambda,\theta_i)$ We have

$$\bar{Z}_{ijk}(\lambda,\theta_i) = \max_{\mu_{\theta_i}\ge 0}[\min_{\theta_j}\min_Q[D(Q \| P_{\theta_i}) + \sum_{\theta_i}\mu_{\theta_i} D(Q \| P_{\theta_j}) - \sum_{\theta_i}\mu_{\theta_i} D(Q \| P_{\theta_i})] - \sum_{\theta_i}\mu_{\theta_i}\lambda]$$

By simple algebra, we have

$$\bar{Z}_{ijk}(\lambda,\theta_i) = \max_{\mu_{\theta_i}\ge 0}[\min_{\theta_j}\min_Q[\sum_{r=1}^{|\Sigma|} Q(\sigma_r)\log(Q(\sigma_r)A(\sigma_r))] - \sum_{\theta_i}\mu_{\theta_i}\lambda] \tag{16}$$

where $A(\sigma_r) = \frac{1}{P_{Y^{(k)}|\theta_i}(\sigma_r)} \cdot \prod_{\theta_i} \left(\frac{P_{Y^{(k)}|\theta_i}(\sigma_r)}{P_{Y^{(k)}|\theta_j}(\sigma_r)} \right)^{\mu_{\theta_i}}$.

Note that the optimization over Q is convex and the optimization over μ_{θ_i} is concave, thus, this problem can be solved efficiently. In fact, the optimization over Q can be solved analytically yielding

$$Q(\sigma_l) = \frac{1}{A(\sigma_l)} \Bigg/ \left(\sum_{r=1}^{|\Sigma|} \frac{1}{A(\sigma_r)} \right), l = 1, \ldots, |\Sigma|.$$

$\bar{Z}_{ijk}(\lambda, \theta_i)$ is convex and nonincreasing in λ for all θ_i. Furthermore, the exponent of the type II GLRT error probability is no smaller than $\bar{Z}_{ijk}(\lambda) = \min_{\theta_i} \bar{Z}_{ijk}(\lambda, \theta_i)$. Note that $\bar{Z}_{ijk}(\lambda)$ is also nonincreasing in λ, and $\lim_{\lambda \to \infty} \bar{Z}_{ijk}(\lambda) = 0$. Assuming that $\bar{Z}_{ijk}(\lambda) > 0$, there exists a $\bar{\lambda}^*_{ijk} > 0$ such that $\bar{Z}_{ijk}(\bar{\lambda}^*_{ijk}) = \bar{\lambda}^*_{ijk}$. Furthermore, both error probability exponents in (11) and (12) are no smaller than $\bar{\lambda}^*_{ijk}$

Following the same line of development as before, set

$$\bar{d}_{ijk} = \max\{\bar{\lambda}^*_{ijk}, \bar{\lambda}^*_{jik}\} \tag{17}$$

and define $\bar{i}, \bar{j}$, and $P^{(e)}_{ijk,n}$ in the same way as earlier. It can be seen that $\hat{d}_{ijk} \geq \bar{d}_{ijk}$. We arrive at the following proposition which provides a weaker but more easily computable probabilistic guarantee on the probability of error.

Proposition IV.7 *Suppose that the GLRT at clusterhead B_k compares $X_{\bar{i}\bar{j}k}(y^{(k),n})$ to $\bar{d}_{ijk}$. The maximum probability of error satisfies*

$$\limsup_{n \to \infty} \frac{1}{n} P^{(e)}_{ijk,n} \leq -\bar{d}_{ijk}.$$

OPTIMUM CLUSTERHEAD PLACEMENT

In this section, we focus on how to place the $K \leq M$ clusterheads at positions in B to facilitate localization. We start by considering the multiple hypothesis testing problem of identifying the region $l \in L$ in which the sensor we seek resides.

Multiple Composite Hypothesis Testing

We assume, without loss of generality, that we have placed clusterheads in positions $B_1, \ldots, B_K$, each one making n i.i.d. observations $y^{(k),n} = (y_1^{(k)}, \ldots y_n^{(k)})$. Let d_{ijk} be the GLRT threshold obtained in the previous section for each region pair (i, j), $i < j$, and clusterhead k. Specifically, d_{ijk} is the Chernoff distance if we compare regions each with a single pdf associated to observations at B_k; in that case as we explained earlier we apply the LRT decision rule. If however, one of the hypothesis is composite, i.e., there is a pdf family associated with at least one of the two regions, then we apply the GLRT and the error exponent is obtained from either (15), or (17), depending on which optimization problem we elect to solve.

Figure 1. Clusterhead placement MILP formulation

$$\max \quad \epsilon \tag{18}$$

$$\text{s.t.} \quad \sum_{k=1}^{M} x_k = K \tag{19}$$

$$\sum_{k=1}^{M} y_{ijk} = 1, \ i, j = 1, \ldots, N, i < j, \tag{20}$$

$$y_{ijk} \leq x_k, \ \forall i, j, i < j, k = 1, \ldots, M, \tag{21}$$

$$\epsilon \leq \sum_{k=1}^{M} d_{ijk} y_{ijk}, \ \forall i, j, i < j, \tag{22}$$

$$y_{ijk} \geq 0, \ \forall i, j, i < j, \forall k, \tag{23}$$

$$x_k \in \{0, 1\}, \ \forall k. \tag{24}$$

We make $N-1$ binary decisions with the LRT or GLRT rule to arrive at a final decision. Specifically, we first compare L_1 with L_2 to accept one hypothesis, then compare the accepted hypothesis with L_3, and so on and so forth. For each one of these L_i vs. L_j decisions we use a single clusterhead B_k as detailed in the previous section and the exponent of the corresponding maximum probability of error is bounded by d_{ijk}. All in all we make $N-1$ binary hypothesis decisions.

Clusterhead Placement

Our objective is to minimize the worst case probability of error. To that end, for every pair of regions L_i and L_j we need to find a clusterhead that can discriminate between them with a probability of error exponent larger than some ε and then maximize ε . This is accomplished by the mixed integer linear programming problem (MILP) formulation of Figure 1.

In this formulation, the decision variables are x_k, y_{ijk}, and ε where $k = 1,...,M$, $i, j = 1,...,N$, $i < j$. x_k is the indicator function of a clusterhead been placed at position B_k. Equation (19) represents the constraint that K clusterheads are to be placed. Constraint (22) enforces that for every region pair there exist a clusterhead k with d_{ijk} larger than ε . Let x_k^*, y_{ijk}^*, and ε^* $(k = 1, \ldots, M, i, j = 1, \ldots, N, i < j)$ be an optimal solution of this MILP. Although this problem is NP-hard (Ray et al. 2006), it can be solved efficiently for sites with more than 100 locations by using a special purpose algorithm proposed in the sequel.

The next proposition establishes a useful property for the optimal solution and value of the MILP in Figure 1. In preparation for that result consider an arbitrary placement of K clusterheads. More specifically, let Y be any subset of the set of potential clusterhead positions B with cardinality K. Let $x(Y) = (x_1(Y), \ldots, x_M(Y))$ where $x_k(Y)$ is the indicator function of B_k being in Y. Define:

$$\varepsilon(Y) = \min_{\substack{i,j=1,\ldots,N \\ i<j}} \max_{k|x_k(Y)=1} d_{ijk} \tag{25}$$

We can interpret $\max_{k|x_k(Y)=1} d_{ijk}$ as the best decay rate for the probability of error in distinguishing between locations L_i and L_j from some clusterhead in Y. Then $\varepsilon(Y)$ is simply the worst pairwise decay rate.

Proposition V.1 *For any clusterhead placement Y we have*

$$\varepsilon^* \geq \varepsilon(Y) \tag{26}$$

Moreover, the selected placement achieves equality; i.e.,

$$\varepsilon^* = \min_{\substack{i,j=1,\ldots,N \\ i<j}} \max_{k|x_k^*=1} d_{ijk} \tag{27}$$

Proof: Consider the placement Y and let

$$y_{ijk} = \begin{cases} 1, \text{ if } k = \arg\max_{k|x_k(Y)=1} d_{ijk}, \\ 0, \text{ otherwise,} \end{cases} \quad \forall i\ j, i<j, \forall k.$$

If more than one y_{ijk} are 1 for a given pair (i, j), we arbitrarily set all but one of them to 0 to satisfy Eq. (20). Then

$$\min_{\substack{i,j \\ i<j}} \sum_{k=1}^{M} d_{ijk} y_{ijk}^* = \min_{\substack{i,j \\ i<j}} \max_{k|x_k(Y)=1} d_{ijk} = \varepsilon(Y).$$

Observe that $x(Y)$, y_{ijk}'s (as defined above), and $\varepsilon(Y)$ form a feasible solution of the MILP in Figure 1. Clearly, the value of this feasible solution can be no more than the optimal ε^*, which establishes (26).

Next note that (22) is the only constraint on ε. So, we have

$$\varepsilon^* = \min_{\substack{i,j \\ i<j}} \sum_{k=1}^{M} d_{ijk} y_{ijk}^* = \min_{\substack{i,j \\ i<j}} \sum_{k|x_k^*=1}^{M} d_{ijk} y_{ijk}^* . \tag{28}$$

The second equality is due to (21). The final observation is that the right hand side of the above is maximized when

$$y_{ijk}^* = \begin{cases} 1, \text{ if } k = \arg\max_{k|x_k^*=1} d_{ijk}, \\ 0, \text{ otherwise,} \end{cases} \quad \forall i\ j, i<j, \forall k.$$

(Again, at most one y_{ijk}^* is set to 1 for a given (i, j) pair.) Thus, an optimal solution satisfies the above. This, along with (28) establishes (27). ∎

As before, $\max_{k|x_k^*=1} d_{ijk}$ is the best decay rate for the probability of error in distinguishing between locations L_i and L_j from some clusterhead in the set Y^*. Then ε^* is simply the worst such decay rate over all pairs of locations. Moreover, according to Proposition V.1, this worst decay rate is no worse than the corresponding quantity $\varepsilon(Y)$ achieved by any other clusterhead placement Y.

Efficient Computation of the Proposed MILP

In this section, we propose an algorithm that solves the MILP presented in Figure 1 faster than a general purpose MILP-solver such as CPLEX (CPLEX 8.0 2002). Our approach is to construct an alternate formulation of the proposed MILP first, and then solve it using an iterative algorithm. The computational advantage of this approach lies in the fact that we solve a *feasibility* problem in each iteration that contains only $O(M)$ variables and $O(N^2)$ constraints instead of $O(N^2M)$ variables and $O(N^2M)$ constraints that appear in the formulation in Figure 1, and thus can be solved much faster.

Alternate Formulation

Let us sort the d_{ijk}'s, in nonincreasing order, and let b_{ijk} denote the index of d_{ijk}. We let equal distances have the same index. Note that b_{ijk} is a positive integer upper bounded by $MN(N-1)/2$. Now consider the MILP problem shown in Figure 2. This problem is actually the MILP formulation of the *vertex K -center problem* (Daskin 1995). The following proposition establishes that the formulations of Figure 1 and Figure 2 are indeed equivalent.

Proposition V.2 *Suppose* (s^*, t^*, π^*) *is an optimal solution to the problem in Figure 2. Then* $(x^* = s^*, y^* = t^*, \varepsilon^* = d_{i^*j^*k^*})$ *is an optimal solution to the MILP problem in Figure 1, where* (i^*, j^*, k^*) *is such that* $b_{i^*j^*k^*} = \pi^*$.

Proof: A proof analogous to the one of Prop. V.1 establishes

$$\pi^* = \max_{\substack{i,j \\ i<j}} \min_{k|s_k^*=1} b_{ijk} \tag{29}$$

Let (i^*, j^*, k^*) be such that $\pi^* = b_{i^*j^*k^*}$. Then, $(x^*, y^*, \varepsilon^*) = (s^*, t^*, d_{i^*j^*k^*})$ is an optimal solution of the MILP problem in Figure 1. To that end, observe that x^*, y^* satisfy constraints (19)–(21), (23) and (24). Moreover, since b_{ijk} was defined as the index of d_{ijk}, Eqs. (29) and (27) imply the optimality of $(x^*, y^*, \varepsilon^*)$; namely, the min-max of the d_{ijk}'s is equivalent to the max-min of their rank. ■

Figure 2. Equivalent formulation of the MILP of Figure 1

$$\min \quad \pi \tag{30}$$

$$\text{s. t.} \quad \sum_{k=1}^{M} s_k = K, \tag{31}$$

$$\sum_{k=1}^{M} t_{ijk} = 1, \quad \forall i,j = 1,\ldots,N, i<j, \tag{32}$$

$$t_{ijk} \le s_k, \quad \forall i<j,\ k = 1,\ldots,M, \tag{33}$$

$$\pi \ge \sum_{k=1}^{M} b_{ijk} t_{ijk}, \quad \forall i,j,i<j, \tag{34}$$

$$t_{ijk} \ge 0, \quad \forall i,j,k,i<j, \tag{35}$$

$$s_k \in \{0,1\}, \quad k = 1,\ldots,M \tag{36}$$

Figure 3. The feasibility problem. c_{ijk}'s are defined by Eq. (41)

$$\max \quad 0 \tag{37}$$

$$\text{s.t.} \quad \sum_{k=1}^{M} c_{ijk} w_k \geq 1, \ \forall i, j = 1, \ldots, N, i < j, \tag{38}$$

$$\sum_{k=1}^{M} w_k = K, \tag{39}$$

$$w_k \in \{0, 1\}, \quad \forall k = 1, \ldots, M, \tag{40}$$

We remark that it is also true that there is a corresponding optimal solution to the problem of Figure 2 for every optimal solution to the problem of Figure 1.

Iterative Algorithm

Proposition V.2 allows us to solve the problem of Figure 2 instead of the problem of Figure 1. So we will concentrate on the former. Our approach is to solve this problem by an iterative *feasibility* algorithm along the lines proposed in (Daskin 1995). In particular, we use a slightly modified version of a *two-phase* algorithm proposed in (Ilhan et al. 2006, Ozsoy et al. 2005).

The core idea of the iterative algorithm is to solve the feasibility problem shown in Figure 3. The problem of Figure 3 depends on a parameter θ by the following equation:

$$c_{ijk} = \begin{cases} 1, & \text{if } b_{ijk} \leq \theta, \\ 0, & \text{otherwise.} \end{cases} \tag{41}$$

Intuitively, θ represents the index of some d_{ijk} distance in the nonincreasingly sorted list that we initially created, and the feasibility problem checks whether all pairs of locations can be distinguished (by at least one clusterhead) with an error exponent greater than or equal to the d_{ijk} distance pointed by θ. If not, θ is increased, which means that it now points to a smaller d_{ijk} distance, and the process is repeated. At termination, θ, which corresponds to the largest feasible d_{ijk} distance, provides the optimal value of the problem in Figure 2. The formal iterative algorithm is shown in Figure 4. It is clear that this algorithm terminates in a finite number of steps. In particular, if $\theta = MN(N-1)/2$, we see that c_{ijk} = 1 for all i, j, k, and the feasibility conditions of problem of Figure 3 are trivially satisfied. Next we show that at termination, we obtain the optimal solution to the problem of Figure 2.

Proposition V.3 *Let θ^* be the value of θ when the algorithm of Figure 4 terminates and w^* the optimal solution to problem of Figure 3 at the last iteration. Then $s^* = w^*$ induces an optimal solution to the problem of Figure 2 with optimal objective function value $\pi^* = \theta^*$.*

Proof: First note that at the last iteration, for any i, j ($i < j$), there exists at least one k such that $b_{ijk} = \theta^*$ with $w^*_k = 1$; otherwise the problem is infeasible. Next we construct a feasible solution (s^*, t^*, π^*) to the problem of Figure 2 as follows. Given a pair (i, j), we select one k such that $b_{ijk} = \theta^*$ with $w^*_k = 1$. Then we set $t^*_{ijk} = 1$ and $t^*_{ijl} = 0$ for any $l \neq k$. We repeat this process for all pairs (i, j). Finally, we set s^* =

Figure 4. Iterative feasibility algorithm

1) Set $\theta = 1$.
2) Determine c_{ijk}, $\forall i, j, k$ and solve the problem shown in Fig. 3.
   a) If the problem of Fig. 3 is infeasible, set $\theta := \theta + 1$ and go to step 2.
   b) Else, if the problem is feasible, stop.

w^* and $\pi^* = \theta^*$. Then, the triplet (s^*, t^*, π^*) satisfies all the constraints of the problem of Figure 2 and is therefore a feasible solution.

Next we prove the optimality of (s^*, t^*, π^*) by contradiction. Suppose that there exists a feasible solution $(\tilde{s}, \tilde{t}, \tilde{\pi})$ to the problem of Figure 2 such that $\tilde{\pi} < \pi^*$. Then according to the algorithm in Figure 4, there is a step where $\theta = \tilde{\pi}$. This implies that the corresponding problem of Figure 3 was infeasible; otherwise the algorithm would not have increased the value of θ beyond $\tilde{\pi}$. However, since $(\tilde{s}, \tilde{t}, \tilde{\pi})$ is feasible for the problem of Figure 2 we have

$$\tilde{\pi} \geq \sum_{k|\tilde{s}_k=1} b_{ijk} \tilde{t}_{ijk} \quad \forall i\ j, i < j,$$

which implies that for all (i, j) with $i < j$ there exists at least one k with $\tilde{s}_k = 1$ such that $b_{ijk} \leq \tilde{\pi}$. Hence, $w = \tilde{s}$ is feasible for the problem of Figure 3 when $\theta = \tilde{\pi}$. We arrived at a contradiction. ∎

To expedite the convergence in the actual implementation, we use the two-phase algorithm shown in Figure 5. In the first part (*LP phase*), we construct the linear programming (LP) relaxation of the problem of Figure 3 by replacing the binary constraint (40) by

$$w_k \in [0,1], \quad \forall k = 1, \ldots, M \tag{42}$$

Then we solve the relaxed problem and compute the smallest integer $\theta = \theta_0$ such that the LP relaxation is feasible. In the second part (*IP phase*), we compute θ^* by executing the iterative algorithm starting from $\theta = \theta_0$, but this time we solve the integer programming problem instead of the LP relaxation. The important difference between our algorithm and the algorithm proposed in (Ilhan et al. 2006, Ozsoy et al. 2005) is that we employ binary search both in the LP-phase and IP-phase whereas the authors of (Ilhan et al. 2006, Ozsoy et al. 2005) use linear search in the LP-phase. Our use of binary search in the IP-phase further decreases the computation time for large problem instances almost by an order of magnitude.

Performance Guarantee

We will use the decision rule outlined in the beginning of this section and for every region pair (i, j) we will rely on the best positioned clusterhead $B_{k_{ij}^*}$ to make the corresponding decision. The following theorem establishes a performance guarantee.

Proposition V.4 *Let x^*, y^* be an optimal solution of the MILP in Figure 1 with corresponding optimal value ε^*. Place clusterheads according $Y^* = \{B_k \mid x_k^* = 1\}$ and for every (i, j) select one clusterhead with*

Figure 5. Two-phase iterative feasibility algorithm

1) (LP phase) Set $l := 1$, $u := \frac{MN(N-1)}{2}$.
2) Let $\theta := \lceil \frac{l+u}{2} \rceil$, determine c_{ijk} $\forall i, j, k$ and solve the relaxed version of the problem of Fig. 3.
    a) If the relaxed version of the problem of Fig. 3 is feasible, set $u := \theta$.
    b) Else, set $l := \theta$.
3) If $u - l \leq 1$, go to step 4. Else go to step 2.
4) (IP phase) If the relaxed version of the problem of Fig. 3 is feasible, set $\theta := l$, else set $\theta := u$. Then set $l := \theta, u := \frac{MN(N-1)}{2}$.
5) Determine c_{ijk}, $\forall i, j, k$ and solve the problem of Fig. 3.
    a) If the problem of Fig. 3 is infeasible, set $l := \theta, \theta := \lceil \frac{l+u}{2} \rceil$ and go to step 5.
    b) Else, if the problem is feasible
        i) If $\theta \neq u$, set $u := \theta, \theta := \lceil \frac{l+u}{2} \rceil$ and go to step 5.
        ii) Otherwise, stop.

index k_{ij}^* so that $y_{ijk_{ij}^*}^* = 1$. Then, the worst case probability of error for the decision rule described in the beginning of this section, $P_n^{(e),opt}$, satisfies

$$\limsup_{n\to\infty} \frac{1}{n} \log P_n^{(e),opt} \leq -\varepsilon^* . \tag{43}$$

Proof: Recall the results of Theorem. IV.1 and Propositions IV.6 and IV.7 for the case where d_{ijk} is defined either by (4), (15), or by (17), respectively. Define $(\bar{i}, \bar{j})$ as in the previous section. The clusterhead with index k_{ij}^* will use the GLRT which compares $X_{\bar{i}\,\bar{j}k_{ij}^*}(y^{(k_{ij}^*),n})$ to $d_{ijk_{ij}^*}$, thus, achieving a maximum probability of error with exponent no smaller than $d_{ijk_{ij}^*}$. Now, for every i and $j \neq i$ define $E_n(i, j)$ as the event that the GLRT employed by the clusterhead at $B_{k_{ij}^*}$ will decide L_j under P_{θ_i}. For all $\delta_n > 0$ and large enough n we have

$$P_{\theta_i}[\text{error}] \leq P_{\theta_i}[\bigcup_{j\neq i} E_n(i, j)] \leq \sum_{j\neq i} e^{-n(d_{ijk_{ij}^*}+\delta_n)} \leq (N-1)e^{-n(\varepsilon^*+\delta_n)} .$$

The 2nd inequality above is due to Thm. IV.1 or Props. IV.6, or IV.7 and the last inequality above is due to (27). Since the bound above holds for all I we obtain (43). ■

LOCALIZATION DECISIONS

In this section we consider the implementation of the decision rule described in the previous section. We assume that the WSN has a single gateway. We seek to devise a distributed localization algorithm in order to minimize the information that needs to be exchanged between clusterheads and the gateway. The primary motivation is that in WSNs communication is, in general, more expensive than processing. For the remainder of this section we will assume that the clusterheads and the gateway form a connected network. Otherwise, one can simply add a sufficient number of relays.

Centralized Approach

We first describe a naive, centralized, approach. Every clusterhead observes $y^{(k),n} = (y_1^{(k)}, \ldots, y_n^{(k)})$ and transmits this information to the gateway. The clusterheads do not need to store anything and perform no processing; they are simple sensors that transmit their measurements. Letting S_l the message size (in bits) needed to encode the measurement $y_l^{(k)}$, for some l, the total amount of information that needs to be transported is $O(S_l nK)$ bits. Each one of these bits has to be sent over multiple hops to reach the gateway; in the worst case over K hops. Thus, the worst case communication cost is $O(S_l nK^2)$ bits. Once this information is received, the gateway can apply the decision rule discussed in the previous section to identify the region at which the sensor in question resides.

Distributed Approach

In this subsection we describe a distributed implementation for the decision rule. We start with an arbitrary pair of regions, say L_1 vs. L_2. The clusterhead at $B_{k^*_{1,2}}$ based on the observations $y^{(k^*_{12}),n}$ uses the GLRT to make the decision; let L_{l_1} be the hypothesis accepted. The clusterhead at $B_{k^*_{1,2}}$ sends the information that l_1 is accepted to the clusterhead at $B_{k^*_{l_1,3}}$ which follows up with the decision L_{l_1} vs. L_3, and so on and so forth. Let now L_{l_i} denote the hypothesis accepted at stage i of the algorithm, for $i = 1, \ldots, N-1$, where we set $l_0 = 1$. At the i-th stage, the clusterhead at $B_{k^*_{l_{i-1},(i+1)}}$ makes the decision $L_{l_{i-1}}$ vs. L_{i+1} and sends the result to the clusterhead at $B_{k^*_{l_i,(i+2)}}$, where the clusterhead at $B_{k^*_{l_{N-1},(N+1)}}$ is the gateway. All in all this procedure takes $N-1$ stages and $L_{l_{N-1}}$ is the final accepted hypothesis.

Each clusterhead is responsible for a set of region pairs and needs to store the corresponding pdfs and thresholds d_{ijk} as well as the necessary information to decide where to forward its decision. At every stage $i = 1, \ldots, N-1$ it takes $O(n)$ work to perform the GLRT, yielding an overall $O(nN)$ processing effort distributed to the K clusterheads. In terms of communication cost, $N-1$ messages get exchanged each consisting of $O(\log N)$ bits needed to encode the decision. Each of these messages can, in the worst case, be sent over $O(K)$ hops if two distant clusterheads need to communicate, yielding an overall worst case communication cost of $O(KN \log N)$. However, one can sequence the regions in such a way that geographically close regions are close in the sequence. As a result, it will often be the case that clusterheads responsible for region pairs close in the sequence will be geographically close resulting in messages between clusterheads traveling a few hops. It follows that the overall communication cost will often be $O(N \log N)$.

Based on the preceding analysis, Table 1 compares the centralized and distributed approaches. In the distributed case we report both the best and worst case in terms of the communication cost based on the discussion in this subsection. Some observations are in order. The total processing cost is the same for both approaches but in the distributed case the work is distributed among the K clusterheads. To compare the communication costs note that typically $K = O(N)$ to ensure reasonable performance (e.g., one clusterhead for a fixed number of regions). Moreover, S_l is the message size for the raw measurements at a clusterhead corresponding to a packet sent from the transmitting sensor, while n can be large enough (e.g., 20-30) so that the probability of error becomes small enough. It follows that $O(N \log N)$ is much preferable to $O(S_l nK^2)$.

Note that both the centralized and the distributed approach guarantee the performance of the system obtained in Prop. V.4, i.e., the savings from the distributed approach come with no performance loss.

Table 1. Comparing the centralized and distributed approaches

	Communication cost (bits)	Processing cost
Centralized	$O(S_1 n K^2)$	$O(nN)$ at the gateway
Distributed	worst: $O(KN \log N)$ best: $O(N \log N)$	$O(nN)$ at the K clusterheads

TESTBED AND EXPERIMENTAL RESULTS

Next, we provide experimental results from a localization testbed we have installed at Boston University (BU) see Figure 6. We have appropriately named our system the *Boston University Statistical Localization System (BLoc)* (http://pythagoras.bu.edu/bloc/index.html). The testbed has a web interface through which one can poll a specific WSN node (identified by an ID). The system responds with a building floor map like the one shown in Figure 7 highlighting the room number and the region where the node was found.

The testbed uses MICAz motes manufactured by Crossbow Inc. We covered 16 rooms and corridors and defined 60 regions. Within each region we placed a mote on some furniture or on the wall. These 60 positions make up the set B of possible clusterhead positions. Hence, in our testbed $N = M = 60$ and B_j can be thought as the center of L_j. All 60 motes are connected to a base MICAz through a mesh network. The base mote is docked on a programming board which is connected to a laptop acting as a server.

The experimental validation of our localization approach can be divided into the five phases outlined in Figure 8. Phase 1 can be carried out automatically by scheduling the motes so that when one is broadcasting the others are listening. For Phase 2 we construct our pdf databases by measuring 200 packets for each pair of motes sent over two frequency channels and with two different power levels. The pdfs

Figure 6. Floor plan for the bestbed

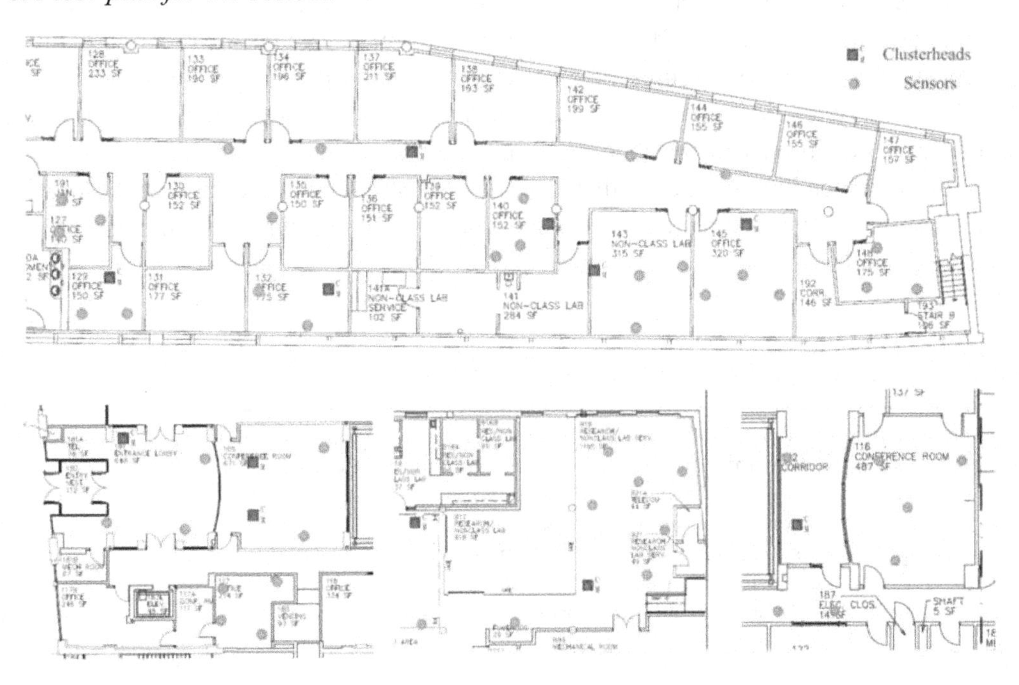

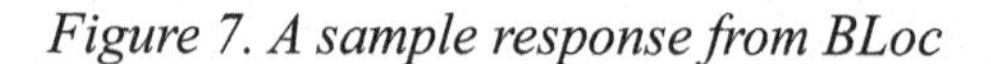

Figure 7. A sample response from BLoc

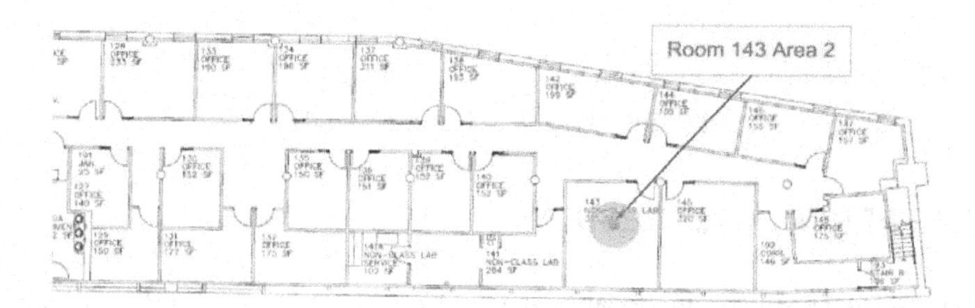

were estimated by histograms, and for each pdf we define an interval $[m_{jk} - \hat{m}_{jk}, m_{jk} + m_{jk}]$ and select points $\theta_{j,1}, \ldots, \theta_{j,R}$ in this interval. We construct the family $\{P_{Y^{(k)}|\theta_j}(y), \theta_j \in \Omega_j\}$ so that the l-th member has the same shape as $P_{Y^{(k)}|B_j}(y)$ but a mean equal to $\theta_{j,l}$, for $l = 1,\ldots,R$. $\hat{m}_{jk}$ is selected appropriately so that the union over j, k of the intervals $[m_{jk} - \hat{m}_{jk}, m_{jk} + m_{jk}]$ is maximized and there is no overlap. In the optimal placement obtained in Phase 4 we used 12 clusterheads to achieve a small enough probability of error and have some built-in redundancy in the clusterhead network.

We obtained results for three versions of the localization system. We made 100 localization tests in positions spread within the covered area. Each test used 20 packets (RSSI measurements) broadcasted by the mote to be located (5 over each channel and power level pair for the 2F2P cases described below). In Version 1 the mote we want to locate transmits packets at a single frequency and a single power level and the system uses the GLRT (we write 1F1P – G to indicate Ver. 1 in Figure 9) to determine the region where the mote resides based on RSSI observations at the clusterheads. In Version 2 (denoted by 2F2P –G) RSSI observations are made for packets transmitted in two different frequencies and two different power levels and the GLRT is again used. Version 3 (denoted by 2F2P – L) is identical to Version 2 but the LRT rather than the GLRT is used where every region is represented by just the pdf observed in Phase 1 (rather than a pdf family). For each Version 1–3 results are reported in Figure 9(a)–(c), respectively. In each of these figures we plot the histogram of the error distance (in inches) based on 100 trials. If the system identifies region L_j as the one where the transmitting mote is located then the error distance is defined as the distance between the transmitting mote and B_j. For each system we also report the corresponding mean error distance ($\bar{D}_e$). We stress that for each trial the location of the transmitting mote is randomly selected and is almost never the one at which RSSI measurements have been made in Phase 1.

Figure 8. Phases of the experimental validation

1) For each pair of positions (B_k, B_j) estimate the pdf $p_{\mathbf{Y}^{(k)}|B_j}(\mathbf{y})$ of RSSI at B_k when the mote at B_j is transmitting. Let m_{jk} the corresponding mean.
2) For each (B_k, B_j) construct a pdf or a pdf family $\{p_{\mathbf{Y}^{(k)}|\theta_j}(\mathbf{y}), \theta_j \in \Omega_j\}$ to characterize transmissions from positions within L_j.
3) Compute the exponent d_{ijk} as described in Sec. IV.
4) Determine the clusterhead placement by the algorithm in Sec. V-B.
5) Determine the location of any mote in the coverage area by the decision rule of Sec. V-A.

Figure 9. Results for various versions of the system

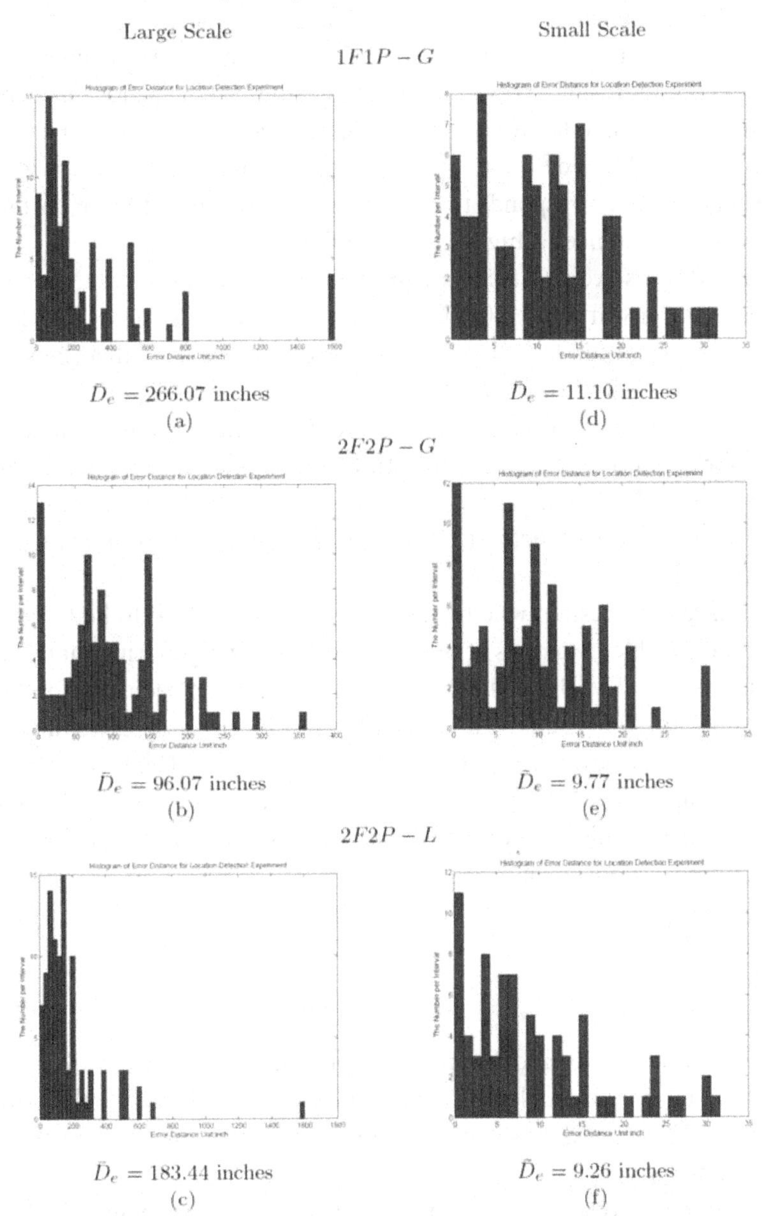

The results show that the 2F2P−G system, which exploits frequency and power diversity, outperforms the 1F1P−G system. Clearly, RSSI measurements at multiple power and frequency levels contain more information about the transmitter location. Also, the 2F2P−G system outperforms the 2F2P −L system which uses the standard LRT decision rule. This demonstrates that, as envisioned, the GLRT provides robustness leading to better performance. The issue with the LRT is that a single pdf can not adequately represent a relatively large region. We also note that the total coverage area was 5258 feet^2, that is, about 87 feet^2 per region. With a mean error distance of $\overline{D}_e$= 8 feet the mean area of "confusion"

was $8^2 = 64$ feet^2. From these results it is evident that we were able to achieve accuracy on the same order of magnitude as the mean area of a region. That is, the system was identifying the correct or a neighboring region most of the time. Put differently, we can say that the achieved mean error distance is about the same as the radius of a region, defined as radius = $\sqrt{\text{area}}$ (for our experiments $\sqrt{87} = 9.3$ feet which is in fact larger than the mean error distance of 8 feet). We used a clusterhead density of 1 clusterhead per 5258/12 = 438 feet^2. Note that our system is *not* localizing based on "proximity" to a clusterhead; one clusterhead corresponds to about 5 regions thus resulting in cost savings compared to proximity-based systems that need a higher density of observers.

For comparison purposes, we also used the same testbed and the exactly same tests with the stochastic trilateration method of (Patwari et al. 2003). (Patwari et al. 2003) assumes that the RSSI (in db) at B_k when the mote at B_j is transmitting, say $Y(k) \mid B_j$, is a random variable with a Gaussian distribution. The mean of RSSI satisfies the path loss formula $\overline{Y}^{(k)} \mid B_j = Y_0 - 10n_p \log_{10}(\zeta_{kj} / \zeta_0)$, where ζ_{ij} is the distance between B_k and B_j and ζ_0 is a normalizing constant. From prior measurements we obtained $n_p = 3.65$ and $Y_0 = -48.62$ dBm for $\zeta_0 = 3$ feet. The location estimation is obtained by maximum likelihood estimation. Applying this method and using our clusterheads in the exactly same position as before resulted in a mean error distance of 341.72 inches (29 feet) which is much larger (a factor of 3.6) than the 8 feet obtained by our method.

These results raised the question whether smaller regions can lead to better accuracy. To that end, we placed 12 motes on a table (two rows of 6 motes each). Two neighboring motes in one row (or in one column) were 6 inches apart. We defined a 36 inches^2 region around each mote and followed the exactly same procedure as before. The results of this "small scale" localization experiment are in Figure 9(d)–(f). As before frequency and power diversity improve performance. Here, however, the GLRT does not make a difference compared to LRT and this is because every region is small enough. With the LRT we can achieve a mean error distance of 9.26 inches, that is, we can again achieve an accuracy on the same order of magnitude as the mean area of a region.

CONCLUSION

We have presented a unified robust and distributed approach for locating the area (region) where sensors of a WSN reside. We posed the problem of localization as a multiple hypothesis testing problem and proposed a combined LRT- or GLRT-based decision rule depending on the appropriate probabilistic characterization of a region.

We developed asymptotic results on the type I and type II error exponents which are critical in posing and solving the problem of optimally placing a given number of clusterheads to minimize the probability of error. We devised a mixed integer linear programming formulation to determine the optimal clusterhead placement, and a fast algorithm for solving it. We evaluated the scalability of the proposed MILP as well as the quality of the resultant placement. Our implementation of the proposed fast algorithm shows that the proposed MILP is capable of solving realistic problems within reasonable time-frames. Furthermore, we proposed a distributed approach to implement our localization algorithm and demonstrated that this can lead to savings in the communication cost compared to a centralized approach.

We validated our approach using testbed implementations involving MICAz motes manufactured by Crossbow. Our experimental results demonstrate that a combined LRT- and GLRT-based system provides significant robustness (and improved performance) compared to a simpler LRT-based system.

Furthermore, our approach leads to significantly improved accuracy compared to a stochastic trilateration technique like the one in (Patwari et al. 2003). We showed that we can achieve an accuracy on the same order of magnitude as the mean radius, which is the best possible accuracy one can achieve with a discrete system. As a result, smaller regions (and more clusterheads) lead to better accuracy but at the expense of more initial measurements (training) and a higher equipment cost. This provides a rule of thumb for practical systems: define as small regions as possible given a tolerable amount of initial measurements and cost.

ACKNOWLEDGMENT

We would like to thank Binbin Li for implementing the stochastic trilateration approach which was compared to ours.

REFERENCES

Bahl, P., & Padmanabhan, V. (2000, March). RADAR: An in-building RF-based user location and tracking system. *In Proceedings of IEEE INFOCOM, 2,* 775-784. Tel-Aviv, Israel.

Battiti, R., Brunato, M., & Villani, A. (2003, January). *Statistical learning theory for location fingerprinting in wireless LANs*, University of Trento, Department of Information and Communication Technology, Trento, Italy, DIT 02-086.

Castro, P., Chiu, P., Kremenek, T., & Muntz, R. (2001, September). A Probabilistic Room Location Service for Wireless Networked Environments. *Proceedings of the 3rd international conference on Ubiquitous Computing* (pp.18-34), Atlanta, Georgia, USA

Chernoff, H. (1952). A measure of asymptotic efficiency for tests of a hypothesis based on the sum of observations. Ann. Math. Statist., *23*, 493-507.

Daskin, M. (1995). *Network and Discrete Location.* New York: Wiley.

Dembo, A., & Zeitouni, O. (1998). *Large Deviations Techniques and Applications*, 2nd ed. NY: Springer-Verlag.

Hightower, J., Want, R., & Borriello, G. (2000, February). *SpotON: An indoor 3d location sensing technology based on RF signal strength.* University of Washington, Department of Computer Science and Engineering, Seattle, WA, UW CSE 00-02-02.

Hodes, T. D., Katx, R. H., Schreiber, E. S., & Rowe, L. (1997, September). Composable ad hoc mobile services for universal interaction. *Proceedings of the 3rd annual ACM/IEEE international conference on Mobile computing and networking* (pp. 1-12) Budapest, Hungary.

Hoeffding, W. (1965). Asymptotically optimal tests for multinomial distributions. *Ann. Math. Statist., 36*, 369.401.

Hofmann-Wellenhof, B., Lichtenegger, H., & Collins, J. (1997). *Global Positioning System: Theory and Practice.* 4th ed. Springer-Verlag.

Ilhan, T., & Pinar, M. (2001). *An efficient exact algorithm for the vertex p-center problem.* http://www.optimization-online.org/ DB-HTML/2001/09/376.html

ILOG CPLEX 8.0, ILOG, Inc., Mountain View, California, July 2002, http://www.ilog.com.

Klingbeil, L., & Wark, T. (2008). A wireless sensor netwok for real-time indoor localisation and motion monitoring. *Proceedings of 2008 International Conference on Information Processing in Sensor Networks* (pp. 39-50).

Meissner, A., Luckenbach, T., Risse, T., Kirste, T., & Kirchner, H. (2002). Design challenges for an integrated disaster management communication and information system. *1st IEEE Workshop on Disaster Recovery Networks.* New York, USA

Ozsoy, F. A., & Pinar, M. C. (2004, November). An exact algorithm for the capacitated vertex p-center problem. *Computers and Operations Research, 33*(5), 1420-1436.

Paschalids, I. C., & Guo, D. (2007, December). Robust and distributed localization in sensor networks. *Proceedings of 46th IEEE Conference on Decision and Control,* (pp. 933-938), New Orleans, Louisiana

Patwari, N., Hero, A. O., Perkins, M., Correal, N. S., & O'Dea,R. J. (2003). Relative location estimation in wireless sensor networks. *IEEE Transactions on signal processing, 51*(8), 2137-2148.

Patwari, N., & Agrawal, P. (2008). Effects of correlated shadowing: connectivity, localization, and RF Tomography. *In 2008 International Conference on IPSN* (pp. 82-93).

Prasithsangaree, P., Krishnamurthy, P., & Chrysanthis, P. K. (September 2002). On indoor position location with wireless LANs. *Proceedings of 13th IEEE International Symposium on Personal, Indoor, and Mobile Radio Communications* (pp. 720-724).

Priyantha, N. B., Chakraborty, A., & Balakrishnan, H. (2000). The cricket location-support system. *In Mobile Computing and Networking* (pp. 32-43).

Ray, S., Lai, W., & Paschalidis, I. C. (2006). Statistical location detection with sensor networks. *IEEE Transactions on Information Theory, Joint special issue with IEEE/ACM Transactions on Networking on Networking and Information Theory, 52*(6), 2670-2683.

Wu, Y., Hu, J. B., & Chen, Z. (2007). Radio map filter for sensor network indoor localization systems. *5th IEEE International Conference on INdustrial Informatics* (pp. 63-68).

Youssef, M. (2008). *Collection about Location Determination Papers available online* http://www.cs.umd.edu/~moustafa/location_papers.htm

Zeitouni, O., Ziv, J., & Merhav, N. (1992 September). When is the generalized likelihood ratio test optimal. *IEEE Transactions on Information Theory, 38*(2), 1597-1602.

APPENDIX A: UPPER BOUND FOR BINARY CASE

Theorem A.1 $P_n^{(e)} \leq e^{-nd_{ijk}}$ *for all n.*

Proof: From our assumption that $y_1^{(k)},\ldots,y_n^{(k)}$ are i.i.d, we have

$$X_{ijk}(y_1^{(k)},\ldots,y_n^{(k)}) = \sum_{i=1}^{n} X_{ijk}(y_i^{(k)}) \tag{44}$$

Since ML rule rejects L_j if the likelihood ratio is greater than 0, for any n,

$$\begin{aligned}
\beta_n &= P_{L_j}[X_{ijk}(y_1^{(k)},\ldots,y_n^{(k)}) > 0] \\
&= P_{L_j}[\lambda \cdot X_{ijk}(y_1^{(k)},\ldots,y_n^{(k)}) > \lambda \cdot 0] \\
&= P_{L_j}[e^{\lambda \sum_{i=1}^{n} X_{ijk}(y_i^{(k)})} > e^{\lambda \cdot 0}] \\
&\leq E_{L_j}[e^{\lambda \sum_{i=1}^{n} X_{ijk}(y_i^{(k)})}] \\
&= \left[E_{L_j}[e^{\lambda X_{ijk}(y_i^{(k)})}] \right]^n \\
&= e^{n \log\left(E_{L_j}[e^{\lambda X_{ijk}(y_i^{(k)})}] \right)}
\end{aligned}$$

Optimizing with respect to λ gives $\beta_n \leq e^{-n \cdot d_{ijk}}$. Noting that d_{ijk} is symmetric with respect to the conditional distributions, $\alpha_n \leq e^{-n \cdot d_{ijk}}$, and thus $P_n^{(e)} \leq e^{-n \cdot d_{ijk}}$. ∎

APPENDIX B: PROOF OF LEMMA IV.3

***Proof*:** For all $\theta_j \in \Omega_j$ we have

$$\begin{aligned}
\alpha_{ijk,n}^{S^*}(\theta_j) &= P_{\theta_j}[y^{(k),n} \in S_{ijk,n}^*] \\
&= \sum_{\{L_{y^{(k)},n} | T_n(L_{y^{(k)},n} \subseteq S_{ijk,n}^*)\}} |T_n(L_{y^{(k)},n})| P_{Y^{(k)}|\theta_j}(y^{(k),n}) \\
&\leq \sum_{\{L_{y^{(k)},n} | T_n(L_{y^{(k)},n} \subseteq S_{ijk,n}^*)\}} e^{nH(L_{y^{(k)},n})} e^{-n[H(L_{y^{(k)},n}) + D(L_{y^{(k)},n} \| P_{\theta_j})]} \\
&= \sum_{\{L_{y^{(k)},n} | T_n(L_{y^{(k)},n} \subseteq S_{ijk,n}^*)\}} e^{-nD(L_{y^{(k)},n} \| P_{\theta_j})} \\
&\leq (n+1)^{|\Sigma|} e^{-n\lambda},
\end{aligned}$$

which establishes (6). For the first inequality above note that the size of the type class of $L_{y^{(k)},n}$ is upper bounded by $e^{nH(L_{y^{(k)},n})}$ and that the probability of a sequence can be written in terms of the entropy and the relative entropy of its type (see Dembo and Zeitouni 1998, Chap. 2). In the last inequality above we used the definition of $S_{ijk,n}^*$ and the fact that the set of all possible types, L_n, has cardinality upper bounded by $(n+1)^{|\Sigma|}$ (Dembo and Zeitouni 1998, Chap. 2).

Let now $S_{ijk,n}$ be some other decision rule satisfying constraint (6), hence, for all $\varepsilon > 0$ and all large enough n

$$\alpha_{ijk,n}^{S}(\theta_j) \le e^{-n(\lambda+\varepsilon)} \tag{45}$$

Meanwhile for all $\varepsilon > 0$, all large enough *n*, and any $y^{(k),n} \in S_{ijk,n}$

$$\begin{aligned}\alpha_{ijk,n}^{S}(\theta_j) &= \sum_{\{L_{y^{(k)},n} | T_n(L_{y^{(k)},n} \subseteq S_{ijk,n})\}} |T_n(L_{y^{(k)},n})| P_{Y^{(k)}|\theta_j}(y^{(k),n}) \\ &\ge \sum_{\{L_{y^{(k)},n} | T_n(L_{y^{(k)},n} \subseteq S_{ijk,n})\}} (n+1)^{-|\Sigma|} e^{-nD(L_{y^{(k)},n} \| P_{\theta_j})} \\ &\ge e^{-n[D(L_{y^{(k)},n} \| P_{\theta_j})+\varepsilon]},\end{aligned}$$

where the first inequality above uses (Dembo and Zeitouni 1998, Lemma 2.1.8). Comparing the above with (45) it follows that if $y^{(k),n} \in S_{ijk,n}$ then for all θ_j $D(L_{y^{(k)},n} \| P_{\theta_j}) \ge \lambda$, hence, $y^{(k),n} \in S^{*}_{ijk,n}$ and $S_{ijk,n} \subset S^{*}_{ijk,n}$. Consequently, for all θ_i $\beta^{S}_{ijk,n}(\theta_i) \ge \beta^{S^*}_{ijk,n}(\theta_i)$ which establishes that the generalized Hoeffding test maximizes the exponent of the type II error probability. We conclude that it satisfies the GNP criterion. ■

Chapter X
Theory and Practice of Signal Strength-Based Localization in Indoor Environments

A. S. Krishnakumar
Avaya Labs Research, USA

P. Krishnan
Avaya Labs Research, USA

ABSTRACT

In this chapter, the authors concentrate on signal strength-based localization in indoor wireless networks, with emphasis on 802.11 networks. The authors briefly summarize some architectures and approaches researchers have taken to address this problem. They then present some insight into theoretical limits to location accuracy, and identify that the issues driving research work in this area will not only be location accuracy but other factors like deployment ease, management simplicity, adaptability, and cost of ownership and maintenance. With this insight, they present the LEASE architecture for localization that allows easy adaptability of localization models. The chapter discusses the use of Bayesian networks for localization and presents a zero-configuration Bayesian localization algorithm that simplifies the maintenance of the model. Although presented in the context of signal strength-based localization in indoor environments, the concepts are general enough to be applicable to sensor, ad hoc, mesh, and infrastructure-based deployments. They conclude with some open issues.

INTRODUCTION

Indoor wireless networks, especially 802.11-based wireless systems, are increasingly being deployed. Looking beyond simple untethered network access, services based on end-user location information provide compelling benefits, and in some cases satisfy regulatory concerns. Examples of such services include location-aware content delivery, emergency location, presence-enabled applications, services using location-based resource management, and location-based access control.

The techniques used for localization will depend on the constraints imposed on the problem, and the underlying technology being used. For example, base stations in outdoor wireless (e.g., cellular) networks are controlled by the service provider and have specialized hardware and software. The endpoints may have service provider-specific software as well. In contrast, indoor wireless networks are built using off-the-shelf components (e.g., access points) and endpoints that are more open (e.g., laptops). Clearly, these two environments present different constraints for localization, and hence the architecture and techniques needed for localization in these environments would differ. Sensor networks usually consist of low-power, low-bandwidth components and these impose additional constraints on the techniques employed.

Localization is of value in both wired and wireless environments. In this chapter, we concentrate on wireless localization, and more specifically on indoor wireless environments. A commercially attractive option for localization in some scenarios is the Global Positioning System (GPS). Usually, GPS technology works well outdoors but has problems in indoor environments. Furthermore, GPS receivers form a closed platform and are co-resident with the device being located. In typical indoor environments, the devices used are general-purpose and do not necessarily have GPS receivers. Therefore, we look at localization aided by the wireless technology itself, namely the classes of techniques that can be used in radio networks. We then concentrate on localization in indoor wireless networks, specifically networks based on IEEE 802.11.

In radio networks, four classes of techniques have generally been used for localization, as depicted in Figure 1. These techniques are based on different features of the radio signal: angle of arrival, time of arrival, time difference of arrival, and received signal strength. The first is a technique using angles or an *angulation* technique. The other three are based on distances and thus are *lateration* techniques. With each technique, the location may be obtained directly by employing geometry, by using scene analysis techniques, or by probabilistic methods. These techniques have been used in many application contexts, e.g., navigation, radar, cellular communication systems, and robotics. An overview of the application of these techniques and others for indoor localization can be found in Hightower and Borriello (2001) and Pahlavan et al. (2002). An overview of localization in CDMA cellular systems is available in Caffrey and Stuber (1998).

Some of the techniques mentioned above require capabilities not typically found in off-the-shelf components used in indoor wireless communications. In general, time-of-arrival and time-difference-of-arrival techniques require an accurate time reference. This is usually available in systems such as cellular communications since an accurate time reference is needed for proper communication as well. Special equipment – such as multiple directional antennas or an antenna with a steerable beam – is needed to measure the angle of arrival. Other systems such as *Cricket* described in Priyantha et al. (2000) require special co-located radio and ultrasound transceivers.

Our emphasis in this chapter is on localization in indoor wireless networks, and specifically, techniques for localization in 802.11-based wireless networks. As mentioned earlier, the capabilities mentioned above are not available in typical off-the-shelf components used for indoor wireless networks. For example, with 802.11 systems the typical time reference available is of the order of 100ns[1], which is insufficient for accurate location based on time of arrival or time difference of arrival. The special requirements for localization via angle of arrival, time of arrival, and time difference of arrival methods make signal

Figure 1. Different techniques used for wireless location

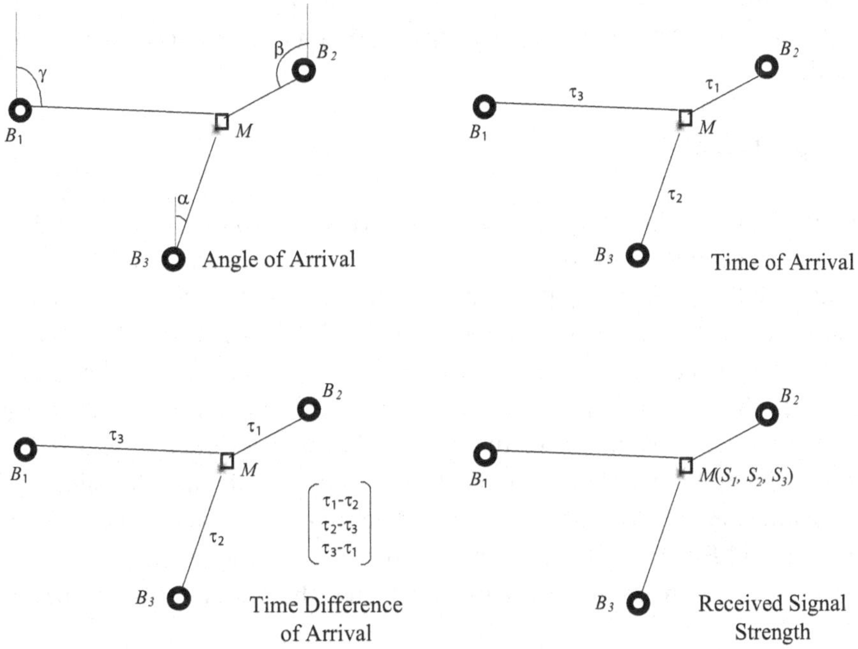

strength-based approaches more attractive for indoor wireless localization. In signal strength-based localization, the received signal strength measurements from several transmitters are used as the basis for localization. The topic of signal strength-based localization has seen a lot of interesting work and many techniques have been proposed, experimentally verified (e.g., in Bahl and Padmanabhan (2000a), Krishnan et al. (2004), Ladd et al. (2002), Prasithsangaree et al. (2002), Roos et al. (2002), Saha et al. (2003), Youssef et al. (2003)), and analytically studied (e.g., in Elnahrawy et al. (2004), Krishnakumar and Krishnan (2005), Malaney (2004)).

The rest of the chapter is organized as follows. First, we provide an overview of the various techniques used for wireless localization. We then concentrate on signal strength-based localization, and discuss both experimental techniques and theoretical work in this area. Other issues that are important in practice, including deployment and maintenance cost, and research work addressing these concerns are considered next. We conclude with open issues.

LOCALIZATION IN WIRELESS NETWORKS: A BRIEF OVERVIEW

Localization is a much-studied topic in the context of radar, cellular WAN and many other technologies. Even if we restrict ourselves to indoor localization, there are many approaches to the problem, as surveyed in Hightower and Borriello (2001), Nerguizian et al. (2001) and Pahlavan et al. (2002). While most of this chapter concentrates on signal-strength based localization techniques, it is instructive to briefly understand systems based on other approaches: specifically, time of arrival, angle of arrival, time difference of arrival, and proximity and we present them in this section.

All the work described in subsequent sections focuses on localization in a plane. In many practical situations, one would like to have at least a quantized estimate of location in the third dimension. Although this issue has not been studied extensively, we outline some initial attempts at dealing with this in a later section.

Systems Using Time of Flight

Global Positioning Satellite (GPS) navigation system is a very well-known example of localization systems. This system uses the time-of-flight of radio signals to perform lateration. Its performance indoors can be problematic due to signal propagation issues. When used outdoors, GPS provides an accuracy of 1-5 meters (only with differential correction; see e.g., USDOT (2002) and Moore et al. (2002)) over 95% of the time. In this chapter sequel, we use "accuracy of x units (y %)" to indicate an accuracy of x units y % of the time. The time of flight approach is difficult to use in indoor wireless systems due to a lack of an accurate time reference in commercially available network interface cards. For example, as pointed out earlier, with 802.11 wireless systems, the typical time reference available is of the order of 100ns, which is insufficient for accurate location based on time of arrival or time difference of arrival. However, in Günther and Hoene (2005), the authors present a technique to increase the resolution by using multiple delay measurements and applying statistical techniques to improve the accuracy.

Systems Using Angle of Arrival

Obtaining angle-of-arrival information requires directional antennas or advanced signal processing. This is usually not available in typical off-the-shelf mobile terminals. The use of angle-of-arrival information indoors is also problematic due to the effects of multipath propagation. There may be situations where the signal that is registered is not the direct, line-of-sight signal, but a reflected one. This could introduce significant error in the location measurement. However, this technique can be and is used outdoors with satisfactory results. Some examples are emergency E911 service in North American wide-area cellular systems and VHF omnidirectional ranging used in flight navigation. An indoor positioning system using angle of arrival is described in Niculescu and Nath (2003). Based on simulations, they report positioning error as a fraction of communication range and as such it cannot be directly compared with experimental results reported for other methods.

Systems Using Time Difference of Arrival

A good example of a system using time-difference-of-arrival information is the Long Range Navigation (LORAN) system (see Sonnenberg (1988)) used by ships and aircraft. This was the main navigation system used by marine craft before the advent of GPS navigation systems. If the position of two LORAN transmitters is known, then the receiver is positioned somewhere on a hyperbolic curve between the transmitters where the time difference of received signals is constant. Using another pair of transmitters with at least one different transmitter, the position can be determined as the intersection of two hyperbolic curves. The Active Bat system described in Harter et al. (1999) employs time difference of arrival in a different fashion than above to compute ranges and perform multi-lateration. In this system, the entities to be tracked carry tags that emit an ultrasonic pulse to a grid of ceiling mounted receivers. This pulse is sent as a response to a request sent by a controller using radio frequency signals. When the controller sends a request to the tag, it simultaneously sends a reset signal to the ceiling receivers using a wired network. The ceiling receivers use the time between the reset signal from the controller and the ultrasound signal from the tag to compute a range value to the tag. These are then sent to the controller, which then performs multi-lateration. This system has an accuracy of 9cm (95%). The Cricket system described in Priyantha et al. (2000) is similar, but does not require ceiling-mounted receivers or a central controller. The mobile terminals act as receivers and also perform timing and computation functions. In their evaluation, the Cricket system estimates location within a 4ft x 4ft (1.2m x 1.2m) cell every time.

Systems Using Proximity

One of the earliest reported indoor location systems is the Active Badge system described in Want et al. (1992). This system used mobile diffuse infrared transmitters and a fixed grid of receivers. There is no angulation or lateration involved; location is determined via proximity. It locates objects at the granularity of the size of a typical room. Many commercially available avalanche transceivers (see e.g., EN282:1997 (1997)) used by backcountry skiers are also based on proximity information. Another example of a proximity-based location system is the RFID system, which is becoming increasingly popular. These systems use an RFID tag, active or passive, on the entity being tracked. An active tag can transmit its ID when in the vicinity of an RFID reader. When a passive tag is in the vicinity of reader, it reflects the incident RF energy with its own ID information superimposed on it. This information can be used to infer the location of the tagged terminal. These tags can be usually read from a range of 3ft (0.9m) for passive tags to 20ft (1.6m) for active ones.

SYSTEMS FOR AD HOC AND WIRELESS SENSOR NETWORKS

It is instructive to understand, at a high level, the general principles used in localization in wireless sensor networks. Wireless sensor networks are a special class of ad hoc networks and have been an active area of investigation and continue to be so. The interested reader is referred to the survey in Akyildiz et al. (2002) for an introduction. These networks typically operate in environments where there is no available infrastructure and it is desirable to minimize centralized processing and control. In these systems, all mobile terminals have similar sensors and processing capabilities. In such environments, localization techniques that do not depend on an infrastructure are needed. By exchanging information with their neighbors and engaging in distributed computing, the mobile terminals converge to an estimate of nearby objects' positions. These techniques use some combination of proximity, triangulation, and scene analysis.

There is a wealth of literature on localization in wireless sensor networks that is beyond the scope of this chapter. What follows is a brief introduction to this substantial literature.

Systems using proximity have been described in Bulusu et al. (2000). In that paper, it is assumed that there are a fixed number of reference nodes arranged in a regular mesh which transmit a periodic beacon signal with period T. These nodes have an overlapping coverage region. The localization technique is based on connectivity. The beacon transmissions of the reference nodes are synchronized such that there is exactly one beacon from each reference node in any time interval T. A node listens for a specified time interval t and computes a connectivity metric that is the ratio of the number of beacon transmissions heard from a reference node to the total number of beacons transmitted by that node in that interval. If this ratio exceeds a threshold, then that reference node is considered *connected* to the node being localized. The location of the node is then computed as the centroid of the locations of all the reference nodes that are connected to this node.

A more general method using connectivity-induced constraints is due to Doherty et al. (2001). This method removes the constraint of a regular mesh for reference nodes and introduces a more generic constraint model. As before, the locations of a few reference nodes are assumed to be known. If two nodes are in communication, the distance between them is constrained to be less than or equal to the radio range of the terminals. Given the known locations and the communication graph (i.e., who can hear whom), the problem of localization is converted to a feasibility problem of satisfying all the constraints imposed by the communication graph. This formulation requires centralized processing.

Unlike the systems considered above, the SpotON system described in Hightower et al. (2000) uses lateration based on measured distance between low-cost radio tags where the measurement is based on received signal strength information. The use of received signal strength for localization will be elaborated in the rest of the chapter. Some other examples of localization systems for wireless sensor networks may be found in Albowicz et al. (2001), He et al. (2003), and Savvides et al. (2001).

USING RECEIVED SIGNAL STRENGTH FOR LOCALIZATION

Our emphasis in this chapter, as described earlier, is signal strength-based localization. We now outline the general principles used in signal strength-based localization and categorize techniques to help in understanding them. It is to be noted that these principles apply as well to systems using signal strength for localization in wireless sensor networks. In this case, some of the sensor nodes are at known locations, thus taking on the role of access points in indoor WLAN environments. In the sequel, it is to be understood that wherever there is a reference to access points, it could be substituted by transmitters in a sensor network.

Please refer to Figure 2 for an explanation of terms used in this section. The access points (or sensor transceivers) are marked as AP_i. The profiled signal strength vector at location *i* is s_i whose components are the received signal strengths (a deterministic value or a random variable with a specified distribution) from the different access points. For example, $s_{i,1}$ is the received signal strength at location *i* from the access point AP_1.

We classify localization techniques along two dimensions: one is based on the nature of the computational technique used and the other is based on where the measurements are collected. More specifically, computational techniques are classified as *deterministic* or *probabilistic* while measurements may be collected at the *client* or by the *infrastructure*. We elaborate on these classifications below.

Deterministic and Probabilistic Techniques

In one approach, the received signal strength measurements can be used to determine the range to the transmitter (based on propagation models) and then for localization via multi-lateration. Another approach is to build an *a priori* radio signal strength map of the region under consideration and determine location based on the measured signal strengths using some optimization criterion. For example, the minimum euclidean distance in the signal-strength vector space can be used as the optimization criterion as described in Bahl and Padmanabhan (2000a). We call such techniques *deterministic*. As described in Rappaport (1996), the received signal strength is a random process that may be modeled by a log-normal distribution. Based on this fact, many probabilistic approaches have been proposed. In these approaches, a signal strength probability distribution based on profiling measurements is determined on a lattice of locations superimposed upon the area under consideration. When attempting to determine the unknown location of a terminal, the received signal strength measurement is estimated using probabilistic methods such as maximum likelihood estimation. Instead of profiling, it is possible to postulate a propagation model and estimate the posterior distribution of the parameters and the location based on the measurements and a specified prior distribution. All such techniques may be termed *probabilistic*.

1. **Deterministic techniques:** One example of deterministic techniques is RADAR described in Bahl and Padmanabhan (2000a). Earlier, Christ and Godwin (1993) proposed a conceptually similar technique using custom hardware and non-802.11 radios. In RADAR, the area under consideration is divided into cells and a radio profile is obtained. The radio profile assigns a signal strength vector to each cell; the components of the vector are the measured received signal strengths from fixed radio access points. To determine the location, various flavors of nearest neighbor algorithms are employed. In Bahl and Padmanabhan (2000a), a multi-lateration approach is also described. In this approach, the profiling data is used to compute the parameters of a propagation model. The computed model is used to convert received signal strength measurements into range values. Subsequently, multi-lateration is used to estimate the location. The authors report an accuracy of 2.9-4.3m (50%).

Figure 2. Basic elements of analysis based on received signal strength

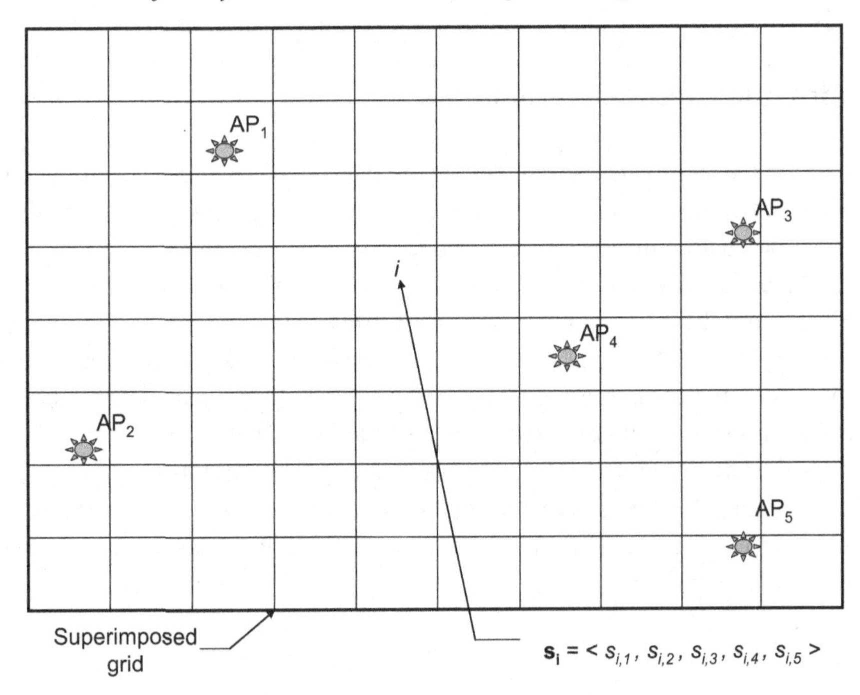

2. **Probabilistic techniques:** Several types of probabilistic techniques have been proposed in the literature. One class of techniques requires profiling; see for example, Ladd et al. (2002). In this approach, signal strength measurements are made for profiling purposes. However, each cell is assigned a probability distribution for the signal strength values instead of a single vector value. The measured signal strength value and the prior distributions are then used to compute a maximum likelihood estimate of the location. Other estimators could also be used, e.g., the mean value of a Bayesian posterior distribution (based on an assumed prior distribution for the location). Other probabilistic approaches may be found in Abnizova et al. (2001) and Myllymäki et al. (2001). A system using probabilistic techniques is the HORUS system described in Youssef and Agrawala (2005), Youssef et al. (2003), and Youssef and Agrawala (2004a), which has as its goals high accuracy and low computation. Location clustering is used to reduce computational burden. This system addresses the causes of wireless channel variations to improve accuracy. Signal strength distributions instead of single values are used in its radio map. The reported accuracy of the system in Youssef et al. (2003) is 7ft (2.1m) (90%) and 3.5ft (1.1m) (50%). In Youssef and Agrawala (2004b), an argument is provided to show that probabilistic techniques can provide better accuracy than deterministic ones.

Another class of probabilistic techniques does not require any prior profiling. In this case, measurements from multiple terminals are used to estimate all their locations simultaneously. This is done by assuming a parameterized propagation model and estimating both the parameters of the model and the locations of the terminals simultaneously. This technique has the added advantage that it automatically adapts to changing radio environments. Recently, the use of Bayesian networks to perform this computation has been reported in Madigan et al. (2005). An earlier example of the use of Bayesian networks for location is the Nibble system described in Castro et al. (2001). Nibble was used only to localize the radio terminal at room-level granularity as opposed to a more accurate estimate of the actual location described in Madigan et al. (2005). Also, Nibble locates terminals individually whereas the method in Madigan et al. (2005) estimates the location of multiple terminals simultaneously.

Client- vs. Infrastructure-Based Systems

Another way to group the techniques is based on where the signal strength measurements are made: at the terminal being located or at other terminals in the network, e.g. sensors, sniffers or access points. We call these client-based and infrastructure-based techniques respectively.

In this grouping, techniques such as RADAR and those described in Ladd et al. (2002), Myllymäki et al. (2001) will fall into the client-based group. An infrastructure-based system called LEASE was described in Krishnan et al. (2004) and further developed in Ganu et al. (2004), and is outlined in a later section. Note that most infrastructure-based techniques can be deployed as client-based techniques. Similarly, several client-based techniques can be adapted to be infrastructure-based. The main issue in such adaptations is the amount of data needed by the technique to build its models.

In Table 1, we present a 2 x 2 matrix that groups some of the reported techniques based on received signal strength measurements according to these two classifications: deterministic/probabilistic and client-based/infrastructure-based.

Table 1. Grouping of localization techniques

	Client-based	Infrastructure-based
Deterministic	Bahl and Padmanabhan (2000a) – *RADAR* Prasithsangaree et al. (2002)	Krishnan et al. (2004) – *LEASE*
Probabilistic	Youssef et al. (2003) Youssef and Agrawala (2004a) – *HORUS* Madigan et al. (2005) – *Bayesian Nets* Abnizova et al. (2001)	Madigan et al. (2005) – *Bayesian Nets* Castro et al. (2001) - *Nibble*

LOCALIZATION ACCURACY

As pointed out earlier, the fact that received signal strength (RSS)-based techniques can be implemented using existing hardware (off-the-shelf) makes them extremely attractive, leading to significant research in RSS-based localization. Even though signal strength measurements are available for "free," they are not the easiest to use. Multi-path propagation in indoor environments makes working with signal strength measurements challenging. The terminals and associated network cards are also heterogeneous, adding to the complexity of the problem. There is also the issue of the correct metrics to use when analyzing location accuracy. For example, is distance error the correct metric? Or, is it a room level accuracy that is desired? Notwithstanding these issues, the median estimation error has traditionally been popular amongst researchers in experimental studies of localization.

Several techniques have been proposed and experimentally evaluated for localization using signal strength. For example, as mentioned earlier, the RADAR system from Bahl and Padmanabhan (2000a) exhibited a localization accuracy of 2.9-4.3m (50%) and Youssef et al. (2003) reported an accuracy of 7ft (2.1m) (90%). The accuracy of the LEASE system from Krishnan et al. (2004) was 7-15ft (2.1-4.5m) (50%). However, these studies were done in different experimental environments.

Recently, some interesting studies have compared various techniques on the same experimental test bed (see Elnahrawy et al. (2004) and references therein). A noteworthy observation was presented in Elnahrawy et al. (2004): over a range of algorithms, approaches, and environments, there appeared to be limits to achievable localization accuracy. In particular, a median localization error of 10ft (3m) and 97th percentile of 30ft (9.1m) was generally observed. More specifically, the focus in Elnahrawy et al. (2004) is on area-based algorithms that trade accuracy (the likelihood that an object is within an area) for precision (the size of the returned area). The authors present and evaluate three area-based algorithms, determining accuracy and precision at the distance and room level. Using data from two sites (a university building and an industrial setting), they show that a wide range of area-based algorithms have similar fundamental performance. They then compare against point-based algorithms – specifically choosing the algorithms from Bahl and Padmanabhan (2000a) and Roos et al. (2002) (with some possible variants) as representative of deterministic and probabilistic strategies respectively. A key result was that the proposed area-based and chosen point-based techniques had "striking similarity" in performance graphs (apart from some Bayesian techniques that had lesser localization accuracy).

To understand if there was something more fundamental that dictated the similarity in localization accuracy results, they studied uncertainty probability density functions along the x and y axes generated from their Bayesian network. The wide distributions observed, especially in the industrial data set, were indicative of a high degree of uncertainty, which they concluded was more fundamental rather than an artifact of the techniques alone. They also allude to results from Battiti et al. (2002) that found that a host of learning approaches had similar performance to maximum likelihood estimation (as in Roos et al. (2002)).

An interesting question raised by these studies is: "Is there any theoretical limit to the localization accuracy using techniques based on signal strength?" To understand this, we present an analytical framework that tries to ascertain the attainable accuracy of such techniques.

ANALYTICAL UNDERSTANDING OF SIGNAL STRENGTH-BASED LOCALIZATION

There is limited analytical work attempting to understand the fundamental issues governing localization accuracy. We present below a synopsis of technical work dealing with this issue.

In Youssef and Agrawala (2004b), the authors developed an analytical framework for calculating the average distance error and the probability of error in location. They showed that probabilistic decision techniques can provide more accurate localization than deterministic ones, since unlike deterministic techniques, probabilistic methods can take into account that the signal strength vector is not always symmetric and identical at all locations. In Krishnakumar and Krishnan (2005), the authors analyzed the fundamental limits of the accuracy of localization using signal strength measurements. The received signal strength is a stochastic variable due to the effect of multi-path propagation. The main intuition in Krishnakumar and Krishnan (2005) was to recognize that a variation in measured signal strength due to change in location is indistinguishable from a variation due to shadowing. Hence, any decision rule will map a set of locations in the neighborhood of a point (x,y) to the point (x,y). Intuitively, this is the *uncertainty* in the location estimate caused by signal variance. The main idea was to define a quantity that captured this uncertainty in localization and derive an expression for this quantity under very general assumptions by mapping uncertainty in signal strength space to location uncertainty in the (x,y) plane. Various features of uncertainty were then analyzed, and also specifically for a log-linear radio propagation model. The lower bound derived matched favorably with experimental work providing an analytical explanation for the observations in Elnahrawy et al. (2004). We now provide more details of this work.

Theoretical Accuracy Limits

Assume that a signal strength vector $\vec{s}$ of dimension n is used to locate a terminal. This vector can be determined, for example, from n access points, sniffers or sensors, depending on whether a client-based or infrastructure-based deployment is used. The received signal strength is a stochastic variable due to multi-path propagation. As explained in Rappaport (1996), the logarithm of the received signal strength can be modeled as a normal distribution around a mean value with variance σ_i^2, $1 \leq i \leq n$. The σ_i can be different from the shadowing variance if filtered signal strength measurements are used. A common method to deal with short-term variations in signal strength due to fast fading is to use the median of a few uncorrelated measurements. It is reasonable to assume that the individual components of the signal strength vector $\vec{s}$ are independent and that the mean signal strength at a location is a differentiable function over a region of interest.

The approach in Krishnakumar and Krishnan (2005), summarized in Figure 3, is to define an α-region in location space such that the total probability that the observed signal strength is due to an emitter located at some point in the region is α. (There may be more than one such region satisfying the condition.) As shown in the figure, let $\mathcal{T}$ be a mapping from location to mean signal strength. Restricting attention to the cases when $\mathcal{T}$ is one-to-one (e.g., when propagation loss is a monotonic function of distance from the emitter as with inverse exponential propagation functions) and when signal variance distribution is symmetric, we can compute the characteristics of the α-region in location space by mapping it to the signal space and then back to the location space.

To do this, one first computes the characteristics of the hypervolume in signal strength space centered on the mean signal strength vector that encloses a probability mass of α. It is shown that the hypervolume is a hyperellipsoid with semi-axes $R_n\sigma_i$, where R_n is a scaling factor related to the confidence level α. This relationship is given by

$$\alpha = \frac{\Gamma(n/2, R_n^2/2)}{\Gamma(n/2)},$$

where Γ(· , ·) is the incomplete gamma function. Given α and n, the equation above can be used to compute R_n. (Details of this derivation appear in Krishnakumar and Krishnan (2005), Appendix II.)

Clearly, there is a mapping between location and mean signal strength. In practice, for signal strength estimation techniques to work, this mapping has some properties; e.g., each location maps to a mean signal strength vector. Using the nature of the region in signal strength space as derived earlier, the analysis then derives the structure of the uncertainty in location. In particular, it is shown that the uncertainty region in the (x, y) plane is an ellipse whose equation is $ax^2 + by^2 + cxy + d = 0$, where

Figure 3. Mapping uncertainty in signal strength space to uncertainty in location

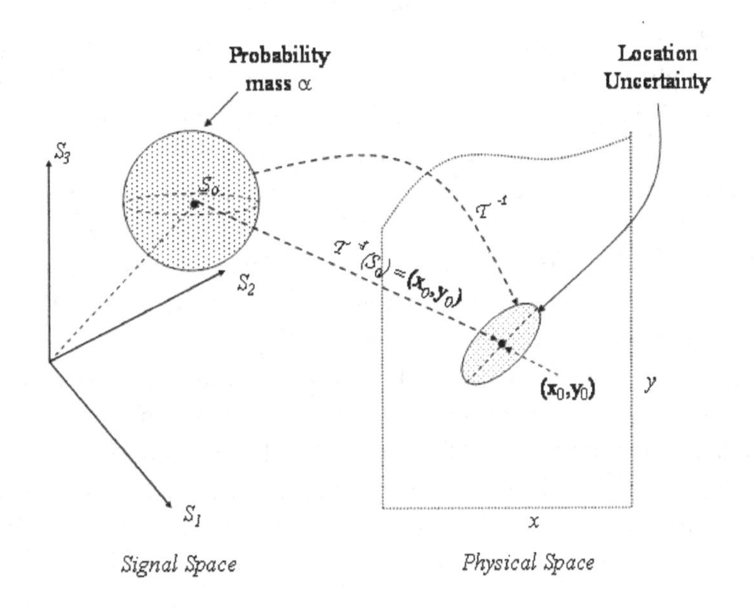

$$a = \sum_{i=1}^{n} \frac{t_{i1}^2}{\sigma_i^2} \quad , \quad b = \sum_{i=1}^{n} \frac{t_{i2}^2}{\sigma_i^2}$$

$$c = \sum_{i-1}^{n} \frac{2t_{i1}t_{i2}}{\sigma_i^2} \quad , \quad d = -R_n^2$$

and $T = \{t_{i,j}\}$ is the Jacobian of the mapping $\mathcal{T}$ from location to mean signal strength measurements. The mapping T is an n x 2 matrix. In particular, $t_{i,1} = \partial s_i / \partial x$, and $t_{i,2} = \partial s_i / \partial y$, where s_i is the ith component of the signal strength vector $\vec{s}$.

Properties of the Uncertainty Region

Several interesting properties of the uncertainty region can be determined once its structure is known. The semi-major axis, semi-minor axis and their geometric mean are quantities of interest. In Krishnakumar and Krishnan (2005), they are defined to be the upper, lower, and mean uncertainty, respectively, and shown to be bounded quantities. The maximum uncertainty over the convex hull bounded by the access points for various configurations was studied assuming a log-linear radio propagation model. Several interesting characteristics that were observed are summarized here. For example, the variation in uncertainty when the APs (transmitters) are at the vertices of an equilateral triangle or a regular square, both inscribed in a circle of radius 100 units, is shown in Figure 4. The figures show the contours of equal uncertainty on the surface and their projections on the (x, y)-plane.

There are several open problems in understanding and analytically determining the properties of uncertainty in the region of the convex hull bounded by the APs. For example, it appears that there is at least one location of locally minimum uncertainty in the convex hull of the AP locations. Is this always so, and is an analytical expression for the location and values of such minima obtainable? Is there a provable number of such minima in the convex hull region? What happens when the APs are not uniformly located? And so on.

The expressions also led to other fundamental observations. For example, for confidence levels above 0.8 (i.e., $\alpha > 0.8$) while keeping all other quantities unchanged, the uncertainty increases disproportionately. Under simplifying assumptions, it is shown that uncertainty is proportional to the variance in signal strength. This dependence is important to understand because it is a factor that can be influenced by the localization algorithm. Several algorithms effectively reduce this variance to improve the localization performance. This may be achieved by using multiple samples as described in Bahl and Padmanabhan (2000a) and Krishnan et al. (2004), probabilistic techniques as described in Ladd et al. (2002) and Youssef et al. (2003), or autoregressive models as described in Youssef and Agrawala (2004a). The tradeoff between computational complexity and the method used for variance reduction must be understood when considering the use of a technique.

The deployment of APs or sniffers affects the attainable accuracy in determining location. It is, therefore, important to understand this relationship. The analysis example in Krishnakumar and Krishnan (2005) shows that if APs are added to the same area, the uncertainty decreases. However, if the area of coverage is increased while increasing the number of APs, the minimum uncertainty increases or remains stable in some cases. The evaluation of the variation of minimum uncertainty for different AP/sniffer placement strategies is an interesting open issue. Uncertainty is also related to the circular probability; the reader is referred to Krishnakumar and Krishnan (2005) for more details omitted from this summarization.

Experimental Validation

It is instructive to use actual measurements from an indoor 802.11b network to calculate the parameters needed to compute minimum uncertainty. This was done in Krishnakumar and Krishnan (2005), where the computed minimum uncertainty was compared to the reported median errors in the experimental literature. Details may be found in the original reference. Considering an 802.11b network where 3 APs are located at the vertices of an equilateral triangle of side 150 ft (45.7m), the minimum uncertainty location is at the centroid of the triangle and is computed to be approximately 4.5ft (1.4m), using non-preprocessed

Figure 4. Max. uncertainty (in ft) in the region enclosed by n APs; $\alpha = 0.75$, and $\sigma_i = 0.707$, $\forall i$ (reprinted with permission from Krishnakumar, A.S., and Krishnan, P., On the Accuracy of Signal Strength-Based Location Estimation Techniques. Proceedings of the 2005 IEEE Infocom Conference, © 2005 IEEE).

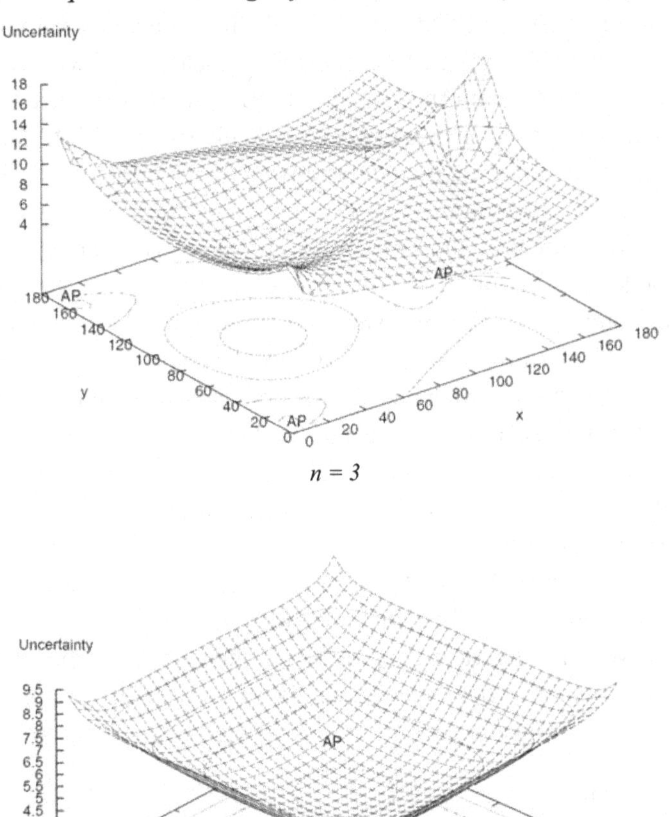

signal measurements to compute variance. This computed minimum uncertainty compares favorably with the reported median error values in the experimental literature, e.g., RADAR (≈ 2.9m) (Bahl and Padmanabhan (2000a)), LEASE (2.1 – 4.5m) (Krishnan et al. (2004)), HORUS (≈ 1.1m) (Youssef et al. (2003)). Therefore, the uncertainty appears to be a realistic indicator of the median error. The lower value using the probabilistic technique in Youssef et al. (2003) could be attributed to the preprocessing of the measurements leading to a lower signal variance and to the smaller distance between the APs. This corroborates in a different way some of the observations in Elnahrawy et al. (2004).

Experimental results obtained from systems with widely varying parameters are difficult to compare. Elnahrawy et al. (2004) address this problem by comparing them on the same platform. An alternative is to evaluate the effect of the various algorithms on the raw parameters that feed into the uncertainty computation and determine how the lower bound will be affected as was done in Krishnakumar and Krishnan (2005).

An Analysis Using Cramér-Rao Bounds

Motivated by a security aspect of localization, a lower-bound on the variance of the position of a node in the 2-dimensional plane is derived in Malaney (2004). In the application considered by Malaney, terminals are equipped with a GPS system and report their location. However, the terminals may not be trusted to report accurate location information. The problem here is to verify if the supplied location information seems trustworthy, based on the network's own internal signal strength measurements. The problem becomes one of verifying rather than determining the location of a terminal.

In trying to assess if a claimed location by a terminal is indeed feasible given the network's signal strength measurements, Malaney used the Cramér-Rao bound to compute the lower bound on the localization variance. The location reported by the terminal, signal strength measurements made by the network, and the lower-bound on localization variance are then used to compute the confidence level of the location reported by the terminal. While the form of the analysis in Krishnakumar and Krishnan (2005) has parallels to the analysis in Malaney (2004) and Malaney's analysis effectively computes a lower-bound on the localization accuracy, the use of the technique in Malaney (2004) is for different purposes and the two analyses are not the same. Malaney's analysis technique is summarized here in the spirit of being another method to measure and use localization accuracy estimates that can also be used to evaluate localization estimators.

Specifically, in Malaney (2004), the author assumes the log-normal distribution for received signal strength. By then writing the expression for distribution of signal strength received at a point (x_0, y_0) from a node at position (x_i, y_i), Malaney derives an expression to calculate the terms of the Fisher information matrix. If b denotes $(10k/(\sigma \ln 10))^2$, where σ is the standard deviation of the shadowing (in dB), and k is the environment-dependent path-loss exponent, the Fisher matrix turns out to be:

$$\begin{pmatrix} b\sum_{i=1}^{n} \frac{\sin^2 \varphi_i}{d_i^2} & \frac{b}{2}\sum_{i=1}^{n} \frac{\sin 2\varphi_i}{d_i^2} \\ \frac{b}{2}\sum_{i=1}^{n} \frac{\sin 2\varphi_i}{d_i^2} & b\sum_{i=1}^{n} \frac{\sin^2 \varphi_i}{d_i^2} \end{pmatrix}$$

where d_i refers to the distance from the node of unknown position (x_0, y_0) to a node of known position (x_i, y_i), n is the number of dimensions in the signal vector, $\sin\varphi_i = (x_i - x_0)/d_i$ and $\cos\varphi_i = (y_i - y_0)/d_i$

Malaney then computes the inverse of the Fisher matrix, and the sum of the diagonal terms of this matrix provides the Cramér-Rao bound on the variance of the position of a node in the 2-dimensional plane. This quantity, σ^2_{CR}, is derived to be

$$\sigma^2_{CR} = \frac{\sum_{i=1}^{n} \frac{1}{d_i^2}}{\left(\frac{10k}{\sigma \ln 10}\right)^2 \left(\sum_{i=1}^{n-1}\sum_{j=i+1}^{n} \frac{\sin^2(\varphi_i - \varphi_j)}{d_i^2 d_j^2}\right)}.$$

The estimate computed above is used to determine if a terminal is misrepresenting its location by verifying the best estimate of the actual position using the derived Cramér-Rao bound, and, more importantly, figuring out how many standard deviations the terminal is away from its claimed position. It is assumed that authorized nodes could pick up the signal strength readings from the terminal and these could be employed by the algorithm for detecting violations.

MANAGING THE COMPLEXITY OF DEPLOYMENT AND OPERATION

In addition to the aspects previously considered that focused on localization accuracy, there are other issues to be taken into account. One such issue is the effect of model variation with environmental changes and the need to adapt the models to preserve localization accuracy. Another is the cost and topology of deployment to obtain the best coverage at the least cost. The sections that follow expand upon these topics.

Adaptation to Environmental Changes

In this subsection, the cost and complexity of building and maintaining the model in changing environments is considered. Even in normal office environments, changing environmental, building, and occupancy conditions could have an effect on signal propagation models as observed in Bahl et al. (2000b). This variation could be due to environmental changes such as rearranged furniture, seasonal occupancy changes, or structural changes such as addition or removal of temporary walls. More rapid changes may be expected in dynamic environments such as an operational warehouse with moving forklifts. It is a challenge to keep the model adapted so that the results remain reasonably accurate. Moves, additions, and changes of transmitters may also require the model to be rebuilt. The models are difficult to maintain and update if purely static techniques are used. We discuss techniques to deal with this problem in this section. Additionally, as pointed out in Smailagic et al. (2001), profiling involves an upfront cost and effort to deployment, and adds to the complexity of maintaining the model. In this context, one seeks simple non-parametric models that can be built with little or no profiling and achieve localization accuracy comparable to techniques that profile the site extensively. This is discussed in the next subsection on zero profiling techniques.

The complexity of building and maintaining the model was identified in Bahl et al. (2000b) and Smailagic et al. (2001). In indoor environments, multi-path propagation is accommodated by using an appropriate exponent in the inverse power law propagation model. This exponent has to be determined experimentally. The exponent could change over time due to changes in the physical environment of the site. Even if a propagation model is not used explicitly, an equivalent underlying radio map has to be determined. This map is also time-variant as with the propagation model. Some techniques have been attempted to address this problem. In Bahl et al. (2000b), an appropriate model from a database of models was chosen based on the reference signal strength seen between access points. The model building problem was tackled to some extent in Smailagic et al. (2001) where a specific functional relationship between signal strength and distance was generated empirically for their site. In practice, the measured signal strength contours are usually anisotropic, unlike the circularly symmetric functions used in Smailagic et al. (2001). In Pandey et al. (2005), a client-assisted data collection scheme to address the issue of dynamically updating the information needed for the radio profile is presented. Their system uses sniffers and client software to build a database of signal strength maps that are continually updated.

The LEASE system to address the model adaptation issue was presented in Krishnan et al. (2004). The LEASE system comprises three main components: stationary emitters (SEs), sniffers, and a location estimation engine (LEE). In Figure 5 we show a possible office site floor with some access points (APs/transmitters), SEs, and sniffers. The LEE can be located anywhere in the network.

The SEs in the LEASE system are standard, inexpensive wireless transmitters that emit a few packets occasionally. The SEs are of small form factor and usually battery-powered. The sniffers sniff on the wireless medium, cycling through a set of specified frequencies and listen for all communication from wireless clients and SEs. They record the received signal strength (RSS) from them. This information is sent to the LEE. The LEE also needs the coordinates of the SEs which could be broadcast by the SE. Using the SEs as "fixed points" of known location, a localization model is built by the LEE as needed, based on the RSS from the SEs. In a sensor network, different sensors could take on the roles of SEs or sniffers.

The LEE builds a model for each sniffer as follows. First, it smoothes the data points, e.g., using a generalized additive model (GAM) (see Hastie and Tibshirani (1990)). Second, a synthetic model is generated. The site is divided into small grids (e.g., grids of 3ft × 3ft (0.9m × 0.9m) cells). Using Akima splines (see Akima (1996a), Akima (1996b) and Akima (1970)) the smoothed values obtained from the GAM are interpolated to estimate the RSS at each grid center. The synthetic model for the specific sniffer is the generated RSS-grid information with an estimated RSS for each grid point. Repeating the above technique for each of the n deployed sniffers, gives a set of grids for the site, where each grid has an associated n-vector of estimated RSS. This n-vector corresponds to the profiled RSS from each AP as seen at each grid point, assuming the APs and sniffers are co-located. The LEE also uses absence of a signal in localization. This is useful since the absence narrows the search area by indicating that the point in question is far away from the sniffer. A variation of the nearest-neighbor algorithm is used for matching the received client RSS vector to the model.

One main result in Krishnan et al. (2004) was that for an office site of size 30,000 sq ft. (2,787 sq. m), with only 12 SEs, a median error of 15ft (4.5m) can be obtained. Increasing the number of SEs does reduce the error further; e.g., with 104 SEs, the median error was just 7ft (2.1m). Different sites also exhibited different accuracies. The authors also motivated a normalized error metric that takes into account the work done in building the model, the localization errors, how dynamic the signal environment is, etc. They showed that the LEASE technique is efficient in terms of the normalized error metric.

Figure 5. A possible office site with components of the LEASE system

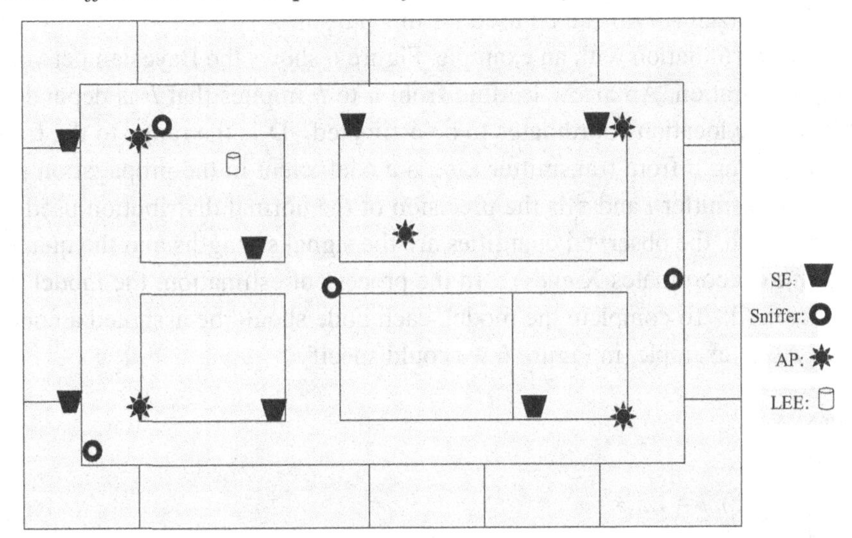

Zero-Profiling Techniques

Localization systems that make use of an explicit propagation model, or a radio map of the region of coverage, require initial profiling. This can be a labor-intensive process. Although automated systems to perform this measurement have been reported in the literature (e.g., Hills and Schlegel (2004)), it still represents an additional cost for deployment. This brings up the feasibility of zero-profiling techniques. As the name implies, these techniques will require no profiling and can be deployed out-of-the-box. As such, they can be used to advantage for the following purposes: a) adaptation in dynamic environments, and b) lowering the cost of deployment. One such technique making use of Bayesian networks was described in Madigan et al. (2005). While this approach does not require profiling, it is computationally expensive and needs further work to be suitable for real-time operation. This remains an interesting area of research. We briefly describe the Bayesian approach below.

In Madigan et al. (2005), instead of trying to locate a single terminal, the model tries to *simultaneously* locate a set of terminals. By appropriately exploiting signal strength information from a collection of terminals, it is shown that the localization for the entire set can be improved. The model is particularly relevant as the number of wireless terminals increases. The methodology uses hierarchical Bayesian graphical models (see Gelman et al. (2003) and Spiegelhalter and Lauritzen (1990)) for wireless localization. The study demonstrates that a hierarchical Bayesian approach, incorporating physical knowledge about the nature of Wi-Fi signals, can provide accurate location estimates without *any* location information in the training data, leading to a truly adaptive, zero-profiling technique for localization.

A *graphical model* is a multivariate statistical model embodying a set of conditional independence relationships. A graph displays the independence relationships. The vertices of the graph correspond to random variables and the edges encode the relationships. In the Bayesian framework, model parameters are random variables and appear as vertices in the graph. When some variables are discrete and others continuous, or when some of the variables are latent or have missing values, a closed-form Bayesian analysis generally does not exist. Analysis then requires either analytic approximations of some kind or simulation methods. The Markov chain Monte Carlo (MCMC) simulation method as described in Spiegelhalter and Lauritzen (1990) can be used for this purpose.

We illustrate this formulation with an example. Figure 6 shows the Bayesian network graph for the problem under consideration. An arrow leading from *a* to *b* implies that *b* is dependent on *a*. In this model, X and Y are the location coordinates to be estimated. D_i is the range to the transmitter *i*, S_i is the received signal strength from transmitter *i*, b_{ij} is a coefficient in the propagation model for signal transmission from transmitter *i* and τ_i is the precision of the normal distribution used to model signal strength variation. Here, the observed quantities are the signal strengths and the quantities to be estimated are the position coordinates X and Y. In the process of estimation, the model variables b_{ij} and D_i are estimated as well. To complete the model, each node should be assigned a conditional density given its parents. As an example, in Figure 6 we could specify[2]:

$X \sim \text{uniform}(0,L)$
$Y \sim \text{uniform}(0,W)$
$S_i \sim N(b_{i0} + b_{i1} \log D_i , \tau_i), i = 1,2,3,$
$b_{i0} \sim N(0,0.001), i = 1,2,3,$
$b_{i1} \sim N(0,0.001), i = 1,2,3.$

Here, L and W are the length and width of the rectangular area under consideration. The signal strength (measured on a logarithmic scale) decays approximately linearly with the logarithm of distance. The notation $N(\mu, \tau)$ indicates a normal distribution with mean μ and precision τ. If we have prior information about the values of b_{i0} and b_{i1}, they could be incorporated via the mean value.

The model shown in Figure 6 treats the model parameter set for each transmitter as independent from each other. While this is quite general, it can also be computationally expensive. Although the propagation model of each transmitter has different model parameters, we could model them all as stochastic variables that are identically distributed with a common mean and variance (or equivalently, precision). This is a hierarchical model and can simplify the computational burden without sacrificing predictive accuracy. The Bayesian network graph for the hierarchical model is shown in Figure 7. For additional details and experimental results, the reader is referred to Madigan et al. (2005).

Deployment for Coverage and Localization

The location of the transmitters and the powers at which they transmit are chosen to satisfy certain objectives in any deployment. Most common is a minimum-cost deployment that ensures signal strength above a certain threshold over the service area, i.e. coverage. Another criterion could be the minimization of communication cost as in Kasetkasem and Varshney (2001). Other considerations may come into play. For example, for adequate coverage it may be desired that at any given point in the coverage area at least two transmitters with adequate signal strength be visible. Maximum coverage at minimum cost has been the main metric in wireless network deployment. There is a wealth of literature on placement optimization for coverage in wireless sensor networks, see for example, Meguerdichian et al. (2001) and Zou and Chakrabarty (2004).

However, a deployment that satisfies this criterion may not be optimal for localization, and may in fact be inadequate. As an example, visibility of three transmitters at any point is required for localization. Consider a long, rectangular area to be covered, as in Figure 8. From a coverage perspective, a line of

Figure 6. Bayesian network graph for localization

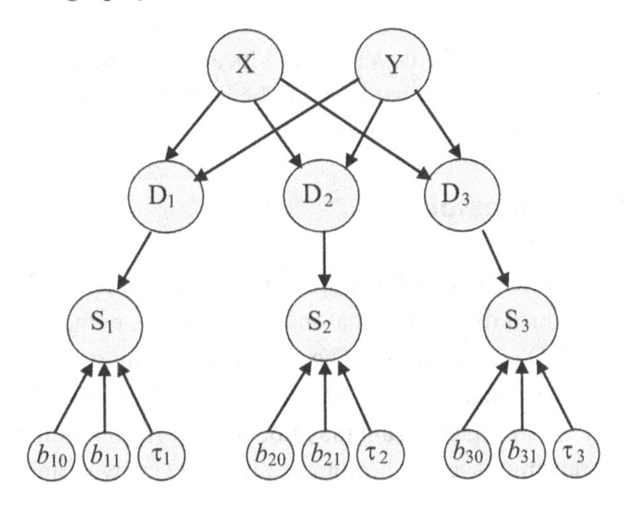

Figure 7. Graph of a hierarchical Bayesian network for the localization problem

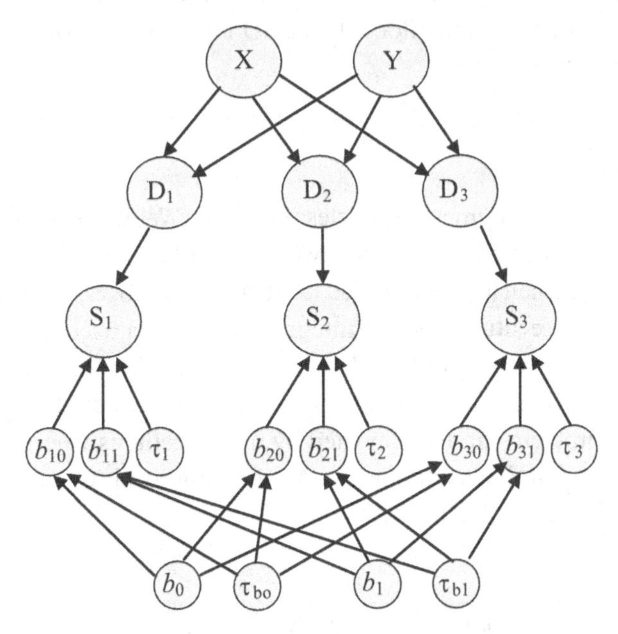

transmitters running down the middle parallel to the long side would be optimal, and will minimize the amount of signal leaked to the outside. However, this is not a good configuration for localization, since there is no way to disambiguate locations that are symmetrically situated about this axis of transmitters. For example, the location of the two terminals shown in Figure 8 can not be disambiguated. Placement optimization for localization has not been studied as extensively as that for coverage. Joint optimization for coverage and localization is even less studied. An example can be found in Chakrabarty et al. (2002). This remains an interesting topic for further research.

OTHER ISSUES IN LOCALIZATION

Apart from the problem of deployment for coverage and localization described earlier, there are other issues in signal strength-based localization that have not been studied extensively and present opportunities for research. Three categories of issues are summarized below.

Dealing with the Third Dimension

While most researchers have concentrated on locating a terminal on a building floor (i.e., two dimensional localization), there is in fact a third dimension that may be discrete, namely, which floor the terminal is on. One could call this "2.5-D" localization. The common assumption is that a floor attenuates a signal significantly and that the strongest signal will be from the floor on which the terminal resides. However, the problem is that the strongest signal may not be seen from a transmitter on the same floor. As an example, consider a transmitter at the bottom of an open stairwell and a terminal at the top of the stairwell.

A heuristic for floor estimation was described in Krishnan et al. (2004). The main idea was to take into account all visible transmitters (or sniffers that can see transmissions from a terminal). The heuristic uses modified majority logic to estimate the floor. The 2.5D localization problem is an interesting one that the reader may want to pursue.

Security Considerations

Security is a major concern in the deployment of wireless networks. While many of the security problems with WEP encryption etc. have been addressed by newer standards such as IEEE 802.11i (2005), there are a class of problems that arise when location is used for security purposes. For example, access control may be based on location. In these situations, the integrity of the data used to perform the estimation becomes very important. In the case of RSS-based methods, this means the integrity of RSS measurements. If client-based reporting is used, it is possible for a malicious terminal to report false values and thus spoof its location. This problem can be alleviated by using methods where the measurements are obtained by sniffers, sensors or similar components. Another approach to evaluating the integrity of client-reported data may be found in Malaney (2004).

Figure 8. Deployment for coverage leading to ambiguity in location

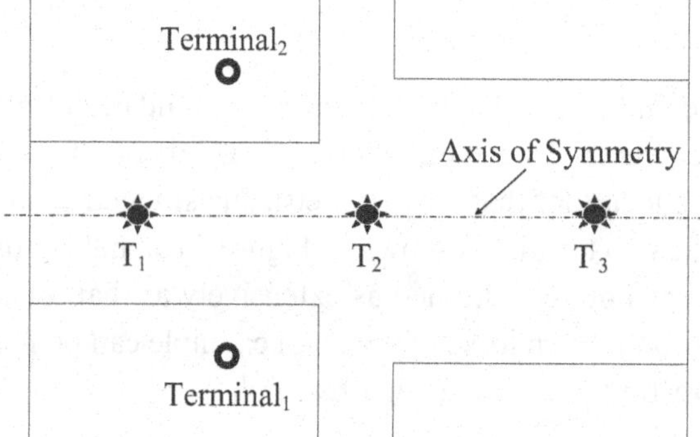

Even if such methods are used, a client terminal using directional antennas can throw off the localization algorithm. A malicious terminal could use a directional antenna to project itself into a secure space. While it can be easy to determine that a directional antenna is being used inside the region of coverage, such use is difficult to determine without additional equipment when the malicious terminal is located outside the coverage area. The use of a small number of reflector antennas located on the periphery of the building, facing outside, could address this issue. To our knowledge, this has not been investigated.

Management

While management is also a cost of ownership issue, there are some interesting questions in management that need to be noted. Does the architecture require additional components to be deployed (hardware and software)? If so, what is the method to manage these components? Does the scheme depend on fail-safe operation of these components? Is the security of the solution adequately addressed? We believe that several of these questions have not been adequately addressed in current work and represent an opportunity for future research in this area.

CONCLUSION

In this chapter, we have given a brief overview of indoor localization techniques. We focused in particular on received signal strength-based techniques for indoor wireless deployments using 802.11. Some of the techniques may be applied to other radio technologies also, e.g., Bluetooth. We summarized work that dealt with theoretical limits to accuracy of localization using received signal strength, including some experimental results. We presented techniques used to adapt to changing environments, and a zero-profiling Bayesian approach for localization. We have listed several other research issues in this area, including deployment, floor estimation, security and management.

This chapter is intended as an overview of signal strength-based localization, the research issues and some results obtained by various researchers. It is not intended to be a comprehensive survey of the vast literature in this area.

REFERENCES

Abnizova, I., Cullen, P., & Taherian, S. (2001). Mobile Terminal Location in Indoor Cellular Multi-path Environment. *Proceedings of the Third IEEE Workshop on Wireless LANs*, retrieved October 13, 2008, from http://www.wlan01.wpi.edu/proceedings/wlan69d.pdf.

Akima, H. (1996a). Algorithm 760: Rectangular-Grid-Data Surface Fitting that has the Accuracy of Bicubic Polynomial. *ACM Transactions on Mathematical Software, 22*(3), 357-361.

Akima, H. (1996b). Algorithm 761: Scattered-Data Surface Fitting that has the Accuracy of Cubic Polynomial. *ACM Transactions on Mathematical Software, 22*(3), 362-371.

Akima, H. (1970). A new method of interpolation and smooth curve fitting based on local properties. *Journal of the ACM, 17*(4), 589-602.

Akyildiz, I. F., Su, W., Sankarasubramaniam, Y., & Cayirci (2002). E. Wireless sensor networks: a survey. *Computer Networks, 38*, 393-422.

Albowicz, J., Chen, A., & Zhang, L (2001). Recursive Position Estimation in Sensor Networks. *Proceedings of the 9th International Conference on Network Protocols*, Riverside, CA, (pp. 35-41).

Bahl, P., & Padmanabhan, V. N. (2000a). RADAR: An In-Building RF-based User Location and Tracking System. *Proceedings of IEEE Infocom, 2*, Tel Aviv, Israel, (pp. 775-784).

Bahl, P., Padmanabhan V. N., & Balachandran, A. (2000b). Enhancements to the RADAR User Location and Tracking System. *Microsoft Research Technical Report*, MSR-TR-2000-12.

Battiti, R., Brunato, M., & Villani, A. (2002). Statistical Learning Theory for Location Fingerprinting in Wireless LANs. *University of Trento, Informatica e Telecomunicazioni*, Technical Report DIT-02-086.

Bulusu, N., Heidemann, J., & Estrin, D. (2000). GPS-less Low Cost Outdoor Localization for Very Small Devices. *IEEE Personal Communications Magazine*, *7*(5), 28-34.

Caffrey, J. J., & Stuber, G. L. (1998). Overview of Radio Location in CDMA Cellular Systems. *IEEE Communications Magazine, 36*(4), 38-45.

Castro, P., Chiu, P., Kremenek, T., & Muntz, R. (2001). A Probabilistic Location Service for Wireless Network Environments. *Proceedings of Ubicomp 2001*, (pp. 18-24). Springer Verlag.

Chakrabarty K., Iyengar, S. S., Qi, H., & Cho, E. (2002). Grid Coverage for Surveillance and Target Location in Distributed Sensor Networks. *IEEE Transactions on Computers*, *51*(12), 1448-1453.

Christ, T. W., & Godwin, P. A. (1993). A Prison Guard Duress Alarm Location System. *Proceedings of the IEEE International Carnahan Conference on Security Technology,* (pp. 106-116).

Doherty, L., Ghaoui, L., & Pister, K. (2001). Convex position estimation in wireless sensor networks. *Proceedings of IEEE Infocom 2001*, (pp. 1655-1663).

Elnahrawy, E., Li, X., & Martin, R. (2004). The Limits of Localization Using Signal Strength: A Comparative Study. *IEEE SECON 2004*, Santa Clara, California, USA, (pp. 406-414).

EN282:1997 (1997). *Avalanche beacons.* Transmitter/receiver systems. Safety requirements and testing, ISBN 0580268233.

Ganu, S., Krishnakumar, A. S., & Krishnan, P. (2004). Infrastructure-based Location Estimation in WLAN Networks. *Proceedings of the IEEE Wireless Communications and Networking Conference (WCNC)*, (pp. 465-470).

Gelman, A., Carlin, J. B., Stern, H. S., & Rubin, D. B. (2003). *Bayesian Data Analysis* (2nd ed.). Chapman and Hall.

Günther, A., & Hoene, C. (2005). Measuring Round trip Times to Determine the Distance between WLAN Nodes. *Proceedings of Networking 2005*, Waterloo, Canada (pp. 768-779).

Harter, A., Hopper, A., Steggles, P., Ward, A., & Webster, P. (1999). The anatomy of a context-aware application. *Proceedings of the 5th annual ACM/IEEE International Conference on Mobile Computing and Networking*, Seattle, WA, USA, (pp. 59-68).

Hastie, T., & Tibshirani, R. (1990). *Generalized Additive Models.* Chapman and Hall.

He, T., Huang, C., Blum, B. M., Stankovic, J. A., & Abdelzaher, T. (2003). Range-Free Localization Schemes for Large Scale Sensor Networks. *Proceedings of MobiCom '03*, San Diego, CA, (pp. 81-95).

Hightower, J., & Borriello, G. (2001). Location Systems for Ubiquitous Computing. *IEEE Computer, 34*(8), 57-66.

Hightower, J., Want, R., & Borriello, G. (2000). SpotON: An Indoor 3D Location Sensing Technology Based on RF Signal Strength. *University of Washington Technical Report*, Seattle, WA, UWCSE 2000-02-02.

Hills, A., & Schlegel, J. (2004). Rollabout: A Wireless Design Tool. *IEEE Communications, 42*(2), 132-138.

IEEE 802.11i Standard (2004). *The IEEE 802.11i Standard,* retrieved October 13, 2008, from http://standards.ieee.org/getieee802/download/802.11i-2004.pdf.

Kasetkasem, T., & Varshney, P. K. (2001). Communication Structure Planning for Multisensor Detection Systems. *IEEe Proceedings on Radar, Sonar and Navigation, 148*, 2-8.

Krishnakumar, A. S., & Krishnan, P. (2005). On the Accuracy of Signal Strength-based Location Estimation Techniques. *Proceedings of the 2005 IEEE Infocom Conference*, Miami, FL, (pp. 642-650).

Krishnan, P., Krishnakumar, A. S., Ju, W. H., Mallows, C., & Ganu, S. (2004). A System for LEASE: System for Location Estimation Assisted by Stationary Emitters for Indoor RF Wireless Networks. *Proceedings of 2004 IEEE Infocom Conference* (pp. 1001-1011).

Ladd, A. M., Bekris, K. E., Rudys, A., Marceau, G., Kavraki, L. E., & Wallach, D. S. (2002). Robotics-Based Location Sensing using Wireless Ethernet. *Proceedings of the Eighth ACM International Conference on Mobile Computing and Networking (MOBICOM)*, (pp. 227-238).

Madigan, D., Elnahrawy, E., Martin, R., Ju, W. H., Krishnan, P., & Krishnakumar, A. S. (2005). Bayesian Indoor Positioning Systems. *Proceedings of the 2005 IEEE Infocom Conference*, Miami, Florida, USA, (pp. 1217-1227).

Malaney, R. A. (2004). A Location Enabled Wireless Security System. *Proceedings of IEEE Globecom*, Dallas, TX, *4*, 2196-2200.

Meguerdichian, S., Koushanfar, F., Potkonjak, M., & Srivastava, M. B. (2001). Coverage problems in wireless ad-hoc sensor networks. *Proceedings of IEEE Infocom Conference*, (pp. 1380-1387).

Moore, T., Hill, C., & Monteiro, L. S. (2002). Maritime DGPS: Ensuring the best availability and continuity. *Journal of Navigation, 55*(3), 485-494.

Myllymäki, P., Roos, T., Tirri, H., Misikangas, P., & Sievanen, J. (2001). A Probabilistic Approach to WLAN User Location Estimation. *Proc. of the Third IEEE Workshop on Wireless LANs,* retrieved October 13, 2008, from http://www.wlan01.wpi.edu/proceedings/wlan18d.pdf.

Nerguizian, C., Despins, C., & Affes, S. (2001). Framework for Indoor Geolocation Using an Intelligent System. *Proceedings of the Third IEEE Workshop on Wireless LANs,* retrieved October 13, 2008, from http://www.wlan01.wpi.edu/proceedings/wlan44d.pdf.

Niculescu, D., & Nath, B., (2003). Ad-hoc positioning system (APS) using AoA. *Proceedings of the 2003 IEEE INFOCOM Conference, 3,* 1734-1743, San Francisco, CA.

Pahlavan, K., Li, X., & Makela, J. (2002). Indoor Geolocation Science and Technology. *IEEE Communications Society Magazine, 40*(2), 112-118.

Pandey, S., Kim, B., Anjum, F., & Agarwal, P. (2005). Client Assisted Location Data Acquisition Scheme for Secure Enterprise Wireless Networks. *Proceedings of the IEEE Wireless Communications and Networking Conference (WCNC),* New Orleans, LA, USA, *2,* 1174-1179.

Prasithsangaree, P., Krishnamurthy, P., & Chrysanthis, P. K. (2002). On Indoor Position Location with Wireless LANs. *Proceedings of the 13th IEEE International Symposium on Personal, Indoor, and Mobile Radio Communications (PIMRC), 2,* 720-724.

Priyantha, N. B., Chakraborty, A., & Balakrishnan, H. (2000). The Cricket Location-Support System. *Proceedings of the 6th annual ACM/IEEE International Conference on Mobile Computing and Networking,* Boston, MA, USA, (pp. 32-43).

Rappaport, T. S. (1996). *Wireless Communication: Principles & Practice.* Prentice Hall.

Roos, T., Myllymäki P., & Tirri, H., (2002). A Statistical Modeling Approach to Location Estimation. *IEEE Transactions on Mobile Computing, 1*(1), 59-69.

Saha, S., Chaudhuri, K., Sanghi, D., & Bhagwat, P. (2003). Location Determination of a Mobile Device Using IEEE 802.11b Access Point Signals. *Proceedings of the 2003 IEEE Wireless Communications and Networking Conference (WCNC), 3,* 1987-1992.

Savvides, A., Han, C., & Srivastava, M. B. (2001). Dynamic fine-grained localization in ad-hoc networks of sensors. *Proceedings of the 7th annual international conference on mobile computing and networking,* Rome, Italy, (pp. 166-179).

Smailagic, A., Siewiorek, D. P., Anhalt, J., Kogan, D., & Wang, Y. (2001). Location Sensing and Privacy in a Context Aware Computing Environment, *Pervasive Computing Conference,* available online from http://citeseerx.ist.psu.edu/viewdoc/summary?doi=10.1.1.24.791, as of October 13, 2008.

Sonnenberg, G. (1988). *Radar and Electronic Navigation.* Butterworths.

Spiegelhalter, D. J. (1998). Bayesian graphical modeling: A case-study in monitoring health outcomes. *Applied Statistics, 47*(1), 115-133.

USDOT (US Department of Transportation) – Federal Highway Administration (2002). Phase I High Accuracy-Nationwide Differential Global Positioning System Report. FHWA-RD-02-110.

Want, R., Hopper, A., Falcao, V., & Gibbons, J. (1992). The active badge location system. *ACM Transactions on Information Systems, 10*(1), 91-102.

Youssef, M., & Agrawala, A. (2005). The Horus WLAN Location Determination System. *Proceedings of the Third International Conference on Mobile Systems, Applications, and Services (MobiSys 2005)*, Seattle, WA, USA, (pp. 205-218).

Youssef, M., Agrawala, A., & Udaya Shankar, A. (2003). WLAN Location Determination via Clustering and Probability Distributions. *Proceedings of the IEEE International Conference on Pervasive Computing and Communications (PerCom)*, (pp. 143-150).

Youssef, M., & Agrawala, A. (2004a). Handling Samples Correlation in the Horus System. *Proceedings of the IEEE Infocom Conference*, Hong Kong, *2*, 1023-1031.

Youssef, M., & Agrawala, A. (2004b). On the Optimality of WLAN Location Determination Systems. *Proceedings of the Communication Networks and Distributed Systems Modeling and Simulation Conference.*

Zou, Y., & Chakrabarty K. (2004). Sensor deployment and target localization in distributed sensor networks. *ACM Transactions on Embedded Computing Systems*, *3*(1), 61-91.

ENDNOTES

1. This applies to 802.11b/g. The chip rate used by 802.11b is 11 times the symbol rate due to the use of 11-chip Barker codes. Given that the symbol rate is 1 MHz, this leads to a chip rate of 11 MHz and hence a clock period of about 91ns. Hence the limitation. Since commercially available chips support both b and g, we use the 100ns figure.
2. Although b_{i0} is non-zero, the distribution assumes a zero mean value indicating a total lack of knowledge about the location of the terminal. As the measurement data are incorporated and posterior distributions are calculated, the distribution will become centered on the actual mean value. If prior knowledge is available, as in tracking situations, it can be incorporated by using an appropriate mean value in the prior distribution for b_{i0}.

Chapter XI
On a Class of Localization Algorithms Using Received Signal Strength

Eiman Elnahrawy
Rutgers University, USA

Richard P. Martin
Rutgers University, USA

ABSTRACT

This chapter discusses radio-based positioning. It surveys and compares several received signal strength localization approaches from two broad categories: point-based and area-based. It also explores their performance and means to improve it. It describes GRAIL - a sample positioning system. It finally concludes with a brief discussion of sensor applications that utilize location information.

INTRODUCTION

Location information is essential for many emerging applications, ranging from a diverse set of areas including asset tracking, workflow management, and physical security. Sensor networks offer an unprecedented potential for realizing many of these applications. Combined with a localization system, sensor nodes can be attached to objects and people and continuously track their locations. Those locations, when communicated back to a network backend, can then be utilized for functions such as controlling access to spaces, making decisions for workflow, or managing inventory.

Outdoors, the location information can be easily obtained using Global Positioning System (GPS) units. However, often it is not feasible to attach a GPS unit to each sensor, because of the additional cost to the sensor node, or because localization with GPS consumes considerable power. Also, GPS does not work well indoors because there is no clear line of sight to the satellites, and many applications must run in indoor environments.

For stationary sensors, a straightforward approach to localization is to simply store their positions during deployment. However, in many situations the node is mobile and the entire network is dynamic. Thus a localization system is necessary to track the positions of the sensor nodes and objects they are attached to.

This chapter will survey research on positioning the sensor nodes using the received signal strength (RSS) of wireless packet transmissions. Given that all modern radio chipsets include the hardware necessary to measure and report the signal strength of received packets, there is a tremendous cost and deployment advantage to re-using the existing RSS infrastructure of the communication network for localization purposes. However, additional hardware, such as ultrasound, can also be added to sensor nodes. The cost/performance tradeoffs and the impacts of additional localization resources in the sensor network are not well understood, and are the subjects of ongoing investigation.

We begin with a broad survey of existing localization approaches and algorithms. We then briefly discuss the causes of positioning uncertainty and how such uncertainty can be expressed in a meaningful and useful way for the higher-level applications. We will then describe two methods of location presentation along with their pros and cons: single-point localization and area-based localization. In the former, a single (x,y) spatial location is returned while in the latter a regular or irregular area is returned, for example an ellipse versus a set of tiles. We will elaborate on sample representative algorithms from each class.

We include a brief evaluation of the various approaches. We show how random and systematic variations in the signal strength affect the performance. Our basic performance metric is the accuracy of the returned position, i.e., how far it is from the true location of the sensor, along with variants of this metric. We also discuss methods and guidelines to improve the localization performance, such as optimally placing anchors in the environment.

We then describe a general infrastructure for indoor localization called the GRAIL system and show how the different approaches we discuss fit into such a system. We will briefly list some development and deployment issues. Finally, we conclude by discussing current and emerging applications of sensors that leverage the location information.

LOCALIZATION APPROACHES

The numerous approaches to localization defy a simple taxonomy. However, there are only a handful of overall strategies and approaches. In all cases *anchors* or *landmarks*, i.e., sensor or gateway nodes with known locations are needed at some point in the process.

Aggregate approaches position sensors using a collection of measurements from a large number of nodes. In contrast, *individual* approaches use information between a single sensor and a set of landmarks.

Orthogonal to the number of sensors participating in the process is the algorithmic approach used. *Lateration* approaches use some function of distance between the sensors and the landmarks. In contrast, *Scene matching* approaches match sensor observations to known maps and do not require any concept of direct physical distance in the algorithm.

When using lateration, the distances could be derived directly from a signal strength decay function, or more indirectly through hop counts (Niculescu et al., 2001). Using the actual sensed data, however, the resulting set of distance equations often has no exact solution, and so approximations must be found. Finding the best approximation is often difficult. A classic approach minimized the residual using least squares, as described in (Patwari et al., 2005). However, the problem can be generalized to viewing the system of distance constraints as an optimization problem, and then applying a range of optimization solvers (Dohertyl et al., 2001). Another approach views the sensor observations as existing in a high dimensional observational space and uses multi-dimensional scaling to estimate the positions in the lower-dimensional physical space (Shang et al., 2003). A further set of lateration approaches averages the coordinates of the landmarks observed by a sensor, either averaging the entire set, or selectively averaging overlapping regions (He et al., 2003; Stolero et al., 2004).

Matching algorithms can be generalized to a classic matching learning problem: given a known signal map and a set of observations, the localization system must derive the position on the map that best fits the observed data. Characterizing localization as matching thus opens the door to a wealth of machine learning approaches, including neural networks, Bayesian matching, and maximum likelihood estimation (Elnahrawy et al., 2004).

Figure 1 and Equation 1 show an example aggregate lateration approach that uses hop count. Node C computes its distance to landmark A, d, by multiplying the average hop distance, l_{avg}, over the network by 2 hops ($n = 2$). It repeats for all the landmarks (grey nodes) then uses lateration to compute its position.

$$d = n \times l_{avg} \tag{1}$$

Rather than using hop-counts, *individual* lateration must use another form of distance estimation to the landmarks. The following equation is a sample propagation model that translates signal strength, ss_{ij}, in dBm units, from sensor i to landmark j, to the distance between them, d_{ij}. The propagation parameters, a_j, and b_j, are unique for each environment and may even vary from one landmark j to another.

Figure 1. An aggregate localization scenario. Landmark nodes are plotted in grey. Node C is two hops away from node A.

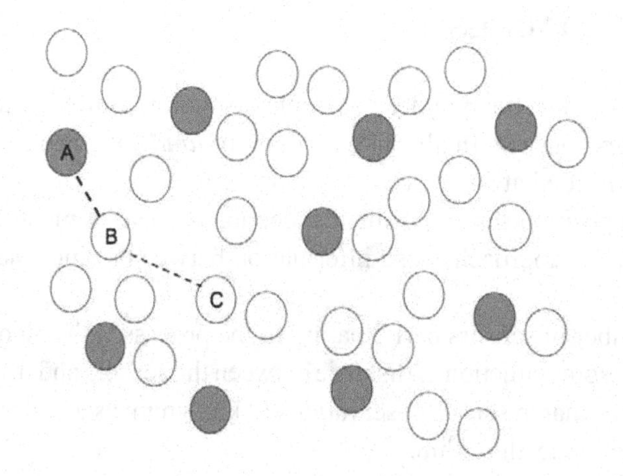

$$ss_{ij} = a_i + b_i \times \log d_{ij} \tag{2}$$

Where

$$d_{ij} = \sqrt{(x_i - x_j)^2 + (y_i - y_j)^2} \tag{3}$$

Typically, least squares estimation is then used to compute the unknown coordinates (x_i, y_i). Once at least 3 landmarks in two dimensions (4 in 3 dimensions) are known, a sensor node can trilaterate (i.e., use 3 landmarks) its position. Figure 2 shows an example network. Node B computes its distance to each landmark A_i using the measured signal to each landmark and Equation 2 with corresponding propagation parameters.

Aggregate approaches are practical only in dense networks, because they are often built on explicit proximity and isotropy assumptions (Niculescu et al., 2001). In most scenarios those assumptions do not hold. The sensor network can be sparse following irregular shape, which yields large errors in the position estimates. In the remainder of the chapter, we will therefore explore additional individual scene matching and lateration-based approaches.

UNCERTAINTIES AND OFFLINE TRAINING

Following Equations 2, 3 and assuming that signals travel with no obstructions, the unknown coordinates of any sensor node can be computed precisely and directly by translating radio ranges to distances using lateration. The environments where sensor networks are often deployed are however challenging, especially indoors. The radio signal suffers attenuations and distortions because of obstructions that absorb, reflect, refract or scatter the signals causing multipath effects. The distances to signals computed using equations 2, 3 are therefore inaccurate. Using lateration on those estimates directly will yield highly erroneous positions.

Scientists have thus opted for machine learning theory to map a set of radio signal strengths to a spatial position with high confidence along with a level of uncertainty. Next section describes an array of representative approaches.

Broadly speaking, there are two families of strategies. The first family uses a *signal strength to distance* function. These rely on estimating the electromagnetic wave properties to compute the distance to known anchors from the observed signal strength, similar to Equation 3. The second family assumes the mapping can be completely arbitrary in that the received signal strength has no direct relationship with neighboring points or distance to the transmitting sensors, *i.e.*, strictly scene matching.

Both families follow what is so called *offline training* and *online localization* phases. Training data is collected during the offline phase and then applied during the online phase to infer the unknown position of the mobile sensor. The training data generally correlates the distance from the mobile sensor node to the anchor node (*i.e.*, landmark) with the strength of the (received/transmitted) signal, *i.e.*, the radio range. It can be in the form of a discrete training set or a continuous gridded map.

Training Set

A training set, T, consists of a set of empirically measured signal strength *fingerprints* from the n anchors in the network along with the m locations where they were collected. A fingerprint at a location,

Figure 2. An individual localization scenario. Landmarks are placed around the network, plotted in grey. Node B computes the distance d_i from each landmark A_i using the range to distance function.

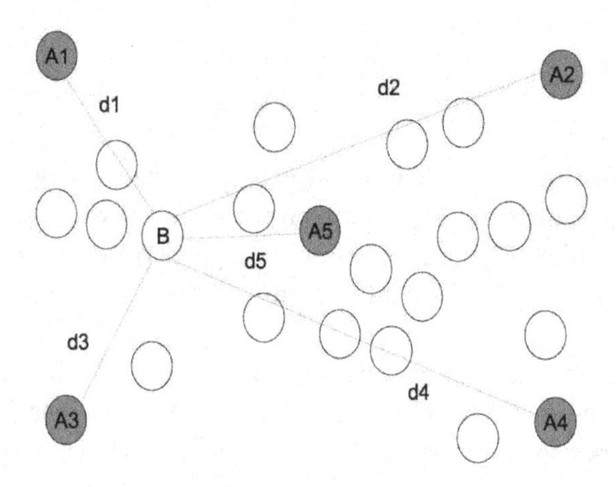

i, with coordinates (x_i, y_i), is the set of average received signal strength, s_{ij}, from each anchor j. Typically, it is computed from a series of k signal strength *samples* collected at that location. A default value for s_{ij} is usually assigned in a fingerprint if no signal is received from anchor j at a location i. That is, $T = \{[(x_i, y_i), s_{i1}, s_{i2,...,} s_{in}]\}$, $i = 1 ... m$.

The sensor node to be localized collects a set of *received signal strengths* (RSS). An RSS is similar to a fingerprint in that it contains a mean signal strength for each anchor j, $j = 1... n$. An RSS may also maintain a standard deviation of the sample set at each location i and anchor j, σ_{ij}. The collection of training points forms the training set.

Gridded Map

Some localization approaches require a continuous map of fingerprints over the two-dimensional localization space. An approximation is to build a grid of regular simple shapes such as tiles that describe the expected fingerprint for the area described by the tile. The tiles can be tiny or coarse. A direct measurement of the fingerprint for each tile is expensive in terms of time and labor costs. Additional signal strength to location points may be interpolated, however, from a set of discrete training points in order to form the gridded map.

Specifically, a surface fitting approach is used to interpolate a fingerprint at each tile from a training set that would be similar to an observed one. Several approaches in the literature can be utilized for the interpolation such as splines. A map M_i, for each anchor node, i , $i = 1... n$, is built independently using the discrete observed training fingerprints for that anchor. It was found that the observed training points' spacing need only to follow a uniform distribution rather than have precise spacing (Elnahrawy et al., 2004).

Figure 3 shows a sample gridded map. This map was built using triangle-based linear interpolation in an indoor office building. In this approach the two-dimensional area is divided into triangular regions. The locations of the observed training points serve as the triangles' vertices. The expected signal strengths in intermediate locations (or tile) are then linearly interpolated using the "height" of the triangle at the center of the tile. This approach also naturally extends to volumes. Notice how the

Figure 3. A sample interpolated radio map for an anchor node. The dots show the actual observed training point locations while the square in the middle show the true location of the anchor.

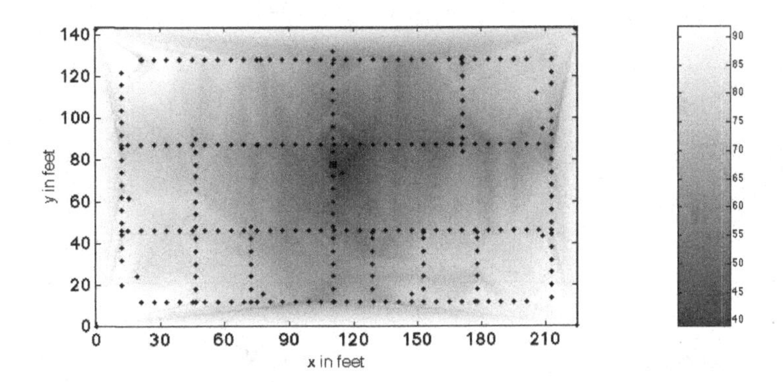

signal is distorted and does not follow regular circular shapes as would be expected in a free space area where the signal decays uniformly with distance in all the directions.

ALGORITHMS

Localization algorithms are classified based on their output into two classes: point-based and area-based. In *point-based* localization, the goal is to return a single point for the sensor node while in *area-based* the goal is to return the *possible locations* of the sensor as an area or volume (areas and volumes are interchangeable from our perspective) (Elnahrawy et al., 2004). The area could be a regular shape such as a circle or an ellipse, or an irregular shape such as a set of tiles or a cloud of points. Figure 4 shows the various representations. The true location of the node is marked as a star.

This section gives an overview of various individual localization approaches from each class. There is generally a myriad of algorithms in literature on this topic, which stemmed out of long research over the past years. The algorithms selected here span broad techniques that proved practical and useful from surveying the references (Elnahrawy et al., 2004; Bahl et al., 2000; Youssef et al., 2003; Youssef et al., 2004; Battiti et al., 2002; Moore et al., 2004; Fang et al., 2005; Savvides et al., 2001; Lorincz et al., 2006; Hazas et al., 2003; Priyantha et al., 2000; Want et al., 1992; Krishnan et al., 2004; Ladd et al., 2002; Roos et al., 2002; Smailagic et al., 2002; Lim et al., 2006). They are intended to overview strategies rather than drill down into detail. The reader is therefore encouraged to pursue the references for further explanation.

Point-Based Algorithms

Point-based approaches can be further categorized as deterministic or statistical. In deterministic localization the sensor's RSS is matched against the training data set using a systematic deterministic strategy. The localization output is the coordinates of the closest matching point in the set. For statistical localization the matching is done probabilistically. This section sketches two major point-based algorithms, one from each category.

Figure 4. Different localization outputs: Point (top left), ellipse or circle (top right), tiles (bottom left), and cloud (bottom right). The ground truth is marked as a black star.

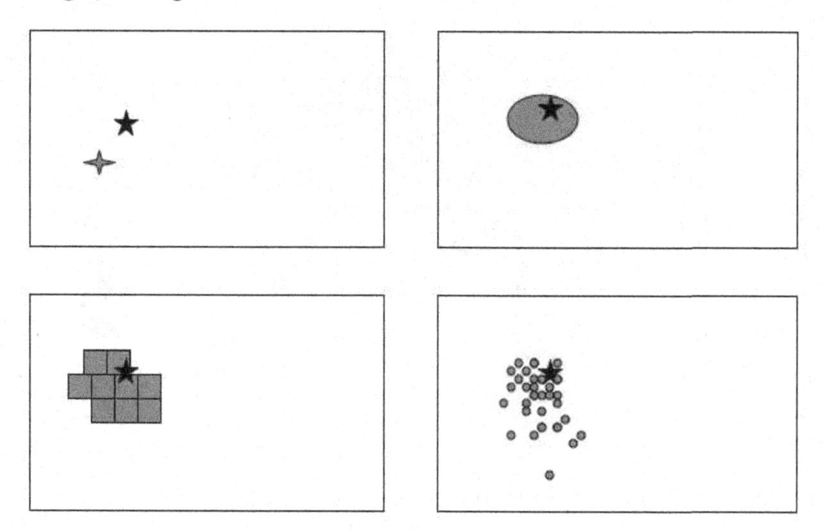

Deterministic Localization

A sample deterministic point-based algorithm we discuss here is the nearest neighbor approach. The strategy is to return the location of the closest fingerprint to the RSS fingerprints in the training data. It uses Euclidean distance in "signal space" as the deterministic measurement function. It views the fingerprints as points in an N-dimensions space, where each anchor node forms a dimension. Specifically, it computes the signal distance between each fingerprint vector in the radio signal map and the measured fingerprint for the localized sensor. It then picks the vector with the minimum distance and returns its coordinates as the estimated sensor's location (Bahl et al., 2000).

This approach is sometimes referred to in literature as the RADAR approach. Other versions of this approach return the average position (centroid) of the top k closest vectors; i.e., averaged RADAR algorithm. For example if $k = 2$, it takes the closest two candidates and returns the mid-point on the floor between them. A disadvantage of deterministic approaches is that they require a large number of training points to perform adequately. To compensate for a small training set gridded signal maps are used as discussed before, e.g., as in a variant of RADAR called gridded RADAR (Bahl et al., 2000; Elnahrawy et al., 2004).

Statistical Localization

Many statistical approaches have been devised within the context of point-based localization. They range from applying simple Bayes' rule and maximum-likelihood estimation to sophisticated support vector machines (Battiti et al., 2002; Youssef et al., 2003; Ladd et al., 2002; Roos et al., 2002). They generally map the problem of localizing sensors to a probability inference problem. First, a statistical model of the RSS as well as some system and environment parameters is constructed during the training phase. The model is then used during the online phase to infer the unknown sensor's position.

Let us elaborate on an example probabilistic approach that applies Bayes' rule. Signals received from the different anchors are assumed to be independent. For each anchor j, j = 1... n, the received signal strengths at each (x_i, y_i) in the training data is s_{ij}. Using Bayes' rule, the probability of being at each fingerprint's location in the training data given the received signal vector of the sensor, $\overline{S_l} = (s_{lj})$, is computed as follows.

$$P(L_i \setminus \overline{S_l}) = \frac{P(\overline{S_l} \setminus L_i) \times P(L_i)}{P(\overline{S_l})} \tag{4}$$

However, $P(\overline{S_l})$ is a constant c. Moreover, given there is no prior knowledge about the exact sensor's location, it is assumed that it is equally likely to be at any location, i.e., $P(L_i) = P(L_j), \forall i, j$. Therefore, Equation 4 is rewritten as:

$$P(L_i \setminus \overline{S_l}) = c \times P(\overline{S_l} \setminus L_i) \tag{5}$$

Without having to know the value c, the location in the training $L_{\max}, L_{\max = \arg\max_i}(P(\overline{S_l} \setminus L_i))$ is returned. Specifically, $P(\overline{S_l} \setminus L_i)$ is computed for every fingerprint i in the training set and the location of the highest probability candidate is returned as the predicted location. This approach hence inherently requires large enough training sets. Variants of the approach may return the midpoint of the top two or an average of top k candidates.

Area-Based Algorithms

We discussed that environmental effects impact localization and introduce fundamental uncertainty in the estimated position. Area-based approaches are better able to utilize and describe this uncertainty as compared to point-based approaches (Elnahrawy et al., 2004). Specifically, they provide an understanding of the localization confidence in a more natural and intuitive manner, where the term confidence is used loosely to refer to the positioning certainty. Hence, the larger the returned area, the less confidence we have in placement of the sensor in a particular location because many probable locations are included in the returned result. These approaches are also able to adjust the localization confidence by controlling the size of the returned area. Point-based approaches have difficulty describing such a trade-off systematically to the higher-level applications or users. A second advantage is that an area can naturally be mapped into a set of directions to search for the sensor in relation to the likelihood of its presence in the area, for example by beginning the search in the most likely area then continually expanding to the next most likely area and so on.

Figure 5 shows two example returned areas for a floor. The areas are shown by a dark color. The true location of the sensor is shown as a "*". The smallest circumscribing circles and rectangles are also shown. Figure 5(a) shows the localization can contain the sensor to an area the size of a single room while in Figure 5(b); the localization is more diffuse, in this case spanning two rooms.

The circumscribing circles show that augmenting a point with a distance to describe the uncertainty, in point-based localization, would likely return a much larger area than a strictly area-based approach. Returning rectangles, while reducing the inaccuracy of circles, no longer fits the definition of a point-based approach, however.

Figure 5. Sample areas returned by area-based presentation, specifically SPM, versus single-point based presentation. The true location is marked by a "".*

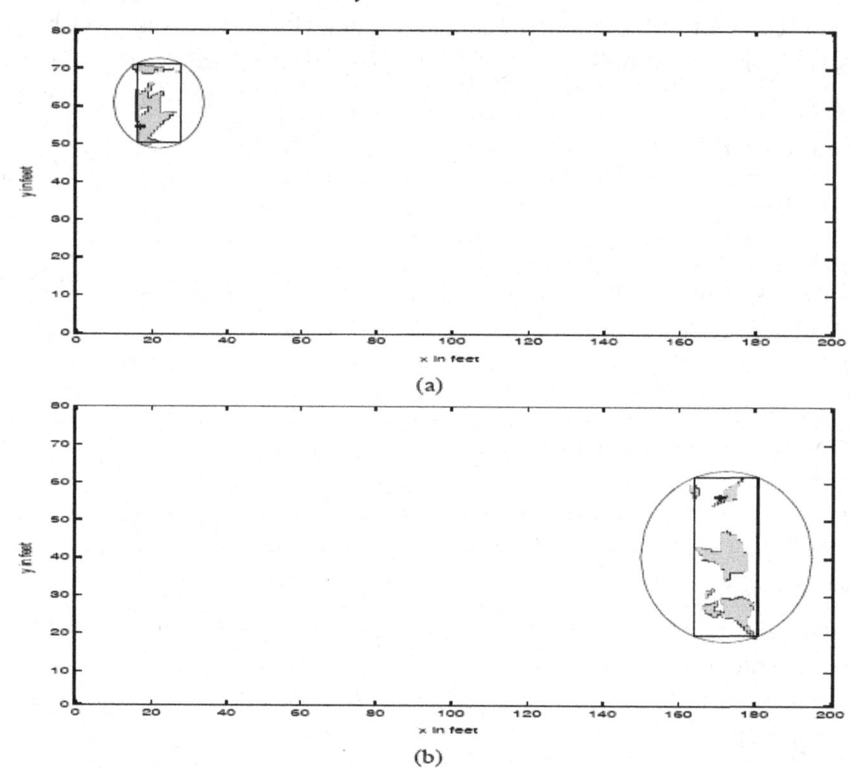

Simple Point Matching

Simple Point Matching (SPM) finds the set of those tiles whose signal strengths fall within a threshold of the received signal for each anchor node independently. Specifically, for each anchor j, $j = 1... n$, it "matches" all fingerprints $\overline{S_l} = (\overline{s_{l1}},...,\overline{s_{1n}})$ from the floor tiles (Elnahrawy et al., 2004). The matching tiles for each anchor j are computed by adding an expected "noise" level q to its received signal $\overline{S_{lj}}$, and then returning all the area tiles that fall within the expected threshold $\pm q$.

It then returns the area formed by intersecting all matched tiles from the individual anchors' tile sets. A gridded training map is used (Elnahrawy et al., 2004). The approach is eager, *i.e.*, it finds the fewest most probable tiles by starting with a very low q. It then incrementally increases it by trying $2q$, $3q$, .., until an intersection is found. Even in the worst case, a non-empty intersection will result, although q may expand to the dynamic range of signal readings. The value q is usually bounded by the standard deviation of the signals in the localization environment.

SPM is a Maximum Likelihood Estimation approach that assumes the anchor nodes are totally independent. Specifically, to localize a sensor i, the set of received signals between each anchor j and the sensor during a time window is estimated by a Gaussian distribution centered around the average measured signal $\overline{S_{lj}}$ with variance equals to $(\sigma_{lj})^2$.

Therefore, a $(1-\alpha)$ 100% confidence interval for the estimator is as follows, where $z_{\frac{\alpha}{2}}$ is a constant that depends on α, *e.g.*, it equals 1.96 for a 95% confidence interval of the estimator

$\overline{s_{lj}} \pm z_{\frac{\alpha}{2}} \times \sigma_{lj}$.

For single-mode distributions, such as Gaussian, increasing the confidence level, $(1 - \alpha)$, increases the width of the estimator's interval at the cost of adding less probable values to it. That is, less confidence is indeed better in our context; since higher probability values are the only ones included in the interval. Although the Gaussian approximation assumption may not be true in general, it has been proven effective in practice.

In SPM, the noise level $z_{\frac{\alpha}{2}} \times \sigma_{lj}$ corresponds to SPM eagerly attempts to find the appropriate (lowest) confidence level $(1-\alpha)$ for each anchor j that yields an overall non-empty area. It assumes that, for each j, the collected fingerprint follows a Gaussian distributions with a standard deviation q, equals to the highest σ_{ij}, among all the fingerprints in T. Therefore, it starts searching by adding a noise level of q_j, then $2q_j$, and so on, till a non-empty overall area is found.

Area Based Probability

The Area-Based Probability (ABP- α) algorithm returns a set of tiles bounded by a probability, α that the object is within the returned area. The probability is also called the *confidence*, and it is an adjustable parameter. ABP's approach to finding the tile set is to compute the likelihood of a received signal vector matching a fingerprint for each tile, and then normalizing these likelihoods given the prior conditions: (1) the sensor must be in the localization floor, and (2) all tiles are a-priori equally likely. ABP then returns the top probability tiles whose sum matches the desired confidence. The confidence controls the accuracy (error) versus the precision (size of returned area) tradeoff, both terms will be defined in more detail later in this chapter. ABP thus stands on a more formal mathematical foundation than SPM (Elnahrawy et al., 2004).

Similar to SPM, signals received from different anchors are assumed to be independent. Using Bayes' rule, ABP computes the probability of being at each tile's location, L_i, on the floor given the fingerprint vector of the sensor using Equations 4, 5 as before, $P(\overline{S_l} \setminus L_i)$. The exact probability is then computed for every tile/location rather than returning the location (tile) $L_{\max}, L_{\max=\arg\max_i} (P(\overline{S_l} \setminus L_i))$.

ABP extends the statistical point based approach discussed above by its final step where it computes the actual probability density of the sensor for each tile given that the sensor must be at exactly one tile, i.e., $\sum_{i=1}^{L} P(L_i \setminus \overline{S_l}) = 1$. Using the resulting density, ABP returns the top probability tiles up to its confidence, α, i.e., the top probability tiles/locations such that their overall probability is $\geq$ *confidence*. Useful values of α have a wide dynamic range between 0.5 and less than 1. While a confidence of 1 returns all the tiles on the floor, picking a useful α is not difficult because in practice some tiles have a much higher probability than the others, while at the same time the difference between these high-probability tiles is small.

Bayesian Networks

Bayes nets are graphical models that encode dependencies and relationships among a set of random variables. The vertices of the graph correspond to the variables and the edges represent dependencies (Gelman et al., 2004). A Bayes net can be utilized to encode the relationship between the received signal and its location based on the signal-versus-distance propagation model described above. The initial parameters of the model are assumed to be unknown, and the training data is then used to compute a probabilistic model for each of the specific parameters. Various Bayesian networks have been designed and tested for localization. They differ in their complexity and assumptions. Here we describe a basic simple model as in Figure 6.

Each random variable s_j, $j = 1... n$ denotes the expected signal strength from the corresponding anchor node or landmark j. The values of these random variables depend on the Euclidean distance D_j between the landmark's location, (x_j, y_j), and the location where the signal s_j is measured (x,y). The baseline expected value of s_j follows a signal propagation model $s_j = b_{0j} + b_{1j} \times logD_j$, where b_{0j}, b_{1j} are the parameters specific to each landmark j in that environment. The distance $D_j = \sqrt{(x-x_j)^2 + (y-y_j)^2}$ in turn depends on the location (x,y) of the measured signal. The network accounts for noise and outliers by modeling the expected value, s_j, as a probabilistic distribution around the above propagation model, with some variance, τ_j.

Using the training fingerprints T and the fingerprint vector of the sensor, the network then learns the specific values for all the unknown parameters b_{0j}, b_{1j}, τ_j and the joint distribution of the (x,y) location of the sensor. In general, there is no closed form solution for the returned joint distribution of the (x,y) location. A simulator such as Markov Chain Monte Carlo is used to draw samples from the joint density for (x,y) (Madigan et al., 2005; Kleisouris et al., 2006; Heckerman et al., 1995). Samples that give, e.g., a 95% confidence on the density are mapped to tiles and returned as the estimated area. A substantive drawback of this approach is that it yields a large number of disconnected tiles. Although the tiles are concentrated around the most likely location, the scatter is substantial and can interfere with higher-level functions (Madigan et al., 2005; Elnahrawy et al., 2004).

LOCALIZATION PERFORMANCE

There have been many experiments that compare and contrast different individual localization algorithms and how they perform. Detail and in depth evaluations can be found in the references, e.g., (Elnahrawy et al., 2004; Battiti et al., 2002). Due to the limited space, the goal of this section is rather to conclude the performance one would expect when using such a positioning approach based on those extensive studies. It first describes the evaluation metrics usually used for assessment.

Figure 6. A simple Bayesian network used for localization

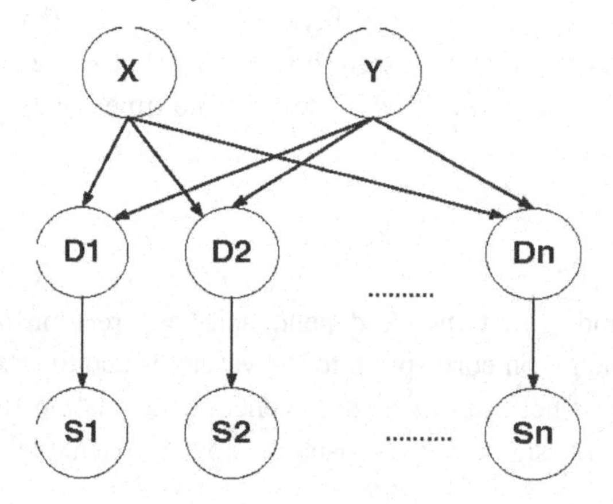

Evaluation Metrics

A traditional metric to assess the performance of point-based localization is *the localization error*. This is the distance from the true location, *i.e.*, the ground truth, of the sensor and the estimated location. There are many ways to express this metric. Scientists usually use the error CDF (Cumulative Density Function) to plot the error along with the corresponding probability of obtaining it. Figure 7 depicts a sample error CDF curve for a hypothetical approach. As shown the approach yields an error less than 10 feet 80% of the time. For area-based systems two metrics are used; accuracy and precision. *The accuracy* is a generalization of *localization error* to areas. It is the distance between the true position of the sensor and the returned area. It is quantified using CDFs of the order statistics such as the median distance. *Precision* describes the size of the area. A point is hence infinitely precise, but may not be very accurate. On the other hand, the area containing the entire scope of the localization system (e.g., a whole building) would have a high accuracy but poor precision. Accuracy and precision are useful utilities to quantitatively describe the performance of different localization approaches by observing the impact of increased precision (*i.e.*, less area) on accuracy (Bahl et al., 2000; Battiti et al., 2002; Madigan et al., 2005; Hand et al., 2001).

Localization Error

Outdoors, signals travel with no or little obstruction following the free space model. The variance and the attenuations are minimal and the localization error is negligible. Indoor environments, on the other hand are more challenging with obstacles everywhere. Researchers have hence focused on studying the performance of indoor localization rigorously. Figure 8 shows sample performance of those individual approaches described earlier in the Algorithms Section along with their variants in an indoor environment. The error CDFs are plotted. For area-based approaches the median accuracy CDF is used for the comparison. The individual curves are not labeled, as the goal is to show that they have similar performance. Although area-based approaches are better at describing uncertainty, their absolute performance is similar to point-based approaches. No existing approach has a substantial advantage in terms of localization performance.

A general rule of thumb is that using radio received signal strengths with much sampling one can expect a median error of roughly 10 feet and with relatively sparse sampling, every 20 feet, one can still get median errors of 15 feet[1]. Researchers therefore concluded that there are fundamental limitations in indoor localization's performance that cannot be transcended without qualitatively more complex models of the indoor environment, for example by modeling every wall, desk or shelf, or by adding extra hardware in the sensor node above that required for communication, for example, very high frequency clocks to measure the time of arrival (Battiti et al., 2002; Elnahrawy et al., 2004; Kaemarungsi et al., 2004).

Anchor Placement

Placement of the anchor nodes (*i.e.*, landmarks) in the environment also has an impact on the localization performance (Chen et al., 2006; Krishnakumar et al., 2005). The problem is generally an optimization problem with the goal of finding the anchor placement that minimizes the error between the true positions of the sensors and the estimated positions.

Figure 7. Sample Error CDF. The x-axis is the error while the y-axis is the probability. The crossing lines mean that 80% of the time the distance error is less than or equals to10 feet.

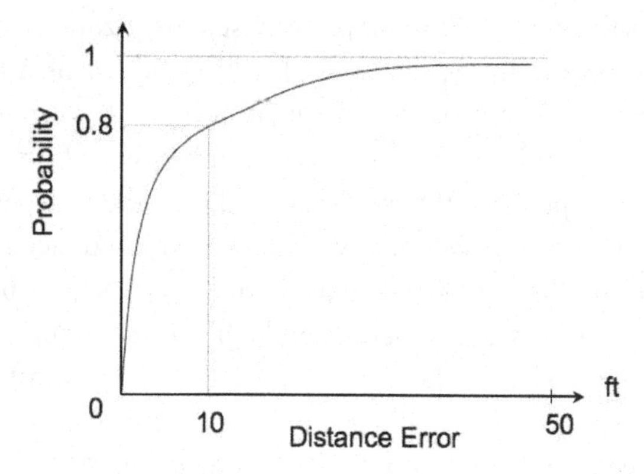

Figure 8. Error across a wide variety of point-and area-based approaches. The CDFs are clustered which shows similar performance.

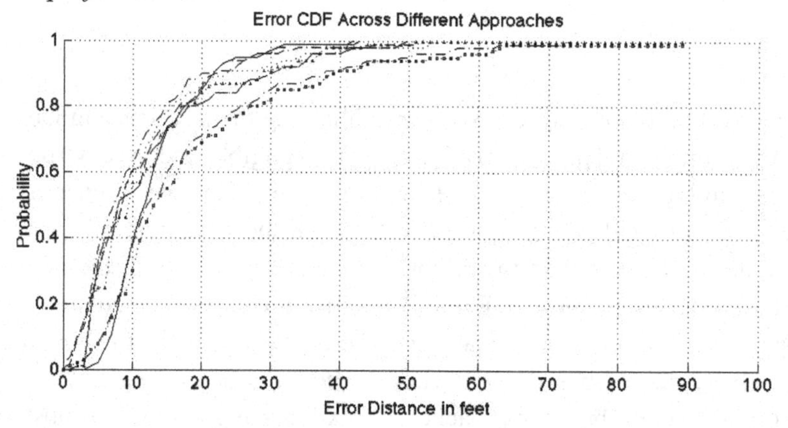

Linear placement of anchors is the worst layout because the confusion along the other dimension will never get resolved, causing large errors. A uniform deployment in contrast is intuitive and practical. Comparing the error between the two scenarios showed that the latter improves the localization error and makes indoor localization less sensitive to the environment noise and biases. In terms of localization metrics the error CDFs, *e.g.*, similar to the one shown in Figure 7, shift up and to the left compared to their counterparts when using linear anchor placement.

Researchers found that optimal anchor placement follows some simple and symmetric patterns that are easy to achieve. They derived some guidelines that are easy to follow when deciding about placing anchors. The placement patterns should follow simple shapes such as squares and equilateral triangles, or enclosing of them. Complex shapes such as pentagons or hexagons have been shown not to be optimal. Figure 9 shows the patterns for 3, 4, 5, 6, and 7 anchors. Generalization to higher number of anchors is straightforward. It is important to take into consideration the physical constraints of the environment where the network is deployed. A slight deviation from the guidelines in the form of stretching or shrinking the shapes has been shown to be tolerable.

A SAMPLE LOCALIZATION SYSTEM

We showed that individual radio range-based localization approaches have similar limited performance with respect to the localization error. They generally differ however in their applicability to a certain localization scenario or an environment. For example, they vary in how much training they require to achieve a decent performance, how long it takes to compute the estimates (*i.e.*, complexity), how many sensors can be localized simultaneously (*i.e.*, scalability), the ability to compute consistent positions (*i.e.*, outliers), and tolerance to measurement errors and biases. In order to understand these deployment issues it helps to think of the core localization approach as a piece of the higher-level applications (LaMarca et al., 2005; Savvides et al., 2001; Chen et al., 2008). We therefore briefly describe a sample core localization system called GRAIL (General purpose Real-time Adaptable Localization) (SourceForge, 2008; Chen et al., 2008). GRAIL can be integrated seamlessly into any application that utilizes radio positioning via simple Application Program Interfaces (APIs). It has been used to simultaneously localize multiple devices running 802.11 (WiFi), 802.15.4 (ZigBee) and special customized RollCall™ radios (InPoint Systems, 2008).

GRAIL has the following key properties: (1) General Purpose, it supports positioning of a variety of physical modalities, networks or radios, devices and algorithms. Specifically, it localizes any wireless device that transmits packet data. It adopts a centralized approach in order to be able to localize a diverse set of radios. Specifically, the inherent anchor-based localization strategy eliminates the need to install special software on the devices to be localized, e.g., sensors, and therefore enables rapid integration of any radio device. (2) Real-time, it can localize stationary and mobile sensors in real time. (3) Adaptable to indoor noise and multi-path effects. (4) Indoors, GRAIL was originally designed to scale in indoor environments. It naturally works in outdoor environments however.

Figure 9. Layout for optimal anchor placement. As the number of anchors increases from 3 up the placement takes simple triangle or square shapes and then enclosing of them.

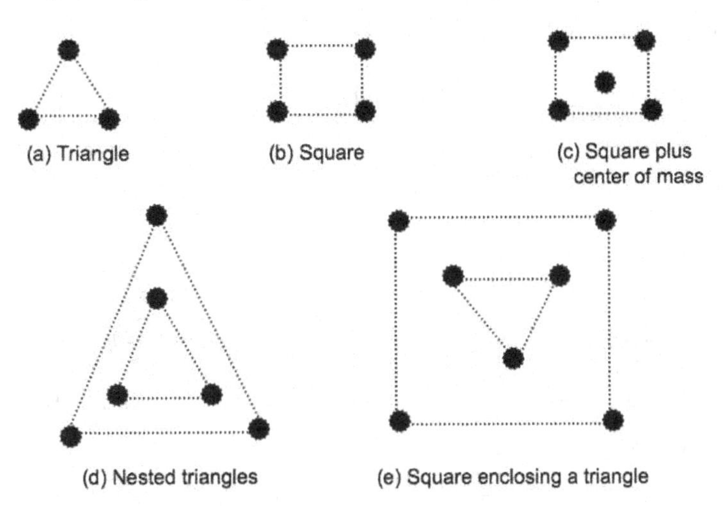

A Localization Scenario

To understand how a specific localization algorithm fits into a system let us describe a typical localization scenario within GRAIL. The localization process starts once a transmitter (*i.e.*, a sensor) in the network transmits packets. The anchor nodes continuously monitor the radio traffic at the packet-level. They timestamp the observed packets and extract the value of the observed physical property, i.e., the received signal strength in this case. The anchors then forward these values along with other header information to a central entity called the server. The server collects traffic data from all the anchor nodes in real time and aggregates those values into fingerprints. It then sends those aggregates to an instance of a solver. The solver entity utilizes a "localization approach" to estimate the locations. It uses an implementation of a localization approach along with some training data and information about the localization environment (anchor placement, for example). Once the localization estimates are computed it sends them to the server. The process ends when the locations are stored in the database or disseminated back to the network nodes.

GRAIL Components

Figure 10 shows the main components of GRAIL: transmitters, landmarks, the server, solvers, the database, and the web server. The transmitters, landmarks and solvers correspond to the sensor nodes, anchors, and localization approaches in our context. We give a brief overview of each of the components and their functionality next.

Much like the Hypertext Transfer Protocol (HTTP) used on the web and the Transaction Language 1 (TL1) used as a standard protocol in the telecommunications equipment industry, all communications between the system components use a simple text-based protocol over TCP sockets. The reader may refer to the references for a detailed description.

Figure 10. Overall architecture for the GRAIL system

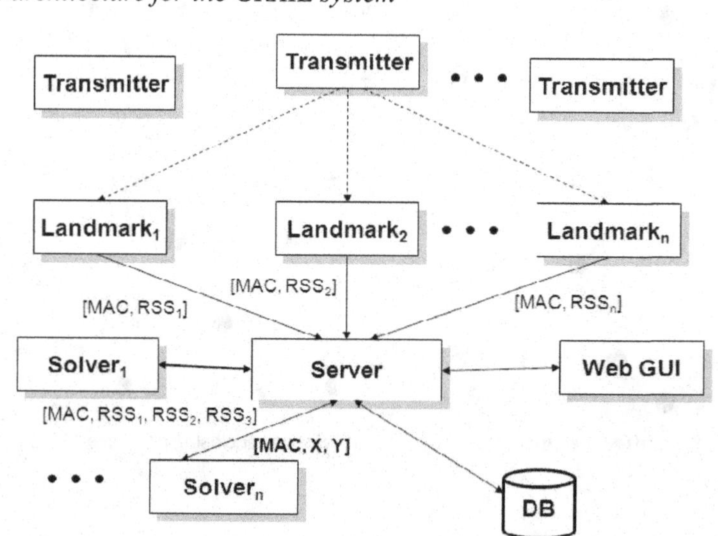

- **Transmitter:** Any device transmitting radio packets that needs to be localized.
- **Landmark:** These are the anchor nodes. They passively monitor the existing network traffic and forward the raw data or a summary to the server. Limiting the role of the anchors in the system or network to traffic monitoring enables scalability because regular unsophisticated nodes can easily work as anchors.
- **Database:** The database is a repository for storing the hard state of the GRAIL system. Specifically, it maintains the localization results, fingerprints computed from data samples, anchor information, environment information, and the transmitters in the network.
- **Web Server:** The web server provides a front-end to the GRAIL system. It provides simple authentication and means to interact with users, e.g., to view location estimates, adjust the system parameters or settings. A set of APIs is also provided to support any potential higher-layer application built on top of the core localization.
- **Server:** The localization server is a centralized moderator that collects data samples from the landmarks, summarizes and cleans them, and then passes the data to solvers to compute the unknown positions. The server interacts with a web server as a user interface. It is responsible for storing all the related traffic information and the estimated positions in the database.
- **Solver:** A solver is an implementation of one or more localization approach. It computes the location estimates and sends the results back to the server. GRAIL's architecture is flexible in that multiple different solvers can run simultaneously against a single server, which can improve the overall localization accuracy and allow for load balancing. The server or the user can control which approach to use at every localization attempt depending on the environment, the number of nodes, the training data and so on.

CONCLUSION AND APPLICATIONS

In this chapter we surveyed a set of localization approaches and algorithms for sensor networks. An important conclusion we can draw is that no existing approach or algorithm has been shown to be the best, or even good enough, for most applications. The primary reason is that the location-based services and applications built on top of the localization system are still in their infancy. Until there are more widespread and longer deployments of applications using sensor networks, the performance requirements and resulting cost/performance tradeoffs will not be well understood.

Helping to fill the applications gap are commercial products and solutions that have recently emerged targeting different markets, including wireless security, access control, and workflow management in health-care and industrial plants (Aeroscout, 2008; Ekahau Inc., 2008; Kordinate, 2008; Newbury Networks, 2008; Airtight Networks, 2008). While the mapping from the spatial location to application function is straightforward in security-related applications it is not as intuitive in the latter ones. The general idea is to improve the cost of operation by attaching sensors to employees, inventory and equipment in factories, and additionally caregivers and patients in hospitals. Activities are detected using proximity information from the estimated position and higher-level decisions are taken or actions are made accordingly. For example, if a doctor and a nurse are both localized in the same room as a patient then it will be concluded that this patient is getting treated. If a high traffic of workers has moved within proximity of a factory machine then it might be an indication of a machinery breakdown, and so on.

As an example illustrating the unknown cost-performance tradeoffs, we can consider a healthcare application measuring the productivity of caregivers by measuring activity using location as the base input. In this application, sensors attached to caregivers are mobile and may not form a dense network. Individual lateration approaches are hence very applicable and have better cost-performance, than say, an aggregate approach requiring a high sensor density. Also, the sensor lifetime might only be on the order of a few hours, for example, the length of one shift. However, sensor cost and form factors are critical variables to obtaining good data, as people misplace sensors, or fail to wear them if they are too bulky or look strange.

We are still in exciting times with regards to sensor networks and their applications. Location is one critical piece of the puzzle that has yet to be solved.

REFERENCES

Aeroscout. (2008). *Aeroscout enterprise visibility solutions, white paper.* Retrieved October 10, 2008, from http://www.aeroscout.com

Airtight Networks. (2008). *Airtight networks, white paper.* Retrieved October 10, 2008, from http://www.airtightnetworks.net

Bahl, P., & Padmanabhan, V. (2000). RADAR: An in-building RF-based user location and tracking system. *In proceedings of Nineteenth Annual Joint Conference of the IEEE Computer and Communications Societies* (pp. 775-784).

Bahl, P., & Padmanabhan, V., & Balachandran, A. (2000*). Enhancements to the RADAR user location and tracking system.* Microsoft Research Technical Report.

Battiti, R., & Brunato, M., & Villani, A. (2002). *Statistical Learning Theory for Location Fingerprinting in Wireless LANs* (Tech. Rep. No. DIT-02086). University of Trento, Informatica e Telecomunicazioni.

Chen, Y., Chandrasekaran, G., Elnahrawy, E., Francisco, J. A., Kleisouris, K., Li, X., Martin, R. P., Moore, R. S., & Turgut, B. (2008). GRAIL: A general purpose localization system. *Sensor Review* (pp. 115-124).

Chen, Y., Francisco, J., Trappe, W., & Martin, R. P. (2006). A practical approach to landmark deployment for indoor localization. In *proceedings of the Third Annual IEEE Communications Society Conference on Sensor, Mesh and Ad Hoc Communications and Network* (pp. 365-373).

Dohertyl, L., Pister, K. S. J., & ElGhaoui, L. (2001). Convex position estimation in wireless sensor networks. In *proceedings of the IEEE International Conference on Computer Communications* (pp. 1655–1663).

Ekahau, Inc. (2008). *The Ekahau Positioning Engine.* Retrieved October 10, 2008, from http://www.ekahau.com

Elnahrawy, E., Li, X., & Martin, R. P. (2004). The Limits of Localization Using Signal Strength: A Comparative Study. *In proceedings of IEEE International Conference on Sensor and Ad hoc Communications and Networks* (pp. 406–414).

Elnahrawy, E., & Li, X., & Martin, R. P. (2004). Using area-based presentations and metrics for localization systems in wireless LANs. *In LCN's Fourth International IEEE Workshop on Wireless Local Networks* (pp. 650–657).

Fang, L., Du, W., & Ning, P. (2005). A beacon-less location discovery scheme for wireless sensor networks. *In proceedings of Annual Joint Conference of the IEEE Computer and Communications Societies* (pp. 161–171).

Gelman, A., Carlin, J. B., Stern, H. S., & Rubin, D. B. (2004). *Bayesian Data Analysis.* Chapman and Hall.

Hazas, M., & Ward, A. (2003). A high performance privacy-oriented location system. *In proceedings of the IEEE International Conference on Pervasive Computing and Communications* (pp. 216).

Hand, D., Mannila, H., & Smyth, P. (Ed.) (2001). *Principles of Data Mining.* The MIT Press.

He, T., & Huang, C., & Blum, B., & Stankovic, J., & Abdelzaher, T. (2003). Range-free localization schemes in large scale sensor networks. *In proceedings of the ACM International Conference on Mobile Computing and Networking* (pp. 81–95).

Heckerman, D. (1995). *A tutorial on learning with Bayesian networks* (Tech. Rep. No. TR-95-06). Microsoft Research.

InPoint Systems. (2008). *White paper.* Retrieved October 10, 2008, from http://inpointsys.com

Kaemarungsi, K., & Krishnamurthy, P. (2004). Modeling of indoor positioning systems based on location fingerprinting. *In proceedings of Twenty Third Annual Joint Conference of the IEEE Computer and Communications Societies* (pp. 1012-1022).

Kleisouris, K., & Martin, R. P. (2006). Reducing the computational cost of Bayesian indoor positioning systems. *In proceedings of the Third IEEE International Conference on Sensor and Ad hoc Communications and Networks* (pp. 555-564).

Kordinate LLC. (2008). http://www.kordinate.com

Krishnakumar, A., & Krishnan, P. (2005). On the accuracy of signal strength-based location estimation techniques. In *proceedings of Twenty Fourth Annual Joint Conference of the IEEE Computer and Communications Societies* (pp. 642-650).

Krishnan, P., & Krishnakumar, A. S., & Ju, W., & Mallows, C., & Ganu, S. (2004). A system for LEASE: Location estimation assisted by stationary emitters for indoor RF wireless networks. *In proceedings of the IEEE International Conference on Computer Communications* (pp. 1001–1011).

Ladd, A. M., Bekris, K. E., Rudys, A., Marceau, G., Kavraki, L. E., & Wallach, D. S. (2002). Robotics-based location sensing using wireless Ethernet. *In proceedings of The Eighth ACM International Conference on Mobile Computing and Networking* (pp. 227–238).

LaMarca, A., Chawathe, Y., Consolvo, S., Hightower, J., Smith, I., Scott, J., Sohn, T., Howard, J., Hughes, J., Potter, F., Tabert, J., Powledge, P., Borriello, G., & Schilit, B. (2005). Place lab: Device positioning using radio beacons in the wild. *In proceedings of Pervasive Computing* (pp. 116-133).

Lim, H., Kung, L., Hou, J., & Luo, H. (2006). Zero-Configuration, robust indoor localization: Theory and experimentation. *In proceedings of the IEEE International Conference on Computer Communications* (pp. 1-12).

Lorincz, K., & Welsh, M. (2006). Motetrack: A robust, decentralized approach to RF-based location tracking. *Springer Personal and Ubiquitous Computing, Special Issue on Location and Context-Awareness.*

Madigan, D., Elnahrawy, E., Martin, R. P., Ju, W. H., Krishnan, P., & Krishnakumar, A. S. (2005). Bayesian indoor positioning systems. *In proceedings of Twenty Fourth Annual Joint Conference of the IEEE Computer and Communications Societies* (pp. 1217-1227).

Moore, D., Leonard, J., Rus, D., & Teller, S. (2004). Robust distributed network localization with range measurements. *In proceedings of Conference on Embedded Networked Sensor Systems* (pp. 50–61).

Newbury Networks. (2008). *Newbury networks, white paper.* Retrieved October 10, 2008, from http://www.newburynetworks.com

Niculescu, D., & Nath, B. (2001). Ad hoc positioning system (APS). In *proceedings of the IEEE Global Telecommunications Conference* (pp. 2926–2931).

Patwari, N., Ash, J. N., Kyperountas, S., Hero, A. O., Moses, R. L., & Correal, N. S. (2005). Locating the nodes. *IEEE Signal Processing Magazine.*

Priyantha, N., Chakraborty, A., & Balakrishnan, H. (2000). The cricket location-support system. *In proceedings of the ACM International Conference on Mobile Computing and Networking* (pp. 32–43).

Roos, T., Myllymaki, P., & Tirri, H. (2002). A Statistical Modeling Approach to Location Estimation. *IEEE Transactions on Mobile Computing* (pp. 59–69).

Savvides, A., Han, C. C., & Srivastava, M. (2001). Dynamic Fine-Grained Localization in Ad-Hoc Networks of Sensors. In *proceedings of the Seventh Annual ACM International Conference on Mobile Computing and Networking* (pp. 166–179).

Shang, Y., W. Ruml, W., Zhang, Y., & Fromherz, M. P. J. (2003). Localization from mere connectivity. In *proceedings of the ACM International Symposium on Mobile Ad-Hoc Networking and Computing* (pp. 201–212).

Smailagic, A., & Kogan, D. (2002). Location sensing and privacy in a context aware computing environment. *IEEE Wireless Communications* (pp. 10-17).

SourceForge.net. (2008). The GRAIL Real Time Location Service, documentation and source code. Retrieved October 10, 2008, from http://grailrtls.sourceforge.net.

Stoleru, R., & Stankovic, J. (2004). Probability grid: A location estimation scheme for wireless sensor networks. *In proceedings of IEEE International Conference on Sensor and Ad hoc Communications and Networks.*

Want, R., Hopper, A., Falcao, V., & Gibbons, J. (1992). The active badge location system. In *proceedings of ACM Transactions on Information Systems* (pp. 91–102).

Youssef, M., & Agrawala, A. (2004). Handling samples correlation in the HORUS system. *In proceedings of Annual Joint Conference of the IEEE Computer and Communications Societies* (pp. 1023–1031).

Youssef, M., Agrawal, A., & Shankar, A. U. (2003). WLAN location determination via clustering and probability distributions. *In proceedings of IEEE International Conference on Pervasive Computing and Communications* (pp. 143–150).

ENDNOTE

[1] The 802.11 Wireless Local Area Network (WLAN) technology was used for the evaluations, however the results apply to all radio-based localization.

Chapter XII
Machine Learning Based Localization

Duc A. Tran
University of Massachusetts, USA

XuanLong Nguyen
Duke University, USA

Thinh Nguyen
Oregon State University, USA

ABSTRACT

A vast majority of localization techniques proposed for sensor networks are based on triangulation methods in Euclidean geometry. They utilize the geometrical properties of the sensor network to infer the sensor locations. A fundamentally different approach is presented in this chapter. This approach is based on machine learning, in which the authors work directly on the natural (non-Euclidean) coordinate systems provided by the sensor devices. The known locations of a few nodes in the network and the sensor readings can be exploited to construct signal-strength or hop-count based function spaces that are useful for learning unknown sensor locations, as well as other extrinsic quantities of interest. They discuss the applicability of two learning methods: the classification method and the regression method. They show that these methods are especially suitable for target tracking applications

INTRODUCTION

A sensor node knows its location either via a built-in GPS-like device or a localization technique. A straightforward localization approach is to gather the information (e.g., connectivity, pair-wise distance measure) about the entire network into one place, where the collected information is processed centrally to estimate the nodes' locations using mathematical algorithms such as Semidefinite Programming (Doherty et al. (2001)) and Multidimensional Scaling (Shang et al. (2003)).

Many techniques attempt localization in a distributed manner. The relaxation-based techniques (Savarese et al. (2001), Priyantha et al. (2003)) start with all the nodes in initially random positions and keep refining their positions using algorithms such as local neighborhood multilateration and convex optimization. The coordinate-system stitching techniques (Capkun et al. (2001), Meertens & Fitzpatrick (2004), Moore et al. (2004)) divide the network into overlapping regions, nodes in each region being positioned relatively to the region's local coordinate system (a centralized algorithm may be used here). The local coordinate systems are then merged, or "stitched", together to form a global coordinate system. Localization accuracy can be improved by using a set of nodes with known locations, called the beacon nodes, and extrapolate unknown node locations from the beacon locations (Bulusu et al. (2002), Savvides et al. (2001), Savvides et al. (2002), Niculescu & Nath (2003a), Nagpal et al. (2003), He et al. (2003)).

Most current techniques assume that the distance between two neighbor nodes can be measured, typically via a ranging procedure. In this procedure, various information can be used to help estimate pair-wise distance, such as Received Signal Strength Indication (RSSI) (Whitehouse (2002)), Time Difference of Arrival (TDoA) (Priyantha (2005), Kwon et al. (2004)), or Angle of Arrival (AoA) (Priyantha et al. (2001), Niculescu & Nath (2003a)). Other range measurement methods can be found in (Priyantha (2001b), Savvides et al. (2001b), Priyantha (2005b), Lee & Scholtz (2002), Gezici et al. (2005)).

To avoid the cost of ranging, range-free techniques have been proposed (Bulusu et al. (2002), Meertens & Fitzpatrick (2004), He et al. (2003), Stoleru et al. (2005), Priyantha et al. (2005)). APIT (He et al. (2003)) assumes that a node can hear from a large number of beacons. Spotlight (Stoleru et al. (2005)) requires an aerial vehicle to generate light onto the sensor field. (Priyantha et al. (2005)) uses a mobile node to assist pair-wise distance measurements until a "global rigid" state can be reached where the sensor locations can be uniquely determined. DV-Hop (Niculescu & Nath (2003b)) and Diffusion (Bulusu et al. (2002), Meertens & Fitzpatrick (2004)) are localization techniques requiring neither ranging nor external assisting devices.

All the aforementioned techniques use Euclidean geometrical properties to infer the sensor nodes' locations. Recently, a number of techniques that employ the concepts from machine learning have been proposed (Brunato & Battiti (2005), Nguyen et al. (2005), Pan et al. (2006), Tran & Nguyen (2006), Tran & Nguyen (2008), Tran & Nguyen (2008b)). The main insight of these methods is that the topology implicit in sets of sensor readings and locations can be exploited in the construction of possibly non-Euclidean function spaces that are useful for the estimation of unknown sensor locations, as well as other extrinsic quantities of interest. Specifically, one can assume a set of beacon nodes and use them as the training data for a learning procedure. The result of this procedure is a prediction model that will be used to localize the sensor nodes of previously unknown positions.

Consider a sensor node S whose true (unknown) location is (x, y) on a 2-D field. There are more than one way we can learn the location of this node. For example, we can model the localization problem as a classification problem (Nguyen et al. (2005), Tran & Nguyen (2006), Tran & Nguyen (2008)). Indeed, we can define a set of classes (e.g., A, B, and C as in Figure 1) that represent geographic regions chosen appropriately in the sensor network area. We then run a classification procedure to decide the membership of S in these classes. Based on these memberships, we can localize S. For example, in Figure 1, if the output of the classification procedure is that S is a member of class A, of B, and of C, then S must be in the intersectional area $A \cap B \cap C$.

We can also solve the localization problem as a regression problem (Pan et al. (2006), Tran & Nguyen (2008b)). We can use a regression tool to infer the Euclidean distances between *S* and the beacon nodes based on the signal strengths that *S* receives from these nodes, or when *S* cannot hear directly from them, based on the hop-count distances between *S* and these nodes. After these distances are learned, trilateration can be used to estimate the location of *S*. Alternatively, we can apply a regression tool that maps the signal strengths *S* receives from the beacon nodes *directly* to a location. One such a tool was proposed by (Pan et al. (2006)), which is based on Kernel Canonical Correlation Analysis (Hardoon et al. (2004)).

Compared to geometric-based localization techniques, the requirements for the learning-based techniques to work are modest. Neither ranging measurements nor external assisting devices are needed. The only assumption is the existence of a set of beacon nodes at known locations. The information serving as input to the learning can be signal strengths (Nguyen et al. (2005), Pan et al. (2006)) or hop-count information (Tran & Nguyen (2006), Tran & Nguyen (2008)), which can be obtained easily at little cost.

The correlation between the signal-strength (and/or hop-count) space and the physical location space is generally non-linear. It is also usually not possible to know *a priori*, given a sensor node, the exact features that uniquely identify its location. A versatile and productive approach for learning correlations of this kind is based on the kernel methods for statistical classification and regression (Scholkopf & Smola (2002)). Central to this methodology is the notion of a *kernel function*, which provides a generalized measure of similarity for any pair of entities (e.g., sensor locations, sensor signals, hop-counts). The functions that are produced by the kernel methods (such as support vector machines and kernel canonical correlation analysis) are sums of kernel functions, with the number of terms in the sum equal to the number of data points. Kernel methods are examples of nonparametric statistical procedures – procedures that aim to capture large, open-ended classes of functions.

Given that the raw signal readings in a sensor network implicitly capture topological relations among sensor nodes, kernel methods would seem to be particularly natural in the sensor network setting. In the simplest case, the signal strength/hop-count would itself be a kernel function. More generally, and more realistically, derived kernels can be defined based on the signal strength/hop-count matrix. In particular, inner products between vectors of received signal strengths/hop-counts can be used in kernel methods. Alternatively, generalized inner products of these vectors can be computed – this simply involves the use of higher-level kernels whose arguments are transformations induced by lower-level kernels. In general, hierarchies of kernels can be defined to convert the initial topology provided by the

Figure 1. If we can define a set of classes that represent geographic regions, a sensor node's location can be estimated based on its memberships in these classes

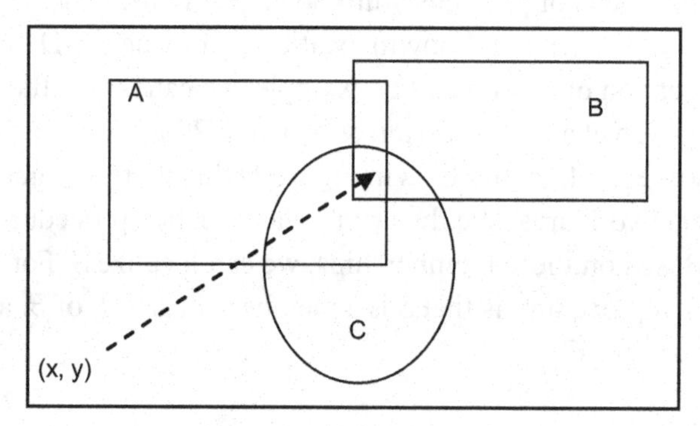

raw sensor readings into a topology more appropriate for the classification or regression task at hand. This can be done with little or no knowledge of the physical sensor model.

In this chapter, we describe localization techniques that build on kernel-based learning methods for classification and regression/correlation analysis.

NOTATIONS AND ASSUMPTIONS

We consider a wireless sensor network of N nodes $\{S_1, S_2, ..., S_N\}$ deployed in a 2-D geographic area $[0, D]^2$ $(D > 0)$. (Here, we assume two dimensions for simplicity, though the techniques to be presented can work with any dimensionality.) We assume the existence of k beacon nodes $\{S_1, S_2, ..., S_k\}$ with known location $(k < N)$. We will devise learning-based algorithms where an estimate can be made for the location of each remaining node $\{S_{k+1}, S_{k+2}, ..., S_N\}$.

We assume that the network is connected and an underlying routing protocol exists to provide a path $path(S_i, S_j)$ to navigate from any sensor node S_i to any other S_j, whose hop-count distance (or distance, *in short*) is denoted by $hc(S_i, S_j)$. If the routing protocol defines this path to be the shortest path in hop-count from S_i to S_j, the distance $hc(S_i, S_j)$ is the least number of hops between them. The sensor coverage is not necessarily uniform; hence, $path(S_i, S_j)$ may not equal $path(S_j, S_i)$ and $hc(S_i, S_j)$ may not equal $hc(S_j, S_i)$. Also, we denote by $ss(S_i, S)$ the signal strength a sensor node S receives from each beacon S_i.

If the network is small enough that any sensor node can hear directly from a majority of the beacons, we can use signal-strength information to estimate the locations for sensor nodes. In practice, however, there is a large class of sensor networks where a node may hear directly from just a few beacons and there may be nodes that do not hear directly from any beacon node. For this type of networks, we learn to estimate the locations based on hop-count information rather than signal-strength information.

Before we present the details in the next sections, the localization procedure is summarized as follows:

1. The beacon nodes communicate with each other so that for each beacon node S_i we can obtain the following k-dimensional distance vector

   $$h_i = (\, hc(S_1, S_i) \quad hc(S_2, S_i) \quad \ldots \quad hc(S_k, S_i)\,)$$

   or, for the case of a small network, the k-dimensional signal-strength vector

   $$s_i = (\, ss(S_1, S_i) \quad ss(S_2, S_i) \quad \ldots \quad ss(S_k, S_i)\,)$$

2. One beacon node is chosen, called the head beacon, to collect all these vectors from the beacon nodes and run a learning procedure (regression or classification). After the learning procedure, the prediction model is broadcasted to all the nodes in the network. Furthermore, each beacon node broadcasts a HELLO message to the network also.
3. As a result of receiving the HELLO message from each beacon, each sensor node $S_j \in \{S_{k+1}, S_{k+2}, ..., S_N\}$ computes the following k-dimensional distance vector

   $$h_j = (\, hc(S_1, S_j) \quad hc(S_2, S_j) \quad \ldots \quad hc(S_k, S_j)\,)$$

or, for the case of a small network, the k-dimensional signal-strength vector

$$s_j = (\ ss(S_1, S_j) \quad ss(S_2, S_j) \quad \ldots \quad ss(S_k, S_j)\ )$$

The sensor node then applies the prediction model it has obtained previously to this distance (or signal-strength) vector to estimate the node's location.

LOCALIZATION BASED ON CLASSIFICATION

As we mentioned in the Introduction section, the localization problem can be modeled as a classification problem. The idea was initiated in (Nguyen et al. (2005)). Generally, the first two steps are as follows:

- **Class definition:** Define a set of classes $\{C_1, C_2, \ldots\}$, with each class C_i being a geographical region in the sensor network area
- **Training data:** Because the beacon locations are known, the membership of each beacon node in each class C_i is known. The distance (or signal-strength) vector of each beacon node serves as its feature vector. The feature vector and membership information serves as the training data for the classification procedure on class C_i

We then run the classification procedure to obtain a prediction model. This model is used to estimate for each given sensor node S and class C_i the membership of S in class C_i. As a result, we can determine the area in which S is located. To solve the classification problem, it is proposed in (Nguyen et al. (2005), Tran & Nguyen (2006), Tran & Nguyen (2008)) that we use Support Vector Machines (SVM), a popular and efficient machine learning method (Cortes & Vapnik (1995)). Specifically, these techniques use binary SVM classification methods – the traditional form of SVM. A brief background on binary SVM classification is presented below and then how it is used for sensor localization.

Binary SVM Classification

Consider the problem of classifying data in a data space U into a class G or not in class G. Suppose that k data points $u_1, u_2, \ldots, u_k$, are given, called the *training* points, for which the corresponding memberships in class G are known. We need to predict whether a new data point u is in G or not. This problem is called a binary classification problem.

Support Vector Machines (SVM) (Boser et al. (1992), Cortes & Vapnik (1995)) is an efficient method to solve this problem. Central to this method is the notion of a kernel function K: $U \times U \to R$ that provides a measure of similarity between two data points in U. For the case of finite data space (e.g., location data of nodes in a sensor network), this function must be symmetric and the $k \times k$ matrix $[K(u_i, u_j)]$ $(i, j \in \{1, 2, \ldots, k\})$ must be positive semi-definite (i.e., has non-negative eigenvalues).

Given such a kernel function K, according to Mercer's theorem (cf., Scholkopf & Smola (2002)), there must exist a feature space in which the kernel K acts as the inner product, i.e., $K(u,u') = \langle \Phi(u), \Phi(u') \rangle$ for some mapping $\Phi\ (u)$. Suppose that we associate with each training data point u_i a label l_i to represent that $l_i = 1$ if $u_i \in G$ and -1 otherwise. The idea is to find a hyperplane in the feature space, that maximally separates the training points in class G from those not in G. For this purpose, the SVM and related kernel-based algorithms find a linear function $h_K(u) = \langle w, \Phi(u) \rangle - b$ in the feature space, where the vector w and parameter b are chosen to maximize the margin, or distance between the parallel hyperplanes that are as far apart as possible while still separating the training data points. Thus, if the training data points are linearly separable in the feature space, we need to minimize $\|w\|$ subject to $1 - l_i h_K(u_i) \leq 0$ for all $1 \leq i \leq k$.

Solving the above minimization problem requires the knowledge about the feature mapping $\Phi(u)$. Fortunately, by the Representer Theorem (cf., Scholkopf & Smola (2002)), the function h_K can be expressed in terms of the kernel function K only

$$h_K(u) = \sum_{i=1}^{k} \alpha_i l_i K(u, u_i) + b$$

for an optimizing choice of coefficients α_i. Using this dual form, to find the function $h_K(u)$, we solve the following maximization problem:

Maximize $$W(\alpha) = \sum_{i=1}^{k} \alpha_i - \frac{1}{2} \sum_{i,j=1}^{k} l_i l_j \alpha_i \alpha_j K(u_i, u_j)$$

subject to $$\sum_{i=1}^{k} l_i \alpha_i = 0 \text{ and } 0 \leq \alpha_i \text{ for } i \in \{1, 2, \ldots, k\}$$

Suppose that $\{\alpha_1^*, \alpha_2^*, \ldots, \alpha_k^*\}$ is the solution to this optimization problem. We choose $b = b^*$ such that $l_i h_K(u_i) = 1$ for all i with $0 < \alpha_i^*$. The training points corresponding to such (i, α_i^*)'s are called the *support vectors*. The decision rule to classify a data point u is: $u \in G$ iff $sign(h_K(u)) = 1$, where

$$h_K(u) = \sum_{i=1}^{k} \alpha_i^* l_i K(u, u_i) + b^*.$$

Under standard assumptions in statistical learning, SVM is known to yield bounded (and small) classification error when applied to the test data. The SVM method presented above is the 1-norm soft margin version of SVM. There are several extensions to this method, whose details can be found in (Boser et al. (1992), Cortes & Vapnik (1995)).

The main property of the SVM is that it only needs the definition for a kernel function K that represents a similarity measure between two data points. This is a nice property because other classifier tools usually require a known feature vector for every data point, which may not be available or derivable in many applications. In our particular case of a sensor network, it is impossible to find the features for each sensor node that uniquely and accurately identify its location. However, we can provide a similarity measure between two sensor nodes based on their relationships with the beacon nodes. Thus, SVM is highly suitable for the sensor localization problem.

Class Definition

There are more than one way to define the classes $\{C_1, C_2, \ldots\}$. For example, as illustrated in (Nguyen et al. (2005)), each class C_i can be an equi-size disk in the sensor network area such that any point in the sensor field must be covered by at least three such disks. Thus, after the learning procedure, if a sensor node S is found to be a member of three classes C_i, C_j, and C_k, the location of S is approximated as the centroid of the intersectional area $C_i \cap C_j \cap C_k$.

Using the above disk partitioning method, or any method requiring that any point in the sensor network field be covered by two or more regions represented by classes, the number of classes in the learning procedure is dependant on the field dimension and could be very high. Alternatively, (Tran & Nguyen (2006), Tran & Nguyen (2008)) propose the LSVM technique which partitions the sensor network field using a fixed number of classes, thus independent of the network field dimension (LSVM is the abbreviation for **L**ocalization based on **SVM**). Hereafter, unless otherwise mentioned, the technique we describe is LSVM. As illustrated in Figure 2, LSVM defines ($2M$-2) classes as follows, where $M = 2^m$ for some m determined later:

- M-1 classes for the X-dimension $\{cx_1, cx_2, ..., cx_{M-1}\}$, each class cx_i containing nodes with the x-coordinate $x < iD/M$
- M-1 classes for the Y-dimension $\{cy_1, cy_2, ..., cy_{M-1}\}$, each class cy_i containing nodes with the y-coordinate $y < iD/M$

We need to solve ($2M$-2) binary classification problems. Each solution, corresponding to a class cx_i (or cy_i), results in a SVM prediction model that decides whether a sensor node belongs to this class or not. If the SVM learning predicts that a node S is in class cx_{i+1} but not class cx_i, and in class cy_{j+1} but not class cy_j, we conclude that S is inside the square cell $[iD/M, (i+1)D/M] \times [jD/M, (j+1)D/M]$. We then simply use the cell's center point as the estimated position of node S (see Figure 3). If the above prediction is indeed correct, the location error (i.e., Euclidean distance between true location and estimated location) for node S is at most $\frac{D}{M\sqrt{2}}$. However, every SVM is subject to some classification error, and so a challenge is to maximize the probability that S is classified into its *true* cell, and, to minimize the location error in the case that S is classified into a wrong cell (Tran & Nguyen (2008)).

Kernel Function

The kernel function $K(S_i, S_j)$ provides a measure for similarity between two sensor nodes S_i and S_j. We define the kernel function as a Radial Basis Function because of its empirical effectiveness (Chang & Lin (2008)):

Figure 2. Definition of class cx_i ($i = 1, 2, ..., 2^m$-1)

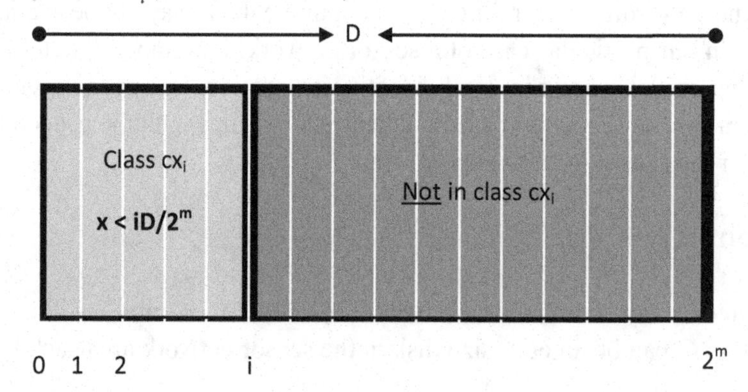

Figure 3. Localization of a node based on its memberships in regions cx_i and cy_j

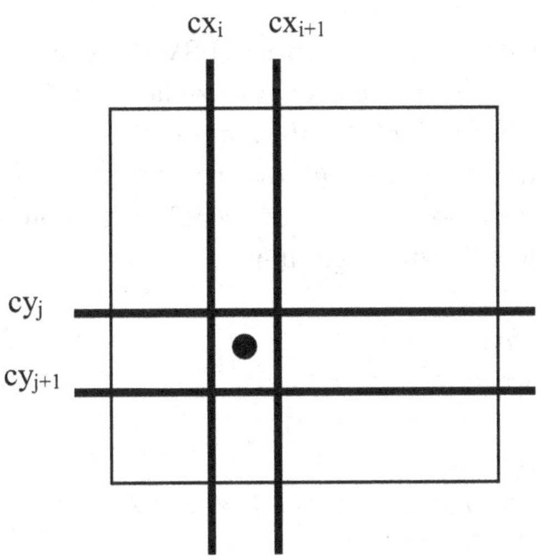

$$K(S_i, S_j) = \exp(-\gamma \left\| h_i - h_j \right\|^2)$$

where γ is a constant to be computed during the cross-validation phase of the training procedure, and h_i the k-dimensional distance vector of sensor node S_i with the j-th entry of the vector representing the hop-count distance from S_i to beacon node S_j. More examples for the kernel function are discussed in (Nguyen et al. (2005)).

Training Data

For each binary classification problem (for a class $c \in \{cx_1, cx_2, ..., cx_{M-1}, cy_1, cy_2, ..., cy_{M-1}\}$), the training data is the set of beacon nodes with corresponding labels $\{l_1, l_2, ..., l_k\}$, where $l_i = 1$ if beacon node S_i belongs to class c and -1 otherwise.

Now that the training data and kernel function have been defined for each class c, we can solve the SVM optimization problem aforementioned to obtain $\{\alpha_1^*, \alpha_2^*, ..., \alpha_k^*\}$ and b^*. We then use the decision function $h_K(.)$ to decide whether a given node S is in class c:

$$h_K(S) = \sum_{i=1}^{k} \alpha_i^* l_i K(S, S_i) + b^*$$

The training procedure is implemented as follows. The head beacon obtains the hop-count vector and location of each beacon. Then, it runs the SVM training procedure (e.g., using a SVM software tool like *libsvm* (Chang & Lin (2008)) on all (2M-2) classes $cx_1, cx_2, ..., cx_{M-1}, cy_1, cy_2, ..., cy_{M-1}$ and, for each class, computes the corresponding b^* and the information $(i, l_i\alpha_i^*)$. This information is called the SVM model information. This model information is used to predict the location of any sensor given its distance vector.

Location Estimation

Let us focus on the classification along the X-dimension. LSVM organizes the cx-classes into a binary decision tree, illustrated in Figure 4. Each tree node is a cx-class and the two outgoing links represent the outcomes (0: "not belong", 1: "belong") of classification on this class. The classes are assigned to the tree nodes such that if the tree is traversed in the *in-order* order {*left-subtree*→ *parent*→ *right-subtree*}, the result is the ordered list $cx_1 \rightarrow cx_2 \rightarrow ... \rightarrow cx_{M-1}$. Given this decision tree, each sensor node *S* can estimate its x-coordinate using the following algorithm:

Algorithm: X-dimension localization
Estimate the x-coordinate of sensor node S:

1. *Initially,* $i = M/2$ *(start at root of the tree* $cx_{M/2}$*)*
2. *IF (SVM predicts S not in class* cx_i*)*
   - *IF (*cx_i *is a leaf node – i.e., having no child decision node)*
     RETURN $x'(S) = (i - 1/2)D/M$
   - *ELSE Move to left-child* cx_j *and set* $i = j$
3. *ELSE*
   - *IF (*cx_i *is a leaf node) RETURN* $x'(S) = (i + 1/2)D/M$
   - *ELSE Move to right-child* cx_t *and set* $i = t$
4. *GOTO Step 2*
5. *END*

Similarly, a decision tree is built for the Y-dimension classes and each sensor node *S* estimates its y-coordinate *y'*(*S*) based on the *Y-dimension localization algorithm* (like the *X-dimension localization algorithm*). The estimated location for node *S*, consequently, is (*x'*(*S*), *y'*(*S*)). Using these algorithms, the localization of a node requires visiting log_2M nodes of each decision tree, after each visit the geographic

Figure 4. Decision tree: $\mathbf{m} = 4$

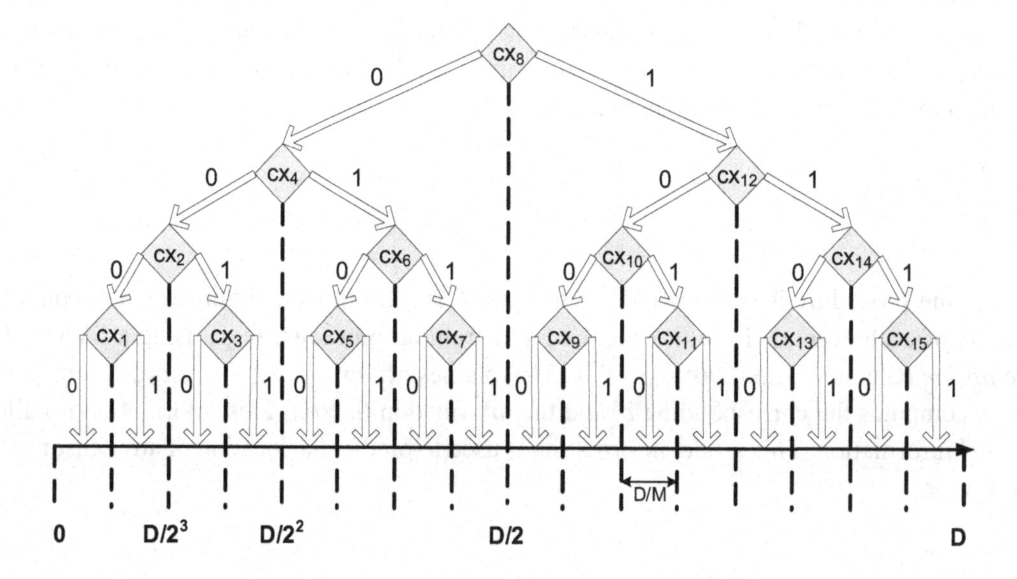

range that contains node S downsizing by a half. The parameter M (or m) controls the precision of the localization.

SVM is subject to error and so is LSVM. Let ε be the worst-case SVM classification error when SVM is applied to solve the *(2M-2)* binary classification problems, each regarding one of the classes $\{cx_1, cx_2, \ldots, cx_{M-1}, cy_1, cy_2, \ldots, cy_{M-1}\}$. For each class c, a misclassification occurs when SVM predicts that a sensor node is in c but in fact the node is not, or when SVM predicts that the node is not in c but the node actually is. The SVM classification error corresponding to class c is the ratio between the number of sensor nodes for which SVM predicts correctly to the total number of all sensor nodes. In (Tran & Nguyen (2008)), it is shown that for a uniformly distributed sensor network field, the location error expected for any node is bounded by

$$E^u = \sqrt{2}D\left(\frac{1}{2^m} + \frac{7}{8} - \frac{(1-\varepsilon)^m}{2^{m+1}} - \frac{(2-\varepsilon)^m}{2^m} + \frac{(4-3\varepsilon)^m}{2^{2m+3}}\right)$$

The location error expectation E^u decreases as the SVM error ε gets smaller. Figure 5 plots the error expectation E^u for various values of ε. There exists a choice for m (no larger than 8) that minimizes the error expectation. In a real-world implementation, it is recommended that we use this optimal m. A nice property of SVM is that ε is typically upper-bounded and under certain assumptions on the choice of the kernel function, the bound diminishes if the training size gets sufficiently large. In the evaluation study of (Tran & Nguyen (2008)), when simulated on a network of 1000 sensors with non-uniform coverage, of which 5% serves as beacon nodes, the error ε is no more than 0.1. This is one example showing that SVM offers a high accuracy when used to classify the sensor nodes into their correct classes. Later in this chapter more evaluation results are presented to demonstrate the localization accuracy of LSVM.

LOCALIZATION BASED ON REGRESSION

Trilateration is a geometrical technique that can locate an object based on its Euclidean distances from three or more other objects. In our case, to locate a sensor node we do not know its true Euclidean distances from the k beacon nodes. We can use a regression tool (e.g., *libsvm* (Chang & Lin (2008))) to learn about these distances using hop-count information. The beacon leader constructs a linear regression function $f: N \rightarrow R$ with the following training data

$$f(hc(S_i, S_j)) = d(S_i, S_j) \text{ for all } i, j \in \{1, 2, \ldots, k\}$$

where $d(S_i, S_j)$ is the Euclidean distance between S_i and S_j. Once this regressor f is computed, it is broadcast to all the sensor nodes. Since each node receives a HELLO message from each beacon, the former can compute its distance vector and apply the regressor f to compute its location (Tran & Nguyen (2008b)). A similar approach, but applied on signal-strength data, was considered by (Kuh & Zhu (2006), Zhu & Kuh (2007), Kuh & Zhu (2008)). Kuh & Zhu uses least squares SVM regression to solve the localization problem with beacon locations as training data. This involves solving a system of linear equations. To achieve sparseness, a procedure is used to choose the support vectors based on training data error.

If the network is sufficiently small, each sensor node can hear from all the beacon nodes. It is observed that if two nodes S_i and S_j receive similar signal strengths from the beacon nodes, and, if the

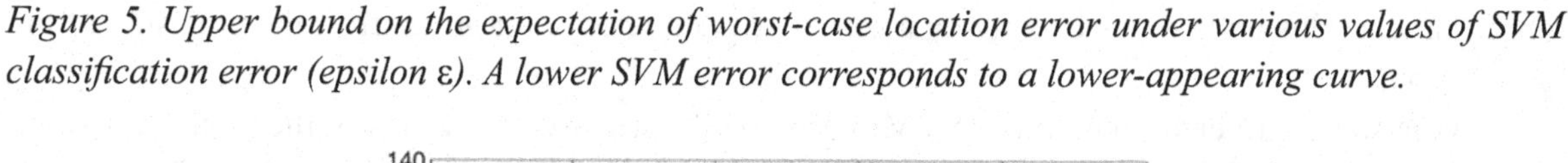

Figure 5. Upper bound on the expectation of worst-case location error under various values of SVM classification error (epsilon ε). A lower SVM error corresponds to a lower-appearing curve.

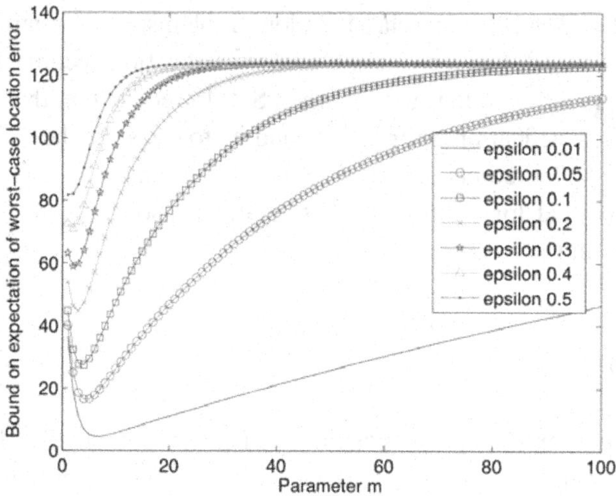

number of beacons is large enough (at least 3), these nodes should be near each other in the physical space. Thus, one could be able to exploit directly the high correlation statistics between the similarity of signal strengths and that of sensor locations. This insight was observed by (Pan et al. (2006)), who proposed to use Kernel Canonical Correlation Analysis (KCCA) (Akaho (2001), Hardoon et al. (2004)) for the regression that maps a vector in the signal-strength space to a location in the physical space. We briefly present KCCA below and then how it is used for the localization problem.

Kernel Canonical Correlation Analysis (KCCA)

KCCA is an efficient non-linear extension of Canonical Correlation Analysis (CCA) (Hotelling (1936), Hardoon et al. (2004)). Suppose that there are two sets of multidimensional variables, $s = (s_1, s_2, \ldots, s_k)$ and $t = (t_1, t_2, \ldots, t_k)$. CCA finds two canonical vectors, w_s and w_t, one for each set such that the correlation between these two sets under the projections, $a = (\langle w_s, s_1 \rangle, \langle w_s, s_2 \rangle, \ldots, \langle w_s, s_k \rangle)$ and $b = (\langle w_t, t_1 \rangle, \langle w_t, t_2 \rangle, \ldots, \langle w_t, t_k \rangle)$ is maximized (the correlation is defined as

$cor(a,b) = \frac{\langle a,b \rangle}{\|a\|\|b\|}$).

While CCA only exploits linear relationship between s and t, its extension using kernels KCCA can work with non-linear relationships. KCCA defines two kernels, K_s for the s space and K_t for the t space. Each kernel K_s (or K_t) represents implicitly a feature vector space Φ_s (or Φ_t) for the corresponding variable s (or t). Then, a mapping that maximizes the correlation between s and t in the feature space is found using the kernel functions only (requiring no knowledge about Φ_s and Φ_t).

KCCA for Localization

(Pan et al. (2006)) applies KCCA to find a correlation-maximizing mapping from the signal-strength space to the physical location space (because the relationship is non-linear, KCCA is more suitable than CCA). Firstly, two kernel functions are defined, a Gaussian kernel K_s for the signal space

$$K_s(s_i,s_j) = \exp(-\gamma \left\|s_i - s_j\right\|^2)$$

and a Matern kernel K_t for the location space

$$K_t(t_i,t_j) = \frac{2(\sqrt{v}w\left\|t_i - t_j\right\|)^v}{\Gamma(v)} K_v(2\sqrt{v}w\left\|t_i - t_j\right\|)$$

where v is a smoothness parameter, $\Gamma(v)$ the gamma function, and $K_v(.)$ the modified Bessel function of the second kind. The signal strengths between the beacon nodes and their location form the training data. In other words, the k instances (s_1, t_1), (s_2, t_2), ..., (s_k, t_k), where (s_i, t_i) represents the signal-strength vector and the location of beacon node S_i, serve as the training data.

After the training is completed, suppose that q pairs of canonical vectors (ws_1, wt_1), (ws_2, wt_2), ..., (ws_q, wt_q) are found. The choice for q is flexible. Technically, more than one pair of canonical vectors can be found recursively in such a way that a newly found pair must be orthogonal to the previous pair and maximally correlate the canonical variates resulted from the previous pair. Thus, q can be chosen as large as we can find a new pair of canonical vectors that improves the correlation according to the previous pair by a *significant* margin (which can be defined by some threshold).

A sensor node $S \in \{S_{k+1}, S_{k+2}, ..., S_N\}$ is localized as follows:

- Compute the signal-strength vector s of sensor node S: $s = (ss(S_1, S), ss(S_2, S), ..., ss(S_k, S))$
- Compute the projection of
  $$P(s) = \left(\langle ws_1, s\rangle, \langle ws_2, s\rangle, ..., \langle ws_q, s\rangle\right)$$
- Choose from the set of beacon nodes m nodes $\{S_i\}$ whose projections $\{P(s_i)\}$ are nearest to $P(s)$. The distance metric used is a weighted Euclidean distance where the weights are obtained from the KCCA training procedure and the canonical vectors wt_1, wt_2, ..., wt_q
- Compute the location for S as the centroid position of these m neighbors

EVALUATION RESULTS

This section presents some evaluation results that demonstrate the effectiveness of the learning-based approach to the sensor localization problem. The main overhead for this approach is the training procedure. It involves communication among the beacon nodes to obtain their distance (or signal-strength) vectors. Then, the head beacon collects this information to run the SVM, resulting in a prediction model which is then broadcast to all the nodes in the network. The location estimation procedure at each node consists of only a small number of comparisons and simple computations. Thus, the approach is fast and simple.

In the following, we show the location error results for LSVM – the classification based technique that uses the hop-count information to learn the sensor nodes' locations. These results are extracted from the evaluation study presented in (Tran & Nguyen (2008)). The evaluation results for the other learning-based techniques can be found in (Tran & Nguyen (2008b)) (regression-based localization using hop-count information), (Nguyen et al. (2005)) (classification-based localization using signal strength) and (Pan et al. (2006)) (regression-based localization using signal strength).

In (Tran & Nguyen (2008)), LSVM is compared to Diffusion (Bulusu et al. (2002), Meertens & Fitzpatrick (2004)). Diffusion is an existing technique that does not require ranging measurements and also uses beacon nodes with known locations. Unlike LSVM, Diffusion is not based on machine learning. In Diffusion, each sensor node's location is initially estimated as a random location in the sensor network area. Each node, a sensor node or a beacon node, then repeatedly exchanges its location estimate with its neighbors and uses the centroid of the neighbors' locations as the new location estimate. This procedure after a number of iterations will converge to a state where each node does not improve its location estimate significantly.

Consider a network of 1000 sensor nodes located in a 100m by 100m 2-D area. The selection of the beacon nodes among the sensor nodes is based on uniform distribution. The communication radius for each node is 10m. Five different beacon populations are considered: 5% of the network size (k = 50 beacons), 10% (k = 100 beacons), 15% (k = 150 beacons), 20% (k = 200 beacons), and 25% (k = 250 beacons). The algorithms in the *libsvm* software kit (Chang & Lin (2008)) are used for SVM classification. The parameter m is set to 7 (i.e., M = 128).

Figure 6 shows that LSVM is more accurate than Diffusion. In this study, node locations are uniformly distributed in the network area. Diffusion converges after 100 iterations (Diff-100). It does not improve when more iterations are run, 1000 iterations (Diff-1000) or 10,000 iterations (Diff-10000). The difference between the two techniques is the most noticeable when the number of beacons is small (k = 50) and decreases as more beacon nodes are used. In any case, even when k = 50 (only 5% of the network serve as beacon nodes), the location error for an average node using LSVM is always less than 6m.

Another nice property of LSVM is that it distributes the error fairly across all the nodes. As an example, Figure 7 shows the localization results for the case k = 50. In this figure, a line connects the

Figure 6. LSVM vs. diffusion: Average location error with various choices for the number of beacons

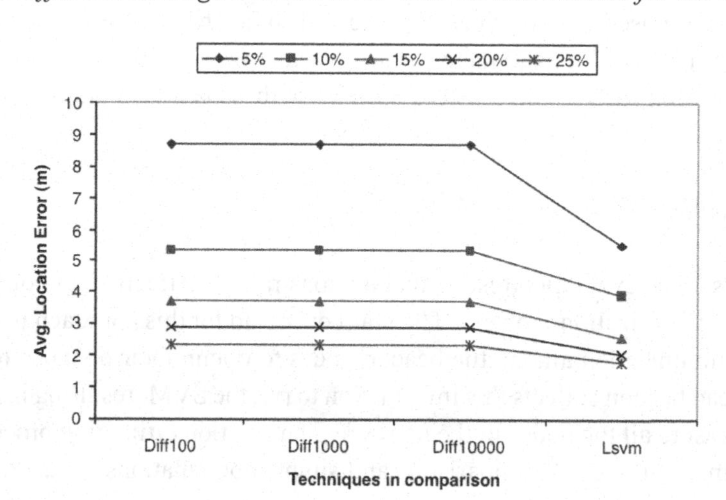

true location and the estimated location for each node. It can be observed that Diffusion suffers severely from the border problem: nodes near the border of the network area are poorly localized. LSVM does not incur this problem.

Many networking protocols such as routing and localization suffer from the existence of coverage holes or obstacles in the sensor network area. (Tran & Nguyen (2008)) also shows that even so, LSVM remains much better than Diffusion. For example, with the sensor network placement shown in Figure 8, where there is a big hole of radius 25m centered at position (50, 50). Table 1 shows that LSVM improves the location error over Diffusion by at least 20% in all measures (average, worst, standard deviation) under every beacon population size.

Figure 7. Diffusion (1000 iterations, left) vs. LSVM (right): A line connects the true location and the estimated location of each sensor node (total 1000 nodes, 50 beacon nodes)

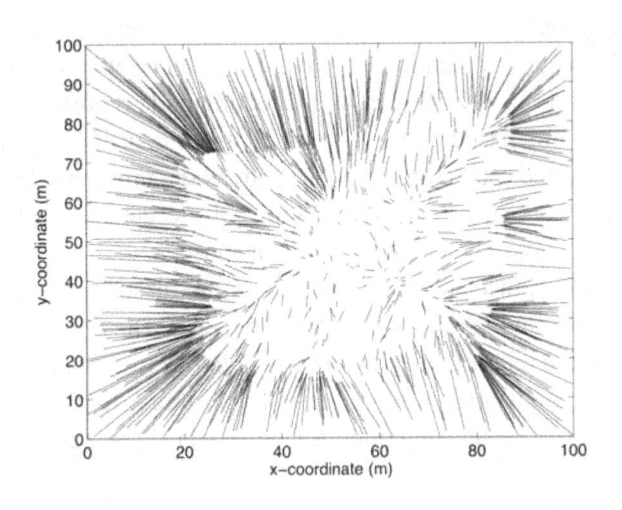

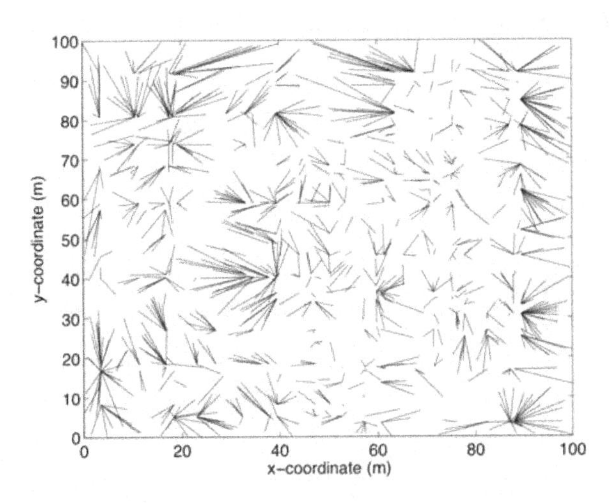

APPLICATION TO TARGET TRACKING

An appealing feature of the presented learning-based approach is that the localization of a sensor node can be done independently from that of another sensor node. The training procedure involves the beacon nodes only, whose result is a prediction model any sensor node can use to localize itself without knowledge about other nodes. This feature is suitable for target tracking in a sensor network where, to save cost, not every sensor node needs to run the localization algorithm; only the target needs to be localized. For example, consider a target tracking system with k beacon nodes deployed at known locations. When a target T occurs in an area and is detected by a sensor node S_T, the detecting node reports the event to the k beacon nodes. The distance vector $[hc(S_T, S_i)]$ $(i = 1, 2, ..., k)$ is forwarded to the sink station who will use the prediction model learned in the training procedure to estimate the location of target T.

An important issue in the learning-based approach is that its accuracy depends on the size of the training data; in our case, the number of beacon nodes. However, in many situations, the beacon nodes are deployed incrementally, starting with a few beacon nodes and gradually with more. In other cases, the set of beacon nodes can also be dynamic. The beacon nodes that are made available to a sensor node (or target) under localization may change depending on the location of this node (or target). We need a solution that learns based on not only the current measurements but also the past. For example, reconsider the target tracking system mentioned above. When a target is detected, sending the event to all the beacon nodes can be very costly. Instead, the detecting node reports the event to a few, possibly random, beacon nodes. Learning based on the current measurements (signal strengths or hop-counts) may be inaccurate because of the sparse training data, but as the target moves, by combining the past learned information with the current, we can better localize the target. Sequential prediction techniques (Cesa-Bianchi & Lugosi (2006)) can be helpful for this purpose.

(Letchner et al. (2005)) propose a localization technique aimed at such dynamism of the beacon nodes. The technique is based on a hierarchical Bayesian model which learns from signal strengths to estimate the target's location. It is able to incorporate new beacon nodes as they appear over time. Alternatively, (Oh et al. (2005)) consider a challenging problem of multiple-target tracking by Markov chain Monte Carlo inference in a hierarchical Bayesian model. Recently, (Pan et al. (2007)) address the problem of

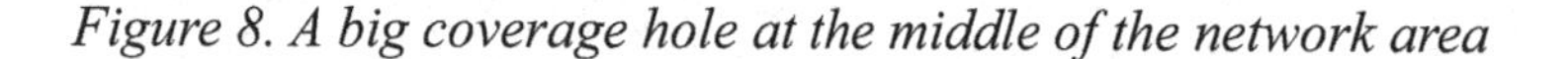

Figure 8. A big coverage hole at the middle of the network area

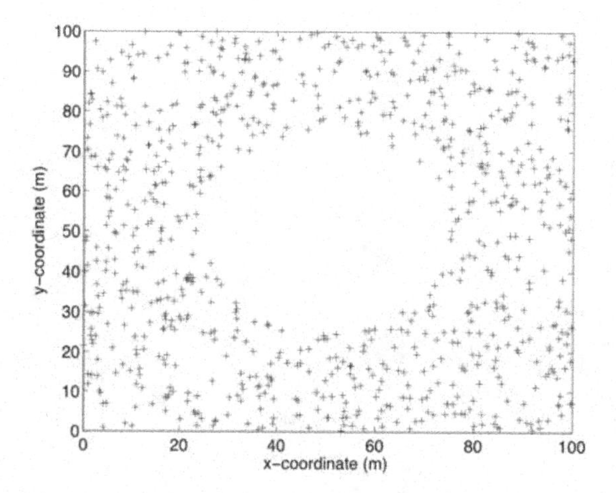

Table 1. Location-error improvement of LSVM over Diffusion for the network with a coverage hole shown in Figure 8

	k = 50	100	150	200	250
Average-case	30.97%	31.30%	34.88%	33.91%	26.50%
Worst-case	27.35%	21.40%	34.35%	33.42%	23.83%
Std. deviation	35.74%	36.29%	37.66%	37.40%	28.34%

not only locating the mobile target but also dynamically located beacon locations. The solution proposed in (Pan et al. (2007)) is based on online and incremental manifold-learning techniques (Law & Jain (2006)) which can utilize both labeled and unlabeled data that come sequentially.

Both (Letchner et al. (2005)) and (Pan et al. (2007)) learn from signal strength information, thus suitable for small networks where measurements of direct signals from the beacons are possible. The ideas could be applicable to a large network where hop-count information is used in the learning procedure rather than signal strengths. The effectiveness, however, has not been evaluated. Investigation in this direction would be an interesting problem for future research.

SUMMARY

This chapter provides a nonconventional perspective to the sensor localization problem. In this perspective, sensor localization can be seen as a classification problem or a regression problem, two popular subjects of Machine Learning. In particular, the presented localization techniques borrow the ideas from kernel methods.

The learning-based approach is favored for its simplicity and modest requirements. The localization of a node is independent from that of others. Also, past information is useful in the learning procedure and, therefore, this approach is highly suitable for target tracking applications where the information about the target at each time instant is partial or sparse, insufficient for geometry-based techniques to work effectively.

Although the localization accuracy can improve as more training data is available, collecting large training data or having many beacon nodes results in significant processing and communication overhead. A challenge for future research is to reduce this overhead. Also, it would be interesting to make one or more beacon nodes mobile and study how learning can be helpful in such an environment.

REFERENCES

Akaho, S. (2001). A kernel method for canonical correlation analysis. *In Proceedings of the International Meeting of the Psychometric Society (IMPS 2001).*

Boser, B. E., Guyon, I. M., & Vapnik V. N. (1992). A training algorithm for optimal margin classifiers. In *5th Annual ACM Workshop on COLT,* (pp. 144-152). ACM Press.

Brunato, M., & Battiti, R. (2005). Statistical learning theory for location fingerprinting in wireless LANs. *Computer Networks, 47*(6), 825-845.

Bulusu, N., Bychkovskiy, V., Estrin, D., & Heidemann, J. (2002). Scalable ad hoc deployable rf-based localization. *In 2002 Grace Hopper Celebration of Women in Computing Conference.* Vancouver, Canada.

Capkun, S., Hamdi, M., & Hubauz, J.-P. (2001). Gps-free positioning in mobile ad hoc networks. In *2001 Hawai International Conference on System Sciences.*

Cesa-Bianchi, N., & Lugosi, G. (2006). *Prediction, Learning, and Games.* Cambridge University Press. ISBN-10 0-521-84108-9.

Cortes, C., & Vapnik, V. (1995). Support-vector networks. *Machine Learning, 20*(3), 273-297.

Chang, C.-C., & Lin, C.-J. (2008). *LIBSVM – A library for Support Vector Machines.* National Taiwan University. URL http://www.csie.ntu.edu.tw/ cjlin/libsvm

Doherty, L., Ghaoui, L. E., & Pister, K. S. J. (2001). Convex position estimation in wireless sensor networks. *In IEEE INFOCOM,* 2001.

Gezici, S., Giannakis, G., Kobayashi, H., Molisch, A., Poor, H., & Sahinoglu, Z. (2005). Localization via ultra-wideband radios: a look at positioning aspects for future sensor networks. *IEEE Signal Processing Magazine, 22*(4), 70-84, 2005.

Hardoon, D. R., Szedmak, S., & Shawe-Taylor, J. (2004). Canonical correlation analysis; an overview with application to learning methods. *Neural Computation, 16,* 2639–2664.

He, T., Huang, C., Blum, B., Stankovic, J., & Abdelzaher, T. (2003). Range-free localization schemes in large scale sensor networks. In *ACM Conference on Mobile Computing and Networking,* (pp. 81-95).

Hotelling, H. (1936). Relations between two sets of variants. *Biometrika, 28,* 321-377.

Kwon, Y., Mechitov, K., Sundresh, S., Kim, W., & Agha, G. (2004). Resilient localization for sensor networks in outdoor environments. Tech. rep., University of Illinois at Urbana-Champaign, 2004.

Kuh, A., Zhu, C., & Mandic, D. P. (2006). Sensor network localization using least squares kernel regression. In *Knowledge-Based Intelligent Information and Engineering Systems,* (pp. 1280-1287).

Kuh, A., & Zhu, C. (2008). Sensor network localization using least squares kernel regression. In D. Mandic et al. (Eds.), *Signal Processing Techniques for Knowledge Extraction and Information Fusion.* (pp. 77-96), Springer, April 2008.

Law, M. H. C., & Jain, A. K. (2006). Incremental nonlinear dimensionality reduction by manifold learning. *IEEE Transactions on Pattern Analysis and Machine Intelligence, 28*(3), 377–391.

Lee, J. –Y., & Scholtz, R. Ranging in a dense multipath environment using an UWB radio link. *IEEE Journal on Selected Areas in Communications, 20*(9), 1677-1683.

Letchner, J., Fox, D., & LaMarca, A. (2005). Large-Scale Localization from Wireless Signal Strength. In *Proc. of the National Conference on Artificial Intelligence (AAAI),* (pp. 15-20).

Meertens, L., & Fitzpatrick, S. (2004). *The distributed construction of a global coordinate system in a network of static computational nodes from inter-node didstances.* Tech. rep., Kestrel Institute, 2004.

Moore, D., Leonard, J., Rus, D., & Teller, S. (2004). Robust distributed network localization with noisy range measurements. In *ACM Sensys*, (pp. 50-61). Baltimore, MA.

Nagpal, R., Shrobe, H., & Bachrach, J. (2003). Organizing a global coordinate system from local information on an ad hoc sensor network. In *International Symposium on Information Processing in Sensor Networks* (pp. 333-348).

Nguyen, X., Jordan, M. I., & Sinopoli, B. (2005). A kernel-based learning approach to ad hoc sensor network localization. *ACM Transactions on Sensor Networks, 1*, 134-152.

Niculescu, D., & Nath, B. (2003a). Ad hoc positioning system (aps) using aoa. *In IEEE INFOCOM.*

Niculescu, D., & Nath, B. (2003b). Dv based positioning in ad hoc networks. *Telecommunication Systems, 22*(1-4), 267–280.

Oh, S., Sastry, S., & Schenato, L. (2005). A Hierarchical Multiple-Target Tracking Algorithm for Sensor Networks. *In Proc. International Conference on Robotics and Automation.*

Pan, J. J., Kwok, J. T., & Chen, Y. (2006). Multidimensional Vector Regression for Accurate and Low-Cost Location Estimation in Pervasive Computing. *IEEE Transactions on Knowledge and Data Engineering, 18*(9), 1181-1193.

Pan, J. J., Yang, Q., & Pan, J. (2007). Online Co-Localization in Indoor Wireless Networks by Dimension Reduction. *In Proceedings of the 22nd National Conference on Artificial Intelligence (AAAI-07),* (pp. 1102-1107).

Priyantha, N., Chakraborty, A., & Balakrishnan, H. (2000). The cricket location-support system. In *ACM International Conference on Mobile Computing and Networking (MOBICOM),* (pp. 32-43).

Priyantha, N., Miu, A., Balakrishnan, H., & Teller, S. (2001). The cricket compass for context-aware mobile applications. *In ACM conference on mobile computing and networking (MOBICOM),* (pp. 1-14).

Priyantha, N. B., Balakrishnan, H., Demaine, E., & Teller, S. (2003). Anchor-free distributed localization in sensor networks. *In ACM Sensys,* (pp. 340-341).

Priyantha, N. B., Balakrishnan, H., Demaine, E., & Teller, S. (2005). Mobile-Assisted Localization in Wireless Sensor Networks. *In IEEE INFOCOM.* Miami, FL.

Priyantha, N. B. (2005b). *The Cricket Indoor Location System.* Ph.D. thesis, Massachussette Institute of Technology.

Savarese, C., Rabaey, J., & Beutel, J. (2001). Locationing in distributed ad-hoc wireless sensor networks. In *IEEE International Conference on Acoustics, Speech, and Signal Processing,* pp. 2037-2040. Salt Lake city, UT.

Savvides, A., Han, C.-C., & Strivastava, M. B. (2001b). Dynamic fine-grained localization in ad hoc networks of sensors. In *ACM International Conference on Mobile Computing and Networking (Mobicom)*, (pp. 166–179). Rome, Italy.

Savvides, A., Park, H., & Srivastava, M. (2002). The bits and flops of the n-hop multilateration primitive for node localization problems. In *Workshop on Wireless Networks and Applications (in conjunction with Mobicom 2002)*, (pp. 112-121). Atlanta, GA.

Shang, Juml, Zhang, & Fromherz (2003). Localization from mere connectivity. In *ACM Mobihoc,* (pp. 201-212).

Scholkopf, B., & Smola, A. (2002). *Learning with kernel.* Cambridge, MA: MIT Press.

Stoleru, R., Stankovic, J. A., & Luebke, D. (2005). A high-accuracy, low-cost localization system for wireless sensor networks. In *ACM Sensys,* (pp. 13-26). San Diego, CA.

Tran, D. A., & Nguyen, T. (2006). Support vector classification strategies for localization in sensor networks. In *IEEE Int'l Conference on Communications and Electronics.*

Tran, D. A., & Nguyen, T. (2008). Localization in Wireless Sensor Networks based on Support Vector Machines. *IEEE Transactions on Parallel and Distributed Systems, 19*(7), 981-994, July.

Tran, D. A., & Nguyen, T. (2008b). Hop-count based learning techniques for passive target tracking in sensor networks. *IEEE Transactions on Systems, Man, and Cybernetics*, submitted, 2008.

Whitehouse, C. (2002). *The design of calamari: an ad hoc localization system for sensor networks.* Master's thesis, University of California at Berkeley.

Zhu, C., & Kuh, A. (2007). Ad hoc sensor network localization using distributed kernel regression algorithms. In *Int'l Conference on Acoustics, Speech, and Signal Processing, 2*, 497-500.

Chapter XIII
Robust Localization Using Identifying Codes

Moshe Laifenfeld
Boston University, USA

Ari Trachtenberg
Boston University, USA

David Starobinski
Boston University, USA

ABSTRACT

Various real-life environments are exceptionally harsh for signal propagation, rendering well-known trilateration techniques (e.g. GPS) unsuitable for localization. Alternative proximity-based techniques, based on placing sensors near every location of interest, can be fairly complicated to set up, and are often sensitive to sensor failures or corruptions. The authors propose a different paradigm for robust localization based on identifying codes, a concept borrowed from the information theory literature. This chapter describes theoretical and practical considerations in designing and implementing such a localization infrastructure, together with experimental data supporting the potential benefits of the proposed technique.

PROBLEM STATEMENT

Dense indoor or urban settings, underwater or underground systems, and many emergency environments typically exhibit signal propagation properties that are extremely difficult to predict. Within these environments, signal strength or time-of-flight measurements do not accurately convey distance information, often due to spurious multi-path effects, occlusions, or noise that is very hard to characterize or model. As a result, traditional trilateration techniques, such as the Global Positioning System (GPS), are very hard to adapt to such systems without significant and often catastrophic error.

In practice, many schemes for localization in such environments are proximity-based, meaning that the location of an object is determined by the closest sensor. Generally, these systems are based on short range sensing techniques such as infrared in the Active Badge system (Want, Hopper, Falcao, & Gibbons, 1992), ultrasound (combined with RF) in the Cricket system (Priyantha, Chakraborty, & Balakrishnan, 2000), Bluetooth in (Aalto, Göthlin, Korhonen, & Ojala, 2004), and radio frequency identification - RFID in the LANDMARK system (Ni, Liu, Lau, & Patil, 2005). Some of these approaches utilize several nearby sensors to make the localization more accurate, but the underlying system organization remains proximity-based, so that the sensing area must be divided into similarly-sized regions, typically with minimal intersection. Properly setting up such systems can be fairly complicated, and the localization provided can be sensitive to sensor failures or corruptions, as well as interferences due to the significant environment changes that might take place.

Another approach to localization involves careful measurement and mapping of signal features within the coverage area. These methods are commonly referred to as fingerprinting, since a specific location region is identified by a unique set of features (i.e. a fingerprint) of the sensed signal. One of the most common fingerprinting techniques is based on the signal strength (or signal to noise ratio) of a received RF signal. Systems such as RADAR (Bahl & Padmanabhan, 2000), SpotOn (Hightower, Borriello, & Want, 2000), and Nibble (Castro, Chiu, Kremenek, & Muntz, 2001) map the signal strength received from several beacons onto a coverage area in order to train a probabilistic localizer. Other similar systems focus on commercially used communication standards to provide localization services, such as Wi-Fi (Ladd et al., 2002), and Groupe Spécial Mobile - GSM (Varshavsky, de Lara, Hightower, LaMarca, & Otsason, 2007). These systems usually perform with relatively high accuracy, although they require careful and complex planning. There are also inherent issues with robustness in such systems, since signal strength or signal to noise ratio are susceptible to the radio frequency - RF propagation channel, and can vary considerably with small changes within the environment, especially in the environments considered.

The performance of a location detection system can be characterized by many measures: resolution, responsiveness (delay until detection), etc. For many applications an important performance measure is the probability of correctly determining the region in which a target is located (i.e. the *correctness* of the system). For example, within the context of emergency response systems, correctness is usually much more important than resolution: to locate a trapped victim, it is usually sufficient to know her general location (e.g. floor and room); on the other hand, sending rescuers to the wrong location in an emergency situation can be deadly.

Motivated by these applications, localization schemes have been proposed that are based on identifying codes (Ray, Starobinski, Trachtenberg, & Ungrangsi, 2004), a concept borrowed from information-theory with links to coverings and superimposed codes. In this approach, only a small subset of sensors is activated as beacons. The subset is chosen so that its sensors have an *identification property*, meaning that a unique (or identifying) collection of these beacons can be detect at any location of interest, with specific regard to physical proximity. As such, a user can identify its location by simply tallying which beacons it can detect.

These systems benefit from heterogeneous sensor placements, and often require significantly fewer sensors to cover a localization area. Moreover such localization can be made robust to spurious connections or sensor failures through the judicious addition of redundant sensors. In general, identifying codes based localization can be viewed as a particular case of fingerprinting where the sensed feature is a binary variable indicating whether a particular beacon is detected. Therefore, identifying codes based localization is expected to be simpler and have much higher robustness than traditional fingerprinting methods, although at the expense of reduced localization accuracy. This tradeoff is advantageous to a number of harsh environments.

Outline

In this chapter we provide a survey of the literature relevant to identifying-code based localization. Specifically, we describe theoretical and practical considerations in the proper design and setup of such a system within a harsh environment, including:

- **Finding good codes:** Presenting provably good approximation algorithms for finding robust identifying codes of arbitrary graphs.
- **Robustness:** Providing means of protecting a localization network against unanticipated changes in topology or signal propagation path.
- **Distributed computation:** Presenting distributed methods of determining robust identifying codes in a sensor network.

Throughout the work, we provide the advantages and shortcomings of identifying-code localization, supported by theoretical results and simulations on Erdos-Renyi graphs and geometric random graphs, and experiments on a testbed on the fourth floor of the Photonics center at Boston University.

BASIC DEFINITIONS

Consider a graph G with vertex set V. A *neighborhood* of a vertex v in V consists of all vertices adjacent to v together with v itself. An *identifying code* on the graph is defined to be a subset $C \subseteq V$ with the property that each neighborhood of the graph consists of a unique collection of vertices in C.

Figure 1 depicts an identifying code of a seven-vertex graph (labeled a through g). The identifying code consists of four highlighted vertices $\{a,b,c,d\}$, which we shall call *codewords*; the unique intersection of each neighborhood with the code is denoted in braces next to the node. For example, vertex g has neighbors $\{a,d,e,f,g\}$, of which a and d are codewords; no other neighborhood contains exactly the set $\{a,d\}$, meaning that, within the parameters of our model, a user that detect beacons from a and d can safely assume that it is at location g.

A *minimum* or an *optimal* identifying code of a given graph is defined to be an identifying code with the smallest possible number of vertices for that graph.

ALGORITHMS

An identifying code based localization system divides the coverage area into a finite set of regions, and reports a point in this region as the location for a given target (see Figure 2).

Figure 1. An example of an identifying code on a graph. The vertices are labeled by letter a...g, with codewords corresponding to highlighted vertices.

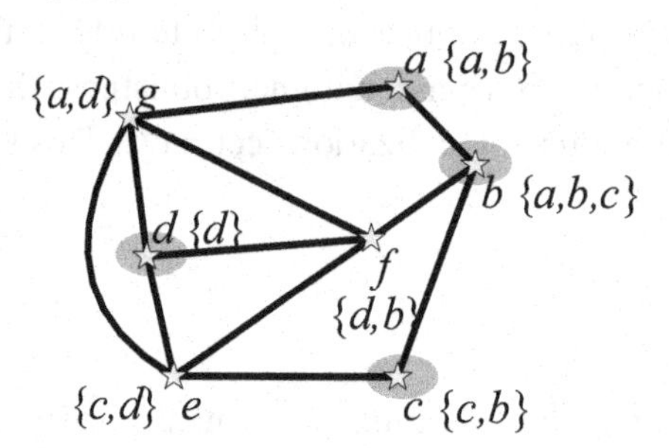

The system operates in either or both of two modes, location service or location tracking, which are equivalent (up to implementational details). In the first mode, the system periodically broadcasts identity (ID) packets from designated beacons, which an observer can harvest to determine his location. In the location tracking mode, an observer transmits his ID and the system determines his location from the sensors receiving the ID.

In the system's design process a set of points is selected for a given coverage area. Then transmitting beacons are placed on a subset of these points, based on their RF-connectivity, in a manner corresponding to an identifying code. This placement guarantees that each point is covered by a unique set of transmitters. Thus, an observer can determine its location from the unique collection of ID packets that it receives.

In summary, identifying codes based localization involves:

1. Choosing a set of discrete points and transmitters in a given region such that each point is covered by a distinct set of transmitters (optimization of the transmitters' locations should be considered as a separate step)

Figure 2. An office location detection plan. The continuous office area is quantized into finite regions a,...,f (left). The RF connectivity between the regions is shown by black arrows (right).

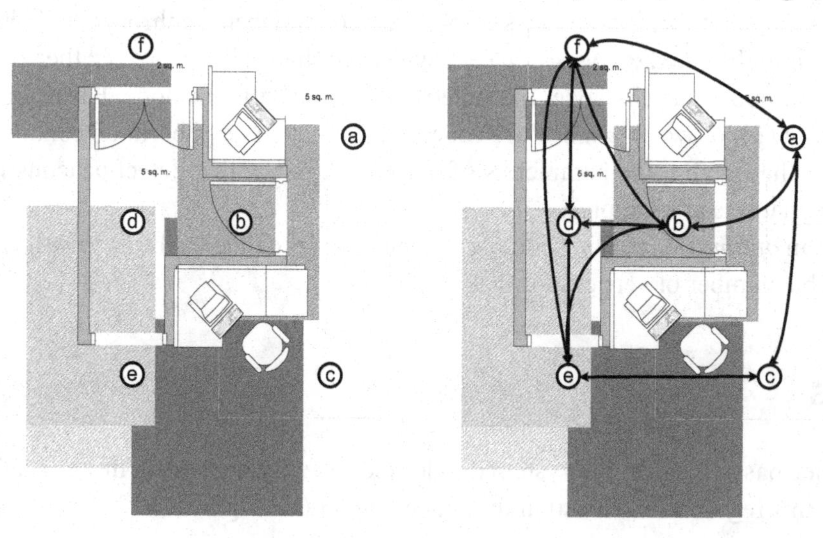

2. Resolving the location of a sensor to be the discrete point covered by the same set of transmitters.

Example

The following example illustrates the identifying-code location-detection approach in more detail. Consider the points *a,...,f* on a simple office floor plan illustrated in Figure 2, and let the RF connectivity among these points be represented by the arrows in Figure 2 (right); in other words, there is an arrow from position *a* to position *b* if and only if *b* can receive an RF signal transmitted by *a*. Given such connectivity information between every pair of points, our objective is to build a system using a minimum number of transmitters that allows an observer to infer his location at any point in {*a,...,f*}. We thus place four wireless transmitters at positions *a,b,c* and *d*, with each transmitter periodically broadcasting a unique ID. We assume that packet collisions are avoided by an appropriate medium access control (*e.g.,* simple randomization or a full-scale protocol) and that the observer collects received packets over a (small) amount of time.

For example, in Figure 1, an observer in the region of point *e* would receive IDs from the transmitters at position *c* and *d*. We shall denote by $I_C(x)$ the set of IDs received at a given position *x* under a set *C* of beacons as the *identifying set* of *x*. If the identifying set of each point in the graph is unique, then targets can be correctly located at these points using a table lookup of the packet IDs received. The reader can simply verify that all identifying sets are unique in our example.

In general, we model a physical environment by a graph of vertices and edges; its vertices model locatable regions and its edges connect regions with RF connectivity. Figure 1 shows the modeling graph for the office floor plan in Figure 2. The problem of finding the minimum number of transmitters is thus translated into the problem of finding the minimum identifying code over the RF connectivity graph.

In the next subsections we provide some algorithms for approximating the minimum identifying codes, with tight performance guarantees.

Identifying Codes in Arbitrary Graphs: First Attempt

In the most general situation, finding a minimum size identifying code for arbitrary undirected and directed graphs was proven to be NP complete in (Charon, Hudry, & Lobstein, 2003). Therefore, rather than looking for an optimal solution, in this section we focus on polynomial time approximations for constructing identifying codes.

An algorithm for generating *irreducible* identifying codes was first suggested in (Ray, Ungrangsi, Pellegrinin, Trachtenberg, & Starobinski, 2003). The irreducibility property means that the deletion of any *codeword* (an element in the identifying code) results in a code that is no longer an identifying code. Since adding codewords to an identifying code results in another identifying code, the algorithm starts with the trivial[1] identifying code - the entire set of vertices of the graph, and at each iteration it removes a vertex in a predetermined order ***a***, until the code is no longer identifying.

Algorithm 1 : ID-CODE($G,\boldsymbol{a}$) (Ray et al., 2003)
Given a graph $G=(V,E)$ and a list of vertices $\boldsymbol{a}$ do:
1) **set** $C = V$ **2)** **for** each $x \in \boldsymbol{a}$, $D = C \setminus \{x\}$ **do** 3) **if** D is an identifying code **then** $C=D$ **4)** **return** C

Clearly, the algorithm ID-CODE generates irreducible codes and therefore it always converges to a local minimum. However, it turns out that these local minima can be arbitrarily far ($O(|V|)$) from a global one, as was shown in (Moncel, 2006). Still, ID-CODE performs well on random graphs, and different heuristics for selecting the ordering list $\boldsymbol{a}$ were suggested in (Ray et al., 2003), showing a benefit in removing first vertices of too low or too high degree.

In the next section we describe a more methodical approach that leads to a polynomial time approximation for the minimum identifying code problem, with provable performance guarantees.

Identifying Codes and the Set Cover Problem

Since their introduction in 1998 identifying codes have been linked to many well studied problems in coding theory and computer science, but these fundamental links have only recently matured into polynomial time approximations. In (Laifenfeld, Trachtenberg, & Berger-Wolf, 2006) the minimum identifying code and the minimum *set cover* problems were first tied together to yield a provably good approximation, which we shall now describe.

Let U be a base set of m elements and let S be a family of subsets of U. A *cover* $C \subseteq S$ is a family of subsets whose union is U. The *set cover problem*, one of the oldest and most studied NP-complete problems, asks to find a cover C of smallest cardinality.

The set cover problem admits the following greedy approximation (GreedySetCover): at each step, and until exhaustion, choose the heretofore unselected set in S that covers the largest number of uncovered elements in the base set. The performance ratio of the greedy set cover algorithm has also been well-studied. The classic results of Lovasz and Johnson (see e.g. (Johnson, 1974)) showed that

$$\frac{s_{\text{greedy}}}{s_{\min}} \leq \ln m, \tag{1}$$

where $s_{\min}$, s_{greedy} are the minimum and the greedy covers sizes, and m is the size of the base set. The example in Figure 3 actually shows an instance of a family of problems that attains this bound.

In order to link the set cover and identifying code problems together we need some additional definitions.

A ball, $B(v)$, is defined as the set of the immediate neighbors of v including v itself. A vertex v is said to *distinguish* between a pair of vertices (u, z) if exactly one of them is a member of the its ball, *i.e.*, $|B(v) \cap \{u, z\}| = 1$. Clearly any code, C, that includes v, satisfies $I_C(u) \neq I_C(z)$; and furthermore (by the definition of an identifying code) a code C is identifying if its members distinguish between all pairs of distinct vertices in the graph[2]. We shall use $\mathbf{U}$ to denote the set of all pairs of distinct vertices, *i.e.*,

Figure 3. A set cover problem example with m=14 elements (dots) and a family of 5 subsets $S=\{S_1,...,S_5\}$ (rectangles). The minimum set cover is of size $s_{min}=2$, $C=\{S_1,S_2\}$, and the greedy set cover is of size $s_{greedy}=3$, $C_{greedy}=\{S_3,S_4,S_5\}$.

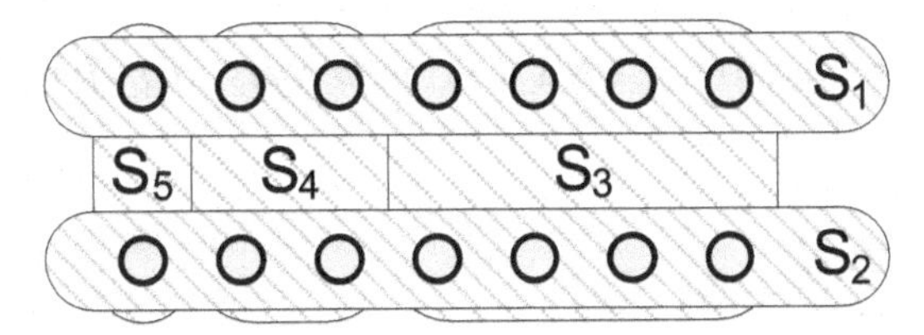

$\mathbf{U}=\{(u,z)\mid u\neq z\in V\}$, and let the distinguishing set of a vertex $v\in V$ be the set of all vertex pairs $(u,z)\in\mathbf{U}$ it distinguishes:

$$\delta_v=\{(u,z)\in\mathbf{U}:\ |B(v)\cap\{u,z\}|=1\}. \quad (2)$$

We are now ready to introduce the ID-CENTRAL algorithm.

Algorithm 2 : C_{greedy} ←ID- CENTRAL (G) (Laifenfeld et al., 2006)
Given a graph $G=(V,E)$ and the set cover heuristic GreedySetCover (U,S) **do**:
1) Compute $\Delta=\{\delta_v \mid v\in V\}$ *2)* C← GreedySetCover ($\mathbf{U}$, Δ) **3) Output** C_{greedy}← $\{v \mid \delta_v\in C\}$

To see why C_{greedy} is an identifying code, we observe that the set cover heuristic of Step 2 produces a set of distinguishing sets which cover all distinct pairs of vertices, or in other words a set of vertices that distinguish between all possible vertex pairs.

Figure 4 shows an example of ID-CENTRAL(G) on a graph G consisting of a 10 node ring; in our case, the algorithm produces the optimum identifying code, although the outcome may vary depending on the node labeling scheme and the manner of breaking ties. In the example of Figure 4, ties are broken in favor of vertices of lower label.

Figure 4. Demonstration of the ID-CENTRAL for 10 nodes ring, starting on the left. Nodes are labeled 1 to 10 clockwise (the labels appear in the inner perimeter). Solid squares represent codewords, and the distinguishing sets sizes, obtained from the greedy set-cover (GreedySetCover) appear in the outer perimeter. The resultant identifying code (right) can be shown to be optimal. (Laifenfeld, Trachtenberg, Cohen, & Starobinski, 2008).

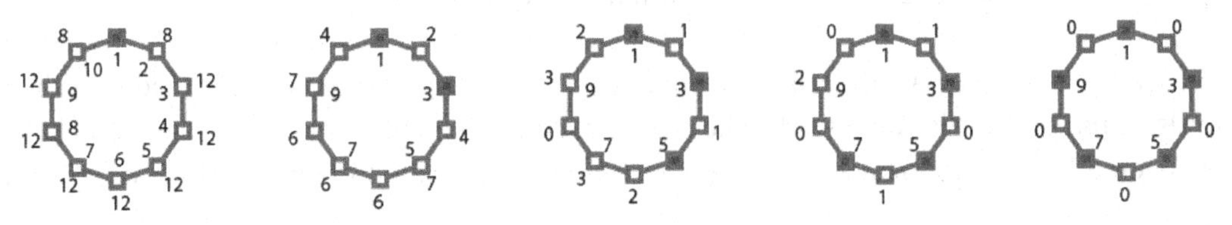

The following theorem follows straightforwardly from Equation (1) and the fact that there are $\binom{n}{2}$ distinct vertex pairs ($n=|V|$).

Theorem 1
For any given graph G of n vertices ID- CENTRAL *generates identifying code of size* c_{greedy} *that satisfies*

$$\frac{c_{\text{greedy}}}{c_{\min}} \leq 2\ln n,$$

where $c_{\min}$ *is the size of a minimum identifying code.*

In fact the work in (Laifenfeld et al., 2006; Laifenfeld & Trachtenberg, 2008) shows that the performance guarantee of Theorem 1 is also tight (up to a small factor), namely:

Theorem 2 (Laifenfeld et al., 2006 ; Laifenfeld & Trachtenberg, 2008)
There exists a family of graphs and arbitrary small $\varepsilon>0$*, for which the* ID- CENTRAL *generates identifying code of size* c_{greedy} *such that*

$$\frac{c_{\text{greedy}}}{c_{\min}} \geq (1-\varepsilon)\ln n.$$

Furthermore, using hardness of approximation results for the set cover problem, these results were extended to show that no polynomial time algorithm can approximate the minimum identifying code with a guarantee better than $(1-\varepsilon)\ln n$ under common complexity assumptions.

Similar but weaker results were later obtained in (Suomela, 2007; Gravier, Klasing, & Moncel, 2006), where the identifying codes problem was also linked to the minimum dominating set problem. Other algorithms for approximating the minimum identifying code were also suggested in the literature. In fact, an algorithm suggested in (Laifenfeld et al., 2006; Laifenfeld & Trachtenberg, 2008) based on a *test cover* approximation achieves the hardness of approximation bound within a small additive factor, guaranteeing performance ratio of $1+\ln n$.

Towards a Practical Location Detection Algorithm: Robust Identifying Codes

Robustness of the location detection system can be critical to many applications. For instance in emergency location detection systems typical corruptions include:

1. Destruction of ID-transmitting beacons (*e.g.,* collapse, fire/water).
2. Radio path changes (*e.g.,* structural disintegration, object movements).
3. Spectrum saturation due to significant communications.

In the previous section we described techniques for constructing identifying codes with as few codewords as possible. This framework inherently provides some amount of robustness, since a point may be covered by sensors located far away, thereby creating spatial diversity. However, in practice, the identifying set received by an observer might fluctuate due to environmental conditions, and thus we seek to guarantee that the scheme works even if the received identifying set differs minimally from the original. In this section we describe a generalization of identifying codes, first suggested in (Ray et al., 2003), that achieves this goal by guaranteeing to be robust in the face of spurious fluctuations in observed identifying sets.

Definition 1

An identifying code C over a given graph $G=(V,E)$ is said to be r-robust if $A \oplus I_C(v) \neq B \oplus I_C(u)$ for all $u,v \in V$, and $A,B \subseteq C$ with $|A|,|B| \leq r$. We use the symbol $\oplus$ to denote the symmetric difference (i.e., $A \oplus B=(A \setminus B) \cup (B \setminus A)$).

Simply stated, an identifying code is r-robust if the addition or deletion of up to r codewords (IDs) around any vertex does not change its identifying capability. Alternatively, we may determine the robustness of a code C by measuring the minimum symmetric difference between the identifying sets of any two vertices, $d_{\min}(C)=\min_{u,v \in V} |I_C(v) \oplus I_C(u)|$.

We thus have the following theorem as a straightforward application of the definitions.

Theorem 3

A code C is r-robust if and only if $d_{\min}(C) \geq 2r+1$

A straightforward extension to ID-CODE can generate r-robust identifying codes, but with the same lack of a performance guarantee. Instead, robust identifying codes were linked to the *set multi-cover problem* in (Laifenfeld, Trachtenberg, Cohen, & Starobinski, 2007). The minimum set k-multicover problem is a natural generalization of the minimum set cover problem, in which one is given a pair $(\mathbf{U}, S)$ and seeks the smallest subset of S that covers every element in $\mathbf{U}$ at least k times. The set multicover problem admits a similar greedy heuristic to the set cover problem, GreedySetMultiCover: in each iteration select the set which covers the maximum number of non k-multicovered elements. It is well known (Vazirani, 2001) that the performance guarantee of this heuristic is upper bounded by $1+\ln\alpha$, where α is the largest set's size.

Algorithm 3 : $C_{central}$ ←rID- CENTRAL (G,r) (Laifenfeld et al., 2007)
Given a graph $G=(V,E)$, a non-negative integer r, and the set multicover heuristic GreedySetMultiCover (U,S,k) **do**:
1) **Compute** $\Delta=\{\delta_v \mid v \in V\}$ 2) $C \leftarrow$ GreedySetMultiCover $(\mathbf{U}, \Delta, 2r+1)$ 3) **Output** $C_{central} \leftarrow \{v \mid \delta_v \in C\}$

The correctness of Algorithm 3 is guaranteed in Step 2) where GreedySetMultiCover generates a multicover of distinguishing sets that distinguish every pair of vertices at least 2r+1 times, or in other words $|I_C(v) \oplus I_C(u)| \geq 2r+1$ for all $(v, u) \in \mathbf{U}$.

The following theorem follows from the well known performance bound of the greedy set multicover heuristic, and the fact that the size of the most distinguishing vertex (a vertex whose identifying set is of maximal size) is $\alpha = \max_{v \in V} \big(B(v)(n - B(v))\big)$.

Theorem 4

For any given graph G of n vertices, rID-CENTRAL *generates an identifying code of size* $c_{central}$ *that satisfies*

$$\frac{c_{central}}{c_{min}} \leq \ln\alpha + 1.$$

rID-CENTRAL requires the knowledge of the entire graph in order to operate. It was observed in (Ray et al., 2003) and (Laifenfeld & Trachtenberg, 2005) that an r-robust identifying code can be built in a localized manner, where each vertex only considers its two-hop neighborhood. The resulting *localized* identifying codes are discussed next, and the approximation algorithm we derive is utilized to construct the distributed algorithm of the next section. In the heart of the localized identifying code lies the observation that nodes, which are the neighbors of a codeword, must differ in their identifying set (by at least one element) from nodes that are not. Therefore if every node is in the neighborhood of some codeword, then it is enough to require that only the nodes in every such neighborhood have unique identifying sets. Equivalently, if the code is a *dominating set*[3], then only the pairs of nodes that are two hops apart need to be considered in Algorithms 2 or 3 to guarantee it to be identifying. In the following we cement this observation with some additional technical definitions and lemmas.

Let $G = (V, E)$ be an undirected graph, we define the distance metric $\rho(u, v)$ to be the number of edges along the shortest path from vertex u to v. The ball of radius l around v is denoted $B(v; l)$ and defined to be $\{w \in V \mid \rho(w, v) \leq l\}$. So far we encountered balls of radius l=1, which we simply denoted by $B(v)$.

Recall that a vertex cover (or dominating set) is a set of vertices, S, such that every vertex in V is in the ball of radius 1 of at least one vertex in S. We extend this notion to define an r-dominating set to be a set of vertices S_r such that every vertex in V is in the ball of radius 1 of at least r vertices in S_r.

Lemma 1

Given a graph G=(V, E), an (r+1)-dominating set C is also an r-robust identifying code if and only if every pair $(u, v) \in U$ *such that* $\rho(u, v) \leq 2$ *is distinguished by at least by* $2r + 1$ *codewords in C.*

Lemma 1 can serve as the basis for an algorithm for a robust identifying code problem based on the greedy set multicover heuristic, similarly to Algorithm 3. The main difference is that we will restrict the basis elements to vertex pairs that are at most two hops apart, and we then need to guarantee that the resulting code is still r-dominating.

Towards this end we define $\mathbf{U}^2 = \{ (u, v) \mid \rho(u, v) \leq 2 \}$, the set of all pairs of vertices (including (v,v)) that are at most two hops apart. Similarly, we will localize the distinguishing set δ_v to $\mathbf{U}^2$ as follows:

$$\delta_v^2 = (\delta_v \cap \mathbf{U}^2) \cup \{(u, u) \mid u \in B(v)\}, \quad (3)$$

The resulting *localized* identifying code approximation is thus given by Algorithm 4 and can be shown in a similar manner to provide an r-robust identifying code for any graph that admits one.

Figure 5. Demonstration of the rID-LOCAL for 10 nodes ring, and r=0, starting on the left. Nodes are labeled 1 to 10 clockwise (the labels appear in the inner perimeter). Solid circles represent codewords, and the distinguishing sets sizes, obtained from the greedy set-multicover (GreedySetMultiCover) appear in the outer perimeter.

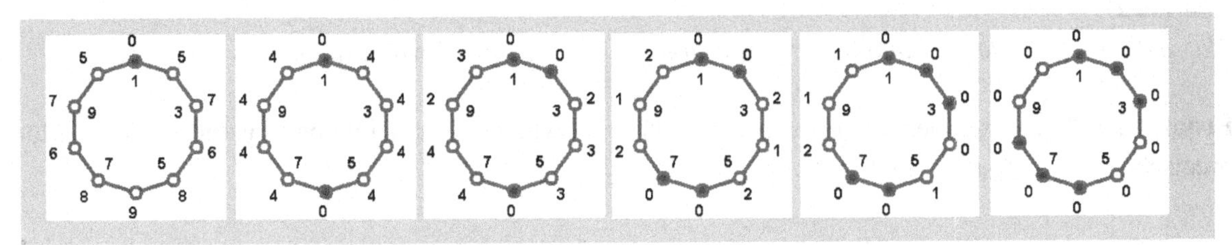

Algorithm 4 : $C_{\text{local}} \leftarrow$rID- LOCAL (G,r) (Laifenfeld et al., 2007)
Given a graph $G=(V,E)$, a non-negative integer r, and the set multicover heuristic GreedySetMultiCover (U,S,k) **do**:
1) **Compute** U^2 and $\Delta^2=\{\delta_v^2 \mid v\in V\}$ 2) $C\leftarrow$ GreedySetMultiCover $(\mathbf{U}^2, \Delta^2, 2r+1)$ 3) **Output** $C_{\text{local}} \leftarrow \{v \mid \delta_v^2 \in C\}$

An example of rID-LOCAL(G,0), where G is a 10 nodes ring is shown in Figure 5 . Comparing to ID-CENTRAL of Figure 4 it can be observed that the two differ in the sizes of their distinguishing sets. rID-LOCAL considers only pairs of nodes that are at most two hops apart, and therefore the initial size of the distinguishing sets is smaller. This also affects the iterations to follow, resulting in a larger identifying code.

Theorem 5

Given a graph $G=(V,E)$ of n vertices, the performance ratio of rID- LOCAL *is upper bounded by:*

$$\frac{c_{\text{local}}}{c_{\min}} \leq \ln\gamma + 1,$$

where $\gamma = \max_{v\in V} |B(v)|(|B(v;3)|-|B(v)|+1)$.

Proof: The proof derives from the performance guarantee of the greedy set multicover algorithm, which is upper bounded by $1+\ln\alpha$ for a maximum set size α.

The size of δ_v^2 is $|B(v)|(|B(v;3)|-|B(v)|+1)$, which, at its maximum, can be applied to this performance guarantee to complete the proof.

Roughly speaking this performance bound is similar to the bound we derived for the centralized algorithm, when the size of the largest $B(v;3)$ is of the order of the number of vertices, n. However, when $|B(v;3)|$ is much smaller, the performance bound of Theorem 5 can be significantly tighter.

Note that although Algorithm 4 "localizes" the set of vertices pairs, $\mathbf{U}^2$, it is not a distributed algorithm, since it still requires a central entity to select the most distinguishing codeword at each iteration. In the next section we present a distributed implementation of the identifying code localized approximation. The following lemma supplements Lemma 1 by providing additional "localization". At the heart of this lemma lies the fact that each codeword distinguishes between its neighbors and the remaining vertices.

Lemma 2

The distinguishing sets δ_v^2 *and* δ_u^2 *are disjoint for every pair (u, v) with* $\rho(u,v)>4$.

Proof: Clearly, δ_v^2 includes all vertex pairs $(x, y) \in \mathbf{U}^2$ where x is a neighbor of v and y is not. More precisely, $(x, y) \in \delta_v^2$ if

$$x \in B(v) \quad \text{and} \quad y \in (B(x;2) \setminus B(v)). \tag{4}$$

Moreover, for all such (x,y), $\rho(x,y)\leq 3$ and $\rho(y,v)\leq 3$. On the other hand, for $(x',y') \in \delta_u^2$ with $\rho(u,v)>4$, either x' or y' must be a neighbor of u, and hence of distance >3 from v. Thus, δ_v^2 and δ_u^2 are disjoint.

Lemma 2 implies that, when applying the localized algorithm, a decision to choose a codeword only affects decisions on vertices within four hops; the algorithm is thus localized to vicinities of radius four.

IMPLEMENTATIONS

Several parallel algorithms exist in the literature for the set cover problem and for more general covering integer programs (e.g. (Rajagopalan & Vazirani, 1998)). There are also numerous distributed algorithms for finding a minimum (connected) dominating set based on set cover and other well-known approximations such as linear programming relaxation (e.g. (Bartal, Byers, & Raz 1997)). Unfortunately, the fundamental assumption of these algorithms is that the elements of the basis set are independent computational entities (i.e. the nodes in the network); this makes it non-trivial to apply them in our case, where elements correspond to pairs of nodes that can be several hops apart. Moreover, we assume that the nodes are energy constrained so that reducing communications is very desirable, even at the expense of longer execution times and reduced performance.

We next provide two distributed algorithms first devised in (Laifenfeld et al., 2007). The first is completely asynchronous, guarantees a performance ratio of at most $1+\ln\gamma$, and requires $\Theta(c_{dist})$ iterations at worst, where c_{dist} is the size of the identifying code returned by the distributed algorithm and $\gamma=\max_{v\in V}|B(v)|(|B(v;3)|-|B(v)|+1)$. The second is a randomized algorithm, which requires a coarse synchronization, guarantees a performance ratio of at most $1+\ln\gamma$, and for some arbitrarily small $\varepsilon>0$ operates within $O\left(\frac{1}{K}\gamma n^{\frac{K+2+\varepsilon}{K-1}}\right)$ time slots (resulting in $O(c_{dist}\max_{v\in V}|B(v;4)|)$ messages). $K\geq 2$ is a design parameter that trades between the size of the resulting r-robust identifying code and the required number of time slots to complete the procedure.

In the next subsection we describe the setup and initialization stages that are common to both distributed algorithms.

Setup and Initialization

We assume that every vertex (node) is pre-assigned a unique serial number and can communicate reliably and collision-free (perhaps using higher-layer protocols) over a shared medium with its immediate neighborhood. Every node can determine its neighborhood from the IDs on received transmissions, and higher radius balls can be determined by distributing this information over several hops. In our case, we will need to know $G(v;4)$ the subgraph induced by all vertices of distance at most four from v.

Our distributed algorithms are based on the fact that, by definition, each node v can distinguish between the pairs of nodes which appear in its corresponding distinguishing set δ_v^2 given in Equation (3). This distinguishing set is updated as new codewords are added to the identifying code being constructed, C; their presence is advertised by flooding their four-hop neighborhood.

The Asynchronous Algorithm rID-ASYNC

The state diagram of the asynchronous distributed algorithm is shown in Figure 6. All nodes are initially in the *unassigned* state, and transitions are effected according to messages received from a node's four-hop neighborhood. Two types of messages can accompany a transition: *assignment* and *declaration* messages, with the former indicating that the initiating node has transitioned to the *assigned* state, and the latter being used to transmit data. Both types of messages also include five fields: the *type*, which is either "assignment" or "declaration", the *ID* identifying the initiating node, the *hop* number, the *iteration* number, and *data*, which contains the size of the distinguishing set in the case of a declaration message.

Following the initialization stage, every node declares its distinguishing set's size. As a node's declaration message propagates through its four hop neighborhood, every forwarding node updates two internal variables, ID_{max} and δ_{max}, representing the ID and size of the *most distinguishing* node (ties are broken in favor of the lowest ID). Hence, when a node aggregates the declaration messages initiated by all its four hop neighbors (we say that the node reached its *end-of-iteration* event), ID_{max} should hold the most distinguishing node in its four hop neighborhood. A node that reaches end-of-iteration event transitions to either the *wait-for-assignment* state or to the final *assigned* state depending on whether it is the most distinguishing node.

The operation of the algorithm is completely asynchronous; nodes take action according to their state and messages received. During the iterations stage, nodes initiate a declaration message only if

Figure 6. Asynchronous distributed algorithm (rID-ASYNC) simplified state diagram in node $v \in V$ (Laifenfeld et al., 2007)

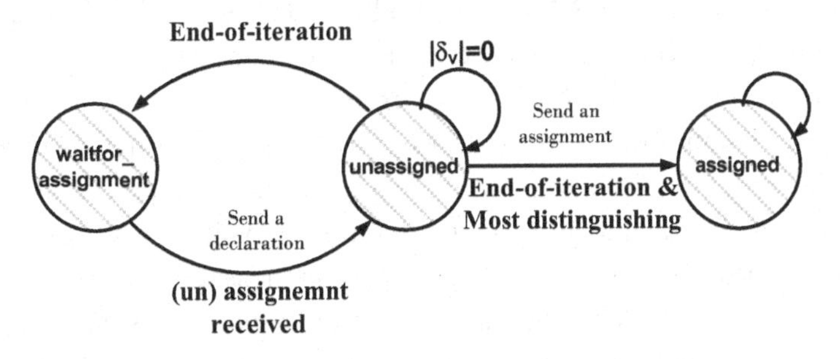

they receive an assignment message or if an updated declaration (called an *unassignment* message) is received from the most distinguishing node of the previous iteration. All messages are forwarded (and their hop number is increased) if the hop number is less than four. To reduce communications load, a mechanism for detecting and eliminating looping messages should be applied.

Every node, v, terminates in either an "unassigned" state with $|\delta_v^2|=0$ or in the "assigned" state. Clearly, nodes that terminate in the "assigned" state constitute a localized r-robust identifying code.

Algorithm 5 : $C_{greedy} \leftarrow$rID-ASYNC (G,r) (Laifenfeld et al., 2007)

Given $G=(V,E)$ with vertices labeled by ID, a non-negative integer r, **do** at every node $v \in V$:

1) **Precompute**
   a. Compute δ_v^2
   b. Initiate a declaration message and set *state* = *unassigned*
   c. Set $ID_{max}=ID(v)$, $\delta_{max}=|\delta_v^2|$.
2) **Iteration**
   a. Increment $hop(ms)$ and forward all messages of $hop(ms)<4$.
   b. if received an *assignment* message with *state≠assigned* then
      i. Update δ_v^2 by removing all pairs covered $2r+1$ times.
      ii. Initiate a declaration message and set *state* = *unassigned.*
      iii. Perform Comp($ID(v)$,$|\delta_v^2|$).
   c. if *state=waitfor-assignment* and received an *unassignment* message then
      i. Initiate a declaration message and set *state* = *unassigned.*
      ii. Perform Comp($ID(v)$,$|\delta_v^2|$).
   d. if received a declaration message *ms* with *state≠assigned* then perform Comp ($ID(ms)$,$data(ms)$)
   e. if *end-of-iteration* reached then,
      i. if $ID_{max}=ID(v)$ and $|\delta_v^2|>0$ then *state=assigned,* initiate an assignment message.
      ii. Otherwise $ID_{max}=ID(v)$, $\delta_{max}=0$, and *state=waitfor-assignment.*

Comp (id, δ): if $\delta_{max}<\delta$ or ($\delta_{max}=\delta$ and $ID_{max}>id$) then $\delta_{max}=\delta$, $ID_{max}=id$.

We illustrate the operation of rID-ASYNC $(0,G)$ over a simple ring topology of 10 nodes in Figure 7. The nodes are labeled from 1 to 10 clockwise. Solid squares represent assigned vertices (or codewords), and the size of the distinguishing sets and the value of ID_{max}, at the end of each iteration, appear in the outer perimeter, separated by a comma. The network is shown at the end of the first iteration in the upper left subfigure, where all nodes have evaluated their distinguishing sizes and communicated them to their 4-hop neighborhoods. We can see that all nodes can distinguish up to 9 pairs, and that all but node 6 have concluded node 1 to be the most distinguishing ($ID_{max}=1$) by the rule that lower labels take precedence. Since node 6 is just outside $B(1;4)$ it concludes that node 2 is the most distinguishing in its 4 hop neighborhood. Note that theoretically node 6 could have assigned itself to be a codeword without loss in performance since it is more than 4 hops away from node 1.

At the start of iteration 2 (subfigure 2a) node 1 transmits an *assignment* message that gets propagated in $B(1;4)$ (shown as solid arrows). The *assignment* message transitions all the nodes in its way from *wait-for-assignment* to *unassigned* state and triggers them to reevaluate their distinguishing set sizes and send *declaration* messages. One half of node 2 *declaration* message is shown by the dashed arrows. This declaration message reaches node 6 when it is still in the *wait-for-assignment* state. Since node 6 is awaiting an *assignment* message from node 2, this declaration message serves as an *unassignment* message and transitions node 6 to the *unassigned* state and invokes a *declaration* message that is shown as dashed arrows in subfigure 2b, which makes the rest of the nodes to conclude that node 6 is the most distinguishing. Iterations 3 to 6 operate in a similar manner and are shown in the bottom of Figure 7. In total rID-ASYNC returns an identifying code of size 6 - only one node more than a minimum one. The outcome of rID-ASYNC is heavily dependent on the way nodes resolve ties and therefore is sensitive to nodes relabeling; however the performance guarantee of the theorem of the next subsection holds for any such arrangement.

Performance Evaluation

Theorem 6

The algorithm rID-ASYNC *requires* $\Theta(c_{dist})$ *iterations and has a performance ratio*

$$\frac{c_{dist}}{c_{min}} \leq \ln\gamma + 1,$$

where $\gamma = \max_{v\in V} |B(v)|(|B(v;3)|-|B(v)|+1)$.

The first part of the Theorem follows from the fact that it performs exactly as the localized identifying code algorithm and the fact that only the most distinguishing set in a four hop neighborhoods is assigned to be a codeword. To see the number of iterations of the algorithm, we first note that in each iteration at least one codeword is assigned. The case of a ring topology (Figure 7) demonstrates that, in the worst case, exactly one node is assigned per iteration. It follows that the amount of communications

Figure 7. Operation of rID-ASYNC over a ring of 10 nodes, labeled from 1 to 10 clockwise (displayed in the inner perimeter), and r=0. Solid squares represent assigned vertices, and the size of the distinguishing set and the value of ID_{max} at the end of each iteration appear in the outer perimeter, separated by a comma. The iteration number appears at the upper left corner of each subfigure. The path of assignment (declaration) messages is shown by solid (dashed) arrows. (Laifenfeld et al., 2008).

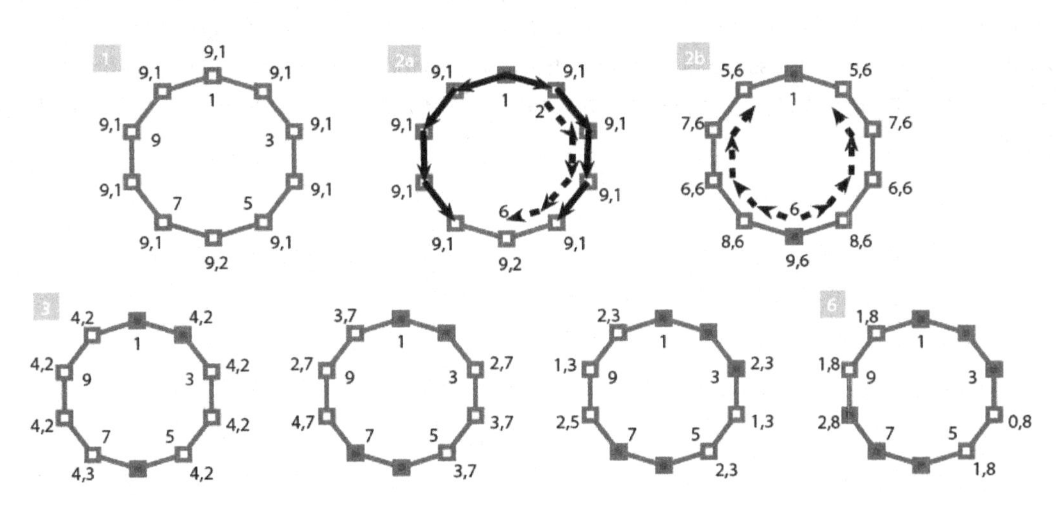

required in the iteration stage is $\Theta\left(c_{\text{dist}} \max_{v \in V} |B(v;4)|\right)$, which can be a significant load for a battery powered sensor network. This can be significantly reduced if some level of synchronization among the nodes is allowed.

In the next section we suggest a synchronized distributed algorithm that eliminates declaration messages altogether.

A Low-Communications Randomized Algorithm rID-SYNC

In this subsection we assume that a coarse time synchronization among vertices within a neighborhood of radius four can be achieved. In particular, we will assume that the vertices maintain a basic time *slot*, which is divided into L *subslots*. Each subslot's duration is longer than the time required for a four hop one-way communication together with synchronization uncertainty and local clock drift. After an initialization phase, the distributed algorithm operates on a time *frame*, which consists of F slots arranged in decreasing fashion from s_F to s_1. In general, F should be at least as large as the largest distinguishing set (e.g. F=n(n-1)/2 will always work). Frame synchronization within a neighborhood of radius four completes the initialization stage.

The frame synchronization enables us to eliminate all the declaration messages of the asynchronous algorithm. Recall that the declaration messages were required to perform two tasks: (i) determine the most distinguishing node in its four hop neighborhood, and (ii) form an iteration boundary, i.e. end-of-iteration event. The second task is naturally fulfilled by maintaining the slot synchronization. The first task is performed using the frame synchronization: every node maintains a synchronized slot counter, which corresponds to the size of the current *most distinguishing* node. If the slot counter reaches the size of a node's distinguishing set, the node assigns itself to the code. The subslots are used to randomly break ties.

Iteration Stage

Each iteration takes place in one time slot, starting from slot s_F. During a slot period, a node may transmit a message *ms* indicating that it is assigning itself as a codeword; the message will have two fields: the identification number of the initiating node, *id*(*ms*), and the hop number, *hop*(*ms*). A node assigns itself to be a codeword if its *assignment time*, which refers to a slot *as* and subslot l, has been reached. Every time an assignment message is received, the assignment slot *as* of a node is updated to match the size of its distinguishing set; the assignment subslot is determined randomly and uniformly at the beginning of every slot.

Algorithm 6 : $C_{greedy} \leftarrow$rID-SYNC (G,r) (Laifenfeld et al., 2007)
Given a graph $G=(V,E)$ with vertices labeled by *ID*, a non-negative integer *r*, **do** at every node $v \in V$:
1) Precompute a. Compute $\delta_v^{\,2}$, set $as=\|\delta_v^{\,2}\|$. b. Set: $slot=s_F$, $subslot = L$, $state = unassigned$ **2) Iteration** while *state≠assigned* and $slot \geq s_1$ do, *a.* l = random({1,...,*L*}) b. if received assignment message, *ms* then update $\delta_v^{\,2}$ by removing all pairs covered $2r+1$ times and set $as=\|\delta_v^{\,2}\|$. c. elseif *subslot=l* and *slot=as* then transmit an assignment message, *state=assigned.*

We illustrate in Figure 8 in the next page the operation of rID-SYNC (0,*G*) over a simple ring topology of 10 nodes. The nodes are labeled from 1 to 10 clockwise. Solid circles represent assigned vertices (or codewords), and the size of the distinguishing sets and the value of *L*, at the end of each iteration, appear in the outer perimeter, separated by a comma. The network is shown at the end of slot 9 in the upper left subfigure, where all nodes have evaluated their distinguishing set sizes and selected randomly a transmission sub-slot from *L*-1 to 0.

We can see that all nodes can distinguish up to 9 pairs, and that node 3 was the first to transmit its assignment message as it selected the largest subslot number. This message is spread through its 4 hop neighborhood preventing other nodes from assigning themselves. However, since node 8 is just outside

Figure 8. Operation of rID-SYNC over a ring of 10 nodes, labeled from 1 to 10 clockwise (displayed in the inner perimeter) for r=0 and L=10 (top) and L=3 (bottom). Solid squares represent assigned vertices, and the size of the distinguishing set and randomly selected subslot, l, at the end of each iteration, appear in the outer perimeter, separated by a comma. The slot number appears at the upper left corner of each subfigure. Dashed arrows represent assignment messages. (Laifenfeld et al., 2008).

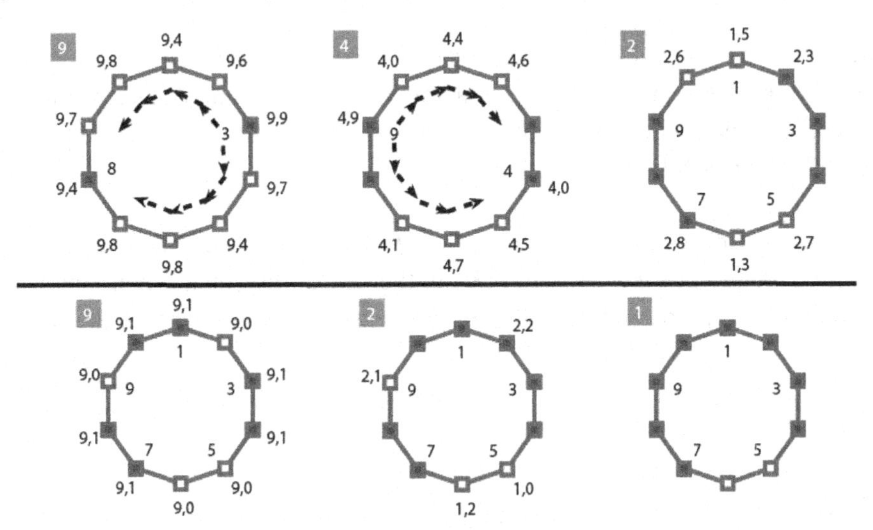

B(3;4) it is free to assign itself as indicated in the subfigure. Note this fundamental difference from rID-ASYNC where in a similar situation node 8 wouldn't be assigned (see Figure 7). rID-SYNC stays idle for the next 5 slots where nodes keep rough synchronization using their internal clocks. The assignment processes resumes at slots 4 and 2 to conclude the procedure returning an identifying code of size 6 - only one node more than a minimum one. Similarly to rID-ASYNC, the outcome of rID-SYNC can vary depending on the way nodes resolve ties and therefore is sensitive to the random selection of subslots. The bottom of Figure 8 shows a run with a small number of subslots (L=3), which due to large amount of over-assignments returns a relatively large code. Nevertheless, the performance guarantee of the theorem below holds for any combination of random selections of subsets and labeling.

Performance Evaluation

Algorithm rID-SYNC requires at most $O(n^2)$ slots ($O(Ln^2)$ subslots), though it can be reduced to $O(L\gamma)$ if the maximum size of a distinguishing set is propagated throughout the network in the precomputation phase. The communications load is low (i.e. $O(c_{\text{dist}} \max_{v\in V} |B(v;4)|)$), and includes only assignment messages, which are propagated to four hop neighborhoods.

In the case of ties, rID-SYNC can provide a larger code than gained from the localized approximation. This is because ties in the distributed algorithm are broken arbitrarily, and there is a positive probability (shrinking as the number of subslots L increases) that more than one node will choose the same subslot within a four hop neighborhood. As such, the L is a design parameter, providing a tradeoff between performance ratio guarantees and the runtime of the algorithm as suggested in the following theorem.

Theorem 7

For asymptotically large graphs, rID-SYNC *guarantees (with high probability) a performance ratio of*

$$\frac{c_{\text{dist}}}{c_{\min}} \leq \ln\gamma + 1,$$

where $\gamma = \max_{v\in V} |B(v)|(|B(v;3)|-|B(v)|+1)$. *The algorithm also requires* $O\left(\frac{1}{K}\gamma n^{\frac{K+2+\varepsilon}{K-1}}\right)$ *subslots to complete for design parameter* $K\geq 2$ *and arbitrarily small* $\varepsilon>0$.

Proof: If no more than K tied nodes assign themselves simultaneously on every assignment slot, then we can upper bound the performance ratio by a factor K of the bound in Theorem 5, as in the theorem statement. We next determine the number of subslots L needed to guarantee theabove assumption asymptotically with high probability.

Let $P(K)$ denote the probability that no more than K tied nodes assign themselves in every assignment slot. Clearly, $P(K)\geq(1-p(K))^{c_{\text{dist}}}$, where $p(K)$ is the probability that, when t nodes are assigned independently and uniformly to L subslots, there are at least $K<t$ assignments to the same subslot. One can see that

$$p(K) \leq Lt\left(\frac{te}{LK}\right)^K$$

for e being the natural logarithm and based on the assumption that $\frac{te}{LK} < 1$.

Let $t=c_{\text{dist}}=n$ (this only loosens the bound) and $L=\frac{e}{K}n^{\frac{K+2+\varepsilon}{K-1}}$. Then,

$$P(K)\geq\left(1-tL\left(\tfrac{te}{LK}\right)^{K}\right)^{c_{\text{dist}}}\geq\left(1-\tfrac{e}{k}\tfrac{1}{n^{1+\varepsilon}}\right)^{n}\to 1$$

SIMULATIONS AND EXPERIMENTS

In this section we provide the advantages and shortcomings of identifying-code localization, through simulations on Erdos-Renyi random graphs and geometric random graphs, and experiments on a test-bed on the fourth floor of the Photonics center at Boston University.

Random Graphs

Erdos-Renyi random graphs and random geometric graphs were studied recently in the context of identifying codes. In (Moncel, Frieze, Martin, Ruszink, & Smyth, 2005) it was shown that for asymptotically large random graphs, any subset of a certain threshold size (roughly logarithmic in the size of the graph) is almost surely an identifying code. It was also shown that the threshold is asymptotically sharp, meaning that the probability of finding an identifying code of slightly smaller size approaches zero. Extension of this result to robust identifying graphs was provided in (Laifenfeld, 2007), where it was further shown that with relatively small addition of codewords (doubly logarithmic in n) identifying codes become r-robust in large random graphs.

Unit disk geometric random graphs (GRGs), in which vertices are placed uniformly at random on a two-dimensional plane and where two vertices are adjacent if their Euclidian distance is less than a unit, were studied in the context of identifying codes in (Muller & Sereni, 2007). GRGs are commonly used to model ad-hoc wireless networks as well as large scale sensor networks. Unlike large Erdos-Renyi random graphs, it has been shown in (Muller & Sereni, 2007) that most of the large unit-disk GRGs do not possess identifying codes.

We have simulated all of the identifying code algorithms described and applied them to both Erdos-Renyi random graphs with different edge probabilities, and to two dimensional GRGs with different nodes densities. We use the average size of the resulting identifying code as a performance measure. For the case of $r=0$ (i.e., simple identifying code) the simulation results are compared to a combinatorial lower bound derived by Karpovsky et al. (1998), and the asymptotic result of Moncel et al. (2005).

Figure 9 compares ID-CENTRAL to ID-CODE and the combinatorial lower bound. It can be observed that ID-CENTRAL demonstrates a significant improvement over ID-CODE. It should also be noted that the curves for basically any algorithm should converge very slowly to Moncel's asymptotic result as n grows, and this is illustrated in Figure 11. This apparently slow convergence rate suggests that there is a lot to gain from using the suggested algorithms, even for reasonably large networks (Moncel et al., 2005). The results of the centralized r-robust identifying code algorithm, rID-CENTRAL, are shown in Figure 10 versus the theoretical results of (Laifenfeld, 2007). The tradeoff between the increase in robustness and the increase in the code's size is evident. Note that there is a pretty close agreement between the theoretical result of (Laifenfeld, 2007) and the simulations. However, as r increases the asymptotic theoretical assumptions no longer hold (for a fixed n) and the two results diverge.

Figure 12 shows the simulation results for the localized (rID-LOCAL) and distributed (rID-SYNC) algorithms compared to the centralized one. Recall that the performance of the asynchronous algorithm,

Figure 9. Average identifying code size generated by ID-CODE and ID-CENTRAL algorithms for 128 nodes random graphs with different edge probabilities (Laifenfeld & Trachtenberg, 2008), in comparison to a combinatorial lower bound of (Karpovsky, Chakrabarty, & Levitin, 1998), and asymptotic bound of (Moncel et al., 2005).

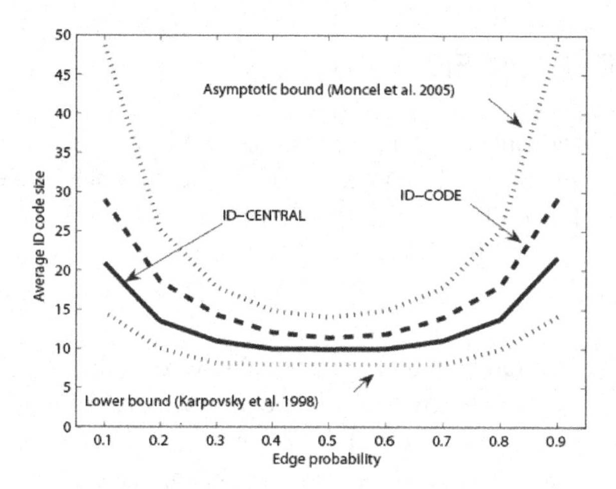

Figure 10. Centralized r-robust identifying codes algorithm, rID-CENTRAL, (solid) and the asymptotic bound (Laifenfeld, 2007) (dotted) for 128 nodes random graphs with different edge probabilities (Laifenfeld, 2007).

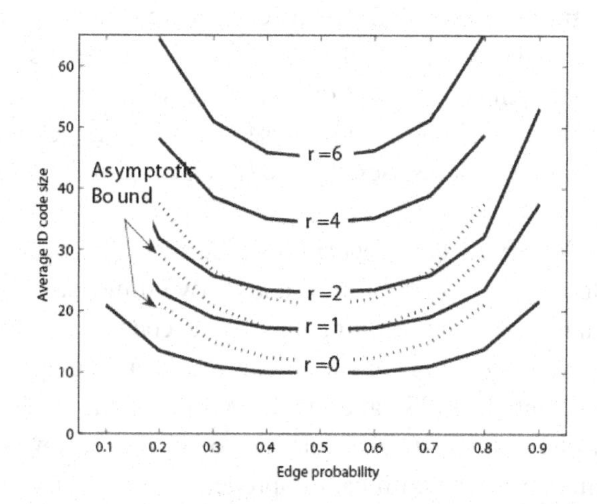

rID − ASYNC, is identical to the localized approximation, and the simulation results of the localized algorithm nearly match the results of the centralized algorithm (rID-CENTRAL). Divergence is evident for low edge probabilities where it is harder to find a dominating set. Recall that there is a tradeoff between performance and the runtime of the synchronized distributed algorithm, rID − SYNC. The smaller the number of subslots parameter, L, the shorter the runtime and the larger the degradation in performance due to unresolved ties. Degradation in performance is also more evident when ties are more likely to happen, i.e., when the edge probability approaches 0.5.

Figure 11. Normalized (by log(n)) average size of the identifying code returned by ID-CODE and ID-CENTRAL for random graphs with edge probability p = 0.1, and various numbers of vertices (Laifenfeld & Trachtenberg, 2008).

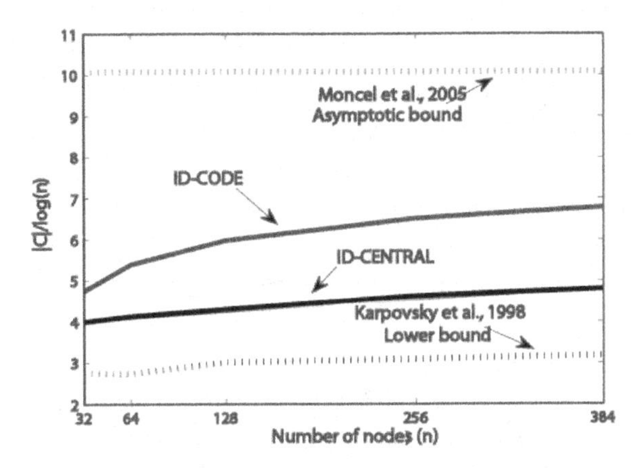

Figure 12. Average identifying code size of the rID − CENTRAL, rID − LOCAL, and rID − SYNC with different number of subslots parameter, L, for 128 nodes random graphs with different edge probabilities (Laifenfeld et al., 2007).

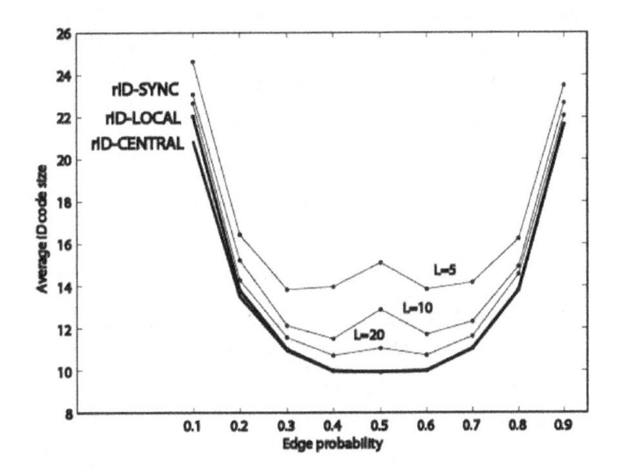

Figure 13 and Figure 14 show the codewords fraction (from the total number of nodes) for GRGs using the localized and distributed approaches, and the fraction of such graphs admitting an identifying code. It also presents the largest fraction of indistinguishable nodes obtained in the simulation. As can be seen the localized and distributed algorithms (with $L = 10$) yield very similar code sizes. The fraction of graphs admitting identifying codes is rather small (less than half the graphs) even for high node densities; an observation that matches the theoretical results of (Muller & Sereni, 2007). However, the sizes of the undistinguishable sets of vertices (undistinguishable set is a set of vertices that have a common identifying set) are relatively small, as indicated in Figure 14, suggesting that the *resolution* of the localization system can still be high, since most of the undistinguishable sets are within a small geometrical proximity with high likelihood. Therefore, from the location detection perspective, identifying codes provide an adequate solution, in spite of the technical fact that most of the geometrical random graphs do not possess identifying codes.

Figure 13. Codeword fraction (out of all nodes) for the localized (rID − LOCAL) and distributed (rID − SYNC) algorithms for GRGs with different nodes densities (nodes per unit area) (Laifenfeld et al., 2007).

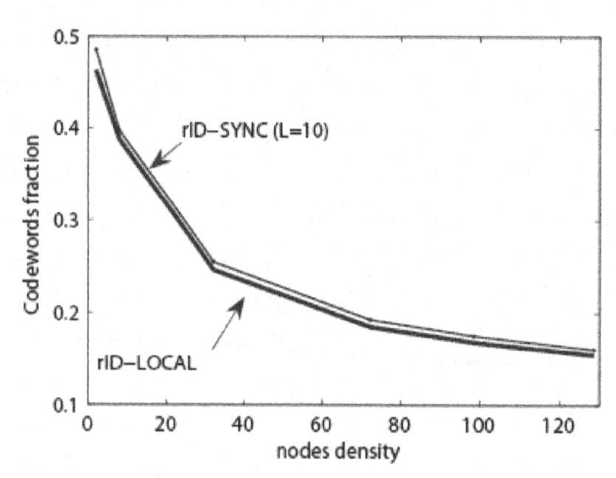

Figure 14. Fraction of graphs admitting an identifying code, and maximum fraction of indistinguishable nodes for GRGs with different node densities (nodes per unit area) (Laifenfeld et al., 2007).

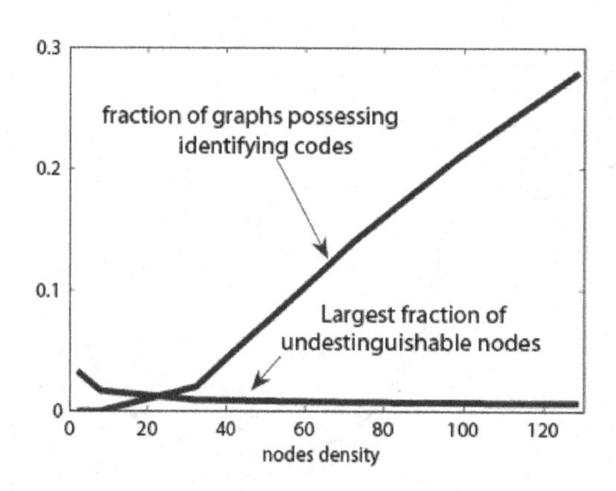

Figure 15. The Experimental testbed – 4th floor of the Photonics building in Boston University. The resolution (0-70 feet) of the location detection system with a 90% confidence level. Stars are transmitters; plain circles are locatable points (Ray et al., 2004).

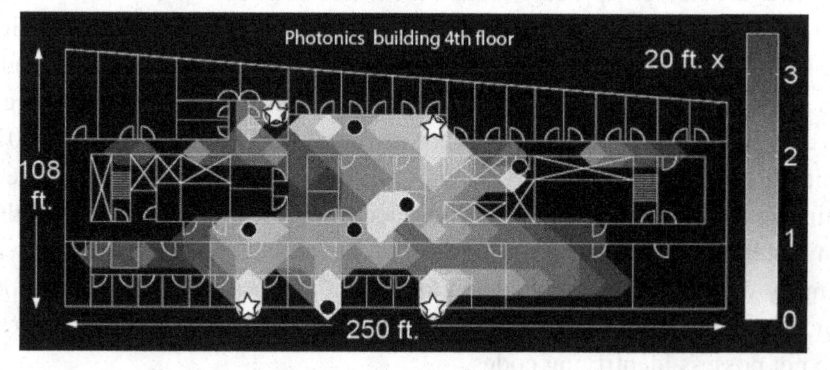

Figure 16. The connectivity graph of the testbed and its identifying code. The bold circles denote the codewords (transmitters) (Ray et al., 2004).

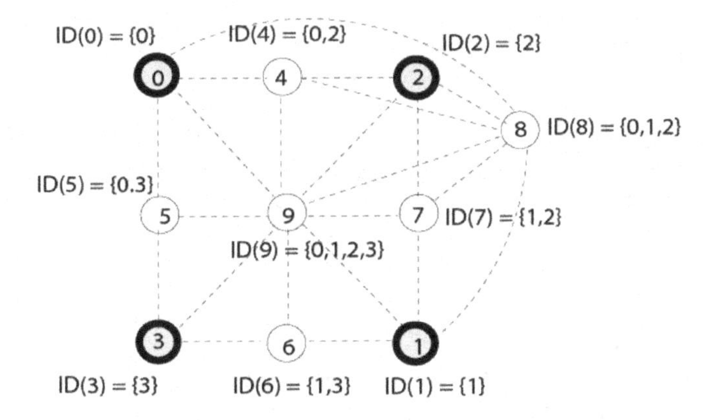

Experimental Testbed

A basic experimental testbed was developed in (Ray et al., 2004) to verify the location detection scheme, and in particular the rID-CODE algorithm. The testbed is located on the 4th floor of the Photonics Building at Boston University and is depicted in Figure 15. The white circles represent 10 discrete positions selected on the floor, and the test bed included five laptop computers equipped with IEEE 802.11 transceivers; four of them as transmitters and the fifth one as the receiver. In order to determine whether two points are connected, a simple thresholding scheme was employed. Specifically, each transmitter transmits 40 packets per second and two points are considered connected if the number of packets received during a sample interval exceeds a certain threshold.

The resultant connectivity graph for the testbed is shown in Figure 16. The solution produced by the ID-CODE algorithm results in a placement of the transmitters at positions (0; 1; 2; 3) (correspondingly identified in Figure 15 by stars). The identifying sets for each position are shown in Figure 16.

The location detection system was evaluated by dividing the floor plan into a grid, where each grid location represents a 10 by 10 sq.ft. area. At each grid location, the packet arrival rate from all transmitters was recorded as a vector of the form $(n_0; n_1; n_2; n_3)$, where n_i represents the number of packets received from transmitter i during a 1 second sample interval. 60 such measurements were taken with different antenna orientations to average out long-term fading variability in the RF channel. The receiver's location was determined to be one of the positions of Figure 16 if its corresponding identifying set matched the received identifying set[4]. If no matching identifying set could be found then the location was determined to be the predefined position whose identifying set was the closest (in terms of Hamming distance) to the received identifying set, where ties are broken arbitrarily.

The resolution achieved with this testbed location detection system is depicted in the contour map shown in Figure 15. Resolution is defined as the (Euclidean) distance between the location resolved by the system and the actual user's location. In the figure, the resolution varies from 0 ft. to 70 ft. A lighter shade corresponds to a higher resolution. The confidence level is 90%, i.e., at each position, at least 90% of the samples achieve the shown resolution. Although the system consists of only four transmitters, it achieves a reasonable resolution, most of the time within 50 ft. As expected, the resolution becomes

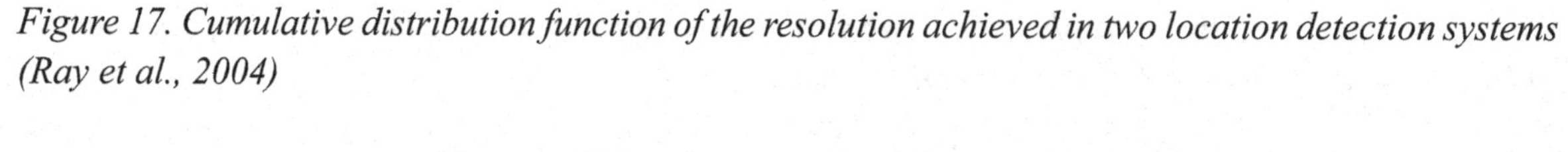

Figure 17. Cumulative distribution function of the resolution achieved in two location detection systems (Ray et al., 2004)

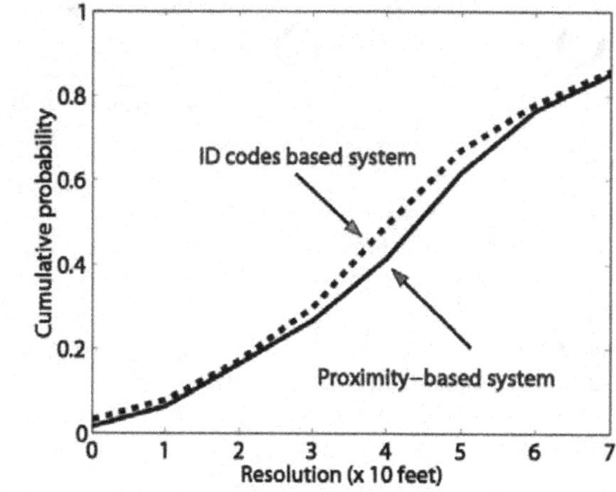

coarser in areas that are distant from any of the discrete measurement points. We note here that experiments recently run on the moteLab testbed at Harvard University (Chao, 2008) reveal similar qualitative results. Moreover, due to the higher density of transmitters at moteLab, the worst-case resolution was found to be about 10 ft.

As a basis of comparison, the resolution obtained with a simple proximity-based scheme was also evaluated; a user resolves his location to be that of his "closest" transmitter, that is, the transmitter from which it correctly receives the largest number of packets. Figure 17 shows the cumulative distribution function (CDF) of the resolution for identifying code based location detection system and the proximity-based system. It can be observed that a larger number of positions are within a given error distance in the identifying codes based system than in the proximity-based system. This non-negligible gain in resolution is achieved through the sole use of identifying code techniques, and, thus, illustrates how judicious use of coding-theoretic approaches can contribute to improving the performance of location-detection systems.

CONCLUSION

We have described a localization method for environments with challenging signal transmission properties (e.g., indoor, urban, or underground areas), based on a new identifying code paradigm. Our work has demonstrated that this fundamental concept, borrowed from information theory and theoretical computer science, can be effectively applied to our problem, and we have provided results, in the form of experiments and simulation, to demonstrate this. We have further provided a survey of existing algorithms for generating identifying codes, making them robust to underlying uncertainty, and distributing their computation throughout a network.

REFERENCES

Aalto, L., Göthlin, N., Korhonen, J., & Ojala, T. (2004). Bluetooth and WAP push based location-aware mobile advertising system. *In Proceedings of the 2nd international Conference on Mobile Systems, Applications, and Services*, (pp. 49-58), Boston, MA.

Bahl, P., & Padmanabhan, V. N. (2000). RADAR: An in-building RF-based user location and tracking system, In *Proceedings of the 19th Annual Joint Conference of the IEEE Computer and Communications Societies*, *2*, 775-784. Tel-Aviv, Israel.

Bartal, Y., Byers, J. W., & Raz D. (1997). Global optimization using local information with applications to flow control. *In IEEE Symposium on Foundations of Computer Science*, (pp. 303–312).

Castro, P., Chiu, P., Kremenek, T., & Muntz, R.R. (2001), A probabilistic room location service for wireless networked environments, In G. D. Abowd, B. Brumitt, and S. A. Shafer, (Eds.), *Proceedings of the 3rd international Conference on Ubiquitous Computing* (pp. 18-34). Springer-Verlag, London.

Chao, S. (2008). *A Comparison of Indoor Location Detection Systems for Wireless Sensor Networks*, MS project report, Boston MA: Boston University.

Charon, I., Hudry, O., & Lobstein, A. (2003). Minimizing the size of an identifying or locating-dominating code in a graph is NP-hard, *Theoretical Computer Science*, *290*(3), 2109–2120.

Gravier, S., Klasing, R., & Moncel, J. (2006). *Hardness results and approximation algorithms for identifying codes and locating-dominating codes in graphs*. (Technical Report RR-1417-06), Bordeaux, France: Laboratoire Bordelais de Recherche en Informatique (LaBRI).

Hightower, J., Borriello, G., & Want R. (2000), *SpotON: An indoor 3D location sensing technology based on RF signal strength*, (Tech. Rep. No. 2000-02-02), University of Washington.

Johnson, D. S. (1974). Approximation algorithms for combinatorial problems. *Journal of Computer and System Sciences*, *9*, 256–278.

Karpovsky, M. G., Chakrabarty, K., & Levitin, L. B. (1998). A new class of codes for identification of vertices in graphs, *IEEE Transactions on Information Theory*, *44*(2), 599–611.

Ladd, A. M., Bekris, K. E., Rudys, A., Marceau, G., Kavraki, L. E., & Wallach, D. S. (2002). Robotics-based location sensing using wireless ethernet. *In Proceedings of the 8th Annual international Conference on Mobile Computing and Networking* (pp. 227-238), Atlanta, Georgia.

Laifenfeld, M. (2007). *Coding for network applications: Robust identification and distributed resource allocation*. Doctoral dissertation, Boston University, MA.

Laifenfeld, M., & Trachtenberg, A. (2005). Disjoint identifying codes for arbitrary graphs. In *IEEE International Symposium on Information Theory*, (pp. 244- 248), Adelaide Australia.

Laifenfeld, M., & Trachtenberg, A. (2008). Identifying codes and covering problems. *IEEE Transaction on Information Theory, 54*(9), 3929-3950.

Laifenfeld, M., Trachtenberg, A., & Berger-Wolf, T. (2006). Identifying codes and the set cover problem. *In Proceedings of the 44th Annual Allerton Conference on Communication, Control, and Computing,* Urbana-Champaign, IL.

Laifenfeld, M., Trachtenberg, A., Cohen, R., & Starobinski, D. (2007). Joint monitoring and routing in wireless sensor networks using robust identifying codes. In *IEEE Proceedings of the 4th international Conference on Broadband Communications, Networks and Systems,* (pp. 197- 206).

Laifenfeld, M., Trachtenberg, A., Cohen, R., & Starobinski, D. (2008) Joint monitoring and routing in wireless sensor networks using robust identifying codes. *In Springer Journal on Mobile Networks and Applications (MONET).* online *http://www.springerlink.com/content/8386234004837647*

Moncel, J. (2006). On graphs of n vertices having an identifying code of cardinality $\lceil \log_2(n+1) \rceil$. *Discrete Applied Mathematics, 154*(14), 2032–2039.

Moncel, J., Frieze, A., Martin, R., Ruszink, M., & Smyth, C. (2005). Identifying codes in random networks. In *IEEE International Symposium on Information Theory,* (pp. 1464- 1467), Adelaide, Australia.

Muller, T. & Sereni, J.-S. (2007). Identifying and locating-dominating codes in (random) geometric networks. 2007, *submitted to Combinatorics, Probability and Computing. ITI Series* 2006-323 and *KAM-DIMATIA Series* 2006-797.

Ni, L., Liu, Y., Lau, Y. C., & Patil, A. P. (2005), LANDMARC: Indoor location sensing using active RFID, *Wireless Networks, 10*(6), 701-710.

Priyantha, N. B., Chakraborty, A., & Balakrishnan, H. (2000). The Cricket location-support system. *In Proceedings of the 6th Annual international Conference on Mobile Computing and Networking,* (pp. 32-43). Boston, Massachusetts.

Rajagopalan, S. & Vazirani, V. (1998). Primal-dual RNC approximation algorithms for set cover and covering integer programs. *SIAM Journal on Computing, 28,* 525–540.

Ray, S., Starobinski, D., Trachtenberg, A., & Ungrangsi, R. (2004). Robust location detection with sensor networks, *IEEE Journal on Selected Areas in Communications (Special Issue on Fundamental Performance Limits of Wireless Sensor Networks), 22*(6), 1016 – 1025.

Ray, S., Ungrangsi, R., Pellegrinin, F. D., Trachtenberg, A., & Starobinski D. (2003). Robust location detection in emergency sensor networks, In *Proceedings of the 22nd Annual Joint Conference of the IEEE Computer and Communications Societies, 2,* 1044–1053.

Suomela, J. (2007), Approximability of identifying codes and locating–dominating codes. *Information Processing Letters, 103*(1), 28–33.

Varshavsky, A., de Lara, E., Hightower, J., LaMarca, A., & Otsason, V. (2007). GSM indoor localization. *Pervasive Mob. Comput., 3*(6), 698-720.

Vazirani V. (Ed.). (2001). *Approximation Algorithms.* Springer-Verlag.

Want, R., Hopper, A., Falcao, V., & Gibbons, J. (1992). The active badge location system. *ACM Trans. on Info. Systems, 10*(1), 91–102.

ENDNOTES

[1] This algorithm applies only to those graphs that admit an identifying code. Some graphs, such as the complete graph, do not.

[2] This condition becomes sufficient if the empty set is a valid identifying set.

[3] A dominating set or a vertex cover is a set of vertices that the union of their balls is equal to the entire set of vertices.

[4] The received identifying set was obtained by thresholding the vector $(n0; n1; n2; n3)$ at that location, namely a transmitter was assumed to be a member of the received identifying set if the number of its received packets exceeded a certain threshold.

Chapter XIV
Evaluation of Localization Algorithms

Michael Allen
Coventry University, UK

Sebnem Baydere
Yeditepe University, Turkey

Elena Gaura
Coventry University, UK

Gurhan Kucuk
Yeditepe University, Turkey

ABSTRACT

This chapter introduces a methodological approach to the evaluation of localization algorithms. The chapter contains a discussion of evaluation criteria and performance metrics followed by statistical/empirical simulation models and parameters that affect the performance of the algorithms and hence their assessment. Two contrasting localization studies are presented and compared with reference to the evaluation criteria discussed throughout the chapter. The chapter concludes with a localization algorithm development cycle overview: from simulation to real deployment. The authors argue that algorithms should be simulated, emulated (on test beds or with empirical data sets) and subsequently implemented in hardware, in a realistic Wireless Sensor Network (WSN) deployment environment, as a complete test of their performance. It is hypothesised that establishing a common development and evaluation cycle for localization algorithms among researchers will lead to more realistic results and viable comparisons.

INTRODUCTION

Evaluating the relative performance of localization algorithms is important for researchers, either when validating a new algorithm against the previous state of the art, or when choosing existing algorithms which best fit the requirements of a given WSN application. However, there is a lack of unification in the WSN field in terms of localization algorithm evaluation and comparison. In addition, no standard methodology exists to take an algorithm through modelling, simulation and emulation stages, and into real deployment. As a result it can be hard to quantify exactly *how* and under what circumstances one algorithm is better than another. Moreover, deciding what performance criteria localization algorithms are to be compared or evaluated against is important for the success of the resulting implementation given that different applications will have differing needs.

Since localization algorithms are expected to be used in real applications, it is not conclusive to verify their performance in simulation only. The authors here argue that algorithms should be emulated (on test beds or with empirical data sets) and subsequently implemented in hardware, in a realistic WSN deployment environment, as a complete test of their performance.

In this chapter, performance evaluation metrics are discussed alongside three criteria – localization accuracy, cost, and coverage. Given that WSNs are typically constrained in terms of node/network lifetime and per-node computational resources, addressing these constraints leads to trade-offs in the performance of localization algorithms. For example, if maximising localization accuracy is the foremost priority, specific hardware may have to be added to each sensor node, increasing node size, cost and weight. Conversely, if the hardware available is already determined, then the application expectations with respect to performance criteria (such as accuracy) must be adjusted accordingly.

The chapter is structured as follows: a discussion of the various performance criteria and evaluation metrics that are readily used in the analysis of localization algorithms is first presented. Next, representative topologies that affect performance criteria are given, followed by simulation models and parameters that affect the performance of localization algorithms. A case study is presented, outlining an acoustic monitoring sensor network with high accuracy constraints enforced by application requirements. This case study is contrasted with an example where scalability and longer network lifetime are required at the expense of complexity and localization accuracy. Finally, the chapter closes with a brief discussion on the development cycle of a localization algorithm, from simulation to real deployment.

It should be noted that although this chapter makes particular emphasis on simulation and comparison of range-based localization algorithms, many of the metrics and techniques described are applicable to other approaches, such as Angle of Arrival (AoA) based algorithms, for example.

EVALUATION CRITERIA

Whilst the intuitive measure of the performance of a localization algorithm may be to show how well it can estimate positions of nodes compared to the known ground truth (to the degree of accuracy required by the WSN application, as discussed further below), localization algorithms are also subject to the general constraints of wireless networked sensing. It follows that a broader set of evaluation criteria for localization algorithms are needed (and are useful to both developers and users of localization algorithms), examples of which are accuracy, cost, coverage, robustness and scalability. These criteria reflect the constraints already mentioned - computational limitations, power constraints, unit cost and network scalability.

Some evaluation criteria are binary in nature: algorithms either have a specific property or they do not (for example, they are self-configuring or not; they are anchor free or not). Classifications and binary criteria can be used by researchers to narrow the set of existing algorithms to evaluate against, or to choose from. For example, one may only consider distributed, anchor-free, range based localization algorithms, immediately limiting the number of algorithms to compare to. Some evaluation criteria and trade-offs however, need quantification and qualification. These are described below in more detail, and questions are posed that might be useful to the algorithm or WSN application designer in establishing a given algorithm's performance.

Scalability

Can the localization algorithm scale from less than ten nodes to hundreds, or even thousands? Moreover, is it necessary from the WSN application standpoint for the algorithm to hold this property?

A centralised localization algorithm will typically aggregate all input data at a central, more capable sink to carry out processing; this represents a single point of error, and potential bottleneck for network communication. In contrast, a distributed localization algorithm's execution is shared throughout the network with no reliance on a central sink. However, centralised algorithms are conceptually simple and easier to implement in cases where it is known that the network will be small and will not increase. By comparison, distributed algorithms are harder to develop and deploy, but may be advantageous for researchers if the network does not have a simple logical topology (i.e. a tree of nodes sending data to a sink), and will need to support a large number of nodes (tens to hundreds). Theoretically, scalability is an important general consideration; however in actual deployment for specific applications this is not necessarily an overriding one (primarily due to the relatively small numbers of nodes that will be deployed and the amount of effort it takes to deploy them).

Accuracy

How well do the positions estimated by the algorithm match the known, ground truth positions? How well has the WSN application been specified in terms of its minimal localization accuracy needs?

One may think that positional accuracy compared to ground truth is the over-riding goal of a good localization algorithm. On reflection, this is largely application-dependent - different WSN applications will have different requirements on the resolution of the accuracy. Consider a tracking application – the estimated positions of nodes in the network directly affect the accuracy of the tracking. The granularity of the required accuracy may be a ratio of the inter-node spacing. For example, if the average node spacing is 100m, up to 1m error may be acceptable. However, if the average node spacing is 0.5 m, the same error level is clearly unacceptable.

Resilience to Error and Noise

How well can the localization algorithm deal with errors and noise in the input data?

It is important to understand how well the localization algorithm will perform without an accurate or full set of input data. Some algorithms, for example classical multi-dimensional scaling, used by Shang et al (2003) assume measurements from every node to every other for the localization algorithm to converge, which is an overbearing assumption given the realities of most deployment environments.

Evaluation should show how measurement noise, bias or uncorrelated error in the input data affects the algorithm's performance, and also establish the number of nodes that can actually be localized. Errors in measurement are particularly important to consider when adapting a localization algorithm that assume 2D to work for 3D applications (a common assumption in the research community). For example, a simple multilateration computation in 3D is far more sensitive to noisy/inaccurate measurements than its 2D counterpart, due to the extra degree of freedom (the Z axis). Convergence in 3D may then result in flips and reflections of the estimated coordinate, as observed by Allen et al. (2006) and shown in Figure 1.

Coverage

How much of the network can be localized by the algorithm, given a specific network topology/deployment?

Some algorithms may have problems localizing the whole network if nodes do not have enough neighbours ("enough" is specific to the details of the algorithm) in terms of connectivity or distance constraints/estimates. Coverage may relate to the physical network density, i.e. one may be more likely to get 100% localization coverage in a densely deployed network. In addition, it is worth considering how easy it is to add another node to the network after the initial localization algorithm has completed.

Cost

How expensive is the algorithm in terms of power consumption, time taken to localize a node, communication and pre-deployment set-up (i.e. need for, and number of anchors)?

There are several parameters which one could classify as "individual costs", such as per-node hardware or software cost, power consumption required to complete node localization, time taken to converge on a network wide localization solution, and amount of communication required (messages/data transmitted). An algorithm which can minimise several cost constraints is likely to be desirable if maximising network lifetime is a primary deployment goal. For example, an algorithm may focus on

Figure 1. Using multilateration to estimate unknown positions in 3D with noisy range measurements

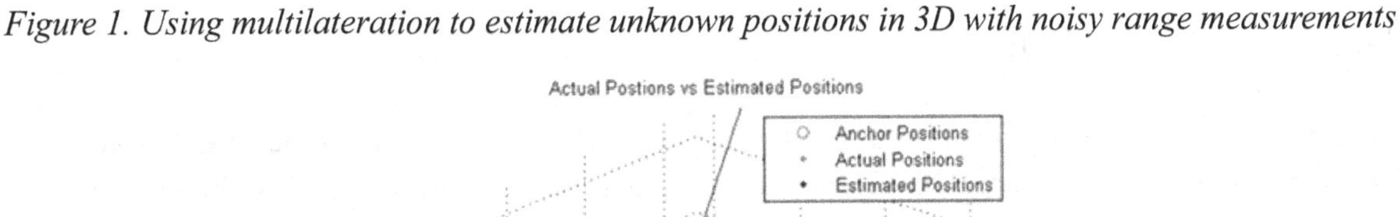

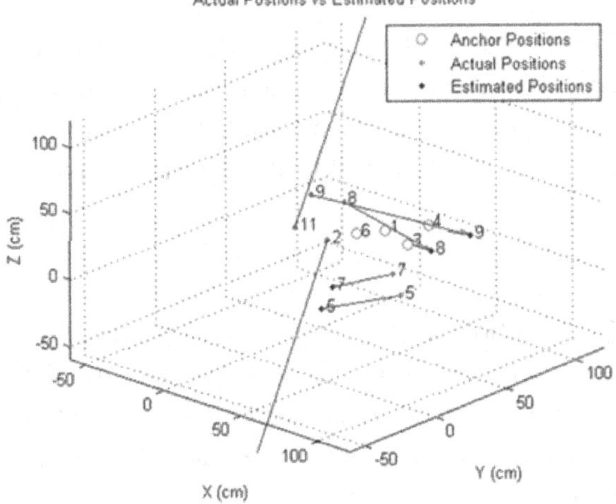

minimising communication and complex processing to achieve quick convergence, but at the expense of the overall accuracy.

Discussion

Clearly, the perfect localization algorithm would provide suitably accurate results (relative to the scale requirements of the application), in a simple and decentralised way, with low communication and processing overhead, whilst allowing incremental addition of nodes, and requiring zero anchor nodes. However, all existing and possibly future algorithms will most likely have to trade these criteria off against one another. For example, is it better to increase memory footprint and processing time for accuracy, or use a simpler algorithm and reduce positional estimation accuracy and time taken? Deployment practice and expertise indicate that trade-offs are best resolved when intimately related to the specifics of the class of applications that the particular WSN is deployed to address. The quantitative measurements required to understand these trade-offs are described in the next section.

LOCALIZATION ALGORITHM EVALUATION, COMPARISON AND METRICS

In order to address quantitatively how well a localization algorithm might perform against the criteria described in the previous section, a set of metrics are available. This section breaks down accuracy, cost and coverage and describes well known metrics or common measures used in their evaluation.

Accuracy Metrics

The basic goal of the localization accuracy metric is to show how well matched the ground truth and estimated positions are. Accuracy is likely to be related to measurement noise, bias, accuracy and precision in the input data provided to the localization algorithm. The accuracy metrics described below are separated into those which use ground truth as comparison, and those which do not. Throughout this section, it is assumed that a WSN is composed of n sensor nodes deployed over a given area.

Metrics with Ground Truth

Globally, the positions determined by a localization algorithm represent a geometrical layout of the physical positions of the sensors. This layout must be compared to the ground truth, or known layout of the sensors. It is important therefore that not only the error between the estimated and real position of each node is minimised, but also that the geometric layout determined by the algorithm matches well the original geometric layout.

Mean Absolute Error

The simplest way to describe localization performance is to determine the residual error between the estimated and actual node positions for every node in the network, sum them and average the result. Broxton et al (2006) do this using the mean absolute error metric (MAE), which, for each of n nodes in the network, calculates the residual between the node's estimated $(\hat{x}_i, \hat{y}_i, \hat{z}_i)$ and actual (x_i, y_i, z_i) coordinates. This is shown in (1):

$$MAE = \frac{\sum_{i=1}^{n} \sqrt{(x_i - \hat{x}_i)^2 + (y_i - \hat{y}_i)^2 + (z_i - \hat{z}_i)^2}}{n} \quad (1)$$

The resulting metric represents the average positional error in the network, aggregating individual residual errors into one statistic. The MAE computation has much similarity to root mean square (RMS) error, a commonly used calculation to measure the difference (or residual) between predicted and observed values. Slijepcevic et al (2002) also note that whilst knowing the mean absolute error is important in some cases, it is also beneficial to know the maximum error exhibited in the position estimation, as shown in (2).

$$MAX\_ERROR = \max_{i=1..N} \sqrt{(x_i - \hat{x}_i)^2 + (y_i - \hat{y}_i)^2 + (z_i - \hat{z}_i)^2} \quad (2)$$

FROB

A slightly different approach is taken by Efrat et al (2006) who use the FROB (Frobenius) metric. In this case, the residual error between all *n* nodes in the network is calculated. It is assumed that the estimated and actual inter-node distances have already been determined. The Frobenius metric is shown in (3), where $\hat{d}_{ij}$ and *dij* are the estimated and ground truth distances respectively and *n* is the number of nodes in the network.

$$FROB = \sqrt{\frac{1}{n^2} \sum_{i=1}^{n} \sum_{j=1}^{n} (\hat{d}_{ij} - d_{ij})^2} \quad (3)$$

FROB essentially determines the RMS of the total residual error, which represents the global quality of the localization algorithm.

GER and GDE

As discussed briefly at the start of this section, it is important for the accuracy metric to reflect not only the positional error in terms of distance, but also in terms of the geometry of the network localization result. If only average node position error is used, there is no sense of the *correctness* of the relative geometry of the network – it is entirely possible that for a given localization result the average error metric is low, but the actual layout created by the algorithm does not match well the physical layout of the network. This problem was identified by Priyantha et al (2003), and addressed by defining the Global Energy Ratio (GER) metric, shown in (4).

$$GER = \frac{1}{n(n-1)/2} \sqrt{\sum_{i=1}^{n} \sum_{j=i+1}^{n} \left(\frac{\hat{d}_{ij} - d_{ij}}{d_{ij}} \right)^2} \quad (4)$$

The distance error between nodes $(\hat{d}_{ij} - d_{ij})$ is normalised by the known distance between the two nodes (d_{ij}), making the error a percentage of the known distance. (One should notice the similarity between GER and FROB – note that FROB does not normalise the distances, and takes the RMS). Ahmed et al (2005) note that the GER metric does not exactly reflect RMS error. They address this by defining an accuracy metric which better reflects the RMS error calculation, called Global Distance Error (GDE), shown in (5).

$$GDE = \frac{1}{R}\sqrt{\frac{\sum_{i=1}^{n}\sum_{j=i+1}^{n}\left(\frac{\hat{d}_{ij} - d_{ij}}{d_{ij}}\right)^2}{n(n-1)/2}} \quad (5)$$

GDE takes the RMS error over the network of n nodes and normalises it using the constant R. In Ahmed et al's context, R represents average radio range, meaning the localization results are represented as a percentage of the average distance nodes can communicate over.

ARD

Gotsman and Koren (2005) derive another quality metric called Average Relative Deviation (ARD), shown in (6). ARD is simply the normalised average of the estimate, rather than the RMS error.

$$ARD = \frac{2}{n(n-1)}\sum_{i<j}\frac{\left|\hat{d}_{ij} - d_{ij}\right|}{\min(\hat{d}_{ij}, d_{ij})} \quad (6)$$

The individual distances are normalised in this case by the shorter of the two distances (either the estimated or ground truth), which may not always be the known distance (as in GDE and GER).

BAR

The BAR metric by Efrat et al (2006) is a measure of how well the estimated positions of nodes that sit on the boundary of the localized network match the actual positions. It is in essence the sum-of-squares normalised error taken from matching the estimated boundary with the actual boundary. This metric may be useful as an alternative to GER, in cases where the topology formed does not seem to match well the actual topology, even though the distance error metrics indicate it should. In this case, the average error is not helpful, as the metric can be diluted by high error variance across the network. BAR is used as the minimisation metric for the Iterative Closest Points (ICP) algorithm which matches estimated and actual boundary points. A BAR metric is computed for each iteration of the ICP algorithm (Zhang, 1992), giving a measurement of how well the two boundaries match. When the change in the BAR metric is negligible, ICP has determined the best alignment possible. The BAR metric therefore represents how well the outer geometry matches, and can potentially give insight into where the problems lie for a particular localization algorithm.

Girod (2005) uses a similar technique to compare the shape of a localized network, irrespective of translation, scale and rotation. He defines a four-step approach influenced by the Procrustes method of characterising shape, and uses it to measure estimation fit with ground truth. Firstly, a scaling factor between the real and estimated topologies is established; the maps are then translated and scaled relative to the origin, which is defined as the node closest to the centroid of the estimated topology. The estimated topology is then rotated according the angular offsets between nodes, and finally translated by the average distance between estimated and ground truth points. Average error can now be taken using any of the metrics we have previously described (MAE, GER, GDE, etc). Whilst both Girod and Efrat's approaches take into account the shape of the network, BAR needs only a subset of the nodes on the boundary to contribute towards the computation; Girod's method uses all nodes in the network.

Metrics without Ground Truth

The accuracy metrics above rely on prior knowledge of the actual node position and physical network topology in order to evaluate the localization quality and error. In realistic, un-positioned WSN deployments, this information is not known, and so measurement of error must be determined relative to what information *is* available. For example – if we assume a range-based localization algorithm where nodes measure distance between one another and their positions are estimated based on this information, a metric not using ground truth must compare the measured ranges with the ranges derived from the estimated positions. Unlike the ground truth metrics above, this means that only the actual measured distances can be compared with localization derived ranges. Toward this aim, Girod (2005) defines an average distance error metric, shown in (7). In this case, the estimated distance between two nodes i and j is subtracted from the observed range R_{ij} between them.

$$error = \frac{2}{n(n-1)} \sum_{i,j,i<j} R_{ij} - \sqrt{(X_i - X_j)^2 + (Y_i - Y_j)^2 + (Z_i - Z_j)^2} \quad (7)$$

Başaran (2006) suggests a FROB similar metric, called SPFROB (shortest-path FROB), based on the shortest path between two nodes, rather than Euclidean distance. This metric is potentially useful for multi-hop localization algorithms which infer distance by the estimated shortest path from a node to an anchor, such as the ad-hoc positioning system (APS) by Niculescu and Nath (2001).

Cost Metrics

Cost metrics relate to how "expensive" it is for localization to be carried out. These costs are related to the traditional constraints of wireless networked sensing devices – low power operation, low computational capability, and redundancy through scale and density. Cost is an important trade-off against accuracy, and is often motivated by realistic application requirements, which are discussed in more detail at the end of the metrics section. As such, cost metrics are typically used to evaluate the trade-offs that are not addressed by positional error and coverage. Several common metrics are described below, along with how they may be determined.

Anchor to Node Ratio

Minimising the number of anchors in the network is desirable from an equipment (cost, power usage) or deployment point of view. For example, using anchors that can estimate position through the Global Positioning System (GPS) will require extra hardware which is both expensive and power-hungry, thus limiting the node lifetime. Similarly, pre-defining anchor positions may be hard if the supposed deployment mechanism is random placement (i.e. nodes being thrown from a vehicle). The *anchor to node ratio* is simply the number of anchors in the network divided by the number of nodes. This metric will typically be used to investigate the trade-off on the accuracy of the localization algorithm, i.e. as the percentage of anchors decreases, how does this affect the accuracy, and the percentage of nodes that can be localized? In anchor based localization algorithms, one must also consider the placement and density of anchors – this is discussed in the Coverage Section further in the Chapter. When using few anchors, Nagpal et al (2003) find that dense networks (on average 15 neighbours per node) are required to provide relatively accurate localization results, but that this accuracy is bounded by the method used to estimate inter–node distance (in their case radio range).

Communication Overhead

Since radio communication is assumed to be a large consumer of power relative to the overall consumption of a wireless sensor node, minimising communication overhead is paramount in maximising the potential network lifetime. Communication overhead will most likely be measured either by actual power consumed or number of packets transmitted to achieve the localization goal. For example, Langendoen et al (2003) use the average number of packets sent per node; power consumption can be derived from this if one knows the cost of sending a single packet (as is discussed in more detail in Modelling Section). This metric will typically be evaluated with respect to the scaling of the network – how does communication cost increase as the network increases in size?

Power Consumption

The proportion of available power that a node spends on localization can affect its lifetime (and the network lifetime). Power consumption will be a combination of the power used to perform local operations and the power used to send and receive messages associated with localization. The more complex the local processing for localization is, the longer it will take the node to process. As above, this metric will also typically be evaluated with respect to scaling of the network – how does power consumption increase as the network increases in size?

Algorithmic Complexity

Standard notions of computational complexity in time and space (i.e. big O notation) can be used as comparison metrics for the relative cost of localization algorithms. For example, as a network increases in size, a localization algorithm with $O(n^3)$ complexity is going to take a longer time to converge than an $O(n^2)$ algorithm. The same is true for space complexity – as the number of nodes increases, the amount of RAM needed (either per node, or centrally) is going to increase at a particular rate; algorithms which require less memory (comparatively) at a given scale may be preferable. This may help motivate a trade-off between centralised and decentralised algorithms – i.e. the centralised approach might be better in some cases if the per-node memory footprint becomes too large as the network scales (this would obviously be offset with the communication overhead).

Convergence Time

Measuring the time taken for both initial measurement gathering and localization algorithm convergence can both provide important comparison metrics. Time taken will most likely be evaluated against network size. For example – how does time taken to gather measurements or localize the network increase as the network increases in size? On the other hand, even for applications with fixed numbers of nodes, a network that takes a long time to localize may be useless if the application requires rapid deployment and processing immediately related to node positions, such as tracking of a moving target. Similarly, if one or more of the nodes in the network are mobile, the time taken to update position may not reflect the current physical state of the network – i.e. positional information may have become stale.

If the localization algorithm is based on non-linear optimisation, there may also be a trade-off to be made between accuracy and convergence time – the extra time taken and energy expended to get a slightly more accurate solution may not be beneficial.

Hybrid Metrics

Hybrid metrics encourage the evaluation of trade-offs in localization algorithms by combining several individual metrics into one composite metric. The way in which the metrics are combined will vary from one hybrid metric to another – one such example is the performance cost metric by Ahmed et al (2005).

Performance Cost Metric (PCM)

The performance cost metric (PCM) is a simple hybrid metric where performance cost C and localization error GDE are weighted by a parameter α, as shown in (8). This weighting is determined by the relative importance the evaluation wishes to place on the relevant components of the metric.

$$PCM = \alpha(GDE) + (1-\alpha)C \qquad (8)$$

Here, GDE (Global Distance Error) localization accuracy metric is a variant of GER, as described in the previous section. The cost aspect C of the PCM metric is described by the average per-node energy required to complete the localization (although one could imagine it being any quantitative cost metric). In deciding whether to use hybrid metrics instead of individual performance metrics, researchers should establish whether the values determined by the hybrid approach represent a fair or meaningful comparison.

Coverage Metrics

Some localization algorithms may not be able to localize all of the nodes in the network. Coverage is simply a measure of the percentage of nodes in the deployed network that can be successfully localized, regardless of the localization accuracy (which is described by previous metrics). However, density of deployment, as well as placement of anchors can have effects on coverage results for different localization algorithms. The effects and their evaluation and are discussed in the following subsections.

Density

The specific approaches that localization algorithms take can directly affect coverage. This can have different implications for anchor based and anchor free localization algorithms. For example, the robust localization algorithm proposed by Moore et al (2004) is an anchor free localization algorithm, based on range estimates between nodes. In order for a node to be considered a candidate for localization, there must be sufficient range estimates between the node and its neighbours to satisfy certain *rigidity* constraints (to protect against positional ambiguities which adversely affect localization results). If the density of the deployment is low, it may be impossible to localize many nodes. Figure 2 shows the relationship between node density, number of anchors and localization error for a multi-hop localization algorithm with random topology (Basaran et al, 2008). As the average node density increases, neighbor nodes generate more information, which can potentially improve localization performance with respect to localization error. Localization algorithms focusing on denser networks should bear in mind that radio traffic, number of message collisions and energy consumption of the nodes will also increase with the increasing average node density in the sensor network.

In anchor free localization algorithms, density is measured simply by the average number of neighbours a node has, as in AFL by Priyantha (2003) and the robust localization algorithm proposed by Moore (2004). Density can be used to determine the *minimum* neighbour density required for 100% localization coverage, or for an acceptable level of accuracy.

With reference to anchor based localization algorithms, Bulusu et al (2001) investigate the effects of anchor placement on localization (discussed further below), evaluating mean and median error improvement against anchor density (or *degree*) per square metre, given a random placement strategy. Similarly, in work on partially localizable networks, Goldenberg et al (2005) examine the percent of localizable nodes in the network as the number anchors increases. The authors measure anchor density in terms of average anchors a node has either in its effective radio communication or measurement region.

Anchor Placement

The position of anchors in the network may have a considerable impact on localization error, especially if the localization algorithm assumes that anchors are uniformly or randomly positioned in fixed locations. Assumptions about a pre-defined anchor placement scheme do not take into account environmental factors, terrain (that can affect placing of anchors), and signal propagation conditions, as well as optimal anchor placement. The geometry of anchor nodes with respect to any un-localized nodes in the network can have a varying effect on the accuracy of resulting position estimates. This effect is notably observed in GPS systems, where positional accuracy is seen to decrease when GPS satellites are closer to one

Figure 2. The change in the Frobenius error on a random grid topology for different node densities and increasing number of anchors

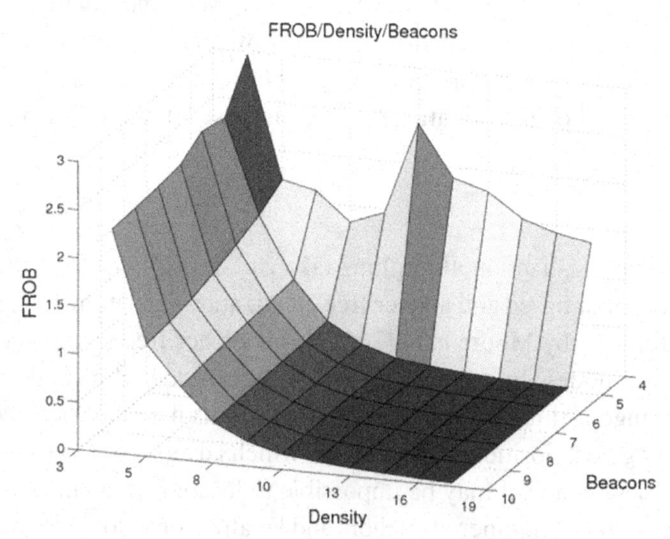

another. The Geometric Dilution Of Precision (GDOP) metric is used in GPS systems to describe the geometric "strength" of the GPS satellites' current positions with respect to the target, and thus can give an indication of whether the accuracy is likely to be good (a small value), or bad (a large value).

Savvides et al (2005) use a GDOP metric to investigate anchor placement in WSNs, using the metric to find the ideal anchor geometry. They conclude that a convex hull of anchors surrounding un-localized nodes is the most favourable configuration for minimising the effects of geometry on localization accuracy.

It is worth noting that some localization algorithms iteratively localize nodes. As a consequence, geometrically significant nodes (i.e. nodes that might allow others to be localized) may not themselves be localized, which could result in low a coverage percentage. Complimentary to this position, Bulusu et al (2002) hypothesise anchor placement needs to be *adaptive* in the face of noisy and unpredictable environmental conditions, proposing and evaluating two simple, mobility based proximity algorithms for incremental anchor placement.

Mobile anchors (or beacons) could also potentially be used to supplement coverage or reduce the number of anchors necessary. A mobile beacon based Bayesian approach to localizing network nodes has been proposed by Sichitiu (2004), and mobility models for simulation are discussed in the Models section of this Chapter.

Evaluating Coverage

In evaluating coverage performance for localization algorithms, researchers must be prepared to try various placement scenarios/strategies for nodes and anchors, as well as various densities. One can evaluate how the accuracy improves as either the number of anchor nodes or neighbours per node increases. Bulusu et al (2002) note that increasing the anchor density does not necessarily guarantee more accurate localization or better coverage; there is essentially a "saturation point" after which no additional gains in accuracy can be made. This is supported by the results shown in Figure 2. Therefore, localization algorithms should be investigated not only with respect to the fewest anchors that can be used, but also the point at which anchors give little or no improvement.

In addition, excessively noisy, biased or missing input data may cause the localization algorithm to behave in unpredictable ways, and may reduce coverage. Therefore as part of understanding coverage, a localization algorithm should also be evaluated with respect to its resilience in the face of varying amounts of measurement noise, as in Langendoen (2003).

Discussion

Accuracy, cost and coverage represent trade-offs for localization algorithms. This is a consequence of localization algorithms usually needing to be optimised toward a set of specific constraints, such as low power operation, speed of localization, scalability or a maximum positional error. A good understanding of trade-offs is important in the context of localization, as it is in general for WSN application design. For example, deploying a network with a large number of anchors is expensive, and requires a large amount of careful placement, especially to guarantee coverage. However, in attempting to minimise or remove entirely the need for anchors, a localization algorithm may compromise its accuracy and simplicity; anchor-free localization algorithms are frequently centralised (even the robust localization algorithm proposed by Moore et al (2004) requires a central phase), and framed as non-linear optimisation or minimisation problems, such as Girod et al (2006), Gotsman and Koren (2005). These approaches may not be tractable to run directly on resource constrained nodes.

It has been shown that accuracy metrics based on average position error may not capture the accuracy of the layout geometry. This is especially true for anchor-free localization algorithms. It has also been shown that the cost of a localization algorithm can take many forms, and can be highly dependent on the application requirements the WSN is designed and deployed to address. Coverage is greatly affected by placement of nodes in the network, be they anchors or regular nodes.

In creating new metrics for algorithm comparison, the designer must carefully consider the performance metrics that need to be addressed. Hybrid metrics can be useful if more than one metric must be analysed at the same time, and it makes sense to evaluate them together. Otherwise, using individual metrics to isolate specific aspects of localization performance is a fine way to evaluate and compare localization algorithms.

The first step toward fully evaluating a localization algorithm is to use the metrics presented in this section and apply them in simulation, along with relevant parameters that best represent the WSN application scenario. These matters are addressed in the next section.

EVALUATING LOCALIZATION PERFORMANCE: REPRESENTATIVE TOPOLOGIES AND SIMULATION MODELS

Evaluation and comparison of localization algorithms can be performed at various scales and using various metrics, as discussed in the previous section. Because real life deployments are expensive and difficult to scale to large numbers, simulation is a relatively easy and highly available tool to validate the performance of localization algorithms. It allows comparative performance evaluation for different environmental models and requirements imposed by the application domain. It also allows researchers to test the robustness of localization algorithms against variable conditions such as ranging error, various network topologies, anchor densities, and numbers of nodes. Simulations can also allow individual characteristics of algorithms to be isolated and evaluated by factoring out or simplifying real-world effects. However, statistical models used in localization simulations can make unrealistic assumptions about the ranging characteristics of deployment environments, which may result in misleading or incorrect results that only come to light during final deployment.

Measuring the performance of a localization algorithm via simulation requires a simulation environment and input parameters that are derived either from statistical models or empirically. The accuracy and achievable precision of localization algorithms strongly depends on the accuracy of the models used in the derivation of these input parameters, making this one of the over-riding limitations of simulation. There are a number of general purpose and localization specific simulators that can be used to evaluate and compare algorithms, including ns-2, OmNet++, and RiST (Reichenbach 2006). Some of these simulators have support for mobility and mobile radio communication, which can aid localization simulation.

This section presents some commonly used component models and building blocks for localization techniques. First, representative network topologies are introduced. This is followed by a presentation of a set of models for: inter-node ranging, noisy radio communication links, and energy consumption. The potential effect of ranging irregularities on localization performance is discussed, as well as other parameters that affect performance (such as node density, anchor/beacon placement, and mobility).

Topologies

Defining ground truth node deployment topologies in simulations can play an important role when comparing the performance of localization algorithms. For example, uniform grid, C-shape and ring-shape topologies can induce effects on localization algorithms that compromise their accuracy. There are essentially two main categories of sensor network topology, *even* and *random*. Even topologies distribute sensor nodes (and anchors) over the deployment area in an exact grid, whilst random topologies perturb individual nodes positions on the grid with random noise (with some predetermined range and variance). Figure 3 shows examples of both topologies. The results collected from the exact grid topology (Figure 3.a) are useful because they are visually simple – it is clear to see deviations in position estimation caused by the localization algorithm. Random topologies, however, better reflect the deployment scenarios in real-world environments (nodes cannot necessarily be placed uniformly). This is also because sensor networks may be deployed in locations where manual placement is either limited (e.g. in a thick forest) or almost impossible (e.g. inside a volcano). In these cases, it is generally assumed that nodes are randomly dropped from some deployment vehicle, and uniform placement cannot be guaranteed.

For these reasons, random network topologies are generally more popular for involved simulation and comparison studies. Topologies can be further sub-classed (*regular* and *irregular* topologies) according to the regularity of their placement densities and shapes.

Regular Topologies

In regular topologies (such as those shown in Figure 3), nodes are typically uniformly distributed over an area as a grid. This has the advantage that average node density in each part of the deployment area is relatively consistent. Many well known multi-hop localization algorithms, such as APS by Niculescu and Nath (2001), estimate the shortest-path distances (in terms of actual distances or number of hops) between sensor nodes and derive an overall Euclidean distance from this to estimate position. Such algorithms, when evaluated in simulation using regular topologies, may appear to be highly accurate, or at least have bounded error. However, this is not sufficient to prove the general effectiveness of a localization algorithm; regular topologies do not necessarily accurately reflect realistic deployment scenarios due to the variety of geographical factors that may restrict placement of sensor nodes.

Irregular Topologies

In these topologies, the shortest-path distances between nodes can deviate greatly from the actual Euclidian distances between nodes, and individual node density in a region may deviate greatly from the average density of the WSN. C-shaped, L-shaped and ring-shaped topologies are typical irregular topology examples, and represent irregular deployment configurations that applications may find themselves constrained by. Therefore, such topologies are generally employed to compare and stress various attributes of localization algorithms. In Figure 4 two types of C-shape topologies are presented. Note that in Figure 4, the difference between the Euclidian distance and the shortest-path distance between certain nodes can be large. As a result, individual errors in the localization algorithm may accumulate, resulting in large overall localization errors.

These simple topologies may be combined to generate either larger or more complex sensor network topologies. Obviously, a localization algorithm is more robust and generally usable when it generates accurate results for these types of topologies.

Figure 3. Even and random topology examples

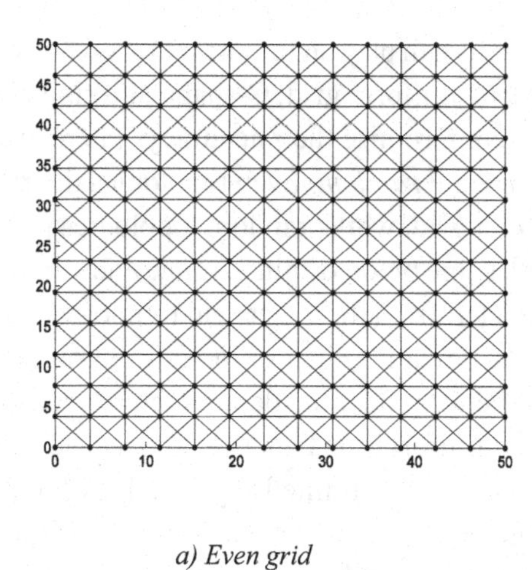

a) Even grid

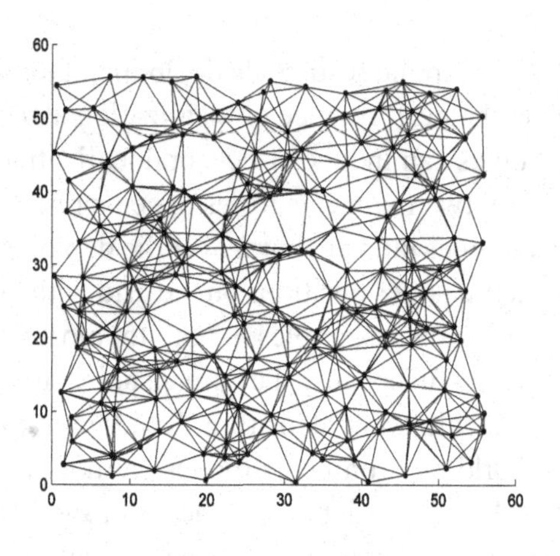

b) Random uniform

Third Dimension in Topologies

Today, most of the localization studies tend to ignore or trivialize the third dimension in topology setup and simulation. However, the third dimension is unavoidable in most real-life deployment scenarios and, unfortunately, introduces additional complexities to the localization algorithms, as examined by Ghosh (2007). For instance, in a network deployed on a hill or mountain, geographical obstacles hinder the radio communication among nodes. In such scenarios, a node may experience better packet reception but worse transmission rates compared with nodes on higher ground. This increases the percentage of asymmetric links, which may therefore affect communication and ranging assumptions.

Ranging Models

Ranging is the process of estimating the inter-node distance or angle using one or more modalities (for example signal strength or acoustic time of flight). In simulation, a widely used ranging modality is radio signal strength, but other modalities include acoustic time of flight (ultrasonic or audible) and ultra-wide band (UWB), as discussed by Yu (2004). All ranging techniques approximate the distance between nodes, therefore error related to measurement accuracy, multi-path effects and non-line of sight is expected.

Noisy Disk Model

An accurate, sensible ranging model is a critical aspect of a range or angle-based localization algorithm. The noisy disk, where a node can emit ranging signals to all neighbors within a maximum range R (the radius of the disk), is a commonly used ranging model in simulation. The model has two components: noise and connectivity. The noise component indicates the distribution of error, which is added to the actual distance (e.g., Gaussian, uniform) to form the estimated distance. The connectivity component indicates the maximum distance d_{max} between two nodes at which a distance estimate can be obtained.

Figure 4. Example irregular topologies

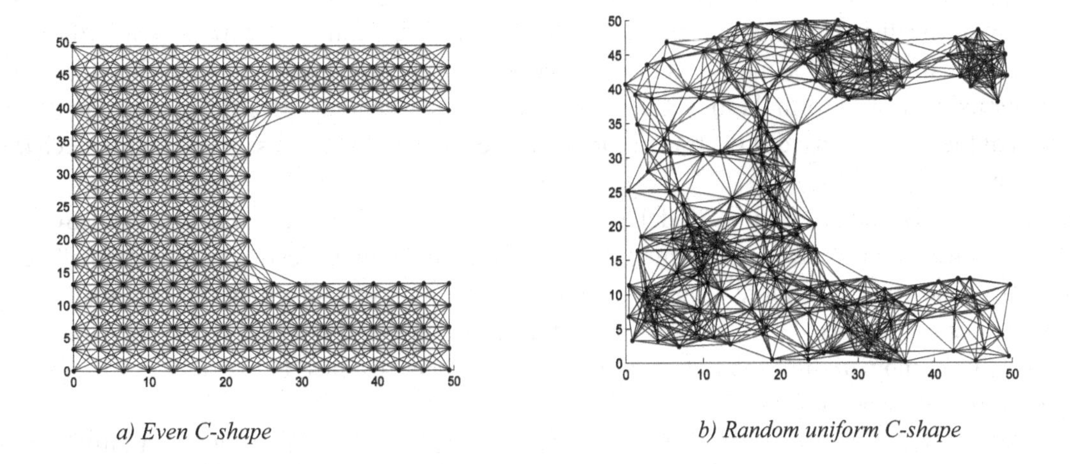

a) Even C-shape *b) Random uniform C-shape*

For example, using Gaussian noise (with variance σ), the Noisy Disk defines the distance estimate $\hat{d}_{ij}$ between nodes i and j in terms of the true distance d_{ij} as:

$$\hat{d}_{ij} = \begin{cases} N(d_{ij}, \sigma) & d_{ij} \le d_{\max} \\ undefined & otherwise. \end{cases} \tag{9}$$

The Noisy Disk model with no noise component (i.e., it only models the connectivity between nodes) is also known as the Unit Disk model. In simulations, range data used in distance estimations are usually generated from a parametric function of a theoretical propagation model. The level of the detail of the propagation model used is particularly important. Insufficient details can produce unrealistic error in estimations. Unfortunately, even detailed theoretical models may be significantly different from estimations made in real life deployments because of hardware and environmental ranging irregularities.

Whitehouse et al (2005) have proposed an alternative simulation technique – *Statistical Emulation*, where data for simulation is generated by randomly drawing measurements from an empirical data set (discussed further in the range irregularity section). Reichenbach et al (2006) have also proposed a tool using defective observations as input thus enabling more realistic simulations. Theoretical propagation models are briefly discussed below.

Radio Propagation

In radio models, the received signal strength is usually represented with the following formula, measured in decibels:

$$\textit{Received Signal Strength} = \textit{Sending Power} - \textit{Path Loss} + \textit{Fading} \tag{10}$$

The *Sending Power* of a node is determined by the battery status and the type of transmitter, amplifier and antenna. *Path Loss* describes the signal's energy loss as it propagates to the receiver. Path loss can be calculated using different physical models. The "Free Space Model" assumes the ideal propagation condition: that there is only one clear line-of-sight path (LOS) between the transmitter and receiver with no obstacles nearby to cause reflection or diffraction. The path loss is modeled as being proportional to the square of the distance between the transmitter and receiver, and also proportional to the square of the frequency of the radio signal. This model accounts for the propagation distance between sender and receiver using a fixed formula for signal loss, and does not include hardware specific factors such as the gain of the antennas used at the transmitter and receiver, nor any loss associated with mechanical imperfections.

The 'Two-Ray-Ground Reflection Model' considers antenna orientation and distance from ground for both the transmitter and receiver, performing detailed radio ray tracing to estimate reflection of signals. This model is known to give more accurate predictions at a long distance than the free space model. However it does not perform as well at short distance due to the oscillation caused by the constructive and destructive combination of the two rays. In the case where distance between nodes is small, the free space model may be preferred.

The effects of reflection, diffraction and scattering of signals as they hit obstacles will influence the free propagation of signals, leading to observation errors at the receiving node. These effects cause an exponential decay on the signal strength with respect to distance. Signal strength is also assumed to be log-normally distributed for a given distance *d*. The log-normal shadowing path loss model which is the most commonly used radio propagation model in WSN simulations is given as follows:

$$P_r(d) / P_r(d0) = -10\beta \log(d / d0) + X \tag{11}$$

Where $P_r(d)$ is the received power for distance d and $P_r(d0)$ is the received power for a reference distance $d0$. β is the path loss exponent (rate at which signal decays). X is a Gaussian random variable with zero mean and standard deviation σ. β and σ are obtained through curve fitting of empirical data. To approximate a communication link with the shadowing model, Ramadurai and Sichitiu (2003) suggest a simple approach to calculate the distance using a certain radio propagation model and introduce a random error E to the calculated distance.

Acoustic Ranging

A useful model for acoustic ranging error (audible or ultrasonic) proposed by Girod (2005) can be given as follows:

$$R_{ij} = \|d_{act} - d_{est}\| + X_{ij} + N_{ij} \tag{12}$$

In (12), d_{act} is the actual distance between nodes i and j, d_{est} is the estimated distance and X_{ij} is a Gaussian random variable with zero mean and standard deviation σ. N_{ij} is a fixed bias, which is present only when line-of-sight is blocked. This model represents the basic error components that one finds in acoustic ranging – a non-line of sight bias component and a Gaussian error component. X_{ij} can be reduced by repeated observations but N_{ij} needs to be filtered at higher layers.

Ultra Wide Band Ranging

An UWB radio ranging system has the ability to resolve multi-path components of the wireless propagation channel with extremely high time resolution. A standardized UWB channel model for IEEE

802.15.3a is claimed to best match the empirical measurements (Lee, 2002; Yu, 2004; Forrester, 2003). Additionally, Shah et al. (2005) has shown UWB ranging can be used to devise algorithms robust for both LOS and NLOS environments. The impulse response of the UWB channel model is:

$$h(t) = X\sum_{l=0}^{L}\sum_{k=0}^{K}\alpha_{k,l}\,\delta(T - T_l - T_{k,l}) \tag{13}$$

where $\alpha_{k,l}$ and $T_{k,l}$ are the multipath gain and delay of the kth ray in the lth cluster, respectively. T_l represents the delay of the lth cluster and X indicates the log-normal shadowing effect. Detailed distribution functions of different variables in can be found in Forrester (2003).

Range Irregularities

Aforementioned range models assume circular propagation ranges, whereas in reality propagation ranges tend to have an irregular shape. Range irregularity is one of the main sources of asymmetric links in WSNs. Irregularities are caused by three main factors, relative to the ranging model: device properties, propagation medium and environmental factors. Device properties include the antenna type, the transmission power, antenna gains, receiver sensitivity, receiver threshold and the Signal-Noise Ratio (SNR). Propagation medium properties include the medium type and background noise and environmental factors include attributes such as the temperature of the environment and obstacles within the deployment area (Zhou et al., 2004). As an example, the radiated pattern of the inverted-F antenna installed in the widely-used ChipCon CC2420 radio (Andersen, 2007) is very obviously non-isotropic. Therefore, it is clear that simple radio models that assume a perfect, spherical radio range cannot accurately predict or describe real-world radio characteristics.

Range irregularity models aim to reduce the discrepancy between the simulation and real-world results; two such examples are described below.

Statistical Emulation: Acoustic or Radio Ranging Irregularities

Whitehouse and Culler (2006) identified four different types of empirical ranging irregularities arising from empirical ultrasound/radio range data, which they use to extend the Noisy Disk model (as mentioned previously in this chapter). They define the ranging irregularities as:

- **Extreme overestimates:** An excess of range estimates that are larger than the true distance by more than two standard deviations.
- **Extreme underestimates:** An excess of range estimates that are smaller than the true distance by more than two standard deviations.
- **Long-range proficiency:** The existence of range estimates between nodes farther than nominal range d_{max}.
- **Short-range deficiency:** The existence of range failures between nodes closer than nominal range d_{max}.

The authors then study five stages of ranging models, each incorporating more ranging irregularity detail than the previous one:

- **Model 1)** Noisy Disk (No irregularities)
- **Model 2)** Model 1 + Extreme Overestimates
- **Model 3)** Model 2 + Extreme Underestimates
- **Model 4)** Model 3 + Long-range proficiency
- **Model 5)** Model 4 + Short-range deficiency

Whitehouse and Culler use their empirical ranging data to generate ranging irregularities in simulation, proposing a technique they call *Statistical Emulation.* The authors find that small variations in ranging model can cause large variations in localization error for several algorithms.

RIM: Radio Irregularity Model

Zhou et al (2004) establish a radio model for simulation, called the Radio Irregularity Model (RIM). From experimental results, they assign the following properties to radio sensing hardware:

- **Non-isotropic:** The radio signal from a transmitter has different path loss in different directions.
- **Continuous variation:** The signal path loss varies continuously with incremental changes of the propagation direction from a transmitter.
- **Heterogeneity:** Differences in hardware calibration and battery status lead to different signal sending powers, hence different received signal strengths.

RIM enhances radio models by approximating these three main properties of radio signals. To reflect the two main properties of radio irregularity, namely non-isotropic and continuous variation, Zhou et al (2004) adjust the previously mentioned path loss formula of (10) with two new parameters: the Degree of Irregularity (DOI) and Variance of Sending Power (VSP). DOI is the maximum received signal strength percentage variation per unit degree change in the direction of radio propagation. When the DOI is set to zero, there is no range variation, and the communication range is a perfect sphere. When it is increased, the communication range becomes more and more irregular. The path loss formula is adjusted as follows:

$$Path\ Loss_{DOI} = K_i\ Path\ Loss \tag{14}$$

K_i is a coefficient to represent the difference in path loss in different directions, and α is a random number between -1 and 1, which is generated according to the Weibull distribution (Devore, 1982). Specifically, K_i is the ith degree coefficient, which is calculated as follows:

$$K_i = \begin{cases} 1 \ , \ i = 0 \\ K_{i-1} + \alpha \ \ DOI \ , \ 0 < i < 360 \ \wedge \ i \in N \end{cases} \tag{15}$$

where $K_0 - K_{359} \leq DOI$

Based on (15), 360 K_i values for 360 different directions can be generated by randomly fixing direction as the starting direction represented by i=0. The second parameter, Variance of Sending Power (VSP) is defined as the maximum percentage variance of the signal sending power among different devices. The signal sending power is adjusted as follows:

$$Sending\ Power_{VSP} = Sending\ Power\ (1 + \alpha\ VSP) \qquad (16)$$

In (16), Zhou et al (2004) assume that the variance of sending power follows a Normal distribution, which is broadly used to measure the variance caused by the hardware, and α is a random number between 0 and 1. With these two new parameters, *DOI* and *VSP*, the RIM model is formulated as follows:

$$Received\ Signal\ Strength = Sending\ Power_{VSP} - Path\ Loss_{DOI} + Fading \qquad (17)$$

The authors implement the RIM model in GloMoSim, discovering that the radio irregularity has a greater impact on the routing layer than the underlying link layer.

The ranging models presented in this section affect the accuracy of the estimated distances between nodes. However, as previously discussed in this chapter, other characteristics of localization algorithms should be evaluated, such as running time, coverage, total energy or communication cost.

Communication Models

Bartelli et al. (2007) stated that many recently proposed localization algorithms have both distributed and range based characteristics. For these classes of localization algorithms, there is a dependency on the reliable communication of local neighborhood information in the network. Therefore, simulation and emulation evaluations of these algorithms require an adequate link abstraction. For example, a node running a distributed localization algorithm may want to collect neighborhood information in order to determine its relative position. Other nodes will most likely be performing the same tasks, causing simultaneous packet transmissions, and therefore collisions. Because the communication overhead of a localization algorithm affects both the running time and energy cost, it is important to model links well in evaluating these metrics. For example, an algorithm which generates a lot of traffic will most likely cause problems in a large network, and may significantly reduce the network lifetime unless properly coordinated.

Packet Reception Ratio (PRR), which is a function of the distance between transmitter and receiver, can be used to model the link, as described by Zuniga et al (2005). An alternative is to use a statistical model. A commonly used packet loss abstraction for wireless link layer simulation is a two state Markov model called the Gilbert-Elliott channel. The loss process is determined by the current state of a discrete time stationary binary Markov process. It is assumed that no packets are lost in a 'good state' S_g while all packets are lost in the 'bad state' S_b. The stationary probability of a channel being in the bad state is:

$$P(S_b) = \alpha / (\alpha + \beta) \qquad (18)$$

where $\alpha = P_{gb}$ and $\beta = P_{bg}$ denote transition probabilities between S_g and S_b, and vice versa, respectively. Thus, the average packet error probability of the channel is:

$$P_s = P_b P(S_b) + P_g (1 - P(S_b)) \qquad (19)$$

where P_b and P_g are the error probabilities in bad and good states respectively. The state transition diagram for a Gilbert-Elliot channel is given in Figure 5. (18) may also be used to model instant node failures in a localization simulation similarly, where S_b denotes the failure state of a node and S_g denotes the non-failure state. In the failure state, all packets sent to the node are lost regardless of the wireless channel state.

Power Consumption Model

Measuring the energy cost of a localization algorithm relies on the battery model used. A commonly used model is referenced by De Marco (2006) – when the sensor transmits k bits, the radio circuitry consumes $kP_{Tx}T_B$ energy, where P_{Tx} is the power required to transmit a bit which lasts in T_B seconds. By adding the radiated power $P_t(d)$, the energy cost E_{Tx}

$$E_{Tx}(k,d) = kP_{Tx}T_B + P_t(d) \tag{20}$$

The model is completed in (21) by adding the term E_{rx} for the reception of packets as well as transmission:

$$E(k,d) = E_{Tx}(k,d) + E_{Rx}(k,d) = kP_{Tx}T_B + P_t(d) + kP_{Rx}T_B \tag{21}$$

P_{Rx} is the power required to correctly receive (demodulate and decode) one bit. In addition to this, the energy consumption model for a single sensor can be enhanced by considering a duty cycle, which may be useful for extremely low power localization algorithms. In this model, a node can be in three operational states each draining different amounts of energy from the battery; *active state* in which the node is either transmitting/receiving/sensing data, *idle state* in which the receiver is on and the node is waiting for an activity to be triggered and *sleep state* in which the node cannot take part in any network activity (from Chiasserini and Garetto 2004). Incorporating these models can help researchers account not only for the number of packets sent by a particular localization algorithm in simulation, but also the power consumption of the packet transmission, and implications of duty cycling.

Simulating Mobility

As mobility can have an impact on the whole network performance in various ways, the model used in simulating the behavior of mobile beacons throughout the simulation is also important. The performance

Figure 5. The Gilbert-Elliott channel model

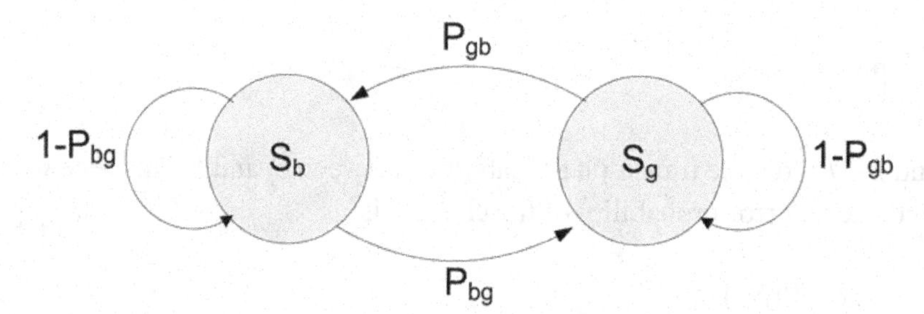

of the algorithm can vary significantly with different mobility models used for mobile entities. It is even possible to get different performance when the same mobility model is used with different parameters. Widely used wireless network simulators such as ns-2, GloMoSim, Qualnet, or Opnet support various mobility models. However, it is important to use a mobility model that most closely matches the expected real-world scenario. In a localization simulation, if the expected real-world mobility scenario is unknown then researchers should experiment with several of the available mobility models. Camp (2002) suggests that if an entity mobility model is desired, this can be well modeled by the Random Waypoint Mobility model, the Random Walk Mobility model or the Gauss-Markov Mobility model.

Discussion

This section has described various models that can be used for the evaluation of localization algorithms in simulation. However, simplistic assumptions made by these models can affect the overall performance and realism of a simulation. Using models derived from empirical results may be useful in addressing these issues, by informing statistical models. In order to evaluate individual aspects of localization algorithms, researchers may wish to iteratively add complexity to simulations. For example evaluation may start with simple radio and ranging models as well as simple deployment topologies, and progress to adding more complex power measurements and a variety of empirically modeled ranging modalities. This allows not only isolation of relevant components (accuracy being paramount initially), but a more thorough validation at several levels. This moves the simulation into a more realistic domain, preparing researchers to implement and evaluate the localization algorithm with realistic hardware in real environments. In the next section, a case study is presented showing how the real-life, application related requirements on localization can lead to tensions in performance criteria.

CASE STUDY

Given that requirements of WSN applications vary to a large degree with respect to localization, it is not easy to choose a general, representative case study for localization algorithm performance evaluation. The case study chosen here describes a localization system developed by Girod et al (2006). The system is anchor-free, highly accurate over relatively large distances (~4cm 2D positional error at tens of metres), and requires low node densities. The system was deployed and demonstrated with 10 nodes across an 80 x 50m area.

The platform requirements and localization techniques employed are described in this section, and the constraints associated with the localization algorithm together with the trade-offs in performance toward fitting application requirements. A different application is considered as a counter point to illustrate the variety of application specific demands that affect localization and localization algorithms. The reader should note that in this section, we refer to the localization of a WSN as *self-localization*, and the localization of non WSN events of interest as *source localization*.

Acoustic Source Localization

The Acoustic Embedded Networked Sensing Box (ENSBox) platform designed by Girod et al (2006) is a capable platform to support distributed acoustic sensing (see Figure 6). Acoustic sensing applications remain a persistent challenge in wireless sensing as they imply high data rates and are a rich source of challenging problems relating to source and self localization, for example Ali (2007) and Allen (2008). Relatively high processing power is needed in order to locally process data and reduce network overhead.

The primary motivator for the design of the ENSBox was a class of scientific localization applications – namely the source localization of animals and birds, based on their vocalizations. Such applications enable species census, classification and behaviour studies to be performed. For source localization, the ENSBox node employs four microphones per node as a local array, in a tetrahedral configuration over a 12cm^2 area. A network of ENSBoxes allows the use of beam crossing techniques for source localization, where nodes individually estimate direction of arrival (DoA) of animal vocalizations using the time-difference of arrival (TDoA) of the signals at each microphone. This is possible as the acoustic signals are coherent across the node's array of four microphones. Network wide position estimations can then be made by triangulating the DoA observations. (The lack of coherency of wide-band acoustic signals over tens of metres means that it is difficult to reliably determine the 'start' of the signal at each node, meaning the use of TDoA for position estimation will not yield accurate enough results).

Using the Acoustic ENSBox for source localization requires self-localization to be performed to a high accuracy. Girod envisages the network being used to surround a target at 30-50m spacing (target to node). As such, he sets the self-localization accuracy requirements to be ±0.5m positional estimation and ±1 degree orientation estimation. This is sufficient to keep the source localization positional error within the bounds of the state of the art (±2.5 degrees for the most comparable system to an ENSBox network). Because DoA estimates are used in source localization, it is important that the geometry of the physical topology estimated by the self-localization algorithm be consistent with the actual physical topology. Actual distance error is not as important, given that for DoA triangulation, it is the angles between nodes that must be accurate, hence the topology need only be correct to a scaling factor. The average node density requirements of such an application are not clear, although a minimum of three nodes is required to remove ambiguity of the location of the acoustic event in two dimensions. Given

Figure 6. The Acoustic ENSBox's compact microphone array. The four microphones are arranged in a tetrahedral configuration over a 12cm^2 area.

that nodes can potentially be deployed over any terrain, it is important that the localization algorithm works well in 3D as well as 2D.

The self-localization solution the Acoustic ENSBox employs is based on acoustic time of flight (ToF) and direction of arrival (DoA) estimation, employing an iterative non-linear least squares minimisation multilateration algorithm (NLLS). The ENSBox nodes are equipped with omni-directional speakers to emit pseudo-noise ranging chirps from; nodes chirp at known times in a sequence, and estimate ToF ranges from each other. DoA estimates are based on an approximate six way cross correlation of the ranging chirp across the four audio channels. The ToF and DoA estimates are used as constraints in the NLLS algorithm, which is carried out in a centralised manner – all nodes report ranges to one elected leader, who performs the localization computation.

The NLLS self-localization algorithm works best when its system of equations is over constrained (that is, there are many range and angle measurements per node). This means that erroneous measurements can be removed at certain points during the position estimation process through outlier rejection procedures. These rejections are based on heuristics such as residual error between two nodes' range estimates and residual error between estimated position and estimated range. Node orientations are iteratively estimated between NLLS iterations by averaging the error between observed DoA and angle based on the NLLS result. Convergence is assumed when residual error for different aspects (yaw, pitch, roll, range) falls below an empirically determined threshold. Sometimes, this means that under constrained systems do not converge.

Girod notes that raw residual error is not sufficient to detect outliers from the linear system formed as part of the NLLS localization algorithm. Therefore, in order to remove outliers, the localization algorithm makes use of studentized residuals (where residual error is divided by an estimate of its standard deviation), a common method of detecting outliers in statistics. Outlier detection is performed after the algorithm has converged, so that the most outlying residual can be removed as a constraint from the linear system. This will potentially enhance the overall localization result, and can be iterated while the algorithm still converges. Girod observes that: 1) average residual error itself is not a good metric to determine a bad fit of coordinates (when ground truth is not available), and in his experiments, 2) that there was not an obvious relationship between average residual error and average positional error. However, there was seemingly a relationship between under-constrained nodes and positional error, pointing to a potential metric which can account for average residual error and under-constrained nodes, although average constraint density is not likely to be sufficient on its own.

Over several experiments in different, semi-obstructed environments, ten nodes were localized with an average 2D error of between 4.4cm and 11.1cm over an 80 x 50m area. The average 3D error was between 26.0cm and 57.3cm – this difference was due to a lack of variation in the Z axis for localization experiments. In practice, it is sometimes possible to make use of a 2D solution by adjusting the pith and roll of the nodes such that their local arrays are approximately planar. This is useful if the user's confidence in the 3D solution is low.

Evaluation

The self-localization system alone will now be examined with respect to the performance criteria established earlier in this chapter – scalability, accuracy, cost and coverage. It has been established at the start of this section that the dominating requirements with respect to the application under discussion here are geometrical accuracy and robustness to ranging error, and that the system meets these requirements by taking advantage of the hardware required for the application. In terms of scalability, although the algorithm is anchor free, the processing it performs is centralised, and comes at a large computational cost. The assumption in this case is that the number of nodes deployed will not be so large to take an unreasonable amount of time for the algorithm to converge on a solution (order of minutes). The algorithmic complexity in this case is $O(N^3)$, which precludes the use of this algorithm for large networks.

In terms of cost, the localization system is expensive – requiring high sample rate audio. The platform has plentiful resources (64MB RAM, 400 MHz ARM CPU), use of which comes at the expense of a shorter battery life. The hardware expense is understandable in the context of the application – acoustic source localization requires multiple microphones, computationally expensive signal processing techniques and data sampled at high rates. The components that aid the localization – time synchronisation and node-to-node state sharing – require constant communication (at least 1 packet every 4 seconds per node), which is not conducive to low power operation. The system is highly accurate, more than meeting its positional requirement in 2D (worst case 10cm error) and just going over 0.5m error in 3D, due to the local array configuration (as previously noted). Special care is given to robust behaviour, but the cost for this is a high number of measurements for each node – the localization algorithm requires an over constrained linear system to remove outliers. In a topology where the number of range measurements per node is low, outliers are likely to become difficult to remove, or even identify; this is likely to be encountered in larger networks. Because the localization algorithm is computed centrally with all measurements, coverage is either 0 or 100%; the algorithm either converges on a result or it does not. This is clearly a problem for scalability.

To conclude, in maximising the accuracy and resilience to measurement noise, the localization system becomes constrained in scalability and unconstrained in cost (power usage, message sending, and computational complexity). This is intuitive if one imagines the criteria in tension – one cannot be maximised without affecting the others. This would seem to limit the generality of the localization approach, but one could argue that any self-localization motivated by a specific application (rather than application class) will make similar optimisations to maximise performance.

Counterpoint

As a simple, brief counterpoint, and with the aim to bring the points discussed so far into a sharper focus, the requirements of a different application are compared to see to what extent the previous localization procedure would suit them. In this motivating example, a WSN network is deployed over a forest in order to monitor it for potential fire events. Nodes in the network acquire temperature and humidity data as part of the calculation of the Fire Weather Index, to help predict dangerous areas for fires. This prediction is intended as an "early warning" system and the network will localize areas in the forest which are highly likely to have fires (as well as detecting fires when they occur).

Forest fires usually occur in summer, and it is envisaged that the network will be deployed before and removed (or replaced) after summer, hence needing a minimum lifetime of at least 6 months continuous operation. Since the deployment area is not likely to be dangerous at deployment times, it can be assumed that nodes will be manually deployed, but that terrain surveying processed are too expensive for the size of the network. The individual constraints on the localization performance are discussed below.

Scale and Density

This network is likely to be far larger than the acoustic sensing network, in terms of number of nodes required and area to be covered. The network is required to be dense in terms of communication – between 10 and 20 neighbours on average is ideal to ensure reliable multi-hop communication paths and allow for duty cycling. Deploying 20 nodes over every 100m by 100m area is likely to be sufficient to maintain at least an average degree of 10 per node.

The fire must be related to an actual physical position, so there must be at least some nodes in the network which are GPS-enabled. However, it is unreasonable for each node to be equipped with GPS, as there is a strong likelihood that it will be rendered useless under the forest canopy. Therefore nodes equipped with GPS could be deployed around the edges of the forest, acting as anchors when required. These nodes would not necessarily have to have the same sensing capabilities as the general network, and as such could be used only when required for localization.

Cost

Network life-time must also be maximised, meaning that radio communication must be kept to a minimum. Ideally, nodes will be duty cycled to take advantage of the deployment density. Additionally, when considering concrete solutions to the forest fire application, hardware cost becomes a factor, meaning that it is not only the power consumption cost that must be considered, but also the per-unit cost. The overall cost of the network will limit how many nodes can be purchased, and so accurate ranging hardware will most likely have to be traded off for simpler, cheaper ranging approaches which are not extra to the application functionality of the system, such as RSSI ranging for example.

Accuracy

Fire event localization is unlikely to be performed in the same way as acoustic localization. The resolution requirement of a fire's geographical location is related to the type of material on the forest floor and how flammable it is. Estimating the fire position could be as coarse as the nearest 100m, and still acceptable.

Coverage

Attaining 100% coverage is important for this application. If any nodes exist in the network which are capable of flagging fire events, but that have not been localized, the network is not meeting its application goals.

Summary

To summarise, the overriding constraints in this application are cost (node price and power consumption), network lifetime and scalability. In order to meet these constraints, it is likely that the network will have to compromise on accuracy. This accuracy trade-off is likely to be manifested in a simple ranging mechanism – highly accurate ranging approaches such as audible acoustic or ultrasonic time of flight represent an extra expense which the nodes cannot justify. In this case, a distributed algorithm would seem to be the best approach. It would not have to be anchor free, although the anchor density would most likely be low, and at a low duty cycle.

It is clear that the approach that the Acoustic ENSBox network uses would not work for the forest fire application – the hardware is too heavy weight to deal with the constraints of the application, and the battery life is not suitable for a long-lived application. The NLLS localization algorithm is too specific to apply to this network, where no angle of arrival measurements could be taken. Also, the computation of the algorithm is not scalable without modification.

A LOCALIZATION ALGORITHM DEVELOPMENT CYCLE

The development and evaluation of a localization algorithm should be considered in its entirety – this implies theoretical modelling and simulation as well as real-life validation of the algorithm. Each stage of the development should characterise and validate a specific aspect of the algorithm. Simulation validates how the algorithm can operate under controlled, simulated conditions – this verifies that the algorithm *functions* correctly. Emulation verifies that the algorithm can work correctly using *empirical data* that reveals conditions which are hard to simulate. Realistic validation shows that the algorithm can work in target environments and with the hardware platforms which are being targeted to support it.

Whitehouse et al (2004) propose that whilst simulation is different from real-world performance, one would expect it to be *indicative* (within some error bound of empirical results) and *decisive* (an algorithm which performs best in simulation should perform best in reality). Therefore, when one is evaluating a localization algorithm against others, one must make sure it performs better in both simulation and realistic deployment.

The verification and validation of a localization algorithm at each of the four stages (modelling, simulation, emulation and deployment) becomes more expensive in terms of (at least) time and cost as we approach real-life deployment. The value of simulation/emulation comes forth with respect to scalability and low cost of entry for researchers – there are no embedded hardware requirements.

Simulation

Researchers can use simulation to simplify some of the difficulties of real deployment (time synchronisation, for example) such that any algorithmic flaws can be isolated at an early stage. For this reason, it is not sensible to try to start with in-situ deployment without simulation verification. Environments such as Matlab, ns-2, OmNet++, Ptolemy and EmStar would be used to simulate the performance of localization algorithms. Different simulation environments allow lesser or greater control over node and network parameters relevant to localization. Simulators such as ns-2 and OmNet++ aim to provide the user with accurate models of wireless propagation and protocol performance, providing a high level language in which to implement simulations. Their wide academic use is desirable for consistency between institutions in a way custom simulators cannot guarantee. Custom simulators can be designed in a variety of languages (Java, C and its variants). Ptolemy provides a hugely powerful framework for modelling, simulation and design of embedded systems using graphical techniques to create state machines, akin to Matlab's Simulink. Development frameworks like EmStar allow researchers to develop end-to-end wireless sensing systems, allowing the same code to be used for simulation, emulation and deployment. Hardware specific simulators, such as TOSSIM and AVRORA can be used when accurate profiling is required (in power consumption analysis, for example). There also exist localization specific simulators, such as Silhouette by Whitehouse (2004, 2006) and SeNeLEx, RiST by Reichenbach (2006).

These environments do not need to be used in isolation, of course – measurements and observations derived from one could be used as set-up parameters in another, or to help inform custom simulation software.

Emulation

Using empirical data to inform simulation parameter values, rather than purely calculating them (for example, ranging or communication data) represents an addition to the realism of a simulation. Empirical data sets can capture some of the environment-specific effects that simple models cannot. The *Statistical Emulation* method proposed by Whitehouse et al (2004), is an example of gathering a data trace in-situ, and using it to power a realistic localization simulation (thus making it an emulation). Part of the challenge of performing this type of emulation is gathering a data set which represents the environment in sufficient detail. Whitehouse (2004) gathered range data using 20 ultrasound enabled nodes that have been arranged in such a way that all ranges between 0.5m and 4.5m (at 0.25cm intervals) can be measured. This captures environmental specific problems, such as non-estimates (range could not be measured) and node-to-node ranging variations (induced by electronic or mechanical differences between nodes). Whitehouse uses this range data set in his Matlab based Silhouette localization software to investigate its effects on the performance of several localization algorithms, comparing the results with pure simulation and finding a disparity between the two. Similarly, real connectivity data can be gathered from a test bed and pushed into a simulation, creating an emulated system.

One of the most powerful emulation frameworks to date is EmStar (Girod et al., 2007). EmStar allows the user to perform simulation, emulation and real deployment using the same framework. This means code developed and simulated can be cross-compiled and tested on real embedded hardware. This approach is advantageous as there is a reduction in the amount of porting required. EmStar allows network connectivity to be emulated in real-time using test bed data, making it a powerful tool for transitioning to real hardware from simulation through emulation.

Real Life Deployment

The strength of using test beds lays in actually being able to run algorithms on real hardware, and gather non-simulated data. This can be particularly useful for testing radio communication, for example. However, creating localization test beds can often be difficult because algorithms are affected by environmental context. Ranging mechanisms will most likely work differently indoors and outdoors, for example if signal strength is being used to determine range or location. Evaluating an algorithm on a test bed in a different environment than the application targets may give an incorrect indication of the algorithm's performance.

Real life deployment of a localization algorithm on hardware in an indicative environment (i.e. similar to where the real network will be deployed) is the most important evaluation of a localization algorithm. Unfortunately, it is also the most time consuming, costly, and error prone aspect of localization evaluation. An in-situ evaluation of a localization algorithm will most likely be as demanding as a real deployment of the network in terms of planning, deployment equipment and time taken to deploy.

The deployment phase of localization algorithm evaluation is also the most error prone and unpredictable, so researchers should have a detailed plan of how and what data needs to be gathered. The aim should not be to perform a large amount of testing, but to have well directed and easily planned experimentation. Software will most likely need to be adapted to work correctly in the field, and worst case scenarios (what to do if pretty much everything fails) should be planned for. Several days should be set aside for deployment, with the understanding that the likelihood is high that things will *not* work as expected first time.

CONCLUSION

When evaluating localization algorithms, it is difficult to separate the issues arising from actual deployments from theoretical drawbacks and constraints of various algorithms. From a theoretical perspective, it is desirable to have an algorithm that is independent of the ranging technique used and platform capability, as well as being robust to the deployment environment and generic with respect to application requirements.

Given that a WSN is deployed for some realistic, physical monitoring and processing aim, the localization algorithm designer should always have some set of motivating applications in mind, throughout the design process. These can be general classes of applications such as tracking and location awareness or very specific, clearly specified applications such as forest fire monitoring and animal call localization. Different applications will place different weightings on the various criteria discussed at the start of this chapter – scalability, accuracy, coverage and cost.

In conclusion, evaluating localization algorithms is not to be underestimated by researchers. In order to fully evaluate a localization algorithm, its performance must be tested in simulation, emulation and realistic environments. Both the design and development process for new localization algorithms and the process of selecting a "best fit" algorithm for a particular application requires consideration of the trade-offs between accuracy, cost, coverage and scalability the localization system needs to achieve. Although simulation is the least costly and most used tool for evaluating algorithms within the WSN domain, with respect to localization researchers must be aware of the limitations of purely simulated models, especially for radio communication and inter-node distance estimation.

The use of metrics to describe the quality of localization is important for all evaluation criteria, but possibly most notably for accuracy evaluation. Using Euclidean error is the simplest, but not always the most telling way of measuring how well a localization solution "fits" ground truth. Also, when ground truth is not available, an equivalent metric must be found which tells the user how well the localization estimate matches the initial constraints (such as inter-node spatial estimates).

Considering the domain's state-of-the-art, being able to instantiate a specific localization algorithm is still not an easy thing to do. Even after choosing a localization algorithm that is most suitable for the motivating application, it is likely that researchers will still have to implement it on specific hardware (with relevant ranging measurement mechanisms, if applicable) before being able to evaluate its performance.

REFERENCES

Ahmed, A. A., Shi, H., & Shang, Y. (2005). SHARP: A New Approach to Relative Localization in Wireless Sensor Networks. *Proceedings of the 25th IEEE International Conference on Distributed Computing Systems Workshops (ICDCSW'05)* (pp 892-888).

Ali, A. M., Yao, K., Collier, T. C., Taylor, C. E., Blumstein, D. T., & Girod, L. (2007). An empirical study of collaborative acoustic source localization. *In Proceedings of the 6th international Conference on information Processing in Sensor Networks (IPSN '07)*, (pp 41-50).

Allen, M., Girod, L., Newton, R., Madden, S., Blumstein, D.T., Estrin, D. (2008). VoxNet: An Interactive, Rapidly-Deployable Acoustic Monitoring Platform. *In Proceedings of the 7th International Conference on Information Processing in Sensor Networks (IPSN '08)*, (pp. 371-382).

Allen, M., Gaura, E., Newman, R., Mount, S., (2006). Experimental Localization with MICA2 Motes. *In Proceedings of NSTI Nanotech 2006*, (pp. 435-440).

Andersen, A., (2007). 2.4 GHz Inverted F Antenna. *Texas Instruments Design Note DN0007*

Başaran, C. (2007). A Hybrid Localization Algorithm for Wireless Sensor Networks. *Master's Thesis, Yeditepe University,* Turkey.

Başaran, C., Baydere S., & Kucuk, G. (2008). RH+: A Hybrid Localization Algorithm for Wireless Sensor Networks, *IEICE Transactions on Comm., E91-B*(No.06), 1852-1861.

Battelli, M., & Basagni S. (2007). Localization for wireless sensor networks: Protocols and perspectives. *In Proceedings of IEEE CCECE 2007,* Vancouver, Canada, April 22-26 2007, (pp. 1074-1077).

Bergamo, P., & Mazzini, G. (2002). Localization in Sensor Networks with Fading and Mobility, *In Proceedings of IEEE PIMRC 2002*, Sept 2002, Lisboa, Portugal, (pp. 750-754).

Broxton, M., Lifton, J., & Paradiso, J. A. (2006). Localization on the pushpin computing sensor network using spectral graph drawing and mesh relaxation. *SIGMOBILE Mob. Comput. Commun. Rev. 10*(1) (Jan. 2006), 1-12.

Bulusu, N., Heidemann, J., & Estrin, D. (2001). Adaptive beacon placement. *21st International Conference on Distributed Computing Systems (ICDCS-21),* (pp. 489-498).

Bulusu N., Heidemann J., Bychkovskiy V., & Estrin D. (2002). *Density adaptive beacon placement algorithms for localization in ad hoc wireless networks.* UCLA Computer Science Department Technical Report UCLA-CS-TR-010013, July 2001.

Camp T., Boleng J., & Davies V. (2002). A Survey of Mobility Models for Ad Hoc Network Research. *Wireless Communication & Mobile Computing (WCMC): Special issue on Mobile Ad Hoc Networking: Research, Trends and Applications*, *2*(5), 483-502.

Cerpa, A., Wong, J. L., Kuang, L., Potkonjak, M., & Estrin D. (2005). Statistical Model of Lossy Links in Wireless Sensor Networks. *International Conference on Information Processing in Sensor Networks.*

Chiasserini C.-F., & Garetto, M. (2004). Modeling the Performance of Wireless Sensor Networks. *IEEE INFOCOM* 2004, (pp. 220-231).

Efrat, A., Erten, C. Forrester, D., Iyer, A., & Kobourov, S.G. (2006). Force-Directed Approaches to Sensor Localization. *Proceedings of the 8th SIAM Workshop on Algorithm Engineering and Experiments (ALENEX)*, (pp. 108-118).

Forester, J. (2003). Channel modeling sub-committee report final. *IEEE802.15-02/490r1-SG3a*, Feb 2003.

Ghosh, A., Wang, Y., Krishnamachari, B., & Hsieh, M. (2007). Efficient Distributed Topology Control in 3-Dimensional Wireless Networks. *4th Annual IEEE Communication Society Conference on Sensor, Mesh and Ad Hoc Communications and Networks, SECON 2007*, 2007, (pp. 91-100).

Girod, L., Lukac, M., Trifa, V., & Estrin, D. (2006). The design and implementation of a self-calibrating distributed acoustic sensing platform. In *Proceedings of the 4th international Conference on Embedded Networked Sensor Systems* (Boulder, Colorado, USA, October 31 - November 03, 2006). SenSys '06. ACM, New York, NY, (pp. 71-84).

Girod, L. (2005). A Self-Calibrating System of Distributed Acoustic Arrays. *Ph.D. Thesis, UCLA, USA.*

Girod, L., Ramanathan, N., Elson, J., Stathopoulos, T., Lukac, M., & Estrin, D. 2007. Emstar: A software environment for developing and deploying heterogeneous sensor-actuator networks. *ACM Trans. Sen. Netw., 3*(3), (Aug. 2007), 13.

Goldenberg, D. K., Krishnamurthy, A., Maness, W. C., Yang, Y. R., Young, A., Morse, A. S., & Savvides, A. (2005). Network localization in partially localizable networks. *INFOCOM 2005. 24th Annual Joint Conference of the IEEE Computer and Communications Societies.* IEEE 1(13-17) (pp. 313-326).

Gotsman, C., & Koren, Y. (2005). Distributed Graph Layout for Sensor Networks. *Lecture Notes in Computer Science*, 3383/2005, Springer Berlin/Heidelberg, (pp. 273-284).

Heidemann, J., Bulusu, N., Elson, J., Intanagonwiwat, C., Lan, K., & Xu, Y., et al. (2001). Effects of Detail in Wireless Network Simulation. *Proceedings of the SCS Multiconference on Distributed Simulation.* Phoenix, Arizona, USA, USC/Information Sciences Institute, Society for Computer Simulation. January, 2001, (pp. 3-11).

Krishnamurthi N., Jay Yang S., & Seidman M. (2004). Modular Topology Control and Energy Model for Wireless Ad Hoc Sensor Networks. *Proceedings of OPNETWORK '04.*

Langendoen, K., & Reijers, . (2003). Distributed localization in wireless sensor networks: a quantitative comparison. *Comput. Netw., 43*(4) (Nov. 2003), 499-518.

Lee, J. Y., & Scholtz, R. A. (2002). Ranging in a dense multipath environment using an UWB Radio Link. *IEEE J. Select. Areas Commun., 20,* December 2002, (pp. 1677-1683).

Lin G., Noubir G., & Rajaraman R. (2004). Mobility Models for Ad hoc Network Simulation. *IEEE INFOCOM* 2004, (pp. 463).

De Marco, G., Yang, T., & Barolli, L. (2006). Impact of Radio Irregularities on Topology Tradeoffs of WSNs. *Proceedings of the 17th International Conference on Database and Expert Systems Applications (DEXA'06),* (pp. 50-54).

Moore, D., Leonard, J., Rus, D., & Teller, S. (2004). Robust distributed network localization with noisy range measurements. In *Proceedings of the 2nd international Conference on Embedded Networked Sensor Systems* (Baltimore, MD, USA, November 03 - 05, 2004). SenSys '04. ACM, New York, NY, (pp 50-61).

Nagpal, R., Shrobe, H., & Bachrach, J. (2003). Organizing a Global Coordinate System from Local Information on an Ad Hoc Sensor Network. *In the 2nd International Workshop on Information Processing in Sensor Networks (IPSN '03), Palo Alto, April, 2003, published as Lecture Notes in Computer Science LNCS 2634.*

Niculescu, D., & Nath, B. (2001). Ad hoc positioning system (APS). *Global Telecommunications Conference, 2001. GLOBECOM IEEE, 5,* 2926-2931.

Priyantha, N. B., Balakrishnan, H., Demaine, E., & Teller, S. (2003). Anchor-Free Distributed Localization in Sensor Networks. *LCS Tech. Report #892, MIT, USA.*

Ramadurai V., & Sichitiu M. L. (2003). Simulation-based Analysis of a Localization Algorithm for Wireless Ad-Hoc Sensor Networks, *OPNETWORK* 2003.

Reichenbach F., Koch M., & Timmermann D. (2006). Closer to Reality - Simulating Localization Algorithms Considering Defective Observations in Wireless Sensor Networks. *Proceedings of the 3rd Workshop on Positioning, Navigation and Communication (WPNC'06),* 2006, (pp. 59-65).

Savvides, A., & Garber, W. L. (2005). An Analysis of Error Inducing Parameters in Multihop Sensor Node Localization. *IEEE Transactions on Mobile Computing, 4*(6) (Nov. 2005), 567-577.

Shah, S. F. A., & Tewfik, A. H. (2005). Enhanced position location with UWB in obstructed LOS and NLOS multipath environments. *In Proc.of European Signal Processing Conf. (EUSIPCO).*

Shang, Y., Ruml, W., Zhang, Y., & Fromherz, M. P. (2003). Localization from mere connectivity. *In Proceedings of the 4th ACM international Symposium on Mobile Ad Hoc Networking & Computing* (Annapolis, Maryland, USA, June 01 - 03, 2003). MobiHoc '03. ACM, New York, NY, (pp. 201-212).

Sichitiu, M. L., & Ramadurai, V. (2004). Localization of wireless sensor networks with a mobile beacon. *In Proc. of the First IEEE Conference on Mobile Ad-hoc and Sensor Systems (MASS 2004),* Fort Lauderdale, FL, Oct. 2004, (pp. 174-183).

Slijepcevic, S., Megerian, S., & Potkonjak, M. (2002). Location errors in wireless embedded sensor networks: sources, models, and effects on applications. *SIGMOBILE Mob. Comput. Commun. Rev., 6*(3) (Jun. 2002), 67-78.

Wang, Y., Li, F., & Dahlberg, T. A (2006). Power Efficient 3-Dimensional Topology Control for Ad Hoc and Sensor Networks. *IEEE Global Communications Conference,* (pp. 1-5).

Whitehouse, K., & Culler, D. (2006). A robustness analysis of multi-hop ranging-based localization approximations. *In Proceedings of the Fifth international Conference on information Processing in Sensor Networks* (Nashville, Tennessee, USA, April 19 - 21, 2006). IPSN '06. ACM, New York, NY, (pp. 317-325).

Whitehouse, K., Karlof, C., & Culler, D. (2007). A practical evaluation of radio signal strength for ranging-based localization. *SIGMOBILE Mob. Comput. Commun. Rev., 11*(1) (Jan. 2007), (pp. 41-52).

Whitehouse, K., Karlof, C., Woo, A., Jiang, F., & Culler, D. (2005). The Effects of Ranging Noise on Multihop Localization: An Empirical Study. *IPSN '05,* (pp. 73-80).

Yu, J., & Oppermann, I. (2004). UWB positioning for wireless embedded networks. *In Proc. of IEEE Radio and Wireless Conference,* (pp. 459–462).

Zhang, Z. (1992). Iterative Point Matching for Registration of Free-Form Curves and Surfaces. *International Journal of Computer Vision, 13*(2), 119-152.

Zuniga, M., & Krishnamachari, B. (2004). Exploring the predictability of network metrics in the presence of unreliable wireless links. *SenSys 2004,* (pp. 275-276).

Chapter XV
Accuracy Bounds for Wireless Localization Methods

Michael L. McGuire
University of Victoria, Canada

Konstantinos N. Plataniotis
University of Toronto, Canada

ABSTRACT

Node localization is an important issue for wireless sensor networks to provide context for collected sensory data. Sensor network designers need to determine if the desired level of localization accuracy is achievable from their network configuration and available measurements. The Cramér-Rao lower bound is used extensively for this purpose. This bound is loose since it uses only information from measurements in its calculations. Information, such as that from the sensor selection process, is not considered. In addition, non-line-of-sight radio propagation causes the regularity conditions of the Cramér-Rao lower bound to be violated. This chapter demonstrates the Weinstein-Weiss and extended Ziv-Zakai lower bounds for localization error which remain valid with non-line-of-sight propagation. These bounds also use all available information for bound calculations. It is demonstrated that these bounds are tight to actual estimator performance and may be used determine the available accuracy of location estimation from survey data collected in the network area.

INTRODUCTION

To provide context for data collected by wireless sensor networks, it is necessary for the sensor network to supply accurate location information for its component sensor nodes (Sheu et al. 2006). To this end, several algorithms and sensor types have been developed for sensor node localization in these networks (Patwari et al. 2003; Ray et al. 2006). These proposed localization systems have been shown to provide excellent localization accuracy for the sensor nodes and mobile terminals in these networks.

The remainder of this book describes the design and use of several of these algorithms. However, an important issue for a network designer is to determine what sensors and network topologies are required to achieve the necessary level of localization accuracy for their application. To make these design decisions, tools are required for analytically evaluating the performance of different localization systems with different sensor positions.

The purpose of this chapter is to describe tools for the accuracy analysis of localization systems for wireless sensor networks. The chapter will focus on localization systems based on base stations at known positions making measurements of the radio signals from the sensor nodes. It should be noted that while this chapter describes only radio-based measurements for localization within sensor networks, the mathematical tools are easily applied to other measurements such as acoustic-based distance measurements.

Evaluation methods for localization systems serve two purposes. First, they allow a network designer, prior to the creation of the senor network, to obtain a quantitative bound on how well the localization of sensor nodes can be performed with given types of localization measurements and with different geometric arrangements of the measuring base stations. A network designer can then determine which of a set of possible network designs will achieve the required localization accuracy for their application. Second, these tools can be used to evaluate the performance of an existing localization system to see if all the potential location accuracy is being achieved or if further improvements are possible. The tools help to quantify the cost and accuracy tradeoffs of different component choices in a localization system's design.

In the radiolocation literature, there have been several figures of merit proposed for localization accuracy such as the Circular Error Probable (CEP) and the Geometric Dilution of Precision (GDOP) (Torrieri 1984; Tekinay et al. 1998). These figures of merit provide useful information for the analysis of the performance of location systems, but these values are difficult to calculate for localization systems coping with multipath or non-line-of-sight (NLoS) radio propagation. In Line-of-Sight (LoS) radio propagation, radio signals travel directly on the shortest straight line path from the node to be located to the measuring base stations, whereas during NLoS radio propagation this path is obstructed and the signal is reflected and diffracted during propagation from the target node to the measuring base stations. NLoS propagation complicates the localization problem since the signal characteristics are not only a function of the node and base station locations but also a function of the location of obstructions in the propagation environment.

To provide accuracy information for localization in the presence of multipath and NLoS propagation, figures of merit have been derived in the localization literature such as the Cramér-Rao lower bound on the mean square error of the location estimates. The local Cramér-Rao lower bound has been derived for localization in the presence of multipath and random NLoS radio propagation and used to evaluate the performance of many localization systems (Qi et al. 2002; Botteron et al. 2004). This bound provides an excellent method of evaluating the effects of the locations of the base stations and measurement noise levels on localization accuracy. A difficulty with the use of the Cramér-Rao lower bound as a general evaluation tool for localization accuracy is that it considers the current radio signal measurements as the only source of information on node location. In other words, the Cramér-Rao lower bound assumes that the node location is a deterministic value and the localization system has no other information about the node location prior to the measurements. Other sources of information, such as the sensor selection procedure or the measurements taken in the past, are not considered, so the Cramér-Rao lower bound is no longer a valid lower bound for localization systems where this information is available.

An extension of the Cramér-Rao lower bound, known as the Bayesian Cramér-Rao lower bound, has been developed to manage information other than that contained in the measurements about the node's location. The additional information is modeled as generating a probability density function for a node's location prior to the availability of the measurements. Unfortunately, the Bayesian Cramér-Rao lower bound has several regularity conditions that are violated for the node localization scenarios of greatest interest. This chapter addresses this problem by demonstrating how the more general Weinstein-Weiss lower bound (Weinstein and Weiss 1988) and the Extended Ziv-Zakai lower bound (Bell et al. 1997) are applied to the node localization problem. The chapter demonstrates the use of these bounds for considering the effect of sensor selection on the localization error bounds. It is demonstrated how, with the use of a motion model, a lower bound is calculated for the localization of nodes that are in motion. This chapter also shows how the Extended Ziv-Zakai bound is used with a measurement survey data set collected in a wireless sensor network's environment to determine a measure of what localization accuracy is attainable for the wireless sensor network at different noise levels.

The rest of this chapter is organized as follows. The next section gives a brief summary of the signal models used in this chapter. The mathematical notation for the chapter is also presented. The third section of the chapter contains a review of the evaluation methods for localization of nodes when the nodes have deterministic locations. The concepts of the CEP and GDOP are reviewed and explained. The fourth section of the chapter describes the Cramér-Rao lower bound on terminal localization from radio measurements. The effects of sensor geometry on propagation distance measurements and the received signal measurements are discussed. The fifth section of the chapter describes the integration of prior information on node location into the bound calculations. The Bayesian Cramér-Rao lower bound is described and the difficulties with its application to node localization are noted. The more general bounds on terminal location, the Weinstein-Weiss and extended Ziv-Zakai bounds, are then introduced. The sixth section of the chapter provides examples of the lower bound calculations with a summary of how the bounds described in previous sections are calculated for a sample network. The last section of the chapter presents the conclusions of the chapter with an overview of the results.

MEASUREMENT AND SIGNAL MODEL

In this chapter, vectors are denoted with bold lower case letters and matrices are denoted with bold upper case letters. Subscripts are used to index the entries of matrices and vectors so $\mathbf{v}_i$ is referring to the ith entry of vector **v**, while $C_{i,j}$ is the jth entry of the ith row of the matrix **C**. Many of the variables in this chapter are random and they are specified in terms of their probability density functions. The function $f(\mathrm{x})$ is the probability density function of the random vector **x**, and $f(\mathrm{x} \mid \mathrm{y})$ is the conditional probability density function of random vector **x** given the value of random vector **y**.

This chapter demonstrates how to calculate bounds for node localization from radio received signal strength (RSS), time of arrival (ToA), time difference of arrival (TDoA), or angle of arrival (AoA) measurements. In this paper, the terms node and terminal are used interchangeably. These measurements were selected since they are the most popular radio signal measurements for wireless node localization. The methods described for calculations of lower bound on localization error are applicable to other measurements that have been proposed for node localization, such as acoustic distance measurements or radio impulse response matching.

The node location at sample time k is specified by the vector $\boldsymbol{\theta}(k) = [p_x(k)\ p_y(k)]^T$ where $(p_x(k), p_y(k))$ are the x and y coordinates of the node of interest. This chapter concentrates on two dimensional localizations. The presented bound calculations are easily generalized to three dimensional localizations, if required. The measurement vector for sample time k is denoted as $\mathbf{z}(k)$. The node localization is performed with measurements from m fixed location base stations at known locations. In wireless sensor networks, the base stations are either localized by measurements made at the time the sensor network is setup or these nodes are equipped with Global Position System (GPS) receivers.

For RSS, ToA, or AoA measurements, the ith entry of the measurement vector is given by

$$\mathbf{z}_i(k) = \mathrm{m}[\boldsymbol{\theta}(k), \mathbf{b}^i] + \mathbf{n}_i(k) \tag{1}$$

where $\mathrm{m}[\boldsymbol{\theta}(k), \mathbf{b}^i]$ gives the noise free measurement for radio propagation from location $\boldsymbol{\theta}(k)$ to the ith base station's location specified by the vector $\mathbf{b}^i$, and $\mathbf{n}_i(k)$ is the ith entry of the measurement noise vector $\mathbf{n}(k)$. To simplify later calculations, the time measurement for the ToA measurement is converted to a distance measurement by multiplication of the measurement by the radio signal propagation speed.

The measurement function for unobstructed LoS propagation is specified by

$$\mathrm{m}[\boldsymbol{\theta}(k), \mathbf{b}^i] = \begin{cases} 10\alpha \log_{10} \left\| \boldsymbol{\theta}(k) - \mathbf{b}^i \right\| & \text{RSS measurement (dB)} \\ \left\| \boldsymbol{\theta}(k) - \mathbf{b}^i \right\| & \text{ToA measurement (m)} \\ \angle[\boldsymbol{\theta}(k) - \mathbf{b}^i] & \text{AoA mesurement (radians)} \end{cases} \tag{2}$$

with $\left\|\mathbf{v}\right\|$ being the Euclidean length of vector $\mathbf{v}$, and α being the radio pathloss propagation constant varying from 2 to 4 in urban environments, and the $\angle[\boldsymbol{\theta}(k) - \mathbf{b}^i]$ operator gives the angle of the difference vector $\mathbf{Q}(k)$-$\mathbf{b}^i$ (Steele 92). It is assumed for ToA measurements that the time of signal transmission from the source is known. For TDoA measurements, each entry of the measurement vector is the difference between the propagation times for the measuring base station and the reference base station. Therefore, the TDoA measurement vector is calculated from a ToA measurement vector as $\mathbf{z}^{\text{TDoA}}(k) = \mathbf{F}\,\mathbf{z}^{\text{ToA}}(k)$ where, without loss of generality, if base station 1 is the reference base station then $\mathbf{F} = [-\mathbf{1}^{m-1} \quad \mathbf{I}^{m-1}]$ with $\mathbf{1}^{m-1}$ being an $(m-1)\times 1$ vector with all one entries and $\mathbf{I}^{m-1}$ being an $(m-1)\times(m-1)$ identity matrix.

The measurement noise for RSS, ToA, and AoA measurements is usually specified as a zero mean Gaussian random vector with a covariance matrix given by $\mathrm{Var}[\mathbf{n}(k)] = \mathbf{C}$. If this is the case, the measurement noise vector for TDoA measurements is a zero mean vector of length m-1 with a covariance given by $\mathbf{C}^{\text{TDoA}} = \mathbf{F}\mathbf{C}^{\text{ToA}}\mathbf{F}^T$.

These measurement equations are only provided to assist with the description of the examples later in the chapter. The lower bound methods described in this chapter are not dependent on these propagation equations.

EVALUATION OF LOCALIZATION ACCURACY FOR NODES WITH DETERMINISTIC LOCATIONS

The node localization problem is specified as computing an estimate of the mobile terminal location at time k based on the measurements taken at time k:

$$\hat{\boldsymbol{\theta}}(k) = \mathrm{e}[\mathbf{z}(k)]\cdot \tag{3}$$

$\hat{\boldsymbol{\theta}}(k) = [\hat{p}_x(k)\ \hat{p}_y(k)]^T$ is the estimated location at sample interval k. $e[\mathbf{z}(k)]$ is an estimator function which maps from measurements to estimated locations. The reader is referred to the remainder of this volume for more details on how to implement these functions. This chapter instead focuses on figures of merit and bounds on the accuracy of these functions for several measurement types.

Figures of merit that have been proposed in the previous literature for stationary target localization are the Mean Distance Error (MDE), the Mean Square Error (MSE), the Root Mean Square Error (RMSE), the Geometric Dilution of Precision (GDOP), and the Circular Error Probable (CEP) (Torrieri 1984; Tekinay et al. 1998). These figures of merit quantify the uncertainty in the localization when the mobile terminal location is at a given point. The most commonly used figures or merit for stationary target location accuracy are the Mean Distance Error (MDE), Mean Square Error (MSE), and Root Mean Square Error (RMSE). These figures of merit give quantitative values to specify the magnitude of localization errors. The MDE is the mean distance of the estimated terminal location from the true mobile terminal location: $\text{MDE}(k) = \text{E}\left\{\sqrt{[p_x(k) - \hat{p}_x(k)]^2 + [p_y(k) - \hat{p}_y(k)]^2}\right\}$ where $\text{E}[\bullet]$ is the statistical expectation operator. The MSE is the mean squared distance of the estimated terminal location from the true mobile terminal location: $\text{MSE}(k) = \text{E}\{[p_x(k) - \hat{p}_x(k)]^2 + [p_y(k) - \hat{p}_y(k)]^2\}$. RMSE is simply the square root of MSE: $\text{RMSE}(k) = \sqrt{\text{MSE}(k)}$. The advantage of MDE is that its value is easily mapped to useful performance measures in the localization application domain. However, MDE is unfortunately difficult to calculate in analysis of estimation algorithms. Conversely, the MSE is easily calculated in theoretical work but has less correspondence to real world distances. As a compromise, RMSE is often used as figure of merit. RMSE is easily calculated and its value is significant to field applications since it can be easily shown that RMSE is always greater than MDE:

$$\begin{array}{rccc} & \text{Var}[\text{MDE}(k)] & \geq & 0 \\ \Rightarrow & \text{MSE}(k) - [\text{MDE}(k)]^2 & \geq & 0 \\ \Rightarrow & \text{MSE}(k) & \geq & [\text{MDE}(k)]^2 \\ \Rightarrow & \text{RMSE}(k) & \geq & \text{MDE}(k). \end{array} \tag{4}$$

where $\text{Var}[\bullet]$ is the statistical variance operator. This is a useful result for performance bound purposes, since bounds on RMSE are easily calculated from bounds on MSE.

Due to non-linearity in the relationship between the mobile terminal locations and the available measurement vectors, the magnitude of the location error is dependent on the relative location of the mobile terminal to the measuring base stations (Spirito 2001). The Geometric Dilution of Precision (GDOP) figure of merit is useful in the analysis of the location dependence in the localization error. GDOP is defined as the ratio of RMSE error over the standard deviation of the measurement errors, given by, using the definition from (Torrieri 1984):

$$\text{GDOP} = \frac{\text{RMSE}}{\sqrt{\text{Var(measurement noise)}}}. \tag{5}$$

GDOP allows for the uncertainty when the mobile is at different positions relative to the base stations to be specified relative to the variance of the available measurements. GDOP is useful when evaluating the choices of different measuring nodes for a given location system. A high GDOP indicates the geometry of measuring base station positions is inappropriate for accurate localization.

The mobile terminal localization error can be decomposed into two parts: a bias which is a fixed localization error vector resulting from the non-linearity in the relationship from measurement to location, and a random localization error vector. Circular Error Probable (CEP) provides quantitative values for the magnitude of the random portion of the localization error. CEP is defined as the radius of the circle which, for a given true location of the mobile terminal, contains half of the estimated locations of the mobile terminal (Torrieri 1984; Tekinay et al. 1998). This is illustrated in Figure 1. These figures of merit have been extensively studied in the case for line-of-sight (LoS), single path radio propagation and several useful results are available (Torrieri 1984). However, in more general cases, other figures of merit are required. The next subsection describes the use of lower bounds on localization error as figures of merit.

THE CRAMÉR-RAO BOUND ON TERMINAL LOCALIZATION ERROR

One factor that is lacking in the evaluation methods described in the previous section is that they do not give an indication of how much improvement is possible in a given mobile terminal localization system. This section describes methods for calculating bounds on the localization errors. The performance of any localization system cannot be better than these lower bound values. These bounds can be used as an indication of how much improvement can be made to a given estimator or how close a localization system is to providing optimal performance.

Another purpose for deriving these bounds is that they give a quantifiable measurement of how much information a single measurement vector, $\mathbf{z}(k)$, contains about the location of the mobile terminal. This information measure is needed for the derivation of the bounds on the localization error for time filtering of localization measurements provided later in this chapter.

A classification of some commonly used bounds for parametric estimation problems is given in Table 1. The Cramér-Rao bound gives a lower bound on the MSE of estimators of a deterministic parameter (Kay 1993). Cramér-Rao bounds have been derived for these performance measures for location estimates using ToA and TDoA measurements (Spirito 2001; Qi and Kobayashi 2002; Botteron et al. 2004). It should be noted that the bound in (Spirito 2001) is identical to the standard Cramér-Rao lower bound without using the standard Cramér-Rao lower bound derivation (Kay 1993). The standard Cramér-Rao bound gives a lower bound on the MSE of unbiased localization when the node is at a given location $\theta(k)$:

Figure 1. Circular error probable (CEP) definition

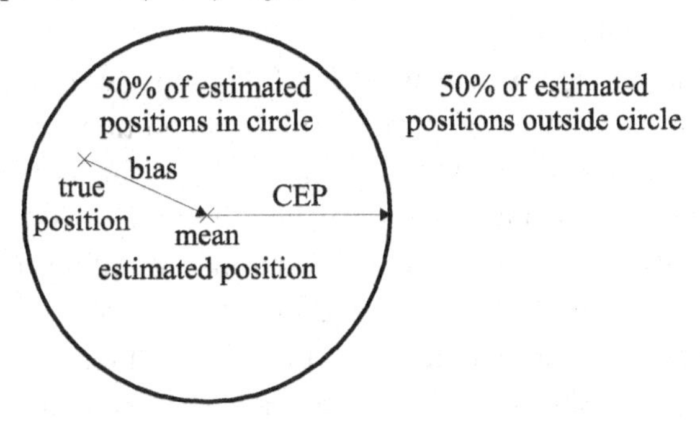

Table 1. Partial taxonomy of estimator lower bounds

Deterministic θ(*k*)		Random θ(*k*) (Bayesian Estimation)	
Restrictions on *f*(**z**(k) | θ(*k*))	Bound	Restrictions on *f*(**z**(k), θ(*k*))	Bound
None	Barankin (Rife et al., 1975)	None	Weiss-Weinstein (Weinstein and Weiss1988) Extended Ziv-Zakai (Bell et al. 1997)
Twice differentiable w.r.t. elements of θ(*k*)	Cramér-Rao (Kay 1993)	Twice differentiable w.r.t. elements of θ(*k*)	Bayesian Cramér-Rao (Van Trees 2001)

$$\mathrm{E}\{[p_x(k)-\hat{p}_x(k)]^2+[p_y(k)-\hat{p}_y(k)]^2 \mid \boldsymbol{\theta}(k)\} \geq \{\mathbf{J}[\boldsymbol{\theta}(k)]\}^{-1}{}_{1,1}+\{\mathbf{J}[\boldsymbol{\theta}(k)]\}^{-1}{}_{2,2}, \qquad (6)$$

where $\mathbf{J}[\boldsymbol{\theta}(k)]$ is defined as

$$\mathbf{J}[\boldsymbol{\theta}(k)] = -\mathrm{E}\left\{\nabla_{\boldsymbol{\theta}(k)}\left[\nabla_{\boldsymbol{\theta}(k)}\log\left(f\left(\mathbf{z}(k)\middle|\boldsymbol{\theta}(k)\right)\right)\right]^T \middle|\boldsymbol{\theta}(k)\right\} \qquad (7)$$

with $f(\mathbf{z}(k)\mid\boldsymbol{\theta}(k))$ being the conditional probability density function of the measurement vector $\mathbf{z}(k)$ given the true location $\boldsymbol{\theta}(k)$ during sample interval k. E[x | y] is the conditional expectation operator. The ∇ operator is defined as

$$\nabla_{\mathbf{v}} = \left[\frac{\partial}{\partial \mathbf{v}_1},\frac{\partial}{\partial \mathbf{v}_2},\ldots,\frac{\partial}{\partial \mathbf{v}_n}\right]. \qquad (8)$$

The matrix $\mathbf{J}[\boldsymbol{\theta}(k)]$ is the Fisher information matrix of the locations given the measurements.

For the Cramér-Rao lower bound to be valid, certain necessary 'regularity' conditions must be satisfied. These 'regularity' conditions are that the partial derivatives within (7) exist and that the expectations are bounded. For the localization problem, this is equivalent to the requirement that

$$\mathrm{E}\{\nabla_{\boldsymbol{\theta}(k)}\log[f(\mathbf{z}(k)\mid\boldsymbol{\theta}(k))]\mid\boldsymbol{\theta}(k)\}^T = \begin{bmatrix}0\\0\end{bmatrix} \qquad (9)$$

for all $\boldsymbol{\theta}(k)$ (Kay, 1993). Because of these conditions, many of the bounds presented in the localization literature are based on three assumptions that are not unconditionally true in many environments of interest (Spirito 2001; Botteron et al. 2004). Each of these assumptions is addressed below.

1. **Radio propagation is line-of-sight to the node.** The LoS propagation assumption is used to derive $f(\mathbf{z}(k)\mid\boldsymbol{\theta}(k))$ functions. The LoS assumption means the conditional probability density function value for $\mathbf{z}(k)$ is dependent only on mobile terminal and measuring base station positions. The assumption of LoS propagation is not always satisfied in urban or indoor environments. NLoS propagation measurements are dependent on the position of buildings and other geographic features in the propagation environment as well as terminal and base stations positions. In order to calculate bounds on the localization error during NLoS propagation, not only must the locations of measuring base stations and mobile terminal be known but also the geometry of obstacles to radio

propagation. When the conditional distribution of measurements is not continuous, the 'regularity' conditions of the Cramér-Rao bound are not satisfied. NLoS propagation creates discontinuities in the derivatives of the conditional probability density functions of the measurements on the border of NLoS regions where the propagation switches from LoS to NLoS.

Prior work on the development of localization error bounds in the presence of both NLoS and LoS propagation has demonstrated that where some base stations have LoS propagation and the other base stations have NLoS propagation, only the measurements from LoS base stations need to be considered to calculate a valid Cramér-Rao lower bound (Qi and Kobayashi, 2002, Botteron et al., 2004). This work modeled NLoS propagation as a random effect with no correlation to a node's location. In this case, the occurrence of NLoS propagation only degrades localization accuracy. However, in practice, NLoS propagation at some locations is deterministic as the LoS propagation paths to measuring base stations are blocked by large immobile objects such as walls or buildings.

2. **The localizations are locally unbiased.** For a locally unbiased estimator, $\mathrm{E}[\hat{\boldsymbol{\theta}}(k) \mid \boldsymbol{\theta}(k)] = \boldsymbol{\theta}(k)$. Terminal localization made using ToA or TDoA measurements are rarely unbiased because of the non-linear relationship of the measurement vector with the terminal location (Kim et al. 2001). If an expression for the bias can be calculated, it is possible to calculate a modified form of (7) which takes bias into account (Kay, 1993). Unfortunately, analysis of zero memory estimators has shown that closed form expressions for this bias are extremely difficult, if not impossible, to obtain (Torrieri 1984). Such a closed form expression would be equivalent to a closed form solution to (3) which is non-trivial for any $f(\mathbf{z}(k) \mid \boldsymbol{\theta}(k))$ conditional probability density function of reasonable complexity. It is known, that for some geometries of base station and node locations, the bias can be quite large in proportion to the total estimation error (Torrieri 1984; Spirito 2001). An example of a base station geometry resulting in a large bias is when the base stations' locations are all collinear with the node location.

   This condition is separate from the condition that localization is globally unbiased, which is the condition that $\mathrm{E}[\hat{\boldsymbol{\theta}}(k)] = \mathrm{E}[\boldsymbol{\theta}(k)]$. Clearly, a locally unbiased estimator is also globally unbiased but the converse is not always true. For example, the trivial localization system that $\hat{\boldsymbol{\theta}}(k) = \mathrm{E}[\boldsymbol{\theta}(k)]$ for all $\mathbf{z}(k)$ is globally unbiased but is not locally unbiased.
3. **No information is available on the node location aside from the measurement vector.** This assumption is required for the Cramér-Rao bound to be a true lower bound. In an urban or indoor region, base stations are located throughout the network area. As a mobile terminal moves through the network area, the hand off algorithm ensures that the mobile terminal is always communicating with a base station that is close to its location. Furthermore, base stations close to the mobile terminal are more likely to be making measurements for localization than base stations further away from the mobile terminal. Therefore, the base station selection for localization is not independent of terminal location. These two factors indicate that the wireless sensor network has prior statistical knowledge of the terminal location before any measurements are made. This knowledge is, in fact, the basis of Phase I of the FCC's E911 wireless location requirement for cellular telephone localization, where the network can identify from which cell a user is making a cellular telephone call (Federal Communications Commission, 1996).

The two largest problems for the calculation of localization error bounds are the unknown bias in the location estimates, and the existence of information on mobile terminal position available prior to the localization. The next section will describe how lower bounds on localization error are calculated in this case.

LOWER BOUNDS ON LOCATION ESTIMATION ERROR WITH PRIOR INFORMATION

If information on the node location is available before the localization procedure begins, then a Bayesian localization procedure with localization MSE that is lower than the Cramér-Rao bound described above is available. To bound this estimator, a Bayesian lower bound on the localization error is required. The most common Bayesian bound is the Bayesian Cramér-Rao Bound (BCRB) given by

$$\mathrm{E}\{[\boldsymbol{\theta}(k)-\hat{\boldsymbol{\theta}}(k)][\boldsymbol{\theta}(k)-\hat{\boldsymbol{\theta}}(k)]^T\} \geq \{\tilde{\mathbf{J}}[\boldsymbol{\theta}(k)]\}^{-1} \tag{10}$$

where the matrix $\tilde{\mathbf{J}}[\boldsymbol{\theta}(k)]$ is defined as

$$\tilde{\mathbf{J}}[\boldsymbol{\theta}(k)] = -\mathrm{E}\{\nabla_{\boldsymbol{\theta}(k)}[\nabla_{\boldsymbol{\theta}(k)} \log f(\mathbf{z}(k),\boldsymbol{\theta}(k))]^T\} \tag{11}$$

when $f(\mathbf{z}(k),\boldsymbol{\theta}(k))$ is the joint probability density function of $\mathbf{z}(k)$ and $\theta(k)$ with no restrictions on the bias of the estimator for bound validity (Van Trees, 2001). The inequality of $\mathbf{A} \geq \mathbf{B}$ indicates that the matrix given by $\mathbf{D} = \mathbf{A} - \mathbf{B}$ is non-negative definite. Using standard Bayesian theory, the joint probability density function is given by $f(\mathbf{z}(k),\boldsymbol{\theta}(k)) = f(\mathbf{z}(k)\,|\,\boldsymbol{\theta}(k)) f(\boldsymbol{\theta}(k))$ where $f(\mathbf{z}(k)\,|\,\boldsymbol{\theta}(k))$ is the conditional probability density function used to calculate the standard Cramér-Rao bound above. The probability density function $f(\boldsymbol{\theta}(k))$ represents the prior probability density function for the mobile terminal location given the selection of measuring base stations or other information.

A useful property of the BCRB is that the effects of the measurements and the prior information are easily separated in terms of computation. This separation is obtained by factoring the measurement conditional probability density function from the joint probability density function of the measurements and location:

$$\tilde{\mathbf{J}}[\boldsymbol{\theta}(k)] = \tilde{\mathbf{J}}_D[\boldsymbol{\theta}(k)] + \tilde{\mathbf{J}}_P[\boldsymbol{\theta}(k)] \tag{12}$$

where if $\mathrm{I}\{g[\boldsymbol{\theta}(k)]\} = -\mathrm{E}\{\nabla_{\boldsymbol{\theta}(k)}[\nabla_{\boldsymbol{\theta}(k)} \log \mathrm{g}(\boldsymbol{\theta}(k))]^T\}$ then $\tilde{\mathbf{J}}_D[\boldsymbol{\theta}(k)] = \mathrm{I}[f(\mathbf{z}(k)\,|\,\boldsymbol{\theta}(k))]$ and $\tilde{\mathbf{J}}_P[\boldsymbol{\theta}(k)] = \mathrm{I}[f(\boldsymbol{\theta}(k))]$. The information matrix $\tilde{\mathbf{J}}_D[\boldsymbol{\theta}(k)]$ is the statistical expectation of the standard Cramér-Rao bound over all possible node locations and quantifies the information from the measurement vector on the localization error bound and the information matrix $\tilde{\mathbf{J}}_P[\boldsymbol{\theta}(k)]$ quantifies the effect of the prior information on the localization error bound. For stationary nodes, the source of the prior information represents information from the selection of base stations used for localization. When non-stationary nodes are located, this prior information can also represent information from measurements in previous sample intervals. More information on this latter case is provided in a later section of this chapter. The computation of the BCRB for localization error using Monte Carlo integration is summarized as Algorithm 1. This calculation gives better approximations to the true bound as $N \to \infty$.

Unfortunately, the BCRB does not exist for all estimation problems. For the BCRB to be valid, two conditions are required (Van Trees 2001). The first condition is that the first and second order partial derivatives of $f(\mathbf{z}(k),\boldsymbol{\theta}(k))$ with respect to all entries of $\theta(k)$ must be absolutely integrable with respect to the entries of $\mathbf{z}(k)$ and $\boldsymbol{\theta}(k)$. The second condition is that the matrices $\tilde{\mathbf{J}}_D[\boldsymbol{\theta}(k)]$ and $\tilde{\mathbf{J}}_P[\boldsymbol{\theta}(k)]$ must be invertible.

The existence of NLoS propagation creates discontinuities in the first and second partial derivatives of $f(\mathbf{z}(k),\boldsymbol{\theta}(k))$ with respect to the entries of $\boldsymbol{\theta}(k)$. The discontinuities occur at the boundaries between regions of LoS and NLoS propagation creating impulses in the partial derivatives of the joint probability density function. The second order partial derivatives in (11) become unbounded, which makes the BCRB invalid. In addition, if the second partial derivatives of $f(\boldsymbol{\theta}(k))$ with respect to the entries of $\boldsymbol{\theta}(k)$ are unbounded then the BCRB is invalidated. This condition is created for many common prior density functions. For example, if the node location is uniformly distributed over a finite area, the second partial derivatives of $f(\boldsymbol{\theta}(k))$ are infinite at the boundaries of the region of support for $f(\boldsymbol{\theta}(k))$ invalidating the BCRB.

The BCRB provides a perfect calculation of the optimal MSE when the conditional probability density function $f\left(\boldsymbol{\theta}(k) \mid \mathbf{z}(k)\right)$ is a multivariate Gaussian probability density function for all $\mathbf{z}(k)$ (Van Trees 2001). The non-linearity of the relationship between the measurements and terminals locations means that this is rarely the case for the localization problem. However, the BCRB provides MSE bounds close to the actual optimal MSE when $f\left(\boldsymbol{\theta}(k) \mid \mathbf{z}(k)\right)$ is well approximated by a multivariate Gaussian probability density function in that there exists a multivariate Gaussian probability density function, $f_G(\boldsymbol{\theta}(k))$ such that for some small constant ε, $\left|f(\boldsymbol{\theta}(k) \mid \mathbf{z}(k)) - f_G(\boldsymbol{\theta}(k))\right| < \varepsilon$ for all $\boldsymbol{\theta}(k)$ and $\mathbf{z}(k)$. When this is not true, then the BCRB MSE bound value can be significantly lower than the true MSE.

Because of these problems, other bounds are needed to calculate a bound on the MSE for localization in the presence of NLOS propagation. The discontinuities created by the presence of both LOS and NLOS propagation in the mobile terminal environment can be handled by either the Barankin bound, the Weinstein-Weiss bound or the extended Ziv-Zakai bound, as can be seen in Table 1. The Barankin bound is valid for localization MSE if the mobile terminal location is deterministic. If there exists a prior probability density function for $\theta(k)$, then the Weinstein-Weiss or extended Ziv-Zakai lower bounds are applicable.

Algorithm 1. Monte Carlo integration calculation of Bayesian Cramér-Rao bound

1. Generate N independent samples of $\boldsymbol{\theta}(k)$: $\mathbf{t}^1, \mathbf{t}^2, \ldots, \mathbf{t}^N$
   $\mathbf{t}^k$ has probability density function of $f(\boldsymbol{\theta}(k))$
2. Calculate $\tilde{\mathbf{J}}_D(\boldsymbol{\theta}(k))$:
   $$\tilde{\mathbf{J}}_D(\boldsymbol{\theta}(k)) = \frac{1}{N}\sum_{k=1}^{N}(\mathbf{D}^k)^T\mathbf{C}^{-1}(\mathbf{D}^k) \text{ where } \mathbf{D}^k = \nabla_{\boldsymbol{\theta}(k)} \mathrm{E}[\mathbf{z}(k) \mid \boldsymbol{\theta}(k)]\Big|_{\boldsymbol{\theta}(k)=\mathbf{t}^k}$$
3. Calculate $\tilde{\mathbf{J}}_P(\boldsymbol{\theta}(k))$:
   $$\tilde{\mathbf{J}}_P(\boldsymbol{\theta}(_k)) = -\frac{1}{N}\sum_{k=1}^{N}\mathbf{B}^k \text{ where } \mathbf{B}^k = \nabla_{\boldsymbol{\theta}(k)}\left[\nabla_{\boldsymbol{\theta}(k)} \log{}_f(\boldsymbol{\theta}(k))\right]^T\Big|_{\boldsymbol{\theta}(k)=\mathbf{t}^k}$$
4. Calculate bound on localization MSE from (10).

The Weinstein-Weiss Lower Bound

The Weinstein-Weiss lower bound gives a lower bound on the MSE of any estimator (Weinstein and Weiss, 1988). The basis of the Weinstein-Weiss bound is that for any vector function e($\mathbf{z}(k)$), the following inequality holds:

$$E\{[\boldsymbol{\theta}(k)-e(\mathbf{z}(k))][\boldsymbol{\theta}(k)-e(\mathbf{z}(k))]^T\}\geq \mathbf{H}\mathbf{G}^{-1}\mathbf{H}^{\mathrm{T}} \tag{13}$$

where $\mathbf{G}$ is the matrix with entries given by

$$\mathbf{G}_{i,j}=\frac{E\{M[s_i,\mathbf{z}(k),\mathbf{h}_i]M[s_j,\mathbf{z}(k),\mathbf{h}_i]\}}{E\{L^{s_i}[\mathbf{z}(k),\boldsymbol{\theta}(k)+\mathbf{h}_i,\boldsymbol{\theta}(k)]\}E\{L^{s_j}[\mathbf{z}(k),\boldsymbol{\theta}(k)+\mathbf{h}_j,\boldsymbol{\theta}(k)]\}} \tag{14}$$

with $\mathbf{H}=[\mathbf{h}_1\mathbf{h}_2]$ for any vectors ($\mathbf{h}_1$,$\mathbf{h}_2$) and scalars (s_1,s_2). The functions $L^s[\mathbf{z}(k),\boldsymbol{\theta}(k)+\mathbf{h},\boldsymbol{\theta}(k)]$ are defined as

$$L^s[\mathbf{z}(k),\boldsymbol{\theta}(k)+\mathbf{h},\boldsymbol{\theta}(k)]=\left[\frac{f(\mathbf{z}(k),\boldsymbol{\theta}(k)+\mathbf{h})}{f(\mathbf{z}(k),\boldsymbol{\theta}(k))}\right]^s, \tag{15}$$

and the function $M[s,\mathbf{z}(k),\boldsymbol{\theta}(k),\mathbf{h}]$ is defined as

$$M[s,\mathbf{z}(k),\boldsymbol{\theta}(k),\mathbf{h}]=L^s[\mathbf{z}(k),\boldsymbol{\theta}(k)+\mathbf{h},\boldsymbol{\theta}(k)]-L^{1-s}[\mathbf{z}(k),\boldsymbol{\theta}(k)-\mathbf{h},\boldsymbol{\theta}(k)]. \tag{16}$$

All expectations are taken with respect to $f(\mathbf{z}(k),\boldsymbol{\theta}(k))$; that is, with regards to both the measurements and the mobile terminal location. Since for any localization system $\hat{\boldsymbol{\theta}}(k)=e(\mathbf{z}(k))$, the bound in (13) is also a bound on the MSE of any localization system. The tightest Weinstein-Weiss lower bound is achieved by calculating the values of s_1, s_2, $\mathbf{h}_1$, and $\mathbf{h}_2$ that maximize the right hand side of (13). Finding the tightest bound is a computationally expensive task.

Fortunately, any set of values of s_1, s_2, $\mathbf{h}_1$, and $\mathbf{h}_2$ bound the estimator performance. Selecting $s_1 = s_2 = 1/2$ and

$$\mathbf{H}=h\begin{bmatrix}1&0\\0&1\end{bmatrix} \tag{17}$$

results in a bound with a very practical cost. For these values, the elements of $\mathbf{G}$ can be rewritten as

$$\mathbf{G}_{i,j}=2\frac{\mu\left(\frac{1}{2},\mathbf{h}_i-\mathbf{h}_j\right)-\mu\left(\frac{1}{2},\mathbf{h}_i+\mathbf{h}_j\right)}{\mu\left(\frac{1}{2},\mathbf{h}_i\right)\mu\left(\frac{1}{2},\mathbf{h}_j\right)} \tag{18}$$

with $\mu(s,\mathbf{h})=E\{L^s[\mathbf{z}(k),\boldsymbol{\theta}(k)+\mathbf{h},\boldsymbol{\theta}(k)]\}$ where $\mathbf{h}_i$ and $\mathbf{h}_j$ designate the ith and j^{th} respective columns of the matrix $\mathbf{H}$. A tight bound can be found by finding the value h which maximizes (13) with the other variables set as above. The limit of this case for $h \to 0$ results in the BCRB, provided the BCRB's conditions are satisfied. As stated above, the BCRB conditions are rarely satisfied in the wireless terminal localization application, but this shows that the best Weinstein-Weiss bound will be at least as tight to optimal localization performance as the BCRB in cases where both bounds are valid.

The special case of $\mu(1/2,\mathbf{h})$ is simplified as

$$\begin{aligned}
\mu\left(\frac{1}{2},\mathbf{h}\right) &= \mathrm{E}\left\{\left[\frac{f(\mathbf{z}(k),\boldsymbol{\theta}(k)+\mathbf{h})}{f(\mathbf{z}(k),\boldsymbol{\theta}(k))}\right]^{\frac{1}{2}}\right\} \qquad (19)\\
&= \int_{\mathbf{S}(k)}\int_{\Re^l}\left[\frac{f(\mathbf{z}(k),\boldsymbol{\theta}(k)+\mathbf{h})}{f(\mathbf{z}(k),\boldsymbol{\theta}(k))}\right]^{\frac{1}{2}} f(\mathbf{z}(k),\boldsymbol{\theta}(k))\mathrm{d}\mathbf{z}(k)\mathrm{d}\boldsymbol{\theta}(k)\\
&= \int_{\mathbf{S}(k)}\int_{\Re^l}\left[f(\mathbf{z}(k),\boldsymbol{\theta}(k)+\mathbf{h})\right]^{\frac{1}{2}}\left[f(\mathbf{z}(k),\boldsymbol{\theta}(k))\right]^{\frac{1}{2}}\mathrm{d}\mathbf{z}(k)\mathrm{d}\boldsymbol{\theta}(k)\\
&= \int_{\mathbf{S}(k)}\int_{\Re^l}\left[f(\mathbf{z}(k)\,|\,\boldsymbol{\theta}(k)+\mathbf{h})f(\boldsymbol{\theta}(k)+\mathbf{h})\right]^{\frac{1}{2}}\left[f(\mathbf{z}(k)\,|\,\boldsymbol{\theta}(k))f(\boldsymbol{\theta}(k))\right]^{\frac{1}{2}}\mathrm{d}\mathbf{z}(k)\mathrm{d}\boldsymbol{\theta}(k)
\end{aligned}$$

where $f(\boldsymbol{\theta}(k))$ is the prior probability density function for the node location with support $\mathbf{S}$(k), and the measurement vectors are of length l. If the measurement vectors given the node locations are Gaussian the conditional probability density function of the measurement vector $\mathbf{z}(k)$ given the location $\boldsymbol{\theta}(k)$ is

$$f(\mathbf{z}(k)\,|\,\boldsymbol{\theta}(k)) = (2\pi)^{-l/2}\left|\mathbf{C}\right|^{-1/2}\exp\left(-\frac{1}{2}\left\|\tilde{\mathbf{z}}[\boldsymbol{\theta}(k)]\right\|^2_{\mathbf{C}^{-1}}\right) \qquad (20)$$

with the signal difference vector defined as $\tilde{\mathbf{z}}[\boldsymbol{\theta}(k)] = \mathbf{z}(k) - \mathrm{E}[\mathbf{z}(k)\,|\,\boldsymbol{\theta}(k)]$ where $\mathrm{E}[\mathbf{z}(k)\,|\,\boldsymbol{\theta}(k)]$ is the expected measurement vector given the wireless node is at position $\boldsymbol{\theta}(k)$, and $\mathbf{C}$ is the covariance matrix of the measurement vector given the node position $\boldsymbol{\theta}(k)$. To simplify the expressions for the Gaussian probability density functions, we make use of the Mahalanobis quadratic distance function defined as $\left\|\mathbf{x}\right\|^2_{\mathbf{C}^{-1}} = \mathbf{x}^T\mathbf{C}^{-1}\mathbf{x}$ (Duda et al. 01). Substituting (20) into (19), the $\mu(1/2,\mathbf{h})$ is defined as

$$\mu\left(\frac{1}{2},\mathbf{h}\right) = \int_{\mathbf{S}(k)}\int_{\Re^l}(2\pi)^{-\frac{l}{2}}\left|\mathbf{C}\right|^{-\frac{1}{2}}\exp\left\{-\frac{1}{4}\left[\left\|\tilde{\mathbf{z}}[\boldsymbol{\theta}(k)]\right\|^2_{\mathbf{C}^{-1}} + \left\|\tilde{\mathbf{z}}[\boldsymbol{\theta}(k)+\mathbf{h}]\right\|^2_{\mathbf{C}^{-1}}\right]\right\}$$

$$\times[f(\boldsymbol{\theta}(k))f(\boldsymbol{\theta}(k)+\mathbf{h})]^{\frac{1}{2}}\mathrm{d}\mathbf{z}(k)\mathrm{d}\boldsymbol{\theta}(k) \qquad (21)$$

To simplify (21), we expand the Gaussian density functions, complete the squares, and then integrate to obtain

$$\mu\left(\frac{1}{2},\mathbf{h}\right)=\int_{\mathbf{S}(k)}\exp\left\{-\frac{1}{8}\left\|\mathrm{E}[\mathbf{z}(k)\,|\,\boldsymbol{\theta}(k)]-\mathrm{E}[\mathbf{z}(k)\,|\,\boldsymbol{\theta}(k)+\mathbf{h}]\right\|_{\mathbf{C}^{-1}}^{2}\right\}[f(\boldsymbol{\theta}(k))f(\boldsymbol{\theta}(k)+\mathbf{h})]^{\frac{1}{2}}\,d\boldsymbol{\theta}(k). \tag{22}$$

In general, except for very simple node arrangements, (22) can only be integrated numerically. For the two dimensional localization problem, the matrix **G** from (18) is given by

$$\mathbf{G}=2\begin{bmatrix}\dfrac{\mu\left(\frac{1}{2},\begin{bmatrix}0\\0\end{bmatrix}\right)-\mu\left(\frac{1}{2},\begin{bmatrix}2h\\0\end{bmatrix}\right)}{\mu\left(\frac{1}{2},\begin{bmatrix}h\\0\end{bmatrix}\right)^{2}} & \dfrac{\mu\left(\frac{1}{2},\begin{bmatrix}h\\-h\end{bmatrix}\right)-\mu\left(\frac{1}{2},\begin{bmatrix}h\\h\end{bmatrix}\right)}{\mu\left(\frac{1}{2},\begin{bmatrix}h\\0\end{bmatrix}\right)\mu\left(\frac{1}{2},\begin{bmatrix}0\\h\end{bmatrix}\right)}\\[2ex] \dfrac{\mu\left(\frac{1}{2},\begin{bmatrix}-h\\h\end{bmatrix}\right)-\mu\left(\frac{1}{2},\begin{bmatrix}h\\h\end{bmatrix}\right)}{\mu\left(\frac{1}{2},\begin{bmatrix}h\\0\end{bmatrix}\right)\mu\left(\frac{1}{2},\begin{bmatrix}0\\h\end{bmatrix}\right)} & \dfrac{\mu\left(\frac{1}{2},\begin{bmatrix}0\\0\end{bmatrix}\right)-\mu\left(\frac{1}{2},\begin{bmatrix}0\\2h\end{bmatrix}\right)}{\mu\left(\frac{1}{2},\begin{bmatrix}0\\h\end{bmatrix}\right)^{2}}\end{bmatrix}. \tag{23}$$

The matrix **G** from (23) is then substituted into (13) to obtain the final Weinstein-Weiss bound matrix. From this matrix, it is possible to calculate a lower bound on the MSE for the localization of stationary nodes. The calculation of the Weinstein-Weiss lower bound for localization using Monte Carlo integration is summarized in Algorithm 2.

The Weinstein-Weiss bound's primary advantage compared to the BCRB is that it handles a wider variety of measurement conditional and location prior probability density functions. It provides MSE bounds close to the optimal MSE values when the conditional probability density function $f(\boldsymbol{\theta}(k)\,|\,\mathbf{z}(k))$ is well approximated as a multivariate Gaussian probability density function (Van Trees and Bell 2007). When this approximation is not good, the Weinstein-Weiss bound is significantly less than the optimal MSE. This is similar to the BCRB, but the Weinstein-Weiss bound is still valid when the prior probability density function $f(\boldsymbol{\theta}(k))$ has only finite support while the BCRB becomes invalid in this case.

While the Weinstein-Weiss lower bound works well in most cases, there are some cases where it can be difficult to calculate, such as when a closed form for $f(\mathbf{z}(k),\boldsymbol{\theta}(k))$ is not available or it is difficult to integrate. In these cases, the Extended Ziv-Zakai bound may be easier to compute.

Algorithm 2: Monte Carlo Integration of Weinstein Weiss bound

The Extended Ziv-Zakai Lower Bound for Location

The Extended Ziv-Zakai lower bound provides a lower bound on the estimation error for Bayesian estimation based on binary detection theory (Bell et al. 1997). The calculation provides a lower bound on the value of $\mathbf{a}^{T}\mathbf{R}_{\mathrm{e}}\mathbf{a}$, where $\mathbf{R}_{\mathrm{e}}=\mathrm{E}\{[\boldsymbol{\theta}(k)-\hat{\boldsymbol{\theta}}(k)][\boldsymbol{\theta}(k)-\hat{\boldsymbol{\theta}}(k)]^{T}\}$. By suitable selections of the vector **a**, bounds can be calculated for all entries of the matrix $\mathbf{R}_{\mathrm{e}}$.

The bound is based on the binary detection problem of deciding whether a mobile terminal is located at position $\theta(k)$ or at position $\boldsymbol{\theta}(k)+\boldsymbol{\delta}$ given the measurement vector $\mathbf{z}(k)$. If the greatest lower bound

Algorithm 2. Monte Carlo Integration of Weinstein Weiss bound

1. Compute Weinstein-Weiss bound for a sweep of h values.
   a. Set $\mathbf{H}$ according to (17) for selected h value.
   b. Compute $\mu(1/2,\bullet)$ function values for entries of $\mathbf{G}$ from (23) using sub-algorithm below.
   c. Compute bound from $\mathbf{G}$ and $\mathbf{H}$ using (13)
2. For the MSE bounds with different h values calculated in step 1, select the highest MSE as the Weinstein-Weiss bound.

Sub-algorithm: Calculation of $\mu(1/2,\mathbf{h})$

a) Generate N independent samples of $\boldsymbol{\theta}(k)$: $\mathbf{t}^1, \mathbf{t}^2, \ldots, \mathbf{t}^N$

$\mathbf{t}^k$ has the probability density function $f(\boldsymbol{\theta}(k))$

b) Compute:

$$\mathbf{c}_k = \exp\left\{-\frac{1}{8}\left\|\mathrm{E}[\mathbf{z}(k)\,|\,\boldsymbol{\theta}(k)=\mathbf{t}^k]-\mathrm{E}[\mathbf{z}(k)\,|\,\boldsymbol{\theta}(k)=\mathbf{t}^k+\mathbf{h}]\right\|^2_{\mathbf{C}^{-1}}\left[f(\boldsymbol{\theta}(k)=\mathbf{t}^k)f(\boldsymbol{\theta}(k)=\mathbf{t}^k+\mathbf{h})\right]^{\frac{1}{2}}\right\}$$

c) Compute $\mu(1/2,\mathbf{h})=\frac{1}{N}\sum_{k=1}^{N}\mathbf{c}_k$

on the probability on this decision error is given by $\mathrm{P}_{\min}[\boldsymbol{\theta}(k),\boldsymbol{\theta}(k)+\boldsymbol{\delta}]$, then the extended Ziv-Zakai bound is written as

$$\mathbf{a}^T\mathbf{R}_e\mathbf{a} \geq \frac{1}{2}\int_0^\infty h\,\mathrm{V}\left\{\max_{\boldsymbol{\delta}:\mathbf{a}^T\boldsymbol{\delta}=\mathbf{h}}\left[\int_{\mathbf{S}(k)}[f(\boldsymbol{\theta}(k))+f(\boldsymbol{\theta}(k)+\boldsymbol{\delta})]\mathrm{P}_{\min}[\boldsymbol{\theta}(k),\boldsymbol{\theta}(k)+\boldsymbol{\delta}]\mathrm{d}\boldsymbol{\theta}(k)\right]\right\}\mathrm{d}h. \quad (24)$$

The V[$g(h)$]is the so-called valley-filling function defined as $\mathrm{V}\left[g(h)\right]=\max_{x\geq h} g(x)$. The tightest bound is found by performing the maximization to find the best value of δ for each value of h. However, the use of any vector δ subject to $\mathbf{a}^T\boldsymbol{\delta}=h$ still results in valid bound, available at a reduced computational cost.

Standard Bayesian detection theory is used to calculate the greatest lower bound on the decision error (Van Trees 2001). If the prior $f(\boldsymbol{\theta}(k))$ is uniform over a finite region and the measurement vector $\mathbf{z}(k)$ given the node location $\boldsymbol{\theta}(k)$ is a Gaussian random vector with mean $\mathrm{E}[\mathbf{z}(k)\,|\,\boldsymbol{\theta}(k)]$ and a covariance matrix of $\mathbf{C}=\mathbf{I}^m\sigma^2$ for all locations then the minimum error probability is

$$\mathrm{P}_{\min}[\boldsymbol{\theta}(k),\boldsymbol{\theta}(k)+\boldsymbol{\delta}]=\mathrm{P}^{el}_{\min}[\boldsymbol{\theta}(k),\boldsymbol{\theta}(k)+\boldsymbol{\delta}]=\mathrm{erfc}\left(\frac{\left\|\mathrm{E}[\mathbf{z}(k)\,|\,\boldsymbol{\theta}(k)]-\mathrm{E}[\mathbf{z}(k)\,|\,\boldsymbol{\theta}(k)+\boldsymbol{\delta}]\right\|}{2\sqrt{2}\sigma}\right) \quad (25)$$

with $\|\mathbf{v}\|$ being the Euclidean length of vector $\mathbf{v}$ (Van Trees 2001). Substituting (25) into (24), and then using a numerical integration technique, such as Monte Carlo integration, provides a general method for bounding localization MSE.

For a general prior probability density functions, the lower bound of (24) can be difficult to compute. In these cases, a looser lower bound is obtained at a much lower cost from the computation of

$$\mathbf{a}^T\mathbf{R}_e\mathbf{a} \geq \int_0^\infty h\,\mathrm{V}\left\{\max_{\boldsymbol{\delta}:\mathbf{a}^T\boldsymbol{\delta}=h}\left[\int_{\mathbf{S}(k)}\min[f(\boldsymbol{\theta}(k)),f(\boldsymbol{\theta}(k)+\boldsymbol{\delta})]\mathrm{P}^{el}_{\min}[\boldsymbol{\theta}(k),\boldsymbol{\theta}(k)+\boldsymbol{\delta}]\mathrm{d}\boldsymbol{\theta}(k)\right]\right\}\mathrm{d}h \quad (26)$$

where $P^{el}_{\min}[\boldsymbol{\theta}(k), \boldsymbol{\theta}(k)+\boldsymbol{\delta}]$ is the optimum decision error between $\theta(k)$ and $\boldsymbol{\theta}(k)+\boldsymbol{\delta}$ calculated using the assumption that both locations are equally likely, so $P^{el}_{\min}[\boldsymbol{\theta}(k), \boldsymbol{\theta}(k)+\boldsymbol{\delta}] = P_{\min}[\boldsymbol{\theta}(k), \boldsymbol{\theta}(k)+\boldsymbol{\delta}]$ with $P_{\min}[\boldsymbol{\theta}(k), \boldsymbol{\theta}(k)+\boldsymbol{\delta}]$ from (25) (Bell et al., 1997). The extended Ziv-Zakai bound calculation using Monte Carlo integration is summarized in Algorithm 3.

Bounds for Location Accuracy from Survey Set Information

For complex propagation environments, such as those encountered in indoor or dense urban locations, simple analytical propagation models, such as those described in the second section of this chapter, do not adequately describe the measurement model to determine the terminal localization accuracy. In these cases, the main recourse for the network designer is to collect survey points in the environment for the implementation of node localization. This section shows a simple calculation that allows a network designer to solve for MSE bounds on the accuracy of localization in their network using a propagation survey set collected in the network area. This calculation is based upon the extended Ziv-Zakai bound presented in the previous section with some modifications for the discrete nature of the survey set information.

Two assumptions are made about the survey set for this procedure to be valid. It is first assumed that the survey set measurements are noise free, in that the survey measurement for a given location is the expected measurement for that location. It is also assumed that the survey set locations are dense enough in the network environment to provide a good representation of the expected variation of the signal measurements over the network area. In other words, these two assumptions state that it is possible to reconstruct all major features of the mean signal function $E[\mathbf{z}(k) \mid \boldsymbol{\theta}(k)]$ from some interpolation of the survey set. Both of these assumptions are standard requirements for the construction of good localization systems based on survey data so are not additional requirements over those of standard survey-based radio localization (McGuire et al., 2003a).

This bounding procedure assumes that the measurement noise distribution is identical for all locations. This assumption is well met if a sensor selection system is used to ensure that only sensors with

Algorithm 3. Calculation of extended Ziv-Zakai bound

1. Numerically integrate $\int_0^\infty h \mathrm{V}[g(h)]dh$ for lower bound on $\mathbf{a}^T\mathbf{R}_e\mathbf{a}$ for $\mathbf{a}=[1\ 0]^T$
   a. $g(h) = \nu(\boldsymbol{\delta})$ where $\boldsymbol{\delta}$ is selected to maximize $\nu(\boldsymbol{\delta})$ subject to $\mathbf{a}^T\boldsymbol{\delta} = h$
      i. Option: for a quick but looser bound, select $\boldsymbol{\delta} = h\mathbf{a}$ where $|\mathbf{a}| = 1$
2. Numerically integrate $\int_0^\infty h \mathrm{V}[g(h)]dh$ for lower bound on $\mathbf{a}^T\mathbf{R}_e\mathbf{a}$ for $\mathbf{a}=[0\ 1]^T$
3. Combine results from step 1 and 2 to obtain extended Ziv-Zakai bound.

Sub-algorithm: Calculation of $\nu(\boldsymbol{\delta})$

a) Generate N independent samples of $\boldsymbol{\theta}(k)$: $\mathbf{t}^1, \mathbf{t}^2, \ldots, \mathbf{t}^N$
b) Compute:
$$\mathbf{c}_k = \min[f(\boldsymbol{\theta}(k)=\mathbf{t}^k), f(\boldsymbol{\theta}(k)=\mathbf{t}^k+\boldsymbol{\delta})]P^{el}_{\min}(\boldsymbol{\theta}(k)=\mathbf{t}^k, \boldsymbol{\theta}(k)+\boldsymbol{\delta}=\mathbf{t}^k+\boldsymbol{\delta})$$
where $P^{el}_{\min}(\boldsymbol{\theta}(k)=\mathbf{t}^k, \boldsymbol{\theta}(k)+\boldsymbol{\delta}=\mathbf{t}^k+\boldsymbol{\delta})$ is computed using (25).
c) Compute $\nu(\boldsymbol{\delta}) = \frac{1}{N}\sum_{k=1}^{N}\mathbf{c}_k$

high signal-to-noise ratios (SNRs) are used for the localization procedure. Since using other sensors will contribute little to the accuracy of the system and incur a communications cost penalty, this restriction will be satisfied by most real world localization systems.

By defining the non-negative random variable $\varepsilon = |\hat{\boldsymbol{\theta}}(k) - \boldsymbol{\theta}(k)|$, the MSE of the estimation error is computed with the expression

$$E[\varepsilon^2] = \frac{1}{2}\int_0^{\infty} \Pr\left(\varepsilon \geq \frac{h}{2}\right) h\,\mathrm{d}h. \tag{27}$$

To use the survey points to compute a bound on the MSE, it is assumed that the survey points represent all possible locations for the nodes, and that all survey point locations are equally likely. This is equivalent to the expression that

$$\Pr(\boldsymbol{\theta}(k) = \tilde{\boldsymbol{\theta}}^i) = 1/N \tag{28}$$

with $\tilde{\boldsymbol{\theta}}^i$ being the location of the ith survey point and N being the total number of survey points. The probability of the error distance is bounded given the node location, assuming the measurement noise distribution is identical for all $\theta(k)$, as

$$\Pr[\varepsilon \geq e \,|\, \boldsymbol{\theta}(k) = \tilde{\boldsymbol{\theta}}^i] \geq \max_{j:\left\|\tilde{\boldsymbol{\theta}}^i - \tilde{\boldsymbol{\theta}}^j\right\| \geq e} \Pr\left(\left\|\mathbf{z}(k) - \tilde{\mathbf{z}}^j\right\| \leq \left\|\mathbf{z}(k) - \tilde{\mathbf{z}}^i\right\| \,\middle|\, \boldsymbol{\theta}(k) = \tilde{\boldsymbol{\theta}}^i\right) \tag{29}$$

where $\tilde{\mathbf{z}}^i$and $\tilde{\mathbf{z}}^j$being the measured signals for survey points i and j, respectively. Combining (28) with (29), the following probability bound is created:

$$\Pr(\varepsilon \geq e) \geq \frac{1}{N}\sum_{i=1}^{N} \max_{j:\left\|\tilde{\boldsymbol{\theta}}^i - \tilde{\boldsymbol{\theta}}^j\right\| \geq e} \Pr\left(\left\|\mathbf{z}(k) - \tilde{\mathbf{z}}^j\right\| \leq \left\|\mathbf{z}(k) - \tilde{\mathbf{z}}^i\right\| \,\middle|\, \boldsymbol{\theta}(k) = \tilde{\boldsymbol{\theta}}^i\right). \tag{30}$$

To obtain the final bound, equation (30) is substituted into (27) to obtain

$$E\left[\varepsilon^2\right] \geq \frac{1}{2}\int_0^{\infty} h \frac{1}{N}\sum_{i=1}^{N} \max_{j:\left\|\tilde{\boldsymbol{\theta}}^i - \tilde{\boldsymbol{\theta}}^j\right\| \geq \frac{h}{2}} \Pr\left(\left\|\mathbf{z}(k) - \tilde{\mathbf{z}}^j\right\| \leq \left\|\mathbf{z}(k) - \tilde{\mathbf{z}}^i\right\| \,\middle|\, \boldsymbol{\theta}(k) = \tilde{\boldsymbol{\theta}}^i\right) \mathrm{d}h. \tag{31}$$

For the simplifying assumption of the Gaussian measurement noise of (20) with $\mathbf{C} = \mathbf{I}^m\sigma^2$ then the probability from (30) is given by

$$\Pr\left(\left\|\mathbf{z}(k) - \tilde{\mathbf{z}}^j\right\| \leq \left\|\mathbf{z}(k) - \tilde{\mathbf{z}}^i\right\| \,\middle|\, \boldsymbol{\theta}(k) = \tilde{\boldsymbol{\theta}}^i\right) = \frac{1}{2}\operatorname{erfc}\left(\frac{\left\|\tilde{\mathbf{z}}^i - \tilde{\mathbf{z}}^j\right\|}{2\sqrt{2\sigma^2}}\right). \tag{32}$$

To bound localization error using a survey set, the value of the noise covariance σ^2 is determined from measurements taken at a single location and then by substituting the survey set point locations and measurements into (31) using the probability from (32), a lower bound on the localization MSE is calculated.

To calculate the covariance bound, a discrete version of (24) is derived to obtain

$$\mathbf{a}^T \mathbf{R}_e \mathbf{a} \geq \frac{1}{2}\int_0^\infty h \frac{1}{N}\sum_{i=1}^{N} \max_{j:\left\|\mathbf{a}^T\left(\tilde{\boldsymbol{\theta}}^j - \tilde{\boldsymbol{\theta}}^i\right)\right\| \geq \frac{h}{2}} \Pr\left(\left\|\mathbf{z}(k) - \tilde{\mathbf{z}}^j\right\| \leq \left\|\mathbf{z}(k) - \tilde{\mathbf{z}}^i\right\| \middle| \boldsymbol{\theta}(k) = \tilde{\boldsymbol{\theta}}^i\right) \mathrm{d}h. \quad (33)$$

From (33) it is possible, with suitable selections of $\mathbf{a}$, to calculate bounds on all entries of the squared error matrix, $\mathbf{R}_e$.

Bounds on Location for Multiple Measurements

The previous sections provide bounds on the localization accuracy from a single measurement vector. To consider additional measurements on the localization error bound, the obvious approach is to just extend the measurement vector, $\mathbf{z}(k)$, to include the new measurements. However, an easier approach is available if the measurement error for the new measurements, $\mathbf{z}'(k)$, is independent of the old measurement vector, $\mathbf{z}(k)$; i.e. $f(\mathbf{z}(k), \mathbf{z}'(k) \mid \boldsymbol{\theta}(k)) = f(\mathbf{z}(k) \mid \boldsymbol{\theta}(k)) f(\mathbf{z}'(k) \mid \boldsymbol{\theta}(k))$. In this case, the joint density of the measurements and location are

$$f(\mathbf{z}'(k), \mathbf{z}(k), \boldsymbol{\theta}(k)) = f(\mathbf{z}'(k) \mid \boldsymbol{\theta}(k)) f(\mathbf{z}(k) \mid \boldsymbol{\theta}(k)) f(\boldsymbol{\theta}(k)) = f(\mathbf{z}'(k) \mid \boldsymbol{\theta}(k)) f(\mathbf{z}(k), \boldsymbol{\theta}(k)) \quad (34)$$

so that we can consider the distribution of the location given both measurements as

$$f(\boldsymbol{\theta}(k) \mid \mathbf{z}(k), \mathbf{z}'(k)) = \frac{f(\mathbf{z}'(k) \mid \boldsymbol{\theta}(k)) f(\boldsymbol{\theta}(k) \mid \mathbf{z}(k))}{f(\mathbf{z}'(k) \mid \mathbf{z}(k))} \text{ with } f(\boldsymbol{\theta}(k) \mid \mathbf{z}(k)) = \frac{f(\boldsymbol{\theta}(k), \mathbf{z}(k))}{f(\mathbf{z}(k))}. \quad (35)$$

The denominators of the ratios in (35) are normalization constants and do not affect the localization. From (35), it can be seen that the conditional probability density function of $\boldsymbol{\theta}(k)$ given $\mathbf{z}(k)$ and $\mathbf{z}'(k)$ is a normalized product of the conditional probability density functions of $f(\boldsymbol{\theta}(k) \mid \mathbf{z}(k))$ and $f(\mathbf{z}'(k) \mid \boldsymbol{\theta}(k))$. This motivates the use of a sequential estimation system where the prior probability density function $f(\boldsymbol{\theta}(k) \mid \mathbf{z}(k))$ is first calculated and then $\boldsymbol{\theta}(k)$ is estimated from $\mathbf{z}'(k)$ using the prior distribution of $f(\boldsymbol{\theta}(k) \mid \mathbf{z}(k))$. This is the basis of the Kalman filter and all other sequential estimation algorithms (Mendel, 1995).

For bound computations using the BCRB considering additional measurements, the decomposition in (12) is extended using partial derivatives of the logarithm of (35) creating the BCRB matrix:

$$\tilde{\mathbf{J}}(\boldsymbol{\theta}(k)) = \tilde{\mathbf{J}}'_D(\boldsymbol{\theta}(k)) + \tilde{\mathbf{J}}_D(\boldsymbol{\theta}(k)) + \tilde{\mathbf{J}}_P(\boldsymbol{\theta}(k)) \quad (36)$$

with the information matrices $\tilde{\mathbf{J}}'_D(\boldsymbol{\theta}(k)) = \mathrm{I}[f(\mathbf{z}'(k) \mid \boldsymbol{\theta}(k))]$, $\tilde{\mathbf{J}}_D(\boldsymbol{\theta}(k)) = \mathrm{I}[f(\mathbf{z}(k) \mid \boldsymbol{\theta}(k))]$, and $\tilde{\mathbf{J}}_P(\boldsymbol{\theta}(k)) = \mathrm{I}[f(\boldsymbol{\theta}(k))]$. Additional measurements with independent measurement errors contribute extra terms to the sum in (36). The information bound for the parameter $\boldsymbol{\theta}(k)$ imposed by a set of measurements is just the sum of the information matrix created by the prior before measurements plus the information matrix for each of the measurement vectors. The bound on the MSE is then calculated by inverting this matrix.

An important consideration for the use of bounds is how to account for the effect of measurements made at different time intervals on the lower bound of localization error. The use of measurements made at different times to track moving nodes is the basis of several proposed time-filtering location algorithms. These filtering algorithms include the Kalman filter (Chen, 1999, Hellebrandt and Mathar, 1999), the Extended Kalman Filter (Liu et al., 1998), and an extended version of the Interactive Multiple Model (IMM) filters (McGuire et al., 2003b).

Bounds on localization error with time filtering are dependent on a description of the time evolution of the node location using a model of the node motion. The location state of the node at sample interval k is given by $\mathbf{x}(k)$. For example, if the location state includes both the location and velocity of the mobile terminal, $\mathbf{x}(k)$ is then defined as $\mathbf{x}(k) = [p_x(k)\ v_x(k)\ p_y(k)\ v_y(k)]^T$ where $(p_x(k), p_y(k))$ is the terminal position at sample time k, where $(v_x(k), v_y(k))$ is the node velocity at sample time k. If the estimated location state for sample interval k given measurements up to sample interval $k\text{-}d$ is denoted as $\hat{\mathbf{x}}(k \mid k-d)$, the estimation bound is denoted as

$$\tilde{\mathbf{P}}(k \mid k-d) \le \mathrm{E}\left\{[\mathbf{x}(k) - \hat{\mathbf{x}}(k \mid k-d)][\mathbf{x}(k) - \hat{\mathbf{x}}(k \mid k-d)]^T\right\}. \tag{37}$$

If the motion model is Markovian so that if $\mathbf{x}(k)$ is known then $\mathbf{x}(k+1)$ is independent of all $\mathbf{x}(k-d)$ for d>0, then the bound on the MSE of the filtering is given by (Šimandl et al. 2001):

$$[\tilde{\mathbf{P}}(k+1 \mid k+1)]^{-1} = [\tilde{\mathbf{P}}(k+1 \mid k)]^{-1} + \mathbf{L}_k^k, \text{ and} \tag{38}$$

$$[\tilde{\mathbf{P}}(k+1 \mid k)]^{-1} = \mathbf{O}_{k+1}^{k+1} - \mathbf{O}_{k+1}^{k,k+1}[\tilde{\mathbf{P}}(k \mid k)^{-1} + \mathbf{O}_{k+1}^{k}]\mathbf{O}_{k+1}^{k,k+1}. \tag{39}$$

The $\mathbf{O}$ matrices are defined as

$$\mathbf{O}_{k+1}^{k} = -\mathrm{E}\{\nabla_{\mathbf{x}(k)}[\nabla_{\mathbf{x}(k)} \ln f(\mathbf{x}(k+1) \mid \mathbf{x}(k))]^T\}, \tag{40}$$

$$\mathbf{O}_{k+1}^{k,k+1} = -\mathrm{E}\{\nabla_{\mathbf{x}(k+1)}[\nabla_{\mathbf{x}(k)} \ln f(\mathbf{x}(k+1) \mid \mathbf{x}(k))]^T\}, \tag{41}$$

$$\mathbf{O}_{k+1}^{k+1} = -\mathrm{E}\{\nabla_{\mathbf{x}(k+1)}[\nabla_{\mathbf{x}(k+1)} \ln f(\mathbf{x}(k+1) \mid \mathbf{x}(k))]^T\}, \text{ and } \mathbf{O}_{k+1}^{k+1,k} = [\mathbf{O}_{k+1}^{k,k+1}]^T. \tag{42}$$

The $\mathbf{L}_k^k$ is defined as

$$\mathbf{L}_k^k = -\mathrm{E}\{\nabla_{\mathbf{x}(k)}[\nabla_{\mathbf{x}(k)} \ln f(\mathbf{z}(k) \mid \mathbf{x}(k))]^T\} \tag{43}$$

with entries that can be taken from $\tilde{\mathbf{J}}_D(\boldsymbol{\theta}(k))$. If the motion model is a linear Markovian model with measurement and process noise given by independent Gaussian vector processes with no time dependencies then equations (38)-(43) will simplify down to the well known Ricatti equations for Kalman filters(Mendel 1995, Brookner, 1998). This derivation assumes that all the derivatives exist and all the expectations are bounded. The basic requirement is that the BCRB exists for localization error. If this condition is not satisfied, then more sophisticated bound calculations are required (Bobrovsky and Zakai, 1975; Kerr, 1989; Bobrovsky et al., 1990; Doerschuk, 1995; Tichavský et al., 1998, Van Trees and Bell, 2007). A recursion similar to (38)-(43) is also available for Weinstein-Weiss lower bounds (Rapoport and Oshman, 2004). These calculations are considerably more complex and well beyond the scope of this chapter.

EXAMPLES OF LOWER BOUND COMPUTATIONS

This section presents examples for calculation of the lower bound on two dimensional localization RMSE using ToA, TDoA, RSS, and AoA measurements. The bounds are calculated for a simple network scenario where the optimum Minimum Mean Square Error (MMSE) localization, $\hat{\boldsymbol{\theta}}(k) = \mathrm{E}[\boldsymbol{\theta}(k) \mid \mathbf{z}(k)]$ is available and easily calculated (Mendel, 1995). It is demonstrated that the Weinstein-Weiss and extended Ziv-Zakai bound calculations provide excellent measures of estimator performance for ToA, TDoA, AoA, and RSS measurements for two different base stations configurations with differing levels of GDOP. Bounds are also calculated for a mobile terminals located in a dense urban environment. The successful use of the localization accuracy bounds under NLoS radio propagation conditions for ToA and TDoA measurements is demonstrated. The last set of simulations demonstrate the use of the accuracy bounds to measure the accuracy of estimation possible for an actual indoor wireless sensor network using RSS measurements from an IEEE 802.11 WLAN base stations.

For the ToA, RSS, and AoA measurements, the measurement probability density function from (20) is assumed with the covariance given as $\mathbf{C} = \mathbf{I}^m\sigma^2$. The value of σ is varied to see the estimators' performance under different noise levels.

Bounds for a Simple Localization Scenario

The simulated network environment for the first three sets of localization is shown in Figure 2. For these simulations, the node locations are uniformly distributed within a disk of radius R_{area}. The node localization is performed from measurements made by *m=3* base stations uniformly located on a circle of radius R_{BS} with the same center as the disc for the terminal locations. The local Cramér-Rao bound is not valid for this localization because of the existence of a prior probability density function for node location. The Cramér-Rao bound would give a bound on localization error only for localization performed without the use of prior information. The BCRB is not valid for this localization either because the finite support for the prior probability density function of the node location causes the regularity conditions of the bound to be violated.

Localization is performed based on RSS, ToA, or AoA measurements made by the base stations with the LoS propagation measurement model from (2). The first sets of simulations are ToA measurements performed with $R_{area} = R_{BS} = 15$m, which results in localization with low GDOP (Spirito, 2001). The RMSE for the ToA MMSE localization is compared with bounds calculated with the Ziv-Zakai and Weinstein-Weiss lower bound in Figure 3. It can be seen that the Ziv-Zakai lower bound provides the tightest bound to the optimal localization MMSE performance.

To show the danger of inappropriate use of the BCRB, the calculated BCRB bound values are also shown in Figure 3. The danger of the BCRB calculation is the calculations can still provide finite values, even though the bound is invalid, such as in this case. As can be seen in Figure 3, the BCRB MSE 'bound' exceeds the optimal MSE as the mean noise power increases. The BCRB calculations for each $\boldsymbol{\theta}(k)$ value use only the local curvature of $f(\mathbf{z}(k),\boldsymbol{\theta}(k))$ and do not include the boundary conditions. For the same reason, local Cramér-Rao bounds calculated near the boundaries of the region of support for $f(\boldsymbol{\theta}(k))$ are also invalid as they do not exclude estimated locations where $f(\boldsymbol{\theta}(k)) = 0$. The BCRB in this case is a lower bound on localization performed without use of the prior probability density function $f(\boldsymbol{\theta}(k))$ also known as Maximum Likelihood Estimation (Van Trees and Bell, 2007).

Figure 2. Simulation set-up

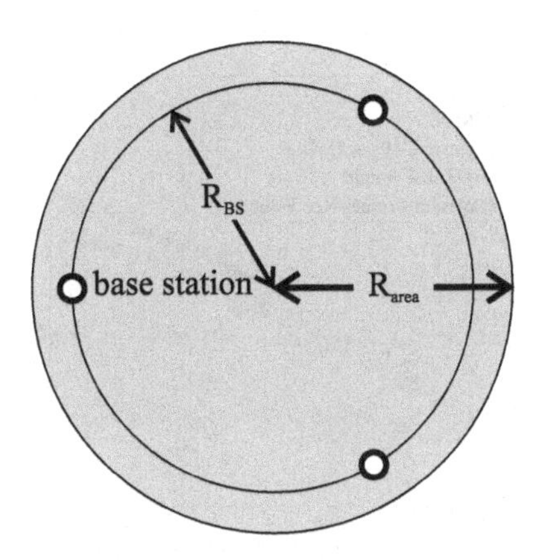

When R_{BS}=15 m for ToA location estimation, the posterior probability density function, $f(\boldsymbol{\theta}(k) \mid \mathbf{z}(k))$ is well approximated by a Gaussian probability density function for most values of $\mathbf{z}(k)$ which results in the Weinstein-Weiss bounds giving good approximations to the optimal MSE for low noise power. As the noise power increases, the effect of the boundary conditions on estimation increases and the Gaussian approximation to the posterior probability density function fails due to truncation at the boundary of the disc, causing the Weinstein-Weiss bounds to become loose with respect to the optimal MSE. However, the extended Ziv-Zakai bound, through its optimizations, is able to better incorporate the effect of boundary conditions which results in a better MSE bound for higher noise levels.

Simulations are performed with ToA measurements with R_{area} = 15m and R_{BS} = 5m which results in a higher GDOP than the previous simulation set for nodes located outside of the base station ring. The RMSE of the optimal MMSE localization and the Ziv-Zakai and Weinstein-Weiss lower bounds are shown in Figure 4. It can be seen that the Ziv-Zakai lower bounds is again tighter to the optimal localization performance than the Weinstein-Weiss lower bound. A comparison of Figure 3 and Figure 4 reveals that the Weinstein-Weiss lower bound is nearly identical for both of these sets. This shows that the Ziv-Zakai provides a better estimate of the lower bound in scenarios with localization with significant GDOP. For ToA location estimation when R_{BS}=5 m, the posterior probability density function is not well approximated as a Gaussian probability density function causing the performance of the Weinstein-Weiss bound to be loose with respect to the optimal MSE.

The last sets of simple localization scenario calculations were performed with LoS RSS propagation and LoS AoA measurements with R_{area} = 15m , R_{BS} = 5m, and α = 3. The results of the RMSE calculations are shown in Figure 5 and Figure 6. These results show that the Ziv-Zakai lower bound and the Weinstein-Weiss lower bound are fairly tight for RSS-based and AoA-based node localization errors as well as for the ToA localization errors. The poor approximation of $f(\boldsymbol{\theta}(k) \mid \mathbf{z}(k))$ as a Gaussian probability density function for RSS and AoA measurements is the chief explanation for the better performance of the extended Ziv-Zakai bound relative to the Weinstein-Weiss bound for these cases.

Bounds for Localization Error in a Dense Urban Environment

This section shows examples of the calculation for the Weinstein-Weiss lower bound calculation for the localization using ToA or TDoA measurements for street locations in a simulated dense urban en-

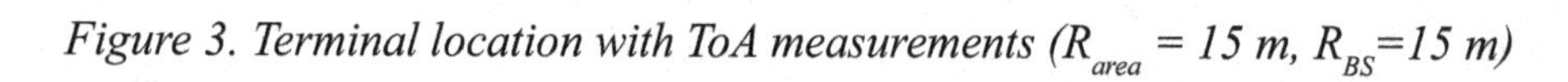

Figure 3. Terminal location with ToA measurements (R_{area} = 15 m, R_{BS}=15 m)

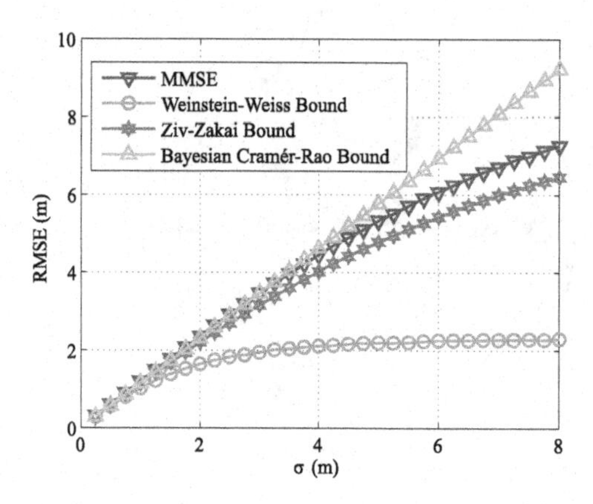

Figure 4. Terminal location with ToA measurements (R_{area}=15 m, R_{BS}=5 m)

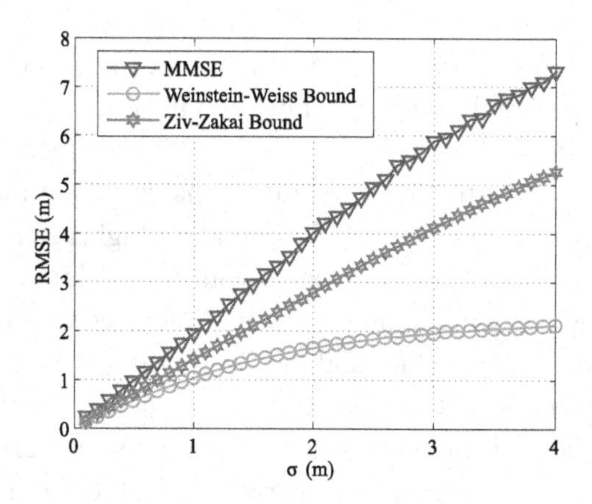

Figure 5. Terminal location with RSS measurements (R_{area}=15 m, R_{BS}=5 m)

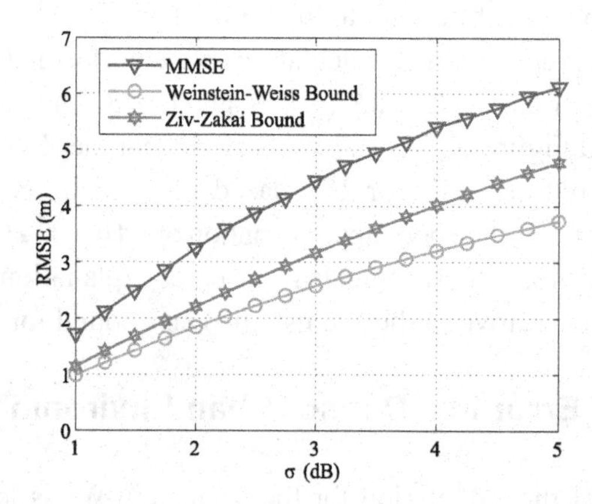

Figure 6. Terminal location with AoA measurements (R_{area}=15 m, R_{BS}=5 m)

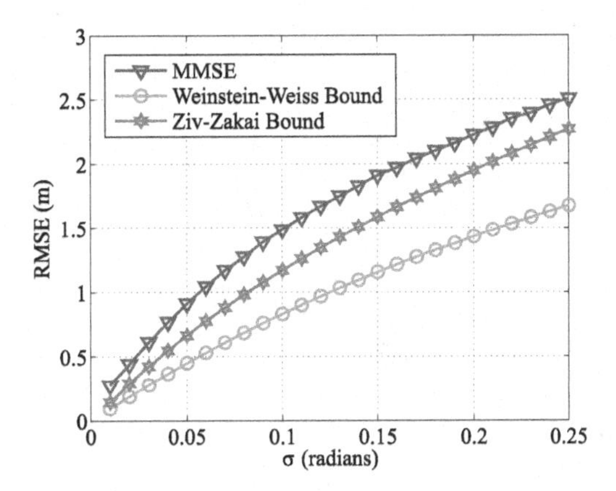

Figure 7. Dense urban environment

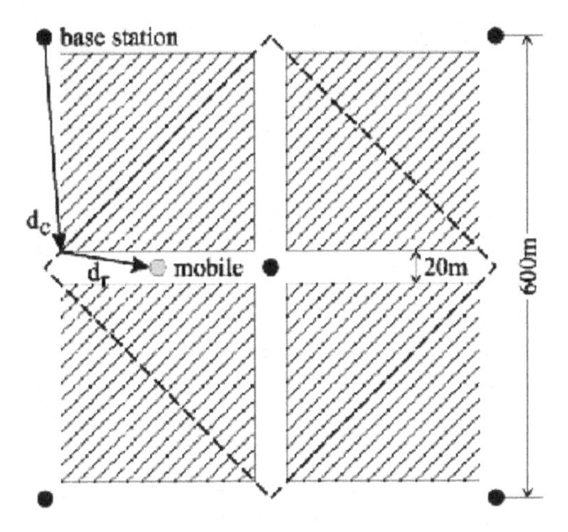

vironment, shown in Figure 7. Each street block is 300 m long with the streets being 20 m wide. The simulated node location is uniformly distributed over all street locations. The node is localized with measurements obtained from the m = 3 nearest base stations to the mobile terminal. Radio propagation is either LoS or NLoS with the building blocking the LoS propagation path if the shortest distance straight line path between the node and the measuring base stations passes through a building block. In this case, the signal diffracts around corners with the noise free propagation distance for a single base station being d_C+d_r as seen in Figure 7.

The RMSE results for the approximate MMSE localization from (McGuire et al., 2003a) using ToA and TDoA measurements in this environment are compared with the Weinstein-Weiss lower bound in Figure 8. For TDoA measurements, the nearest base station is used as the reference base station. These results show that the Weinstein-Weiss lower bound is fairly tight to the actual localization error performance. These results also show that excellent localization accuracy performance is available for the case when NLoS radio propagation is occurring, since when the 3 nearest base stations measurements are used for localization, only the nearest base station has LoS propagation with the other two measuring base stations experiencing NLoS propagation. Note that localization error from TDoA measurements

Figure 8. Weinstein-Weiss lower bound calculations

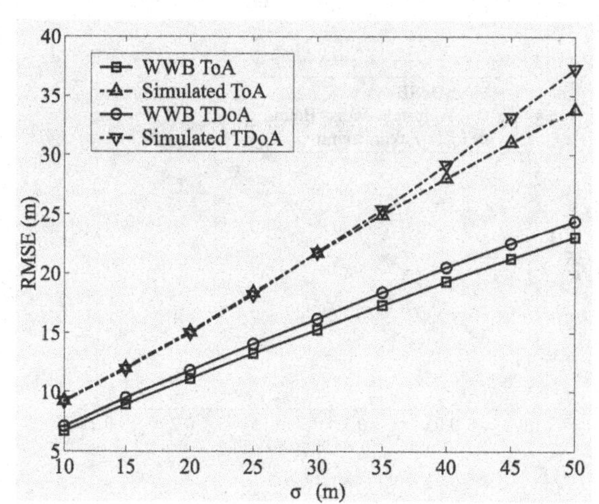

Figure 9. RSS survey set bound calculation

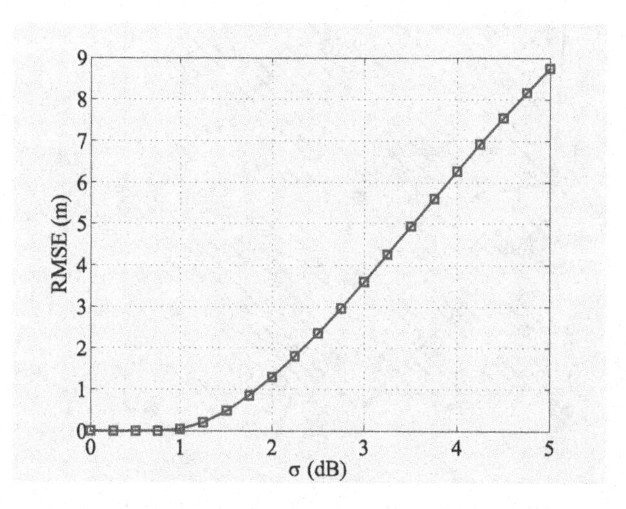

is slightly worse than the ToA localization error. This is due to the fact that the TDoA measurements contain less information than ToA measurements since the full TDoA measurement vector can be calculated from the ToA measurement vector but not vice versa(Shin and Sung 2002).

Survey Set Estimation Lower Bound

In this section, the results of the calculations for the lower bound on the localization RMSE based on an estimator constructed from survey points is presented. For this experiment, RSS survey measurements were collected for each of the 117 seat locations of a 10 m by 13 m lecture hall at the University of Victoria in British Columbia, Canada. At each location, the RSS from 12 of the IEEE wireless local area network base stations for the engineering network were recorded. It should be noted this is a pure

NLoS propagation environment since none of these base stations are located in the lecture hall, and only 3 base stations are located on the same floor as the lecture hall.

The bound on the estimation RMSE for different standard deviations of the measurement noise is shown in Figure 9. The actual standard deviation for the measurement noise for the RSS measurements was 2.1 dB with the resulting RMSE of the estimated location being 2.7 m and the bound on the RMSE is 1.49 m. This shows that the bound is fairly tight to the localization error.

CONCLUSION

This chapter has presented several methods for calculating performance bounds for node localization in wireless sensor networks. These bounds account for information available on the node location from the sensor selection process and consider complicating effects such as radio propagation including both LoS and NLoS propagation paths within the network environment. It is also demonstrated how these bounds can be used to determine the location accuracy available from a given survey set.

Several bounds on localization error have been presented. When the prior probability density function for node location has infinite support and there is only LoS propagation in the network area, the Bayesian Cramér-Rao bound on the localization error is valid. This bound requires only a simple addition of two information matrices, each calculated via simple expectations. However, the Bayesian Cramér-Rao lower bound becomes invalid when either the prior probability density function of the node location has only finite support or the wireless sensor network's environment contains both LoS and NLoS radio propagation. The chapter presents two bounds to handle these cases, the Weinstein-Weiss bound and the extended Ziv-Zakai lower bound. The Weinstein-Weiss lower bound is the easiest to apply, calculated from expectations of functions derived from the measurement and location densities with the optimization performed with respect to a single scalar parameter. The extended Ziv-Zakai bound is calculated from expectations but optimization is performed with respect to a vector of parameters requiring considerably more computational effort.

REFERENCES

Bell, K. L., Steinberg, Y., Ephraim, Y., & Van Trees, H. L. (1997). Extended Ziv–Zakai lower bound for vector parameter estimation. *IEEE Transactions on Information Theory, 43*(2), 624-637.

Bobrovsky, B. Z., & Zakai, M. (1975). A lower bound on the estimation error for Markov processes. *IEEE Transactions on Automatic Control, 20*(6), 785-788.

Bobrovsky, B. Z., Zakai, M., & Kerr, T. H. (1990). Comments on Status of CR-like lower bounds for nonlinear filtering. *IEEE Transactions on Aerospace and Electronic Systems, 26*(5), 895-898.

Botteron, C., Fattouche, M., & Høst-Madsen, A. (2004). Cramer–Rao Bounds for the Estimation of Multipath Parameters and Mobiles' Positions in Asynchronous DS-CDMA Systems. *IEEE Transactions on Signal Processing, 52*(4), 862-875.

Brookner, E. (1998). *Tracking and Kalman filtering made easy.* Toronto, ON: John Wiley & Sons.

Caffery, J. J., & Stüber, G. L. (1998). Overview of radiolocation in CDMA cellular systems. *IEEE Communications Magazine, 36*(4), 38-45.

Chen, P.-C. (1999). A cellular based mobile location tracking system. *In Proceedings of the IEEE Vehicular Technology Conference* (pp. 1979-1983), Houston, TX, USA.

Doerschuk, P. C. (1995). Cramer-Rao bounds for discrete-time nonlinear filtering problems, *IEEE Transacations on Automatic Control, 40*(8), 1465-1469.

Federal Communications Commission. (2006). *FCC Docket 94-102: Report and order and further notice of proposed rulemaking in the matter of revision of the commission's rules to ensure compatibility with enhanced 911 emergency calling systems.*

Duda, R. O., Hart, P. E., & Stork, D. G. (2001). *Pattern Classification.* Second Edition, Toronto, ON: John Wiley & Sons.

Hellebrandt, M., & Mathar, R. (1999). Location tracking of mobiles in cellular radio networks. *IEEE Transactions on Vehicular Technology, 48*(5), 1558-1562.

Kay, S. M. (1993). *Fundamentals of Statistical Signal Processing: Estimation Theory.* Upper Saddle River, NJ: Prentice Hall.

Kerr, T. H. (1989). Status of CR-like lower bounds for nonlinear filtering. *IEEE Transactions on Aerospace and Electronic Systems, 25*(5), 590-600.

Kim, W., Jee, G.-I., & Lee, J. G. (2001). Wireless location with NLOS error mitigation in Korean CDMA system. In *Proceedings of the Second International Conference on 3G Mobile Communication Technologies* (pp. 134-138). London, U.K.

Leon-Garcia, A. (1994). *Probability and Random Processes for Electrical Engineering.* Second Edition, Don Mills, Ontario: Addison-Wesley Publishing Company.

Liu, T., Bahl, P., & Chlamtac, I. (1998). Mobility modeling, location tracking, and trajectory prediction in wireless ATM networks. *IEEE Journal on Selected Areas in Communications, 16*(6), 922-936.

Mendel, J. M. (1995). *Lessons in Estimation Theory for Signal Processing, Communications, and Control.* New Jersey: Prentice-Hall.

McGuire, M., Plataniotis, K. N. & Venetsanopoulos, A. N. (2003a) Location of mobile terminals using time measurements and survey points. *IEEE Transactions on Vehicular Technology, 52*(4), 999-1011.

McGuire, M., & Plataniotis, K. N. (2003b). Dynamic model-based filtering for mobile terminal location estimation. *IEEE Transactions on Vehicular Technology, 52*(4), 1012-1031.

Patwari, N., Hero, A. O., Perkins, M., Correal, N. S., & O'Dea, R. J. (2003). Relative location estimation in wireless sensor networks. *IEEE Transactions on Signal Processing, 51*(8), 2137-2148.

Ray, S., Lai, W., & Paschalidis, I. C. (2006). Statistical location detection with sensor networks, *IEEE Transactions on Information Theory, 52*(6), 2670-2683.

Qi, Y., & Kobayashi, H. (2002). Cramér-Rao lower bound for geolocation in non-line-of-sight environment. In *Proceedings of the 2002 IEEE International Conference on Acoustics, Speech, and Signal Processing, 3*, 2473-2476.

Rapoport, I., & Oshman, Y. (2004). Recursive Weiss-Weinstein lower bounds for discrete-time nonlinear filtering. *In Proceedings of the 43rd IEEE Conference on Decision and Control, 3*, 2662-2667).

Rife, D. C., Goldstein, M., & Boorstyn, R. R. (1975). A unification of Cramér-Rao type bounds. *IEEE Transactions on Information Theory, IT-21*(3), 330-332.

Sheu, J.-P., Li, J.-M., & Hsu, C.-S. (2006). A distributed location estimating algorithm for wireless sensor networks. *In Proceedings of the IEEE International Conference on Sensors Networks, Ubiquitous, and Trustworthy Computing* (pp. 218-225).

Shin, D.-H., & Sung, T.-K. (2002) Comparisons of error characteristics between TOA and TDOA Positioning. *IEEE Transactions on Aerospace and Electronic Systems, 38*(1), 307-311.

Šimandl, M., Královec, & Tichavský, P. (2001). Filtering, predictive, and smoothing Cramér-Rao bounds for discrete-time nonlinear dynamic systems. *Automatica, 37*(11), 1703-1706.

Spirito, M. A. (2001). On the accuracy of cellular mobile station location estimates. *IEEE Transactions on Vehicular Technology, 50*(3), 674-685.

Steele, R. (1992). *Mobile Radio Communications*. Piscataway, NJ: IEEE Press.

Tekinay, S., Chao, E., & Richton, R. (1998). Performance benchmarking for wireless location systems. *IEEE Communications Magazine, 36*(4), 72-76.

Tichavský, P., Muravchik, C. H., & Nehorai, A. (1998). Posterior Cramér-Rao bounds for discrete-time nonlinear filtering. *IEEE Transacations on Signal Processing, 46*(5), 1386-1396.

Torrieri, D. J. (1984). Statistical theory of passive location systems. *IEEE Transactions on Aerospace and Electronic Systems, AES-20*(2), 183-198.

Van Trees, H. L. (2001). *Detection, Estimation, and Modulation Theory.* Volume I, Toronto, Ontario: John Wiley & Sons.

Van Trees, H. L., & Bell, K. L. (Ed.) (2007). *Bayesian Bounds for Parameter Estimation and Nonlinear Filtering/Tracking.* Piscataway, NJ: IEEE Press.

Weinstein, E., & Weiss, A. J. (1988). A general class of lower bounds in parameter estimation. *IEEE Transactions on Information Theory, 34*(2), 338-342.

Wylie, M. P., & Holtzman, J. (1996). The non-line of sight problem in mobile location estimation. In *Proceedings of the International Conference on Universal Personal Communications* (pp. 827-831), Cambridge, MA, USA.

Chapter XVI
Experiences in Data Processing and Bayesian Filtering Applied to Localization and Tracking in Wireless Sensor Networks

Junaid Ansari
RWTH Aachen University, Germany

Janne Riihijärvi
RWTH Aachen University, Germany

Petri Mähönen
RWTH Aachen University, Germany

ABSTRACT

The authors discuss algorithms and solutions for signal processing and filtering for localization and tracking applications in Wireless Sensor Networks. Their focus is on the experiences gained from implementation and deployment of several such systems. In particular, they comment on the data processing solutions found appropriate for commonly used sensor types, and discuss at some length the use of Bayesian filtering for solving the tracking problem. They specifically recommend the use of particle filters as a flexible solution appropriate for tracking in non-linear systems with non-Gaussian measurement errors. They also discuss in detail the design of some of the indoor and outdoor tracking systems they have implemented, highlighting major design decisions and experiences gained from test deployments.

INTRODUCTION

In this chapter we focus on the practical aspects of implementing Wireless Sensor Network (WSN) based localization and tracking solutions. We discuss different design decisions, such as the types of signals used for localization and how to preprocess and filter the sensor readings before applying the complete localization and tracking algorithms. Mobility tracking and localization are multifaceted problems, which have been studied for a long time in different contexts. Many potential applications in the domain of WSNs require such capabilities. The need for mobility tracking is inherent in many surveillance, security and logistic applications. Hence, the development of robust and low-cost techniques has also high practical interest in the industrial context. Vast literature exists on these topics, and especially theoretical underpinnings are relatively well established by now. In this chapter we will focus on explaining some of the practical issues for engineers who are interested in implementing tracking solutions. In particular, we also introduce particle filters, which are often better suited than Kalman filters and their variants to the real-world situations where non-linearities and/or non-Gaussian errors are relatively commonly encountered.

We shall use the following terminology throughout this chapter. By *localization,* we mean the determination of the location of the object at a single time instant either in relative or in absolute coordinates. Sometimes this is performed using *ranging,* that is, by obtaining distance estimates to nodes or devices at known locations. In applications where localization in terms of a global coordinate system is necessary we assume the presence of, for example, differential GPS enabled anchor nodes, which then provide global coordinates for the sensor nodes which use local ranging or localization techniques. Finally, by *tracking* we mean the estimation of the trajectory of an object based on sequential measurements.

Due to space limitations, this chapter is naturally providing only a starting point for someone being interested in deploying wireless sensor based tracking systems. Our aim is to provide enough theoretical background and references to enable similar work by others, but not to provide a comprehensive survey of the field. Moreover, we have provided a number of examples and less emphasized practical lessons from our own work with *deployed experimental networks.* We have drawn our knowledge from a number of networks and deployments, and particularly from a large outdoor vehicular tracking network that was semi-commercially deployed with ca. 100 surveillance nodes, and from indoor tracking networks which have also been using inertial tracking methods. The outdoor network was designed for target tracking and surveillance in the context of SMAUG-project (Ansari *et al.* 2007a) which considered tactical purpose networks. The designed indoor network, which has also common components with the SMAUG-network, was designed for asset tracking and ubiquitous computing purposes, tracking assets and doctors in hospitals being a good example application. Hence, our selection of topics has been heavily influenced by practical projects and from our experience on what are the key issues to be highlighted and take into account when building real deployed systems.

The rest of this chapter is structured as follows: We begin by a short general discussion on major localization techniques followed by an overview of different signal types used in localization. We also discuss different sensor types available for detecting such signals. After these preliminaries, we discuss at some length different data processing and filtering solutions, specifically highlighting common application scenarios and implementation considerations. The techniques considered range from filtering that can be carried out on individual sensor nodes all the way to Bayesian filters for data fusion and for solving the tracking problem using individual localizations as inputs. We then give rather detailed accounts on some of the systems we have developed applying the principles outlined. We discuss both the systems for indoor localization (for asset tracking, localization of terminals or tracking users) as well as for tracking vehicles.

TECHNIQUES FOR LOCALIZATION AND RANGING

Before going into the discussion on data processing and filtering, we shall briefly recall the basic techniques for ranging and localization. For further details of these techniques the reader is referred to the previous chapters of the present volume. Our focus here will be on approaches with relevance for the tracking systems discussed in-depth later in this chapter.

In general localization and ranging can be performed either actively or passively. In active schemes, the listener or the target object transmits a signal solely for the detection process. The listener associates this signal with the position of the object in a given coordinate space. On the other hand, passive localization is performed based on the already present signal in the environment. Active schemes provide more control and flexibility in positioning and generally result in higher accuracy than the passive schemes. The passive systems have their relevance in environments where the detection process is to be done without being noticed, for example in hostile environments or where the localization is to be performed without altering the already existing infrastructure. Examples of systems with such constraints include traffic monitoring and surveillance systems.

For ranging, the main techniques considered in the following are Time of Arrival (ToA), Time Difference of Arrival (TDoA), and techniques based on signal strengths. In Time of Arrival method, also known as Time-of-Flight (ToF), the time duration for a particular signal in propagating from a transmitter to a receiver is measured. Using the propagation speed and the traveling time, the distance between the transmitter and the receiver is computed. The main limitation of ToA is that it requires strict synchronization between the sender and the receiver clocks. TDoA removes this restriction by employing two different types of signals with different propagation speeds. Most commonly used combination of signals is the Radio Frequency (RF) and acoustic signal pair. If t_d is the time between the arrival of a RF and an acoustic signal, then the Line-Of-Sight (LOS) distance d can be computed as

$$d = \frac{t_d}{\frac{1}{speed_{RF}} - \frac{1}{speed_{acoustic}}}.$$

TDoA requires two different kinds of transceivers and is applicable only to active localization systems. The primary limiting factors in ranging accuracy with TDoA are the accuracy of estimation of these delays, clock resolution of the receiver, and changes in signal propagation speeds caused by environmental effects. Examples of ToA-based schemes utilizing audible sound have been developed by Simon *et al.* (2004) and Kuckertz *et al.* (2007) whereas an example of a TDoA-based scheme has been given by Whitehouse and Culler (2003). However, in TDoA-based schemes, ultrasound is usually preferred because of its highly directional properties and due to the fact that it causes no irritation to human ears. The MIT Cricket system is the most famous example here (Priyantha, 2005). Since radio waves cannot travel in water, ultrasonic based positioning systems are very promising in underwater sensor networks as well. Classically, ultrasonic based localization and tracking has been used for over half a century in Sound Navigation and Ranging (SONAR) (Drumheller, 1987).

Techniques based on received signal strength form another important family of passive localization and ranging techniques. Radio waves are one of the popular methods used in signal strength based schemes as virtually all the devices communicating over wireless use radio waves. The distance may be computed from the received signal strength or techniques utilizing signal strength maps can be used for localization. For examples of systems based on these approaches see Bahl and Padmanabhan (2000) and Lorincz and Welsh (2007). The major limiting factor for the accuracy of ranging and localization systems based on radio waves is the complexity of the radio propagation environment. However, radio based signal strength methods are quite popular in sensor networks because no additional hardware is required for localization purpose as every sensor node is already equipped with a radio for communication.

A localization technique not directly relying on ranging is Angle of Arrival (AoA). Instead of distance estimates, triangulation is used to determine the position of the target object in the coordinate space based on the angle of the received signal using an antenna array. Since this method requires a calibrated antenna array, it is usually too expensive and complicated to be used in sensor networks. However, in Enhanced-911 system (Federal Communications Commission), the mobile phone uses RF time of flight and AoA of the signals from the mobile phone to the base station towers in order to find the mobile phone's location.

Another passive localization technique of importance especially in vehicular tracking is obtained by observing local strength of the magnetic field. A magnetic object or an object with significant permeability disturbs the Earth's magnetic lines of forces. This property is used in the magnetometers to estimate the direction and the "magnetic content" in the object to be localized. Typical magnetometers are based on flux-gate, magneto-resistivity and Hall Effect. These are able to detect a changing magnetic field caused by a moving ferromagnetic object, rotation of a magneto-sensitive object or due to a changing electric current. These characteristics can be employed passively in a variety of applications like traffic monitoring, border surveillance applications etc. Magnetic sensors are not only used in detecting the presence of ferromagnetic objects for proximity but also for distance ranging in highly calibrated systems (Arora *et al.*, 2006).

Finally, different optical and infrared (IR) sensors can be effectively used for localization purposes as well. Such techniques are especially appropriate for various surveillance applications and military usage. Both passive and active approaches are possible, depending on the requirements and the constraints of the scenario being considered.

DATA PRE-PROCESSING

In the previous section we described the basic localization and ranging techniques together with their characteristic features and requirements relevant for our applications. We shall now briefly cover the main approaches to data processing on sensor nodes used in various parts of the signal processing chain. Again, our focus is on methods we have found most useful when implementing tracking solutions.

Data Calibration

Sensor readings generally depend on environmental factors such as light intensity, temperature, humidity, etc. The dependencies can be linear as well as non-linear. The process of compensating these effects is known as calibration. Depending on the type of sensor and system requirements, calibration can take place in various stages of the system lifecycle. Often calibration is part of the assembly process, but especially if the environmental factors affect the performance significantly, online calibration becomes necessary. As an example, we consider the TDoA based distance ranging system using a pair of radio frequency and ultrasonic signals. Since the ultrasonic waves are significantly affected by temperature, necessary adjustments are required in the speed of the sound waves for accurate distance ranging. It is convenient if the dependency relationship can be approximated through a mathematical relationship as it allows a sensor node to simply perform a calculation to calibrate the data instead of maintaining large statistical tables, which may not be possible because of the limited available memory. For instance, the temperature dependency of the speed of ultrasonic signal can be expressed as

$$v_{ultrasound} = 331.4 + 0.6T_c \text{ [m/s]},$$

where T_c is the temperature on Celsius scale.

Furthermore, as we will explain in the later sections, the ranging error grows over the real distance in the case of TDoA based ranging using the pair of radio and ultrasonic signals. In this case, appropriate re-adjustments in the distance estimates are performed, which are referred to as post-calibration.

Compensation of Unwanted Constants and Long-Term Drift

In many situations, sensor readings show a constant signal level and the signal of interest overrides this constant level. In these situations, either a differential reading is useful which automatically nullifies the DC level or the constant offset needs to be subtracted explicitly. Occasionally it is more useful to subtract the constant offset before the digitization of the signal so that the constant offset does not reduce the dynamic range of the ADC. In many of the situations, the constant offset depends on the sensor surroundings or post deployment conditions for a sensor network. For instance, magnetometer readings always possess some DC offsets depending upon the ferromagnetic content in the surroundings or the readings from the infrared sensors have a constant offset because of the ambient temperature.

Figure 1. The deviation of the measured ranging distance over real distance, obtained from Cricket System. A post-calibration scheme can be applied to compensate the distance dependence errors.

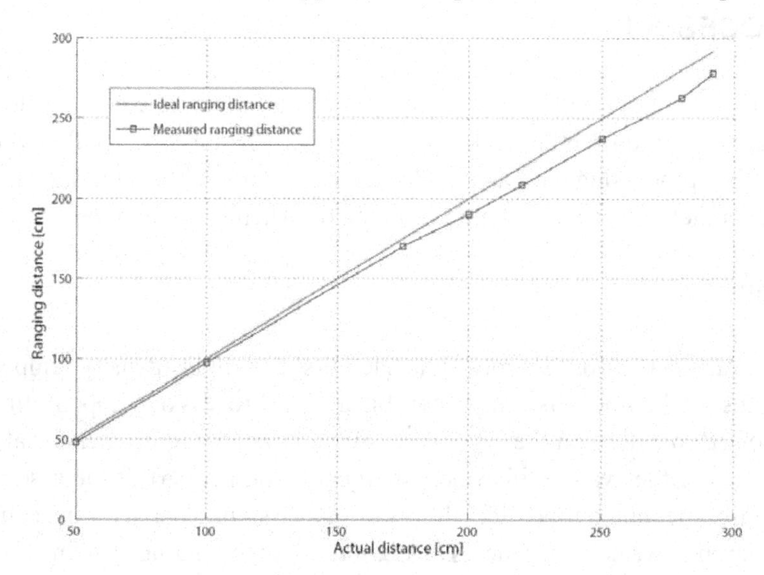

All the magnetometer based compasses require some kind of offset compensation and calibration. A good practice in sensor board design is to include a signal conditioning circuitry that can dynamically compensate the environmental effects. This can be achieved by introducing software controllable circuit elements. A common approach is to use a digital potentiometer based voltage divider to generate similar offset at an input of a differential amplifier so that the resulting static differential signal becomes zero. This differential signal is then amplified.

In many sensors, a drift is developed over long periods of time. This is due to the changes in environmental conditions or the degradation of sensors. Changes in environmental conditions may, for instance, include the day and night effects, temperature and humidity variations and pressure differences, the changes in the Earth's magnetic field strength due to its rotation etc. Long-term drifts restrict the capabilities of sensors completely or partially and need to be eliminated. In an outdoor vehicle tracking scenario, using the Honeywell's HMC1022 magnetometer sensors (Honeywell, 2008a) it was observed that over periods of eight hours, the sensors gradually develop a high enough drift that the output signal starts to saturate. This drift can be compensated by using an exponentially moving average filter (Ansari *et al.*, 2007). The coefficient of the filter needs to be set according to the dynamic characteristics of the signal.

Finally, it is important to note that certain sensors develop anomalies over extended periods of time and lose their sensitivity. Some of the effects are due to the mechanical characteristics and others are due to the intrinsic measuring property. For example, the anisotropic magnetometer sensors have magneto-resistive elements that remain identical in the absence of magnetic field and get aligned in the presence of an external magnetic field. When the sensor is exposed to strong external magnetic field over extended periods of time, the sensor gets magnetized and loses its sensitivity. This can be de-magnetized using a high current pulse, using a circuitry as described in (Honeywell, 2008a).

Filtering of Data

In many cases, the data obtained from the sensors has more information content than desired. This unwanted signal is regarded as noise and is normally filtered out before amplification and digitization. The choice of the filter is highly related to the type of noise in the sensor and the signal characteristics like bandwidth, variance, amplitude, phase etc. Filter selection is also dependent upon the degree of the required accuracy, the computational ability and the available memory at a sensor node. There is a filtering delay associated with each type of filter, which is generally dependent upon the computational complexity and the filter taps, i.e., number of samples required to generate an output. In this context, iterative filtering algorithms are very attractive because of their easy realization with low computational overhead, especially in resource constrained embedded devices like sensor nodes. In the following, we describe some of the commonly used filters in sensor networks and highlight the noise types the filters are effective in dealing with.

In many cases, the sensor readings contain spurious noise samples, which are uncorrelated with their adjacent samples. These noise samples can easily be filtered out using *median filters*. For instance, Cricket system results in some spurious distance estimates bearing no relationship with their neighbours. An effective way to get rid of the unwanted samples is by applying a moving median filter (Priyantha, 2005). *Lowpass filtering* is a popular smoothing technique which can be used to attenuate the noise inherent in many types of waveforms in the measurement dataset. It lets through the lower frequencies and attenuates the higher frequencies. The cut-off frequency is chosen to be compatible with the sampling rate of the ADC and the desired band of frequencies of measured signal. Lowpass filters can be categorized into two main classes based on how they operate on their impulse response:

- **Finite impulse response (FIR):** An FIR filter is usually implemented by using a series of delays, multipliers, and adders to create the filter's output. FIR filters do not use any feedback and are generally easy to implement.
- **Infinite impulse response (IIR):** An IIR filter uses feedback to keep more historical information active in the calculation. Its impulse response is infinite in duration. IIR filter might not be stable compared to FIR filters due to the feedback.

Owing to the limited memory and computational constraints, the filter coefficients used on the sensor nodes should be integer numbers and the order/complexity of the filter should be low. The operation of an 8-tap FIR filter is illustrated in the Figure 2, where the continuous line is the original measurement and the dashed line is the smoothed data.

A *moving average filter* is used to smooth out fluctuations in the data. It is implemented simply by calculating the sum of the measurements over a time window divided by the number of samples within the window. The *Exponentially Weighted Moving Average (EWMA)* approximates an arithmetic moving average by shifting the current estimate for the average by a constant multiple of the latest measurement. This can be expressed as

$$y \leftarrow \alpha z + (1-\alpha)y,$$

where z is the new measurement, y is the variable containing the approximation to the average, and α is a parameter between zero and one. Note that every past value of z in the time series is contained in

Figure 2. Results of applying low pass filtering. The blue curve (solid line) represents the raw data sampled from the ADC and the red curve (dashed line) shows the smoothened version after applying a low-pass-filter.

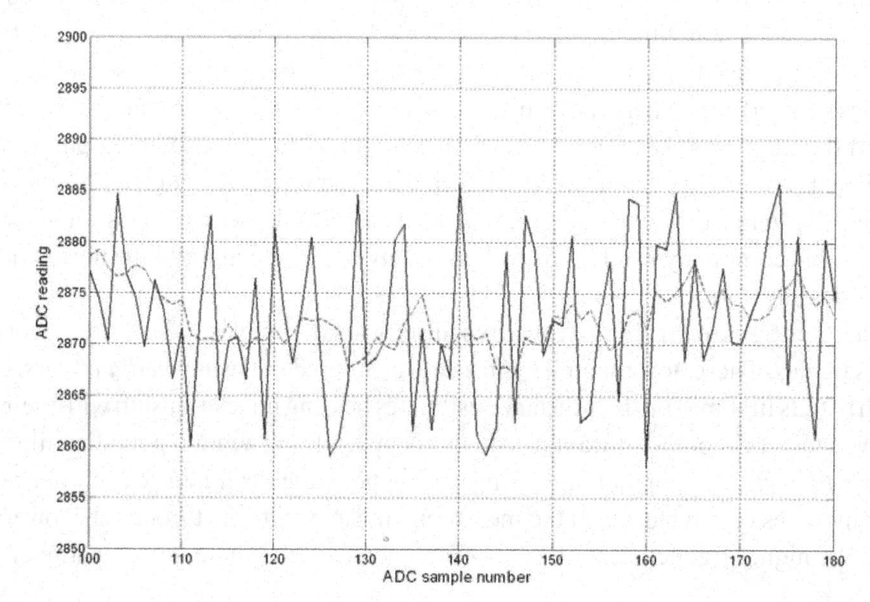

each new result y, but older measurement values get exponentially weighted to insignificance as the series progresses. The advantage of EWMA filter over the regular moving average one is obviously that memory for storing only a single variable is needed.

BAYESIAN FILTERING AND DATA FUSION

The techniques presented in the previous section are absolutely necessary for obtaining accurate individual samples from sensors of various types. We shall now focus on techniques which can be utilized to estimate the behavior of a *system* being measured based on individual localization or ranging samples. We begin with a rather general introduction to sequential Bayesian filtering as a probabilistic tool to reason about the time-evolution of the system. We focus specifically on *particle filters* due to their wide range of applications, and capability to cope with non-linearities and non-Gaussian measurement noises. Our experience with practical tracking systems has led to the conclusion that there are many situations where non-linearities and/or non-Gaussian errors are quite significant. Later on we shall discuss on the applications of these principles for various localization and tracking systems. For further details, we refer the reader to Arulampalam *et al.* (2002) and Van Trees and Bell (2007), on which the following exposition is also based. Another insightful survey on Bayesian filtering in location estimation is given in Fox *et al.* (2003).

Basic Framework

We denote the state of the system under study at time t_k by x_k, where k indexes the time instances at which either measurements become available or prediction of the system state is desired. In general x can be an arbitrary vector, although in localization and tracking applications it is almost invariably the location of the object being tracked (or a Cartesian product of multiple location vectors in case of multi-object localization and tracking is performed). We focus only on the discrete-time case since this seems to cover all the major applications in the present field. We assume that a *system model* of the form

$$x_t = f(x_{t-1}, s_{t-1})$$

is known, where f is a function of the state vector and the so-called *process noise* s. In general f can be a non-linear function of its arguments, although the linear case is obviously of interest as well. We further assume that *measurements* on the system behavior are given by

$$z_k = h(x_k, n_k),$$

where h can again be non-linear, and n is the model for *measurement noise*. The noise processes are usually assumed to be independent and identically distributed, and we shall do so here as well.

Bayesian Solution

We shall now apply Bayesian approach to the system and measurement models to obtain a recursive set of equations as a general solution to the state estimation problem. We assume that the initial state of the system is known in the form of a probability density $p(x_0)$. Our interest is on estimating the posterior probability of the state of the system given all the measurement information, that is, obtaining a law for $p(x_k \mid z_{1:k})$, where the shorthand $z_{1:k} = \{z_i \mid i = 1,...,k\}$ is used.

The usual way of writing down the solution for the Bayesian filtering problem is by means of two steps performed at each update time: *prediction* and *update*. In the prediction phase, the so-called Chapman-Kolmogorov equation is used to write the probability density function in the form

$$p(x_k \mid z_{1:k-1}) = \int p(x_k \mid x_{k-1}) p(x_{k-1} \mid z_{1:k-1}) dx_{k-1}.$$

The first term of the integrand is known from the system model and the second term is known from the previous time step. As the kth measurement is performed, the estimate of the state is updated by applying the Bayes rule, namely by computing

$$p(x_k \mid z_{1:k}) = \frac{p(z_k \mid x_k) p(x_k \mid z_{1:k-1})}{p(z_k \mid z_{1:k-1})}.$$

The likelihood function for the measurements is obtained from the measurement model, and the normalizing constant can be calculated from

$$p(z_k \mid z_{1:k-1}) = \int p(z_k \mid x_k) p(x_k \mid z_{1:k-1}) dx_k.$$

The above equations give the optimal Bayesian solution to the filtering problem in a form of recursive formulae. Unfortunately, the exact computation of the above integrals is not feasible in the most general case. We shall outline two approaches that have been found especially useful in practice. First is the simplification of the above equations by assuming simple form of system and measurement models. This leads to the famous *Kalman filter* and its variants. The second approach is driven by Monte Carlo techniques. The probability densities in the above equations can be approximated by a large sum of delta functions ("particles") turning the integrals into finite sums. This leads to very powerful approximate solutions of the filtering problem by means of *particle filters*.

Kalman Filter and its Major Variants

The simplest special case of the filtering problem is obtained by assuming that both the system model and the measurement model are linear functions, and that both noise models in the system description are Gaussian. The system model thus becomes

$$x_k = F_k x_{k-1} + s_{k-1},$$

where F is now a matrix. Similarly, the measurement model becomes

$$z_k = H_k x_k + n_k.$$

Let now $N(x \mid m, Q)$ denote the Gaussian density with mean m and covariance matrix Q. It can be shown that in this case the Bayesian filtering equations are reduced to

$$p(x_{k-1} \mid z_{1:k-1}) = N(x_{k-1} \mid m_{k-1|k-1}, P_{k-1|k-1}),$$
$$p(x_k \mid z_{1:k-1}) = N(x_k \mid m_{k|k-1}, P_{k|k-1}),$$
$$p(x_k \mid z_{1:k}) = N(x_k \mid m_{k|k}, P_{k|k}), \qquad \text{and}$$

where

$$m_{k|k-1} = F_k m_{k-1|k-1}$$
$$P_{k|k-1} = Q_{k-1} + F_k P_{k-1|k-1} F_k^T$$
$$m_{k|k} = m_{k|k-1} + K_k (z_k - H_k m_{k|k-1})$$
$$P_{k|k} = P_{k|k-1} - K_k H_k P_{k|k-1}$$
$$S_k = H_k P_{k|k-1} H_k^T + R_k$$
$$K_k = P_{k|k-1} H_k^T S_k^{-1}$$

and Q and R denote the covariances of the process and measurement noises, respectively. These are the equations defining the Kalman filter, yielding the optimal solution in the case of linear system with Gaussian noises. These equations can be used to approximate the non-linear case as well by linearization (that is, by approximating the system and measurement models by their Taylor series and discarding the non-linear terms). This results in the *Extended Kalman filter* (EKF). The accuracy of the results obtained from EKF depends heavily on the approximation error made, and the structure of the true shape of the posterior density. As the Kalman filter always results in Gaussian posterior, even the EKF will have limited validity in case of highly skewed or multimodal true posteriors. Another limitation of Kalman filter is the assumption of Gaussian process and measurement noises. As we have seen above the measurement noise for many real-world sensors is non-Gaussian, calling for more advanced solutions to the filtering problem.

Particle Filters

The key observation needed in the following is that any probability density can be approximated by a sum of delta-functions of the form

$$p(x) \approx \sum_{i=1}^{N} w^i \delta(x - x^i),$$

where $\{x^i\}$ are called *particles* and $\{w^i\}$ are *weights* summing to one. The accuracy of the approximation is mainly dependent on the shape of p and the total number of particles N. The weights are introduced to facilitate sampling of the points $\{x^i\}$. Usually one chooses an *importance density* q for which generating samples is easy and makes weights proportional to the ratio of the true distribution and q. In the case of Bayesian filtering using the above representation in the recursive formulation for the density

$$p(x_k \mid z_{1:k}) \approx \sum_{i=1}^{N} w_k^i \delta(x_k - x_k^i)$$

leads to the update rule

$$w_k^i \propto w_{k-1}^i \frac{p(z_k \mid x_k^i) p(x_k^i \mid x_{k-1}^i)}{q(x_k^i \mid x_{k-1}^i, z_k)}$$

for the weights, where q is a suitably chosen importance density (we shall return to the question of choosing q momentarily). This approach, called *Sequential Importance Sampling* (SIS) can be chosen to lead to the exact solution to the filtering equations in the limit of large N.

The SIS algorithm is relatively simple to implement, and has turned out to be practical in a number of applications. It does have, however, certain shortcomings the awareness of which is important. First of these is the so-called *degeneracy problem* manifesting itself in the majority of particles having weights approaching zero. This can be solved by *resampling*, that is, generating a new collection of particles and weights by sampling from the particle cloud approximation of the posterior distribution with weights giving the probabilities of the individual particles occurring in the sample.

Let us now briefly discuss the selection of the importance density q. The simplest choice would be to put

$$q(x_k \mid x_{k-1}^i, z_k) = p(x_k \mid x_{k-1}^i),$$

that is, choosing the importance density to simply be the prior density. The update step for weights is then obviously reduced to multiplication by the likelihood of the measurements, which is not only simple conceptually, but also implementation-wise. The drawback of this choice is that if the prior is highly concentrated, most of the particles will obtain very small weights and the system becomes rapidly degenerate. It also does not take into account the latest measurement, potentially accentuating the degeneracy problem. The remedies for this degeneracy are either substantial increase in the number of particles used, or selection of a more appropriate importance density. Choice between these alternatives has to be ultimately made based on the particulars of the problem at hand, and the computational resources available. For a survey and comparison of different importance densities, see Simandl and Straka (2007).

Numerous variants of the basic particle filter concept discussed above have been presented in the literature. For an overview of some of the most promising of these, see Arulampalam *et al.* (2002). However, for the applications in WSN-based tracking systems we have found the presented solutions quite satisfactory. In the following sections we discuss in more detail two applications of particle filters, namely on improving accuracy of Cricket-type TDoA systems as well as performing data fusion in vehicular tracking. For examples of use of particle filters in solving tracking problems in other systems (such as Ad Hoc and cellular networks), see, for example, Mihaylova *et al.* (2007), Gustafsson *et al.* (2002), Thrun (2002) and Olama *et al.* (2006). Relevant techniques from robotics research are given in Howard (2006). For an earlier application of Bayesian filtering to Cricket-system using Extended Kalman Filters see Smith *et al.* (2004).

INDOOR SYSTEMS FOR LOCALIZATION

Many WSN-based localization systems have been developed for indoor applications as well as for outdoor scenarios. These use one of the basic principles for localization as explained in the previous sections and have their own application-specific design requirements in terms of functionality, scalability, performance and hardware. Representative examples for indoor localization systems include IR based systems like Active Badge (Want *et al.*, 1992), radio signal based systems like MoteTrack (Lorincz and Welsh, 2007), ultrasonic based systems like Active Bat (Ward, Jones and Hopper, 1997) and combination of radio and sonic waves like Cricket system (Priyantha, 2005). Active Badge uses pulse width modulated IR signals and can give an accuracy of 6m. Although, IR signals inherently suffer from ambient interferences, Versus Information Systems Ltd. developed commercial tags for locating doctors in hospitals and medical centers. Active Bat sender-tags transmit an ultrasonic pulse, which is received by a mesh of receivers mounted on the walls and ceilings in order to perform distance ranging. Trilateration is applied on the distance estimates to obtain position of the transmitting tag. MoteTrack uses radio signal signature to estimate the position of an object. It first requires establishing a calibrated signal signature model of the environment, which is obtained by measurements at a number of nodes from known geographical points. In the following we present Cricket System (Priyantha, 2005) in detail as a practical example of an indoor localization system. In the later sub-sections, we would discuss how improved accuracy can be obtained from Cricket System by applying Bayesian filtering framework and by complementing it with inertial sensors.

Cricket System

Cricket system is an indoor localization system, which uses inexpensive wireless sensor nodes (c.f. Figure 3) mounted at known positions. These nodes act as active beacons and transmit a pair of RF and ultrasonic signals. A Cricket node mounted on the target object, receives these signals from various beacons and applies TDoA based distance ranging. Since the atmospheric temperature has a significant influence to the speed of sound, the distance estimates obtained through the TDoA scheme must be calibrated to the ambient temperature conditions as described in the earlier section on Data Calibration. These distances are then used to find the spatial position of the object using lateration. Cricket system is able to determine the location of the target object within a few centimeters of accuracy.

Limitations of the Cricket System

Let us look in detail what are the different constraints posed by the Cricket system:

- Cricket nodes work correctly only if a LOS path between a listener and a beacon node exists. In many sparsely deployed cases, this results in localization blind spots.
- Cricket nodes have a limited range of approximately 11m. The range depends upon the sensitivity of the ultrasonic receiver module, the detection threshold and most importantly on the power level of the transmitted ultrasonic pulse.
- Cricket distance ranging error increases as the actual distance between the listener and the beacon node is increased. The error also grows-up if the angle between the faces of listener and the beacon nodes increases. This has to do with the radiation characteristics of the ultrasonic transmitter and receiver. Beyond certain angles at a particular distance, Cricket nodes cannot compute distance estimates. The coverage over higher angles is limited to smaller distances. Please refer to Figure 4 for distance ranging dependency of the Cricket system on the actual distance and the angle between a transmitting and receiving node.

Figure 3. Cricket node

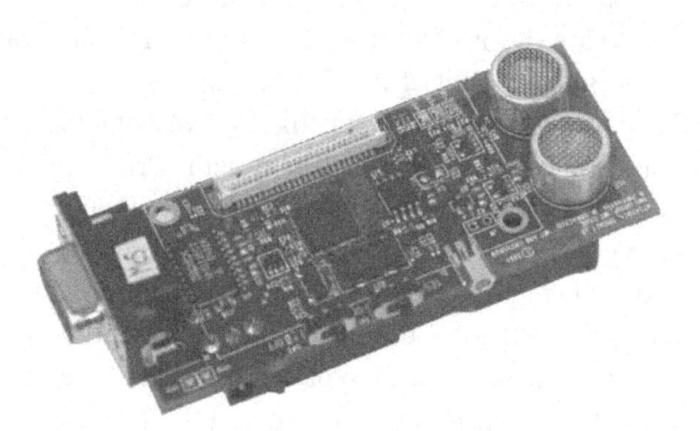

The localization performance of the Cricket system is dominated by the performance of the deployed ultrasonic sensors and their directivity and gain characteristics. When beaconing and listening nodes are faced directly to each other, the system can work properly up to 10 or 11 m. However, if the angle between beaconing and listening nodes increases, the measurements get worse and worse before it ceases to work at all. The larger the angle, the smaller is the operating range of the Cricket system. Near this limit of the maximum operating range, the measurements also get less stable and less accurate. However, for shorter distances, the system provides accurate and reliable distance measurements.

The system design includes various types of delays besides the signal propagation time, which limits the number of beacons transmitted per second and so the possible position updates per second. These delays are inserted in order to avoid radio and ultrasonic mutual interferences from different nodes. The constraints on the maximum number of beacon signals that can be transmitted per second are mostly imposed by the initial delay that has been put to let any stray ultrasonic pulse die down before the transmission of the beacon signal.

At a particular time instant, there needs to be at least four distance estimates to uniquely compute the position of the object in 3-dimentional space. From the ranged distance alone, the listener node is unable to determine the angle to the target Cricket node and therefore cannot calibrate it. A set of Cricket nodes may first apply lateration to determine the position of the target node as well as the angles from each listener node to the target node. Depending upon the angles, post-calibration can be applied on the ranged distances and localization can be performed again. The post-calibration process attempts to compensate the ranging anomalies due to the varying strengths of the ultrasonic transmission pattern at different real distances and angles. Despite the computational effort induced by the process, the post-calibration mechanism can in practice be iterated a few times to achieve higher accuracies. We also enhanced the Cricket system by filtering out non-coherent and spurious distance estimates known as outliers by using a moving median filter (c.f. Filtering of Data above). We found out that besides the low computational complexity, the choice of a moving median filter is very appropriate because of its high robustness, making the likelihood of sporadic noisy distance estimates very low. Furthermore, because of mobility the distance estimates change and can easily be adapted by a moving median filter. The pre-processing leads to better location estimates from the Cricket system (Ansari *et al., 2007b)* and as a consequence results in more accurate tracking outputs.

Figure 4. Plot of the distance ranging error for distances between 50 cm and 1050 cm and angles between 0 degree and 90 degrees

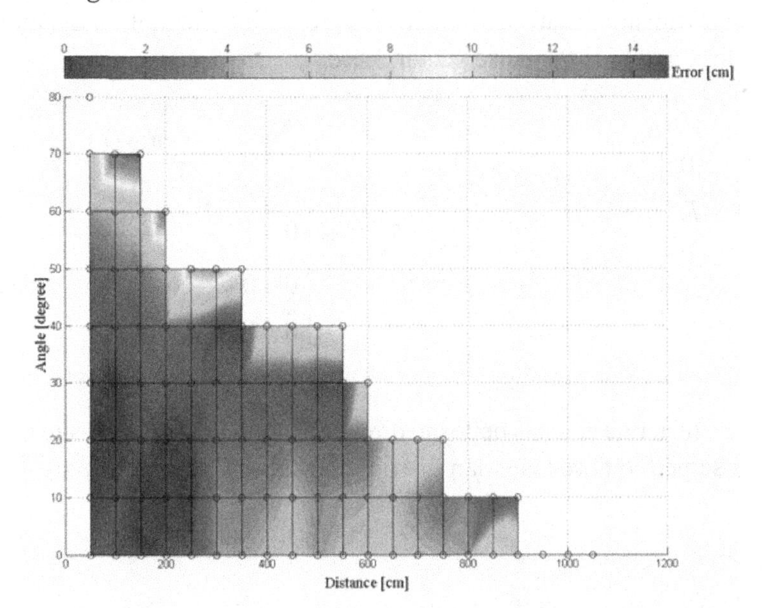

Applying Bayesian Framework on Cricket Localization

As described in the previous section, Bayesian techniques can be applied on localization system to achieve performance gains. In the following we will describe how particle filtering can be combined to the Cricket system for minimizing the positioning error (Ansari *et al., 2007b).*

In the particle filtering setup, the system maintains a set of parameters as a state vector. By modeling the system dynamics appropriately, a given state vector can be used to predict the state vector at the next measurement time step. The filtering framework carries out new measurements to correct the uncertainties/anomalies in the predicted state vector. Appropriate noise models are introduced to incorporate the uncertainties in the measurements and the system models. The system and measurement models may also consist of multiple models and depending upon the case, switching from one to other can be made.

In the following, we will give an example how particle filtering framework be applied on the Cricket system based localization of a toy train. We use this particular example due to its repeatability as well as non-trivial dynamics. However, the framework developed is general, and other dynamics and noise distributions for different motion dynamics can easily be included for applications in other scenarios. The toy train rail track is shown in Figure 5. The motion consists of two types of dynamics and is modeled by Constant Velocity (CV) and Constant Turn (CT) models along the straight and curved paths, respectively.

The state vector at any time instant T_k consists of the coordinate position (x_k, y_k) and the corresponding velocity components $(x^{\cdot}_k, y^{\cdot}_k)$ of the object. The state vector, $\mathbf{X}_k$, can be expressed as $\mathbf{X}_k = [x_k\ y_k\ x^{\cdot}_k\ y^{\cdot}_k]^T$. The system model can be written as

$$\mathbf{X}_{k+1} = \mathbf{F}_k \mathbf{X}_k + \mathbf{G}_k \mathbf{p}_{k,}$$

where $\mathbf{F}_k$ represents the transition state space matrix representing the motion dynamics of the train, $\mathbf{G}_k$ is the noise input matrix and $\mathbf{p}_k$ is the noise vector. Here,

$$F_k^{(CV)} = \begin{bmatrix} 1 & 0 & T_k & 0 \\ 0 & 1 & 0 & T_k \\ 0 & 0 & 1 & 0 \\ 0 & 0 & 0 & 1 \end{bmatrix} \qquad F_k^{(CT)} = \begin{bmatrix} 1 & 0 & \frac{\sin(\omega T_k)}{\omega} & \frac{\cos(\omega T_n) - 1}{\omega} \\ 0 & 1 & \frac{1-\cos(\omega T_k)}{\omega} & \frac{\sin(\omega T_n)}{\omega} \\ 0 & 0 & \cos(\omega T_k) & -\sin(\omega T_k) \\ 0 & 0 & \sin(\omega T_k) & \cos(\omega T_k) \end{bmatrix}$$

where ω represents the turn rate and has opposite direction along the two opposite curved sections. The noise process $\mathbf{p}_k$ is assumed to have Gaussian distribution. The $\mathbf{G}_k$ is given by

$$G_k = \begin{bmatrix} \frac{T_k^2}{2} & 0 \\ 0 & \frac{T_k^2}{2} \\ T_k & 0 \\ 0 & T_k \end{bmatrix}.$$

The measurement model is given by

$$\mathbf{Z}_\mathbf{k} = \mathbf{H}\mathbf{X}_k + \mathbf{n}_{k,}$$

where $\mathbf{Z}_k$ is the measurement vector, $\mathbf{H}$ is the measurement matrix and $\mathbf{n}_k$ represents the measurement noise process. The measurement system provides the coordinate position of the object and the angle information obtained from the digital compass, CMPS03. The digital compass is calibrated and gives

Figure 5. Trajectory of the toy train which exhibits two kinds of motion dynamics. It can be represented by Constant Velocity (CV) model along the straight sections while Constant Turn (CT) model along the curved sections of the path.

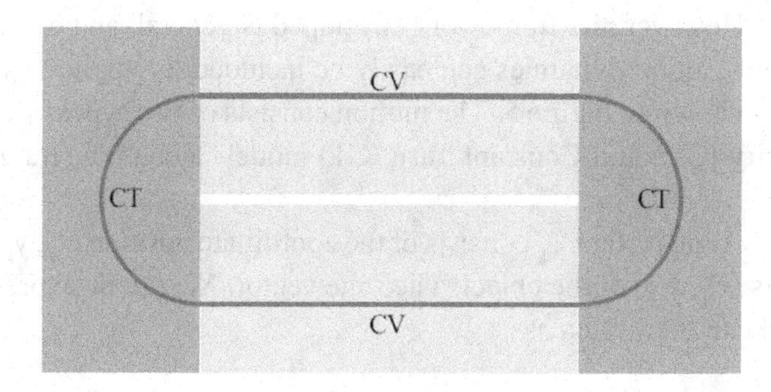

the absolute angle information. The measurement model is therefore expressed as $\mathbf{H}(\mathbf{X}_k) = [x_k\ y_k\ _\theta]^T$. The measurement noise characteristics $\mathbf{n}_k$ are found using approximation on a statistically big enough set of localization and angle measurements. In the above framework, there exist two motion models and it is necessary to switch from one model to the other appropriately. This is generally achieved by maintaining a mode transition or regime switching matrix. We refer the reader to Arulampalam *et al.* (2002) and Ristic *et al.* (2004) for a detailed discussion on multi-modal Bayesian filtering. This approach can be in general visualized as parallel banks of particle filters as shown in Figure 6.

In our simplified case with the availability of absolute angle information from the digital compass, we could easily decide to switch from one motion model to another by observing the change in the absolute angle. More complicated model-switching logic would, of course, be necessary if such absolute information were not available, or the readings obtained from the sensor had significant inaccuracies. Especially in the latter case the accuracy of the models employed becomes critical. Noisy measurements with incorrect noise model can cause significant problems or delays in identifying correct dynamics. Also, if the behavior of the system deviates significantly from the dynamics selected the performance of the framework will, of course, be poor.

The particle filtering framework now estimates the posterior probability $p(\mathbf{X}_k \mid \mathbf{Z}_k)$ by Monte Carlo integration. After applying the above described framework on the toy train, we obtained the tracking results as shown in Figure 7. The plus signs in the figure indicate the location measurements and the continuous line indicates the trajectory estimated by the tracking algorithm for repeated number of rounds. For 358 location estimates in three complete rounds, the average RMSE (Root Mean Squared Error) is found to be approximately 2.8 cm, which is certainly an improvement over the localization obtained from Cricket system alone (A. Smith *et. al.*, 2004). Bayesian filtering can be applied to other systems as well for indoor localization in sensor networks and has shown to result in improved accuracy.

Figure 6. Multiple motion models and switching logic

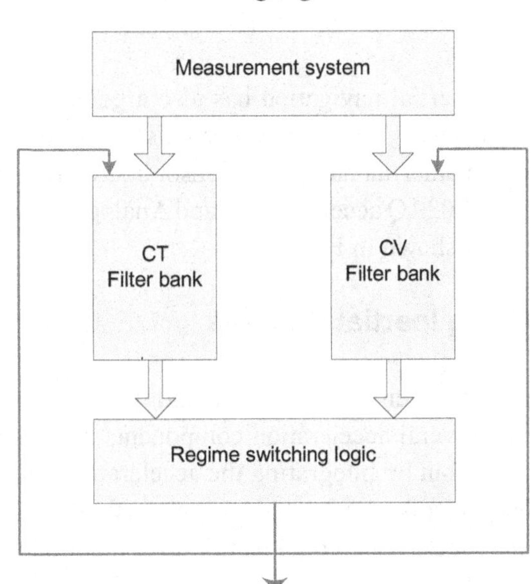

Figure 7. Tracking performance of the toy train after applying particle filtering based technique on the Cricket localization system (adapted from Ansari et al., 2007b; © 2007 IEEE).

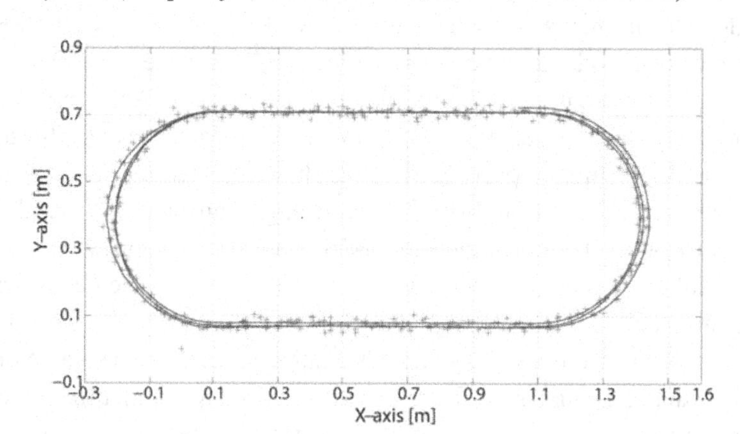

Inertial Navigation

Inertial navigation can be utilized for tracking the position and orientation of a moving object without requiring any external frame of reference. Inertial navigation has outdoor as well as indoor applications. It has been widely used in the field of aviation, military systems and robotics. An inertial navigation system essentially consists of gyroscopes and accelerometers. By simply applying principles of Newtonian Physics on the accelerometer and gyroscope readings, relative position and orientation information of the moving object can be achieved. Accelerometer readings are used to estimate the displacement along different axes and the gyroscope readings give the relative change in the orientation along different directions. By combining the information from the accelerometers and gyroscopes, smoothened trajectory of the moving object can be obtained. The relative positioning information can be mapped to an appropriate reference scale by applying a single rigid transformation. One of the biggest problems of the inertial navigation is the error accumulation. Therefore, the error needs to be compensated from time to time.

Besides other application areas, inertial navigation has also a good potential in sensor networks. This is mainly owing to the design of very low power inertial sensors that can easily be attached to sensor nodes. One such example is the inertial navigation sensor board prototype (Popa *et al.*, 2008). It consists of STMicroelectronics' LIS3L02AQ accelerometer and Analog Devices' ADIS16100 gyroscope connected to a TelosB sensor node as shown in Figure 8.

Indoor Human Tracking Using Inertial Sensors

In order to track a person using inertial sensors, the speed and direction of the motion are required. Human motion is characterized by several acceleration components along various directions and the displacement cannot simply be found out by integrating the acceleration components twice. However, there is a unique vertical acceleration spike associated with each single foot-step as a person walks. For slower and smoother steps, the acceleration components along other axes are less visible. Another peculiar characteristic of human motion is the foot-step size, which remains constant for a particular

Figure 8. Inertial Navigation sensor node platform (from Popa et al., 2008; © 2008 IEEE)

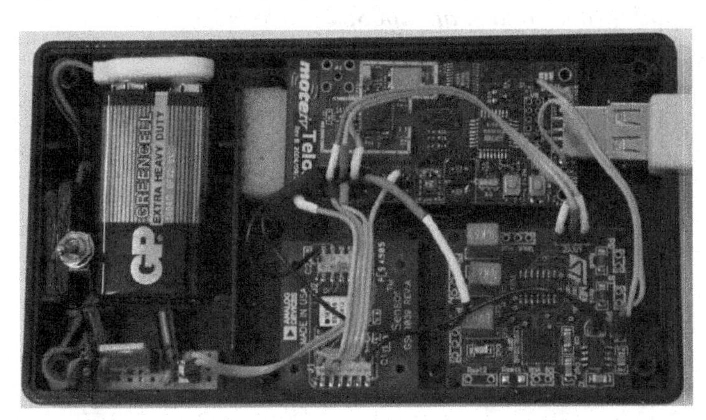

individual in his/her normal motion. Based on the inter-peak distance of the acceleration signature and the step-size for a particular person, his/her speed and hence displacement can be computed.

One of the advantages of inertial navigation is its independence from external factors. However, inertial navigation systems tend to accumulate errors over a period of time and therefore, periodic corrections are necessary. The position corrections can be applied using another system.

Cricket system has coverage problems owing to the line-of-sight requirements but it has relatively high accuracy. One logical approach is to combine Cricket system with inertial navigation. The combined solution can solve the coverage problem by relying on the inertial navigation and the error accumulation can be eliminated by periodically correcting the position using the Cricket system measurements. Figure 9 shows the localization output for Cricket system and a hybrid system, which combines inertial navigation with Cricket system. It can be observed that Cricket system alone has a high accuracy but it suffers from coverage problem owing to the strict line-of-sight requirements. Inertial navigation accumulates errors over time and the output for an arbitrary human motion is not very accurate. The combination of the Cricket system and inertial navigation has better accuracy than achieved by inertial navigation alone and the problem of blind spots is also eliminated. The improved coverage of the combined system is illustrated by the larger number of combined localization estimates compared to the ones obtained from the Cricket system alone.

LOCALIZING AND TRACKING VEHICLES

We shall now move from indoor to outdoor localization systems, focusing especially on vehicular localization and tracking. In particular we report on our work on the design and implementation of a complete, large-scale modular sensor network hardware/software platform for target tracking applications (Ansari *et al.*, 2007a). In this work, the objective was to design a flexible software platform that could be adapted into a variety of scenarios, and to prototype it on a likewise adaptable hardware platform targeted for passive tracking scenarios. In the resulting design the software platform is carefully separated from the hardware by various abstraction layers, and due to the modular design various filtering, data processing and communication solutions can be used in the platform according to the needs of the particular application. The overall hardware design is flexible as well, in the sense of not being confined to tracking applications with a particular target object in mind. Combination of magnetometers and

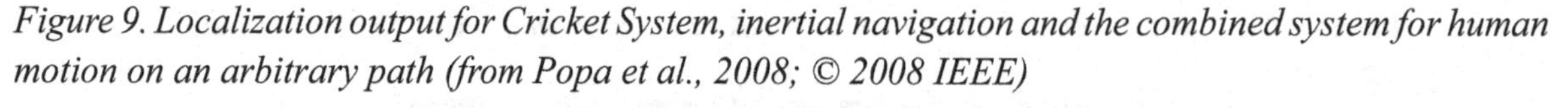

Figure 9. Localization output for Cricket System, inertial navigation and the combined system for human motion on an arbitrary path (from Popa et al., 2008; © 2008 IEEE)

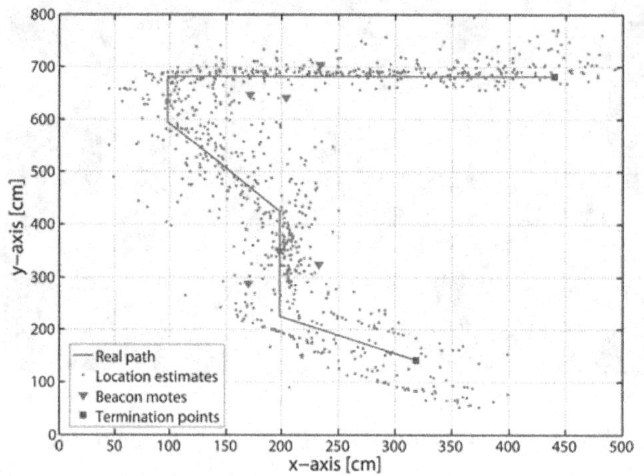

Figure 10. Sensor node platform for vehicular tracking consisting of the passive infrared sensors and magnetometer attached to TelosB (from Ansari et al., 2007a; © 2007 IEEE)

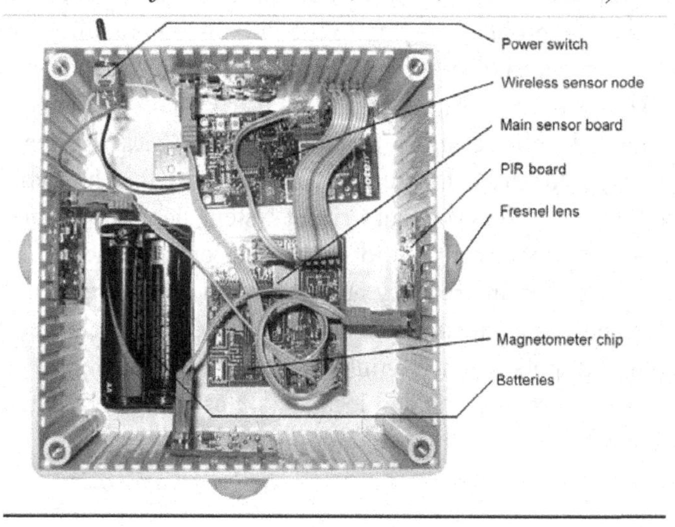

passive infrared sensors makes the platform very versatile, and we enhanced the platform further by developing automatic calibration features to remove the influence of environmental changes.

The sensor nodes used in the tracking system were built using the TelosB platform from Moteiv Inc. extended with a substantial amount of customized hardware. In particular, the nodes feature a sensor board consisting of two types of sensors namely passive infrared (PIR) sensors and anisotropic magneto-resistive (AMR) sensors, as shown in Figure 10. PIR sensors detect the differential thermal energy signal rather than absolute values and are therefore highly suitable for tracking applications. AMR sensors generate an output voltage proportional to the magnetic field strength. A moving ferromagnetic object disturbs Earth's magnetic field and causes the AMR sensors to generate an output signal, which is used for detection purposes as explained below. Low power consumption is one of the key objectives in the

Figure 11. Circuit block diagram of the sensor node platform used for vehicular tracking (from Ansari et al., 2007a; © 2007 IEEE)

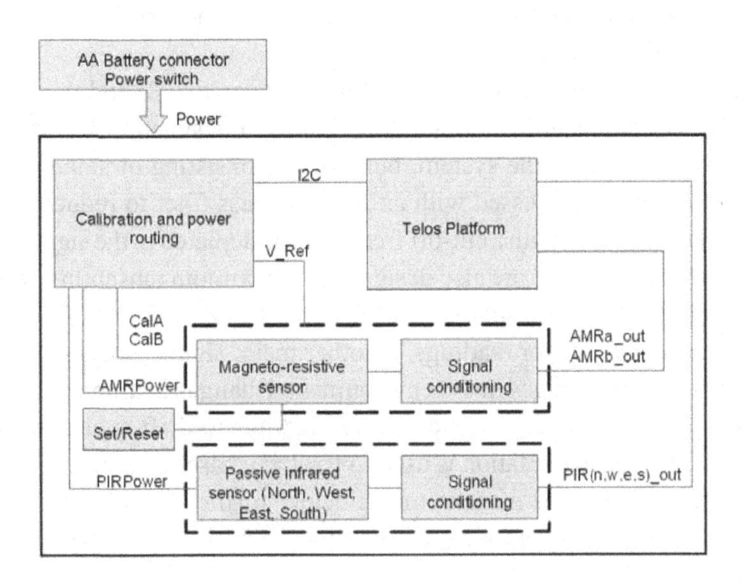

design. The energy efficient design includes individual power control of the PIR and AMR sensors. The data pre-processing includes multiple phases: The output signal of the sensors is amplified using two-stage instrumentation amplifiers before feeding it to the *ADC* of the TelosB platform. The circuit also includes high frequency noise suppression filters as part of the signal pre-processing implemented in the circuitry. In order to demagnetize the AMR sensors, we included a provision for external set/reset circuitry, which we always used before any large scale deployment setups in order to compensate the long term drift and loss of sensitivity of the AMR sensors. The loss of sensitivity and drifts are developed owing to the alignment of ferro-sensitive cells inside the sensor chip when exposed to high magnetic field over extended periods of time as described in the above discussion on Compensation of Unwanted constants and Long-term Drift.

The packaged sensor node is shown in Figure 10 together with the block diagram of its architecture (c.f. Figure 11). A set of four PIR sensors (called as North-West-South-East) are mounted orthogonally for a complete 360-degree field of view. Fresnel lenses are used to increase the sensing range but at the same time not losing the beam-width below 90 degrees per PIR sensor. A combination of two axis magnetometer (AMRa and AMRb) enables the sensor node to detect moving ferromagnetic objects in a field. The amplitude level from the AMR sensors is highly dependent on its orientation with respect to Earth's magnetic field, the ferromagnetic material content in the surroundings and obviously on the strength of Earth's magnetic lines of force. For the optimum swing of the signal, the amplified output of the magnetometer to be fed to the *ADC* should be at the midscale (around 1.5V). This is done by tuning the resistance of a digital potentiometer, and hence adjusting the voltage levels at one of the inputs of the second stage amplifier. This calibration phase (c.f. section on Data Calibration) is a recursive process. Firstly, a set of 10 AMR samples are taken and the mode value is calculated. The underlying reason is that the probability of an outlier is very low. The mode value is converted into voltage and is checked whether it lies within a small window around the mid-scale voltage. Otherwise, the value of the potentiometer is increased or decreased accordingly by sending appropriate commands to the digital potentiometer over the I^2C bus. The process is repeated till the voltage assumes a mid-scale value. This process not only prevents clipping of the signal but also enables the sensor node to calibrate automatically to any environmental condition.

The software developed for the platform also follows a layered and modular design to ensure flexibility. Transfer of the measurement data to gateway nodes, time synchronization, calibration etc. are handled in middleware tailored for the system, but mainly consisting of standard protocol solutions. Sensor readings themselves are processed with an FIR low-pass filter to reduce noise prior to further processing. The FIR filter is chosen with a cut-off frequency as depicted in the signal of interest. The filter components implemented in software are also designed for maximum reusability, with filter parameters such as type, order, cut-off frequency and filter coefficients all being adjustable according to the characteristics of the incoming raw sensor readings. Another major signal processing aspect implemented in the software is adaptation to changes in the environment. Changes in ambient conditions and Earth's magnetic field cause, for example, the magnetometer readings to drift over time. A computationally less intensive EWMA filter implementation is used to track the baseline of the magnetometer readings, which is subtracted from the filtered measurements before further processing is applied. This way a calibrated, zero-offset and non-clipped signal is obtained, which can be processed for the potential object detection purposes.

After applying the data processing techniques, the readings from AMR and PIR sensors are given as inputs to the vehicle detection algorithm running on each node. The algorithm relies on impulse integration with adjustable thresholds as illustrated in Figure 12 for the AMR sensor. The filled area between this curve and the thresholds (dotted lines) represents the integrated impulse value. When the AMR readings return to the area between the thresholds, a dwell timer is started (horizontal line). If the readings cross the threshold before this timer expires, the timer is reset and the integration of impulse continued (crosses terminating the thick timer line). When the dwell timer expires (upward arrow), the impulse value is compared against a pre-set value. Too small impulses are considered as noise, and detection is not signaled (cross over the upward arrow).

Figure 12. Illustration of the vehicle detection algorithm using a magnetometer (from Ansari et al., 2007a; © 2007 IEEE)

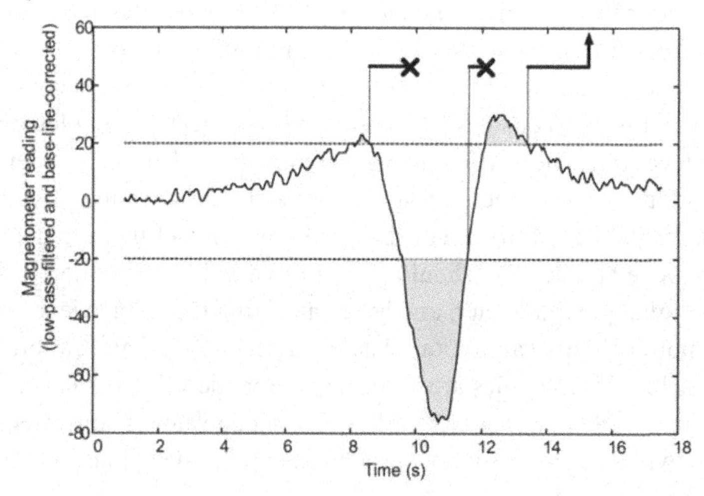

Figure 13. Vehicle tracking output (from Ansari et al., 2007a; © 2007 IEEE)

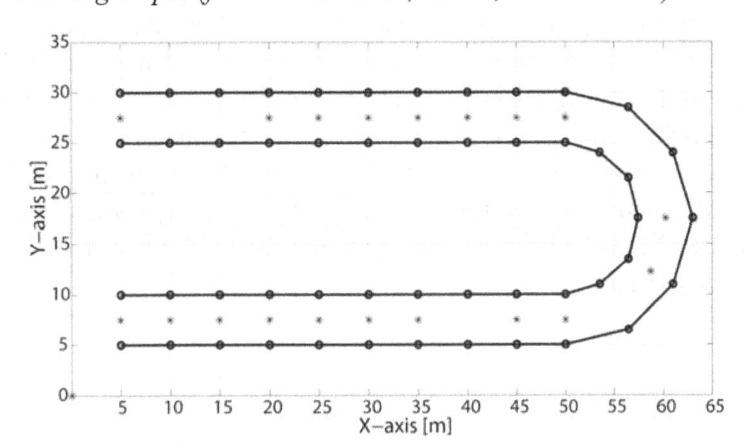

All the detection events from the individual sensor nodes are gathered on a powerful node responsible for further data processing. Rejection of false positives and data fusion in general is performed at this stage. Even though extensive filtering is applied in various parts of the system occasional noise spikes unavoidably still cause false detections on individual nodes, as will sudden changes in environmental conditions. Also detections from PIR sensors are not necessarily all related to vehicles, but might also arise from animals or humans passing by. Neighborhood-based fusion of the readings from both the AMR and PIR sensors from the whole sensor field results, however, in accurate detection. Figure 13 illustrates the performance of the system on a small test track where a single vehicle was moving at a speed of approximately 50 km/h. The dots indicating localized detections were confirmed to be highly accurate. The figure also shows that the system did not, however, reach 100% accuracy as some of the events were missed. The underlying cause of this appears to be the residual unreliability in the communications stack rather than the algorithms described above. With a more aggressive data protection in the network (such as multipath routing) we expect our design to yield highly accurate vehicle detection and tracking performance even in very dynamic environmental conditions.

We have also worked on distributed localization and tracking systems for vehicles utilizing particle filters for data fusion. Our focus has been especially on fusing tracking information obtained from radars with local positioning estimates of the vehicles obtained via vehicle-to-vehicle communications. Particle filters are an especially appropriate solution for such a scenario due to the non-linearities in the system model, highly non-Gaussian measurement noises of radars, as well as due to the need to fuse sensor readings coming from diverse sources. For further details we refer the reader to Riihijärvi *et al.*, 2005.

CONCLUSION

In this chapter we discussed some of the algorithms and solutions that can be applied for data processing and filtering in wireless sensor networks deployed for localization and tracking applications. Following a short overview of the different basic localization and ranging solutions commonly applied in the field, we gave a short account of the basic data filtering algorithms, such as median and lowpass filters, as well as different flavors of averaging. We then focused on Bayesian filtering as a method of choice for solving the tracking problem based on individual measurements. We advocate the use of particle filters due to their ability to handle non-linearities and non-Gaussian measurement noises, both of which appear to be common features in tracking scenarios. The increase in computational complexity induced by particle filters is not very high compared to, for example, usual Kalman filters, and the improvement in accuracy and robustness is significant. They are also not methodologically more difficult to implement or understand than other major Bayesian filter types. Finally, we have discussed the design of indoor and outdoor localization and tracking systems we have implemented based on the data processing and filtering solutions discussed above.

REFERENCES

Ansari, J. *et al.* (2007a). Flexible Hardware/Software Platform for Tracking Applications. In *Proceeding of VTC 2007*, Dublin, Ireland.

Ansari, J., Riihijärvi, J., & Mähönen, P. (2007b). Combining Particle Filtering with Cricket System for Indoor Localization and Tracking Services. In *Proceeding of PIMRC 2007*, Athens, Greece.

Arora, A., Dutta, P., Bapat, S., Kulathumani, V., Zhang, H., Naik, V., Mittal, V., Cao, H., Demirbas, M., Mohamed, G., Choi, Y-R., Herman, T., Kulkarni, S., Arumugam, U., Nesterenko, M., Vora, A., & Miyashita, M. (2006). A line in the sand: A wireless sensor network for target detection, classification and tracking. *Computer Networks*, *46*(5), 605–634.

Arulampalam, S., Maskell, S., Gordon, N., & T. Clapp, (2002). A Tutorial on Particle Filters for On-line Non-linear/Non-Gaussian Bayesian Tracking. *IEEE Transactions on Signal Processing*, 50(2), 174-188.

Bahl, P., & Padmanabhan, V. (2000). RADAR: An In-Building RF Based User Location and Tracking System. In *Proceedings of the IEEE INFOCOM 2000*,Tel Aviv, Israel.

Drumheller, M. (1987). Mobile robot localization using sonar. *IEEE Transactions on Pattern Analysis and Machine Intelligence*, *9*(2), 325–332.

Federal Communications Commision's Enhanced 911 Wireless Services. http://www.fcc.gov/pshs/services/911-services/. Retrieved: 08.10.2008.

Fox, D., Hightower, J., Liao, L., Schulz, D., & Borriello, G. (2003). Bayesian filters for location estimation. *IEEE Pervasive Computing*, *2*(3), 24–33.

Gustafsson, F., Gunnarsson, F., Bergman, N., Forssell, U., Jansson, J., Karlsson, R., & Nordlund, P.-J. (2002). Particle Filters for Positioning, Navigation, and Tracking. *IEEE Transactions on Signal Processing*, *50*(2), 425-437.

Honeywell (2008). *HMC1022 data sheet.* www.ssec.honeywell.com/magnetic/datasheets/hmc1001-2_1021-2.pdf,", Retrieved: 08.10.2008.

Howard, A. (2006). Multi-robot Simultaneous Localization and Mapping using Particle Filters. *The International Journal of Robotics Research*, *25*(12), 1243-1256.

Kuckertz, P., Ansari, J., Riihijärvi, J., & Mähönen, P. (2007). Sniper Fire Localization using Wireless Sensor Networks and Genetic Algorithm based Data Fusion. In *Proceedings of IEEE MILCOM 2007*, Orlando, Florida, USA.

Lorincz, K., & Welsh, M. (2007). Motetrack: A robust, decentralized approach to RF-based location tracking. In *Personal and Ubiquitous Computing, Special Issue on Location and Context-Awareness, 11*(6), 489-503.

Mihaylova, L., Angelova, D., Honary, S., Bull, D., Canagarajah, C., & Ristic, B. (2007). Mobility Tracking in Cellular Networks Using Particle Filtering. *IEEE Transactions on Wireless Communications, 6*(10), 3589-3599.

Olama, M., Djouadi, S., & Pendley, C. (2006). Position and Velocity Tracking in Mobile Cellular Networks Using the Particle Filter. *In Proceedings of IEEE WCNC 2006*, Las Vegas, USA.

Polastre, J., Szewczyk, J., & Culler, D. (2005). Telos: enabling ultra-low power wireless research. *In Proceedings of Information Processing In Sensor Networks 2005*, Los Angeles, USA.

Popa, M., Ansari, J., Riihijärvi, J., & Mähönen, P. (2008). Combining Cricket System and Inertial Navigation for Indoor Human Tracking. *In Proceeding of WCNC 2008*, Las Vegas, USA.

Priyantha, N. (2005). *The Cricket Indoor Location System*. doctoral dissertation. Massachusetts Institute of Technology, Boston, USA.

Riihijärvi, J., Vázquez Martí, C., & Mähönen, P. (2005). Enhancing Radar Tracking with Car-to-Car Communication. In *Proceedings of IEEE SCVT'05*, Enschede, the Netherlands.

Ristic, B., Arulampalam, S., & Gordon, N. (2004). *Beyond the Kalman Filter: Particle Filters for Tracking Applications.* Artech House, Norwood, Massachusetts, USA.

Simandl, M., & Straka, O. (2007). Sampling Densities of Particle Filter: A Survey and Comparison. *In Proceedings of the 2007 American Control Conference*, New York, USA.

Simon, G., Maroti, M., Ledeczi, A., Balogh, G., Kusy, B., Nadas, A., Pap, G., Sallai, J., & Frampton, K. (2004). Sensor Network-Based Countersniper System. In *Proceedings of the 2nd international conference on Embedded Networked Sensor Systems*, Baltimore, USA.

Smith, A., Balakrishnan, H., Goraczko, M., & Priyantha, N. (2004). Tracking Moving Devices with the Cricket Location System. *In Proceedings of the ACM MobiSys 2004*, Boston, USA.

Thrun, S. (2002). Particle Filters in Robotics. In *Proceeding of Uncertainty in AI.*

Van Trees, H., & Bell, K. L. (Eds.) (2007, August). *Bayesian Bounds for Parameter Estimation and Nonlinear Filtering/Tracking.* Wiley-IEEE Press.

Want, R., Hopper, A., Falcao, V., & Gibbons, J. (1992). The active badge location system. *ACM Transactions on Information Systems, 10*(1), 91-102.

Ward, A., Jones, A., & Hopper, A. (1997). A new location technique for the active office. *IEEE Personnel Communications, 4*(5), 42–47.

Whitehouse, K., & Culler, D. (2003). Macro-calibration in Sensor/Actuator Networks. *Mobile Networks and Applications Journal (MONET), Special Issue on Wireless Sensor Networks, 8*(4), 463-472.

Chapter XVII
A Wireless Mesh Network Platform for Vehicle Positioning and Location Tracking

Mohamed EL-Darieby
University of Regina, Canada

Hazem Ahmed
University of Regina, Canada

Mahmoud Halfawy
National Research Council NRC-CSIR, Canada

Ahmed Amer
Zagazig University, Egypt

Baher Abdulhai
Toronto Intelligent Transportation Systems Centre, Dept. of Civil Engineering, Canada

ABSTRACT

Large urban areas in North America as well as many other parts of the world are experiencing unprecedented and soaring congestion problems. It is imperative that modern societies upgrade their transportation systems in order to remain competitive, and maintain the high quality of life and social wellbeing. Current practices in Intelligent Transportation Systems (ITS) data gathering are dominated by the use of point detectors for surveillance, and wire-line communication networks for data transmission. Reliance on point detectors is losing appeal due to detector reliability issues, the cost of building and maintaining detector networks, and potential traffic disruption during construction and maintenance of these networks. This chapter describes a novel wireless mesh network platform for traffic monitoring. The platform uses traveling cars as data collection probes and uses wireless municipal mesh networks to transport sensed data. The platform assumes that cars or drivers' mobile devices are equipped with the widely adopted low-cost Bluetooth wireless technology. Field trials of the proposed platform demonstrated its capability to track cars traveling at speeds of 0 to 70 km/hour. The platform was able to track cars as they travel and turn on a typical road network. In addition, the platform was used to approximate car speeds through determining the change in position in a time period. The preliminary results indicated an accuracy of ± 10%- 15%. The chapter describes the architecture, implementation, and field-testing of the proposed platform. It also discusses aspects of large-scale deployment of the proposed platform to cover large geographic areas.

INTRODUCTION

Large urban areas in North America as well as many other parts of the world are experiencing unprecedented and soaring congestion problems. It is imperative that modern societies upgrade their transportation systems in order to remain competitive and to maintain the high quality of life and social well being that we rightly prize so highly. Transportation management agencies are under increasing pressure to adopt more innovative approaches to enhance the efficiency of existing transportation networks. Solutions in the form of building more roads are neither desirable nor feasible in many cases. A more feasible approach would be to maximize the use of the capacity already afforded by existing networks before expansions can be justified.

Over the past two decades, numerous technologies and methods have been developed and deployed to support real-time monitoring of transportation systems (Zheng, Winstanley, Yan, & Fotheringham, 2008). However, the installation and maintenance costs as well as the inherent limitations (e.g., power consumption, telemetry) of existing technologies constitute a major impediment towards implementing continuous real-time monitoring in a cost-effective manner. The efficiency and economic viability of current monitoring practices not only have limited the deployment of such technologies, but many transportation agencies still do not have an effective or systematic strategy for traffic monitoring.

The "heart" of traffic monitoring lies in gathering and using real-time system information to enable proactive management and control of the network. Current practices in monitoring traffic systems are dominated by the use of point detectors for surveillance, and wire-line communication networks for data transmission. In most large metropolitan areas, major freeways and arteries are covered by pavement-embedded induction loop detector stations to measure traffic volumes and speeds. Gathered information are typically aggregated over 20-30 seconds then transmitted over copper or fiber optic wire lines to the nearest operations centre. This approach is losing appeal due to detector reliability issues, the cost of building and maintaining detector networks, and potential traffic disruption during construction and maintenance operations. Modern off-road detector technologies have improved significantly over the past decade, resulting in new and more mature detector types based on radar, ultrasound, infrared, and acoustic technologies. With the inherent limitations of existing technologies, a new technology that allows for cost-effective real-time and continuous monitoring of traffic systems is urgently needed.

In this chapter, we propose a novel wireless and cost-effective platform for ITS monitoring. The novelty of the platform lies in using traveling cars equipped with Bluetooth devices as probes for collecting raw traffic data. The platform employs municipal wireless mesh network (WMN) infrastructure to gather and transport real-time traffic data to a centralized ITS server. The platform does not require installation of infrastructures which results in further cost-effectiveness by exploiting any existing WMN infrastructure and the wide spread use of Bluetooth devices. It is estimated that 80% of the cars by 2009 will be Bluetooth enabled (Bluetooth SIG, 2006). In case a car is not Bluetooth-enabled, mobile Bluetooth devices in the car such as a driver's cell phone can be used. The cost-effectiveness of the platform is achieved by using unlicensed wireless technologies in addition to common hardware and open source software for building the platform.

The use of location-based sensing technologies and wireless communication devices has been steadily gaining grounds in the industry because of their obvious advantages relative to point detector surveillance technologies. An emerging category of solutions that promises cheaper and broader network coverage involves the use of traveling vehicles as probes transmitting information about the surrounding traffic environment as they progress through the road network. As vehicles travel through major and minor roads, they can serve to collect and transmit valuable traffic information.

Wireless technologies used to collect traffic data can be classified into satellite-based (e.g., Global Positioning System (GPS)) and terrestrial-based technologies (e.g., cellular networks, IEEE 802.11). Selecting a feasible technology for a particular application would involve evaluating trade-offs between the cost of building data collection system, accuracy of collected data, bandwidth available for transmission, system capacity, and ubiquity of the technology. An overview of the most commonly used technologies is provided in the next section.

This chapter describes the architecture, implementation, and field-testing of a novel wireless platform for traffic network monitoring. Performance results of two small-scale field deployments are presented. In addition, considerations for large-scale deployment of the platform and the main technological, economical, and operational factors that affect such deployments are also discussed.

RELATED WORK

This section summarizes the relevant literature on using wireless technologies for vehicle tracking, and provides a brief overview of the WMN technology as it pertains to the proposed platform.

Wireless Vehicle Tracking

In most large metropolitan areas, in order to measure traffic volumes and speeds, major freeways and arterials are covered by induction loop sensors that are embedded into pavements. Gathered information are typically aggregated over 20-30 seconds and transmitted over copper or fiber optic wire lines to the nearest operations centre. At a typically centralized operations centre, ITS software processes the gathered information and produces recommendation on how to, for example, divert traffic to avoid congestions. These systems face many problems. Using induction loop sensors is losing appeal due to sensor reliability issues, cost of building and maintaining detector networks, and potential traffic disruption during construction and maintenance operations. Modern off-road detector technologies have improved significantly over the past decade, resulting in new and more mature detector types based on radar, ultrasound, infrared, and acoustic technologies.

In general, the wireless technologies used to collect traffic data can be classified as satellite-base such as GPS or terrestrial-based such as cellular networks and IEEE 802.11 (i.e., Wi-Fi) technologies. There is typically a trade-off between cost of building a data collection system, accuracy of collected data, bandwidth available for transmission, system capacity and ubiquity of the technology among all these wireless technologies. The trade-offs indicate that no one technology is suitable for all applications.

Recently, there has been an increasing trend towards the use of terrestrial wireless systems for vehicle tracking applications. Satellite-based systems, while providing high positional accuracy, require relatively expensive equipment to locate and communicate vehicle positions (e.g. using GPS and cellular communication). Terrestrial wireless technologies have the advantage of providing more bandwidth and two-way communication, which potentially enables richer applications and information exchange.

GPS position information is very accurate with an error in the range of meters. Changes of the position within an interval of time give velocity information. However, GPS communication requires line-of-sight and consequently it cannot be used inside tunnels and urban areas with tall buildings (the urban canyon effect).

In order to provide tracking information, GPS is typically integrated with wireless communication systems. GPS can be coupled with Short Message Service (SMS) wireless technology to provide vehicle monitoring information to a monitoring server (Al-Rousan, Al-Ali, & Darwish, 2004; Young & Skobla, 2003). The system periodically sends location information each 5 or 10 seconds. However, the time taken to send an SMS message is dependent on the status of the cellular network (e.g. congestions). The expected massive amount of exchanged data makes the use of SMS-based systems both expensive and unreliable in most cases.

GPS may also be integrated with General Packet Radio Service (GPS) or Global System for Mobile communications (GSM) location services to support vehicle monitoring systems. In Zhang et al. (2005), the authors provide a comparison of using the Transmission Control Protocol and User Datagram Protocol to send the position information of the vehicle's on-board GPS module to the monitoring server via vehicle on-board General Packet Radio Service (GPRS) module. The GPRS-based systems can provide accurate position of the vehicle and real time monitoring, and are generally cheaper than the SMS-based systems. The disadvantages of this system include the requirement for installation of GPS modules in the vehicles, and the high operating costs for GPRS subscription and data transmission.

GSM-based location services were introduced in 1995 (Spirito, 2001; Broida, 2003). In general, two standard positioning methods can be used: (1) time of arrival and (2) enhanced observed time difference (E-OTD). The time of arrival method calculates the propagation period of a known signal sent by the mobile station (MS). This requires installation of location measurement units at each base transceiver station (BTS). However, this method does not require modifications to cellular handsets. The readings of three base transceiver stations are used to determine the location by triangulation algorithms. This method is known to be time-sensitive because one microsecond of timing error may result in approximately 300 meters of location error. To reduce such errors, the time difference of arrival (TDOA) method was proposed to enhance the accuracy of the TOA method.

The E-OTD method has three measurement parameters: observed time difference (OTD), real time difference (RTD), and geographical time difference (GTD). The OTD relies on the measurement of the TDOA between two BTS, the RTD is the synchronization error between two BTS (i.e. synchronization difference between two stations), and the GTD is the difference between the OTD and the RTD. This method was found to achieve errors in the range of 100 to 300 meters but it requires modifications to the MS to enable the OTD measurement.

Amongst the rapidly emerging communication technologies is Dedicated Short Range Communication (DSRC) described in IEEE P1609.3/D18 (2005). DSRC systems are being designed to provide short-range, wireless links to transfer information between vehicles and roadside units, other vehicles, or portable roadside units. DSRC is anticipated to be essential to many ITS applications that improve traveler safety, and decrease traffic congestion. Examples of such information transfer include: traffic light control, traffic monitoring, traveler alerts, automatic toll collection, traffic congestion detection, emergency vehicle traffic signal pre-emption and electronic inspection of moving trucks through data transmissions with roadside inspection facilities.

In addition, Bluetooth has been used for indoor object tracking. Two methods that use the Received Signal Strength Indicator (RSSI) to track objects are described in Huang et al. (2006). Bluetooth is a wireless cable replacement technology. It operates in the unlicensed Industrial Scientific Medical (ISM) Frequency range of 2.4 GHZ. Bluetooth is designed to be a low power and low cost wireless technology. Bluetooth is found in many electronic devices such as cellular phones, laptops, headphone, keyboard, and printers. Bluetooth devices can be implemented internally in these devices or can be added as a separate USB dongle. There are three classes of Bluetooth devices classified based on power level which is directly associated with the device communication range. Class 1 has a range of 100 meters where class 3 only has 10 meters of range.

Several research prototypes employed alternative wireless technologies. The Place Lab project, created by Hightower et al. (2006), had shown that using Wi-Fi hotspots and GSM-based cellular phones would efficiently provide vehicle location tracking in downtown of cities. Position error of the system was in the range of 20-30 meters even for different weather conditions. The main disadvantage cited was the dependence on Wi-Fi technology, which is not commonly used in portable devices, mainly due to the high power requirements of Wi-Fi devices and their relatively high cost compared to Bluetooth.

Hull et al. (2006), the authors of the MIT CarTel project, developed a computing system for collecting and processing information from mobile sensors mounted on automobiles. An embedded system on the automotives interfaces with different sensors in the car and transmits the sensor information to a server for processing. CarTel focused on handling intermittent network connectivity inherent in WMNs. For that purpose, a special network stack was developed. However, since CarTel relies on GPS to collect location information, the platform may not perform reliably within "urban canyons."

Wireless Mesh Networks

Wireless Mesh Networks (WMN), also known as municipal wireless networks (Lee, Jianliang, Young-Bae, & Shrestha, 2006; Akyildiza, Wang, & Wang, 2005; Farkas & Plattner, 2005), are posed to be a key infrastructure for enabling new applications in public safety, business, and entertainment. They are typically deployed in a quasi-stationary manner, where some mesh routers are stationary. This wireless infrastructure enables routing of information in a multi-hop manner. WMN nodes comprise mesh clients and mesh routers. Mesh clients, also known as On-Board Units (OBU), can be desktops, laptops, cellular phones. Mesh routers, also known as WMN Access Points (AP), can self-configure themselves to automatically build the wireless infrastructure that establishes and maintains mesh connectivity. Mesh routers typically provide access to a fixed structure network or to the Internet. Each mesh router has a domain of wireless coverage. An OBU is associated with one AP at a time, as long as it is in the router coverage domain. OBUs can move freely and associate themselves with different mesh routers, and may use Bluetooth or Wi-Fi for communication with the WMN APs. Although WMNs can be used for a large number of applications (Spirito, 2001; Khemapech, Duncan, & Miller, 2005), a number of general requirements and characteristics are shared among most of these applications. These characteristics include geographic coverage, cost-effectiveness, scalability, fault resilience, and privacy and security (Ilyas & Mahgoub, 2005; Karl & Willig, 2005; Akyildiz, Su, Sankarasubramaniam, & Cayirci, 2002).

The effectiveness of real-world WMNs for monitoring applications is determined largely by its ability to reliably cover larger areas for longer durations (Hać, 2003). WMNs may be formed by deploying hundreds or thousands of APs across large geographical areas (Ahmed, Shi, & Shang, 2003). APs are generally designed to have sufficient intelligence to gather and disseminate data, and to exchange information and cooperate in processing gathered data, thus enabling the monitoring of a large geographical area through distributed data processing and communication. WMN scalability involves ability to increase the size of coverage area in a manner that does not adversely affect WMN performance (Stoianov, Nachman, & Madden, 2007; Khemapech et al., 2005). Robustness involves ability of sensors and WMNs to tolerate faults or errors in operations (Cardell-Oliver, Smettem, Kranz, & Mayer, 2004). This requires WMN APs to have autonomous capabilities such as self-testing, self-configuring and self-healing (Yu, Prasanna, & Krishnamachari, 2006). The design of WMN must also incorporate security mechanisms in order to prevent unauthorized access and attacks (Wu & Tseng, 2007).

The autonomy of APs simplifies WMN installation and maintenance. Routers exchange data packets to update their routing tables. WMNs typically implement two types of routing protocols: proactive and on-demand routing protocols (Huhtonen, 2004; IETF RFC3626, 2003; IETF RFC3561, 2003). Proactive routing protocols are table-driven. Optimized link state routing (OLSR) is a proactive and table driven routing protocol where APs exchange OLSR "hello" messages periodically to build and maintain the routing table. This dynamic method of building the table enables self-configuring of the AP's. "Hello" messages advertise the one-hop interfaces of each AP. This enables each AP to find information about its neighboring nodes and hence allows the AP to build and maintain its routing table. Since OLSR is a multi-hop protocol, each node forward a message to its immediate neighbor based on the contents of this routing table. The periodic exchange of hello messages also enables the WMN to recover from a failed link or node (IETF RFC3626, 2003). Whenever a change happens in the topology of the WMN, control messages are propagated through the network to announce this change, and update the routing tables maintained in various routers. The flooding of these messages, may potentially span a very large part of the entire network, may be disadvantageous.

On-demand routing protocols, or reactive protocols, do not require this flooding of update messages. The ad-hoc on-demand distance vector (AODV) routing protocol is an example of the reactive routing protocols. These protocols do not maintain routing tables for the entire network, but only requested routes are maintained. That is, routes are calculated only when there is a request to send data from a source node to a destination node. AODV maintains vectors of destinations' routes and costs to use. This renders AODV as a better alternative for more static networks. In addition, AODV requires lower memory and processing power.

ARCHITECTURE AND OPERATIONS OF THE PROPOSED TRAFFIC MONITORING PLATFORM

The proposed platform uses traveling cars as probes transmitting information about traffic as they progress through the road network. The proposed platform adopts a hierarchical networking architecture, shown in Figure 1. With this hierarchical architecture, a geographic area is divided into adjacent but distinct hexagonal clusters/cell. Each cell is controlled by a centralized "head." A WMN AP is configured to operate as a cluster head. It communicates with cars travelling in its geographical cell in order to gather application data. OBUs are sensors in the sense that they generate and transmit data to nearby WMN AP. APs will run software programs for gathering, and pre-processing of raw data. Only a few of WMN APs, called gateways, are connected to the Internet. APs exchange information to identify

Figure 1, A reference architecture for the proposed platform

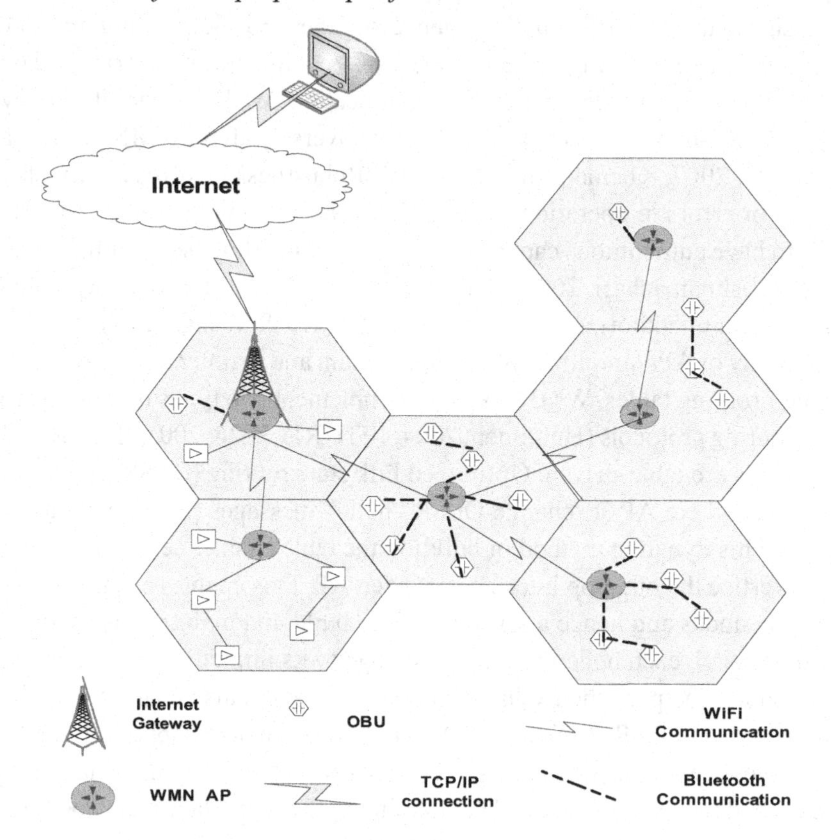

their neighbors and identify gateway AP. WMN APs collaborate and forward collected data towards gateways. The WMN is used to transfer monitoring information, generated by OBU, to a centralized server typically at headquarters offices over the Internet. The set of OBUs represent the lowest level in the hierarchical network. WMN APs represent the next higher level of the hierarchy. The third level in the hierarchy consists of Gateways. The top most level of the hierarchy consists of the servers at the city headquarters. This architecture enables network as a whole to monitor a larger geographical area through large-scale distributed processing and communication of data performed by many devices.

In this hierarchical architecture, different devices at different levels perform different functions. At the lowest level in the hierarchy, traffic and travelling cars data required for applications are gathered. Travelling cars equipped with Bluetooth OBU devices communicate with WMN AP. The middle layers in the hierarchy consist of WMN APs and Gateways that process, aggregate, and forward data, typically for longer distances using Wi-Fi and the Internet. The upper layers of the hierarchy consist of centralized servers that perform decision-support functions. A particular function may be carried out by more than one layer, for instance, each layer could perform a specialized role in computation (Stoianov et al., 2007; Yu, Mokhtar, & Merabti, 2006; Toumpis & Tassiulas, 2006).

This hierarchical architecture provides operational scalability and allows for phased deployment of a network and for node upgrades. However, imposing a logical structure on an existing flat network may result in potential inefficiency. For example, organizing nodes into a hierarchy typically introduces overhead (e.g., execution of clustering algorithm) into the network.

The network model for our platform consists of WMN AP mounted on light posts. The APs can be laid out at arbitrary distances depending on deployment conditions. Each AP is configured to run the OpenWRT Linux distribution (http://www.openwrt.org) for embedded devices. Some APs are connected to the Internet and are configured as gateways. In addition, all APs are configured to run OLSR. OLSR enables AP to automatically configure the WMN and enables self-healing in case of the failure of one AP. Bluetooth dongles are attached to AP through USB ports. We integrated BlueZ (http://www.bluez.org) open-source Linux-based implementation of the Bluetooth stack in each AP. With this setup, each AP can communicate with other nodes using Wi-Fi and/or Bluetooth.

BlueZ is controlled to scan the wireless medium for Bluetooth devices in proximity. We assume monitored cars are equipped with Bluetooth devices. We developed a Linux shell script that is capable of retrieving the Bluetooth (BT) address of near-by Bluetooth devices. The script gathers other information about the detected device including the time a device is detected. The gathered information is then relayed to car tracking server on the Internet using TCP/IP. The route from the AP that detects the Bluetooth Device to the server is controlled by the WMN routing tables maintained at each AP and configured automatically by OLSR.

Figure 2 illustrates the communication model within the platform between Bluetooth OBU device, different APs, gateway and server. The OBU device in the car communicates with the Bluetooth dongle on the AP "Wi-Fi router with BT." Information about this communication is extracted and saved at the AP using a Linux shell script. The AP collects information such as the MAC address of the detected device, the RSSI, the time the device was detected, and the ID of the AP that detected the device. This information is packaged and sent to the Internet server via the WMN. AP "Wi-Fi router with BT" sends an OLSR message to a neighboring AP "Wi-Fi router" that, in turn, forwards it to a neighboring AP "Wi-Fi gateway." The gateway AP forwards the information using TCP/IP on the Internet where packets are rerouted to reach the Internet Server. The server correlates the information it receives from different devices and determines the location of the Bluetooth device at different times.

IMPLEMENTATION AND FIELD TESTING OF THE PLATFORM

In this section, we describe a small-scale deployment of the proposed platform and describe the experiments carried out in order to demonstrate the capability of tracking cars. We carried out two deployments one at the borders of the city to avoid interferences and achieve near-ideal line of sight between sender

Figure 2. Network protocol interactions

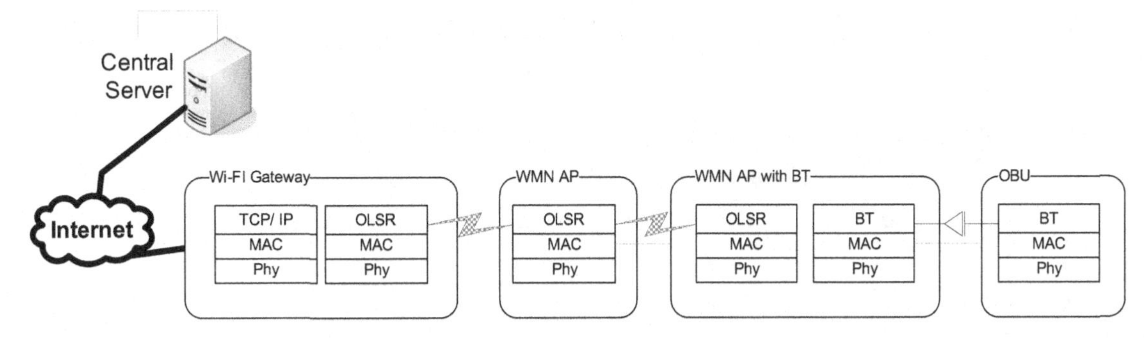

and receiver; and the second deployment was in a residential area with interference and reflection from passing-by cars.

APs were implemented using ASUS® WL-500g Premium AP, with a 266 MHz and 8M flash CPU, 32M RAM, an external dipole antenna, and 2 USB 2.0 ports. Travelling cars were equipped with ultra-slim Bluetooth V1.2 dongles with USB 2.0 and operation range of 100m with built-in antenna and maximum data rate of 3MB. We gather the results on the AP by running a Linux shell script. To enable OLSR message exchange, the OpenWRT firewall was opened at port 698 and forwarding rules were added to the firewall configuration files. For the purposes of preliminary experiments, we do not allow Bluetooth devices to establish connections with AP in order to avoid security and bluejacking issues.

In general, a major issue with Bluetooth detection of devices is the lengthy scanning period of Bluetooth devices. A Bluetooth device may take up to 10 seconds in order to fully detect all the Bluetooth devices in range. We avoided this problem by storing (caching) the information of the to-be-detected Bluetooth module in APs. Caching this information allows Bluetooth to detect a device without performing the standard time-consuming (inquiry and scan) processes. Algorithms for storing, caching, sharing, and managing Bluetooth address is out of the scope of this paper. It has been proven that caching Bluetooth information can reduce detection time by up to 90% (Sang-Hun et al., 2002).

The signal strength, i.e. RSSI received at the AP from the Bluetooth device in the travelling car was measured. RSSI is a measurement of how well the device is receiving a signal and is typically measured in dBm. Generally, the closer the RSSI to zero the stronger the signal level received. Figure 3 and 4 show the RSSI levels (vertical axis) versus time (horizontal axis) as a single car travels towards, by, and away from an AP. In the figures, the horizontal axis represents the second at which we measured the signal strength. An RSSI = 0 dBm indicates that the Bluetooth device is at the closest distance from the AP. In contrast, an RSSI = -13 dBm indicates the Bluetooth device is out of range of the AP.

Figure 3 shows results for the near-ideal deployment in an environment with almost no interference at edges of the city. We realize that the AP started to detect the Bluetooth device at the 64th second, but the car actually entered into range of the AP at the 51st second – an offset of 13 seconds. This delay can be attributed to the slow response of Bluetooth devices in detecting new vehicles. Since the car was approaching the AP, the RSSI started to rise towards 0. The car became at the nearest point to the AP at the 81st- 83rd second. After that, the RSSI level started to decline indicating the car is travelling away from the AP. At the 100th second, the car physically left the Bluetooth range coverage, however the AP

Figure 3. The signal strength of a traveling car in a near-ideal environment

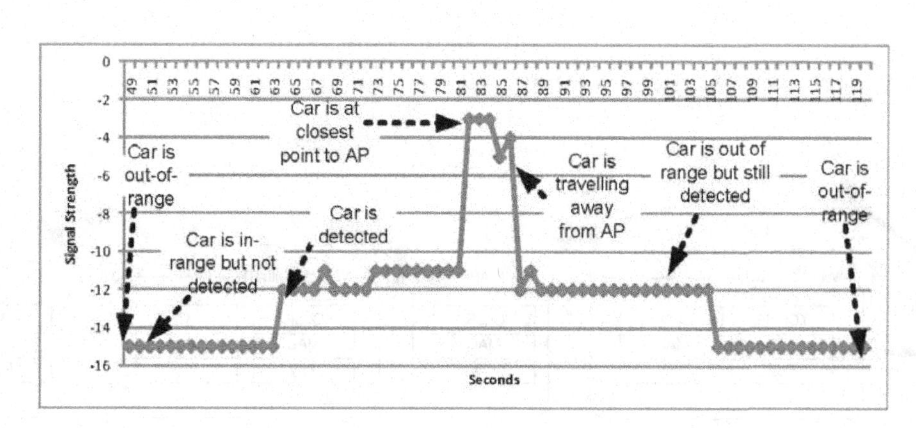

still detected its signal. At the 117th second, the Bluetooth device became completely out of range from AP as indicated by RSSI level, an offset of 17 seconds. This offset is used to correct the measured data for each AP. Repeated experiments on different areas, show that the offset depends on the geographical area surrounding the AP.

In Figure 4, we report the results in the residential area deployment with interference from passing-by cars and signal reflection from houses and parked cars. We realize that the AP started to detect the Bluetooth device at the 30th second. Before that the car was out of range of the AP because RSSI was about -15 dBm. Since the car was approaching the AP, the RSSI started to rise towards 0 dBm. The car became at the nearest point to the AP at the 33rd second. After that, the RSSI level started to decline indicating the car to be travelling away from the AP. We realize that for the following 7 - 8 seconds the Bluetooth device was in the range of the AP. In the figure, we note that Bluetooth RSSI is affected by radio interference from other devices and from signal reflection and refraction from cars and other objects in proximity. For example, the curve at the 34th, 35th, and 36th second indicates that the car is roughly at the same distance from the AP, which is not the case because the car was in constant movement. At the 37th second we realize that the signal strength indicates that the car became closer to the AP which was not the case. The car was actually travelling away from the AP at that instant. It is important to note that not all the Bluetooth dongles that we used showed such inaccuracies. We attribute this to the accuracy in manufacturing the dongles, in addition to the conventional signal fluctuation in complex environments.

Correlating information from the deployed 4 AP, the proposed platform can be used to identify the path of travelling cars and approximate car speeds. The central server maintains a database of the location and identity of each AP. During the experiments, APs detect passing-by cars and notify the central server. Information about which AP the car was close to at what time is gathered at the central server. The central server processes this information to track travelling cars based on the time and location of each Bluetooth device detected by the WMN. In Figure 5 we show the map of our actual AP deployment in the residential area. The squares indicate the locations of the APs. The directed thick line indicates the path and direction followed by the car. This line is constructed at the central server by connecting the squares in the figure. At this stage of development, we used straight lines to connect the triangles, which can be extended to follow actual roads.

Our platform enables the calculation of approximate speed of the car based on the time and location information collected from different APs. We parsed the information collected from each AP to find out the time the car was closest to each AP (the second in time the RSSI received from the car was at

Figure 4: Signal Strength detected for a Car in a residential area

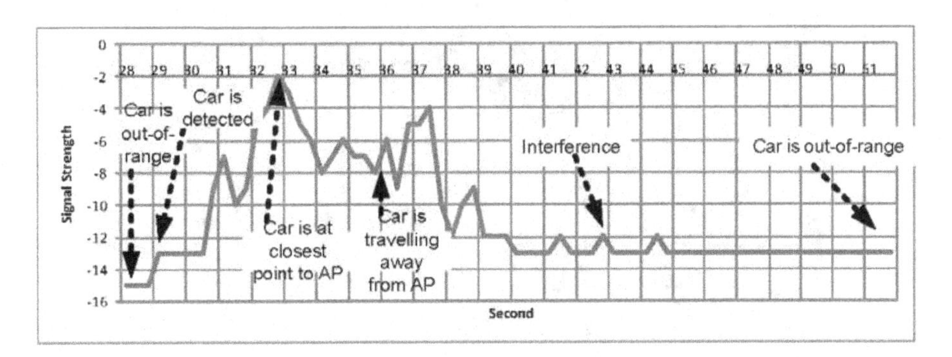

highest value). Before deployment, the time at each AP is synchronized with a centralized server and the distance between APs is measured and reported to the central server. The server calculates the average of the travelling speeds between each two APs and considers this average to be the traveling speed of the car in the area of deployment. Comparing the calculated speed against the actual speed the car traveled at we found out a 10- 15% of difference. We had the car traveling at 20, 30, 40 and 50 km/hour, and repeated for three times, with approximately the same approximated percentage.

DISCUSSION ON LARGE-SCALE DEPLOYMENT OF THE PLATFORM

Here, we report on aspects affecting large-scale deployment of the platform, which provides a roadmap for expanding this work. Large-scale deployment involves units at the lowest two levels in the hierarchy, that is the OBU and the WMN AP levels.

At the lowest level in the hierarchy, OBUs can transmit traffic information using standard wireless communication protocols, of which IEEE 802.15 (Bluetooth), 802.15.4 (ZigBee) and 802.11 (Wi-Fi) are the most common. Of the aspects affecting the design of OBU communication, transmission range of an OBU is one of the most important factors of large-scale deployment. The higher the transmission range of an OBU, the smaller the number of WMN AP required to cover larger geographic areas. A smaller transmission range is typically associated with a multi-hop architecture versus a single-hop architecture.

Another factor that affects OBU design is its transmission rate. The higher the data rate offered by a transceiver device, the smaller the time needed to transmit a given amount of data. Higher data transmission rates are mandatory for vehicle tracking applications (Sarangapani, 2007) because vehicles typically travel at relatively higher speeds. This will allow only a few seconds for OBU-RSU data exchange. OBUs report data to RSU either continuously or in response to an event. With the first model, traffic is continuously monitored and data gathered is continuously reported to the centralized server. In the event-based reporting, the OBU report data if they detect the occurrence of an event or in response

Figure 5. Tracking a traveling car (map courtesy of Google Maps)

to receiving a query request. A hybrid model is also possible using a combination of continuous, event-driven and query-driven data delivery (Akkaya & Younis, 2005).

At the next higher level of the hierarchy, large-scale deployment of WMN AP is controlled by a set of economical and operational factors. Economical factors include the number of AP, the unit price for each AP, and AP installation and maintenance cost. Operational factors include traffic flow characteristics, coverage quality, and surrounding environment.

The number of WMN AP is affected by several factors such as area of deployment region, nature of deployment region, and fault tolerance requirements. The WMN cell size is a design factor for the platform that depends on many factors such as AP transmission range, application type, and required coverage accuracy. In addition the general characteristics of the geographic cell affect WMN deployment. For example, downtown areas typically have higher density deployment of WMN APs as compared to suburban or rural areas because of the density of cars and interference from other wireless devices. The same applies for highways when compared to small city streets.

In addition, WMN deployment always considers a trade-off between AP capabilities and unit price, and power consumption. Since a WMN contains a large number of APs, the cost of a single AP is very important to maintain the overall cost of the network within acceptable limits. The price of an AP is affected by capabilities of its components, such as communication devices, power supplies, and processing units. For the communication device, for example, enhanced features include distinct communication address, enhanced data rates, power consumption efficiency, wider communications range, precise receiver sensitivity, carrier sense capabilities, RSSI, wake up radio, ultra wide band communication, and dynamic modulation scaling (Krishnamachari, 2005; Cheekiralla & Engels, 2005; Khemapech et al., 2005).

A third economic factor to be considered is WMN installation and maintenance costs which largely depend on the number of nodes, mobility of sensors, and type of deployment. Deterministic deployment is almost always expected to have higher installation costs than random deployment. Another aspect of cost is whether the installation is automated or manual. The installation process may also involve costs for licenses, permits, insurance and labor. WMN maintenance and fault tolerance are of the most important network management issues. A WMN is required to provide reliable monitoring in severe circumstances and even if some AP fail. WMNs are, therefore, required to automatically recover and reconfigure themselves. WMNs are typically designed with redundancy in AP to enhance tolerance to faults. The number of WMN AP is affected by the redundancy level required to fulfill QoS or fault tolerance requirements. The higher the fault tolerance level required, the higher the redundancy level and the larger the number of sensors that will be required. In Gao et al. (2004), the problem of evaluating redundant sensing areas among adjacent wireless sensors was analysed and recommendations on the minimum and maximum number of neighbours required to provide complete redundancy were presented.

WMN deployment also depend on operational factors such as interference from unwanted wireless signals available in the environment and weather conditions affecting the WMN operation. Interference may come from other transmitters sending in the same frequency band at the same time. There exist two kinds of interference: co-channel and adjacent-channel (Khemapech et al., 2005). Weather also affects the deployment of WMNs. Snow and rain have an effect on packet loss with the effect of rain being with less severity than of snow (Stojmenovic, 2005).

CONCLUSION

In this chapter, we described how the integration of Wi-Fi based wireless mesh networks and Bluetooth technologies can be used for detecting and tracking travelling cars as well as measuring their speeds. We described our proof-of-concept implementation and deployment of a wireless platform for enabling this. The platform was able to track cars travelling at speeds of 0 to 70 km/hour. We did not test the platform at higher speeds which require unavailable (at the time of the experiment) WMN setups on highways. The tested platform was able to track cars on roads and as they travel through and make turns on different streets. The platform calculated car speed by correlating information gathered at different synchronized AP. Preliminary results indicated that speeds can be measured with ± 10%-15% accuracy. We plan to incorporate more advanced algorithms to enhance speed calculation accuracy in the future.

The proposed platform is cost-effective for three reasons: 1) the platform uses unlicensed wireless technologies; 2) it leverages investments made in municipal WMNs; 3) the platform was built on common hardware and open-source software. Using open source elements makes the developed platform flexible and easily modifiable by us or others. To decrease the number of units required to cover required parts of the city, one can use longer-range Bluetooth devices. The developed platform can be extended to provide many applications and services such as congestion identification and quantification, traveler information systems and navigation and route guidance services. This system has the potential to contribute to reducing fuel consumption and air pollution by reducing traffic congestions.

We identified the following sources of inaccuracy in this experiment due to: 1) difference in manufacturing of Bluetooth dongles and implementations of the Bluetooth stack; and 2) we also realized that signal refractions and interference can affect RSSI measurements. There are different versions of Bluetooth each with a different coverage area and data rates. The availability of metal objects (such as other cars) and other wireless signals in the spectrum affects these measurements.

With the steep growth and expansion of WMNs and the increasing popularity of Bluetooth and Wi-Fi mobile devices, it is logical to predict extensions to this research. The developed system has the potential to use widely available and rapidly expanding components to enable new ITS services in a cost effective manner. Future work will include investigating more accurate algorithms for tracking cars and calculating their speeds, assessing the impact of device quality on and tracking cars and as well measure their speeds as a proxy to congestion level. We will use more complicated algorithms and Geographic Information Systems (GIS) information to draw more accurate tracking of the path that the car traveled. We believe that a large-scale deployment of the platform can track cars in urban areas such as downtowns where other wireless technologies are more expensive (such as GSM) or cannot operate at all (e.g. GPS). Our future studies will include characterizing the effect of differences in manufacturing of Bluetooth dongles on the accuracy of our measurements. We will also study how Bluetooth limits the performance of this infrastructure. In particular, we will study the effect of the delay in detecting a car on the ability of the infrastructure to perform as the speed of travelling car changes. We will also try to quantify the maximum number of cars that can be detected in a second. We also would like to investigate the potential of the platform for two way communications with the mobile devices. This will enable the gathering of traffic information from vehicles as probes and using this information to provide navigational services and traveler information to traveling cars, i.e. enabling each device to be a contributor and beneficiary at the same time. In the future, wider WMNs can be established using private citizens' routers at residences and offices. Consequently, users can join online communities with their AP devices and benefit from the resulting information and services, provided by WMNs.

ACKNOWLEDGMENT

The authors would like to thank the Canadian Natural Science and Engineering Research Council and Canada research Program for their generous funds supporting this work. Parts of this work are based on our previous work published in [Ahmed 2008].

REFERENCES

Ahmed, A. A., Shi, H., & Shang, Y. (2003). A survey on network protocols for wireless sensor networks. *IEEE International Conference on Information Technology: Research and Education, Proceedings. ITRE2003.* (pp. 301-305).

Ahmed H., EL-Darieby M., Abdulhai B. and Morgan Y (2008) *"A Bluetooth and WiFi Based MESH Networks Platform for Traffic Monitoring"*, in the Proceedings of the National Academies Transportation Research Board 87th Annual Meeting, TRB-87, Washington DC, January.

Akkaya, K., & Younis, M. (2005). A survey on routing protocols for wireless sensor networks. *Elsevier Journal of Ad Hoc Networks.* (pp. 325-349).

Akyildiz, I. F., Su, W., Sankarasubramaniam, Y., & Cayirci, E. (2002, November 7). A survey on sensor networks. *Communications Magazine, IEEE,* (pp. 102-114).

Akyildiza, I. F., Wang, X., &Wang, Y., (2005).Wireless Mesh Networks: A Survey. *Elsevier Journal of Computer Networks.* (pp. 445-487)

Al-Rousan, A., Al-Ali, R., & Darwish, K. (2004). GSM-Based Mobile Tele-Monitoring and Management System for Inter-Cities Public Transportations. *IEEE International Conference on Industrial Technology.* (pp. 859- 862)

Bluetooth SIG (2006). *Bluetooth Wireless Technology Becoming Standard in Cars.* Retreived October, 2007, from http:// www.bluetooth.com/Bluetooth/Press/SIG/Test_1.htm

Broida, R. (2003). *How to Do Everything with Your GPS.* McGraw-Hill.

Cardell-Oliver, R., Smettem, K., Kranz, M., & Mayer, K. (2004). Field testing a wireless sensor network for reactive environmental monitoring, *In Proceedings of the 2004 Intelligent Sensors, Sensor Networks and Information Processing Conference* (pp. 7-12)

Cheekiralla, S., & Engels, D. W. (2005, October). *A functional taxonomy of wireless sensor network devices.* Cambridge, MA : IEEE Explore

Farkas, K., & Plattner, B. (2005).Supporting Real-Time Applications in Mobile Mesh Networks, *In Proceedings of the MeshNets 2005 Workshop,* (pp. 540-550).

Gao, Y., Wu, K., & Li, F. (2004). Analysis on the redundancy of wireless sensor networks. *In Proc.2nd ACM Intl. Workshop on Wireless Sensor Networks and Applications (WSNA),* San Diego, CA: ACM (pp. 108-114).

Hać, A. (2003). *Wireless sensor network designs.* West Sussex: John Wiley & Sons Ltd.

Hightower, J., LaMarca, A., & Smith, I. (2006) Practical Lessons from Place Lab, *IEEE Pervasive Computing*, 5(3), 32- 39.

Huang, H. C., Huang, Y. M., & Ding, J. W. (2006). An implementation of battery-aware wireless sensor network using ZigBee for multimedia service. *In Proceedings of Intenrnational Conference on Consumer Electronics.* (pp. 369-370)

Huhtonen, A. (2004).Comparing AODV and OLSR Routing Protocols, *InTelecommunications Software and Multimedia.* Helsenki : Helsinki University of Technology.

Hull, B., Bychkovsky, V., Zhang, Y., Chen, K., Goraczko, M., Miu, A., et al. (2006). CarTel: A Distributed Mobile Sensor Computing System, *The 4th ACM Conference on Embedded Networked Sensor Systems (SenSys).* (pp. 125- 138)

IEEE P1609.3/D18 standard (2005)- IEEE P1609.3 Standard for Wireless Access in Vehicular Environments (WAVE)- Networking Services.

IETF RFC3626 (2003). *Optimized Link State Routing Protocol.* Retreived October 2007, from http:// www.ietf.org/rfc/rfc3626.txt.

IETF RFC3561 (2003). *Ad hoc On-Demand Distance Vector Routing.* Retreived October 2007, from http:// www.ietf.org/rfc/rfc3561.txt.

Ilyas, M., & Mahgoub, I. (Eds.). (2005). *Handbook of sensor networks: Compact wireless and wired sensing systems.* CRC Press LLC.

Karl, H., & Willig, A. (2005). *Protocols and architectures for wireless sensor networks.* Chichester: John Wiley & Sons Ltd.

Khemapech, I., Duncan, I., & Miller, A. (2005). A survey of wireless sensor networks technology. *In Proceedings of the 6th Annual PostGraduate Symposium on the Convergence of Telecommunications, Networking & Broadcasting (PGNET'05), Liverpool, U.K.*

Krishnamachari, B. (2005). *Networking wireless sensors.* New York: Cambridge University Press.

Lee, M., Jianliang, Z., Young-Bae, K., & Shrestha, D. (2006). Emerging Standards For Wireless Mesh Technology, *IEEE Wireless Communications*, *13*(2), 56- 63.

Sang-Hun, C., Hyunsoo, Y., & Jung-Wan, C. (2002). A Fast Handoff Scheme For IP over Bluetooth. *In the Proceedings of the International Conference on Parallel Processing Workshops* (pp. 51).

Sarangapani, J. (2007). *Wireless ad hoc and sensor networks: protocols, performance, and control.* F. L. Lewis (Ed.), Broken Sound Parkway NW: Taylor & Francis Group, LLC.

Spirito, M. A. (2001). On the Accuracy of Cellular Mobile Station Location Estimation. *IEEE Transactions On Vehicular Technology*, 50(3), 674-685.

Stoianov, I., Nachman, L., & Madden, S. (2007). PIPENET: A wireless sensor network for pipeline monitoring. *In 6th International Symposium on Information Processing in Sensor Networks*, Cambridge, MA. (pp. 264 – 273)

Stojmenovic, I. (Ed.). (2005). *Handbook of sensor networks : algorithms and architectures.* Hoboken, New Jersey: John Wiley & Sons, Inc.

Toumpis, S., & Tassiulas, L. (2006, July). Optimal deployment of large wireless sensor networks. *IEEE Transactions On Information Theory*, 52(7), 2935-2953.

Wu, S. L., & Tseng., Y. C. (Eds.). (2007). *Wireless ad hoc networking : personal-area, local-area, and the sensory area networks.* Broken Sound Parkway NW: Taylor & Francis Group, LLC.

Yu, M., Mokhtar, H., & Merabti, M. (2006). A survey of network management architecture in wireless sensor network. *In Proceedings of the Sixth Annual PostGraduate Symposium on The Convergence of Telecommunications*, Liverpool, UK.

Yu, Y., Prasanna, V. K., & Krishnamachari, B. (2006). *Introduction to wireless sensor networks.* In Y. Yu, V. K. Prasanna, & B. Krishnamachari (Eds.), *Information processing and routing in wireless sensor networks* (pp. 1-21). Singabore: World Scientific Publishing Co. Pte. Ltd.

Young, A. S., & Skobla, J. (2003). Robust GPS- SMS communication channel for the AVL system. *Proceedings of the Aerospace Conference.* (pp. 4_1957- 4_1965).

Zhang, P., Shi, Z., & Xu, M. (2005). Design and implementation of vehicle monitoring system based on GPRS. *Proceedings of the Fourth International Conference on Machine Learning and Cybernetics*, Guangzhou. (pp. 3574 - 3578).

Zheng, J., Winstanley, W. C., Yan, L., & Fotheringham, A. S. (2008). Economical LBS for Public Transport: Real-time Monitoring and Dynamic Scheduling Service. *The 3rd International Conference on Grid and Pervasive Computing Worksh*ops, (pp. 184 -188).

Chapter XVIII Beyond Localization: Communicating Using Virtual Coordinates

Thomas Watteyne
Orange Labs & CITI Lab, University of Lyon, France

Mischa Dohler
Centre Tecnològic de Telecomunicacions de Catalunya (CTTC), Spain

Isabelle Augé-Blum
CITI Lab, University of Lyon, France

Dominique Barthel
Orange Labs, France

ABSTRACT

This chapter deals with self-organization and communication for Wireless Sensor Networks (WSNs). It shows that nodes do not always need to know their true physical coordinates to be able to communicate in an energy-efficient manner. They can be replaced by coordinates which are not related to their geographical position, yet are easier to obtain and more efficient when used by routing protocols. The authors start by analyzing the techniques used by a node to infer its geographical location from a small number of location-aware anchor nodes. They describe how nodes can use their geographical locations to self-organize the network. The authors then present an anchor-free positioning algorithm in which nodes acquire virtual coordinates. Through a continuous updating process, virtual coordinates of neighbor nodes are brought close together. Although not related to the nodes' geographical location, routing using these coordinates outperforms routing using true physical coordinates. This chapter hence shows that localization algorithms are not per se required when considering communication in a WSN. A better strategy is to use geographic routing protocols over non-physical virtual coordinates which are easier to obtain.

SELF-ORGANIZING WIRELESS SENSOR NETWORKS: A PARADIGM SHIFT

Self-organization can be defined as "the emergence of system-wide functionality from simple local interactions between individual entities" (Prehofer & Bettstetter, 2005). As we will describe in this section, self-organization principles can be applied to any collection of individual entities, be it a group of economic agents, individual bacteria, a school of fishes, or a wireless multi-hop network.

The goal of self-organization in WSNs is to create a fully-autonomic network, which can be used without human intervention after deployment. From a networking point of view, it includes enabling network-wide communication from local simple interactions between nodes. This is, in fact, the definition of self-organization given above. (Mills, 2007) extended this definition by describing the design strategies of self-organizing systems. In the following paragraphs, we give examples of emergent behavior in economics and biological systems.

Emergent behavior principles apply to economics. Every economic agent uses only local information to decide how to behave. Buyers know only their own preferences and their own budget constraints, sellers know only their own costs. Their buying and selling on markets generate market prices, containing and transmitting all information about preferences, resources and production techniques. This way, market prices guide economic agents in making the best use of the resources available. Adam Smith called the market price "the invisible hand" which leads people to behave in the interest of society even when they seek only their self-interest (McMillan, 2002).

Emergent behavior also applies to much simpler systems such as a colony of Escherichia coli, a type of bacteria. Each bacterium is provided with flagella enabling it to move. In the presence of succinate (a chemical component), each bacterium excretes chemical substances which serve as attractants for other bacteria. Whereas these unicellular beings follow simple rules, these local interactions between individual entities yield chemotactic pattern formation: the bacteria organize into swarm rings and aggregates (Brenner, Levitov, & Budrene, 1998).

In "migrating groups of fish, ungulates, insects and birds, crowding limits the range over which individuals can detect one another" (Couzin, Krause, Franks, & Levin, 2005). Despite the local knowledge of each bird, a flock of birds moves in a coherent way (see Figure 1). Moreover, as detailed in (Prehofer & Bettstetter, 2005), bird flocks exhibit all the advantageous properties of a self-organized system, namely adaptability (the flock changes when attacked by a bigger bird), robustness (the flock is still coherent even when a bird gets killed) and scalability.

The previous paragraphs have shown examples of self-organizing entities and emergent behavior in economics and biological systems. We will see that WSNs have a lot in common with those systems. Because of the potentially very high number of nodes creating a wireless multi-hop network, the manufacturing cost of each individual node needs to be kept low. As a consequence, each node is capable of fulfilling only a limited set of tasks, and can only communicate with a limited number of close neighbor nodes. Hence, the concepts of emergent behavior can apply to large scale WSNs in a fashion similar to what the described biological systems achieve. This emergence enables extraordinary accomplishments by the network as a whole.

The ultimate goal of a self-organizing network is to be fully autonomic: to be deployed and used without any human intervention. The challenge of self-organizing a wireless multi-hop network is exemplified in Figure 2. Each small white circle represents a node and edges interconnect nodes capable of communicating. Self-organization in such a network consists of enabling node C to send a message to node X, by only having nodes communicate locally with their neighbor nodes (the ones within com-

Figure 1. (left) Illustration of the main principles of a self-organizing system borrowed from (Prehofer & Bettstetter, 2005); (right) picture of a starlings flock in Denmark (by Bjarne Winkler)

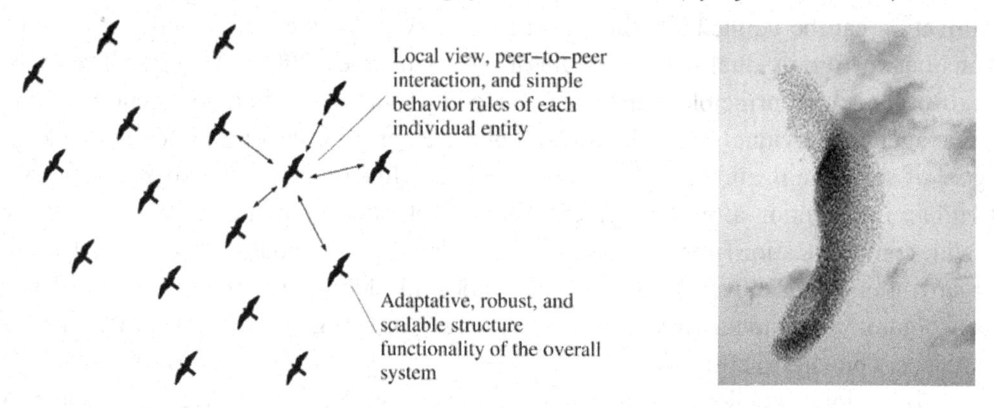

Figure 2. Depicting the problem of self-organizing a wireless multi-hop network

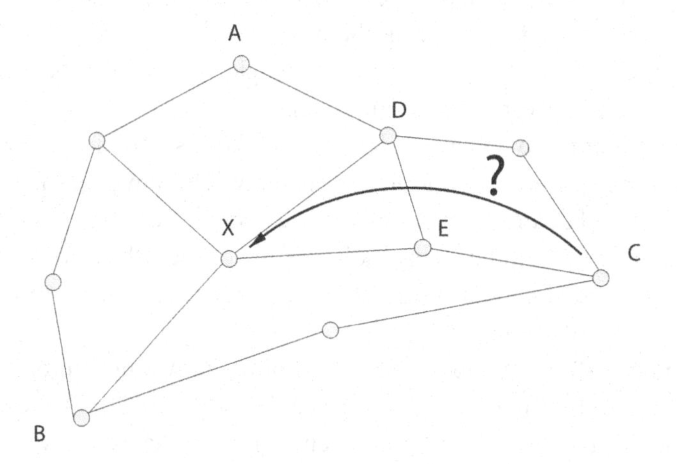

munication range). We call routing the process of finding a sequence of nodes to relay the message from C to X. This process needs to happen in an energy-efficient and robust manner. Energy-efficiency guarantees a long network lifetime; robustness implies that communication is still possible even under lossy links, or when nodes move and/or (dis)appear.

Research on wireless ad-hoc networks (Tonguz & Ferrari, 2006) has yielded a number of self-organization concepts such as clustering and virtual backbones. Clustering refers to grouping nodes together and electing a leader node in each cluster. Routing is thereby simplified as it can be done hierarchically: inside a cluster on a local scale and between a small number of clusterheads on a global scale. Ideas developed are largely inspired by wired networks where routers are grouped into Autonomous Systems and IP addresses are assigned hierarchically. An excellent overview of the concepts of self-organization for ad-hoc networks can be found in (Theoleyre & Valois, 2007).

As stressed by (Karl & Willig, 2005) and (Dohler et al., 2007), while Wireless Sensor Networks and ad-hoc networks are both wireless multi-hop networks, they are different in mainly three aspects: (1)

energy-efficiency is a primary goal for WSNs, (2) in most envisioned applications, the amount of data transported by a WSN is low and (3) all the information flows towards a limited number of destination nodes in WSNs. Clustering does not really answer any of these three specific WSN constraints, mainly because building and maintaining such a structure costs energy. A paradigm shift is thus needed when considering self-organization for WSNs.

In the biological examples described above, all involved entities have a notion of movement and position. A bird in a flock knows where its neighbor birds are, and knows their relative position and heading. In this chapter, we describe how WSN protocols exploit location information to enable network-wide communication. At the end of this chapter, we propose a self-organizing protocol for WSNs which mimics the behavior of a swarm of biological entities.

The remainder of this chapter is organized as follows. We will first describe how location information is used for routing in WSNs. As acquiring location information is expensive, we will detail how (estimated) physical coordinates can be determined, relative to a set of anchor nodes. In case the anchor nodes do not know their true physical coordinates, the other nodes determine non-physical coordinates. We will present a self-organization technique inspired by geographic routing, which uses entirely virtual coordinates in an anchor-free setting.

LOCATION-BASED COMMUNICATION PROTOCOLS

Applications for WSNs are foreseen in a large range of domains (Culler, Estrin, & Srivastava, 2004). In the example case of a city-wide automated water meter reading WSN, nodes are attached to each home's water meter and report the daily consumption to the local water supplier. Knowledge of the physical location of the water meter is not useful as long as the latter can be identified. On the other hand, when considering a WSN used for tracking the location of lions in a National Park, having the location of the sending node in a reported message is essential.

If the application requires the nodes to know their location, there is no overhead to reuse this location information for communication purposes. This is the philosophy behind geographic routing, which uses the knowledge of a node's position together with the positions of its neighbors and the destination node (called 'sink node') to elect the next hop node.

Greedy Geographic Routing Protocols

Greedy geographic routing is the simplest geographic routing protocol (Stojmenovic & Olariu, 2005). When a node receives a message, it relays the message to its neighbor geographically closest to the sink. Several definitions of proximity to the destination exist. We will use Figure 3(a) as a basis for our description, where node *S* wants to send a message to node *D*. Most-forward within radius considers the position of a node's projection on a line between the source and the destination. In Figure 3(a), node *S* would choose *A* as the neighbor node closest to *D*. Another definition considers the Euclidian distance to the destination (in this case, *S* would choose *B*). Finally, a last variant, sometimes referred to as myopic forwarding, chooses the node with the smallest deviation from the line interconnecting the source and the destination (node *C* in Figure 3(a)).

Irrespective of the definition of proximity, greedy routing can fail. In Figure 3(b), if a message is sent from node *A* to *X*, it reaches *X* with a number of hops close to optimal. Consider now the message

Figure 3. Greedy geographic routing. (a) Different ways of defining distance to the destination; (b) Geographic routing may fail

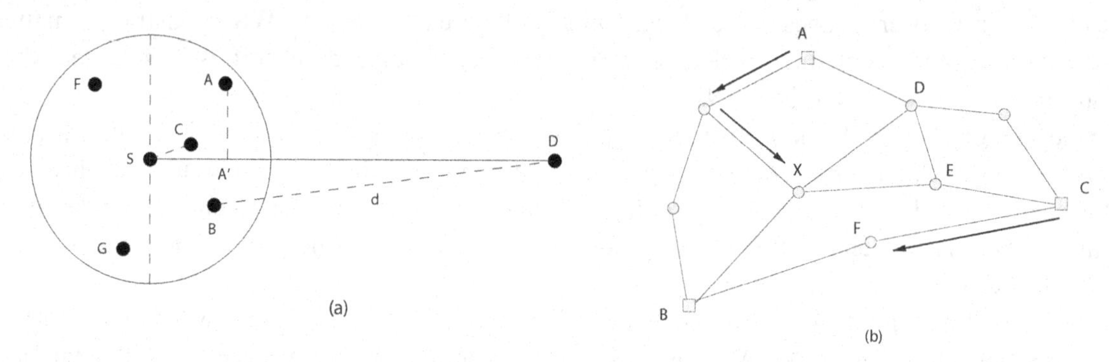

Figure 4. An example of a void area. The plain circle depicts the communication range of node S. The dotted circle is shown for readability only, it is centered at D and has a radius ||DS||. It shows that no neighbor node of S is closer than S to D.

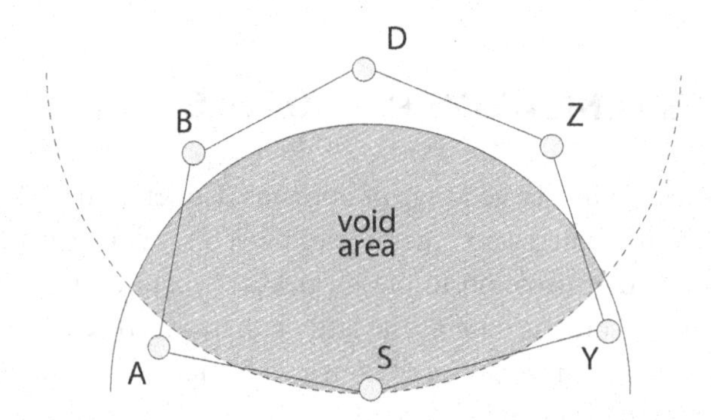

is sent from *C* to *X*. *C* will send it to *F*, it's neighbor closest to *X*. *F*, however, has no neighbor closer to *X* than itself; the message ends up at a local minimum, or a void area. A void area (or simply void) is depicted in Figure 4. It appears when a node has no neighbor closer than itself to the destination. A greedy geographic routing algorithm fails when it reaches a void.

The occurrence of such failures depends on the topology used. In Figure 5, we present simulation results obtained by randomly scattering nodes in a 1000x1000 area. Each node has a circular communication area of radius 200. We tune the number of nodes to obtain the desired average node degree (the average number of neighbors of all nodes in the network) and measure the delivery ratio. Results are averaged over 10^5 runs. For our simulations, the source and the sink nodes are chosen randomly - among connected nodes - and change at each run. A ratio equal to 1 means that all sent messages are received. Note that Figure 5 also shows results for other protocols which will be described later.

Delivery ratio is close to 1 for very high densities because the probability of having void areas decreases as the number of nodes increases. For typical WSN densities (5-10 neighbors), over 20% of sent messages are not received because of this flaw in the routing protocol.

Figure 5. Delivery ratio for different routing protocols when using true physical coordinates, assuming a physically connected network. Note that results for the GFG and 3rule protocols (which will be presented later) coincide at 1, which is the best possible case.

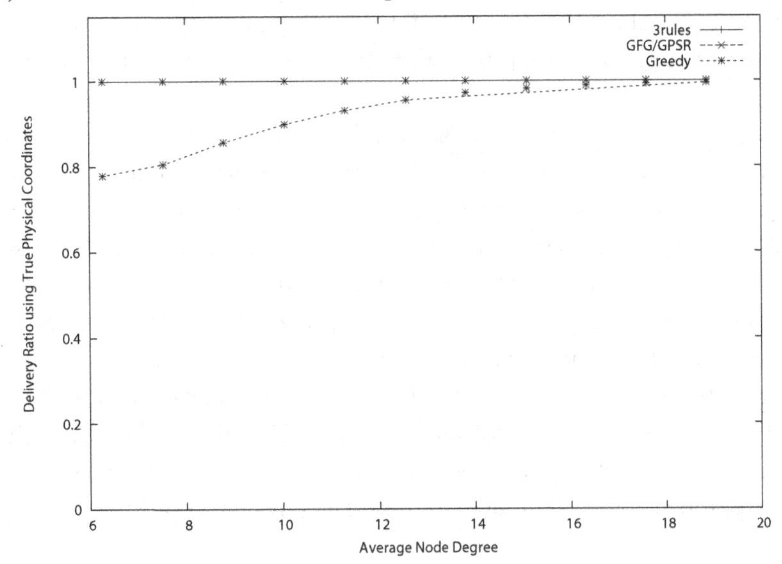

Geographic Routing with Guaranteed Delivery

Some geographic routing protocols guarantee delivery under the assumption of reliable links and nodes. The key idea of these protocols is to switch between two modes. The default mode uses the greedy approach described above. In case this mode fails, a second mode is used to circumnavigate the void area. Once on the other side of this void area, the greedy mode can be resumed.

Greedy-Face-Greedy (GFG) and Greedy Perimeter Stateless Routing (GPSR) (Frey & Stojmenovic, 2006) use exactly this principle. They have been proven to guarantee delivery, which is verified in Figure 5. Although details on geographic routing protocols are out of this chapter's scope, the interested reader is referred to the excellent overview provided in (Stojmenovic & Olariu, 2005).

Note, however, that this protocol fails when the unit disk graph assumption does not hold (i.e. the communication areas of the nodes are not perfect circles with the same radius). We evaluate this effect by simulation later in this chapter.

We have seen that some applications require the nodes to know their locations. Geographic communication protocols take advantage of this knowledge to perform some tasks which would be more expensive otherwise, such as routing. Yet, having a node know its position is expensive. The position of a node can be programmed manually during deployment. This, however, removes the possibility of randomly deploying a large number of nodes.

Another solution is to equip each node with a positioning device (e.g. GPS). However, GPS-like systems have been reported to be "cost and energy prohibitive for many applications, not sufficiently robust to jamming for military applications, and limited to outdoor applications" (Patwari et al., 2005). While not completely solving the problem, reducing the portion of location-aware nodes in a network is a step forward.

INFERRING LOCATION FROM A SET OF ANCHOR NODES

The idea behind using anchor nodes is to only have a subset of nodes be location aware. The cost of location-awareness can be monetary (e.g. the cost of a GPS chip), energy-related (e.g. to power a GPS chip), related to man-power (e.g. manually programming a node's position during deployment) or any combination thereof.

Regardless of the technique used, each anchor node is assumed to know its position (e.g. a set of {x,y} coordinates in a two-dimensional deployment). Non-anchor nodes will need to infer their own coordinates from the anchors using local measurements and localization protocols. When using anchor nodes, there is a clear distinction between localization (i.e. determining the physical positions in space/plane of the nodes) and routing. The nodes in the network typically determine their coordinates first; the geographic routing protocol then uses this information to send a message from any node to the sink.

There are two cases. In the first one, anchor nodes are location aware, meaning that they know their **true physical coordinates** (e.g. by means of GPS). As a result, non-anchor nodes will determine **(estimated) physical coordinates**, as close as possible to their true physical ones. In the second case, anchor nodes do not know their true physical coordinates. Nodes will thus have **relative coordinates**, a concept defined later in the chapter, not related to their true physical coordinates.

Location-Aware Anchors

With anchor nodes knowing their true physical position, the goal of a node is to determine coordinates which are as close as possible to its true physical coordinates. We call these coordinates "(estimated) physical coordinates". Multi-lateration may be used: if each node knows its distance to a set of anchor nodes, it determines its position as the intersection of the circles centered at each anchor node and with radius equal to the distance to this anchor node.

Whereas it is essentially the same idea as the one used by the GPS system, the main difficulty is to determine distances. As WSNs are multi-hop, a first approximation to the distance to an anchor node is the sum of distances of the individual links constituting the multi-hop shortest path. There are a number of techniques to measure these one-hop distances, including received signal strength (RSS) and time of arrival (TOA) measurement. Niculescu and Nath show that angle-of-arrival (AOA) is another valid technique for positioning in a wireless multi-hop network (Niculescu & Nath, 2003). Readers interested in positioning techniques are referred to (Patwari et al., 2005).

In a GPS-like system (Figure 6(a)), localization precision depends on the number of anchors (i.e. satellites), their relative positions and the precision of distance measurements. Things are more complicated when applying trilateration to WSNs. First, distance measurement errors add up on a multi-hop link. Moreover, localization precision depends also on the alignment of nodes on this multi-hop link. As shown in Figure 6(b), $|AX| \neq |AD| + |DX|$ because nodes *A*, *D* and *X* are not aligned. This localization technique is used by the GPS-Free-Free (Benbadis, Friedman, Amorim, & Fdida, 2005) protocol. Localization accuracies of about 40m are reported on networks with an average node degree of 10 neighbors (results are worse with sparser networks).

Benbadis et al. (Benbadis, Obraczka, Cortes, & Brandwajn, 2007) extend these results with simulations showing that the success ratio of greedy routing when using (estimated) physical coordinates is lower than when using true physical coordinates.

Figure 6. The concept of trilateration applied to a GPS-like system (a) and to a multi-hop wireless network (b)

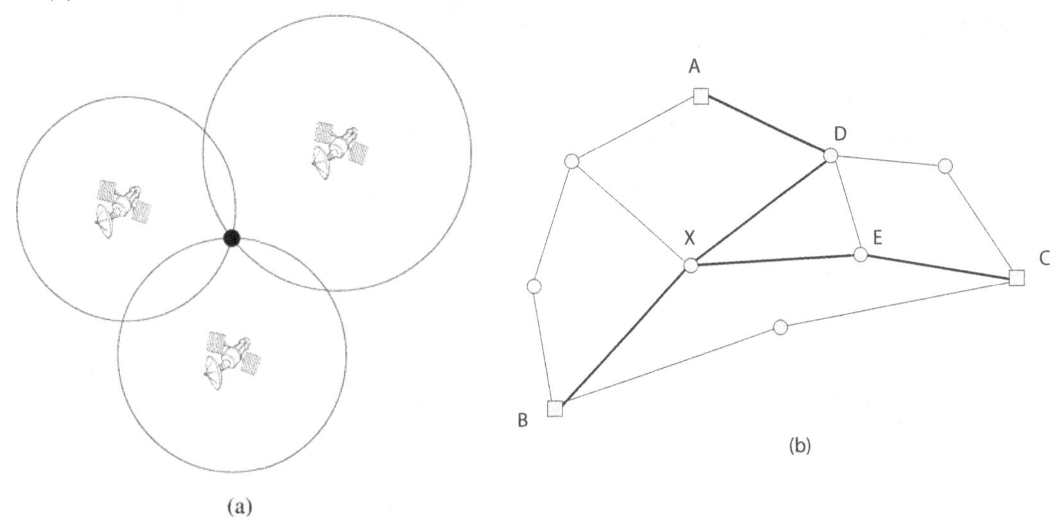

The most critical drawback of using true or estimated physical coordinates for routing is that geographic proximity is not synonymous with electromagnetic proximity. In other words: geographically close nodes can not always communicate, and nodes which can communicate are not always geographically close. This rule by itself annihilates all geographic routing protocol solutions, and has been largely overseen. Most of the proposed protocols are evaluated by simulation. For most of them, the simulated propagation model is the over-simplified on/off link model. In this model, the communication area of each node is a perfect circle, there is no interference outside this circle, and the radius of this communication area is the same for all nodes.

Routing protocols perform well under these assumptions; yet, when confronted with a real propagation model, they fail dramatically. This is shown in (Kim, Govindan, Karp, & Shenker, 2005) for the GFG and GPSR routing protocols. The same observation applies to all routing protocols based only on true or estimated physical coordinates.

In some applications, a node needs to know its physical position in order to report to the sink node where the sensed event is located. Nevertheless, the idea of using this geographical position alone for routing purposes does not hold in the general case because of the over-simplified assumptions on the propagation model it conveys. Physical coordinates (determined by GPS-like hardware, manually programmed or determined relatively to anchor nodes) can not be used directly for routing purposes. A new localization system is needed in this case, which is related to the topology of the network.

Location-Unaware Anchors

Using location-aware anchor nodes is useful for determining (estimated) physical coordinates; while essential to some applications, these coordinates cannot be used as such for routing. New coordinates are needed, which reflect the topology of the network. We will call these coordinates "relative coordinates". These can be determined by using a set of location-unaware anchor nodes.

Relative coordinates of node V are defined as a vector $\{V_1,V_2,\ldots,V_N\}$ where V_i is the hop distance from the current node to anchor node I; N is the number of anchor nodes. A simple way of assigning

Figure 7. An example topology where each node is assigned relative coordinates. Each small white circle represents a node and edges interconnect nodes capable of communicating. A small white square represents an anchor node.

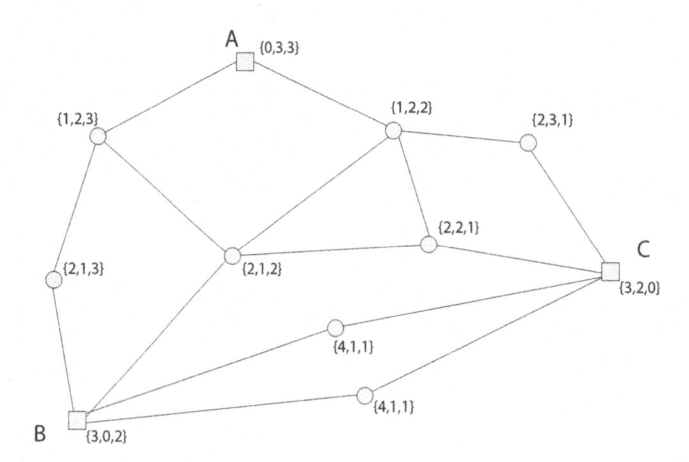

these relative coordinates is to ask each anchor node to periodically broadcast a message containing a counter which is incremented at each hop as it propagates through the network. Note that nodes can learn how many anchor nodes there are by listening to these broadcasted messages. Relative coordinates are not related to true physical coordinates. An example topology where each node is assigned relative coordinates is presented in Figure 7.

Geographic routing needs a notion of distance to be functional. As will be discussed later, note that the resulting relative distance is not directly related to physical distance. Cao et al. (Cao & Abdelzaher, 2004) proposed Euclidian distance as a metric of distance. In their proposal, relative distance $||D||$ between nodes $V=\{V_1,V_2,\ldots,V_N\}$ and $W=\{W_1,W_2,\ldots,W_N\}$ is calculated as

$$\|D\| = \sqrt{\sum_{i=1}^{N} (V_i - W_i)^2}$$

Several aspects of the relative coordinates need to be clarified. First, several distinct nodes may end up having the same coordinates. We call a group of nodes with the same coordinates a "zone". Furthermore, because coordinates are not orthogonal (i.e. having more than three anchor nodes introduces redundancy), $||D||$ is not directly related to physical distance.

Despite these specificities, using relative coordinates is a promising approach to routing in WSNs. Simulation results in (Cao & Abdelzaher, 2004) show that, when using relative rather than physical coordinates, less voids are encountered. This means that the success ratio of greedy geographic routing when using relative coordinates is higher than when using true physical coordinates, and hence more energy in the network is conserved. These results are confirmed experimentally by (Fonseca et al., 2005). This work serves as a proof-of-concept experiment for relative coordinate routing in WSNs.

The difficulty when using anchor nodes is to select those anchor nodes. The Virtual Coordinate Assignment Protocol ("VCap", Caruso, Chessa, De, & Urpi, 2006) elects anchor nodes dynamically during an initialization phase. A distributed protocol is designed to elect a predefined number of anchor nodes, evenly distributed around the edge of the network. This obviates the need for manual selection.

As said above, "zones" refer to a group of nodes which have the same relative coordinates. As the routing protocol bases its decision on these coordinates, ties may appear inside a zone, and the protocol may make the wrong decision. This can cause the multi-hop transmission to fail. Liu and Abu-Ghazaleh address this problem (Liu & Abu-Ghazaleh, 2006) by turning each virtual coordinate into a floating point value, and slightly changing these coordinates as a function of the nodes' neighborhood. The occurrence of ties and inconsistencies in the distances used for routing is hereby drastically reduced. To our knowledge, this is the first paper where a routing process using relative coordinates outperforms a routing process using true physical coordinates, in terms of hop count.

True physical coordinates represent the nodes' geographical positions; relative coordinates represent the topological position of the nodes, i.e. their position in the connectivity graph of the network. Routing using true physical coordinates suffers from void areas which makes greedy geographic routing fail. Some geographic routing protocols can deal with void areas, but they discover paths which are potentially very long. When using relative coordinates, there are less void areas. As a result, routing paths can be shorter than when using true physical coordinates, provided the problem of "zones" is addressed.

So far, relative coordinates were obtained by counting the number of hops separating each node from each anchor node. The GSpring protocol (Leong, Liskov, & Morris, 2007) takes this concept one step further by introducing the spring model. Each link connecting two nodes is considered as a spring. These abstract springs have a rest length which is a function of the node's neighborhood. If two nodes are closer to each other than this rest length (using the distance calculated as a function of the nodes' relative coordinates), the repulsion force of the spring causes their relative coordinates to part away. Inversely, if the length of the abstract spring is larger than its rest length, an attraction force brings the nodes relatively closer together.

During initialization of GSpring, an algorithm identifies a predefined number of anchor nodes on the edge of the network, and initializes their relative coordinates. The relative coordinates of these nodes will not change, and they appear as anchors to the spring system. An iterative process causes the abstract springs to be elongated and shortened until the spring system converges. Simulation results show that using this coordinate system yields better performance (in terms of number of hops) than using true physical coordinates.

Using relative coordinates for routing in WSNs is a very promising approach. Because the coordinate system is related to the topology of the network (and not to the physical location of the nodes), using routing protocols on top of relative coordinates yields better performances than using true physical coordinates. Moreover, relative coordinates avoid the cost of acquiring (estimated) physical coordinates.

Relative coordinates do require either a human operator to manually select the location of the anchor nodes, or a time-consuming and costly election protocol to perform the same task. Moreover, rotating anchor nodes is costly. None of the cited works answers the questions related to network dynamics. During the lifetime of the network, nodes – including anchor nodes – may disappear, and new nodes may appear. Moreover, wireless links are dynamic. The usual answer to these problems is to periodically rebuild the relative coordinate system. This is not satisfactory as coordinates may continuously become outdated, and periodic rebuilding may be unnecessary when there is no traffic. New solutions are needed.

ANCHOR-FREE VIRTUAL COORDINATE-BASED SOLUTIONS

Motivation and Theoretical Basis

To sum up the previous parts, some applications require each node to know (an approximation of) its **true physical coordinates**. These obtained **estimated physical coordinates** (using GPS, manual programming or localization protocols) can not be used as such for routing because they are not related to the network topology. To answer this, a node can determine coordinates relatively to a set of location-unaware anchor nodes. These **relative coordinates** can be used for routing in WSNs (outperforming solutions with true physical coordinates). Nevertheless, the use of relative coordinates suffers from the cost of electing a set of anchor nodes, and from network dynamics.

In this section, we introduce **virtual coordinates**. Like relative coordinates, they are not related to the node's true physical coordinates, but are used for routing in WSNs. They offer solutions which perform significantly better than using true physical coordinates. Unlike relative coordinates, no anchor nodes are required for setting up virtual coordinates, and the solution elegantly copes with network dynamics.

Research on applying non-physical coordinates to wireless multi-hop nodes has been driven by the quest for a greedy embedding. A graph is defined as a set of vertices interconnected by edges. A greedy embedding of a graph is composed of the same edges interconnecting the same vertices, only the vertices have been placed at coordinates such that greedy routing always functions when sending a message between arbitrarily chosen nodes (i.e. there are no void areas).

The notion of greedy embedding was developed by Papadimitriou and Ratajczak (Papadimitriou & Ratajczak, 2004), who studied the special case of the Euclidian space. They provided examples of graphs which do not admit a greedy embedding in the Euclidean plane, yet they conjectured that every 3-connected planar graph admits a greedy embedding in the Euclidean plane.

Kleinberg has extended this work and shown that every connected finite graph has a greedy embedding in the hyperbolic plane (Kleinberg, 2007). The underlying algorithm, however, assumes that the network is capable of computing a spanning tree rooted at some node. Although a fair assumption (distributed protocols for computing a spanning tree are abundant in the literature and in practice), using a spanning tree requires the network to maintain this structure, which may be hard and costly. Moreover, in theory, the worst-case path stretch (the ratio of the number of hops on a greedy route to the number of hops on the shortest route between the same pair of nodes) is linear in the network size.

The solution we propose does not require an initialization phase. This means it is functional as soon as the network is deployed. The nodes use virtual coordinates which are updated throughout the network lifetime. No network-wide periodic updates are needed, and the system is extremely robust against nodes (dis)appearing and link dynamics. The path stretch is small, typically a few percents above 1.

As the nodes' virtual coordinates are constantly updated, there is no distinct localization phase followed by a routing phase, as it is the case when using physical or relative coordinates. This significantly increases network robustness as any topological change will be reflected into the nodes' virtual coordinates immediately. This also means that localization (i.e. nodes determine their virtual coordinates) and routing (i.e. a geographic routing protocol uses these virtual coordinates to find a path to the destination) are intertwined and happen at the same time.

Initialization and Iterative Convergence Process

Let's assume we have a planar 2-D network. Each node has two virtual coordinates (i.e. $\{x,y\}$, x and y being real numbers). When a node is switched on, it chooses its initial virtual coordinates randomly within a common given range, e.g. [0 1000]. The sink node always chooses the virtual coordinates $\{0,0\}$.

Each time a node sends a message, it replaces its virtual coordinates with the average of its neighbors'. The sink node is an exception to this rule as it never changes it virtual coordinates from $\{0,0\}$. Learning the virtual coordinates of its neighbors can easily be implemented as an on-demand service provided by the Medium Access Control (MAC) layer. As an example, such a protocol is proposed in (Watteyne, Bachir, Dohler, Barthel, & Augé-Blum, 2006). After updating its virtual coordinates, the current node appends these new virtual coordinates to the message it is about to send. Other than sending these coordinates along with the message, there is no additional overhead to our approach, i.e. no signaling messages at the network layer.

As a message is sent over the wireless medium, all neighbors will hear the current node's new virtual coordinates. If a neighbor node finds out that it is virtually closer to the current node than a minimal "safety distance", it updates it own virtual coordinates in order to be at a threshold distance *MinVirtual*. After this step, no neighbor table is maintained, i.e. no long-term information is kept.

The complete process is depicted in Figure 8, which represents four nodes placed at their virtual coordinates. Edges interconnect nodes capable of communicating with each other. Node V has three neighbor nodes W, X and Y. In Figure 8(a), it wants to send a message. It learns the virtual coordinates of its neighbors thanks to the MAC layer and replaces its virtual coordinates with the average value of its neighbors' virtual coordinates (Figure 8(b)). Node V now sends its message, appending its updated virtual coordinates. Node X finds out it is virtually closer to V than the threshold virtual distance *MinVirtual*, represented by a dashed circle. Node X thus updates its own virtual coordinates so as to virtually "slide" away from node V, until it is at virtual distance *MinVirtual* from it (Figure 8(c)). Note that when sliding away, node X remains on the same axis XV.

The nodes in the network know that the sink always chooses virtual coordinates $\{0,0\}$. As a result, the sink does not need to broadcast its coordinates to the entire network. This characteristic can be especially helpful when the sink node is relocated to another place (we will detail this case further in the text).

Figure 8. The updating process when using virtual coordinates

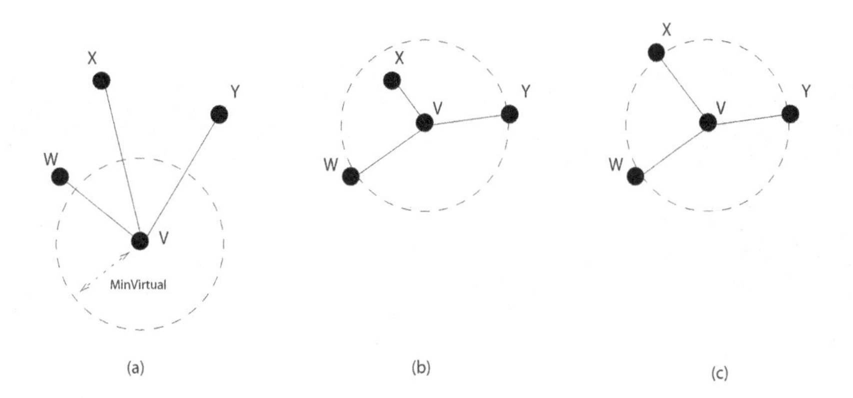

Explaining Network Convergence

Before presenting the performance results when using the presented virtual coordinates, we believe it is important to discuss the intuition behind it. As detailed in the introductory part of this chapter, self-organization in WSN is shifting from complex to lightweight protocols. The functionality of the latter comes from the emergence of a network-wide behavior as a consequence of the (simple) interactions between neighbor nodes.

Analogies can be seen between the presented protocol and the behavior of animal swarms. In a bird flock, a single bird can only see its neighbor birds, and knows where they are. To keep the flock together, a bird moves equally close to each of its neighbors; yet, to avoid collision, it stays at a safety distance. The parameter *MinVirtual* represents this safety distance. Without it, as nodes update their virtual coordinates, they would get virtually closer to one another and closer to the sink. Virtual coordinates would take infinitely small values, which are hard to handle by the fixed-point computation unit typically found in the microprocessors/microcontrollers at the heart of wireless sensors.

The flock forms a homogeneous structure, all bird following a leader in the front (Couzin, Krause, Franks, & Levin, 2005). Our system adopts the same strategy. As described in the next paragraphs, the virtual coordinates of the nodes align, the "leader" role being played by the sink node.

Figure 9. Witnessing network convergence

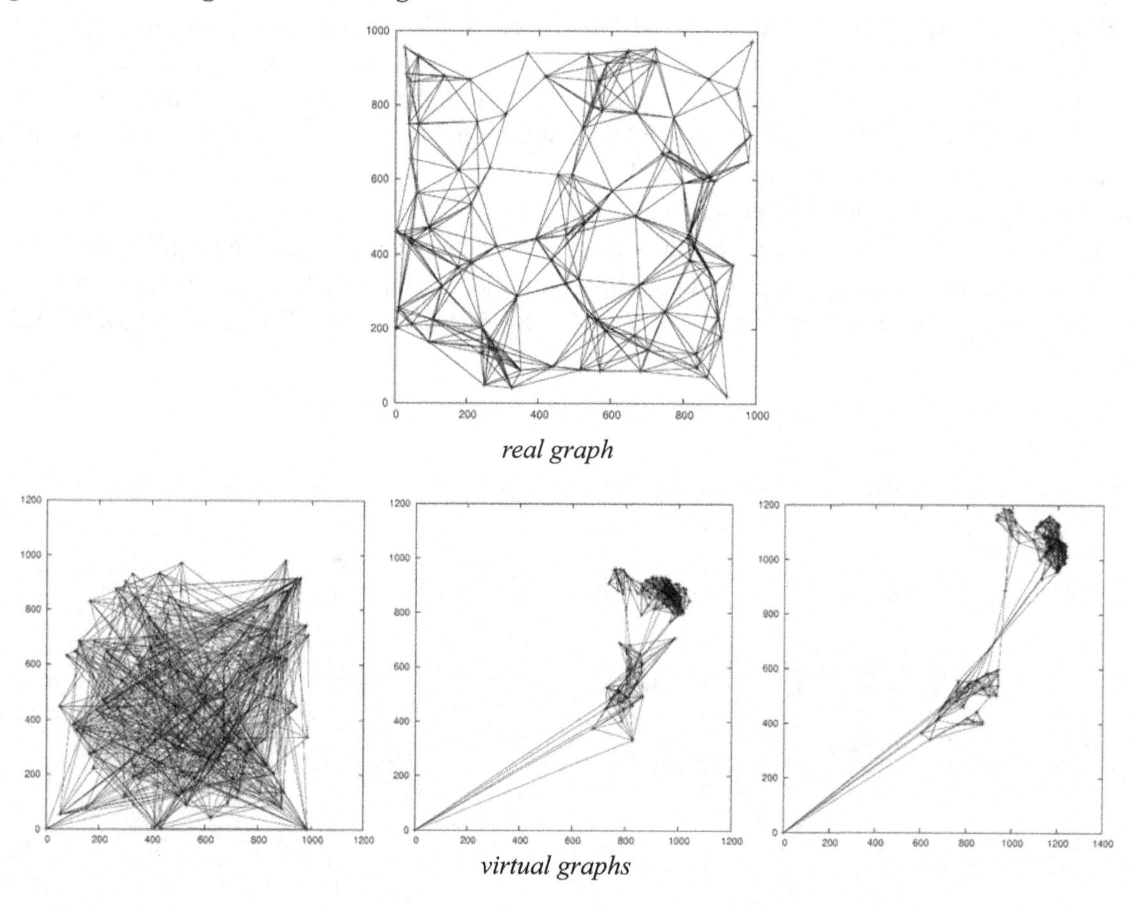

To help the reader visualize the emergent behavior of our WSN, we refer to Figure 9. This figure was obtained by simulating the behavior of 100 nodes, randomly scattered within a two dimensional square space of dimensions 1000×1000; each node has a 200 unit radio range. The upper part shows the real graph, i.e. nodes are positioned at their true physical coordinates with edges interconnecting nodes able to communicate. The lower part represents the virtual graph, i.e. the same vertices and edges as in the upper drawing, only nodes are positioned at their virtual coordinates. We show snapshots of the virtual graph after 0, 100 and 500 messages have been sent (from left to right). Each of these messages is sent from a randomly chosen connected node (different for each message) to the sink node.

The initial virtual graph (Figure 9, lower left) looks erratic as virtual coordinates are initially chosen randomly. As the number of sent messages increases, the virtual coordinates of the nodes align. While messages flow through the network, neighbor nodes are brought virtually closer to one another. In the resulting linear structure, nodes topologically close to the sink node are also virtually close, and vice-versa. Once the virtual coordinates have converged, virtual distance to the sink is hence closely related to the minimum number of hops to the sink. As we will see in the next paragraphs, using geographic routing protocols on top of these virtual coordinates yields near-optimal path length. In these simulations, we have used *MinVirtual=40*. This guard distance causes the virtual coordinates to expand, i.e. after 500 messages, nodes are on average virtually farther away from the sink than after 100 messages.

Note that our solution supports the use of multiple sinks. Without loss of generality, let us assume we have two sinks in the network. Each node would now have two pairs of virtual coordinates, one for each sink. A sink would have fixed virtual coordinates only for "its" set of virtual coordinates. To select the destination sink node, a sending node uses its respective set of virtual coordinates. Note that in case a sink node is moved to a different geographical location, the network automatically re-converges after the relocation. This re-convergence does, however, come with an extra energy-expenditure.

Proving Network Convergence

As shown in Figure 9, lower-left corner, the initial graph is erratic. Without loss of generality, we show that under the assumption that *MinVirtual=0*, the virtual graph converges to a linear virtual graph, i.e. the virtual coordinates of all nodes are on a line. We consider the network is composed of N nodes.

As depicted in Figure 10, the lines passing through the sink node and forming angles with the lower side of the network form a cone which contains all the nodes in the network. We define θ_{min} and θ_{max} as follows:

$$\begin{cases} \theta_{min} = \min\left(\arctan\left(\frac{y_i}{x_i}\right)\right), 0 < i < N \\ \theta_{max} = \max\left(\arctan\left(\frac{y_i}{x_i}\right)\right), 0 < i < N \end{cases}$$

Let's consider the updating process. In particular, let's see how θ_{min} and θ_{max} evolve over time. Both values only change if the updating process affects the node defining this angle (in Figure 10, nodes *A* and *B* for angles θ_{min} and θ_{max}, respectively). We call ξ_A and ξ_B the set of neighbor nodes of A and B, respectively.

Figure 10. Definition of the angles used in the proof

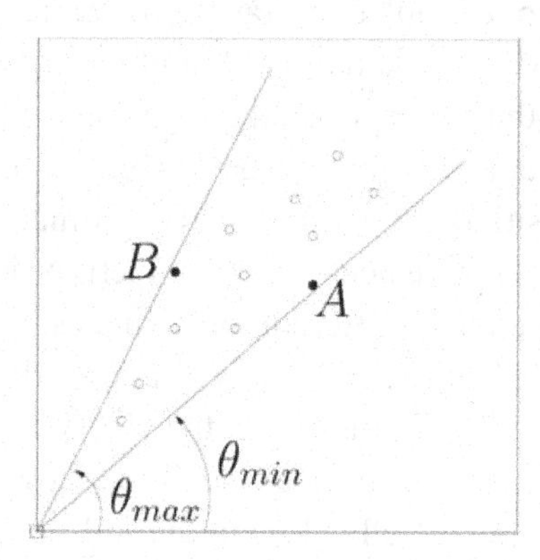

We will focus on θ_{min}, the same analysis applies for θ_{max}. As tan(·) is a strictly increasing function over $\left[0, \frac{\pi}{2}\right]$, we have $\frac{y_A}{x_A} \leq \frac{y_i}{x_i} \forall i \in \xi_A$. If we call $\left[x_A^*, y_A^*\right]$ the new virtual coordinates of A, we have

$$\frac{y_A^*}{x_A^*} = \frac{\Sigma_{i \in \xi_A} y_i}{\Sigma_{i \in \xi_A} x_i} \geq \frac{\Sigma_{i \in \xi_A} x_i \frac{y_A}{x_A}}{\Sigma_{i \in \xi_A} x_i} = \frac{y_A}{x_A}.$$

Therefore, θ_{min} increases. A similar analysis shows that θ_{max} decreases. We have $\lim_{m \to \infty} \theta_{max} - \theta_{min} = 0$, where *m* represents the number of sent messages. Network convergence is hence achieved where the network converges to a linear virtual graph. This analysis still holds with *MinVirtual>0*. The convergence of the system is equivalent, only neighbor nodes are never virtually closer than *MinVirtual*. The cone always stays slightly open.

Note that neighbor nodes always have different coordinates, which is ensured by the use of *MinVirtual*. As a result, the zones problem described previously (in which several neighbor nodes have the same coordinates) does not exist when using virtual coordinates.

Path Stretch and Speed of Convergence

Virtual and relative coordinates were introduced to be used by a geographic routing protocol. To increase the network throughput and reduce the energy expenditure, the multi-hop path discovered by the routing protocol should have the smallest possible number of hops. Hence, to evaluate efficiency of routing protocols, path stretch is a commonly used metric. It is the number of hops obtained by a given protocol divided by the minimum number of hops, using the centralized Dijsktra algorithm (calculating the shortest possible path). A path stretch of 1 is optimal.

We present simulation results in the following paragraphs. These were obtained using the same parameters already described above. Each value is averaged over 10^5 runs and presented with a 95% confidence interval.

In a geographic routing protocol, voids can be met, causing the greedy approach to fail. In this case, a second mode is used to circumnavigate the void until the greedy mode can be resumed. As pointed out in (Kim, Govindan, Karp, & Shenker, 2005), face mode protocols such as GFG or GPSR do not function when the connectivity graph is not a unit disk graph, which is the case of our virtual graph. We therefore use the 3rule routing protocol (Watteyne, Augé-Blum, Dohler, & Barthel, 2007) together with virtual coordinates. In this protocol, each traversed node is asked to append its identifier in the packet's header. Based on a sequence of nodes already traversed, a node can elect the next hop in a way that guarantees delivery.

Performances of the resulting communication architecture are compared with the GFG/GPSR protocols. Note that the performances of GFG/GPSR are extracted assuming all nodes have a perfect knowledge of their true physical coordinates. For our simulations, the source node of each message is chosen randomly among the nodes which are connected to the sink, and changes at each run.

Figure 11 shows how the average path stretch of the virtual coordinate setting decreases as a function of the number of sent messages. As messages flow through the network, virtual coordinates align and the path stretch decreases. After about 100 messages, using virtual coordinates turns out to be more efficient than using true physical coordinates. Although the speed of convergence depends on the topology of the network and on the message generation model, simulations show that the number of messages needed for convergence is roughly proportional to the depth of the network, i.e. the maximum number of hops between any node and the sink. Note that the path stretch of the GFG/GPSR protocols does not depend on the number of messages sent. An optional initialization message could speed up the convergence of the network. Developing such a hybrid solution is relatively straightforward.

Convergence and Energy Efficiency

Figure 12 (left) and Figure 11 have been drawn for sparse networks (average node degree of 4) and dense networks (average node degree of 11), respectively. Because more voids appear as a network gets sparser, GFG/GPSR perform worse on sparse networks than on dense ones. Performances of virtual coordinates degrade only slightly.

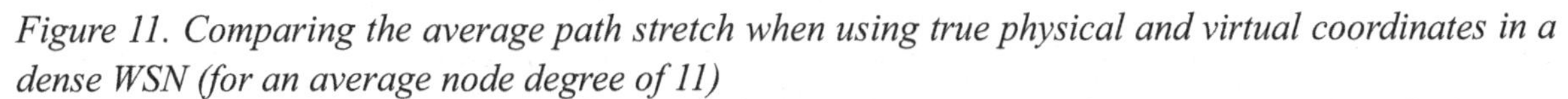

Figure 11. Comparing the average path stretch when using true physical and virtual coordinates in a dense WSN (for an average node degree of 11)

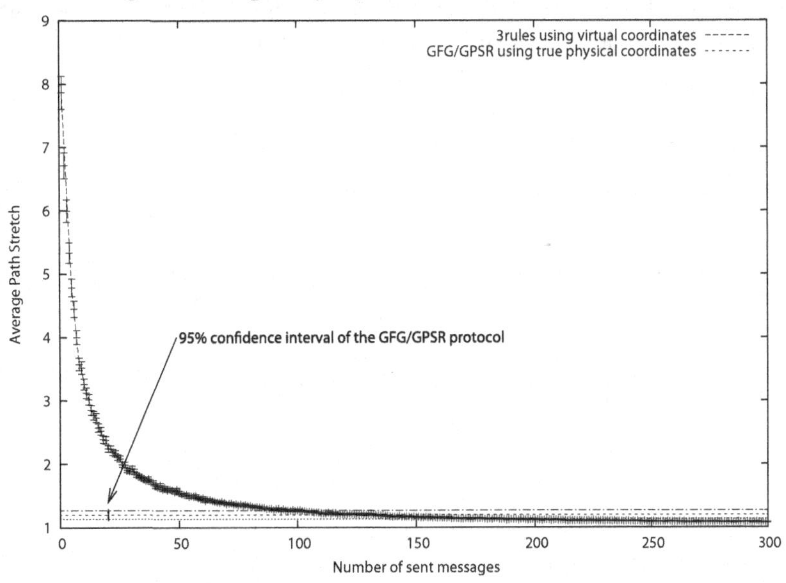

Figure 12. Comparing the average (left) and cumulated (right) path stretch when using true physical and virtual coordinates in a sparse WSN (for an average node degree of 4)

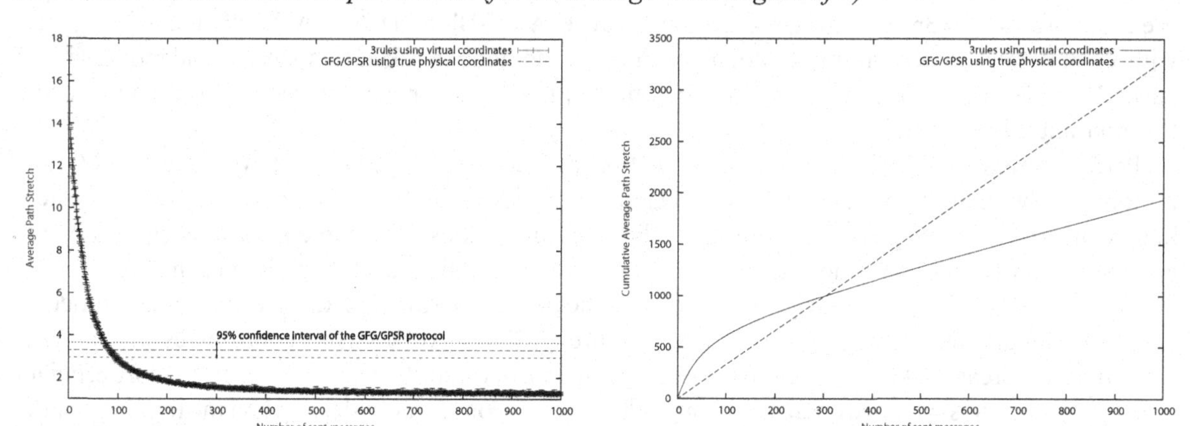

Figure 12 (right) depicts the cumulative average path stretch, i.e. an integration of Figure 12 (left). The cumulative average path stretch is proportional to the total number of messages sent, thus to the total energy consumed. After about 300 messages, it is more energy-efficient to use virtual coordinates than true physical ones, i.e. the gain in using virtual coordinates is larger than the extra cost induced during network ramp up.

Let us take the realistic scenario of a 100-node environmental monitoring WSN where each sensor reports a reading twice a day for 15 years. After 15 years, about 1 million messages will have traversed the network. By using virtual coordinates in such a scenario, the network saves 61.4% of the energy it would spend if using true physical coordinates, as multi-hop paths are shorter. This number is obtained by extrapolating Figure 12 (right) linearly.

Robustness against Nodes (Dis)Appearing

During the lifetime of the network, some nodes will die due to battery exhaustion or hardware failure. In the meantime, the network administrator may decide to add new nodes. These events should be efficiently taken into account when designing self-organization protocols for WSNs.

Figure 13. Robustness against nodes (dis)appearing

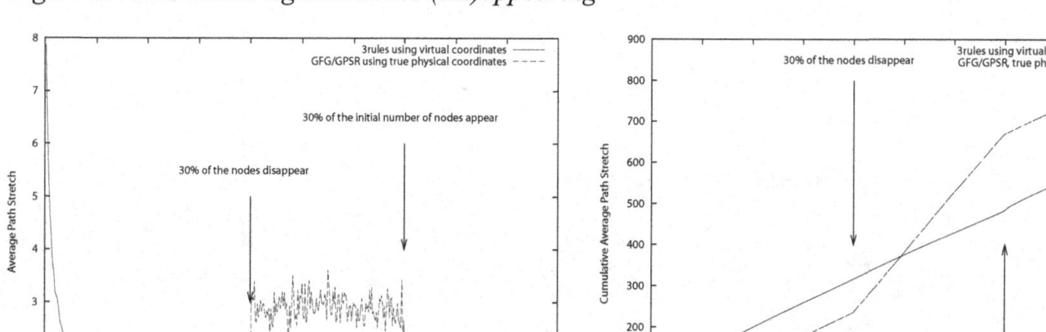

Figure 14. A real graph with 10 obstacles

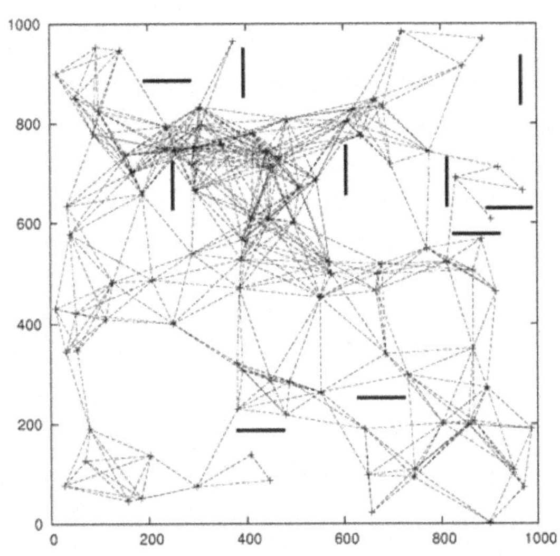

Figure 15. The impact of the number of obstacles (left) and localization accuracy (right) on the delivery ratio of true physical and virtual coordinate-based solutions

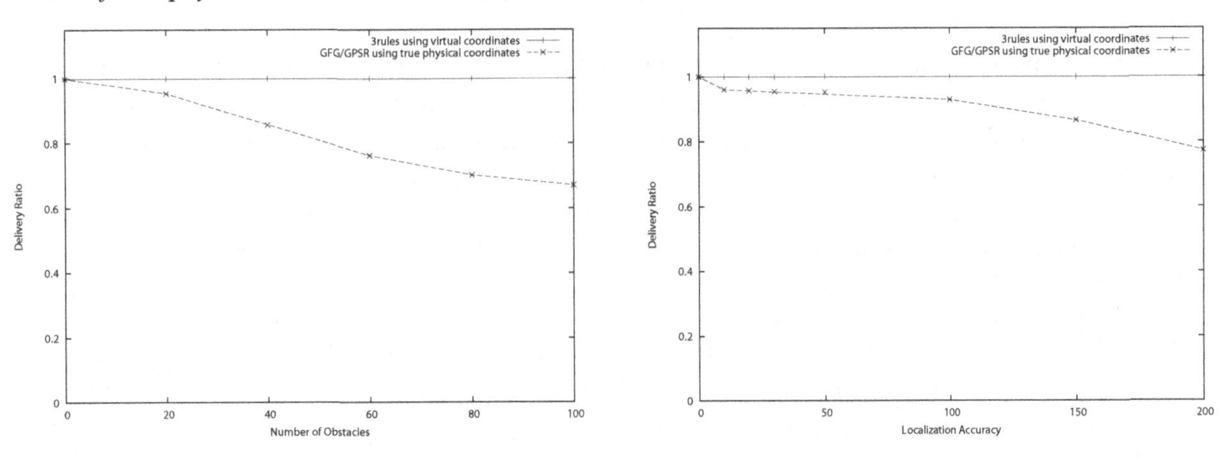

Let us take the extreme case of an earthquake simultaneously destroying 30% of the nodes of a 100-node network. After the tragic event, a helicopter flies over the monitored area and randomly drops new nodes into the network. Simulation results of this scenario are presented in Figure 13. As expected, GFG/GPSR performs worse under low density, which translates into a sharp increase in the energy consumption of the network. Our virtual coordinate-based solution quickly adapts to the new situation, and converges back to a near-optimal state after nodes are removed/added. As can be seen on Figure 13 (right), removing or adding new nodes only has a very limited impact on the network's energy consumption.

Figure 16. The experimental setting used for the proof-of-concept experiment

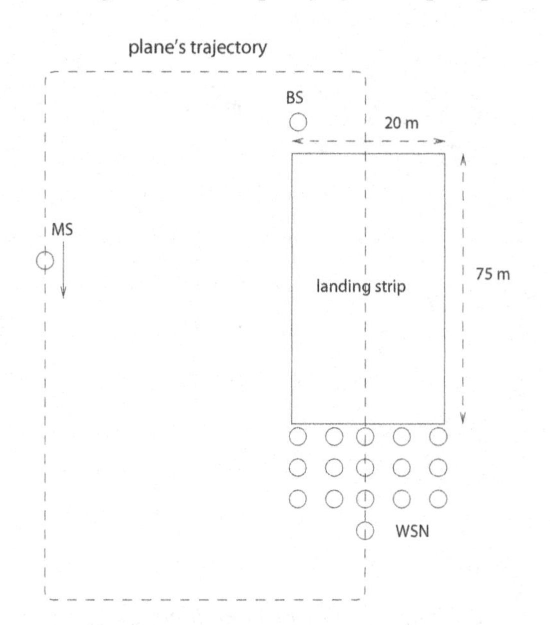

Robustness against a Realistic Transmission Model

So far, all simulations were performed under the unit disk graph assumption, i.e. the communication area of each node is a perfect circle. We break this assumption by randomly deploying linear obstacles of length 100 inside the network. Their positions and orientations (horizontal or vertical) are chosen randomly (see Figure 14).

As shown in Figure 15 (left), the delivery ratio of the true physical coordinate-based solution degrades as the number of obstacles increases, whereas the virtual coordinate-based solution keeps delivering all sent messages.

In Figure 15 (right), we confirm results from (Kim, Govindan, Karp, & Shenker, 2005) stating that GFG/GPSR fails when the nodes do not know their true physical position with sufficient accuracy. Delivery ratio drops with positioning accuracy. We define localization accuracy as follows. When a node believes it is at location $\{x,y\}$ with accuracy a, it means it is somewhere inside the square box with opposite corners at locations $\{x-a,y-a\}$ and $\{x+a,y+a\}$.

Proof-of-Concept Experiment

We have carried out an experiment to prove that virtual coordinates can be implemented on low-end sensor nodes and perform well when facing real world constraints. We have pushed the system to the extreme case of a fast-moving sink node traversing the network.

The experimental setting is shown in Figure 16. A 16-node network is deployed on an airfield, at one end of the landing strip. A base station (denoted *BS*) is installed at the other end and issues requests which can be answered by the WSN. As the WSN and the BS are too far apart to communicate directly, a mobile sink node (denoted *MS*) is mounted on a radio-controlled airplane. After receiving a request from the *BS*, the *MS* circles around and sends the request to the WSN which is broadcasted to all nodes in the WSN. The node which can answer the request sends its reply to the mobile sink using the virtual coordinate routing scheme described in this chapter. After receiving this reply, the *MS* flies to the *BS* and replies to the request. The interested reader is referred to an internal research report (Watteyne, Barthel, Dohler, & Augé-Blum, 2008) which contains all the details about this experiment.

OPEN QUESTIONS AND RESEARCH CHALLENGES

The concept of self-organization has undergone a paradigm shift when applied to WSNs. Because of their stringent energy constraints and low-throughput, building and maintaining complex structures may not be applicable. Inspiration from biological systems such as animal flocks has lead to defining self-organization as an emergent behavior coming from simple interactions between a node and its direct neighbors.

Because of its natural scalability, coordinates have been used as a basis for communication in WSNs. The use of location-aware anchor nodes combined with appropriate localization protocols allows nodes to learn their (estimated) physical coordinates. Although this information is required for some applications, topology-related coordinates are more appropriate for routing. These coordinates can be learned relatively to location-unaware anchor nodes, and are not related to the node's true physical coordinates. To avoid the election of anchor nodes and the use of costly localization protocols, this chapter has introduced virtual coordinates. Using an appropriate updating algorithm triggered each time a message is sent, these coordinates converge to a near-optimal emerging state, which is extremely robust to network dynamics and realistic propagation models.

This proposal constitutes a step towards a fully autonomic network. Being able to cope with dynamics of all kinds, such a network offers a deploy-and-forget experience to the end user. Only with these characteristics will WSNs really get ubiquitous.

A long road still lies ahead, and the work presented opens many perspectives for research. Communication systems, especially self-organizing solutions, largely rely on periodic signaling messages. Such a pro-active system makes sense under high traffic loads. In the context of WSNs, where typical applications require a node to send a message only every now and then, periodically maintaining a structure may be too expensive as the network sits idle most of the time. An important research challenge is to investigate fully on-demand approaches.

True physical coordinates are required by the application; virtual ones are used for communication. As most WSNs cope with both aspects, combining true physical and virtual coordinates is essential. Both types of coordinates could be acquired simultaneously, reducing the signaling required. A challenging approach is to define hybrid coordinates, close enough to the nodes' true physical coordinates, but useful for routing. Depending on the use, virtual coordinates could be extended to more that 2 coordinates. The one dimensional problem could also be considered, although an extension to hybrid coordinates seems harder in such a setting.

ACKNOWLEDGMENT

The authors would like to thank André Watteyne (KULeuven, Campus Kortrijk, Belgium) for his input on self-organizing economics models, Christine Métin (INSERM UMR-S 839, Université Paris 6, France) for her invaluable contribution on the biological aspects of self-organization, and Claude Nicol and Bjarne Winkler for their expert eyes on animal swarms.

REFERENCES

Benbadis, F., Friedman, T., Amorim, M. Dias de, & Fdida, S. (2005). GPS-Free-Free Positioning System for Wireless Sensor Networks. *In Second IFIP International Conference on Wireless and Optical Communications Networks (WOCN)* (p. 541-545). Dubai, United Arab Emirates.

Benbadis, F., Obraczka, K., Cortes, J., & Brandwajn, A. (2007). Exploring Landmark Placement Strategies for Self-Localization in Wireless Sensor Networks. *In 18th Annual International Symposium on Personal, Indoor and Mobile Radio Communications (PIMRC)* (pp. 1-5). Athens, Greece.

Brenner, M. P., Levitov, L. S., & Budrene, E. O. (1998). Physical Mechanisms for Chemotactic Pattern Formation by Bacteria. *Biophysical Journal, 74,* 1677-1693.

Cao, Q., & Abdelzaher, T. (2004). A Scalable Logical Ccoordinates Framework for Routing in Wireless Sensor Networks. *In 25th IEEE International Real-Time Systems Symposium (RTSS)* (pp. 349-358). Lisbon, Portugal.

Caruso, A., Chessa, S., De, S., & Urpi, A. (2006). GPS Free Coordinate Assignment and Routing in Wireless Sensor Networks. *In Annual Joint conference of the Computer and Communication Societies (INFOCOM)* (pp. 150-160). Barcelona, Spain.

Couzin, I. D., Krause, J., Franks, N. R., & Levin, S. A. (2005). Effective Leadership and Decision Making in Animal Groups on the Move. *Nature, 433,* 513-516.

Culler, D., Estrin, D., & Srivastava, M. (2004). Overview of Sensor Networks. *IEEE Computer, Guest Editors' Introduction, 37*(8), 41-49.

Dohler, M., Barthel, D., Maraninchi, F., Mounier, L., Aubert, S., Dugas, C., et al. (2007). The ARESA Project: Facilitating Research, Development and Commercialization of WSNs. *In 4th Annual IEEE Communications Society Conference on Sensor, Mesh and Ad hoc Communications and Networks (SECON)* (pp. 590-599). San Diego, CA, USA.

Fonseca, R., Ratnasamyy, S., Zhao, J., Ee, C. T., Culler, D., Shenker, S., et al. (2005). Beacon Vector Routing: Scalable Point-to-Point Routing in Wireless Sensornets. *In 2nd Symposium on Networked Systems Design & Implementation (NSDI)* (pp. 329-342). Boston, MA, USA.

Frey, H., & Stojmenovic, I. (2006). On Delivery Guarantees of Face and Combined Greedy-Face Routing Algorithms in Ad hoc and Sensor Networks. *In Twelfth ACM Annual International Conference on Mobile Computing and Networking (MOBICOM)* (pp. 390-401). Los Angeles, CA, USA.

Karl, H., & Willig, A. (2005). *Protocols and Architectures for Wireless Sensor Networks*, H. Karl & A. Willig, (Eds.). John Wiley & Sons, Inc., Hoboken, New Jersey.

Kim, Y.-J., Govindan, R., Karp, B., & Shenker, S. (2005). Geographic Routing Made Practical. *In 2nd Symposium on Networked Systems Design & Implementation (NSDI)* (pp. 217-230). Boston, MA, USA.

Kleinberg, R. (2007). Geographic Routing Using Hyperbolic Space. *In 26th International Conference on Computer Communications (INFOCOM)* (pp. 1902-1909). Anchorage , AL, USA.

Leong, B., Liskov, B., & Morris, R. (2007). Greedy Virtual Coordinates for Geographic Routing. *In IEEE International Conference on Network Protocols (ICNP)* (pp. 71-80). Beijing, China.

Liu, K., & Abu-Ghazaleh, N. (2006). Aligned Virtual Coordinates for Greedy Routing in WSNs. *In International Conference on Mobile Adhoc and Sensor Systems (MASS)* (pp. 377-386). Vancouver, Canada.

McMillan, J. (2002). Reinventing the Bazaar. *A Natural History of Markets.* J. McMillan (Ed.). W. W. Norton & Company.

Mills, K. L. (2007). A Brief Survey of Self-Organization in Wireless Sensor Networks. *Wireless Communication and Mobile Computing, 7,* 823-834.

Niculescu, D., & Nath, B. (2003). Ad Hoc Positioning System (APS) Using AOA. *Annual Joint Conference of the Computer and Communication Societies (INFOCOM)* (pp. 1734-1743). San Francisco, CA, USA.

Papadimitriou, C. H., & Ratajczak, D. (2004). *On a Conjecture Related to Geometric Routing (Algorithmic Aspects of Wireless Sensor Networks).* In SpringerLink (Ed.), (Vol. 3121/2004, pp. 9-17). Springer Berlin / Heidelberg.

Patwari, N., Ash, J. N., Kyperountas, S., Hero, A. O. I., Moses, R. L., & Correal, N. S. (2005). Locating the Nodes - Cooperative Localization in Wireless Sensor Networks. *IEEE Signal Processing Magazine, 1,* 54-69.

Prehofer, C., & Bettstetter, C. (2005). Self-Organization in Communication Networks: Principles and Design Paradigms. *IEEE Communications Magazine, 43*(7), 78-85.

Stojmenovic, I., & Olariu, S. (2005). Geographic and Energy-Aware Routing in Sensor Networks. In Stojmenovic, (Ed.), *Handbook of sensor networks: Algorithms and architectures,* (p. 381-416). Wiley.

Theoleyre, F., & Valois, F. (2007). Self-Organization of Ad Hoc Networks: Concepts and Impacts. In H. Labiod (Ed.*), Wireless Ad hoc and Sensor Networks*, (pp. 101-128). ISTE.

Tonguz, O. K., & Ferrari, G. (2006). *Ad hoc Wireless Networks: A Communication-theoretic Perspective,* O. K. Tonguz & G. Ferrari (Eds.). Wiley.

Watteyne, T., Augé-Blum, I., Dohler, M., & Barthel, D. (2007). Geographic Forwarding in Wireless Sensor Networks with Loose Position-Awareness. *In 18th Annual International Symposium on Personal, Indoor and Mobile Radio Communications (PIMRC)* (pp. 1-5). Athens, Greece.

Watteyne, T., Bachir, A., Dohler, M., Barthel, D., & Augé-Blum, I. (2006). 1-hopMAC: An Energy-Efficient MAC Protocol for Avoiding 1-hop Neighborhood Knowledge. *In International Workshop on Wireless Ad-hoc and Sensor Networks (IWWAN)* (pp. 639-644). New York, NY, USA.

Watteyne, T., Barthel, D., Dohler, M., & Augé-Blum, I. (2008). *WiFly: Experimenting with Wireless Sensor Networks and Virtual Coordinates* (Research Report RR-6471). INRIA.

Compilation of References

Aalto, L., Göthlin, N., Korhonen, J., & Ojala, T. (2004). Bluetooth and WAP push based location-aware mobile advertising system. *In Proceedings of the 2nd international Conference on Mobile Systems, Applications, and Services*, (pp. 49-58), Boston, MA.

Abel, J. S. (1990). A divide and conquer approach to least-squares estimation. *IEEE Transactions on Aerospace and Electronic Systems, 26*(2), 423-427.

Abnizova, I., Cullen, P., & Taherian, S. (2001). Mobile Terminal Location in Indoor Cellular Multi-path Environment. *Proceedings of the Third IEEE Workshop on Wireless LANs*, retrieved October 13, 2008, from http://www.wlan01.wpi.edu/proceedings/wlan69d.pdf.

Aeroscout. (2008). *Aeroscout enterprise visibility solutions, white paper.* Retrieved October 10, 2008, from http://www.aeroscout.com

Agee, B. G. (1991). Copy/DF approaches for signal specific emitter location. *the Twenty-Fifth Asilomar Conference on Signals, Systems and Computers* (pp. 994-999).

Agrawal, P., & Patwari, N. (2008). *Correlated Link Shadow Fading in Multi-hop Wireless Networks*, (Tech Report arXiv:0804.2708v2), arXiv.org. Retrieved 18 Apr 2008 from http://arxiv.org/abs/ 0804.2708v2

Ahmed, A. A., Shi, H., & Shang, Y. (2003). A survey on network protocols for wireless sensor networks. *IEEE International Conference on Information Technology: Research and Education, Proceedings. ITRE2003.* (pp. 301-305).

Ahmed, A. A., Shi, H., & Shang, Y. (2005). SHARP: A New Approach to Relative Localization in Wireless Sensor Networks. *Proceedings of the 25th IEEE International Conference on Distributed Computing Systems Workshops (ICDCSW'05)* (pp 892-888).

Aiello, G. R., & Rogerson, G. D. (2003). Ultra-wideband wireless systems. *IEEE Microwave Magazine, 4*(2), 36-47.

Airtight Networks. (2008). *Airtight networks, white paper.* Retrieved October 10, 2008, from http://www.airtightnetworks.net

Akaho, S. (2001). A kernel method for canonical correlation analysis. *In Proceedings of the International Meeting of the Psychometric Society (IMPS 2001).*

Akgul, F. O., & Pahlavan, K. (2007). AOA Assisted NLOS Error Mitigation for TOA-based Indoor Positioning Systems. *IEEE MILCOM.* (pp. 1-5). Orlando, FL.

Akima, H. (1970). A new method of interpolation and smooth curve fitting based on local properties. *Journal of the ACM, 17*(4), 589-602.

Akima, H. (1996). Algorithm 760: Rectangular-Grid-Data Surface Fitting that has the Accuracy of Bicubic Polynomial. *ACM Transactions on Mathematical Software, 22*(3), 357-361.

Akima, H. (1996). Algorithm 761: Scattered-Data Surface Fitting that has the Accuracy of Cubic Polynomial. *ACM Transactions on Mathematical Software, 22*(3), 362-371.

Akkaya, K., & Younis, M. (2005). A survey on routing protocols for wireless sensor networks. *Elsevier Journal of Ad Hoc Networks.* (pp. 325-349).

Akyildiz, I. F., Su, W., Sankarasubramaniam, Y., & Cayirci (2002). E. Wireless sensor networks: A survey. *Computer Networks, 38*, 393-422.

Akyildiz, I. F., Su, W., Sankarasubramaniam, Y., & Cayirci, E. (2002, November 7). A survey on sensor networks. *Communications Magazine, IEEE*, (pp. 102-114).

Akyildiz, I., & Wang, X. (2005). A survey on wireless mesh networks. *IEEE Communicatios Magazine*, 43(9).

Akyildiza, I. F., Wang, X., &Wang, Y., (2005).Wireless Mesh Networks: A Survey. *Elsevier Journal of Computer Networks*. (pp. 445-487)

Alavi, B., & Pahlavan, K. (2003). Bandwidth effect of distance error modeling for indoor geolocation. In *IEEE Personal Indoor Mobile Radio Communications Conference (PIMRC), 3*, 2198-2202.

Alavi, B., & Pahlavan, K. (2006). Modeling of the TOA based Distance Measurement Error Using UWB Indoor Radio Measurements. *IEEE Communication Letters, 10*(4), 275-277.

Alavi, B., & Pahlavan, K. (2006). Studying the effect of bandwidth on performance of UWB positioning systems. In *Proceedings of the IEEE Wireless Communications and Networking Conference (WCNC), 2*, 884-885.

Alavi, B., Pahlavan, K., Alsindi, N., & Li, X. (2005). Indoor geolocation distance error modeling with UWB channel measurements. In *IEEE Personal Indoor Mobile Radio Communications Conference (PIMRC), 1*, 418-485.

Albowicz, J., Chen, A., & Zhang, L (2001). Recursive Position Estimation in Sensor Networks. *Proceedings of the 9th International Conference on Network Protocols*, Riverside, CA, (pp. 35-41).

Ali, A. M., Yao, K., Collier, T. C., Taylor, C. E., Blumstein, D. T., & Girod, L. (2007). An empirical study of collaborative acoustic source localization. *In Proceedings of the 6th international Conference on information Processing in Sensor Networks (IPSN '07)*, (pp 41-50).

Allen, M., Gaura, E., Newman, R., Mount, S., (2006). Experimental Localization with MICA2 Motes. *In Proceedings of NSTI Nanotech 2006*, (pp. 435-440).

Allen, M., Girod, L., Newton, R., Madden, S., Blumstein, D.T., Estrin, D. (2008). VoxNet: An Interactive, Rapidly-Deployable Acoustic Monitoring Platform. *In Proceedings of the 7th International Conference on Information Processing in Sensor Networks (IPSN '08)*, (pp. 371-382).

Al-Rousan, A., Al-Ali, R., & Darwish, K. (2004). GSM-Based Mobile Tele-Monitoring and Management System for Inter-Cities Public Transportations. *IEEE International Conference on Industrial Technology*. (pp. 859- 862)

Alsindi, N., & Pahlavan, K. (2008). Cooperative localization bounds for indoor ultra wideband wireless sensor networks. *EURASIP Journal on Applied Signal Processing (ASP)*,2008. article id 852809. (pp.1-13).

Andersen, A., (2007). 2.4 GHz Inverted F Antenna. *Texas Instruments Design Note DN0007*

Anderson, B. D. O., Belhumeur, P. N., Eren, T., Goldenberg, D. K., Morse, A. S., Whiteley, W., & Yang, Y. R. (2007). Graphical properties of easily localizable sensor networks. *Wireless Networks*. Thre Netherlands: Springer.

Anjum, F., Pandey, S., & Agrawal, P. (2005). Secure Localization in Sensor networks using transmission range variation. *Proceedings of the IEEE Mobile Adhoc and Sensor Systems Conference*.

Ansari, J. *et al.* (2007). Flexible Hardware/Software Platform for Tracking Applications. In *Proceeding of VTC 2007*, Dublin, Ireland.

Ansari, J., Riihijärvi, J., & Mähönen, P. (2007). Combining Particle Filtering with Cricket System for Indoor Localization and Tracking Services. In *Proceeding of PIMRC 2007*, Athens, Greece.

Arora, A., Dutta, P., Bapat, S., Kulathumani, V., Zhang, H., Naik, V., Mittal, V., Cao, H., Demirbas, M., Mohamed, G., Choi, Y-R., Herman, T., Kulkarni, S., Arumugam, U., Nesterenko, M., Vora, A., & Miyashita, M. (2006). A line in the sand: A wireless sensor network for target detection, classification and tracking. *Computer Networks, 46*(5), 605–634.

Arulampalam, S., Maskell, S., Gordon, N., & T. Clapp, (2002). A Tutorial on Particle Filters for On-line Nonlinear/Non-Gaussian Bayesian Tracking. *IEEE Transactions on Signal Processing*, 50(2), 174-188.

Aspnes, J., Eren, T., Goldenberg, D. K., Morse, A. S., Whiteley, W., Yang, Y. R., et al. (2006). A theory of network localization. *IEEE Transactions on Mobile Computing, 5*(12), 1663-1678.

Aspnes, J., Goldenberg, D.K. & Yang, Y.R. (2004) On the computational complexity of sensor network localization, Springer Lecture Notes in Computer Science 3121,

Algorithmic Aspects of Wireless Sensor Networks, (pp. 32-44). Berlin, Springer.

Bahl, P., & Padmanabhan, V. N. (2000). RADAR: An In-Building RF-based User Location and Tracking System. *Proceedings of IEEE Infocom, 2,* Tel Aviv, Israel, (pp. 775-784).

Bahl, P., & Padmanabhan, V., & Balachandran, A. (2000). *Enhancements to the RADAR user location and tracking system.* Microsoft Research Technical Report.

Bancroft, S. (1985). Algebraic solution of the GPS equations. *IEEE Transactions on Aerospace and Electronic Systems AES-21*(1), 56-59.

Barabell, A. (1983). Improving the resolution performance of eigenstructure-based direction-finding algorithms. *IEEE International Conference on Acoustics, Speech, and Signal Processing* (pp. 336-339).

Bartal, Y., Byers, J. W., & Raz D. (1997). Global optimization using local information with applications to flow control. *In IEEE Symposium on Foundations of Computer Science*, (pp. 303–312).

Başaran, C. (2007). A Hybrid Localization Algorithm for Wireless Sensor Networks. *Master's Thesis, Yeditepe University,* Turkey.

Başaran, C., Baydere S., & Kucuk, G. (2008). RH+: A Hybrid Localization Algorithm for Wireless Sensor Networks, *IEICE Transactions on Comm., E91-B*(No.06), 1852-1861.

Battelli, M., & Basagni S. (2007). Localization for wireless sensor networks: Protocols and perspectives. *In Proceedings of IEEE CCECE 2007,* Vancouver, Canada, April 22-26 2007, (pp. 1074-1077).

Battiti, R., & Brunato, M., & Villani, A. (2002). *Statistical Learning Theory for Location Fingerprinting in Wireless LANs* (Tech. Rep. No. DIT-02086). University of Trento, Informatica e Telecomunicazioni.

Beauregard, S. (2006). A Helmet-Mounted Pedestrian Dead Reckoning System. In *Proceedings of the 3rd International Forum on Applied Wearable Computing (IFAWC 2006).* Herzog, O., Kenn, H., Lawo, M., Lukowicz, P., & Troster, G. (Eds.), Bremen, Germany: VDE Verlag. (pp. 79–89).

Bell, K. L., Steinberg, Y., Ephraim, Y., & Van Trees, H. L. (1997). Extended Ziv–Zakai lower bound for vector parameter estimation. *IEEE Transactions on Information Theory, 43*(2), 624-637.

Benbadis, F., Friedman, T., Amorim, M. Dias de, & Fdida, S. (2005). GPS-Free-Free Positioning System for Wireless Sensor Networks. *In Second IFIP International Conference on Wireless and Optical Communications Networks (WOCN)* (p. 541-545). Dubai, United Arab Emirates.

Benbadis, F., Obraczka, K., Cortes, J., & Brandwajn, A. (2007). Exploring Landmark Placement Strategies for Self-Localization in Wireless Sensor Networks. *In 18th Annual International Symposium on Personal, Indoor and Mobile Radio Communications (PIMRC)* (pp. 1-5). Athens, Greece.

Berg, A. & Jordán, T. (2003) A proof of Connelly's conjecture on 3-connected circuits of the rigidity matroid, *J. Combinatorial Theory*, Ser. B. 88, 77-97.

Berg, A. & Jordán, T. (2003) Algorithms for graph rigidity and scene analysis, In G. Di Battista & U. Zwick, (Ed), *11th Annual European Symposium on Algorithms (ESA)* (pp.78-89), Springer Lecture Notes in Computer Science 2832. Berlin, Springer.

Bergamo, P., & Mazzini, G. (2002). Localization in sensor networks with fading and mobility. *The 13th IEEE International Symposium on Personal, Indoor and Mobile Radio Communications* (pp. 750-754).

Berger, B., Kleinberg, J. & Leighton, T. (1999) Reconstructing a three-dimensional model with arbitrary errors, *Journal of the ACM,* 46 (2), 212-235.

Biedka, T. E., Reed, J. H., & Woerner, B. D. (1996). Direction finding methods for CDMA systems. *Thirteenth Asilomar Conference on Signals, Systems and Computers* (pp. 637-641).

Biswas, P., & Ye, Y. (2004). Semidefinite programming for ad hoc wireless sensor network localization. *Third International Symposium on Information Processing in Sensor Networks* (pp. 46-54).

Bliss, D. W., & Forsythe, K. W. (2000). Angle of arrival estimation in the presence of multiple access interference for CDMA cellular phone systems. *Proceedings of the 2000 IEEE Sensor Array and Multichannel Signal Processing Workshop* (pp. 408-412).

Blomenhofer, H., Hein, G., Blomenhofer, E., & Werner, W., (1994). Development of a Real-Time DGPS System in the Centimeter Range. *IEEE 1994 Position, Location, and Navigation Symposium*, Las Vegas, NV, (pp. 532–539).

Bluetooth SIG (2006). *Bluetooth Wireless Technology Becoming Standard in Cars.* Retreived October, 2007, from http:// www.bluetooth.com/Bluetooth/Press/SIG/Test_1.htm

Bobrovsky, B. Z., & Zakai, M. (1975). A lower bound on the estimation error for Markov processes. *IEEE Transactions on Automatic Control, 20*(6), 785-788.

Bobrovsky, B. Z., Zakai, M., & Kerr, T. H. (1990). Comments on Status of CR-like lower bounds for nonlinear filtering. *IEEE Transactions on Aerospace and Electronic Systems, 26*(5), 895-898.

Bollobás, B. (1985) *Random graph.* New York: Academic Press.

Bondy, J.A. & Murty, U.S.R. (2008) *Graph theory.* Springer.

Borg, I., & Groenen, P. (1997). *Modern multidimensional scaling, theory and applications.* New York: Springer-Verlag.

Boser, B. E., Guyon, I. M., & Vapnik V. N. (1992). A training algorithm for optimal margin classifiers. In *5th Annual ACM Workshop on COLT,* (pp. 144-152). ACM Press.

Botteron, C., Fattouche, M., & Host-Madsen, A. (2002). Statistical theory of the effects of radio location system design parameters on the position performance. In *Proc. IEEE Vehicular Technology Conference*, Vancouver, Canada.

Botteron, C., Fattouche, M., & Høst-Madsen, A. (2004). Cramer–Rao Bounds for the Estimation of Multipath Parameters and Mobiles' Positions in Asynchronous DS-CDMA Systems. *IEEE Transactions on Signal Processing, 52*(4), 862-875.

Botteron, C., Host-Madsen, A., & Fattouche, M. (2001). Cramer-Rao bound for location estimation of a mobile in asynchronous DS-CDMA systems. In *Proc. IEEE Conference on Acoustics, Speech and Signal Processing*, Salt Lake City, UT, USA.

Botteron, C., Host-Madsen, A., & Fattouche, M. (2004b). Effects of system and environment parameters on the performance of network-based mobile station position estimators. *IEEE Transactions on Vehicular Technology, 53*(1).

Brenner, M. P., Levitov, L. S., & Budrene, E. O. (1998). Physical mechanisms for chemotactic pattern formation by bacteria. *Biophysical Journal, 74,* 1677-1693.

Breu, H. & Kirkpatrick, D.G. (1998) Unit disk graph recognition is NP-hard. *Comput. Geom. 9* (1-2), 3-24.

Broida, R. (2003). *How to do everything with your GPS.* McGraw-Hill.

Brookner, E. (1998). *Tracking and Kalman filtering made easy.* Toronto, ON: John Wiley & Sons.

Broxton, M., Lifton, J., & Paradiso, J. A. (2006). Localization on the pushpin computing sensor network using spectral graph drawing and mesh relaxation. *SIGMOBILE Mob. Comput. Commun. Rev. 10*(1) (Jan. 2006), 1-12.

Brunato, M., & Battiti, R. (2005). Statistical learning theory for location fingerprinting in wireless LANs. *Computer Networks, 47*(6), 825-845.

Bulusu N., Heidemann J., Bychkovskiy V., & Estrin D. (2002). *Density adaptive beacon placement algorithms for localization in ad hoc wireless networks.* UCLA Computer Science Department Technical Report UCLA-CS-TR-010013, July 2001.

Bulusu, N., Heidemann, J., & Estrin, D. (2000). GPS-less low-cost outdoor localization for very small devices. *IEEE Personal Communications, 7*(5), 28-34.

Bulusu, N., Bychkovskiy, V., Estrin, D., & Heidemann, J. (2002). Scalable ad hoc deployable rf-based localization. *In 2002 Grace Hopper Celebration of Women in Computing Conference.* Vancouver, Canada.

Bulusu, N., Heidemann, J., & Estrin, D. (2001). Adaptive beacon placement. *21st International Conference on Distributed Computing Systems (ICDCS-21),* (pp. 489-498).

C.-C. Shen, C., Srisathapornphat, C., & Jaikaeo, C. (2001). Sensor information networking architecture and applications sensor information networking architecture and applications. *IEEE Personal Communications, 8*(4).

Caffrey, J. J., & Stuber, G. L. (1998). Overview of Radio Location in CDMA Cellular Systems. *IEEE Communications Magazine, 36*(4), 38-45.

Camp T., Boleng J., & Davies V. (2002). A Survey of Mobility Models for Ad Hoc Network Research. *Wireless Communication & Mobile Computing (WCMC): Special issue on Mobile Ad Hoc Networking: Research, Trends and Applications, 2*(5), 483-502.

Cao, M., Anderson, B. D. O., & Morse, A. S. (2005). Localization with imprecise distance information in sensor networks. *Proc. Joint IEEE Conf on Decision and Control and European Control Conf.* (pp. 2829-2834).

Cao, Q., & Abdelzaher, T. (2004). A scalable logical coordinates framework for routing in wireless sensor networks. *In 25th IEEE International Real-Time Systems Symposium (RTSS)* (pp. 349-358). Lisbon, Portugal.

Capkun, S., Hamdi, M., & Hubaux, J. (2001). GPS-free positioning in mobile ad-hoc networks. *34th Hawaii International Conference on System Sciences* (pp. 3481-3490).

Cardell-Oliver, R., Smettem, K., Kranz, M., & Mayer, K. (2004). Field testing a wireless sensor network for reactive environmental monitoring, *In Proceedings of the 2004 Intelligent Sensors, Sensor Networks and Information Processing Conference* (pp. 7-12)

Carter, G. (1981). Time delay estimation for passive sonar signal processing. *IEEE Transactions on Acoustics, Speech, and Signal Processing, 29*(3), 463-470.

Carter, G. (1993). *Coherence and time delay estimation.* Piscataway, NJ: IEEE Press.

Carter, M., Jin, H., Saunders, M., & Ye, Y. (2006). SpaseLoc: An adaptive subproblem algorithm for scalable wireless sensor network localization. *SIAM Journal on Optimization. 17*(4), 1102-1128.

Caruso, A., Chessa, S., De, S., & Urpi, A. (2006). GPS Free Coordinate Assignment and Routing in Wireless Sensor Networks. *In Annual Joint conference of the Computer and Communication Societies (INFOCOM)* (pp. 150-160). Barcelona, Spain.

Castro, P., Chiu, P., Kremenek, T., & Muntz, R. (2001, September). A Probabilistic Room Location Service for Wireless Networked Environments. *Proceedings of the 3rd international conference on Ubiquitous Computing* (pp.18-34), Atlanta, Georgia, USA

Cedervall, M., & Moses, R. L. (1997). Efficient maximum likelihood DOA estimation for signals with known waveforms in the presence of multipath. *IEEE Transactions on Signal Processing, 45*(3), 808-811.

Cerpa, A., Wong, J. L., Kuang, L., Potkonjak, M., & Estrin D. (2005). Statistical Model of Lossy Links in Wireless Sensor Networks. *International Conference on Information Processing in Sensor Networks.*

Cesa-Bianchi, N., & Lugosi, G. (2006). *Prediction, Learning, and Games.* Cambridge University Press. ISBN-10 0-521-84108-9.

Chakrabarty K., Iyengar, S. S., Qi, H., & Cho, E. (2002). Grid Coverage for Surveillance and Target Location in Distributed Sensor Networks. *IEEE Transactions on Computers, 51*(12), 1448-1453.

Chan, Y. T., & Ho, K. C. (1994). A simple and efficient estimator for hyperbolic location. *IEEE Transactions on Signal Processing, 42*(8), 1905-1915.

Chang, C., & Sahai, A. (2004). Estimation Bounds for Localization. *IEEE SECON.* (pp. 415-424).

Chang, C., & Sahai, A. (2006). Cramer-Rao type bounds for localization. *EURASIP Journal on Applied Signal Processing, 2006.* article id 94287. (pp. 1-13).

Chang, C.-C., & Lin, C.-J. (2008). *LIBSVM – A library for Support Vector Machines.* National Taiwan University. URL http://www.csie.ntu.edu.tw/ cjlin/libsvm

Chao, S. (2008). *A comparison of indoor location detection systems for wireless sensor networks*, MS project report, MA: Boston University.

Charon, I., Hudry, O., & Lobstein, A. (2003). Minimizing the size of an identifying or locating-dominating code in a graph is NP-hard, *Theoretical Computer Science, 290*(3), 2109–2120.

Cheekiralla, S., & Engels, D. W. (2005, October). *A functional taxonomy of wireless sensor network devices.* Cambridge, MA : IEEE Explore

Chen, C. K., & Gardner, W. A. (1992). Signal-selective time-difference of arrival estimation for passive location of man-made signal sources in highly corruptive environments. Ii. Algorithms and performance. *IEEE Transactions on Signal Processing, 40*(5), 1185-1197.

Chen, D., & Varshney, P. K. (2007). A survey of void handling techniques for geographic routing in wireless networks. *IEEE Communications Surveys & Tutorials, 9*(1), 50-67.

Chen, J. C., Yao, K., & Hudson, R. E. (2002). Source localization and beamforming. *IEEE Signal Processing Magazine, 19*(2), 30-39.

Chen, P.-C. (1999). A cellular based mobile location tracking system. *In Proceedings of the IEEE Vehicular Technology Conference* (pp. 1979-1983), Houston, TX, USA.

Chen, Y., & Kobayashi, H. (2002). Signal strength based indoor geolocation. In *Proc. IEEE International Conference on Communications*, New York, NY, USA.

Chen, Y., Chandrasekaran, G., Elnahrawy, E., Francisco, J. A., Kleisouris, K., Li, X., Martin, R. P., Moore, R. S., & Turgut, B. (2008). GRAIL: A general purpose localization system. *Sensor Review* (pp. 115-124).

Chen, Y., Francisco, J., Trappe, W., & Martin, R. P. (2006). A practical approach to landmark deployment for indoor localization. In *proceedings of the Third Annual IEEE Communications Society Conference on Sensor, Mesh and Ad Hoc Communications and Network* (pp. 365-373).

Cheng , Y., Chawathe, Y., LaMarca, A., & Krumm, J. (2005) Accuracy Characterization for Metropolitan-scale Wi-Fi Localization. *Proceedings of the Third International Conference on Mobile Systems, Applications, and Services.* (pp. 233-245).

Cheng, D. K. (1989). *Field and wave electromagnetics* (2nd ed.): Addison-Wesley Publishing Company, Inc.

Chernoff, H. (1952). A measure of asymptotic efficiency for tests of a hypothesis based on the sum of observations. Ann. Math. Statist., *23*, 493-507.

Cheung, M. & Whiteley, W. (2008) *Transfer of global rigidity results among dimensions: graph powers and coning*, preprint, York University.

Chiasserini C.-F., & Garetto, M. (2004). Modeling the Performance of Wireless Sensor Networks. *IEEE INFOCOM* 2004, (pp. 220-231).

Chintalapudi, K., Dhariwal, A., Govindan, R., & Sukhatme, G. (2004). Ad-hoc localization using ranging and sectoring. In *IEEE INFOCOM, 4*, 2662-2672.

Chong, C.-Y., & Kumar, S. P. (2003). Sensor networks: Evolution, opportunities, and challenges. *Proceedings of the IEEE, 91*(8), 1247-1256.

Christ, T. W., & Godwin, P. A. (1993). A Prison Guard Duress Alarm Location System. *Proceedings of the IEEE International Carnahan Conference on Security Technology,* (pp. 106-116).

Chu, M., Haussecker, H., & Zhao, F. (2002). Scalable information-driven sensor querying and routing for ad hoc heterogeneous sensor networks. *Int. Journal on High Performance Computing Applications*, *16*(3), 90-110.

Coates, M. (2004). Distributed particle filters for sensor networks. In *Proc. IEEE Information Processing in Sensor Networks*, Berkeley, CA, USA.

Connelly, R. (1982) Rigidity and energy, *Invent. Math.*, 66(1), 11-33.

Connelly, R. (1991) On generic global rigidity, Applied geometry and discrete mathematics, 147–155, *DIMACS Ser. Discrete Math. Theoret. Comput. Sci., 4*, Amer. Math. Soc., Providence, RI.

Connelly, R. (2005) Generic global rigidity, *Discrete Comput. Geom.* 33, 549-563.

Cortes, C., & Vapnik, V. (1995). Support-vector networks. *Machine Learning, 20*(3), 273-297.

COST 231. (1991). Urban transmission loss models for mobile radio in the 900- and 1,800 MHz bands (Revision 2). COST 231 TD(90)119 Rev. 2, The Hague, The Netherlands.

Costa, J., Patwari, N., & Hero, A. O. (2006). Distributed weighted-multidimensional scaling for node localization in sensor networks. *ACM Trans. Sensor Networks, 2*(1), 39-64.

Couzin, I. D., Krause, J., Franks, N. R., & Levin, S. A. (2005). Effective Leadership and Decision Making in Animal Groups on the Move. *Nature, 433,* 513-516.

Crapo, H. (1990) *On the generic rigidity of plane frameworks.* INRIA research report No. 1278.

Crippen, G. M., & Havel, T. F. (1988). *Distance geometry and molecular conformation.* New York: John Wiley and Sons Inc.

Culler, D., Estrin, D., & Srivastava, M. (2004). Overview of Sensor Networks. *IEEE Computer, Guest Editors' Introduction, 37*(8), 41-49.

Cypher, D., Chevrollier, N. Montavont, N., & Golmie, N. (2006). Prevailing over wires in healthcare environments: Benefits and challenges. *IEEE Communications Magazine*, *44*(4l), 56-63.

Dasarathy, B. V. (Ed.) (1991). Nearest Neighbor (NN) Norms: NN Pattern Classification Techniques, ISBN 0-8186-8930-7. *IEEE Computer Society.*

Daskin, M. (1995). *Network and Discrete Location.* New York: Wiley.

Davidon, W. C. (1968). Variance algorithm for minimization. *Computer Journal, 10.*

De Marco, G., Yang, T., & Barolli, L. (2006). Impact of Radio Irregularities on Topology Tradeoffs of WSNs. *Proceedings of the 17th International Conference on Database and Expert Systems Applications (DEXA'06)*, (pp. 50-54).

Deblauwe, N. *GSM-based Positioning: Techniques and Application.* PhD Dissertation, Vrije University, Brussels, Belgium, 2008.

Dembo, A., & Zeitouni, O. (1998). *Large Deviations Techniques and Applications*, 2nd ed. NY: Springer-Verlag.

Deng, P., & Fan, P. (2000). An AOA assisted TOA positioning system. In *IEEE WCC-ICCT2000, 2*, 1501-1504.

Denis, B., Keignart, J., & Daniele, N. (2003). Impact of NLOS propagation upon ranging precision in uwb systems. In *IEEE Conference on Ultra Wideband Systems and Technologies.* (pp. 379-383).

Dharamdial, N., Adve, R., & Farha, R. (2003). Multipath Delay Estimations using Matrix Pencil. *Proceedings of the IEEE Wireless Communication and Networking Conference, 1*, 632-635.

Dickey, F., Romero, L., & Doerry, A. (2001). *Super-resolution and Synthetic Aperture Radar. Sandia Report SAND2001-1532.* Retrieved March 14, 2008 from Department of Energy Scientific and Technical Information Bridge. Web site: http://www.osti.gov/bridge/purl.cover.jsp?purl=/782711-Y2uIQp/native/

Doerschuk, P. C. (1995). Cramer-Rao bounds for discrete-time nonlinear filtering problems, *IEEE Transacations on Automatic Control, 40*(8), 1465-1469.

Dogancay, K. (2005). Emitter localization using clustering-based bearing association. *IEEE Transactions on Aerospace and Electronic Systems, 41*(2), 525-536.

Doherty, L., El Ghaoui, L., & Pister, K. (2001). Convex position estimation in wireless sensor networks. In IEEE, *Proc. of INFOCOM* (pp. 1655-1663). Anchorage, AK.

Dohler, M., Barthel, D., Maraninchi, F., Mounier, L., Aubert, S., Dugas, C., et al. (2007). The ARESA Project: Facilitating Research, Development and Commercialization of WSNs. *In 4th Annual IEEE Communications Society Conference on Sensor, Mesh and Ad hoc Communications and Networks (SECON)* (pp. 590-599). San Diego, CA, USA.

Drane, C., Macnaughtan, M., & Scott, C. (1998). Positioning GSM telephones. *IEEE Communications Magazine*, *36*(4).

Drumheller, M. (1987). Mobile robot localization using sonar. *IEEE Transactions on Pattern Analysis and Machine Intelligence*, *9*(2), 325–332.

Duan, C. D., Orlik, P., Sahinoglu, Z., & Molisch, A. F. (2007). A Non-Coherent 802.15.4a UWB Impulse Radio. *IEEE International Conference on Ultra-Wideband.* (pp. 146-151).

Duda, R. O., Hart, P. E., & Stork, D. G. (2001). *Pattern Classification.* Second Edition, Toronto, ON: John Wiley & Sons.

Dumont, L., Fattouche, M., & Morrison, G. (1994). Super-Resolution of Multipath Channels in a Spread Spectrum Location System. *IEE Electronic Letters, 30*(19), 1583-1584.

Durgin, G., Rappaport, T. S., & Xu, H. (1998). Measurements and models for radio path loss and penetration loss in and around homes and trees at 5.85 GHz. *IEEE Trans. Communications, 46*(11), 1484-1496.

Efrat, A., Erten, C. Forrester, D., Iyer, A., & Kobourov, S.G. (2006). Force-Directed Approaches to Sensor Localization. *Proceedings of the 8th SIAM Workshop on Algorithm Engineering and Experiments (ALENEX)*, (pp. 108-118).

Ekahau, Inc. (2008). *The Ekahau Positioning Engine.* Retrieved October 10, 2008, from http://www.ekahau.com

Elnahraway, E., Li, X., & Martin, R. P. (2004). The limits of localization using RSS. *In Proceedings of the 2nd Intl. Conf. on Embedded Networked Sensor Systems* (pp. 283-284), Baltimore, MD.

Elnahrawy, E., & Li, X., & Martin, R. P. (2004). Using area-based presentations and metrics for localization systems in wireless LANs. *In LCN's Fourth International IEEE Workshop on Wireless Local Networks* (pp. 650–657).

Elnahrawy, E., Li, X., & Martin, R. P. (2004). The limits of localization using signal strength: A comparative study. *First Annual IEEE Conference on Sensor and Ad-hoc Communications and Networks* (pp. 406-414).

Elson, J., & Römer, K. (2003). Wireless sensor networks: A new regime for time synchronization. *ACM Computer Communication Review, 33*(1).

EN282:1997 (1997). *Avalanche beacons.* Transmitter/ receiver systems. Safety requirements and testing, ISBN 0580268233.

Eren, T. (n.d.) Using angle of arrival (bearing) information for localization in robot networks, *Turk J Elec Engin, 15*(2), 169-186.

Eren, T., Goldenberg, D., Whiteley, W., Yang, Y.R., Morse, A.S., Anderson, B.D.O. & Belhumeur, P.N. (2004). Rigidity, Computation, and Randomization in Network Localization. In *IEEE INFOCOM Conference*, (pp.2673-2684) Hong Kong.

Eren, T., Whiteley, W., Morse, A.S., Belhumeur, P.N. & Anderson, B:D.O. (2003). Sensor and network topologies of formations with direction, bearing and angle information between agents, In *42nd IEEE Conference on Decision and Control*, (pp. 3064-3069).

Fang, B. T. (1990). Simple solutions for hyperbolic and related position fixes. *IEEE Transactions on Aerospace and Electronic Systems, 26*(5), 748-753.

Fang, J., Cao, M., Morse, A. S., & Anderson, B. D. O. (2006). Localization of sensor networks using sweeps. In *Proceedings of CDC*, (pp. 4645–4650), San Diego, CA.

Fang, J., Cao, M., Morse, A. S., & Anderson, B. D. O. (2006). Sequential localization of networks. *The 17th International Symposium on Mathematical Theory of Networks and Systems- MTNS 2006.*

Fang, L., Du, W., & Ning, P. (2005). A beacon-less location discovery scheme for wireless sensor networks. *In proceedings of Annual Joint Conference of the IEEE Computer and Communications Societies* (pp. 161–171).

Fang, S.-H., Lin, T.-N., & Lin, P.-C. (2008), Location Fingerprinting In A Decorrelated Space. *IEEE Trans. Knowledge and Data Engineering, 20*(5), 685-691.

Farkas, K., & Plattner, B. (2005).Supporting Real-Time Applications in Mobile Mesh Networks, *In Proceedings of the MeshNets 2005 Workshop*, (pp. 540-550).

FCC-US Federal Communications Commission. (1999). *Announcement of Commision Action, FCC Acts to Promote Competition and Public Safety in Enhanced Wireless 911 Services.* [Online]. Available: http://www.fcc.gov/Bureaus/Wireless/News_Releases/1999/nrwl9040.html

FCC-US Federal Communications Commission. (2002). *Revision of part 15 of the commissions rules regarding ultra-wideband transmission systems.* FCC 02-48, First Report & Order.

Federal Communications Commision's Enhanced 911 Wireless Services. http://www.fcc.gov/pshs/services/911-services/. Retrieved: 08.10.2008.

Federal Communications Commission. (2006). *FCC Docket 94-102: Report and order and further notice of proposed rulemaking in the matter of revision of the commission's rules to ensure compatibility with enhanced 911 emergency calling systems.*

Fekete, Z. & Jordán, T. (2006). Uniquely localizable networks with few anchors, In S. Nikoletseas, S. & Rolim, J.D.P. (Ed.) *Algosensors 2006*, (pp. 176-183). Springer Lecture Notes in Computer Science 4240.

Fekete, Z. & Jordán, T. (2008). Algorithms for minimum cost anchor sets in uniquely localizable networks, preprint.

Fekete, Z. (2006). Source location with rigidity and tree packing requirements. *Operations Research Letters 34*(6), 607-612.

Feuerstein, M. J., Blackard, K. L., Rappaport, T. S., Seidel, S. Y., & Xia, H. H. (1994). Path loss, delay spread, and outage models as functions of antenna height for microcellular system design. *IEEE Trans. Vehicular Technology, 43*(3), 487-498.

Fonseca, R., Ratnasamyy, S., Zhao, J., Ee, C. T., Culler, D., Shenker, S., et al. (2005). Beacon Vector Routing: Scalable Point-to-Point Routing in Wireless Sensornets. *In 2nd Symposium on Networked Systems Design & Implementation (NSDI)* (pp. 329-342). Boston, MA, USA.

Forester, J. (2003). Channel modeling sub-committee report final. *IEEE802.15-02/490r1-SG3a*, Feb 2003.

Fox, D., Hightower, J., Liao, L., Schulz, D., & Borriello, G. (2003). Bayesian filters for location estimation. *IEEE Pervasive Computing, 2*(3), 24–33.

Foy, W. H. (1976). Position-location solutions by Taylor-series estimation. *IEEE Trans. Aerospace and Elect. Sys., AES-12*(2), 187-194.

Frey, H., & Stojmenovic, I. (2006). On Delivery Guarantees of Face and Combined Greedy-Face Routing Algorithms in Ad hoc and Sensor Networks. *In Twelfth ACM Annual International Conference on Mobile Computing and Networking (MOBICOM)* (pp. 390-401). Los Angeles, CA, USA.

Gabow, H.N. & Westermann, H.H. (1992) Forests, frames and games: Algorithms for matroid sums and applications, *Algorithmica* 7, 465-497.

Galstyan, A., Krishnamachari, B., Lerman, K., & Pattem, S. (2004). Distributed particle filters for sensor networks. In Proc. *IEEE Information Processing in Sensor Networks*, Berkeley, CA, USA.

Ganu, S., Krishnakumar, A. S., & Krishnan, P. (2004). Infrastructure-based Location Estimation in WLAN Networks. *Proceedings of the IEEE Wireless Communications and Networking Conference (WCNC)*, (pp. 465-470).

Gao, Y., Wu, K., & Li, F. (2004). Analysis on the redundancy of wireless sensor networks. *In Proc. 2nd ACM Intl. Workshop on Wireless Sensor Networks and Applications (WSNA)*, San Diego, CA: ACM (pp. 108-114).

Gavish, M., & Weiss, A. J. (1992). Performance analysis of bearing-only target location algorithms. *IEEE Transactions on Aerospace and Electronic Systems, 28*(3), 817-828.

Gelman, A., Carlin, J. B., Stern, H. S., & Rubin, D. B. (2003). *Bayesian Data Analysis* (2nd ed.). Chapman and Hall.

Gezici, S., Tian, Z., Giannakis, G. B., Kobayashi, H., Molisch, A. F., Poor, H. V., et al. (2005). Localization via ultra-wideband radios: A look at positioning aspects for future sensor networks. *IEEE Signal Processing Magazine, 22*(4), 70-84.

Ghassemzadeh, S.S., Jana, R., Rice, C.W., Turin, W., & Tarokh, V. (2002). A statistical path loss model for in-home UWB channels. In IEEE, *Proc. of the Conference on Ultra Wideband Systems and Technologies* (pp. 59–64).

Ghosh, A., Wang, Y., Krishnamachari, B., & Hsieh, M. (2007). Efficient Distributed Topology Control in 3-Dimensional Wireless Networks. *4th Annual IEEE Communication Society Conference on Sensor, Mesh and Ad Hoc Communications and Networks, SECON 2007*, 2007, (pp. 91-100).

Girod, L. (2005). A Self-Calibrating System of Distributed Acoustic Arrays. *Ph.D. Thesis, UCLA, USA.*

Girod, L., Lukac, M., Trifa, V., & Estrin, D. (2006). The design and implementation of a self-calibrating distributed acoustic sensing platform. In *Proceedings of the 4th international Conference on Embedded Networked Sensor Systems* (Boulder, Colorado, USA, October 31 - November 03, 2006). SenSys '06. ACM, New York, NY, (pp. 71-84).

Girod, L., Ramanathan, N., Elson, J., Stathopoulos, T., Lukac, M., & Estrin, D. (2007). Emstar: A software environment for developing and deploying heterogeneous sensor-actuator networks. *ACM Trans. Sen. Netw., 3*(3), (Aug. 2007), 13.

Gluck, H. (1975) Almost all simply connected closed surfaces are rigid, Geometric topology (Proc. Conf., Park City, Utah, 1974), pp. 225–239. *Lecture Notes in Math., 438*, Springer, Berlin.

Glunt, W., Hayden, T. L., & Raydan, M. (1993). Molecular conformation from distance matrices. *J. Computational Chemistry, 14*, 114-120.

Goldenberg, D. K., Bihler, P., Cao, M., Fang, J., Anderson, B. D. O., Morse, A. S., & Yang, Y. R. (2006). Localization in sparse networks using sweeps. *Proceedings of Mobicom* (pp. 110–121).

Goldenberg, D. K., Krishnamurthy, A., Maness, W. C., Yang, Y. R., Young, A., Morse, A. S., & Savvides, A. (2005). Network localization in partially localizable networks. *INFOCOM 2005. 24th Annual Joint Conference of the IEEE Computer and Communications Societies.* IEEE 1(13-17) (pp. 313-326).

Gortler, S.J., Healy, A.D. & Thurston, D.P. (2007) Characterizing generic global rigidity, arXiv:0710.0926v3.

Gotsman, C., & Koren, Y. (2005). Distributed Graph Layout for Sensor Networks. *Lecture Notes in Computer Science*, 3383/2005, Springer Berlin/Heidelberg, (pp. 273-284).

Gower, J. (1966). Some distance properties of latent root and vector methods used in multivariate analysis. *Biometrika*, *53*(3,4), 325-338.

Graver, J., Servatius, B. & Servatius, H. (2003). Combinatorial Rigidity, *AMS Graduate Studies in Mathematics* 2.

Gravier, S., Klasing, R., & Moncel, J. (2006). *Hardness results and approximation algorithms for identifying codes and locating-dominating codes in graphs.* (Technical Report RR-1417-06), Bordeaux, France: Laboratoire Bordelais de Recherche en Informatique (LaBRI).

Gudmundson, M. (1991). Correlation model for shadow fading in mobile radio systems. *IEE Electronics Letters*, 27(23).

Günther, A., & Hoene, C. (2005). Measuring Round trip Times to Determine the Distance between WLAN Nodes. *Proceedings of Networking 2005*, Waterloo, Canada (pp. 768-779).

Gustafsson, F., & Gunnarsson, F. (2005). Possibilities and fundamental limitations of positioning using wireless communications networks. *IEEE Signal Processing Magazine*, *22*(7).

Gustafsson, F., Gunnarsson, F., Bergman, N., Forssell, U., Jansson, J., Karlsson, R., & Nordlund, P.-J. (2002). Particle Filters for Positioning, Navigation, and Tracking. *IEEE Transactions on Signal Processing*, *50*(2), 425-437.

Hać, A. (2003). *Wireless sensor network designs.* West Sussex: John Wiley & Sons Ltd.

Halder, B., Viberg, M., & Kailath, T. (1993). An efficient non-iterative method for estimating the angles of arrival of known signals. *The Twenty-Seventh Asilomar Conference on Signals, Systems and Computers* (pp. 1396-1400).

Hand, D., Mannila, H., & Smyth, P. (Ed.) (2001). *Principles of Data Mining.* The MIT Press.

Hardoon, D. R., Szedmak, S., & Shawe-Taylor, J. (2004). Canonical correlation analysis; an overview with application to learning methods. *Neural Computation*, *16*, 2639–2664.

Harter, A., Hopper, A., Steggles, P., Ward, A., & Webster, P. (1999). The anatomy of a context-aware application. *Proceedings of the 5th annual ACM/IEEE International Conference on Mobile Computing and Networking*, Seattle, WA, USA, (pp. 59-68).

Hashemi, H. (1993). The indoor radio propagation channel. *Proc. IEEE*, *81*(7), 943–968.

Hastie, T., & Tibshirani, R. (1990). *Generalized Additive Models.* Chapman and Hall.

Hata, M. (1980). Empirical formula for propagation loss in land mobile radio services. *IEEE Trans.Veh. Technology*, *29*, 317-325.

Hatami, A., & Pahlavan, K. (2005) A comparative performance evaluation of RSS-based positioning algorithms used in WLAN networks. In *Proceedings of the IEEE Wireless Communications and Networking Conference (WCNC '05)*, *4*, (pp. 2331–2337), New Orleans, La, USA.

Haykin, S. (2002). *Adaptive Filter Theory – 4th Ed.* Prentice-Hall.

Hazas, M., & Ward, A. (2003). A high performance privacy-oriented location system. *In proceedings of the IEEE International Conference on Pervasive Computing and Communications* (pp. 216).

He, T., & Huang, C., & Blum, B., & Stankovic, J., & Abdelzaher, T. (2003). Range-free localization schemes in large scale sensor networks. *In proceedings of the ACM International Conference on Mobile Computing and Networking* (pp. 81–95).

Heckerman, D. (1995). *A tutorial on learning with Bayesian networks* (Tech. Rep. No. TR-95-06). Microsoft Research.

Heidari, M., Akgul, F. O., & Pahlavan, K. (2007). Identification of the Absence of Direct Path in Indoor Localization System. *IEEE PIMRC 2007.* (pp. 1-6).

Heidari, M., Akgul, F. O., Alsindi, N., & Pahlavan, K. (2007). Neural Network Assisted Identification of the Absence of the Direct Path in Indoor Localization. *IEEE Globecom 2007.* (pp. 387-392).

Heidemann, J., Bulusu, N., Elson, J., Intanagonwiwat, C., Lan, K., & Xu, Y., et al. (2001). Effects of Detail in Wireless Network Simulation. *Proceedings of the SCS Multiconference on Distributed Simulation.* Phoenix, Arizona, USA, USC/Information Sciences Institute, Society for Computer Simulation. January, 2001, (pp. 3-11).

Hellebrandt, M., & Mathar, R. (1999). Location tracking of mobiles in cellular radio networks. *IEEE Transactions on Vehicular Technology, 48*(5), 1558-1562.

Hendrickson, B. & Jacobs, D. (1997) An algorithm for two-dimensional rigidity percolation: the pebble game, *J. Computational Physics* 137, 346-365.

Hendrickson, B. (1992). Conditions for unique graph realizations. *SIAM J. Comput., 21*(1), 65–84.

Henneberg, L. (1911) *Die graphische Statik der starren Systeme*, Leipzig.

Hightower, J., & Borriello, G. (2001). Location Systems for Ubiquitous Computing. *IEEE Computer, 34*(8), 57-66.

Hightower, J., Borriello, G., & Want R. (2000), *SpotON: An indoor 3D location sensing technology based on RF signal strength*, (Tech. Rep. No. 2000-02-02), University of Washington.

Hightower, J., LaMarca, A., & Smith, I. (2006). Practical Lessons from Place Lab, *IEEE Pervasive Computing, 5*(3), 32- 39.

Hills, A., & Schlegel, J. (2004). Rollabout: A Wireless Design Tool. *IEEE Communications, 42*(2), 132-138.

Hodes, T. D., Katx, R. H., Schreiber, E. S., & Rowe, L. (1997, September). Composable ad hoc mobile services for universal interaction. *Proceedings of the 3rd annual ACM/IEEE international conference on Mobile computing and networking* (pp. 1-12) Budapest, Hungary.

Hoeffding, W. (1965). Asymptotically optimal tests for multinomial distributions. *Ann. Math. Statist., 36,* 369.401.

Hoel, P., & Stone, J. (1971). *Introduction to Probability Theory.* Boston: Houghton Mifflin.

Hofmann-Wellenhof, B., Lichtenegger, H., & Collins, J. (1997). *Global Positioning System: Theory and Practice.* 4th ed. Springer-Verlag.

Honeywell (2008). *HMC1022 data sheet.* www.ssec.honeywell.com/magnetic/datasheets/hmc1001-2_1021-2.pdf,", Retrieved: 08.10.2008.

Hopcroft J.E. & Tarjan, R.E. (1973) Dividing a graph into triconnected components, *SIAM J. Comput.* 2, 135–158.

Hotelling, H. (1936). Relations between two sets of variants. *Biometrika, 28*, 321-377.

Howard, A. (2006). Multi-robot Simultaneous Localization and Mapping using Particle Filters. *The International Journal of Robotics Research, 25*(12), 1243-1256.

Huang, H. C., Huang, Y. M., & Ding, J. W. (2006). An implementation of battery-aware wireless sensor network using ZigBee for multimedia service. *In Proceedings of Intenrnational Conference on Consumer Electronics.* (pp. 369-370)

Huhtonen, A. (2004).Comparing AODV and OLSR Routing Protocols, *InTelecommunications Software and Multimedia.* Helsenki : Helsinki University of Technology.

Hull, B., Bychkovsky, V., Zhang, Y., Chen, K., Goraczko, M., Miu, A., et al. (2006). CarTel: A Distributed Mobile Sensor Computing System, *The 4th ACM Conference on Embedded Networked Sensor Systems (SenSys).* (pp. 125- 138)

IEEE 802.11i Standard (2004). *The IEEE 802.11i Standard,* retrieved October 13, 2008, from http://standards.ieee.org/getieee802/download/802.11i-2004.pdf.

IEEE P1609.3/D18 standard (2005)- IEEE P1609.3 Standard for Wireless Access in Vehicular Environments (WAVE)- Networking Services.

IETF RFC3561 (2003). *Ad hoc On-Demand Distance Vector Routing.* Retreived October 2007, from http://www.ietf.org/rfc/rfc3561.txt.

IETF RFC3626 (2003). *Optimized Link State Routing Protocol.* Retreived October 2007, from http://www.ietf.org/rfc/rfc3626.txt

Ihler, A. T., Fisher, J. W., III, Moses, R. L., & Willsky, A. S. (2005). Nonparametric belief propagation for self-localization of sensor networks. *IEEE Journal on Selected Areas in Communications, 23*(4), 809-819.

Ilhan, T., & Pinar, M. (2001). *An efficient exact algorithm for the vertex p-center problem.* http://www.optimization-online.org/ DB-HTML/2001/09/376.html

ILOG CPLEX 8.0, ILOG, Inc., Mountain View, California, July 2002, http://www.ilog.com.

Ilyas, M., & Mahgoub, I. (Eds.). (2005). *Handbook of sensor networks: Compact wireless and wired sensing systems.* CRC Press LLC.

Imai, H. (1985). On combinatorial structures of line drawings of polyhedra, *Discrete Appl. Math.* 10, 79-92.

InPoint Systems. (2008). *White paper.* Retrieved October 10, 2008, from http://inpointsys.com

Intanagonwiwat, C., Govindan, R., & Estrin, D. (2000). Directed diffusion: A scalable and robust communication paradigm forsensor networks. In ACM, *Proc. of the 6th Int'l Conf. on Mobile Computing and Networks (MobiCom)* (pp. 56-67). Boston, MA.

Jackson, B & Jordán, T. (2008). Globally rigid circuits of the two-dimensional direction-length rigidity matroid, J. Combinatorial Theory, Ser. B., to appear.

Jackson, B & Jordán, T. (2008). Operations preserving global rigidity of generic direction-length frameworks, Egerváry Research Group TR-2008-08, submitted.

Jackson, B. & Jordán, T. (2005). Connected rigidity matroids and unique realizations of graphs, *J. Combinatorial Theory, Ser. B.*, 94, 1-29.

Jackson, B. & Jordán, T. (2008). A sufficient connectivity condition for generic rigidity in the plane, Discrete Applied Math, in press.

Jackson, B., Jordán, T. & Szabadka, Z. (2006). Globally linked pairs of vertices in equivalent realizations of graphs, *Discrete and Computational Geometry*, 35, 493-512.

Jackson, B., Servatius, B. & Servatius, H. (2007). The 2-dimensional rigidity of certain families of graphs, *J. Graph Theory* 54 (2), 154–166.

Jacobs, D., & Hendrickson, B. (1997). An algorithm for two-dimensional rigidity percolation: the pebble game. *J. Comput. Phys., 137*(2), 346–365.

Jayasimha, D., Iyengar, S., & Kashyap, R.(1991). Information integration and synchronization in distributed sensor networks. *IEEE Transactions Systems, Man, and Cybernetics, 21*(5).

Ji, X., & Zha, H. (2004). Sensor positioning in wireless ad-hoc sensor networks using multidimensional scaling. *IEEE INFOCOM* (pp. 2652-2661).

Jian, L., Halder, B., Stoica, P., & Viberg, M. (1995). Computationally efficient angle estimation for signals with known waveforms. *IEEE Transactions on Signal Processing, 43*(9), 2154-2163.

Johnson, D. B. & Maltz, D. B. (1996). Dynamic source routing in ad hoc wireless networks. In T. Imielinski and H. Korth (Ed.), *Mobile Computing* (pp. 153-181). The Netherlands: Kluwer Academic Publishers.

Johnson, D. S. (1974). Approximation algorithms for combinatorial problems. *Journal of Computer and System Sciences, 9*, 256–278.

Jordán, T. & Szabadka, Z. (in press) Operations preserving the global rigidity of graphs and frameworks in the plane, *Comput.Geom.*.

JTC (Joint Technical Committee for PCS T1 R1P1.4). (1994). Technical Report on RF Channel Characterization and System Deployment Modeling, JTC (AIR)/94.09.23-065R6.

Julier, S. J., & Uhlmann, J. K. (1997). A new extension of the Kalman filter to nonlinear systems. *Int. Symp. Aerospace/Defense Sensing, Simul. and Controls.*

Kaemarungsi, K., & Krishnamurthy, P. (2004). Modeling of indoor positioning systems based on location fingerprinting. *In proceedings of Twenty Third Annual Joint Conference of the IEEE Computer and Communications Societies* (pp. 1012-1022).

Kahn, J. M., Katz, R. H., & Pister, K. S. J. (1999). Mobile networking for smart dust. In *Proc. ACM/IEEE MobiCom*, Seattle, WA, USA.

Kalman, R. E. (1960). A New Approach to Linear Filtering and Prediction Problems. *Transactions of the ASME - Journal of Basic Engineering, 82*, 35-45.

Kanaan M., Akgul, F. O., Alavi, B., & Pahlavan, K. (2006a). A Study of the Effects of Reference Point Density on TOA-Based UWB Indoor Positioning Systems.

In the *17th Annual IEEE International Symposium on Personal, Indoor and Mobile Radio Communications, PIMRC*, (pp. 1-5). Finland.

Kanaan, M., & Pahlavan, K. (2004). Algorithm for TOA-based Indoor Geolocation. *IEE Electronics Letters, 40*(22).

Kanaan, M., & Pahlavan, K. (2004). A comparison of wireless geolocation algorithms in the indoor environment. In *IEEE Wireless Communications and Networking Conference (WCNC04), 1*, 177-182.

Kanaan, M., Akgul, F. O., Alavi, B., Pahlavan, K. (2006). Performance Benchmarking of TOA-Based UWB Indoor Geolocation Systems Using MSE Profiling. In *IEEE 64th Vehicular Technology Conference* (VTC-2006 Fall). (pp. 1-5).

Kanaan, M., Heidari, M., Akgul, F., & Pahlavan, K. (2006). Technical Aspects of Localization in Indoor Wireless Networks, *Bechtel Telecommunications Technical Journal, 4*(3).

Kannan, A. A., Mao, G., & Vucetic, B. (2005). Simulated annealing based localization in wireless sensor network. *The 30th IEEE Conference on Local Computer Networks* (pp. 513-514).

Kannan, A. A., Mao, G., & Vucetic, B. (2006). Simulated annealing based wireless sensor network localization with flip ambiguity mitigation. *63rd IEEE Vehicular Technology Conference* (pp. 1022- 1026).

Kaplan, E., & Hegarty, C. (2005). *Understanding GPS: Principles and Applications.* Norwood, MA: Artech House Publishers.

Karl, H., & Willig, A. (2005). *Protocols and architectures for wireless sensor networks.* Chichester: John Wiley & Sons Ltd.

Karlsson, R., & Gustafsson, F. (2003). Particle filter and Cramer-Rao lower bound for underwater navigation. In *IEEE Conference on Acoustics, Speech and Signal Processing (ICASSP)*, Hongkong, China.

Karp, B. & Kung, H. T. (2000). GPSR: Greedy perimeter stateless routing for wireless networks. In ACM, *Proc. of the 6th Int'l Conf. on Mobile Computing and Networks (MobiCom)* (pp. 243-254). Boston, MA.

Karpovsky, M. G., Chakrabarty, K., & Levitin, L. B. (1998). A new class of codes for identification of vertices in graphs, *IEEE Transactions on Information Theory, 44*(2), 599–611.

Kasetkasem, T., & Varshney, P. K. (2001). Communication Structure Planning for Multisensor Detection Systems. *IEEe Proceedings on Radar, Sonar and Navigation, 148*, 2-8.

Katz, B., Gaertler, M. & D. Wagner, (2007). Maximum rigid components as means for direction-based localization in sensor networks, In Jan van Leeuwen et al. (Ed) *SOFSEM 2007, LNCS 4362*, (pp. 330-341), Springer, Berlin.

Kaveh, M., & Bassias, A. (1990). Threshold extension based on a new paradigm for music-type estimation. *International Conference on Acoustics, Speech, and Signal Processing* (pp. 2535-2538).

Kay, S. M. (1993). *Fundamentals of Statistical Signal Processing: Estimation Theory.* Upper Saddle River, NJ: Prentice Hall.

Kerr, T. H. (1989). Status of CR-like lower bounds for nonlinear filtering. *IEEE Transactions on Aerospace and Electronic Systems, 25*(5), 590-600.

Khemapech, I., Duncan, I., & Miller, A. (2005). A survey of wireless sensor networks technology. *In Proceedings of the 6th Annual PostGraduate Symposium on the Convergence of Telecommunications, Networking & Broadcasting (PGNET'05), Liverpool, U.K.*

Kim, D., & Langley, R. B. (2000). GPS Ambiguity Resolution and Validation: Methodologies,Trends and Issues. *7th GNSS Workshop-International Symposium on GPS/GNSS*, Seoul, Korea.

Kim, W., Jee, G.-I., & Lee, J. G. (2001). Wireless location with NLOS error mitigation in Korean CDMA system. In *Proceedings of the Second International Conference on 3G Mobile Communication Technologies* (pp. 134-138). London, U.K.

Kim, Y.-J., Govindan, R., Karp, B., & Shenker, S. (2005). Geographic Routing Made Practical. *In 2nd Symposium on Networked Systems Design & Implementation (NSDI)* (pp. 217-230). Boston, MA, USA.

King, T., Kopf, S., Haenselmann, T., Lubberger, C., & Effelsberg, C. W. (2006). COMPASS: A Probabilistic Indoor Positioning System Based on 802.11 and Digital Compasses. *In Proc. 1st ACM Intl. Workshop on Wireless Network Testbeds, Experimental Evaluation & Characterization (WiNTECH)*, (pp. 34-40), Los Angeles, USA.

Kirchner, D. (1991). Two-way time transfer via communication satellites. *Proceedings of the IEEE, 79*(7), 983-990.

Kirkpatrick, S., Gelatt, C. D., & Vecchi, M. P. (1983). Optimization by simulated annealing. *Science, 220*(4598), 671–680.

Kleinberg, R. (2007). Geographic Routing Using Hyperbolic Space. *In 26th International Conference on Computer Communications (INFOCOM)* (pp. 1902-1909). Anchorage , AL, USA.

Kleisouris, K., & Martin, R. P. (2006). Reducing the computational cost of Bayesian indoor positioning systems. *In proceedings of the Third IEEE International Conference on Sensor and Ad hoc Communications and Networks* (pp. 555-564).

Klingbeil, L., & Wark, T. (2008). A wireless sensor netwok for real-time indoor localisation and motion monitoring. *Proceedings of 2008 International Conference on Information Processing in Sensor Networks* (pp. 39-50).

Klukas, R., & Fattouche, M. (1998). Line-of-sight angle of arrival estimation in the outdoor multipath environment. *IEEE Transactions on Vehicular Technology, 47*(1), 342-351.

Knapp, C., & Carter, G. (1976). The generalized correlation method for estimation of time delay. *IEEE Transactions on Acoustics, Speech and Signal Processing, 24*(4), 320-327.

Koks, D. (2005). *Numerical calculations for passive geolocation scenarios* (No. DSTO-RR-0000). Edinburgh, SA, Australiao. Document Number)

Koorapaty, H. (2004). Barankin bound for position estimation using received signal strength measurements. In *Proc. IEEE Vehicular Technology Conference*, Milan, Italy.

Koorapaty, H., Grubeck, H., & Cedervall, M. (1998). Effect of biased measurement errors on accuracy of position location methods. In *Proc. IEEE Global Telecommunications Conference*, Sydney, Australia.

Krim, H., & Viberg, M. (1996). Two decades of array signal processing research: The parametric approach. *IEEE Signal Processing Magazine*. vol. 13, (pp. 67-94).

Krishnakumar, A. S., & Krishnan, P. (2005). On the Accuracy of Signal Strength-based Location Estimation Techniques. *Proceedings of the 2005 IEEE Infocom Conference*, Miami, FL, (pp. 642-650).

Krishnamachari, B. (2005). *Networking wireless sensors.* New York: Cambridge University Press.

Krishnamurthi N., Jay Yang S., & Seidman M. (2004). Modular Topology Control and Energy Model for Wireless Ad Hoc Sensor Networks. *Proceedings of OPNETWORK '04.*

Krishnan, P., & Krishnakumar, A. S., & Ju, W., & Mallows, C., & Ganu, S. (2004). A system for LEASE: Location estimation assisted by stationary emitters for indoor RF wireless networks. *In proceedings of the IEEE International Conference on Computer Communications* (pp. 1001–1011).

Kuchar, A., Tangemann, M., & Bonek, E. (2002). A Real-Time DOA-Based Smart Antenna Processor. *IEEE Transactions on Vehicular Technology, 51*(6), 1279-1293.

Kuckertz, P., Ansari, J., Riihijärvi, J., & Mähönen, P. (2007). Sniper Fire Localization using Wireless Sensor Networks and Genetic Algorithm based Data Fusion. In *Proceedings of IEEE MILCOM 2007*, Orlando, FL.

Kuh, A., Zhu, C., & Mandic, D. P. (2006). Sensor network localization using least squares kernel regression. In *Knowledge-Based Intelligent Information and Engineering Systems*, (pp. 1280-1287).

Kuh, A., & Zhu, C. (2008). Sensor network localization using least squares kernel regression. In D. Mandic et al. (Eds.), *Signal Processing Techniques for Knowledge Extraction and Information Fusion.* (pp. 77-96), Springer, April 2008.

Kumaresan, R., & Tufts, D. W. (1983). Estimating the angles of arrival of multiple plane waves. *IEEE Trans. Aerosp. Electron. Syst., AES-19*(1), 134-139.

Kwok, C., Fox, D., & Meila, M. (2004). Real-time particle filters. *Proceedings of the IEEE, 92*(3), 469-484.

Kwon, Y., Mechitov, K., Sundresh, S., Kim, W., & Agha, G. (2004). *Resilient localization for sensor networks in outdoor environments.* Tech. rep., University of Illinois at Urbana-Champaign, 2004.

Ladd, A. M., Bekris, K. E., Rudys, A., Marceau, G., Kavraki, L. E., & Wallach, D. S. (2002). Robotics-based location sensing using wireless ethernet. *In Proceedings of the 8th Annual international Conference on Mobile Computing and Networking* (pp. 227-238), Atlanta, Georgia.

Laifenfeld, M. (2007). *Coding for network applications: Robust identification and distributed resource allocation.* Doctoral dissertation, Boston University, MA.

Laifenfeld, M., & Trachtenberg, A. (2005). Disjoint identifying codes for arbitrary graphs. In *IEEE International Symposium on Information Theory,* (pp. 244- 248), Adelaide Australia.

Laifenfeld, M., & Trachtenberg, A. (2008). Identifying codes and covering problems. *IEEE Transaction on Information Theory, 54*(9), 3929-3950.

Laifenfeld, M., Trachtenberg, A., & Berger-Wolf, T. (2006). Identifying codes and the set cover problem. *In Proceedings of the 44th Annual Allerton Conference on Communication, Control, and Computing,* Urbana-Champaign, IL.

Laifenfeld, M., Trachtenberg, A., Cohen, R., & Starobinski, D. (2008) Joint monitoring and routing in wireless sensor networks using robust identifying codes. *In Springer Journal on Mobile Networks and Applications (MONET).* Online http://www.springerlink.com/content/8386234004837647

Laman, G. (1970) On graphs and rigidity of plane skeletal structures, *J. Engineering Math.* 4, 331-340.

LaMarca, A., Chawathe, Y., Consolvo, S., Hightower, J., Smith, I., Scott, J., Sohn, T., Howard, J., Hughes, J., Potter, F., Tabert, J., Powledge, P., Borriello, G., & Schilit, B. (2005). Place lab: Device positioning using radio beacons in the wild. *In proceedings of Pervasive Computing* (pp. 116-133).

Langendoen, K., & Reijers, N. (2003). Distributed localization in wireless sensor networks: A quantitative comparison. *Networks: The International Journal of Computer and Telecommunications Networking, 43*(4), 499-518.

Lanzisera, S., & Pister, K. (2008) Burst Mode Two-way Ranging with Cramér-Rao Bound Noise Performance. *Proceedings of the 2008 IEEE Global Communications Conference.*

Lanzisera, S., Lin, D., & Pister, K. (2006). RF Time of Flight Ranging for Wireless Sensor Network Localization. *Proceedings of the IEEE Workshop on Intelligent Solutions in Embedded Systems.*

Larsson, E. G. (2004). Cramer-Rao bound analysis of distributed positioning in sensor networks. *IEEE Signal Processing Letters, 11*(3), 334-337.

Law, M. H. C., & Jain, A. K. (2006). Incremental nonlinear dimensionality reduction by manifold learning. *IEEE Transactions on Pattern Analysis and Machine Intelligence, 28*(3), 377–391.

Lee, J.-Y., & Scholtz, R. A. (2002). Ranging in a dense multipath environment using an UWB radio link. *IEEE Journal on Selected Areas in Communications, 20*(9), 1677-1683.

Lee, M., Jianliang, Z., Young-Bae, K., & Shrestha, D. (2006). Emerging Standards For Wireless Mesh Technology, *IEEE Wireless Communications, 13*(2), 56- 63.

Lehmann, E. (1991). *Theory of point estimation.* Statistical/Probability series. Wadsworth & Brooks/Cole.

Leong, B., Liskov, B., & Morris, R. (2007). Greedy Virtual Coordinates for Geographic Routing. *In IEEE International Conference on Network Protocols (ICNP)* (pp. 71-80). Beijing, China.

Leon-Garcia, A. (1994). *Probability and Random Processes for Electrical Engineering.* Second Edition, Don Mills, Ontario: Addison-Wesley Publishing Company.

Letchner, J., Fox, D., & LaMarca, A. (2005). Large-Scale Localization from Wireless Signal Strength. In *Proc. of the National Conference on Artificial Intelligence (AAAI),* (pp. 15-20).

Li, J., Halder, B., & Stoica, P. (1995). Computationally efficient angle estimation for signals with known waveforms. *IEEE Transactions on Signal Processing,* 43(9), 2154-2163.

Li, X. (2006). RSS-Based Location Estimation with Unknown Pathloss Model. *IEEE Trans. Wireless Communications, 5*(12), 3626-3633.

Li, X., & Pahlavan, K. (2004). Super-resolution TOA estimation with diversity for indoor geolocation. *IEEE Trans. on Wireless Communications, 3*(1), 224-234.

Li, X., Shi, H., & Shang, Y. (2004). A Partial-Range-Aware Localization Algorithm for Ad-hoc Wireless Sensor Networks. In IEEE, *Proc. of the 29th Int. Conf. on Local Computer Networks (LCN)* (pp. 77-83). Tampa, FL.

Li, X., Shi, H., & Shang, Y. (2006). Sensor network localisation based on sorted RSSI quantisation. *Int. Journal of Ad Hoc and Ubiquitous Computing (IJAHUC), 1*(4), 222-229.

Li, X-Y., Wan, P-J., Wang, Y. & Yi, C-W. (2003). Fault tolerant deployment and topology control in wireless networks. In *ACM Symposium on Mobile Ad Hoc Networking and Computing (MobiHoc)* (pp. 117-128), Annapolis, MD.

Liang, T.-C., Wang, T.-C., & Ye, Y. (2004). *A gradient search method to round the semidefinite programming relaxation for ad hoc wireless sensor network localization*: Standford Universityo. Technical Report.

Lim, H., Kung, L., Hou, J., & Luo, H. (2006). Zero-Configuration, robust indoor localization: Theory and experimentation. *In proceedings of the IEEE International Conference on Computer Communications* (pp. 1-12).

Lin G., Noubir G., & Rajaraman R. (2004). Mobility Models for Ad hoc Network Simulation. *IEEE INFOCOM* 2004, (pp. 463).

Liu, C., Wu, K., & He, T. (2004). Sensor localization with ring overlapping based on comparison of received signal strength indicator. *In Proc. IEEE Mobile Ad-hoc and Sensor Systems (MASS)*, (pp. 516–518).

Liu, K., & Abu-Ghazaleh, N. (2006). Aligned Virtual Coordinates for Greedy Routing in WSNs. *In International Conference on Mobile Adhoc and Sensor Systems (MASS)* (pp. 377-386). Vancouver, Canada.

Liu, T., Bahl, P., & Chlamtac, I. (1998). Mobility modeling, location tracking, and trajectory prediction in wireless ATM networks. *IEEE Journal on Selected Areas in Communications, 16*(6), 922-936.

Lorincz, K., & Welsh, M. (2007). Motetrack: A robust, decentralized approach to RF-based location tracking. In *Personal and Ubiquitous Computing, Special Issue on Location and Context-Awareness, 11*(6), 489-503.

Lovász, L. & Yemini, Y. (1982) On generic rigidity in the plane, *SIAM J. Algebraic Discrete Methods* 3 (1), 91–98.

Luo, J., Shukla, H., and Hubaux, J.-P. (2006). Noninteractive location surveying for sensor networks with mobility-differentiated toa. In *Proc. IEEE INFOCOM*, Barcelona, Spain.

Luo, R., & Kay, M. G. (1989). Multisensor integration and fusion in intelligent systems. *IEEE Transactions Systems, Man, and Cybernetics, 19*(5).

Ma, J., Min, G., Zhang, Q., Ni, L.M., & Zhu, W. (2005). Localized Low-Power Topology Control Algorithms in IEEE 802.15.4-Based Sensor Networks. *IEEE International Conference on Distributed Computing Systems.* (pp. 27-36).

Madigan, D., Elnahrawy, E., Martin, R. P., Ju, W. H., Krishnan, P., & Krishnakumar, A. S. (2005). Bayesian indoor positioning systems. *In proceedings of Twenty Fourth Annual Joint Conference of the IEEE Computer and Communications Societies* (pp. 1217-1227).

Malaney, R. A. (2004). A Location Enabled Wireless Security System. *Proceedings of IEEE Globecom*, Dallas, TX, *4*, 2196-2200.

Mao, G., Fidan, B. & Anderson, B.D.O. (2007) Localisation, In N.P. Mahalik (Ed.) *Sensor networks and configuration* (pp. 281-315), Springer, Berlin.

Maroti M., Kusy B., Balogh G., Volgyesi P., Molnar, Karoly, Dora S., & Ledeczi A. (2005). Radio Interferometric Positioning. *Proceedings of the ACM Conference on Embedded Networked Sensor Systems.*

McGuire, M., & Plataniotis, K. N. (2003). Dynamic model-based filtering for mobile terminal location estimation. *IEEE Transactions on Vehicular Technology, 52*(4), 1012-1031.

McGuire, M., Plataniotis, K. N. & Venetsanopoulos, A. N. (2003) Location of mobile terminals using time measurements and survey points. *IEEE Transactions on Vehicular Technology, 52*(4), 999-1011.

McMillan, J. (2002). Reinventing the Bazaar. *A Natural History of Markets.* J. McMillan (Ed.). W. W. Norton & Company.

Meertens, L., & Fitzpatrick, S. (2004). *The distributed construction of a global coordinate system in a network of static computational nodes from inter-node didstances.* Tech. rep., Kestrel Institute.

Meguerdichian, S., Koushanfar, F., Potkonjak, M., & Srivastava, M. B. (2001). Coverage problems in wireless ad-hoc sensor networks. *Proceedings of IEEE Infocom Conference*, (pp. 1380-1387).

Meissner, A., Luckenbach, T., Risse, T., Kirste, T., & Kirchner, H. (2002). Design challenges for an integrated disaster management communication and information system. *1st IEEE Workshop on Disaster Recovery Networks*. New York, USA

Mendel, J. M. (1995). *Lessons in Estimation Theory for Signal Processing, Communications, and Control*. New Jersey: Prentice-Hall.

Merrill, W. M., Newberg, F., Sohrabi, K., Kaiser, W., & Pottie, G. (2003). Collaborative networking requirements for unattended ground sensor systems. *IEEE Aerospace Conference, 5*, 2153-2165.

Mihaylova, L., Angelova, D., Honary, S., Bull, D., Canagarajah, C., & Ristic, B. (2007). Mobility Tracking in Cellular Networks Using Particle Filtering. *IEEE Transactions on Wireless Communications, 6*(10), 3589-3599.

Mills, K. L. (2007). A Brief Survey of Self-Organization in Wireless Sensor Networks. *Wireless Communication and Mobile Computing, 7*, 823-834.

Moncel, J. (2006). On graphs of n vertices having an identifying code of cardinality $\lceil \log_2(n + 1) \rceil$. *Discrete Applied Mathematics, 154*(14), 2032–2039.

Moncel, J., Frieze, A., Martin, R., Ruszink, M., & Smyth, C. (2005). Identifying codes in random networks. In *IEEE International Symposium on Information Theory*, (pp. 1464- 1467), Adelaide, Australia.

Moore, D., Leonard, J., Rus, D., & Teller, S. (2004). Robust distributed network localization with noisy range measurements. *In Proc. 2nd Intl Conf. Embedded Networked Sensor Systems*, (pp. 50-61), Baltimore, MD.

Moore, T., Hill, C., & Monteiro, L. S. (2002). Maritime DGPS: Ensuring the best availability and continuity. *Journal of Navigation, 55*(3), 485-494.

Motley, A. J. (1988). Radio Coverage in Buildings. *Proc. National Communications Forum*. Chicago. (pp. 1722-1730).

Muller, T. & Sereni, J.-S. (2007). Identifying and locating-dominating codes in (random) geometric networks. 2007, *submitted to Combinatorics, Probability and Computing. ITI Series* 2006-323 and *KAM-DIMATIA Series* 2006-797.

Myllymäki, P., Roos, T., Tirri, H., Misikangas, P., & Sievanen, J. (2001). A Probabilistic Approach to WLAN User Location Estimation. *Proc. of the Third IEEE Workshop on Wireless LANs*, retrieved October 13, 2008, from http://www.wlan01.wpi.edu/proceedings/wlan18d.pdf.

Nagpal, R., Shrobe, H., & Bachrach, J. (2003). Organizing a global coordinate system from local information on an ad hoc sensor network. In *International Symposium on Information Processing in Sensor Networks* (pp. 333-348).

Nanotron Technologies nanoLOC TRX Transceiver (NA5TR1) Datasheet Version 1.03 (2008). Retrieved March 14, 2008 from Nanotron Technologies Web site: http://www.nanotron.com/EN/docs/nanoLOC/DS_nanoLOC_TRX_NA5TR1.pdf

Nefedov, N., & Pukkila, M. (2000). Iterative channel estimation for gprs. *Proceedings of the 11th IEEE International Symposium on Personal, Indoor and Mobile Radio Communications, 2*, 999-1003.

Nerguizian, C., Despins, C., & Affès, C. (2006). Geolocation in Mines With an Impulse Fingerprinting Technique and Neural Networks. *IEEE Transactions on Wireless Communications, 5*(3).

Nerguizian, C., Despins, C., & Affes, S. (2001). Framework for Indoor Geolocation Using an Intelligent System. *Proceedings of the Third IEEE Workshop on Wireless LANs*, retrieved October 13, 2008, from http://www.wlan01.wpi.edu/proceedings/wlan44d.pdf

Newbury Networks. (2008). *Newbury networks, white paper*. Retrieved October 10, 2008, from http://www.newburynetworks.com

Nguyen, X., Jordan, M. I., & Sinopoli, B. (2005). A kernel-based learning approach to ad hoc sensor network localization. *ACM Transactions on Sensor Networks, 1*, 134-152.

Ni, L., Liu, Y., Lau, Y. C., & Patil, A. P. (2005), LANDMARC: Indoor location sensing using active RFID, *Wireless Networks, 10*(6), 701-710.

Niculescu, D. & Nath, B. (2001). Ad-hoc positioning system. In IEEE, *Proc. of Global Telecommunications Conference (Globecom)* (pp. 2926-2931). San Antonio, TX.

Niculescu, D. (2004). Positioning in ad hoc sensor networks. *IEEE Network, 50*(4).

Niculescu, D., & Nath, B. (2003). Ad Hoc Positioning System (APS) Using AOA. *Annual Joint Conference of the Computer and Communication Societies (INFOCOM)* (pp. 1734-1743). San Francisco.

Niculescu, D., & Nath, B. (2003). Localized positioning in ad hoc networks. *IEEE International Workshop on Sensor Network Protocols and Applications* (pp. 42-50).

Niculescu, D., & Nath, B. (2003). Dv based positioning in ad hoc networks. *Telecommunication Systems, 22*(1-4), 267–280.

Oh, S., Sastry, S., & Schenato, L. (2005). A Hierarchical Multiple-Target Tracking Algorithm for Sensor Networks. *In Proc. International Conference on Robotics and Automation.*

Oh-Heum, K., & Ha-Joo, S. (2008). Localization through map stitching in wireless sensor networks. *IEEE Transactions on Parallel and Distributed Systems, 19*(1), 93-105.

Okumura,Y., Ohmori, E., Kawano, T., & Fukuda, K. (1968). Field strength and its variability in VHF and UHF land-mobile radio service. *Review of the Electrical Communication Laboratory, 16*(9-10).

Olama, M., Djouadi, S., & Pendley, C. (2006). Position and Velocity Tracking in Mobile Cellular Networks Using the Particle Filter. *In Proceedings of IEEE WCNC 2006*, Las Vegas, USA.

Oppenheim, A., & Schafer, R. (1975). Digital Signal Processing. Englewood Cliffs, N.J.: Prentice-Hall.

Oppermann, I., Hamalainen, M., & Linatti, J. (Eds.) (2004). *UWB Theory and Applications.* England: John Wiley and Sons.

Ottersten, B., Viberg, M., & Kailath, T. (1991). Performance analysis of the total least squares ESPRIT algorithm. *IEEE Trans. on Signal Processing, 39*, 1122-1135.

Ozsoy, F. A., & Pinar, M. C. (2004, November). An exact algorithm for the capacitated vertex p-center problem. *Computers and Operations Research, 33*(5), 1420-1436.

Pahlavan, K., & Levesque, A. H. (2005) *Wireless Information Networks - 2nd Edition.* Wiley – Interscience. ISBN: 0-471-72542-0, Hardcover, 722 pages.

Pahlavan, K., Akgul, F. O., Heidari, M., Hatami, A., Elwell, J. M., & Tingley, R. D. (2006). Indoor geolocation in the absence of direct path. *IEEE Wireless Communications, 13*(6), 50-58.

Pahlavan, K., Krishnamurthy, P., & Beneat, J. (1998). Wideband radio propagation modeling for indoor geolocation applications. *IEEE Communications Magazine.* (pp. 60–65).

Pahlavan, K., Li, X., & Makela, J. (2002). Indoor geolocation science and technology. *IEEE Commun. Mag., 40*(2), (pp. 112-118).

Pan, J. J., Kwok, J. T., & Chen, Y. (2006). Multidimensional Vector Regression for Accurate and Low-Cost Location Estimation in Pervasive Computing. *IEEE Transactions on Knowledge and Data Engineering, 18*(9), 1181-1193.

Pan, J. J., Yang, Q., & Pan, J. (2007). Online Co-Localization in Indoor Wireless Networks by Dimension Reduction. *In Proceedings of the 22nd National Conference on Artificial Intelligence (AAAI-07),* (pp. 1102-1107).

Pandey, S., Kim, B., Anjum, F., & Agarwal, P. (2005). Client Assisted Location Data Acquisition Scheme for Secure Enterprise Wireless Networks. *Proceedings of the IEEE Wireless Communications and Networking Conference (WCNC)*, New Orleans, LA, USA, *2*, 1174-1179.

Papadimitriou, C. H., & Ratajczak, D. (2004). *On a Conjecture Related to Geometric Routing (Algorithmic Aspects of Wireless Sensor Networks).* In SpringerLink (Ed.), (Vol. 3121/2004, pp. 9-17). Springer Berlin / Heidelberg.

Paschalids, I. C., & Guo, D. (2007, December). Robust and distributed localization in sensor networks. *Proceedings of 46th IEEE Conference on Decision and Control*, (pp. 933-938), New Orleans, Louisiana

Pathan, A.-S.K., Choong, S. H., Hyung-Woo, L. (2006). Smartening the environment using wireless sensor networks in a developing country. *The 8th International Conference on Advanced Communication Technology, 1*, 705-709.

Patwari, N., & Agrawal, P. (2008). Effects of correlated shadowing: connectivity, localization, and RF Tomography. *In 2008 International Conference on IPSN* (pp. 82-93).

Patwari, N., & Hero, A. O. (2003). Using proximity and quantized RSS for sensor localization in wireless networks. *In Proc. 2nd ACM Intl. Conf. Wireless Sensor Networks and Applications (WSNA '03)* (pp. 20-29), San Diego, CA.

Patwari, N., & Hero, A.O. (2006). Signal strength localization bounds in ad hoc & sensor networks when transmit powers are random. *In Proceedings of the Fourth IEEE Workshop on Sensor Array and Multichannel Processing* (SAM-2006) (pp. 299-303), July 12-14, 2006, Waltham, MA.

Patwari, N., Agrawal, P., & Hero, A. O. (2006). Demonstrating Distributed Signal Strength Location Estimation. *In Proc. Fourth Intl. Conf. Embedded Networked Sensor Systems (SenSys'06)*, (pp. 353-354), Boulder, CO.

Patwari, N., Ash, J. N., Kyperountas, S., Hero, A. O., III, Moses, R. L., & Correal, N. S. (2005). Locating the nodes: Cooperative localization in wireless sensor networks. *IEEE Signal Processing Magazine, 22*(4), 54-69.

Patwari, N., Hero, A. O., Perkins, M., Correal, N. S., & O'Dea,R. J. (2003). Relative location estimation in wireless sensor networks. *IEEE Transactions on signal processing, 51*(8), 2137-2148.

Patwari, N., Wang, Y., & O'Dea, R. J. (2002). The Importance of the Multipoint-to-Multipoint Indoor Radio Channel in Ad Hoc Networks. *In Proceedings of the IEEE Wireless Communication and Networking Conference (WCNC'02), 2*, 608-612, Orlando FL.

Paulraj, A., Roy, R., & Kailath, T. (1986). A subspace rotation approach to signal parameter estimation. *Proceedings of the IEEE, 74*(7), 1044-1046.

Polastre, J., Szewczyk, J., & Culler, D. (2005). Telos: enabling ultra-low power wireless research. *In Proceedings of Information Processing In Sensor Networks 2005*, Los Angeles, USA.

Poling, T. C., & Zatezalo, A. (2002). Interferometric GPS ambiguity resolution, *Journal of Engineering Mathematics*, 43(2-4), 135-151(17).

Popa, M., Ansari, J., Riihijärvi, J., & Mähönen, P. (2008). Combining Cricket System and Inertial Navigation for Indoor Human Tracking. *In Proceeding of WCNC 2008*, Las Vegas, USA.

Prasithsangaree, P., Krishnamurthy, P., & Chrysanthis, P. (2002). On indoor position location with wireless LANs. *The 13th IEEE International Symposium on Personal, Indoor and Mobile Radio Communications* (pp. 720-724).

Prehofer, C., & Bettstetter, C. (2005). Self-Organization in Communication Networks: Principles and Design Paradigms. *IEEE Communications Magazine, 43*(7), 78-85.

Priyantha, N. B. (2005). *The Cricket Indoor Location System.* Doctoral thesis. Massachussette Institute of Technology.

Priyantha, N. B., Balakrishnan, H., Demaine, E., & Teller, S. (2003). Anchor-Free Distributed Localization in Sensor Networks. *LCS Tech. Report #892, MIT, USA.*

Priyantha, N. B., Balakrishnan, H., Demaine, E., & Teller, S. (2005). Mobile-Assisted Localization in Wireless Sensor Networks. *In IEEE INFOCOM.* Miami, FL.

Priyantha, N. B., Demaine, E., & Teller S. (2003) Poster abstract: Anchor-free distributed localization in sensor networks. In *SenSys '03: Proceedings of the 1st international conference on Embedded networked sensor systems*, (pp. 340–341,) New York. ACM.

Priyantha, N., Miu, A., Balakrishnan, H., & Teller, S. (2001). The cricket compass for context-aware mobile applications. *In ACM conference on mobile computing and networking (MOBICOM)*, (pp. 1-14).

Qi, Y., & Kobayashi, H. (2002). Cramér-Rao lower bound for geolocation in non-line-of-sight environment. In *Proceedings of the 2002 IEEE International Conference on Acoustics, Speech, and Signal Processing, 3*, 2473-2476.

Qi, Y., & Kobayashi, H. (2002). On geolocation accuracy with prior information in non-line-of-sight environment. In *Proc. IEEE Vehicular Technology Conference*, Vancouver, Canada.

Qi, Y., & Kobayashi, H. (2003). On relation among time delay and signal strength based geolocation methods. In *IEEE Global Telecommunications Conference. GLOBECOM '03., 7*, 4079-4083.

Qi, Y., Kobayashi, H., & Suda, H. (2006). Analysis of wireless geolocation in a non-line-of-sight environment. *IEEE Transactions on Wireless Communications, 5*(3), 672-681.

Qi, Y., Suda, H., & Kobayashi, H. (2004). On time-of arrival positioning in a multipath environment. In *Proc. IEEE 60th Vehicular Technology Conf. (VTC 2004-Fall)*. Los Angeles, CA., *5*, 3540–3544.

Rabbat, M., & Nowak, R. (2004). Distributed optimization in sensor networks. *Third International Symposium on Information Processing in Sensor Networks* (pp. 20-27).

Rajagopalan, S. & Vazirani, V. (1998). Primal-dual RNC approximation algorithms for set cover and covering integer programs. *SIAM Journal on Computing, 28*, 525–540.

Ramadurai V., & Sichitiu M. L. (2003). Simulation-based Analysis of a Localization Algorithm for Wireless Ad-Hoc Sensor Networks, *OPNETWORK* 2003.

Randell, C., Djiallis, C., & Muller, H. (2003). Personal position measurement using dead reckoning. In *Proceedings of the Seventh International Symposium on Wearable Computers, IEEE Computer Society*. (pp. 166–173).

Rao, A., Ratnasamy, S., Papadimitriou, C., Shenker, S., & Stoica, I. (2003). Geographical routing without location information. In ACM, *Proc. of the 9th Int. Conf. on Mobil Computing and Networking (MobiCom)* (pp. 96-108). San Diego, CA.

Rao, B. D., & Hari, K. V. S. (1989). Performance analysis of root-music. *IEEE Trans. ASSP-37*(12), 1939-1949.

Rapoport, I., & Oshman, Y. (2004). Recursive Weiss-Weinstein lower bounds for discrete-time nonlinear filtering. *In Proceedings of the 43rd IEEE Conference on Decision and Control, 3*, 2662-2667).

Rappaport, T. S. (1996). *Wireless Communications: Principles and Practice*. Englewood Cliffs, NJ: Prentice-Hall.

Rappaport, T. S., Reed, J. H., & Woerner, B. D. (1996). Position location using wireless communications on highways of the future. *IEEE Communications Magazine, 34*(10), 33-41.

Ray, S., Lai, W., & Paschalidis, I. C. (2005). Deployment optimization of sensornet-based stochastic location-detection systems. *IEEE INFOCOM 2005* (pp. 2279-2289).

Ray, S., Lai, W., & Paschalidis, I. C. (2006). Statistical location detection with sensor networks. *IEEE Transactions on Information Theory, Joint special issue with IEEE/ACM Transactions on Networking on Networking and Information Theory, 52*(6), 2670-2683.

Ray, S., Starobinski, D., Trachtenberg, A., & Ungrangsi, R. (2004). Robust location detection with sensor networks, *IEEE Journal on Selected Areas in Communications (Special Issue on Fundamental Performance Limits of Wireless Sensor Networks), 22*(6), 1016 – 1025.

Ray, S., Ungrangsi, R., Pellegrinin, F. D., Trachtenberg, A., & Starobinski D. (2003). Robust location detection in emergency sensor networks, In *Proceedings of the 22nd Annual Joint Conference of the IEEE Computer and Communications Societies*, *2*, 1044–1053.

Recski, A. (1989) *Matroid theory and its applications in electric network theory and in statics*, Budapest, Akadémiai Kiadó.

Reichenbach F., Koch M., & Timmermann D. (2006). Closer to Reality - Simulating Localization Algorithms Considering Defective Observations in Wireless Sensor Networks. *Proceedings of the 3rd Workshop on Positioning, Navigation and Communication (WPNC'06)*, 2006, (pp. 59-65).

Richards, M. (2005). *Fundamentals of Radar Signal Processing*. New York: McGraw-Hill.

Rife, D. C., Goldstein, M., & Boorstyn, R. R. (1975). A unification of Cramér-Rao type bounds. *IEEE Transactions on Information Theory, IT-21*(3), 330-332.

Riihijärvi, J., Vázquez Martí, C., & Mähönen, P. (2005). Enhancing Radar Tracking with Car-to-Car Communication. In *Proceedings of IEEE SCVT'05*, Enschede, the Netherlands.

Ristic, B., Arulampalam, S., & Gordon, N. (2004). *Beyond the Kalman Filter: Particle Filters for Tracking Applications*. Artech House, Norwood, Massachusetts, USA.

Röhrig, C., & Spieker, S. (2008). Tracking of Transport Vehicles for Warehouse Management Using a Wireless Sensor Network. *IEEE/RSJ International Conference on Intelligent Robots and Systems*.

Romer, K. (2003). The lighthouse location system for smart dust. *Proceedings of MobiSys 2003 (ACM/USENIX Conference on Mobile Systems, Applications, and Services)* (pp. 15-30).

Roos, T., Myllymaki, P., & Tirri, H. (2002). A statistical modeling approach to location estimation. *IEEE Transactions on Mobile Computing, 1*(1), 59-69.

Roy, R., & Kailath, T. (1989). ESPRIT-estimation of signal parameters via rotational invariance techniques. *IEEE Transactions on Acoustics, Speech, and Signal Processing, 37*(7), 984-995.

Royer, E. & Toh, C. (1999). A review of current routing protocols for ad hoc mobile wireless networks. *IEEE Personal Communications, 6*(2), 46-55.

Saha, S., Chaudhuri, K., Sanghi, D., & Bhagwat, P. (2003). Location Determination of a Mobile Device Using IEEE 802.11b Access Point Signals. *Proceedings of the 2003 IEEE Wireless Communications and Networking Conference (WCNC), 3*, 1987-1992.

Sahai, P. (2002). Geolocation on Cellular Networks. In B. Sarikaya (Ed.) *Geographic location in the Internet*. (pp. 13-49). Boston: Kluwer Academic Publishers.

Sang-Hun, C., Hyunsoo, Y., & Jung-Wan, C. (2002). A Fast Handoff Scheme For IP over Bluetooth. *In the Proceedings of the International Conference on Parallel Processing Workshops* (pp. 51).

Sarangapani, J. (2007). *Wireless ad hoc and sensor networks: protocols, performance, and control*. F. L. Lewis (Ed.), Broken Sound Parkway NW: Taylor & Francis Group, LLC.

Savarese, C., & Rabaey, J. (2002). Robust positioning algorithms for distributed ad-hoc wireless sensor networks. *Proceedings of the General Track: 2002 USENIX Annual Technical Conference* (pp. 317-327).

Savarese, C., Rabaey, J., & Beutel, J. (2001). Locationing in distributed ad-hoc wireless sensor networks. In *IEEE International Conference on Acoustics, Speech, and Signal Processing*, pp. 2037-2040. Salt Lake City, UT.

Savvides, A., Garber, W. L., Moses, R. L., & Srivastava, M. B. (2005). An analysis of error inducing parameters in multihop sensor node localization. *IEEE Transactions on Mobile Computing, 4*(6), 567-577.

Savvides, A., Han, C., & Srivastava, M. B. (2001). Dynamic fine-grained localization in ad-hoc networks of sensors. *Proceedings of the 7th annual international conference on mobile computing and networking*, Rome, Italy, (pp. 166-179).

Savvides, A., Park, H., & Srivastava, M. (2002). The bits and flops of the n-hop multilateration primitive for node localization problems. In ACM, *1st Int'l Workshop on Wireless Sensor Networks and Applications (WSNA'02)* (pp. 112–121). Atlanta, GA.

Saxe, J. (1979). Embeddability of weighted graphs in k-space is strongly NP-hard. *17th Allerton Conference in Communications, Control and Computing* (pp. 480-489).

Sayed, A. H., Tarighat, A., & Khajehnouri, N. (2005). Network-based wireless location. *IEEE Signal Processing Magazine, 22*(7).

Schelkunoff, S. A. (1943). A mathematical theory of linear arrays. *Bell System Technical Journal, 22*, 80-107.

Schell, S. V., & Gardner, W. A. (1993). High-resolution direction finding. *Handbook of Statistics, 10*, 755-817.

Schenato, L., & Gamba, G. (2007). A distributed consensus protocol for clock synchronization in wireless sensor network. In *Proc. IEEE Conference on Decision and Control*, New Orleans, LA, USA.

Schmidt, R. (1986). Multiple emitter location and signal parameter estimation. *IEEE Transactions on Antennas and Propagation, 34*(3), 276-280.

Scholkopf, B., & Smola, A. (2002). *Learning with kernel*. Cambridge, MA: MIT Press.

Schrijver, A. (2003) *Combinatorial Optimization*, Berlin, Springer.

Schroer, R. (2003). Navigation and landing [A century of powered flight 1903-2003]. IEEE Aerospace and Electronic Systems Magazine, *18*(7), 27-36.

Seidel, S.Y., & Rappaport, T. S. (1992). 914 MHz path loss prediction models for indoor wireless communications in multifloored buildings , *IEEE Transactions on Antennas and Propagation, 40*(2), 207-217.

Servatius B. & Whiteley, W. (1999) Constraining plane configurations in CAD: Combinatorics of directions and lengths, *SIAM J. Discrete Math.*, 12, 136–153.

Shah, S. F. A., & Tewfik, A. H. (2005). Enhanced position location with UWB in obstructed LOS and NLOS multipath environments. *In Proc.of European Signal Processing Conf. (EUSIPCO).*

Shang, Juml, Zhang, & Fromherz (2003). Localization from mere connectivity. In *ACM Mobihoc,* (pp. 201-212).

Shang, Y., Ruml, W., Zhang, Y., & Fromherz, M. (2004). Localization from connectivity in sensor networks. *IEEE Transactions on Parallel and Distributed Systems, 15*(11), 961-974.

Shang, Y., Ruml, W., Zhang, Y., & Fromherz, M. P. (2003). Localization from mere connectivity. *In Proceedings of the 4th ACM international Symposium on Mobile Ad Hoc Networking & Computing* (Annapolis, Maryland, USA, June 01 - 03, 2003). MobiHoc '03. ACM, New York, NY, (pp. 201-212).

Shang, Y., Shi, H., & Ahmed, A. (2004). Performance study of localization methods for ad-hoc sensor networks. In IEEE, *Proceedings of the 1st IEEE International Conference on Mobile Ad-hoc and Sensor Systems (MASS'04)* (pp. 184-193). Fort Lauderdale, FL.

Shepard, R. N. (1962). Analysis of proximities: Multidimensional scaling with an unknown distance function i & ii. *Psychometrika, 27,* 125–140, 219–246.

Sheu, J.-P., Li, J.-M., & Hsu, C.-S. (2006). A distributed location estimating algorithm for wireless sensor networks. *In Proceedings of the IEEE International Conference on Sensors Networks, Ubiquitous, and Trustworthy Computing* (pp. 218-225).

Shi, H., Li, X., Shang, Y., & Ma, D. (2005). Cramer-Rao Bound Analysis of Quantized RSSI Based Localization in Wireless Sensor Networks. In IEEE, *Proc. of the 11th Int. Conf. on Parallel and Distributed Systems - Workshops (ICPADS'05)* (pp. 32-36). Fukuoka, Japan.

Shin, D.-H., & Sung, T.-K. (2002) Comparisons of error characteristics between TOA and TDOA Positioning. *IEEE Transactions on Aerospace and Electronic Systems, 38*(1), 307-311.

Sichitiu, M. L., & Ramadurai, V. (2004). Localization of wireless sensor networks with a mobile beacon. *In Proc. of the First IEEE Conference on Mobile Ad-hoc and Sensor Systems (MASS 2004),* Fort Lauderdale, FL, Oct. 2004, (pp. 174-183).

Simandl, M., & Straka, O. (2007). Sampling Densities of Particle Filter: A Survey and Comparison. *In Proceedings of the 2007 American Control Conference,* New York, USA.

Šimandl, M., Královec, & Tichavský, P. (2001). Filtering, predictive, and smoothing Cramér-Rao bounds for discrete-time nonlinear dynamic systems. *Automatica, 37*(11), 1703-1706.

Simon, G., Maroti, M., Ledeczi, A., Balogh, G., Kusy, B., Nadas, A., Pap, G., Sallai, J., & Frampton, K. (2004). Sensor Network-Based Countersniper System. In *Proceedings of the 2nd international conference on Embedded Networked Sensor Systems,* Baltimore, USA.

SiRFStarIII Product Insert (2008). Retrieved March 14, 2008 from SiRF Technologies Web site: http://www.sirf.com/products/GSC3LPProductInsert.pdf

Sivrikaya, F., & Yener, B. (2004). Time synchronization in sensor networks: A survey. *IEEE Network, 18*(4).

Slijepcevic, S., Megerian, S., & Potkonjak, M. (2002). Location errors in wireless embedded sensor networks: sources, models, and effects on applications. *SIGMOBILE Mob. Comput. Commun. Rev., 6*(3) (Jun. 2002), 67-78.

Smailagic, A., & Kogan, D. (2002). Location sensing and privacy in a context aware computing environment. *IEEE Wireless Communications* (pp. 10-17).

Smith, A., Balakrishnan, H., Goraczko, M., & Priyantha, N. (2004). Tracking Moving Devices with the Cricket Location System. *In Proceedings of the ACM MobiSys 2004,* Boston, USA.

Smith, J., & Abel, J. (1987). The spherical interpolation method of source localization. *IEEE Journal of Oceanic Engineering, 12*(1), 246-252.

So, A.M. & Ye, Y. (2007). Theory of semidefinite programming for sensor network localization, *Math. Program.* 109 (3) Ser. B, 367–384.

Sohraby, K., Minoli, D., & Znati, T. (2007). Wireless Sensor Networks – Technology, Protocols and Applications. Hoboken, New Jersey: John Wiley & Sons.

Solis, R., Borkar, V., and Kumar, P. (2006). A new distributed time synchronization protocol for multihop wireless networks. In *Proc. IEEE Conference on Decision and Control,* San Diego, CA, USA.

Song, L., Adve, R., & Hatzinakos, D. (2004). Matrix pencil positioning in wireless ad hoc sensor networks. *Proceedings of First European Workshop on Wireless Sensor Networks,* (pp. 18-27).

Sonnenberg, G. (1988). *Radar and Electronic Navigation.* Butterworths.

SourceForge.net. (2008). The GRAIL Real Time Location Service, documentation and source code. Retrieved October 10, 2008, from http://grailrtls.sourceforge.net.

Spencer, Q., Jeffs, B., Jensen, M., Swindlehurst, A. (2000). Modeling the statistical time and angle of arrival characteristics of an indoor multipath channel. *IEEE Journal on Selected Areas in Communications, 18*(3), 347-360.

Spiegelhalter, D. J. (1998). Bayesian graphical modeling: A case-study in monitoring health outcomes. *Applied Statistics, 47*(1), 115-133.

Spirito, M. A. (2001). On the accuracy of cellular mobile station location estimates. *IEEE Transactions on Vehicular Technology, 50*(3), 674-685.

Spirito, M., & Mattioli, A. (1998). On the hyperbolic positioning of GSM mobile stations. In *Proc. International Symposium on Signals, Systems and Electronics.*

Stanfield, R. G. (1947). Statistical theory of DF finding. *Journal of IEE, 94*(5), 762-770.

Steele, R. (1992). *Mobile Radio Communications.* Piscataway, NJ: IEEE Press.

Steiner, C., Althaus, F., Troesch, F., & Wittneben, A. (2008). Ultra-Wideband Geo-Regioning: A Novel Clustering and Localization Technique. *EURASIP Journal on Advances in Signal Processing.* article id 296937.

Stoianov, I., Nachman, L., & Madden, S. (2007). PIPENET: A wireless sensor network for pipeline monitoring. *In 6th International Symposium on Information Processing in Sensor Networks*, Cambridge, MA. (pp. 264 – 273)

Stoica, P., & Sharman, K. (1990). A Novel Eigenanalysis Method for Direction of Arrival Estimation. *Proc. IEE,* (pp. 19-26).

Stojmenovic, I. (Ed.). (2005). *Handbook of sensor networks : algorithms and architectures.* Hoboken, New Jersey: John Wiley & Sons, Inc.

Stojmenovic, I., & Olariu, S. (2005). Geographic and Energy-Aware Routing in Sensor Networks. In Stojmenovic, (Ed.), *Handbook of sensor networks: Algorithms and architectures,* (p. 381-416). Wiley.

Stoleru, R., & Stankovic, J. A. (2004). Probability Grid: A location estimation scheme for wireless sensor networks. In *IEEE SECON.* (pp. 430-438).

Stoleru, R., Stankovic, J. A., & Luebke, D. (2005). A high-accuracy, low-cost localization system for wireless sensor networks. In *ACM Sensys,* (pp. 13-26). San Diego, CA.

Sun, G., Chen, J., Guo, W., & Ray Liu, K. J. (2005). Signal processing techniques in network-aided positioning: A survey *IEEE Signal Processing Magazine, 22*(7).

Suomela, J. (2007). Approximability of identifying codes and locating–dominating codes. *Information Processing Letters, 103*(1), 28–33.

Tay, T.S. & Whiteley, W. (1985) Generating isostatic frameworks, *Structural Topology* 11, pp. 21-69.

Tayem, N., & Kwon, H. M. (2004). Conjugate esprit (C-SPRIT). *IEEE Transactions on Antennas and Propagation, 52*(10), 2618-2624.

Tekinay, S., Chao, E., & Richton, R. (1998). Performance benchmarking for wireless location systems. *IEEE Communications Magazine, 36*(4), 72-76.

Tenenbaum, J., de Silva, V., & Langford, J. (2000). A global geometric framework for nonlinear dimensionality reduction. *Science, 290*(5500), 2319-2323.

Tewfik, A. H., & Hong, W. (1992). On the application of uniform linear array bearing estimation techniques to uniform circular arrays. *IEEE Transactions on Signal Processing, 40,* 1008-1011.

Theoleyre, F., & Valois, F. (2007). Self-Organization of Ad Hoc Networks: Concepts and Impacts. In H. Labiod (Ed.*), Wireless Ad hoc and Sensor Networks,* (pp. 101-128). ISTE.

Thrun, S. (2002). Particle Filters in Robotics. In *Proceeding of Uncertainty in AI.*

Tichavský, P., Muravchik, C. H., & Nehorai, A. (1998). Posterior Cramér-Rao bounds for discrete-time nonlinear filtering. *IEEE Transacations on Signal Processing, 46*(5), 1386-1396.

Tingley, R. (2000). *Time-Space Characteristics of Indoor Radio Channel.* PhD Thesis, WPI.

Tonguz, O. K., & Ferrari, G. (2006). *Ad hoc Wireless Networks: A Communication-theoretic Perspective,* O. K. Tonguz & G. Ferrari (Eds.). Wiley.

Torgerson, W. S. (1952). Multidimensional scaling: I. Theory and method. *Psychometrika, 17*, 401–419.

Torgerson, W. S. (1965). Multidimensional scaling of similarity. *Psychometrika, 30*, 379–393.

Torrieri, D. J. (1984). Statistical theory of passive location systems. *IEEE Transactions on Aerospace and Electronic Systems, AES-20*(2), 183-198.

Toumpis, S., & Tassiulas, L. (2006, July). Optimal deployment of large wireless sensor networks. *IEEE Transactions On Information Theory*, 52(7), 2935-2953.

Tran, D. A., & Nguyen, T. (2006). Support vector classification strategies for localization in sensor networks. In *IEEE Int'l Conference on Communications and Electronics*.

Tran, D. A., & Nguyen, T. (2008). Localization in Wireless Sensor Networks based on Support Vector Machines. *IEEE Transactions on Parallel and Distributed Systems, 19*(7), 981-994, July.

Tran, D. A., & Nguyen, T. (2008). Hop-count based learning techniques for passive target tracking in sensor networks. *IEEE Transactions on Systems, Man, and Cybernetics*, submitted, 2008.

Tse, D., & Viswanath, P. (2005). *Fundamentals of Wireless Communication*. Cambridge, UK: Cambridge University Press.

Ubisense Limited (2008). *Ubisense System Overview*. Retrieved July 29, 2008 from Ubisense Limited Web site: http://www.ubisense.net/media/pdf/Ubisense%20System%20Overview%20V1.1.pdf

Ulaby, F. (1999). *Fundamentals of Applied Electromagnetics*. Upper Saddle River, NJ: Prentice Hall.

Unbehaun, M. (2002). On the deployment of unlicensed wireless infrastructure. Ph.D. Thesis, Royal Institute of Technology, Department of Signals, Sensors & Systems, Stockholm.

USDOT (US Department of Transportation) – Federal Highway Administration (2002). Phase I High Accuracy-Nationwide Differential Global Positioning System Report. FHWA-RD-02-110.

Van Trees, H. L. (2001). *Detection, Estimation, and Modulation Theory*. Volume I, Toronto, Ontario: John Wiley & Sons.

Van Trees, H. L., & Bell, K. L. (Ed.) (2007). *Bayesian Bounds for Parameter Estimation and Nonlinear Filtering/Tracking*. Piscataway, NJ: IEEE Press.

Varshavsky, A., de Lara, E., Hightower, J., LaMarca, A., & Otsason, V. (2007). GSM indoor localization. *Pervasive Mob. Comput., 3*(6), 698-720.

Vazirani V. (Ed.). (2001). *Approximation Algorithms*. Springer-Verlag.

Walden, M. C., & Rowsell, F. J. (2005). Urban propagation measurements and statistical path loss model at 3.5 GHz. In IEEE, *Proc. of the Antennas and Propagation Society International Symposium* (pp. 363-366).

Wang, Y., Li, F., & Dahlberg, T. A (2006). Power Efficient 3-Dimensional Topology Control for Ad Hoc and Sensor Networks. *IEEE Global Communications Conference*, (pp. 1-5).

Want, R., Hopper, A., Falcao, V., & Gibbons, J. (1992). The active badge location system. *ACM Transactions on Information Systems, 10*(1), 91-102.

Ward, A., Jones, A., & Hopper, A. (1997). A new location technique for the active office. *IEEE Personnel Communications, 4*(5), 42–47.

Watteyne, T., Augé-Blum, I., Dohler, M., & Barthel, D. (2007). Geographic Forwarding in Wireless Sensor Networks with Loose Position-Awareness. *In 18th Annual International Symposium on Personal, Indoor and Mobile Radio Communications (PIMRC)* (pp. 1-5). Athens, Greece.

Watteyne, T., Bachir, A., Dohler, M., Barthel, D., & Augé-Blum, I. (2006). 1-hopMAC: An Energy-Efficient MAC Protocol for Avoiding 1-hop Neighborhood Knowledge. *In International Workshop on Wireless Ad-hoc and Sensor Networks (IWWAN)* (pp. 639-644). New York, NY, USA.

Watteyne, T., Barthel, D., Dohler, M., & Augé-Blum, I. (2008). *WiFly: Experimenting with Wireless Sensor Networks and Virtual Coordinates* (Research Report RR-6471). INRIA.

Weinstein, E., & Weiss, A. J. (1988). A general class of lower bounds in parameter estimation. *IEEE Transactions on Information Theory, 34*(2), 338-342.

Weiss, A. J. (2003). On the accuracy of a cellular location system based on received signal strength measurements. *IEEE Transactions on Vehicular Technology, 52*(6),1508–1518.

Werb, J., Newman, M. Berry, V., & Lamb, S. (2005). Improved Quality of Service in IEEE 802.15.4 Mesh Networks. *Proceedings of International Workshop on Wireless and Industrial Automation.*

Wesson, R., Hayes-Roth, F., Burge, J. W., Stasz, C., & Sunshine, C. A. (1981). Network structures for distributed situation assessment. *IEEE Transactions Systems, Man, and Cybernetics, 11*(1).

Wherenet, (2008). *NYK Logistics Case Study.* Retrieved July 29, 2008 from Wherenet Web site: http://www.wherenet.com/NYKLogisticsCaseStudy.shtml

Whitehouse, C. (2002). *The design of calamari: an ad hoc localization system for sensor networks.* Master's thesis, University of California at Berkeley.

Whitehouse, K., & Culler, D. (2002). Calibration as parameter estimation in sensor networks. In ACM, *Proc. of the Int'l Workshop on Wireless Sensor Networks and Applications* (pp. 59-67). Atlanta, GA.

Whitehouse, K., & Culler, D. (2003). Macro-calibration in Sensor/Actuator Networks. *Mobile Networks and Applications Journal (MONET), Special Issue on Wireless Sensor Networks, 8*(4), 463-472.

Whitehouse, K., & Culler, D. (2006). A robustness analysis of multi-hop ranging-based localization approximations. *In Proceedings of the Fifth international Conference on information Processing in Sensor Networks* (Nashville, Tennessee, USA, April 19 - 21, 2006). IPSN '06. ACM, New York, NY, (pp. 317-325).

Whitehouse, K., Karlof, C., & Culler, D. (2007). A practical evaluation of radio signal strength for ranging-based localization. SIGMOBILE Mob. *Comput. Commun. Rev. 11*(1), 41-52.

Whitehouse, K., Karlof, C., Woo, A., Jiang, F., & Culler, D. (2005, April 24 - 27). The effects of ranging noise on multihop localization: an empirical study. *In Proc. 4th Intl. Symp. Information Processing in Sensor Networks (IPSN'05)* (pp. 73-80). Los Angeles, California.

Whiteley, W. (1990). Vertex splitting in isostatic frameworks, *Structural Topology* 16, 23–30.

Whiteley, W. (1996). Some matroids from discrete applied geometry. Matroid theory (Seattle, WA, 1995), 171–311, *Contemp. Math.,* 197, Amer. Math. Soc., Providence, RI.

Whiteley, W. (2004). Rigidity and scene analysis, In J. E. Goodman and J. O'Rourke (Eds.) *Handbook of Discrete and Computational Geometry,* (pp. 1327-1354), CRC Press, Second Edition.

Wu, S. L., & Tseng., Y. C. (Eds.). (2007). *Wireless ad hoc networking : personal-area, local-area, and the sensory area networks.* Broken Sound Parkway NW: Taylor & Francis Group, LLC.

Wu, Y., Hu, J. B., & Chen, Z. (2007). Radio map filter for sensor network indoor localization systems. *5th IEEE International Conference on INdustrial Informatics* (pp. 63-68).

Wylie, M. P., & Holtzman, J. (1996). The non-line of sight problem in mobile location estimation. In *Proceedings of the International Conference on Universal Personal Communications* (pp. 827-831), Cambridge, MA, USA.

Yamasaki, R., Ogino, A., Tamaki, T., Uta, T., Matsuzawa, N., & Kato, T. (2005). TDOA location system for IEEE 802.11b WLAN. *IEEE Wireless Communications and Networking Conference,* 4, 2338-2343.

Yedavalli, K., Krishnamachari, B., Ravula, S., & Srinivasan, B. (2005). Ecolocation: a sequence based technique for RF localization in wireless sensor networks. *In Proc. 4th Intl. Symp. Information Processing in Sensor Networks,* (pp. 285-292), Los Angeles, CA.

Yemini, Y. (1978). Distributed sensors networks (dsn): An attempt to define the issues. In *Proc. Distributed Sensor Networks Workshop,* Carnegie Mellon, Pittsburgh, PA, USA.

Yemini, Y. (1979). Some theoretical aspects of position-location problems, In *20th Annual IEEE Symposium on Foundations of Computer Science,* (pp. 1–8).

Young, A. S., & Skobla, J. (2003). Robust GPS- SMS communication channel for the AVL system. *Proceedings of the Aerospace Conference.* (pp. 4_1957- 4_1965).

Youssef, M. (2008). *Collection about Location Determination Papers available online* http://www.cs.umd.edu/~moustafa/location_papers.htm

Youssef, M., & Agrawala, A. (2004). Handling samples correlation in the HORUS system. *In proceedings of*

Annual Joint Conference of the IEEE Computer and Communications Societies (pp. 1023–1031).

Youssef, M., & Agrawala, A. (2004). On the Optimality of WLAN Location Determination Systems. *Proceedings of the Communication Networks and Distributed Systems Modeling and Simulation Conference.*

Youssef, M., & Agrawala, A. (2005). The Horus WLAN Location Determination System. *Proceedings of the Third International Conference on Mobile Systems, Applications, and Services (MobiSys 2005),* Seattle, WA, USA, (pp. 205-218).

Youssef, M., Agrawala, A., & Udaya Shankar, A. (2003). WLAN Location Determination via Clustering and Probability Distributions. *Proceedings of the IEEE International Conference on Pervasive Computing and Communications (PerCom),* (pp. 143-150).

Yu, J., & Oppermann, I. (2004). UWB positioning for wireless embedded networks. *In Proc. of IEEE Radio and Wireless Conference,* (pp. 459–462).

Yu, M., Mokhtar, H., & Merabti, M. (2006). A survey of network management architecture in wireless sensor network. *In Proceedings of the Sixth Annual PostGraduate Symposium on The Convergence of Telecommunications,* Liverpool, UK.

Yu, Y., Govindan, R., & Estrin, D. (2001). *Geographical and energy aware routing: a recursive data dissemination protocol for wireless sensor networks* (Tech. Rep. No. ucla/csd-tr-01-0023). Los Angeles, CA: University of California-Los Angeles.

Yu, Y., Prasanna, V. K., & Krishnamachari, B. (2006). *Introduction to wireless sensor networks.* In Y. Yu, V. K. Prasanna, & B. Krishnamachari (Eds.), *Information processing and routing in wireless sensor networks* (pp. 1-21). Singabore: World Scientific Publishing Co. Pte. Ltd.

Zaidi, A. S., & Suddle, M. R. (2006). Global Navigation Satellite Systems: A Survey. In *International Conference on Advances in Space Technologies,* (pp. 84-84).

Zeitouni, O., Ziv, J., & Merhav, N. (1992 September). When is the generalized likelihood ratio test optimal. *IEEE Transactions on Information Theory, 38*(2), 1597-1602.

Zhang, P., Shi, Z., & Xu, M. (2005). Design and implementation of vehicle monitoring system based on GPRS. *Proceedings of the Fourth International Conference on Machine Learning and Cybernetics,* Guangzhou. (pp. 3574 - 3578).

Zhang, Z. (1992). Iterative Point Matching for Registration of Free-Form Curves and Surfaces. *International Journal of Computer Vision, 13*(2), 119-152.

Zhao, Y. (2002). Standardization of mobile phone positioning for 3G systems. *IEEE Communications Magazine, 40*(7).

Zheng, J., Winstanley, W. C., Yan, L., & Fotheringham, A. S. (2008). Economical LBS for Public Transport: Real-time Monitoring and Dynamic Scheduling Service. *The 3rd International Conference on Grid and Pervasive Computing Workshops,* (pp. 184 -188).

Zhu, C., & Kuh, A. (2007). Ad hoc sensor network localization using distributed kernel regression algorithms. In *Int'l Conference on Acoustics, Speech, and Signal Processing, 2,* 497-500.

Zhu, J. (2006). *Indoor/Outdoor Location of Cellular Handsets Based on Received Signal Strength.* Doctoral dissertation, Georgia Tech, Atlanta. Retrieved Aug 12, 2008, from http:// etd.gatech.edu/theses/available/etd-05182006-154920/

Ziskind, I., & Wax, M. (1988). Maximum likelihood localization of multiple sources by alternating projection. *IEEE Transactions on Acoustics, Speech, and Signal Processing, 36*(10), 1553-1560.

Zou, Y., & Chakrabarty K. (2004). Sensor deployment and target localization in distributed sensor networks. *ACM Transactions on Embedded Computing Systems, 3*(1), 61-91.

Zuniga, M. Z., & Krishnamachari, B. (2007). An analysis of unreliability and asymmetry in low-power wireless links. *ACM Trans. Sensor Networks, 3*(2), 1-7.

Zuniga, M., & Krishnamachari, B. (2004). Exploring the predictability of network metrics in the presence of unreliable wireless links. *SenSys 2004,* (pp. 275-276).

About the Contributors

Guoqiang Mao received a bachelor degree in electrical engineering from Hubei University of Technology, China, the master degree in engineering from South East University, China and PhD in telecommunications engineering from Edith Cowan University, Australia in 1995, 1998 and 2002 respectively. After graduation from PhD, he worked in the U.S.-based industrial company "Intelligent Pixel Incorporation" as a Senior Research Engineer for one year. He joined the School of Electrical and Information Engineering, the University of Sydney in December 2002 where he is a Senior Lecturer now. He is seconded to National ICT Australia as a Senior Researcher since 2003. He has published over fifty papers in prestigious journals and refereed conference proceedings. He has been a regular reviewer for many of the leading journals in the area. He has served as a program committee member in a number of international conferences and was the publicity co-chair of 2007 ACM Conference on Embedded Networked Sensor Systems. He was listed in the 25th Anniversary Edition of Marquis "Who's Who in the World" (2008) and in the 9th (2007) and 10th (2008) Anniversary Edition of Marquis "Who's Who in Science and Engineering". His research interests include wireless localization techniques, wireless multihop networks, graph theory and its application in networking, telecommunications traffic measurement, analysis and modeling, and network performance analysis.

Barış Fidan received the BS degrees in electrical engineering and mathematics from Middle East Technical University, Turkey in 1996, the MS degree in electrical engineering from Bilkent University, Turkey in 1998, and the PhD degree in electrical engineering at the University of Southern California, Los Angeles, USA in 2003. After working as a postdoctoral research fellow at the University of Southern California for one year, he joined the National ICT Australia and the Research School of Information Sciences and Engineering of the Australian National University, Canberra, Australia in 2005, where he is currently a senior researcher. His research interests include autonomous multi-agent dynamical systems, sensor networks, cooperative localization, adaptive and nonlinear control, switching and hybrid systems, mechatronics, and various control applications including high performance and hypersonic flight control, semiconductor manufacturing process control, and disk-drive servo systems. He is coauthor of more than seventy publications, including the textbook "Adaptive Control Tutorial" (SIAM, 2006).

* * *

Baher Abdulhai, is Canada Research chair professor in Intelligent Transportation Systems, and the Director, Toronto Intelligent Transportation Systems Centre. Dr. Abdulhai is an associate professor of ITS at the Department of Civil Engineering, University of Toronto. Prof. Baher Abdulhai has 20 years of experience in transportation and highway engineering including 15 years experience in Intelligent Transportation Systems (ITS). Abdulhai has a PhD from the University of California at Irvine in 1996, and training experience at the California PATH of the University of California Berkeley, and additional professional training at the Massachusetts Institute of Technology (MIT), the University of Rome, and the Technical University of Crete.

Piyush Agrawal has been pursuing the PhD degree in electrical engineering at the University of Utah in Salt Lake City since 2006, where he is a graduate researcher in the Sensing and Processing Across Network (SPAN) Lab. Piyush received his BS degree at the Vellore Institute of Technology, India, where he received a merit scholarship in 2003 and 2004. He has done internships at the University of Applied Sciences, Esslingen, Germany; the Indian Institute of Sciences, Banglore, India; and the Defense Research and Development Lab, Hyderabad, India. Piyush's current research is in received signal strength (RSS)-based localization and joint channel modeling in ad hoc and sensor networks.

Ahmed A. Ahmed is a postdoctoral research associate with the Department of Computer Science at Texas State University-San Marcos. He is also an assistant professor with the Department of Computer and Systems Engineering at Zagazig University, Egypt. He received his PhD degree from the University of Missouri-Columbia in 2005. He obtained his BS degree in electronics and communication engineering from Zagazig University, Egypt in 1995. Ahmed's research interests include wireless sensor networks and distributed computing.

Hazim Ahmed received his MASc in electonic systems engineering, at the University of Regina Canada, in 2008, and his BSc in computer and control engineering from Zagazig University, Egypt 2003. Mr. Ahmed also holds a graduate diploma in software development from the Information Technology Institute (ITI), Cairo Egypt. He has extensive software development experience using Java. Mr. Ahmed research interests include software engineering, wireless mesh networks and high performance computing. Hazem authored/co-authored several papers and book chapters in ITS field. Currently he is working as an independent software designer and architect for ITS and enterprise systems.

Ferit Ozan Akgül received his BSc degree from Middle East Technical University, Turkey, in 2002 and his MSc degree from Koc University, Turkey, in 2004 both in electrical engineering. During his master's, he studied efficient resource allocation for high speed CDMA data systems, specifically 1xEV-DO (IS-856). He also worked on the integration methodologies for WLAN and CDMA systems using joint resource allocation and routing. Currently he is a PhD candidate at the Center for Wireless Information Network Studies (CWINS) at Worcester Polytechnic Institute (WPI), where he works as a research assistant on RF based indoor geolocation and indoor channel characterization with a particular focus on multipath aided precise indoor ranging. His interests include algorithms, statistical modeling, and performance evaluation of UWB TOA-based ranging/positioning systems.

Michael Allen is a PhD student at the Cogent Applied Research Centre, Coventry University, where he is advised by Dr. Elena Gaura. He received a BSc (Hons) from Coventry University in 2005. His research interests include acoustic source localization and self-localization in wireless sensor networks, as well as design and development of robust end-to-end sensing systems. He is an active collaborator with the Centre for Embedded Networked Sensing, UCLA, where he has worked on distributed acoustic sensing systems under the guidance of prof. Deborah Estrin.

Nayef Alsindi received the B.S.E.E. degree from the University of Michigan, Ann Arbor, in 2000 and the MSc degree in electrical engineering from WPI, Worcester, MA, in 2004. He is currently working toward the PhD degree in electrical and computer engineering with the CWINS at WPI. From 2000 to 2002, he was a technical engineer with Bahrain Telecom. From 2002 to 2004, he was awarded a Fulbright Scholarship to pursue the MS degree at WPI. His research interests include performance limitations of time-of-arrival-based ultra-wide-band ranging in indoor non-line-of-sight (non-LOS) conditions, cooperative localization for indoor wireless sensor networks, and non-LOS/blockage identification and mitigation.

Ahmed Amer received his BSc degree in computer and control engineering from Zagazig University in July 2003, where he is currently working toward the M.S. degree in wireless sensor networks. His research interests include power-aware routing and clustering of wireless sensor nodes, design of operating systems, computer organization and architecture, and database management systems. He worked in the field of programming for two years. In addition, he worked as a visiting lecturer at Information Technology Institute (ITI) and Remote Sensing Center (RSC). He is working as a teaching assistant at Zagazig University from September 2003 till now.

Junaid Ansari is currently a PhD student and research assistant at the Department of Wireless Networks, RWTH Aachen University, Germany. He completed his MSc in communications engineering from RWTH Aachen University, Germany in 2006 and his Bachelor's degree in Electrical Engineering from National University of Science and Technology, Pakistan in 2002. His current research interests include low-power design and energy efficient networking solutions for wireless sensor networks. He has been actively working and managing various large scale research projects related to wireless sensor networks funded by European Union and German government at the Department of Wireless Networks, RWTH Aachen University, Germany.

Isabelle Augé-Blum holds an MSc in industrial computer sciences (1996) and a PhD on Formal Validation of Real-Time Fieldbus (2000), both from LAAS laboratory, University Toulouse III, France. Since September 2000, she is an associate professor at the National Institute of Applied Sciences (INSA) at Lyon, France, Center for Innovations in Telecommunication and Services integration (CITI Lab). She is a member of the INRIA ARES Project (Architecture of Networks of Services). Her research interests include MAC and routing protocols for autonomic wireless sensor and mesh networks, real-time communication and formal validation.

Dominique Barthel got his engineering degrees from Ecole Polytechnique, Palaiseau, France (1985) and Ecole Supérieure d'Electricité, Gif-sur-Yvette, France (1987). After a first career architecting and developing microcontrollers, DSPs, media processors and a scientific supercomputer CPU, he moved on to work on communication protocols for wireless sensor networks with France Telecom R&D at San Francisco, CA, USA then Meylan, France. He is inventor or co-inventor of 8 patents. His research interests include energy-efficient radios and communication protocols.

Sebnem Baydere is a professor in the Computer Engineering Department at Yeditepe University, Istanbul, Turkey. Prof. Baydere received her BSc and MSc degrees in computer engineering from Middle East Technical University(METU), Ankara, Turkey in 1984 and 1987 respectively. She obtained her PhD degree in computer science from University College London(UCL), UK in 1990. Professor Baydere is the director of the Networking Research Group at Yeditepe since 1996. Her recent research interests are wireless sensor networks, cross layer design, distributed operating systems, context awareness.

Mohamed El-Darieby (M'98) is an associate professor and chair of Software Systems Engineering at the Faculty of Engineering at the University of Regina, Regina, Canada. He received BSc and MScin electrical engineering from Zagazig University, Egypt and PhD in systems and computer engineering at Carleton University, Ottawa, Canada. His research is in the areas of backbone networks, grid computing and wireless mesh networks. Dr. El-Darieby is a member of the IEEE and the Association of Professional Engineers and Geoscientists of Saskatchewan (APEGS).

Mischa Dohler is now senior researcher with CTTC in Barcelona. He has published more than 100 technical journal and conference papers at high citation indexes, holds several patents, co-edited and contributed to several books, has given numerous international short-courses, and participated in standardization activities. He has been TPC member and co-chair of various conferences and is editor for the IEEE Communications Letters, the IEEE Transactions on Vehicular Technology, the IEEE Communications Magazine, the IEEE Wireless Communications, the IET Communications, the Elsevier PHYCOM journal, the EURASIP JWCN journal and other journals. He is a senior member of the IEEE.

Dominique Duncan graduated from the University of Chicago in mathematics and Polish literature with a minor in computational neuroscience in 2007. She is currently pursuing a PhD in Electrical Engineering at Yale University. Her research interests include applied mathematics, systems and control, computational neuroscience, data analysis and modeling. She is a member of IEEE, Control Systems Society (CSS), CSS Women in Control, CSS Technical Committee on Control Education, and Association of Women in Mathematics.

Eiman Elnahrawy is an NJCST post-doc fellow. Her current research interests include indoor localization using wireless radios and its applications in asset tracking, healthcare workflow management, network administration, and security. Dr. Elnahrawy awards include two post-doc Fellowships 2007-2008, a WISE-TRUST summer fellowship at Berkeley in 2007 as well as the best paper award a at the 2004 IEEE SECON. She served as an investigator on grants from the National Science Foundation, Rutgers Collaborative Computing Research and Robert Wood Johnson Foundation. Dr. Elnahrawy received a BSc from Alexandria University, an MS from the University of Maryland, College Park, and a PhD from Rutgers University in computer science.

Jia Fang received a BA in computer science with a minor in mathematics from the University of California, Berkeley, and a PhD degree in electrical engineering from Yale University. Currently, she is a post-doctoral fellow at Yale University. Her main interests include sensor networks and graph theory. She is a member of IEEE.

Barış Fidan received the BS degrees in electrical engineering and mathematics from Middle East Technical University, Turkey in 1996, the MS degree in electrical engineering from Bilkent University, Turkey in 1998, and the PhD degree in electrical engineering at the University of Southern California, USA in 2003. He has been with National ICT Australia and the Research School of Information Sciences and Engineering of the Australian National University, Canberra, Australia since 2005, where he is currently a senior researcher. His research interests include autonomous formations, sensor networks, cooperative localization, adaptive and nonlinear control, switching and hybrid systems, mechatronics, and various control applications.

Elena Gaura is currently a reader in pervasive computing and the founding director of the Cogent Computing Research Centre at Coventry University, UK. Dr. Gaura received the BS and MS degrees in electrical engineering from the Technical University of Cluj Napoca, Romania, in 1989 and 1991, respectively. In 2000, she received the PhD degree in electrical engineering from Coventry University, UK. Presently, her research interests pursue the issues of hardware-software integration and design for real life Wireless Sensor Network applications. She is a member of several national and international advisory bodies in the fields of microsystems and wireless sensor networks.

Fredrik Gunnarsson is a senior research engineer at Ericsson Research, and an associate professor at Linköping University, Sweden. He received the MSc degree in 1996, and the PhD degree in 2000, both in electrical engineering and from Linköping University, Sweden. His research interests include radio network management, radio resource management and signal processing for wireless communications.

Fredrik Gustafsson is professor in Sensor Informatics at Department of Electrical Engineering, Linkoping University. His research interests are in stochastic signal processing, adaptive filtering and change detection, with applications to communication, vehicular, airborne, and audio systems. He is a co-founder of the companies NIRA Dynamics and Softube, developing software signal processing solutions for automotive and music industry, respectively. He was an associate editor for IEEE Transactions of Signal Processing 2000-2006 and is currently associate editor for EURASIP *Journal on Applied Signal Processing* and *International Journal of Navigation and Observation.* In 2004, he was awarded the Arnberg prize by the Royal Swedish Academy of Science (KVA) and in 2007 he was elected member of the Royal Academy of Engineering Sciences (IVA).

Dong Guo is a PhD candidate at Center for Information & Systems Engineering, Boston University. He received his ME and BE degrees from Department of Automation, Tsinghua University in 2004 and 2001, respectively.

Mahmoud Halfawy, PhD, PEng. is a research officer at the Centre for Sustainable Infrastructure Research, Institute for Research in Construction, National Research Council of Canada. During the past twenty years, Dr. Halfway's work has been primarily focused on researching and developing information technology solutions to support the construction and infrastructure industry. Prior NRC, Dr. Halfway held several positions at Carnegie Mellon University, Ohio State University, EMH&T, Inc., Engineering Animation, Inc., and the University of British Columbia. Dr. Halfway has published 35 papers in refereed journal and conference proceedings. Dr. Halfway holds a PhD from the University of Iowa, USA and MSc Cairo University and BSc Asiut University, Egypt, all in engineering.

Mohammad Heidari received his MSc degree in communication and computer networking concentrated on WiFi localization from WPI, Worcester, MA. Currently, he is pursuing his PhD studies at CWINS, WPI and his research interests are analysis of radio channel dynamic behavior for indoor geolocation applications, wireless sensor networks, UWB channel measurement and modeling.

Bill Jackson was born in Sunderland, England in 1953. He studied mathematics as an undergraduate at Imperial College, London from 1970-73 and then as a postgraduate at Queens University and the University of Waterloo in Canada. He received his PhD from Waterloo in 1978 and then returned to

England as a postdoctoral research fellow at the University of Reading. He has lectured at the University of London since 1980 and is currently professor of Mathematical Sciences at Queen Mary. He has served on the British Combinatorial Committee and is a member of the Egerváry Research Group at Eötvös University in Budapest. His main research interests are graph theory, matroid theory, and their applications to the rigidity of frameworks and statistical mechanics.

Tibor Jordán received his PhD in mathematics from the Hungarian Academy of Sciences in 1995. In the following years he was employed by CWI, Amsterdam, and the University of Odense in Denmark as a postdoctoral research fellow, and the University of Aarhus as an associate research professor. In 2000 he joined the Department of Operations Research at Eötvös University in Budapest, Hungary, where he is now associate professor. His research interests range from combinatorial optimization to discrete applied geometry, with focus on graph algorithms and combinatorial rigidity. He is an associate editor of SIAM J. Discrete Mathematics and a member of the Egerváry Research Group at Eötvös University.

A.S. Krishnakumar (Krishna) is currently the director of Networked Systems Research at Avaya Labs. Krishna spent 16 years at Bell Labs including running the first overseas research department of Bell Labs during 1993-1997. He has also been part of startup companies. He has a PhD in EE and an M.S. in Statistics from Stanford University (1984), a master's degree in EE from Northwestern University (1980) and a bachelor's degree in EE from IIT Madras (1979). His research interests include security, special purpose architectures, enterprise communication systems and applications - including wireless IP telephony, mobility, location determination, and location-based applications.

P. Krishnan (PK) is currently a research scientist at Avaya Labs. He has worked in the past with Bell Labs Research and has been part of startup companies. PK obtained his PhD in computer science from Brown University and his B. Tech in computer science and Engineering from IIT Delhi. His research interests include the design and implementation of algorithms; IP and converged networking; wireless, mobility and location management; network management; security; and the World Wide Web.

Gurhan Kucuk is an assistant professor in the Department of Computer Engineering at Yeditepe University, Istanbul, Turkey. He received his BSc in computer engineering from Marmara University of Istanbul, Turkey in 1995, followed by his MSc in computer engineering from Yeditepe University in 1998. He received his PhD degree in computer science and also received SUNY Distinguished Dissertation Award in 2004. His current research interests are in wireless sensor networks and in computer architecture, particularly in the optimizations of both high-end and embedded microprocessors, which are used in wireless sensor devices, for energy efficiency.

Wei Lai received the BE degree in automatic control and the ME degree in systems engineering, both from Huazhong University of Science and Technology, China, in 1999 and 2001, respectively, and the PhD degree in systems engineering from Boston University in 2007.

Moshe Laifenfeld was born in Kiev, Ukraine. In 1992 he received the bachelor degree in electrical engineering from the Technion - Israel Institute of Technology, and in 1998 he received the master's degree magna cum laude in electrical and computer engineering from Tel Aviv University. In 2007 Moshe received the PhD in electrical and computer engineering from Boston University. Throughout these years he has been engaged in research and development projects in the field of wireless communications in several companies, including Rafael, the Israeli Armament Development Authority, and other start-ups for which he has been consulting on diverse aspects of signal processing. Since September 2007, Moshe has also been post-doctoral researcher at the Massachusetts Institute of Technology, where he focuses on aspects of coding theory in communications networks.

Steven Lanzisera received a BS degree from the University of Michigan, Ann Arbor, in 2002. He is currently a PhD candidate in electrical engineering at the University of California, Berkeley. From 1999 to 2002 he was an engineer with the Space Physics Research Laboratory at the University of Michigan where he worked on satellite integration and testing. He has served as a consultant for companies developing wireless and location aware systems and has held internships developing spacecraft electronics and biomedical devices. His research interests include wireless embedded systems, communications and integrated circuits.

Xiaoli Li obtained her PhD degree in computer science from the University of Missouri-Columbia, with the Distributed Computing and Sensor Networks (DCSN) Research Laboratory in 2007. Her research interests include wireless sensor networks, distributed computing, and software engineering.

Petri Mähönen is currently a full professor and holds Ericsson Chair of Wireless Networks at the RWTH Aachen University in Germany. Before joining to RWTH Aachen in 2002, he was a research director and professor at the Centre for Wireless Communications and the University of Oulu, Finland. He has studied and worked in the United States, the United Kingdom and Finland. He has been a principal investigator in several international and national multi-million USD research projects. Dr. Mähönen has published ca. 200 papers in international journals and conferences and has been invited to deliver research talks at many universities, companies and conferences. He is a senior member of IEEE and ACM, and fellow of RAS. He is inventor or co-inventor for over 20 patents or patent applications. He is currently also a research area coordinator and one of the principal investigators for a newly formed Ultra High Speed Mobile Information and Communication (UMIC) research cluster at RWTH, which is one of the German national excellence clusters supported by the Federal Government of Germany established in 2006.

Guoqiang Mao received the bachelor degree in electrical engineering from Hubei University of Technology, China, the Master degree in engineering from South East University, China and PhD in telecommunications engineering from Edith Cowan University, Australia in 1995, 1998 and 2002 respectively. After graduation from PhD, he worked in the U.S.-based industrial company "Intelligent Pixel Incorporation" as a senior research engineer for one year. He joined the School of Electrical and Information Engineering, the University of Sydney in December 2002 where he is a Senior Lecturer now. His research interests include wireless localization techniques, wireless multihop networks, graph theory and its application in networking, telecommunications traffic measurement, analysis and modeling, and network performance analysis.

Richard P. Martin is an associate professor at the Department of Computer Science and a member of the Wireless Information Network Laboratory at Rutgers University. His current research interests include localization in wireless sensor networks and human factors in dependable computing. Recent awards include the best paper award at the 2004 IEEE Conference on Sensor and Ad Hoc Communication Networks as well as a CAREER award from the National Science Foundation. Dr. Martin has served as an investigator on grants from the Defense Advanced Research Projects Agency, the National Science Foundation, and IBM. He received a B.A. from Rutgers University and an MS and PhD in computer science from the University of California at Berkeley.

Michael McGuire is an assistant professor with the Department of Electrical and Computer Engineering at the University of Victoria, Victoria, British Columbia, Canada. He obtained his Bachelors of Engineering and Masters of Applied Science in 1995 and 1997 from the University of Victoria. He worked at Lucent Technologies in Holmdel, New Jersey for two years. In 2003, he obtained his PhD from the University of Toronto while investigating the tracking of cellular telephones in dense urban areas. His current research interests are in the development of enabling methods for location-aware computing.

Stephen Morse was born in Mt. Vernon, New York. He received a BSEE degree from Cornell University, MS degree from the University of Arizona, and a PhD degree from Purdue University. From 1967 to 1970 he was associated with the Office of Control Theory and Application {OCTA} at the NASA Electronics Research Center in Cambridge, Mass. Since 1970 he has been with Yale University where he is presently the Chair of the Department of Electrical Engineering, the Dudley Professor of Engineering and a professor of Computer Science. His main interest is in system theory and he has done research in network synthesis, optimal control, multivariable control, adaptive control, urban transportation, vision-based control, hybrid and nonlinear systems, sensor networks, and coordination and control of large grouping of mobile autonomous agents. He is a fellow of the IEEE, a distinguished lecturer of the IEEE Control System Society, and a co-recipient of the Society's 1993 and 2005 George S. Axelby Outstanding Paper Awards. He has twice received the American Automatic Control Council's Best Paper Award and is a co-recipient of the Automatica Theory/Methodology Prize. He is the 1999 recipient of the IEEE Technical Field Award for Control Systems. He is a member of the National Academy of Engineering and the Connecticut Academy of Science and Engineering.

Thinh Nguyen is an assistant professor at the School of Electrical Engineering and Computer Science of the Oregon State University. He received his PhD from the University of California, Berkeley in 2003 and his BS degree from the University of Washington in 1995. He has many years of experience working as an engineer for a variety of high tech companies. He has served in many technical program committees. He is an associate editor of the IEEE Transactions on Circuits and Systems for Video Technology, the IEEE Transactions on Multimedia, the Peer-to-Peer Networking and Applications. His research interests include network coding, multimedia networking and processing, wireless and sensor networks.

XuanLong Nguyen is a postdoctoral researcher at Duke University's Department of Statistical Science. He received his master's degree in statistics and PhD degree in computer science from the University of California, Berkeley in 2007. Dr. Nguyen is interested in learning with large-scale spatial and nonparametric models with applications to distributed and adaptive systems in computer science, and modeling in the environmental sciences. He is a recipient of the 2007 Leon O. Chua Award from UC Berkeley for his PhD research, the 2007 IEEE Signal Processing Society's Young Author best paper award, an outstanding paper award from the ICML-2004 conference.

Kaveh Pahlavan is a professor of Electrical and Computer Engineering (ECE) and Computer Science (CS) and founding director of the CWINS, WPI, Worcester, Massachusetts. Previously, he was a visiting professor at the Telecommunication Laboratory and Center for Wireless Communications, University of Oulu, Finland. Dr. Pahlavan is the editor-in-chief of the *International Journal of Wireless Information Networks*, an advisory board member of the *IEEE Wireless Magazine*, and an Executive Committee member of the IEEE PIMRC. He has been an IEEE fellow since 1996 and was a Nokia fellow in 1999 and a Fulbright-Nokia scholar in 2001. He has served as the general chair and organizer of many IEEE events and has contributed to numerous seminal technical and visionary publications regarding wireless office information networks, home networking, and indoor geolocation science and technology. Dr. Pahlavan is the principal author of "Wireless Information Networks" (Allen Levesque, co-author), John Wiley and Sons, 1995, 2nd Ed. 2005, and "Principles of Wireless Networks – A Unified Approach" (P. Krishnamurthy, co-author), Prentice Hall, 2002. Additional information regarding his work can be found at www.cwins.wpi.edu. Dr. Pahlavan received a PhD from Worcester Polytechnic Institute, Massachusetts, and an MSc degree from the University of Tehran, Iran, both in electrical engineering.

Ioannis Ch. Paschalidis is an associate professor at Boston University with appointments in the Department of Electrical and Computer Engineering and the Systems Engineering Division. He is a co-director of the Center for Information and Systems Engineering (CISE) and the academic director of the Sensor Network Consortium. He completed his graduate education at the Massachusetts Institute of Technology (MIT) receiving an MS (1993) and a PhD (1996), both in electrical engineering and computer science. In September 1996 he joined Boston University where he has been ever since. He has held visiting appointments with MIT, and the Columbia University Business School. His current research interests lie in the fields of systems and control, networking, applied probability, optimization, operations research, and computational biology.

Neal Patwari received his BS(`97) and MS(`99) in electrical engineering from Virginia Tech, in Blacksburg, VA. He was an undergraduate and graduate researcher at the Mobile & Portable Radio Research Group (MPRG). Between 1999 and 2001, he was a research engineer at Motorola Labs, Florida Communications Research Lab, in Plantation, FL. He received his PhD in electrical engineering from the University of Michigan EECS Department in September, 2005. Neal is currently an assistant professor in the University of Utah Department of Electrical and Computer Engineering, which he joined in August, 2006. He has been awarded the National Science Foundation CAREER Award, the NSF's most prestigious award for young faculty. His research interests generally fall in the area of "radio channel signal processing". He directs the Sensing and Processing Across Networks Lab at the University of Utah, has over thirty publications in refereed conferences, journals, and edited volumes, and holds seven U.S. Patents.

Kristofer S. J. Pister received the BA degree in applied physics from the University of California, San Diego, in 1986 and the MS and PhD degrees in electrical engineering from the University of California, Berkeley, in 1989 and 1992. From 1992 to 1997 he was an assistant professor of Electrical Engineering at the University of California, Los Angeles, where he helped developed the graduate microelectromechanical systems (MEMS) curriculum and coined the term Smart Dust. Since 1996, he has been a professor of Electrical Engineering and Computer Sciences at the University of California, Berkeley. In 2003 and 2004, he was on leave from the University of California, Berkeley, as CEO and then CTO of Dust Networks, Hayward, CA, a company he founded to commercialize wireless sensor networks. He has participated in many government science and technology programs, including the DARPA ISAT and Defense Science Study Groups, and he is currently a member of the Jasons. His research interests include MEMS, micro-robotics, and low-power circuits.

Konstantinos N. Plataniotis is a professor with the Department of Electrical and Computer Engineering at the University of Toronto, Toronto, Ontario, Canada. His research interests are in the areas of multimedia systems, biometrics, image & signal processing, communications systems and pattern recognition. Dr. Plataniotis is a registered professional engineer in the province of Ontario.

Saikat Ray is an assistant professor at University of Bridgeport. He received his B.Tech. degree from Indian Institute of Technology, Guwahati, India in electronics and communications engineering in 2000, and MS and PhD degrees from Boston University in Electrical Engineering in 2002 and 2005, respectively. He spent the summers of 2001 and 2003 as intern in Fujitsu Network Communications (Acton, MA, USA) and Microsoft Research (Cambridge, UK), respectively. He was a post-doctoral researcher in the Department of Electrical and Systems Engineering, University of Pennsylvania from September 2005 to December 2006.

Janne Riihijärvi works as a senior research scientist at the Department of Wireless Networks at RWTH Aachen University. Before joining RWTH he worked in a variety of research projects on wireless networks at VTT Electronics and at the Centre for Wireless Communications at University of Oulu. His research interests have lately been in applications of techniques from spatial statistics and stochastic geometry on characterization of wireless networks, embedded intelligence in general, use of metaheuristics in optimizing component-oriented systems, and various frequency assignment and topology control problems. He has also worked on various enabling technologies for cognitive wireless networks, including participating into the development of the Unified Link-Layer API as well as different localization and tracking frameworks. As a part of his research work he has participated extensively into international research projects as well as research projects carried out in collaboration with the industry.

Yi Shang is an associate professor in the Department of Computer Science at the University of Missouri. He received PhD degree from University of Illinois at Urbana-Champaign in 1997. His research interests include wireless sensor networks, intelligent distributed systems, nonlinear optimization, and bioinformatics. His research has been supported by NSF, NIH, DARPA, Microsoft, and Raytheon. He is a senior member of the IEEE and a member of ACM.

Hongchi Shi is a professor and the chair of Computer Science at Texas State University-San Marcos. Before joining Texas State University, he has been an assistant/associate/full professor of Computer Science and Electrical and Computer Engineering at University of Missouri-Columbia. He obtained his BS degree and MS degree in computer science and engineering from Beijing University of Aeronautics and Astronautics in 1983 and 1986, respectively. He obtained his PhD degree in computer and information sciences from University of Florida in 1994. Hongchi Shi's research interests include parallel and distributed computing, wireless sensor networks, neural networks, and image processing. He has served on many organizing and/or technical program committees of international conferences. He established and chaired SPIE/SIAM International Conference on Parallel and Distributed Methods for Image Processing for several years. He is a lifetime member of ACM and a senior member of IEEE.

David Starobinski received the BSc, MSc and PhD degrees, all in electrical engineering, from the Technion-Israel Institute of Technology, in 1993, 1996 and 1999, respectively. In 1999-2000, he was a visiting post-doctoral researcher at UC Berkeley, and in 2007-2008 he was an invited professor at the School of Computer and Communication Sciences at EPFL (Swiss Institute of Technology in Lausanne). Since September 2000, he has been at Boston University, where he is now an associate professor. Starobinski received a U.S. National Science Foundation (NSF) CAREER award and a U.S. Department of Energy (DOE) Early Career award for his work on Quality of Service engineering and network modeling. He also received a fellowship for prospective researchers from the Swiss National Foundation, and awards from the Gutwirth Foundation and Intel Corp. His research interests are in networks performance evaluation, traffic engineering, and high-speed and wireless networking.

Ari Trachtenberg was born in Haifa, Israel. He received his BS degree from the Massachusetts Institute of Technology in 1994 in mathematics with computer science, and the MS and PhD degrees from the Department of Computer Science at the University of Illinois at Urbana/Champaign, in 1996 and 2000, respectively. He is currently an associate professor with the Department of Electrical and Computer Engineering, Boston University, Boston, MA. At the University of Illinois, he was a University fellow and later a Computational Science and Engineering Fellow from 1994 to 1997. In the summer of 1997, he was a research intern at Hewlett Packard Laboratories, Palo Alto, CA, and, in summers of 1998 and 1999, an instructor with the Center for Talented Youth at the Johns Hopkins University, Baltimore, MD. His research interests are centered around coding theory (iterative decoding, rateless codes) and its application to networks (data synchronization and location detection). Dr. Trachtenberg was the recipient of the David J. Kuck Outstanding Thesis award in 2000, the NSF CAREER award in 2002, and the ECE faculty teaching award in 2003.

Duc A. Tran is an assistant professor in the Department of Computer Science at the University of Massachusetts at Boston, where he leads the Network Information Systems Laboratory (NISLab). He received a PhD degree in computer science from the University of Central Florida (Orlando, Florida) in 2003. Dr. Tran's interests are in the areas of computer networks and distributed systems, particularly in support of information systems that can scale with both network size and data size. His current research projects are focused on data management and networking designs for decentralized networks (e.g., P2P networks, sensor networks). Earlier, he had worked extensively on multimedia systems, specializing in scalable overlay techniques for multimedia multicast. The results of his work have led to research grants from the US National Science Foundation, a Best Paper Award at ICCCN 2008, and a Best Paper Recognition at DaWak 1999. Also, his contribution on P2P streaming is widely-cited in this area (approaching 500 citations according to Google Scholar). Dr. Tran has engaged in many professional activities. He has been a guest-editor for two international journals, a workshop chair, a program vice-chair for AINA 2007, a program committee member for 20+ international conferences, and a referee and session chair for numerous journals/conferences.

Thomas Watteyne holds a masters degree, specialized in Telecommunications and an MSc degree in informatics, specialized in networking, telecommunications and services, from INSA Lyon, France (both 2005). He is now a final-year PhD candidate at France Telecom R&D and CITI Laboratory, INSA Lyon, France. His research interests include wireless sensor networks, self-organization principles and energy-efficiency. He is member of the Student Activities Committee of IEEE Region 8. He has published several papers, holds two patents, and has been organizing and technical program committee member of various conferences.

Index

D

E

F

G

H

I

K

L

M

Z